International Seminar on Mathematics, Science, and Computer Science Education (MSCEIS 2016)

Journal of Physics: Conference Series Volume 812

Bandung, Indonesia
15 October 2016

Editors:

Topik Hidayat
Asep Bayu Dani Nandiayanto
Al Jupri
Irma Rahma Suwarma
Lala Septem Riza
Diah Kusumawati
Endi Suhendi
Heli Siti Halimatul Munawaroh
Siti Aisyah
Marthalina Iriani

ISBN: 978-1-5108-3858-1
ISSN: 1742-6588

Printed from e-media with permission by:

Curran Associates, Inc.
57 Morehouse Lane
Red Hook, NY 12571

Some format issues inherent in the e-media version may also appear in this print version.

Printed by Curran Associates, Inc. (2017)

For permission requests, please contact the Institute of Physics
at the address below.

Institute of Physics
Dirac House, Temple Back
Bristol BS1 6BE UK

Phone: 44 1 17 929 7481
Fax: 44 1 17 920 0979

techtracking@iop.org

Additional copies of this publication are available from:

Curran Associates, Inc.
57 Morehouse Lane
Red Hook, NY 12571 USA
Phone: 845-758-0400
Fax: 845-758-2633
Email: curran@proceedings.com
Web: www.proceedings.com

Table of contents

Volume 812

International Seminar on Mathematics, Science, and Computer Science Education

(MSCEIS 2016)15 October 2016, Bandung, Indonesia

Accepted papers received: 15 February 2017Published online: 29 March 2017

Preface

Papers

IOP Conf. Series: Journal of Physics: Conf. Series **812** (2017) 011001 doi:10.1088/1742-6596/812/1/011001

PREFACE

Mathematics, Science, and Computer Science Education International Seminar (MSCEIS) has been started since 2013 as an International Seminar of Mathematics, Science and Computer Science Education. This seminar is motivated by improving the quality of mathematics, science and computer science education. The aims of the seminar are: (1) To bring together the scientists, education experts and practicioners, students,and civil society organization representatives in the scientific forum. (2) To share and to discuss theoretical and practical knowledge about innovation in mathematics, science and education.

MSCEIS 2016 was the 3rd seminar to provide forum for researchers in Mathematics, Science, and Computer Science Education to share new ideas or research result in their field. The theme for this seminar is "Harnessing Local Wisdom to Build Competencies of Excellence in Research and Collaboration in The New Era of The ASEAN Economic Community". This seminar is organized by FPMIPA UPI and SEAMEO QITEP in Science and sponsored by Universitas Pendidikan Indonesia(UPI).

The conference took place in Bandung, Indonesia, on 15 October 2016. More than 880 participants are involved from various topics in mathematics, science, and computer science education. There are 120 selected papers that go through a strict peer-reviewed process, and these papers will be published in the present conference proceedings.

We would like to thank the organizing committee and the members of reviewers for their kind assistance in reviewing the papers. We would also extend our best gratitude to keynote speakers for their invaluable contribution and worthwhile ideas shared in the conference.

The Editors

Topik Hidayat
Asep Bayu Dani Nandiyanto
Al Jupri
Irma Rahma Suwarma
Lala Septem Riza
Diah Kusumawati
Endi Suhendi
Heli Siti Halimatul Munawaroh
Siti Aisyah
Marthalina Iriani

msceis

organized by:
Faculty of Mathematics and Science Education
Universitas Pendidikan Indonesia
supported by:
SEAMEO QITEP in Science
International Seminar
msceis
2016
Mathematics, Science, and
Computer Science Education
Auditorium FPMIPA UPI, October 15, 2016

IOP Conf. Series: Journal of Physics: Conf. Series **812** (2017) 011001 doi:10.1088/1742-6596/812/1/011001

msceis

Chemistry and Chemistry Education
International Conference (SEAMEO QITEP Science)
Ms Irene TAN
Principal Master Teacher (Chemistry)
Academy of Singapore Teachers, MOE (Singapore)

QUESTION
Problem Statement
1. How to develop students' critical thinking skills using Problem based Learning 4 Core Areas (PBL4C) Model?
2. What are the difficulties that students have in developing their critical thinking skills?
msceis
Plenary Session
Chemistry and Chemistry Education

IOP Conf. Series: Journal of Physics: Conf. Series **812** (2017) 011001 doi:10.1088/1742-6596/812/1/011001

LIST OF COMMITTEE OF MSCEIS 2016

A. STEERING COMMITTEE

Chairperson

Siti Fatimah, S.Pd, M.Si, Ph.D

Members

Dr. Nahadi, M.Pd.
Dr. Indarjani
Prof. Dr. Yaya Sukjaya Kusumah
Prof. Dr. Munir, M.IT.
Dr. Andhy Setiawan, M.Si.
Dr. Bambang Avip Priatna
Dr. Bambang Supriatno
Dr. Dadi Rusdiana, M.Si.
Dr. Diana Rochintaniawati, M.Ed.
Dr. Didik Priandoko
Dr. Hendrawan
Dr. Mimin Nurjhani Kusumastuti
Dr. Mudzakir
Dr. Muslim, M.Pd.
Dr. Sjaeful Anwar
Dr. Sufyani Prabawanto, M.Ed.
Entit Puspita, M.Si.
Eddy Prasetyo Nugroho, M.T.
Jajang Kusnendar, M.T.

International Advisory Board

Prof. Dr. Triyanta, Director of SEAMEO QITEP in Science
Prof. Khairurrijal, Institut Teknologi Bandung, Indonesia
Prof. Dr. Hj. Rr. Hertien K. Surtikanti, Universitas Pendidikan Indonesia, Indonesia
Prof. Dr. H. R. Asep Kadarohman, M.Si, Universitas Pendidikan Indonesia, Indonesia
Prof. Dr. H. Moejiono, M.Sc, Universitas Budi Luhur, Indonesia
Prof. Dato' Dr. Aziz bin Dereman, Universiti Malaysia Terengganu, Malaysia
Prof. John Williamson, University of Tasmania, Australia
Prof. Hsin Kai Wu, National Taiwan Normal University, Taiwan
Prof. Dr. Munir, M.IT., UniversitasPendidikan Indonesia, Indonesia
Prof. Bruce Waldrip, University of Tasmania, Australia
Prof. Peter Aubusson, University of Technology Sydney, Australia
Prof. Lim Cher Ping, The Hong Kong Institute of Education
Assoc. Prof. Dr. Agus Setiabudhi, M.Si., Universitas Pendidikan Indonesia, Indonesia
Assoc. Prof. Dr. Turmudi, M.Ed., Universitas Pendidikan Indonesia, Indonesia

Scientific Committee

Prof. Dr. Didi Suryadi,M.Ed,UniversitasPendidikan Indonesia, Indonesia
Assoc. Prof. Rizky Rosjanuardi, Universitas Pendidikan Indonesia, Indonesia
Prof. Liliasari, M.Pd, Universitas Pendidikan Indonesia, Indonesia
Prof. Dr. Hj. Nuryani Y. Rustaman, M.Pd, Universitas Pendidikan Indonesia, Indonesia
Prof. Dr. Hj. Anna Permanasari, M.Si, Universitas Pendidikan Indonesia, Indonesia

IOP Conf. Series: Journal of Physics: Conf. Series **812** (2017) 011001 doi:10.1088/1742-6596/812/1/011001

Marthalina Iriani, M.T.
Yudi Wibisono, M.T.
Reza Setiawan, M.T.

Public Relation and Documentation
Yaya Wihardi, M.Kom.
Duden Saepuzaman, M.Pd, M.Si.
Haidar Helmy, S.Pd.
Iyan Sopian, S.T.
Irfan Taufik, S.Kom.
Agus Prihatna

Accommodation, Equipment, Decoration and Transportation
Dian Hendriyana, M.Pd.
Eki Nugraha, M.Kom.
Saeful Anwar, AMd.
Wawan Waryana
Yudi Isyuntika
Rudi Suyadi, S.Pd.

Food
Wiwi Siswaningsih, M.Si.
Gebi Dwiyanti, M.Si.
Tini Kartini, S.Pd.

Supporting
Drs. Jajang Saefulloh
Nandang, SAP, MAP
Drs. Rindang Supriyatno
Agus Mawan Giarto, S.Pd.
Fitri Ernawati, AMd.

Peer review statement

All papers published in this volume of *Journal of Physics: Conference Series* have been peer reviewed through processes administered by the proceedings Editors. Reviews were conducted by expert referees to the professional and scientific standards expected of a proceedings journal published by IOP Publishing.

IOP Conf. Series: Journal of Physics: Conf. Series **812** (2017) 012001 doi:10.1088/1742-6596/812/1/012001

Influence of Joyful Learning on Elementary School Students'Attitudes Toward Science

S Anggoro[1,2], W Sopandi[1] and M Sholehuddin[1]

[1] Universitas Pendidikan Indonesia, Jl. Dr. Setiabudhi No. 229, Bandung 40154, Indonesia
[2] Universitas Muhammadiyah Purwokerto, Jl. Dukuhwaluh Kembaran, Banyumas 53182, Indonesia
Email : wsopandi@upi.edu

Abstract. This study investigated the effects of joyful learning approach on elementary school students' attitudes toward science. The method used is quasy experiment with the participants were devided into two gropus.Thirty three of 4th grade students volunteered as an experimental group, and the otherforty two actas a control group. The data was collected by questionnaire that are given before and after the lesson, observation sheet, and interview. The effect of joyful learning on students' attitude was obtained by determining the n-gain and independent t-test. Observation and interview results were used to triangulate and support the quantitative findings. The data showed that the gain scores of the experimental group students' attitudes toward science were significantly higher than the gain scores of control group. In addition, the experimental group made significantly greater progress in their cognitive, affective and conative experiences. Interviews and observations indicated that their attitude toward science changed over the intervention. This indicated that joyful learning approach can enhance theelementary school students' attitudes toward science. According to these findings, it can be concluded that joyful learning approach can be used as an alternative approach to imrpove student's attitude toward science.

1. Introduction

Activities developed in science teaching should aim to encourage students to observe and explore their environment, to understand the relationships in nature, the relationship between man and nature, and to learn to understand human beings as an integral part of the life chain. So learn science will be able to be more fun, both for students and teachers, if it is based on real experiences [1]. Beside that, in science learning process, both of hear and see are not enough to learn. Students will improved their understood, if they can do something with the information obtained.

The study of attitudes toward science has received attention in recent decades [2]and this is the key in the concept mastery of Science[3]. Positive attitude toward scienceis constructed with the subject matter being taught, methods and learning media, teacher academic ability and their interpersonal, physical and social environments that support the student enjoyed learning [4]. This is associated with learningpositively, memorizing and higher order thinking skills successfuly[5],[6],[7]. Attitude is a form of evaluation or feelings toward an object that supports does not support or do not take sides (unfavorable)[8],[9]. It definesas a certain regularity in terms of thinking (cognition), feeling (affection), and predisposition action (conation) a person against an aspect in the surrounding environment[10]. This concept is known as triadic schemes (triadic scheme) the view that attitude as a reaction to a combination of cognitive, affective and behavior of an object. The cognitive, affective,

IOP Conf. Series: Journal of Physics: Conf. Series **812** (2017) 012001 doi:10.1088/1742-6596/812/1/012001

and conation (behavior) as a stand-alone factor which would then form the concept of attitude, or better known as the tripartite models[11]. The attitude is affective on positive or negative assessment of an object[12]. Based on these opinions can be concluded that the attitude response consists of cognition (a statement of what is believed), affection response (statement of what is perceived) and the response conation (statement regarding the behavior [13].

Joyful Learning are strategy, concept and practice of synergic learning of meaningful learning [14][15], contextual learning[1],[16],[17], the theory of constructivism [18],[19], active learning [20], and child psychological development [21]. Joyful perception is a positive influence on the motivation of learners[22]. Some of educational games with the Joyful Learning strategy has been developed and is based on the theory and strategy of science education/pedagogic [23].Joyful strategy Classroom Learning System (JCLS) has been used to help children learning multiplication in math. Based on trial results and field tests show that JCLS provide a deeper understanding of the subject matter through hands-on exercises. Besides that, by using JCLS simultaneously increase students' motivation and improve joyful perception during the learning process. JCLS known also supports teacher each student's ability to obtain information quickly and accurately so as to determine in-class instructional strategy and provide after-school assistances [18].The implementation of the joyful activity-based learning (ABJL) strategy has been elaboratedat several elementary schools in Maharastra IndiaJadal [19],[24]. ABJL strategy is a learning strategy maximum student involvement in the learning process. This strategy refers to the principles of learn by playing, learn by doing, learn by enjoying and learn by problem solving. This learning strategy requires the involvement of multiple sensory organs students during the learning process. Based on several studies that use different learning strategies which refers to the joyful learning activities give positive results.The purpose of this study is to examine and analyze the scores increase attitude toward science (cognition, affection and affection)

2. Research method

This research was conducted by quasy experimental design. The Population of this study are two private elementary school students in Bandung city. The sample are all forth student in these schools. The instrument that used was developed by operational definitions and indicators of the attitude toward science. Attitudestoward science is a response to student cognition, affection, and conation to the statements in the questionnaire to accept or reject the science learning which covers the subject matter, learning models, learning media and teacher character shown on the statement contained in the questionnaire attitude of science learning.Cognitive attitudes include students' attitudes toward the usefulness of materials science, the effectivity of learningmedia and model were used and the ability of teachers to teach. Affective attitude include students' attitudes toward material deemed attractive, media and learning model that is exciting and fun and character of the teacher teaches. While the attitude connative include students' attitudes towards material science, media and inspiring learning models, as well as the enthusiasm of teachers in teaching.

Joyful Learning is learning activities undertaken by the students through role-play activities, experiments, and group discussions conducted inside and outside the classroom. The theme refers to the activities performed science concepts regarding Natural Resources and sounds associated with the object in the field of health (doctors and stethoscope), music (musicians and musical instruments) and forestry (erosion and consequently). The indicators of learning toward science questionnaire are cognitive (11 items), affective (9 items) and conative aspects (9 items).The purpose of this study is to examine and analyze the scores increase attitude toward science on experimental group and the control group. The test is performed to determine differences in attitude scores increase learning and mastery of concepts before and after treatment. The increase that occurred before and after learning the views from the index gain score (N-Gain) withthe Meltzer formula [25].

3. Results and discussion

Description of attitude toward science scores before and after treatment with N-Gain value is shown in Figure 1 dan Figure 2.

IOP Conf. Series: Journal of Physics: Conf. Series **812** (2017) 012001 doi:10.1088/1742-6596/812/1/012001

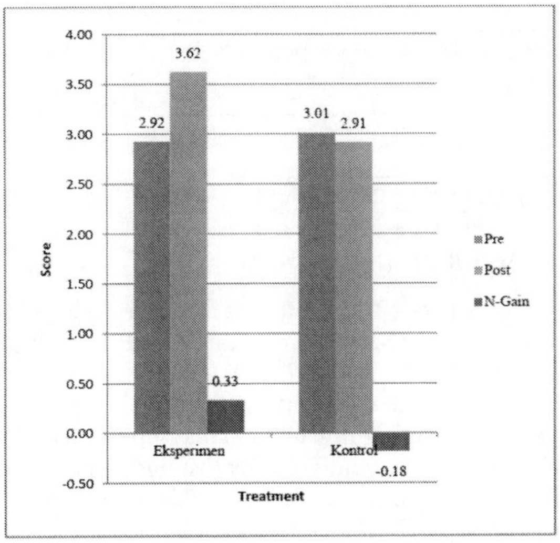

Figure 1. The Attitude toward science performance.

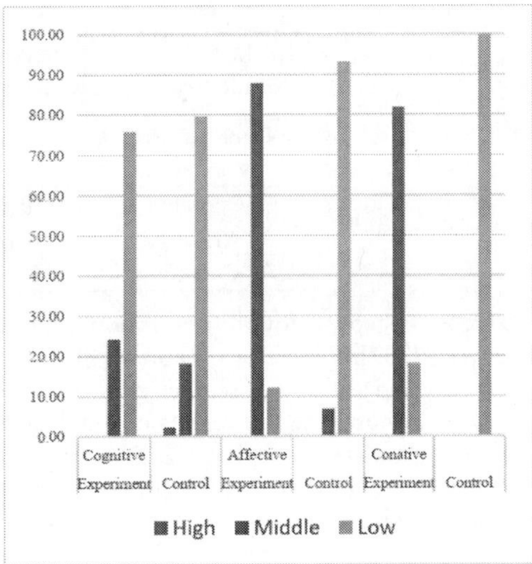

Figure 2. N-Gain categories of attitude toward science.

Figure 1 has described the attitude toward science scores during the study. It has shown increased in experiment class and decreased in control. By Hake category, a score of N-Gain classified into three categories, namely the increase in the category of high, medium and low [25]. Improved attitude toward science score of experimental group is 0.33and including medium category [25], whereas the increase in the control group attitude toward science scores of -0.11 including low category. The results has supportedthat the Joyful Learning gives better results than conventional learning.[4],[17],[19],[26] and also influence on behavior Joyful study or receive information[5],[6],[27],[28].

The number of students in grade experimental and control classes that meet the category of high, medium, and low is shown in Fig. 2.In Figure 2 it appears that the majority of the control class students obtained a score of N-Gain with low category on every aspect of attitude. While the experimental class, most students (over 80%) of students had increased scores on the attitude of learning medium category affective and conative aspects. Based on these results can be drawn a conclusion that the approach Joyful Learning in science learning provides enhanced attitude toward science better than conventional learning. Figure 1 and 2 have shown that are difference in the improvement of attitude toward science score of the second class treatment. These data indicated experiment class was higher than the control.

Based on normality test were using the *Kolmogorov-Smirnov* and *Levene's test* of homogenity, attitude toward science before and after treatment with *N-Gain* have shown normal distributed and homogen. The results of testing the difference improvement and learning behaviors using *Joyful Learning* strategies are presented in Table 1 and 2.

Table 1. Analysis of variance n-gain attitude towards science.

Aspect	Group	Mean	F	Sig.	T	Sig.
Attitude towads science	Experiment	0.338	26. 925	0.000	15. 057	0.000
	Control	0.187				

Table 2. The result analysis of variance of N-Gain aspects of attitude towards science.

Aspects of Attitude toward science	Anova Test Result
Cognitive	(0.20) (-0.09), E (0.025) = 2,026.p <. 05
Affective	(0.42) (- 0.04), E (0.042) = - 10,95.p<. 05
Conative	(0.37) (- 0.26), E (0.023) = 14,372.p <. *05*

Table 1 shows Joyful perception has positive influence on the attitude toward science of learners[9],[22]. Educational games with Joyful Learning strategy has been developed and is based on the theory and strategy of science education/pedagogic[23].

Table 2 shows the test results of the differences in attitudes enhancement treatment. All aspects of attitude toward science have influenced with joyful learning significantly. This shows that the joyfullearning approach can improve the attitude of learning in cognitive, affective and conative in science learning. Joyful Classroom Learning System Strategy (JCLS) can increase students' motivation and improve joyful perception during the learning process[18]. The strategy joyful activity-based learning (ABJL) at several elementary schools in Maharastra India gives better results than using *traditional learning methods*[19][24].

Attitude is a certain regularity in terms of knowledge (cognition), comprehension (affection), and behavior (conation) theperson against an aspect in the surrounding environment[13]. Joyful Learning shown to provide positive influence on the attitudes of students in learning science. In the experimental group, teachers apply learning activities using the principles of the principles of learning by playing, learning by doing, learning by enjoying and learning by problem solving. Students synergy of activities to make tools and play a role like making a stethoscope and checking the heart rate. At other times the students to make a musical instrument from a bottle filled with water in a certain amount and make the tune of their own choosing. The teacher's role in these activities is as a facilitator and evaluator.

Students were enthusiastic in applying such activities, it has been seen from their willingness to bring their own equipment from home and a strong desire to be able to make props and apply it. The most fundamental thing of the invention Piaget's theory of cognitive development is learning on the students do not have to happen just because a teacher teaches something to him. Piaget believe that learning occurs because the student was actively construct knowledge from it, and this is reinforced when students have control and choice about the things learned [29],[30]. This does not negate the teacher factor in the learning process. Teaching by the teacher who invites students to explore, apply manipulation, either in the form of physical or symbolic, ask questions and seek answers, comparing answers from other students will be helping students to learn and understand something.

At the beginning of the learning, the teacher provides introductory material with learning by play, either using tools like puzzle games and science comic books. Beside that, during the learning process, teachers applying intense interaction in students through interactive discussion. At each lesson, the teacher provides time to apply outdoor activities. Activities carried out with regard to the material presented. It turned out to provide a different atmosphere than when done in the classroom. Students can apply to the activities of groups with a wider space. Students are seen more zeal in applying the learning activities as learning activities can be done while playing. According to Vygotsky through play, all aspects of child development can be improved [31]. With children can play freely express and explore to strengthen the things that are already known and discover new things. Through learning by play, children can also develop all their potential optimally, both physical and mental potential intellectual and spiritual. Therefore, children's play is a bridge to the development of physical and psychological [16].

Learning is a process of knowledge development is holistic and sustained through social interaction, based on experience and students adapt to their lives. Through play, social interaction with peers, teachers and parents can be constructed. Interaction with adults (in this case the teachers at school and parents at home) is very important for cognitive development process more quickly achieved. This is consistent with the view of Vygotsky's Zone of Proximal Development concept [32].

At the end of the lesson, the teacher asked us to apply the evaluation activities through competitive activities. Each group presented the results of discussions by the poster they made together, then the other group to provide an assessment. This makes each group tries to show the best. The implications of the cognitive theory of Vygotsky[31], on science learning that are:(1) desired form of cooperative learning classroom setting so that students can interact around tasks and bring mutual problem-solving strategies affective within the zone of proximal development; and (2) In teaching emphasized scaffolding so that students are increasingly responsible for their own learning.

Joyful Learning is a learning approach that involves a sense of fun, happy, and comfortable of the parties who are in the learning process. There is an attachment of love and affection between teachers and learners and among learners,and the learning process will make each party trying to give the best to please others. Teachers with the passionate spirit will seek optimal lead the class in a way that is most attractive, while participants with enthusiasm and compete actively take part in any activity. Thus, Joyful Learning becomes a means that makes teachers and students to be like a session-by-session lesson so that the result will be a maximum [7],[19],[32],[33],[34]

Joyful Learning is done in the experimental class corresponding to the characteristics proposed by [21][27][28]. The existence of a relaxed environment, fun, does not create tension (stress), safe, attractive, and not make students feel free to apply something wrong to achieve success despite high. Availability of the subject matter and the relevant methods, involvement of all the senses and brain activity left and right, the situation challenging learning (challenging) for students to explore the material being studied, as well as learning situations positive emotional when students learn together, to make learning more fun.Joy of Learning indicatorshave been conducted on the experimental class [28]. Some indicators include: (a) find the pleasure in learning; (b) let student create things; (c) show off student work; (d) take time to thinker; (e) make-school spaces inviting; (f) get outside; and (g) read good books. Atmosphere "pleasant discovery" makes students maintain enthusiasm undergo a learning process with joyful feeling [7].Joyful Learning is done in harmony with the experimental class proposed by [27]. Students feel the atmosphere of learning that arouse interest in learning, relax, and attractive so as to make the student spirit and high concentration during lessons. Full involvement in the learning demonstrated by a willingness to provide their own equipment and the division of tasks within the group independently with joy.

4. Conclusion

Based on observations in the experimental class and control class, it can be concluded that the students get to experience positive cognitive learning through learning materials that are useful, methods and appropriate learning media, as well as a good teacher. If it is supported through the affective experience interesting subject matter, methods and media learning engaging and fun as well as enthusiasm and fun, make students have a positive desire or inclination to science learn (conative experience).Students who have an opinion that science learning beneficial to himself, engaging and fun in the learning process, have a tendency to study harder. In contrast, when learning science is considered as a burden, supported by a learning process that is tedious, causing lazy students to learn science. Therefore, creating joyful learning help students to overcome student's barier towards science.

5. References

[1] Hart C, Mulhall P, Berry A, Loughran J, Gunstone R 2000 *J. Res. Sci. Teach* **37** 655-75
[2] Ali M S and Awan A S 2013 *Interdisciplinary Journal of Contemporary Research In Business* **4** 707-18

[3] Nordin A and Ling L H 2011 *Journal of Science & Mathematics Educational* **2** 89-101
[4] Chopra V and Chabra S 2013 *Journal of Unschooling and Alternative Learning* **7** 28-44
[5] Willis J 2007 The neuroscience of joyful education *Educational Leadership* EL 64(2007)
[6] Willis J 2011 Understanding how the brain thinks *Brain Based Learning* (Online: Edutopia)
[7] Kohn A 2004 Feel bad education the cult of rigor and the loss of joy *Feel-Bad Education and Other Contrarian Essays on Children & Schooling Education Week* vol 24 (Boston: Beacon Press) pp 44–45
[8] Kwon J and Vogt C 2008 *Proc. of the 2008 Northeastern Recreation Research Symp* (New York: USDA Forest Service) pp 298-305
[9] Kirikkaya E B 2011 *Educational Research and Reviews Academic Journals* **6** 374-82
[10] Tighezza M 2013 *International Journal of Science and Mathematics Education* **12** 721-40
[11] Eagly A H and Chaiken S 2007 *Guilford Journals* **25** 582-602
[12] Fishbein M and Ajzen I 1975 *Belief, Attitude, Intention, and Behavior: An Introduction to Theory and Research* (MA: Addison-Wesley) p 23
[13] Azwar 1995 *Sikap Manusia Teori dan Pengukurannya* (Yogyakarta: Pustaka Pelajar) pp 20-45.
[14] Vallori A B 2002 *Meaningful Learning in Practice: How to Put Meaningful Learning in The Classroom* (Granada: Seminar on Meaningful Learning)
[15] Morgado P 2010 *Transformative Dialogues: Teaching & Learning Journal* **4** 1-13
[16] Brotherson S 2009 Young children and the important of play *Bright Beginning* vol 23 (North Dakota State University)
[17] Hayes D 2007 *Joyful Teaching and Learning in Primary School,* ed Bell & Bain (Glasgow: Great Britain)
[18] Wei C, Hung I C, Lee L and Chen N 2011 *TOJET: The Turkish Online Journal of Educational Technology* **10**(2) 11-23
[19] Jadal M M 2012 *Indian Streams Research Journal* **2** 1-5
[20] Clark R C and Mayer R E 2008 *Learning by Viewing Versus Learning by Doing: Evidence-Based Guidelines for Principled Learning Environments* vol 47 (Wiley InterScience) pp 5-13
[21] Corbeil P 1999 *Sage Journals* **30** 163-80
[22] Kirikkaya E B, İşeri Ş and Vurkaya G 2010 *The Turkish Online Journal of Educational Technology* **9** 1-13
[23] Kebritchi M and Hirumi A 2008 *Computers & Education* **51** 1729-43
[24] Jadal M M 2012 *Journal of Arts and Culture* **3** 110-4
[25] Meltzer D E 2002 *Am. J. Phys* **70** 1259-68
[26] Hongkong Arts Development Council 2005 *Joyful learning the arts-in education program.* (Hongkong: Hongkong Arts Development Council) pp 118-50
[27] Meier D 2000 *The Accelerated Learning Handbook: A Creative Guide to Designing and Delivering Faster, More Effective Training Programs* (New York: McGraw Hill) p 145
[28] Wolk S 2008 *Joy in school* Educational Leadership EL 66(2008)1
[29] Blake B and Pope T 2008 *Journal of Cross-Disciplinary Perspectives in Education* **1** 59-67.
[30] Simatwa E M W 2010 *Educational Research and Review* **5** 366-71
[31] Tarman B and Tarman I 2011 *Elementary Education* **10** 325-37
[32] Saleh F 2011 *Prosiding Seminar Internasional ke-3 dan Workshop Pedagogik Praktis yang Berkualitas (Bandung)* (Bandung: UPI) pp 13-25
[33] Adodo S O and Gbore L O 2012 *International Journal of psychology and Counselling* **4** 68-72.
[34] Majzub R M, Othman Z and Hasan A 2012 *Research Journal of Applied Sciences* **7** 421-25

IOP Conf. Series: Journal of Physics: Conf. Series **812** (2017) 012002 doi:10.1088/1742-6596/812/1/012002

Seismic Hazard Analysis based on Earthquake Vulnerability and Peak Ground Acceleration using Microseismic Method at Universitas Negeri Semarang

H Sulistiawan*, Supriyadi, and I Yulianti

Departemen Fisika, Fakultas Matematika dan Ilmu Pengetahuan Alam, Universitas Negeri Semarang, Jalan Taman Siswa, Sekaran, Gunung Pati, Sekaran, Gn. Pati, Kota Semarang, Jawa Tengah 50229, Indonesia

Email : hendrisulistiawan@students.unnes.ac.id

Abstract. Microseismic is a harmonic vibration of land that occurs continuously at a low frequency. The characteristics of microseismic represents the characteristics of the soil layer based on the value of its natural frequency. This paper presents the analysis of seismic hazard at Universitas Negeri Semarang using microseismic method. The data acquisition was done at 20 points with distance between points 300 m by using three component's seismometer. The data was processed using Horizontal to Vertical Spectral Ratio (HVSR) method to obtain the natural frequency and amplification value. The value of the natural frequency and amplification used to determine the value of the earthquake vulnerability and peak ground acceleration (PGA). The result shows then the earthquake vulnerability value range from 0.2 to 7.5, while the value of the average peak ground acceleration (PGA) is in the range 10-24 gal. Therefore, the average peak ground acceleration equal to earthquake intensity IV MMI scale.

1. Introduction

The Indonesian archipelago is an archipelago that has a high level of seismicity. Most of the earthquake source in Java, especially in Semarang city comes from active faults in Java. Active faults that affect Semarang city are Opak Fault (Yogyakarta), Lasem Fault, Pati Fault, and Kaligarang Fault.In 1856,Semarang has been shaken by an earthquake with earthquake intensity of VII - VIII MMI, or a magnitude above 5 SR. In 1821 and 1890, Jepara and Pati have also been shaken by an earthquake with earthquake intensity of VII MMI. The most recent earthquake occured in Jepara on October 23, 2015 which was caused by Lasem Fault activity with magnitude of 5 SR.

Semarang city with an area of 373.7 km^2 has several high seismic regions, such as Gunungpati and Mijen districts. Based on the vulnerability of ground movement zone map published by Pusat Vulkanologi Mitigasi Bencana Geologi (PVMBG), Gunung Pati is also consideredas an area that prone to soil movement. The impactcused by earthquake depends on the earthquake intensity, the distance from the source of the earthquake, the scale of the earthquake, the size of the fault zone, and the energy released rock, as well as the condition of the local geology[1].

To determine the seismic hazard zone at Unnes area, geophysical survey should be conducted to determine the vulnerability of the earthquake and the maximum ground acceleration.Earthquake Vulnerability value and maximum ground acceleration can be determined by the microseismic method [2]. Microseismic is a geophysical method that can describe the level of vulnerability of surface soil layers towards deformation during an earthquake [3]. Earthquake Vulnerability useful for predicting a weak zone during earthquake [4], and cracks of the ground due to earthquake [5].

Based on these descriptions, studies and analysis of seismic hazard needs to be done to analyze the risk of earthquakes based on the value of Peak Ground Acceleration (PGA) and earthquake vulnerability with microseismic methods to be used as information for disaster mitigation earthquakes in the Universitas Negeri Semarang.

2. Theory

2.1. Horizontal to Vertical Spectral Ratio (HVSR)

HVSR analysis method was first developed by Nakamura in 1989[2]. HVSR method used to calculate the spectrum ratio of the signal microtremor horizontal component towards the vertical component. Nakamura [6]divided microseismic waves into two types which are Rayleigh wave and body wave. Rayleigh wave is surface wave that propagates in the ground surface,whereas body waves propagates through the bedrock. Under these conditions, the HVSR equation can be written as follows.

$$H_f = A_h\, S_{HB} + S_{HS} \quad (1)$$

$$V_f = A_v\, S_{VB} + S_{VS} \quad (2)$$

so that

$$H/V = \frac{H_f}{V_f} = \frac{A_h S_{HB} + S_{HS}}{A_v S_{VB} + S_{VS}} \quad (3)$$

where H_f and $V_f$are the horizontal and vertical component of microseismic wave, A_h and $A_v$are amplification factor of body wave, S_{HB} and S_{VB} are spectrum of horizontal and verticalmotion in the bedrock, while S_{HS} and S_{VS} are spectrum of horizontal and vertical motion on the soil surface or sediment layers. HVSR curve is obtained by ploting H/V to frequency. The frequency of the highest amplitude in HVSR curve shows the natural frequency of ground layer.Meanwhile, the highest amplitude represents the amplification. The natural frequency and the amplification obtained from the HVSR curve are used to calculate the earthquake vulnerability and also peak ground acceleration.

2.2. Earthquake Vulnerability

Earthquake vulnerability is a value that describes the vulnerability of the surface soil layer due to deformation during the earthquake [3]. Earthquake vulnerability aims to measure the level of vulnerability of land or structures whenearthquake occured [7]which is defined by:

$$kg = \frac{A^2}{fo} \quad (5)$$

where f_o and A are the natural frequency and amplification, respectively. The earthquake vulnerability is essential to predict weak zones during an earthquake [4]. According to Nakamura [3],parameters that affect the vulnerability of the earthquake (Kg) are the value of ground shear-strain (γ) caused by the deformation of the surface layer (δ) and the acceleration of seismic waves on the bedrock (α) which increaseswhen propagates in a low-density medium such as soil layer surface. Ground shear-strain value is written in the following equation:

$$\gamma = \frac{A}{h} \cdot \frac{a}{4\pi^2 fo^2} \quad (6)$$

where A is the amplification, δ is the deformation of the surface soil layer caused by the earthquake $\left(\frac{a}{(2\pi Fo)^2} = \delta\right)$, and h is the thickness of the sediment layer.

2.3. *Peak Ground Acceleration (PGA)*

Peak Ground Acceleration (PGA) is the maximum acceleration of ground vibrations that happened at particular location that caused by earthquake wave. Peak ground accelerationvalue depends on the propagation of seismic waves and the characteristics of the soil layers[8]. Kanai [8] formulated an empirical equation of maximum ground acceleration as follows:

$$PGA=\frac{5}{\sqrt{To}}10^{(0,61M)-\left(1,66+\frac{3,6}{R}\right)\log R+\left(0,167-\frac{1,83}{R}\right)} \quad (7)$$

where*To* is the natural period of ground, *M* is the magnitude of earthquake and *R* is the hypocentre of earthquake.
According Gutenberg and Richter [9] there is a relationship between the value of maximum ground acceleration with earthquake intensity scale (I_{MM}) which is then formulated with the following empirical equation:

$$I_{MM}=3\log PGA+1,5. \quad (8)$$

Earthquake intensity defines the strength of earthquake that were perceived at certain location (on the surface) and is determined from the direct effects of the shake of the earthquake. The value of the maximum ground acceleration and the earthquake intensity depend on the magnitude of the earthquake, the distance from the earthquake source and the geology factor of the region affected by the earthquake.

3. Research Methodology

The research was conducted atUniversitas Negeri Semarang with a research area of 1200 × 750 m. Data collection was done at 20 points as shown in Figure 1 with a recording duration of 30 minutes at each point.

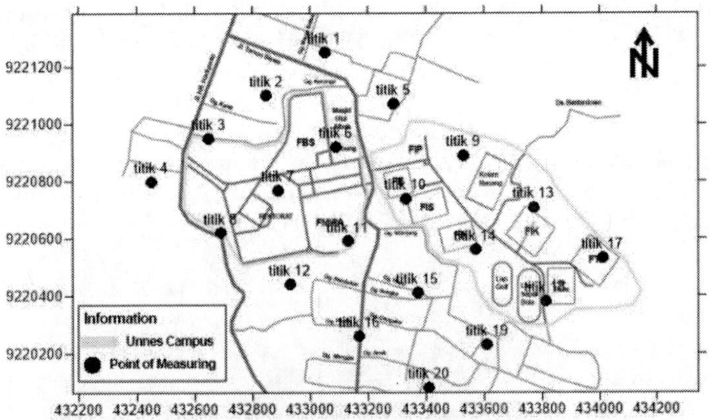

Figure 1. The data collection points at the research location

Data collection were performed by using M.A.E seismometer. The data were then analyzed by using *Geopsys*software to determine the natural frequencies and amplification.

4. Result and Discussion

4.1. *Natural Frequency and Amplification at Unnes Region*

Natural frequency is the dominant frequency that exists in the region,while the amplification is the magnitude of wave amplitude reinforcement when the wave propagates through a particular medium. Herak [10] explains that the value of natural frequency and amplification on the surface associated with physical parameters of the subsurface. The result of HVSR curve analysis using*Geopsy*are the values of natural frequency and the amplification as shown in Figure 2.

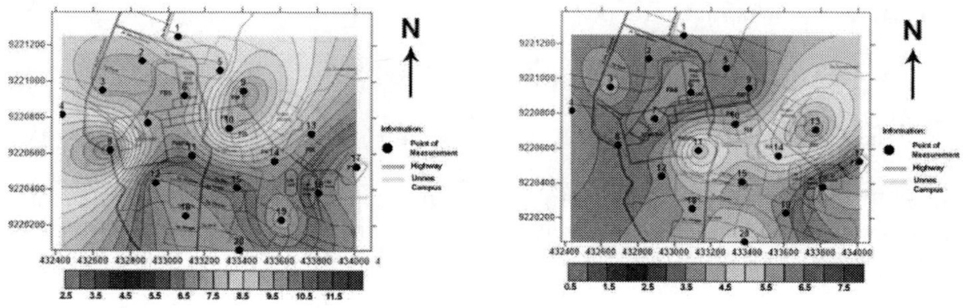

(a) (b)

Figure 2. Values of (a) natural frequency and (b) amplification

It is shown that the natural frequencies at Unnes region is ranging from 2.5 to 12 Hz, with natural frequencies of less than 8 Hz dominatingat theresearch area. It is also found that high natural frequency value (>8 Hz) are located at area around point 8, 9 and 10 which are shown in yellow to red contour. The high natural frequency indicates thin sediment layer, while low natural frequency valuesdescribe thick sediment layer.

In terms of amplification,the results show that the value at Unnes region is ranging from 0.5 to 7.5. In general, almost all research locations have amplification value below 4.5. Therefore, amplification of above 4.5 is considered as high amplification. It is shown in figure 2(b), high amplification value located at around point 11, 13, 14, and 20. The difference of amplification value occured due to deformation (weathering, folding or shifting) of the rocks below ground surface.[11]

4.2. Earthquake Vulnerability at Unnes Region

Earthquake vulnerability shows physical quantities of the impact due to shocks or movement of rock layers. The greater the value of the earthquake vulnerability is, the more vulnerable the affected area by shocks is. The result of the calculation of earthquake vulnerability is shown in Figure 3.

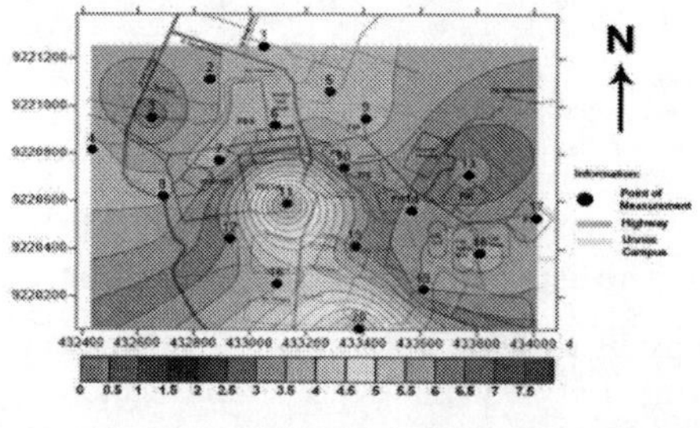

Figure 3. Map of earthquake vulnerability value at Unnes region

The results show that the earthquake vulnerability is in the range of 0.5 up to 7.5. More over, it also can beobserved that more than 50% of the area have earthquake vulnerability of below 4.5. Thus, the area with earthquake vulnerability of above 4.5 is considered as region with high earthquake vunerability which are located at around point 11, 16, and 20. The high earthquake vulnerability value indicates high ground shear-strain value. From equation (6) it is found that the ground-shear strain at point 11, 16, and 20 are 42.7×10^{-4}, 28.9×10^{-4}, and 44.6×10^{-4}, respectively. Based on table of strain dependence of dynamic properties of soil [11], the phenomenon of ground deformation caused by earthquake with those ground shear-strain value are in the form of crack and settlement.

4.3. Peak Ground Acceleration

The parameters used to determine the value of maximum ground acceleration is dominant period of soils derived from the natural frequency value and magnitude and hypocentre of earthquakes that was occured around Semarang as defined by equation (7). The earthquake sourcesthat were used are the earthquakes that happened in the last 10 years with a magnitude of 5 SR or above, which are the earthquake in Yogyakarta on May 27, 2006, earthquake in Jepara on October 23, 2015 and earthquake in Tegal on March 23, 2015. The maximum ground acceleration value are shown in Figure 4.

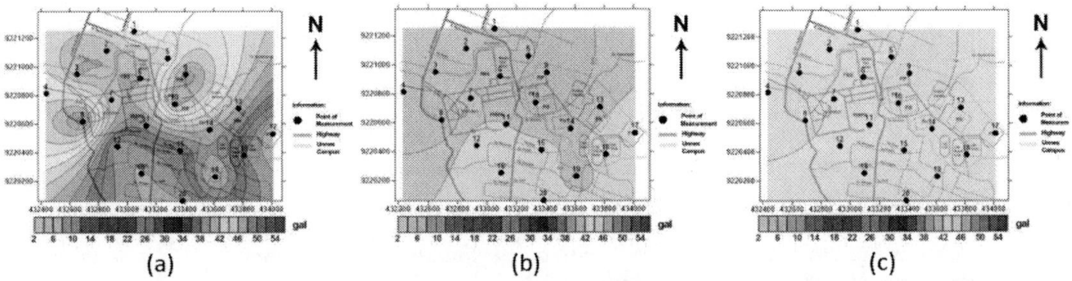

(a) (b) (c)

Figure 4. Map of peak ground acceleration value at Unnes region based on earthquake source in (a) Yogyakarta, (b) Jepara, and (c) Tegal.

Figure 4(a) shows that the peak ground acceleration value at Unnes based on Yogyakarta earthquake are in the range of 26-54 gal. The highest ground acceleration are around point 8, 9 and 10. Based on Jepara earthquake (Figure 4.b), the peak ground acceleration value are in the range of 6-12 gal. Meanwhile, the highest ground acceleration are around point 4, 8, 9 and 10. It is found thatthe peak ground acceleration based on Tegal earthquake (Figure 4.c) are lower compared to peak ground acceleration based on Yogyakarta and Jepara, which in the range of 2-5 gal. The highest ground acceleration based on Tegal earthquake was around point 1, 4, 8, 9 and 10.

The average peak ground acceleration at Unnes are calculated based on the value of the peak ground acceleration caused by the earthquake in Yogyakarta, Jepara and Tegal. The average peak ground acceleration value at Unnes is shown in Figure 5.

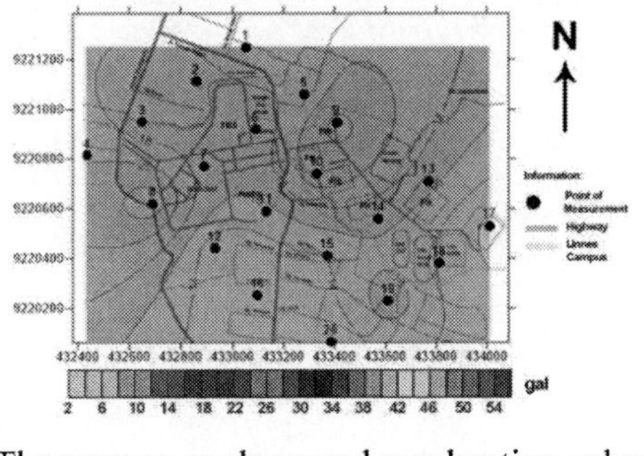

Figure 5. The average peak ground acceleration value at Unnes

Figure 5 shows that the average peak ground acceleration value are in the range of 10-24 gal. The highest ground acceleration value are located around point 8, 9, and 10. From equation (8), it was calculated that the average of earthquake intensity at Unnes region is in the range of IV MMIscale. The impact of earthquake intensity in IV MMI scale areswing of hanging objects and vibration that similar to vibration caused by moving heavy trucks.

4.4. Interpretation of Seismic Hazard

Seismic hazard is a description of a phenomenon caused by earthquakes that have potential to cause damage. Based on earthquake vulnerability and peak grond acceleration obtained in previous subsections, it can be concluded that seismic hazard at Unnes region is relatively low. However, there are several regions with high earthquake vulnerability that should be more alert since these regions can easily deforms. The deformation may cause demage to buildings and infrastructure during an earthquake.Therefore, it is suggested to not build any constructions on these regions. In case it need to build construction, then the construction should be built with deep foundation.

5. Conclusion

The values of earthquake vulnerability at Unnes area are in the range of 0 to 7.5. Meanwhile, the average value of the peak ground acceleration (PGA) at Unnes area are in the range of 10-24 gal. Seismic hazard at Unnes area is generally classified as low seismic hazard. However, in areas with high earthquake vulnerability,it is not advisable to make buildings.

6. References

[1] Nakamura Y 2000 *The 12nd Word Conf. on Earthquake Engineering.*
[2] Nakamura Y 1989 *Q.R. of RTRI* **30**(1) 25-33
[3] Nakamura Y 2008 On The H/V Spectrum ,*The 14th World Conference on Earthquake Engineering*, Beijing, China
[4] Gurler E D, Nakamura Y, Saita J and Sato T 2000 *6th International Conf. on Seismic Zonation, Palm Spring Riviera Resort, California, USA*65
[5] Daryono 2011 *Indeks Kerentanan Seismik Berdasarkan Mikrotremor pada Setiap Satuan Bentuklahan di Zona Graben Bantul, Daerah Istimewa Yogyakarta*, Disertation, Program Pascasarjana Fakultas Geografi (Yogyakarta: Universitas Gadjah Mada)
[6] Nakamura Y 2000 Clear Identification of Fundamental Idea of Nakamura's Technique and Its Application, *The 12nd Word Conference on Earthquake Engineering*, Tokyo, Japan
[7] Nakamura 2001 *4th EQTAP Workshop in Kamakura*
[8] Kanai K 1966 *Proceedings of the Japanese Earthquake Symposium, Not seen, Reported in Trifunac & Brady* (1975) 1-4
[9] Gutenberg B and Richter C F 1942 *Bulletin of the Seismological Society of America* **32** (3)
[10] Herak M 2008*Computers and Geosciences* **34** 1514–26
[11] Nakamura Y 1997 *Seismic Vulnerability Indices for Ground and Structures using Microtremor* (World Congress on Railway Research: Florence)

IOP Conf. Series: Journal of Physics: Conf. Series **812** (2017) 012003 doi:10.1088/1742-6596/812/1/012003

Improving Middle School Students' Critical Thinking Skills Through Reading Infusion-Loaded Discovery Learning Model in the Science Instruction

Nuryakin[1,2] and Riandi[2]

[1] SMP N 4 Pancatengah, Jl. Cayur Km 12, Neglasari, Pancatengah, Tasikmalaya 46194, Indonesia

[2] Science Education, Postgraduate Student, Universitas Pendidikan Indonesia, Jl. Dr. Setiabudi No. 229, Bandung 40154, Indonesia

Email: nuryakin@student.upi.edu

Abstract. A study has been conducted to obtain a depiction of middle school students' critical thinking skills improvement through the implementation of reading infusion-loaded discovery learning model in science instruction. A quasi-experimental study with the pretest-posttest control group design was used to engage 55 eighth-year middle school students in Tasikmalaya, which was divided into the experimental and control group respectively were 28 and 27 students. Critical thinking skills were measured using a critical thinking skills test in multiple-choice with reason format questions that administered before and after a given instruction. The test was 28 items encompassing three essential concepts,vibration, waves and auditory senses. The critical thinking skills improvement was determined by using the normalized gain scoreand statistically analyzed by using Mann-Whitney U test.. The findings showed that the average of students' critical thinking skills normalized gain score of both groups were 59 and 43, respectively for experimental and control group in the medium category. There were significant differences between both group's improvement. Thus, the implementation of reading infusion-loaded discovery learning model could further improve middle school students' critical thinking skills than conventional learning.

1. Introduction

The developedcurriculum in the 21st century learning has been shifted from a trend of merely mastering concept to the higher level [1]. Science instruction in the School-Based Curriculum (known as Kurikulum Tingkat Satuan Pendidikan) should be performed using scientific inquiry to foster thinking skill, process skill and scientific attitude as well as to communicate an important aspect of life skills [2]. Life skills are necessary to address the arisen problems of life, such as globalization, the changing of industry demand and work field, rapidly growing innovation and the trend of using technology and information. One of the life skills required to overcome these challenges is thinking skills. A minimum thinking skill that should be developed to face the globalization and progress in science and technology in the 21st century, according to Costa [3] is the critical thinking skills [4-5].

Critical thinking skills are included in the complex thinking process. Complex thinking is a high-order thinking process which consists of critical thinking, creative thinking, problem solving and decision making [3,6]. Critical thinking skills is one of the essential skills that affect the academic and professional students' success in the future [7]. Based on research in various fields such as social-science said that students who graduate from schools in various countries do not have the ability to compete on a global scale because they do not have the ability to think critically [8]. Thus, the critical thinking skills necessary to be developed at the school level (namely in middle school) as a provision in preparing for student life in the future, particularly on the Natural Sciences subject.

Science learning process that foster thinking skills, process skills and communicating them as an important aspect of life skills can be accomplished through the investigation (discovery / inquiry learning) [9]. Discovery/inquiry learning is a learning process that involves the students perform various activities such as observing, asking questions, seeking referrals from a variety of sources, planning investigations, conducting experiments, analyzing and interpreting the data, answering, explaining and estimating, and communicating results (NRC, 2000 in Ref. [2]). Discovery learning is a student-centered learning which activates students to seek and find a knowledge of natural phenomena that occur around them [10]. Learning by discovery prioritize the activity of reflecting, thinking, experimenting and exploring [11]. In discovery learning, students perform mental operations such as measurement, prediction, observation, inference and groupings [6]. A mental operation involving those intellectual skills can develop students' skills in constructing knowledge and enable them to recognize well their environment with the existing concept or knowledge they learned. However, discovery learning does not entirely provide a positive impact on student achievement [12]. Discovery learning that is only used without providing guidance or assistance do not provide meaningful benefit to their students [13]. Therefore, we need a strategy that supports discovery learning in order to improve student achievement such as critical thinking skills. One strategy used is through reading infusion.

Reading infusion is a reading that is integrated with discovery learning. Through reading infusion students can build scientific knowledge widely because students are provideda learning opportunity to obtain scientific knowledge from various sources as well as from various media information related to the subject matter [14]. Students can also build an argument toclarify an investigation result with the information obtained from reading material [14-15]. In line with these opinions, Muttaqin& Sopandi [1] states that there is a strong relationship between reading with critical thinking skills. In reading process, there should be tactics and strategies to understand literature well [14, 16-17]. A strategy that combines reading and writing during the phase-before-during and after reading [16] is the SQ3Rmethod (Survey, Question, Read, Recite and Review). This reading method can be incorporated with discovery learning on verification phase. At this stage, the students perform reading activities after collecting data through experiments and analyzing the experimental data.

Based on the above rationale, this study was conducted to obtain information about middle school students' critical thinking skills improvement as an effect of the implementation of reading infusion-loaded discovery learning model in science instruction. The learning object in this study is the concept of vibration, waves and auditory senses. Through this study we may provide useful information about an appropriate learning model in teaching science as an effort to improve student learning outcomes, particularly the critical thinking skills.

2. Researh methodology

This study used a quasi-experimental method with the pretest-posttest control group design [18]. Fifty five of eighth-year students in the academic year 2015/2016 at Tasikmalaya regency was involved in this study. They were divided into the experimental group (28 students) who had reading infusion-loaded discovery learning model and control group (27 students) who had conventional learning.

The learning phases of reading infusion-loaded discovery learning model includes the following phases: (1) stimulation, (2) problem statement; (3) data collection; (4) data processing; (5) verification; and (6) generalization. Reading infusion with SQ3R reading technique (survey, question, read, recite and review) performed at the verification phase. Conventional learning is a learning process that applies a verification lab-based direct learning model. Laboratory activities based on the laboratory guide books used in the school.

Students' critical thinking skills are measured using a critical thinking skills test in multiple-choice with reason format questions. The test was 28 items encompassing three essential concepts, vibration, waves and auditory senses, which was preceded by construct and statistical validity. Scoring rubric for the critical thinking skills test was adapted from Stiggins [19] with the minimum and maximum score range from 0 to 4. The critical thinking test refers to the five indicators of critical thinking skills proposed by Ennis [3,6]. Critical Thinking Skills (CTS) measured in this study include indicators skills (sub-indicator): (1) elementary clarification (focusing questions, analyzing arguments, asking and answering questions about an explanation or challange); (2) basic support (considering a source's credibility); (3) advance clarification (defining the term and considering the definitions); (4) inference (making and considering induction, making and considering deduction, and making decisions and considering the results), and (5) strategy and tactics (deciding an action). The critical thinking skills improvement was measured by using students' pretest and posttest score that analyzed using normalized gain score formula proposed by Hake [20].

3. Results and discussion

Critical thinking skills test was administered before and after the given instruction in both groups. Recapitulation of the average pre-test, post-test and n-gain score for experimental and control groups are presented in Figure1.

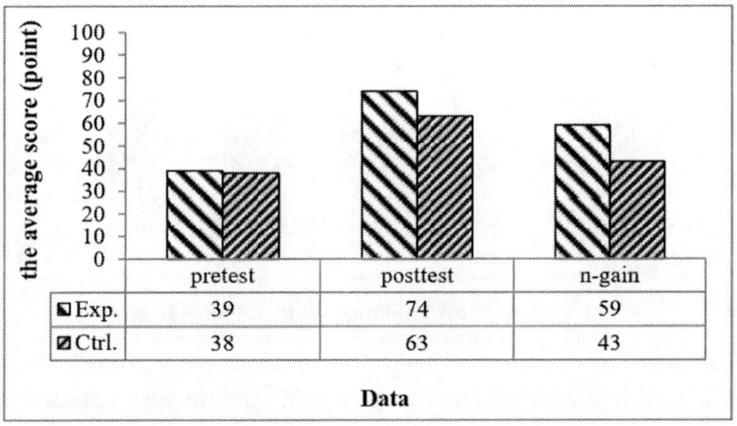

Figure 1. Comparation of the average pretest, posttest, and n-gain score for experimental and control groups.

Figure 1 shows that there is no significant difference on the average of initial critical thinking skills between both groups. The average score of the two groups before the instruction differs only one point, while after instruction differ 11 points. It means that both groups have similar initial critical thinking skill before instruction and there are different changes in critical thinking skills of both groups after instruction. Both groups show medium improvement with the average n-gain score differ 0.11 points, in which n-gain score of the experimental group (59) is higher than n-gain score of the control group (43).

An analysis of the improvement in each critical thinking skills indicators shows that the highest average pre-score for the experimental and control group are respectively 54 and 46 with a difference of 8 points, on the similar advance clarification indicator. While the lowest average prescore is 34 for the experimental group on inference indicator and 35 for the control group on elementary clarification indicator. The highest average post-score is 85 for the experimental group on advance clarification indicator and 73 for the control group on basic support indicator. While the lowest average post-score for experimental and control group are respectively 72 and 58 on the similar inference indicator for both. The highest average n-gain score for the experimental and control groups are respectively 62 on strategy and tactics indicator and 53 on the basic support indicator. Both groups demonstrate a medium improvement. Comparison of the average n-gain score for each critical thinking skills indicators are presented in Figure 2.

Figure 2 shows that the average improvement of each critical thinking skills indicators in the experimental group sequentially from highest to lowest is strategies and tactics, advance clarification, basic support, inferences, and elementary clarification. As for the control groups is basic support, advance clarification, elementary clarification, strategies and tactics, and inference. Overall, the experimental group improvement is higher than the control group.

The difference in critical thinking skill improvement between two groups was analyzed using the mean difference test, which preceded by the normality and homogeneity test. Mean difference test of the students' prescore showed that there was no significance difference in the initial critical thinking skills between the experimental and control groups (sig = 0.446 $> \alpha = 0.05$). While after instruction, there was a significantly different in their critical thinking skills between the experimental and control group ($Z_{count} = -4.767$; sig = 0.000 $< \alpha = 0.05$). The average n-gain score of students' critical thinking skills addressed the significant differences between experimental and control group ($Z_{count} = -5.137$; sig = 0.000 $< \alpha = 0.05$). These mean that the implementation of reading infusion-loaded discovery learning model could further improve students' critical thinking skills than the conventional learning.

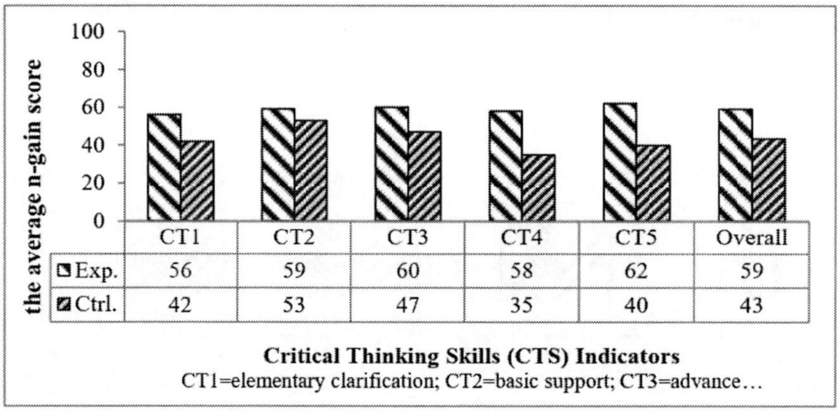

	CT1	CT2	CT3	CT4	CT5	Overall
Exp.	56	59	60	58	62	59
Ctrl.	42	53	47	35	40	43

Figure 2. Comparation of the average n-gain score in each critical thinking skills indicators for both groups.

A higher improvement in the experimental group than control group can be explained by the learning process that provides the opportunity for students to activate critical thinking skills during instruction. It also provides facilities for students to practice their critical thinking skills. Various studies previously mentioned that the discovery learning model can improve student's performance in learning [11, 21-26]. Discovery learning is a process of inquiry-based learning that will practice students' thinking skills gradually, moving from the basic thinking skills towards the higher-order thinking skills [27]. By activating thinking process during the learning process, it affects students' critical thinking skills improvement. Brasford & Donovan [7] states that by practicing critical thinking skills during learning, students are practiced to "how they think," not just to know "what they think". In contrast to conventional learning, students are likely to receive facts through a direct explanation and proven by an experiment. The weakness of verification laboratory that the student is rarely given the opportunity to experience the real scientific efforts in solving new problems. The learning process with discovery learning model provides an opportunity for students, from evoking the problem until a solution is all done by the students themselves [28].

In addition to the learning process, the differences improvement in both groups cannot be separated from reading infusion strategy. Reading infusion is reading activities using the SQ3R technique (survey, question, read, recite and review) conducted at the verification phase. The SQ3R method will make the students studying readings that involve critical thinking skills. It also provides opportunities to determine the important parts that will be the question in understanding the text. These questions then they answer based on the literature they read, so that at the same time the students perform reading and thinking. The SQ3R reading method gives students the opportunity to practice critical thinking skills [29]. This is in line with the opinion of the Syah [30] states that by using SQ3R, student learning outcomes such as critical thinking can be more satisfying, because the students read active and targeted directly at the essence or implicit and explicit contents in the reading text. Reading texts or literature can basically develop critical thinking skills [31]. The previous studies showed that reading can stimulate discussion and inquiry activities [32]. In line with these opinions, reading gave a positive impact on learning outcomes [33]. Hermida [34] and Jones [3] states that the reading activity will stimulate students to use higher order thinking skills. Reading will assist students to fill their own knowledge that will improve their learning outcomes, such as their critical thinking skills [35]. Assigning students to read with SQ3R method that elaborated with innovative learning model will obtain the critical thinking skills improvement. Kaleliogu & Gulbahar [4] states that the innovative learning model, such as reading infusion-loaded discovery learning model, will have an impact on students' critical thinking skills that last a long time and touches all students.

4. Conclusion

The findingsshowed that the average improvement of students' critical thinking skills for the experimental and control group are respectively 59 and 43, in the medium category. There was a significant difference in the improvement of students' critical thinking skills between the experimental and control group (Z_{count} = -5.137; sig = 0.000 $< \alpha$ =0,05). Thus, the implementation of reading infusion-loaded discovery learning model could further improve middle school students' critical thinking skills than conventional learning.

In this study, the designed instruction still requires plenty of time than the usual in the classroom. It need more improvement in designing and planning each learning phase's efficiency. In addition, a continous efforts are important to develope an appropriate learning model by utilizing the potential of discovery learning and reading infusion in science instruction, particularly in different learning objects or themes.

5. References

[1] Muttaqqin A and Sopandi W 2015 *Edusentris: Jurnal Pendidikan dan Pengajaran* **2** 2116-125
[2] Kemdikbud (a) 2013 *Materi Pelatihan Guru Implementasi Kurikulum 2013 SMP/MTs IPA* (Jakarta: Badan Pengembangan Sumber Daya Manusia Pendidikan dan Kebudayaan dan Penjaminan Mutu Pendidikan Kementerian Pendidkan dan Kebudayaan RI) p 213
[3] Costa A L 1985 *Developing Minds: A Resource Book for Teaching Thinking* (Alexandria: Association for Supervision and Curriculum Development) p 45
[4] Kalelioglu F and Gulbahar Y 2014 *Education Technology & Society* **17** 248-58
[5] Kereluik K, Misrha P, Fahnoe C and Terry L 2013 *Journal of Digital Learning in Teacher Education* **29** 127-40
[6] Tawil M and Liliasari 2013 *Berpikir Kompleks dan Impelementasinya dalam Pembelajaran IPA* (Makasar: Badan Penerbit Universitas Negeri Makasar) p 4
[7] Quitadamo I J, Faiola C L, Johnson J E and Kurtz M J 2008 *CBE-Life Science Education* **7** 327-37
[8] Fritjers S Gt D and Gert R 2012 *Learning and Instruction* **18** 66-82
[9] Kemdikbud (b) 2013 *Peraturan Menteri Pendidikandan Kebudayaan Republik Indonesia Nomor 65 Tahun 2013 Tentang Standar Proses Pendidikan Dasar dan Menengah* (Jakarta: Kementerian Pendidikandan Kebudayaan RI)
[10] Sutiadi A 2008 *Jurnal Geliga Sains* **2** 11-6
[11] Anyafulude J C 2013 *Journal of Educational and Social Research* **3** 6105-111
[12] Muttaqin A 2015 *Pengaruh Model Discovery Learning dengan Sisipan Membaca Kritis terhadap Penguasaan Konsep dan Berpikir Kritis Siswa SMP pada Konsep Energi dalam Sistem Kehidupan* (Bandung: Tesis Universitas Pendidikan Indonesia)

[13] Alfieri L, Brooks P J and Aldrich N J 2011 *Journal of Educational Psychology* **102** 11-8
[14] Fang Z and Wei Y 2010 *The Journal of Educational Research* **103** 4262-273
[15] Hanrahan M 2009 *Eurasia Journal of Mathematics, Science & Technology Education* **5** 3289-304
[16] Martin G T 2002 *Reading, writing, and comprehending: encouragingactive active reading in the science classroom* The Science Teacher (Virginia: National Science Teachers Association) pp 56-59
[17] Jhonson B E and Zabrucky K M *International Electronic Journal of Elementary Education* **4** 119-31
[18] Fraenkel J R, Wallen N E and Hyun H H 2012 *How to Design and Evaluate Research in Education 8th edition* (The McGraw Hill Companies. Inc.) p 270
[19] Stiggins R J *Student-Centred Classroom Assesment 2nd edition* (New York: Macmillan College Publishing Company, Inc. Meril) p 246
[20] Hake R R 1998 *American Journal Physics* **66** 164-74
[21] Swaak J, de Jong T and Van J W R 2004 *Journal of Computer Assisted Learning* **20** 225–34
[22] Cohen M T 2008 *NERA Conf. Proc. (Rocky Hill)* vol 30 (Storrs: UCONN Library)
[23] Balim A G 2009 *Eurasian Journal of Educational Research* **35** 1-20
[24] Onevweghede O W 2010 *Journal of Research in Education and Society* **1** 131-39
[25] Widiadnyana I W, Sadia I W and Suastra I W 2014 *E-Journal Program Pascasarjana Universitas Pendidikan Ganesha* **4**
[26] Suryadi S M and Wulan A R 2015 *Proc. of Int. Seminar on Mathematics, Science, and Computer Science Edication (Bandung)*(Bandung: FPMIPA UPI)
[27] Wenning C J 2011 *Journal Physic Teacher Education Online (JPTO)* **6**
[28] Rustaman N Y, Dirdjosoemarti S, Yudianto S A, Yusnani A, Subekti R, Rochintaniawati D and Nurjhani M K 2005 *Strategi Belajar Mengajar Biologi* (Malang: Penerbit Universitas Negeri Malang) p 96
[29] Tomasek T 2009 *International Journal of Teaching and Learning in Higher Education* **2** 1127-132
[30] Syah M 2010 *Psikologi Pendidikan Dengan Pendekatan Baru* (Bandung: PT Remaja Rosdakarya) p 129
[31] Khatib M and Nazari O 2012 *Journal of Comparative Literature And Culture* **1** 29-33
[32] Zmach C C, Sanders, Patrick J D, Dedeoglu H, Charbonnet S, Henkel M, Fang Z, Lame L L and Pringle R 2007 *Infusion Reading Into Science Learning: Researchers and Middle School Teachers Teamed up to Give Students Effective Reading Instruction in Science Class* (Educational Leadership) pp 62-6
[33] Oliveras B, Márquez, C and Sanmartí N 2013 *International Journal Science Education* **35** 6885-905
[34] Hermida J 2009 *The International Journal of Research and Review* **3** 20-30
[35] O'Reilly T and McNamara D 2007 *American Educational Research Journal* **44** 1161-169

Process-based Assignment-Setting Change for Support of Overcoming Bottlenecks in Learning by Problem-Posing in Arithmetic Word Problems

A A Supianto[1, 2], Y Hayashi[1] and T Hirashima[1]

[1] Graduate School of Engineering, Hiroshima University, 1 Chome-3-2 Kagamiyama, Higashihiroshima, Hiroshima Prefecture 739-8511, Japan
[2] Department of Informatics/Computer Science, Universitas Brawijaya, Jl. Veteran No.8t, Malang 65145, Indonesia
Email: afif@lel.hiroshima-u.ac.jp

Abstract. Problem-posing is well known as an effective activity to learn problem-solving methods. Monsakun is an interactive problem-posing learning environment to facilitate arithmetic word problems learning for one operation of addition and subtraction. The characteristic of Monsakun is problem-posing as sentence-integration that lets learners make a problem of three sentences. Monsakun provides learners with five or six sentences including dummies, which are designed through careful considerations by an expert teacher as a meaningful distraction to the learners in order to learn the structure of arithmetic word problems. The results of the practical use of Monsakun in elementary schools show that many learners have difficulties in arranging the proper answer at the high level of assignments. The analysis of the problem-posing process of such learners found that their misconception of arithmetic word problems causes impasses in their thinking and mislead them to use dummies. This study proposes a method of changing assignments as a support for overcoming bottlenecks of thinking. In Monsakun, the bottlenecks are often detected as a frequently repeated use of a specific dummy. If such dummy can be detected, it is the key factor to support learners to overcome their difficulty. This paper discusses how to detect the bottlenecks and to realize such support in learning by problem-posing.

1. Introduction

Problem-posing is recognized as an important component in the nature of mathematical thinking [1]. Problem-posing involves generating new problems and questions aimed at exploring a given situation as well as reformulating a problem during the course of solving a related problem [2]. Recently, technology-enhanced approaches have been used to realize learning by problem-posing in a practical way. An interactive learning environment for problem-posing in arithmetic word problem as sentence-integration has been developed [3], named Monsakun. A long-term evaluation of the system was carried out, and the study confirmed that it was interesting and useful for learning [4]. In problem-posing via sentence integration, several simple sentences are provided to a learner. The learner, then, selects the necessary sentences and arranges them in an appropriate order. This approach makes simple and goal-oriented problem-posing tasks possible even for lower elementary students while maintaining its value as a viable learning method. Nevertheless, in some cases, the environment provides an assignment that makes learners feel difficulty so that it requires figuring out the assignment setting for helping learners posing meaningful problems. Thus, there is a need to support learners in problem-posing activities. This can be achieved by generating problem-posing assignments automatically based on the history of their activities, and hence, the assignments would meet the learners' level of understanding.

In this paper, we propose a method to help learners in problem-posing process on the learning environment called Monsakun by changing the assignment setting based on their bottlenecks of thinking. Supianto et al. [5] conducted detection of trap-states by tracing problem-posing activity sequences from the system's data-logs in a problem-posing learning environment. Moreover, the trap-state was found in the study is conducted by a technology in order to detect the bottlenecks of learners based on their activity, instead of pre-defined errors [6]. Detecting trap-state proposed by [5] was conducted in the group of learners. However, in this study, detecting trap-state is implemented for each learner in order to support individually based on the process of learning by problem-posing in arithmetic word problems. In our proposed method, the log data of learners' activity on Monsakun is collected and formulated. Next, trap-state as learners' bottlenecks in thinking is discovered. Finally, a new setting of the assignment is created according to the trap-state found. The provided assignment is expected to overcome from learners' bottlenecks in problem-posing activity and have a higher level of acceptance by both teachers and learners.

The paper is organized as follows. The next section presents problem-posing activity in Monsakun followed by describes the proposed mechanism support, which includes the detection of bottlenecks and changing assignment setting. Finally, we conclude by summarizing the paper and offer the suggestion for future work in the last section.

2. Problem-posing activity in Monsakun

Monsakun was designed as an interactive learning environment for problem posing as sentence-integration. In the problem-posing activity by using Monsakun, learners do not create their own problem statements, however, they are required to interpret the sentence cards and integrate them into one problem. The interface of Monsakun is shown in Figure 1. A learner is required to integrate three sentence cards from five or six available sentence cards in order to build a problem in the three card slots based on the requirement for each assignment. There are more than three cards provided to the learners, which means that not all the cards included are necessary to complete the required problem. We call such cards "dummy cards," and they are intentionally included to test learners and check for their understanding. For example, an overlooking of the story type or confusion about the formula may lead to problems completing the exercise correctly. The learner selects several sentence cards and arranges them to pose a problem in the correct order. Putting and removing sentence cards are the basic actions of learners, and Monsakun records the result of actions, which are combinations of sentence cards.

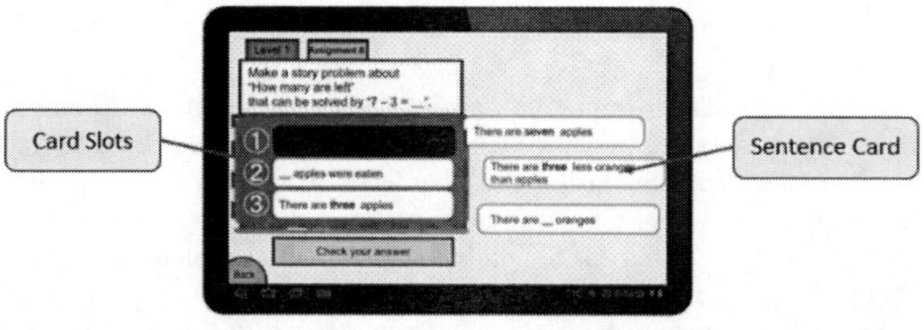

Figure 1. Interface of Monsakun for one operation of addition and subtraction.

The task model of problem posing as sentence integration on Monsakun has been developed based on the consideration of problem types in the triplet structure model [7]. This model is an arithmetical word problem model that is solved by one arithmetical operation composed of three quantities: operand, operand, and result quantity. In this model, all word problems are composed of two "existence sentences" and one "relational sentence." Although an existence sentence can be used in any story type, a relational sentence is used only in one specific story type. There are four types of story in arithmetic word problem of addition and subtraction: 1) combination, 2) increase, 3) decrease, and 4) comparison. Combination and comparison stories are story problems that the order of sentence cards in the slot is not necessary, while increase and decrease are story problem that the order of cards is restricted [8].

3. Method

In this section, we summarize the formulation of the problem before discussing the mechanism of support. Next, the bottlenecks in the process of problem-posing are formulated, and procedure of detection of bottlenecks and support system for overcoming them is elaborated.

3.1. Problem formulation

In this study, each composition of sentence cards is treated as a "state" in the problem-posing process by the learner. All possible states can be obtained by combining all the available sentence cards which divided into three categories: incomplete states, mistaken states, and correct answer state. Each state represents a basic unit of thinking, and a problem state space provides the range of thinking in a problem-posing assignment. Five main constraints to be satisfied by each posed problem have been devised [8], which are calculation, story type, number, objects, and sentence structure. When all five constraints are satisfied, the learner has succeeded in posing a correct problem according to the assignment requirements. When less than five constraints are satisfied, it shows that the learner has acquired a level of understanding in the structure of the arithmetic word problem; however, the final problem does not yet satisfy the requirements. If there are no constraints satisfied by the learner, it shows that the learner is unable to understand the structure of the arithmetic word problem. In order to cover the measurement of the validity of states, we define three values for each constraint: -1, 0, and 1. The value of -1 means the constraint is violated, the value of 0 means the constraint is not violated, and the value of 1 means the constraint is satisfied. The validity of a state is calculated by summing all constraints value, and the number of violated constraints is obtained by counting how many constraints are violated.

Figure 2 depicts an illustrative example of processed log data of a learner's problem-posing activity using Monsakun in Level 5 Assignment 1. The story type of this assignment is about combination problem. The requirement of this assignment was: *Make a word problem about "How many are there in all" that can be solved by "8-3."* There are six sentence cards that could be used by the learner. The sentences for each card, from the first card index to the sixth, are: ***There are 3 white rabbits, There are _ black rabbits, There are 8 white and black rabbits altogether,*** *There are 8 white rabbits, There are 3 more white rabbits than black rabbits,* and *There are 3 brown rabbits*. In this assignment, the correct answer consists of sentence card #1, #2, and #3 (the sentences appearing in bold), while sentence card #4, #5, and #6 are identified as dummy cards. The example of learners' log data of this assignment, which is designed as combination story problem, will be used for explaining the procedure suggested in this paper.

The brown, red, and green rectangles are shown in Figure 2(a) indicate the incomplete states, the mistaken states, and the correct answer state, respectively. In this example, the total number of steps is 67 times and the total number of mistakes is 18 times, which indicate that the learner feels difficult to reach the correct answer, and hence, need to be supported. The blue nodes in Figure 2(b) indicate the states that had already been made by the learner. The yellow lines show the path of the learner's steps and the thickness of the link indicates how many steps from a node to another. Based on this information, it looks the learner got stuck in state 000, 004, 024, 045 and 245. The learner seems struggle to achieve the correct answer, state 123, although in the end, the learner is able to achieve it through state 023. Therefore, in order to help them out of the impasse, a system that can support them according to learner's bottlenecks is conducted. The proposed system presents different assignment setting for the learner based on the detected bottlenecks. This setting expects the learner could overcome the bottlenecks and could accomplish the assignment smoothly.

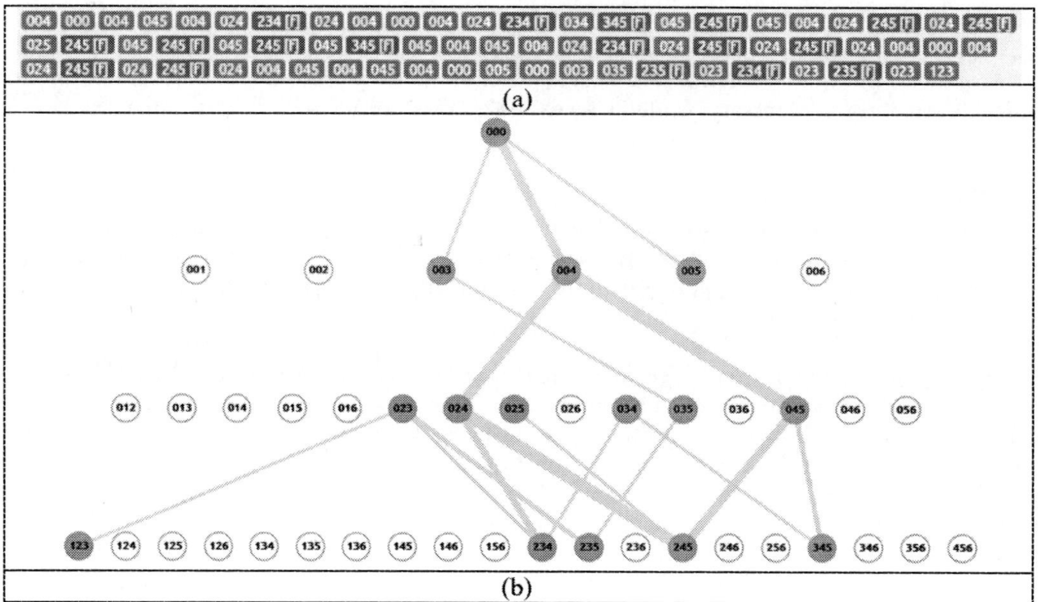

Figure 2. An example of log data of a learner's problem-posing activity in Monsakun. (a) The sequence of states. (b) Graphical representation.

3.2. Proposed procedure of process-based support system

In order to promote adaptive learning, we must constantly detect the learner's bottlenecks and possibly redirect the assignment setting accordingly. Our goal is to provide assignment setting in the system based on previous learner's steps recorded in the log data. These learner's steps are known as states. A state contained at least one dummy card where it could lead the learner to do many steps to get to the correct answer and reached by many actions as a "trap-state." In this paper, such trap-state is defined as learner's bottlenecks.

The proposed support system consists of two functional modules: *detection of bottleneck* and *recommendation setting*. The detection of bottleneck module responsible to find the learner's bottlenecks. It processes the log data of learner's activities and provides trap-state(s). The recommendation setting module generates new assignment setting to a learner based on the trap-state(s) found in the previous stage. As the result, we remove the frequently repeated use of a specific dummy card that makes learner stuck in posing the problem and build new assignment by omitting this dummy. The detail proposed procedure of process-based support system is given as follows:

- Formulation the problem-posing process: the sentence cards are encoded with indexing number. When the slot is still empty, index = 0 is implemented. For instance, when the learner poses the problem by selecting sentence card #1 and arranges it to the second slot, state 010 is generated.
- Tracing learner's activity: Every state stored by the system sequentially produce an order of each state called a "sequence of states." The sequence is then used in the next processes to detect the bottlenecks. An example of a sequence that consists of 67 states is shown in Figure 2(a).
- Finding trap-state: A state found as trap-state when it has the highest trap-value obtained from support value and distance value [5]. When the highest score happens in more than one state, then all states with the highest score will be selected as trap-states.

- Selecting the trap-card (the dummy): the system will select the trap-card in the trap-state(s). If there are more than one trap-cards, the system will sum-up the trap-value of the dummies and select the dummy having the highest score.
- Build new assignment by omitting the dummy obtained from the previous step. This new assignment setting will be inserted to the Monsakun.

4. Changing assignment setting

Removing the dummy, which is from the beginning intended as a distractor and eventually becomes the cause of the bottlenecks, might be able to overcome their impasse. We propose a method to change the characteristic of assignment based on the learners' bottlenecks. The changes of the assignment are generated from the previous one in order to guide the learners. The characteristic of current assignment has caused learners difficult to get the correct answer. Therefore, by changing the characteristic of assignment to be less difficult, having a good understanding of the problem's structure is expected. In this study, the characteristic of the assignment is defined as the violation of the constraints for all states generated in the states space

Figure 3 shows the proportion of the number of states that violate the constraints of Level 5 Assignment 1. It can be seen that the new setting when sentence card #4 is removed will help learners having difficulty in calculation constraint. The number of states in the calculation constraint of this setting is lowest than the other new settings. Thus, this setting is fit for learners that have bottlenecks in states related to the calculation constraints. On the other hand, the new setting when sentence card #5 is removed will help learners having difficulty in the story constraint because of the possibility of the violation of constraint about the story is decreased. While when the new setting in which sentence card #6 is removed will support learners that have difficulty in objects constraint because of the possibility of the violation of constraint about objects is decreased. Therefore, depending on the characteristic of the difficulty that faced by learners, the new setting would be generated.

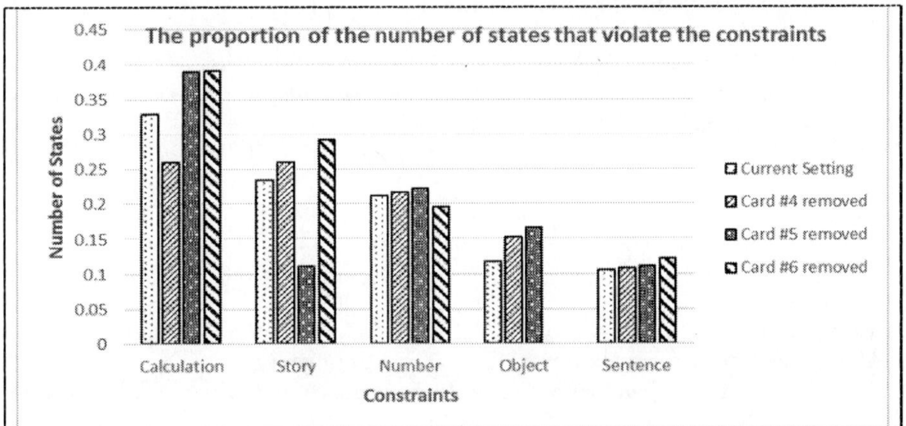

Figure 3. The chart of proportion of the number of states that violate the constraints of Level 5 assignment 1.

Based on the detection of the bottlenecks, the learner is trapped in state 024 during pose the problem in Level 5 Assignment 1. The trap-values of top-5 trap-states are state 024, 004, 245, 045, and 234 with trap-values are 144, 130, 90, 80, and 76, respectively. The difficulty in this assignment is that the learner confused about the gap between the required story type of combination and the numerical expression of subtraction (8–3). Although subtraction implies the story type of decrease and comparison, in this case, the learner must pose a problem of the combination story type. To complete this assignment, the learner needs to transform the numerical expression, "8–3," into the calculation expression representing a combination story, such as "3 + ? = 8," or "? + 3 = 8." Thus, the learner could assign existence sentence cards to "3" and "?". According to the trap-state 024, the learner seems kept trying to use sentence card #4 (*There are 8 white rabbits*). There was a decided tendency based on some sort of thinking that the number "8" in the numerical expression was considered to be an existence sentence, instead of the relational sentence. In this situation, the learner confused and stuck because the correct answer with the number "8" is the number in the relational sentence, sentence card #3 (*There are 8 white and black rabbits altogether*). Such kind of difficulty, it would be helpful for learner when the assignment is changed to the new setting when sentence card #4 is removed because the number of states in the calculation constraint of this setting is lowest than the other new settings.

5. Conclusion and future work

In this paper, we present a method of the detection of learner's bottlenecks in problem-posing activity on the learning environment and adaptively change the assignment setting based on the bottlenecks in order to support for overcoming them. We propose a method to change the characteristic of assignment which generated from the previous one based on the learners' bottlenecks in order to guide the learner to have a good understanding of the problem's structure. The method can generate recommendations of assignment-setting and it would be inserted into an existing problem-posing learning environment. The proposed method responsible for formulating problem-posing process from log data of learner's activities, finding a trap condition that made learners stuck in a condition where they would have to do more steps to reach the correct answer and building new assignments to the learner. For the future work, we plan to implement the proposed setting of assignments in the practical use and analyze the result.

6. References

[1] Kilpatrick J 1987 *Cognitive Science and Mathematics Education* ed A H Schoenfeld and Hillsdale (NJ: Lawrence Erlbaum) pp 123–147
[2] Silver E A 1994 *For the Learning of Mathematics* **14** 19
[3] Hirashima T, Yokoyama T, Okamoto M and Takeuchi A 2007 *Proc. of the 2007 Conf. on Artificial Intelligence in Education: Building Technology Rich Contexts that Work (Los Angeles)* (Amsterdam: IOSPress) p 254-261
[4] Hirashima T, Yokoyama T, Okamoto M and Takeuchi A 2008 *Proc. of ICCE 2008* p 817 *(Taiwan)*
[5] Supianto A A, Hayashi Y and Hirashima T 2015 *Proc. of the 23rd Int. Conf. on Computers in Education (Hangzhou)* (New York: Springer) p 85
[6] Ben-Naim D, Bain M and Marcus N 2009 *Int. Conf. on Educational Data Mining (Raleigh)* p 21
[7] Hirashima T, Yamamoto S and Hayashi Y 2014 *Int. Conf. on Human Interface and the Management of Information (crete)* (New York: Springer) p 42
[8] Supianto A A, Hayashi Y and Hirashima T 2016 *Research and Practice in Technology Enhanced Learning* **11**

Acknowledgments
This work was supported by JSPS KAKENHI Grant Number 15H02931.

IOP Conf. Series: Journal of Physics: Conf. Series **812** (2017) 012005 doi:10.1088/1742-6596/812/1/012005

Development virtual test of reaction rate based visual perceptual skills to measure students' mastery concept

L. Valantika[1], H. Firman[2], Nahadi[2]

[1]*Sekolah Pascasarjana, Universitas Pendidikan Indonesia, Jl. Dr. Setiabudi no 229, Bandung 40154, Indonesia*
[2]*Departemen Pendidika Kimia, Universitas Pendidikan Indonesia, Jl. Dr. Setiabudi no 229, Bandung 40154, Indonesia*
E-mail: *Valantikalutfi@yahoo.com*

Abstract. This study aims to obtain a valid and reliable virtual test of reaction rate based visual perceptual skills to measure students' mastery concept. The test has been developed has the form of virtual test, consists of texts, drawings, graphs, animations, videos, sounds, and images. The method of this study was development and validation method. Data obtained from the result of validation, testing of virtual test, and interview of students which include content validity, reliability, level of difficulty, distinguishing ability, and the transcription of interview students. The results showed that the virtual test is valid from 33 items developed are 32 items that have a CVR value between 0.6 to 1.0 which indicating items both in terms of content validity. Thirty-two items that have been declared valid tested on 177 students of Eleventh-Grade to calculate the reliability of the test, and Cronbach alpha values obtained for the entire items of 0.734 which indicates a high level of consistency of measurement is acceptable. Virtual test of rate reaction based visual perceptual items has distinguishing ability of 0.35, that is indicated good sufficiency category and level of difficulty of 0.45, that is indicated medium category. After being evaluated of distinguishing ability from 32 items of virtual test, three items were rejected. From result of interviews, the students responded positively to the virtual test based visual perceptual on the concept of rate reaction which is developed viewed from sides of performance, grammar of items, feasibility of access and operation, advantages and time of doing.

1. INTRODUCTION

In these '21st century', the assessment as an essential component and affect the learning process, assessment should be changed from the traditional to tecnology assessment. The use of computer technology to conduct assessments to students in schools has managed successfully to drive a high measure of validity in the result, more interactive and wishes in accordance with the measurement objectives to be achieved, in other words more contextual measurement[1]. Recently, information and communication tecnology has been used in assessment. Computer-based systems' specific value can be seen in their efficiency while automatically marking responses, in providing immediate learner feedback on any potential weaknesses, and in being available upon demand and, in the case of larger learning groups, simultaneously accessible by all pupils[2]. The computer is a medium of instruction to visualize different facts, skills, concepts and computer also displays images that move in

IOP Conf. Series: Journal of Physics: Conf. Series **812** (2017) 012005 doi:10.1088/1742-6596/812/1/012005

accordance with its requirements[3]. So that teachers as educators with computer based test can measure the competence of more established than paper and pencil test.

Van Merrienboer[4], one of the leading educational researchers in problem solving, emphasizing the importance of using a computers as an assessment and instructional tool, since the tool can simulate real-world problems, which are ill-structured and complex in nature. Besides assessment by Computer Based Test, has advantages which the tests can explore the computer so that the test form with multimedia can be images, text, graphics, animation, video and others. One of the computer based test that has been and should be developed further is a virtual test. Virtual test is a test form using software that can be carried out both online and offline[5]. From the above definition, it can be concluded that the virtual test is a form of testing by using multimedia computer such as images, text, graphics, diagrams, animations, and video. Mushonev states that the use of visual forms in the test questions will help evaluators to measure students'cognitive abilities were higher compared to just use a statement or just questions and can train simultaneously measuring the ability of the students' science process[6]. The use of media such as text, images, audio, animation, and video better than the media[7]. Multimedia effects provide sufficient retention great for students. video can display a phenomenon which is difficult to be seen directly. This is so that students better understand the statement items are delivered.

Virtual test is developed by eight visual perceptual skills are developed Rocford and archer[8] with the aim to measure the level of student mastery concepts on the reaction rate material. There are several research of virtual tests have been developed, but the virtual development of visual perceptual skills test based on specific chemical content has never been developed. Research that have been developed include Firman and Rusyati (2014) develop a virtual test used to measure critical thinking skills of junior high school students on the theme of human diseases, Saukani (2015) in his research to develop a virtual test used to measure the ability of decision-making high school students on the material solution acid-base, Anggarjati (2015) in his research to develop a virtual test used to measure critical thinking skills of high school students on the material Chemical Equilibrium, and Christian (2009) has developed VSCs (Visual Spatial Chemistry Spesific). VSCs are assessment tools developed to assess visual perceptual skills of the students, where these skills are involved in the formation of a mental image of visual objects. Christian VSCs focuses on the representation of the molecule, and have not focused on a specific chemical content.

Chemistry requires the users, can visualize a world picture macroscopic, microscopic and symbolic. Visual perceptual skills is an integral part of three of the world picture. Visual perceptual skills are fundamental premise that students in learning chemistry is always exposed to view objects in space, where students have formed a mental image of the object[8]. In other words, three level representation of the chemical can be developed in a virtual instrument test. Because the conceptual understanding of chemistry requires the ability to represent and interpret the problems of chemicals in the form of representation macroscopic, microscopic, and symbolic simultaneously[9] .

Visual perceptual capabilities need to be developed on any material, but any material in chemistry has the characteristics of a particular concept. There are several materials that can accommodate the needs of evaluation in measuring students' mastery of concepts in which the material terms with the concepts of real, symbolic and abstract visual perceptual skills require students to understand it, one of which is the rate of reaction. Judging from the characteristics of the concept, the reaction rate has kind of a concept based on the principle of the critical attributes of the abstract but concrete examples. Under these conditions, the reaction rate of the material requires a level of understanding and visualizing high enough, so that it can be used in developing the items that can encourage students to develop visualization.

This study aims to produce a virtual test of visual perceptual skills based on the material reaction rate is valid and reliable to measure students' mastery of concepts. Based on previous research, virtual test based visual perceptual on specific chemical content has never been studied. Though virtual test based visual perceptual is needed to help students understand the statement items that contain high enough understanding and visualization, so that needs to be developed virtual test based visual perceptual test on the chemical content that requires understanding and visualizing them is the concept of a high enough rate of chemical reactions.

2. RESEARCH METHODOLOGY

In general, the research method is defined as a scientific way to obtain valid data with the purpose and usefulness. The method used in this research is Development and Validation Method. Development and validation method is used to assess test developed whether it is feasible to use and can measure the aspects measured[10].

This study seeks to develop and validate a virtual visual-perceptual-based test to measure the level of student mastery of concepts related to the rate of reaction material. Rochfor and Archer theoretical framework consisting of the eight visual-perceptual skills as a basis or foundation (framework) to develop the indicators items. Analysis of the chemical reaction rate of material from the standpoint of visual-perceptual skills done to develop this assessment tool. Where visual-perceptual characteristics of eight visual-perceptual framework Rochford and the Archer, each revealed to be indicator items that collaborated with the indicator reaction rate material. The character of the skills identified clearly delineated, and it is suitable to use in developing the assessment tool to measure the level of mastery of concepts rather overlooked in educational assessment is the process of high-level thinking skills (High order thinking skills-HOTS). With a virtual visual-perceptual-based test can be developed an assessment tool to measure the level of student mastery of concepts included high-level thinking skills consisting of C4 levels, C5 levels, and C6 levels.

Stages of implementation of this study were divided into four main phases namely planning studies (including analysis of competency standard, basic competence, analysis of the concept, representation reaction rate, material analysis reaction rate from the point of view of the skills of visual perceptual), stage of development (including the preparation of the grating and analysis of items virtual test), the validation phase (includes validation and analysis of CVR), and the last stage is the pilot phase of virtual test rate of reaction.

The method used to measure the validity of the content is a method developed by Lawshe[11]. Mechanical CVR (content validity ratio) is used CVR proposed by Lawshe as interpreted by Wilson[12]. Reliability for determining the level of consistency in this study using internal consistency reliability. Reliability index is done for this sort of test, where scoring is done if it is true "1" and if one of "0". This study using coefficient Cronbach alpha developed by using SPSS 16.

3. RESULTS AND DISCUSSION

Virtual test the reaction rate based visual perceptual developed is thirty-three items in the form of simple multiple choice. Based on the assessment validator which consists of 10 people validator (7 lecturers and 3 chemistry teacher), thirty two (32) are valid (accepted) by the CVR value of 0.600 to 1 and one (1) is invalid (rejected) of CVR value (0.400) below

CVR critical value (0.520). Of the 33 items based virtual test of reaction rate based visual perceptual that is validated by 10 expert judgment there are 32 items that valid (acceptable. There is one problem that is invalid (rejected), which is about the number 32 by the value of the CVR (CVR = 0.44) below the critical value of the CVR (CVR = 0.520). CVR analysis results for each item on the virtual test based visual perceptual are shown in Figure 1.

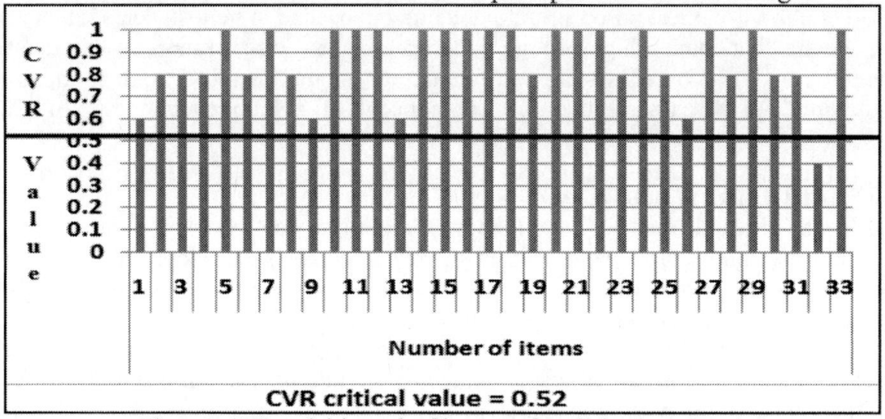

Figure 1. CVR analyisis result of items test based visual perceptual

Virtual test developed has CVI (Content Validity Index) of 0.897, it means that there is a match between the virtual test items based visual perceptual skills developed indicators to measure the level about students' mastery of concepts. Virtual test of reaction rate based visual perceptual that are 32 questions and has been revised subsequently tested ready to pitch. Items that have been validated indicate that the items developed are in accordance with the indicators to measure the level of students' concepts mastery.

In this study, the reliability is used internally consistent reliability. Formulation to determine the reliability index is done to sort of test where scoring is done if it is true "1" and if either "0" by using Cronbach's alpha coefficient developed by Cronbach. Determination of reliability were performed using SPSS 16 by determining the value of Cronbach 's alpha. Cronbach 's alpha value of virtual test visual perceptual reaction rate based on the can at 0.734. Based on Table 1 were developed Jacobs and Chase, this means that this instrument is acceptable and are included in the category of "high".

Table 1. Classification of reliability coefficients

Correlation coefficient	spesification
0.00 – 0.19	very low
0.20 – 0.39	Low
0.40 – 0.59	Sufficient
0.60 – 0.79	High
0.80 – 1.00	very high

Distribution characteristics of matter obtained 35 % items with the characteristics of the visual skills of association, 10 % items with characteristic visual skills constancy, 14 % items with characteristic visual skills discrimation, 7 % items with the skill characteristics of visual figure ground, 17 % items with characteristics of visual memory skills, 7 % items with characteristic visual skills orientation, and 10 % items with characteristic visual sequencing skills.

The advantages of virtual test of reaction rate based visual perceptual is a measurement tool that can be used in the measurement and improvement of skills including visual perceptual skills, and aligned with 21st century learning where feedback is timely, meaningful to learners and teachers. Virtual test of reaction rate based visual perceptual skills displays those items that can visualize various facts related to the concept of reaction rate, and the statement displays a matter involving dynamic images. virtual test of reaction rate based visual perceptual involves multimedia such as text /picture/video/animation is a visualization of the fact concepts of reaction rate. The concept of the macroscopic to the concept of reaction rate as formation of deposits is a concrete concept, then the stem of item it directly demonstrated with video. Such visualization can stimulate more students to achieve a high level of understanding of the concept of chemistry. Figure 2 shows an example of virtual items based visual association test .

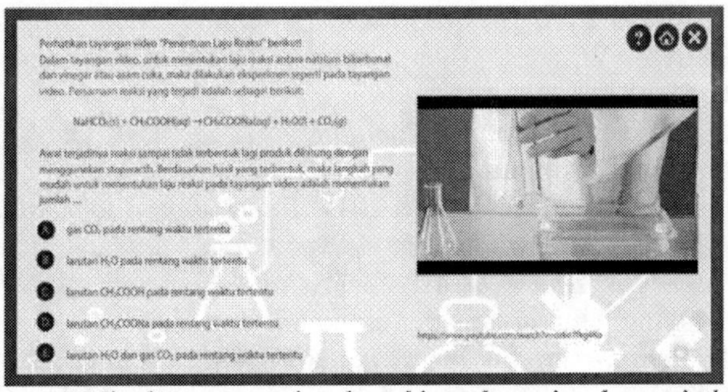

Figure 2. The items were developed based on visual association

Association visual skills are skills that involve abilities related to the concept represented in the image and the written word[13]. Items shown in Figure 1 based visual association, namely the statement items (stem) question about based on the concept that is represented in the image and the written word. While visual sequencing skills reflect the student's ability to remember a sequence of object shape. Examples of items that are developed by visual sequencing is shown in Figure 3.

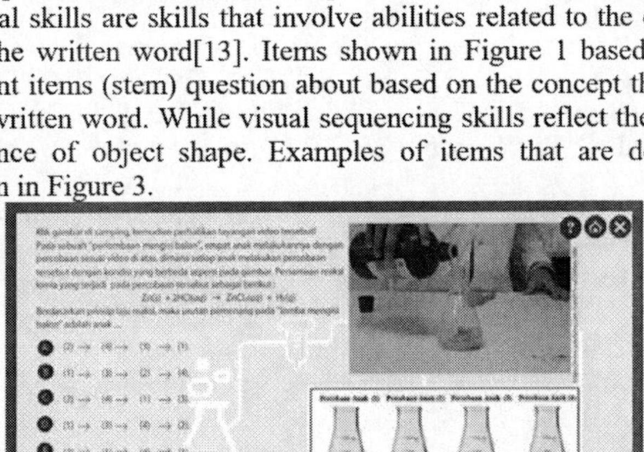

Figure 3. The items were developed based on visual sequencing

Based on the interview can be concluded that the information in the form of video, images, graphics, animation or experimental data contained in items very help students in understanding the statement items. Based on interviews with students, use of multimedia on the virtual test of reaction rate based visual perceptual such as video, images, graphics, animations or interactive experiment data easier for students to understand the items. the use of virtual test can be said to be almost 0 %, for items in the virtual test is made by means of randomized, where each student will get a different order of questions. The use of virtual tests help teachers in the effectiveness and efficiency of test examination time for virtual test software is already equipped with a slide to the results of the evaluation in the form of scores acquisition and analysis results from each answer items .

4. CONCLUSION

The results showed that virtual test of rate reaction based visual perceptual (VTVP) a measure that is valid and reliable. VTVP so it can be used to measure students' mastery of concepts on the concept of rate of reaction. Use of multimedia on the virtual test of reaction rate based visual perceptual such as video, images, graphics, animations or interactive experiment data easier for students to understand the items.

5. REFERENCES

[1] Scalise K and Gifford B 2016 *The Journal of Technology, Learning, and Assessment* **4(6)**

[2] Krause M, Kienast S, Witteckc T and Eilks I 2013 *Chemical Education Research & Practice* **14** 345-353

[3] Coburn P 1985 *Practical Guide to Computer in Education 2nd.* (California: Addison-Wesley Publication Company Inc)

[4] Scherer R, Koppelt J M and Tiemann R 2014 *International journal of STEM Education*

[5] Firman H and Rusyati L 2014 *Virtual Test: Sebuah Student Center Software Sebagai Alat Ukur Berpikir Kritis Siswa SMP Pada Tema Penyakit Manusia* (Laporan PPKBK). Bandung: Universitas Pendidikan Indonesia

[6] Saukani M 2015 *Pengembangan dan Validasi Virtual Test untuk Mengukur Kemampuan Pembuatan Keputusan (Decision Making) siswa pada Materi Larutan Asam Basa* Tesis, Universitas Pendidikan Bandung

[7] Mayer R E 2001 *Multimedia learning (*New york : Cambridge University Press)

[8] Christian C M 2014 *Journal of Research in Science Teaching* **51(8)** 963–81

[9] Russell J W 1997 *Journal of Chemical Education* **74(3)** 330-4

[10] Adam W K and C. E. Wieman C E 2010 *International Joural of Science Education* **33(9)** 1-24

[11] Lawshe C H 1975 *A paper presented at Content Validity* Conference Held at Bowling Green State University **28** 563-575

[12] Wilson R F, Pan W, and Schumsky D A 2012 *Measurement and Evaluation in Counseling and Development* **45(3)** 197-210

[13] Christian C M 2009 *The Development and Validation of a Visual-Spatial Chemistry Specific (VSCS) Assessment Tool* Thesis North Carolina State University

Anodic Stripping Voltammetry with Pencil Graphite Electrode for Determination of Chromium (III)

S Wyantuti*, R A Hafidza, S Ishmayana and Y W Hartati

Department of Chemistry, Faculty of Mathematics and Natural Sciences, Universitas Padjadjaran, Jln. Raya Bandung-Sumedang km. 21 Jatinangor, Sumedang, West Java, 45363, Indonesia
*Corresponding author: santhy.wyantuti@unpad.ac.id

Abstract. Chromium is required as micronutrient that has roles in insulin metabolism and blood glucose level regulation. Chromium (III) deficiency can cause hyperglycemia and glycosuria. However, a high amount of chromium in body can cause allergic reaction, organ damage, and even death because of its toxicity. Chromium is commonly used in steel industries. Simultaneously with the development of industry, the waste disposal that can endanger environment also increased. Therefore, a sensitive and specific analysis method for chromium detection is required. Stripping voltammetry is one of the voltammetric methods that is commonly used for heavy metal analysis due to the very low limit of detection (sub ppb). The present study was conducted to develop an analysis method for chromium (III) determination using pencil graphite electrode. Quantitative determination was performed for chromium (III) which measured at -0.8 to +1.0 V with deposition time for 60 s and 50 mV/s scan rate. Stripping voltammetric analysis of chromium (III) using pencil graphite electrode gave linear range at 12.5 to 75 ppm with limit of detection of 0.31 ppm.

1. Introduction

Chromium (III) is one of micronutrient that is required to maintain normal physiological function. However, when chromium exceeding the permitted limit, it can cause negative effect such as disturbing immune system and oxidative damage to the cellular components [1, 2]. This also confirmed by The Minister of Health of Republic Indonesia decree No. 907/MENKES/SK/VII/2002 which stated that the threshold value for chromium (III) in water is 0.05 ppm. Therefore a method for chromium (III) determination in trace amount need to be developed.

Some methods for chromium (III) determination has been widely developed. Some methods that are commonly used for chromium (III) determination include: inductively coupled plasma-atomic emission spectrometry (ICP-AES) [3], Inductively coupled plasma-mass spectrometry (ICP-MS) [4], flame atomic absorption spectrometry [5], spectrofluorometry [6] and visible spectrophotometry [1]. However, to determine both chromium species, i.e. chromium (III) and chromium (VI), these technique still require preconcentration or speciation stage. It is very important to differentiate between chromium (III) and chromium (VI), because chromium (VI) is easily reduced to chromium (III).

A sensitive and specific analysis method for detection of chromium metal and can differentiate between chromium (III) and chromium (VI) is required [7]. Voltammetric method is one of analysis method that can be applied for various sample. This method has some advantages, including simple preparation of sample and fast analysis [8]. Sample preparation usually only involve dissolution without preconcentration and separation of major components, which eventually reduce error source with detection limit up to 0.1 μg/L and relatively inexpensive instruments. Beside of that, voltammetric method enable investigation of chemical species of heavy metal that can be done with other methods. By finding out the chemical species, toxicity of a sample containing heavy metal can also be determined.

Stripping voltammetric for chromium determination is currently developing. This method has advantages compared to other analytical techniques in that it has very low detection limit [9]. Stripping analysis conducted in two steps, which are deposition of analyte onto the electrode surface (preconcentration) and stripping, which is the release of deposit [10].

The material of working electrode influences the performance of voltammetric method. Carbon based electrode utilization is currently growing in electroanalytical field because it has some advantages including wide electrical potential range, low background current, inexpensive, inert and suitable for wide variety of sensors [10]. Therefore, in the present study, we develop a stripping voltammetric method for chromium metal determination using pencil graphite as electrode.

2. Materials and methods

2.1. Determination of Chromium (III) Solvent

Chromium (III) solution was made from analytical grade chromium chloride ($CrCl_3$) (Merck, ≥98%) diluted in three solvents (double distilled water, 0.01 M hydrochloric acid and 0.33M acetate buffer pH 5.0). The solvents were trialed by measuring the current at potential range of -0.8 to +1.0 V with 90 s deposition time, 10 s cleaning time and 20 mV/s scan rate. Solution of 0.33 M acetate buffer pH 5.0 was used as supporting electrolyte. This supporting electrolyte was used throughout all experiments conducted in this study.

2.2. Electrodeposition of Chromium (III)

To the voltammetric cell containing 3 mL of supporting electrolyte, 3 mL of 50 ppm chromium (III) was inserted followed by measurement of current generated using Metrohm® μAutolab potentiostat at scanning range from -1.5 to +1.0 V with 90 s deposition time, 10 s cleaning time and scan 20 mV/s rate. The reference electrode, counter electrode and working electrode used in the present study were Ag/AgCl, platinum and pencil graphite, respectively. The working geometric area was 0.314 cm^2. The detection of chromium (III) was performed in 0.33 M acetate buffer pH 5.0. All potentials given in this work are with respect to the Ag/AgCl sat. reference electrode.

2.3. Measurement with Cyclic Voltammetry

To the voltammetric cell containing 3 mL of supporting electrolyte, 3 mL of 50 ppm chromium (III) was inserted, followed by measurement of current generated using Metrohm® μAutolab potentiostat at scanning range from -1.000 to +1.000 V with 90 s deposition time, 10 s cleaning time and 50 mV/s scan rate.

2.4. Determination of Optimum Deposition Time and Scan Rate

To the voltammetric cell containing 3 mL of supporting electrolyte, 3 mL of 50 ppm chromium (III) was inserted followed by measurement of current generated using Metrohm® μAutolab potentiostat at scanning range from -1.5 to +1.0 V with 10 s cleaning time and 20 mV/s scan rate. The deposition time was varied at 30, 60, 90, 120 and 180 s at fixed scan rate of 20 mV/s. While for determination of optimum scan rate, the deposition time was fixed at 60 s and the scan rate was varied at 10, 20, 50, and 100 mV/s.

2.5. Determination of Linear Concentration Range

Solutions containing 1 to 500 ppm chromium (III) were measured using Metrohm® μAutolab potentiostat at scanning range from -1.5 to +1.0 V with 10 s cleaning time, optimum deposition time and optimum scan rate.

2.6. Calibration Curve
Solutions of chromium (III) with concentration of 12.5, 25, 50, 62.5 and 75 ppm was measured using Metrohm® μAutolab potentiostat at scanning range from -1.5 to +1.0 V with 10 s cleaning time, at optimum deposition time and optimum scan rate.

2.7. Determination of Measurement Precision
One of standard concentration from chromium (III) standard solution was measured at optimum condition three times and the precision of the measurement was determined.

3. Results and discussion

3.1. Determination of Chromium (III) solvent
Three solvents were trialed in the present study, i.e. 0.01 M hydrochloric acid, acetate buffer pH 5.0 and double distilled water. The measurement was conducted at potential range of -0.8 to +1.0 V with 90 s deposition time, 10 s cleaning time and 20 mV/s scan rate with anodic stripping voltammetry method. The result of the measurement of the solvents is presented on Figure 1. This trial was conducted to investigate the correction factor of solvent background to avoid the presence of undesired current peaks which come from the solvent background.

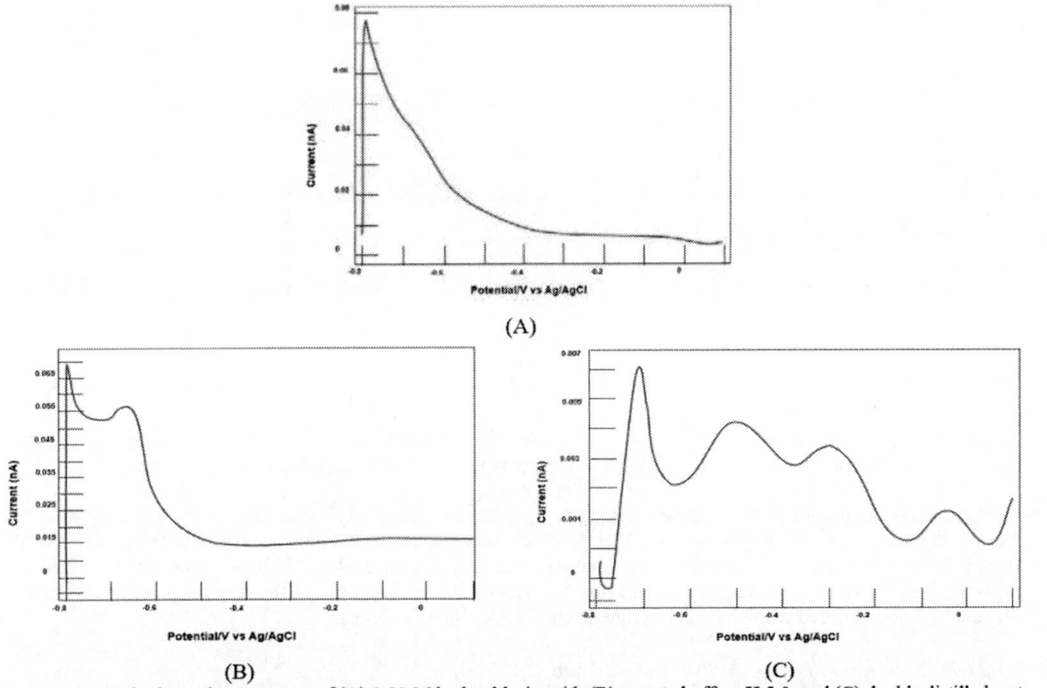

Figure 1. Stripping voltammogram of (A) 0.01 M hydrochloric acid; (B) acetate buffer pH 5.0; and (C) double distilled water. Condition of measurement: scan rate: 20 mV/s, deposition potential: -1.260 V vs Ag/AgCl, deposition time 90 s.

From Figure. 1 it can be observed that no peak detected for 0.01 M hydrochloric acid at potential range -0.8 to +1.0 V. While for acetate buffer pH 5.0 and double distilled water some peaks were observed. Therefore hydrochloric acid was chosen as chromium (III) solvent in the present study. It can be concluded that chromium (III) solution can be made by dissolving chromium (III) chloride in 0.01 M hydrochloric acid pH 2.0 as solvent.

3.2. Characteristics of Chromium (III) Solution with Acetate Buffer pH 5.0 as Supporting Electrolyte

Before performing further measurement, electrodeposition experiment was conducted to identify general characteristics of chromium (III) solution with acetate buffer pH 5.0 as supporting electrolyte using anodic stripping voltamettry and cyclic voltammetry measured at potential range of -1.5 to +1.5 V using pencil graphite electrode. The result is presented on Figure 2 and Figure 3.

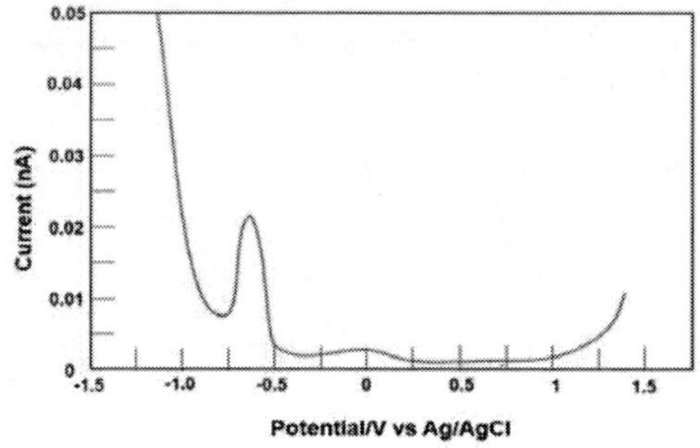

Figure 2. Voltammogram of typical peak of chromium (III) solution using stripping voltammetry. Condition of experiment: concentration of Cr(III) = 50 ppm, scan rate = 20 mV/s, deposition potential 1.260 V vs Ag/AgCl, deposition time= 90 s.

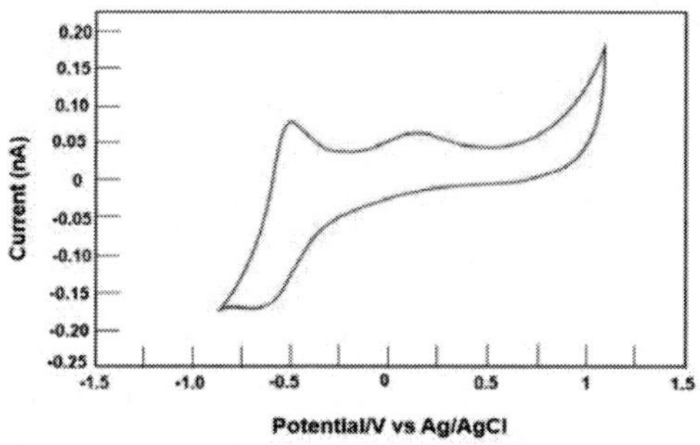

Figure 3. Voltamogram of chromium (III) solution using cyclic voltammetry. Experiment condition: concentration of Cr(III) = 50 ppm, scan rate = 20 mV/s, deposition potential 1.260 V vs Ag/AgCl, deposition time= 90 s.

Fig. 2 shows that the peak current was detected at potential value of -0.634 V. This result was supported by cyclic voltammetry result as presented on Fig. 3, which showed cyclic voltammogram at potential of -0.6 V. Measurement of cyclic voltammetry was conducted to gain additional qualitative information regarding electrochemical reaction of chromium (III). From this results it can be proposed that typical peak of chromium (III) that can be measured using pencil graphite electrode is at -0.634 V peak.

3.3. Determination of Optimum Deposition Time

The effect of deposition time can be observed by comparing voltammogram results of chromium (III) solution measurement with varied deposition time. The results are presented on Fig. 4. Optimum deposition time for determination of chromium (III) using pencil graphite electrode is 60 s, as indicated by the highest peak of current detected.

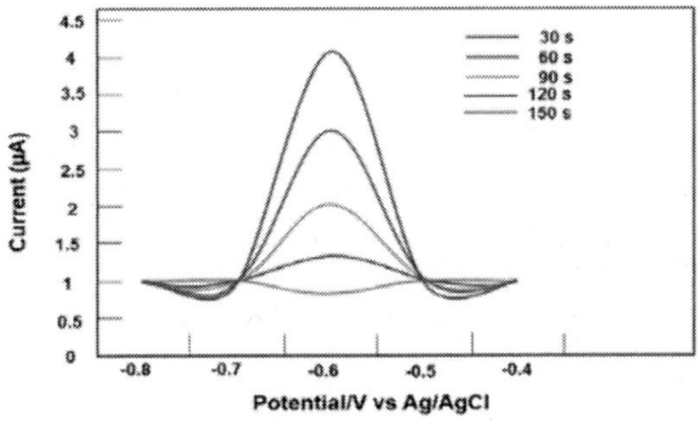

Figure 4. Voltammogram of 50 ppm chromium (III) solution with various deposition time for determination of optimum deposition time.

3.4. Determination of Optimum Scan Rate Condition

On stripping stage, the height and width of peak current can be affected by scan rate. Therefore in the present study, the optimum scan rate was determined. The effect of scan rate was determined by comparing voltammogram of chromium (III) solution with 60 s deposition time and varied scan rate. The results of the experiment is presented on Fig. 5. The results indicate that scan rate affect the height of current peak of chromium (III). The best scan rate was observed at 50 mV/s, because it gave high curve with relatively narrow peak.

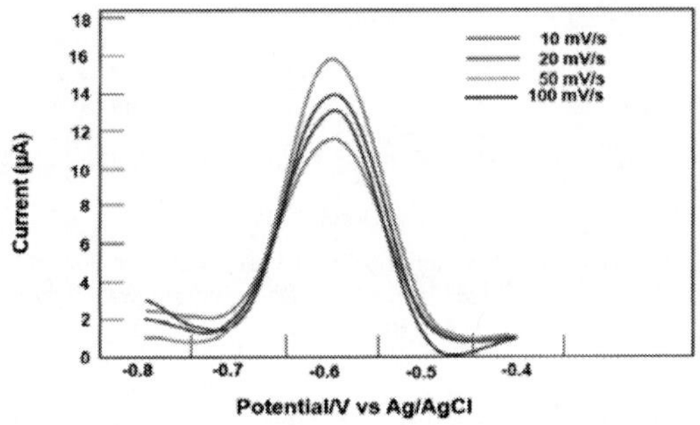

Figure 5. Voltammogram of of 50 ppm chromium (III) solution with various scan rate for determination of optimum scan rate. Experiment condition: deposition time = 90 s, deposition potential = -1.260 V.

3.5. Determination of Linear Concentration Range

Determination of chromium (III) within 1 to 500 ppm concentration range was performed at potential range of -1.5 to +1.0 V with 60 s deposition time, 10 s cleaning time and 50 mV/s scan rate. The results of measurement is presented on Table 1.

Table 1. Determination of chromium (III) in acetate buffer pH 5 with concentration ranging from 1 to 500 ppm.

Concentration (ppm)	Peak Current (µa)
1.0	1.0056
12.5	3.1153
25.0	5.3865
50.0	9.1768
62.5	11.2880
75.0	14.7840
100.0	8.2300

125.0	2.1639
150.0	11.5919
200.0	8.1233
250.0	9.0375
500.0	11.2880

Linear concentration range for chromium (III) determination using pencil graphite electrode was investigated to determine the linear range between height of current peak and concentration of chromium (III). The results of measurement at linear concentration range is presented on Fig. 6, while the data of current peak height against chromium (III) concentration is presented on Table 2. When all data points were used to make calibration curve, the regression coefficient value (R) was very low (0.3860) indicating that it was not linear. We tried to make linear regression using part of the data points. We obtained a good linear regression with five data points, ranging from 12.5 to 75 ppm (R = 0.9936). Therefore, we choose the concentration range of chromium (III) from 12.5 to 75 ppm to determine the analytical parametets.

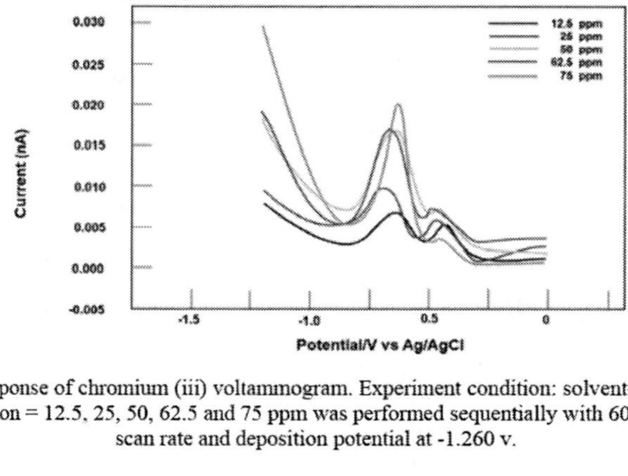

Figure 6. Typical peak response of chromium (iii) voltammogram. Experiment condition: solvent= 0.01 m hydrochloric acid, chromium (iii) concentration = 12.5, 25, 50, 62.5 and 75 ppm was performed sequentially with 60 s deposition time, 50 mv/s scan rate and deposition potential at -1.260 v.

TABLE 2. Results of measurement of chromium (III) at linear range

Concentration (ppm)	Peak current (μA)			Average of peak current (μA)
	1	2	3	
12.5	2.0744	3.5727	3.6988	3.1153
25.0	5.3566	6.3093	4.4936	5.3865
50.0	9.0228	9.8727	8.6350	9.1768
62.5	11.3660	12.5140	9.9840	11.2880
75.0	16.0400	13.5280	-	14.7840

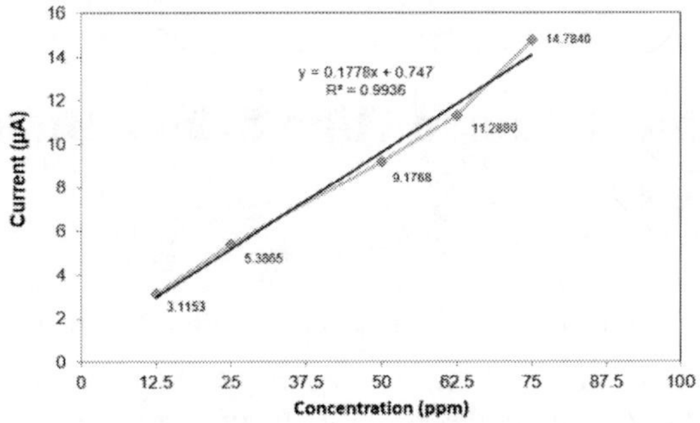

Figure 7. Calibration curve of chromium (iii) with 0.01 m hydrochloric acid as solvent, 60 s deposition time, -1.260 v deposition potential, and 50 mv/s scan rate.

Fig. 7 showed the range of concentration that gave linear correlation against peak current height of chromium (III). The range of linear concentration at 12.5 to 75 ppm gave correlation coefficient of 0.9936, which indicate good relationship with correlation equation of y = 0.1778x + 0.7470.

4. Conclusions

Pencil graphite electrode can be used as a working electrode for determination of chromium (III) with optimum condition achieved at 60 s deposition time and 50 mV/s scan rate. Linear range of chromium (III) measurement with the proposed condition was 12.5 to 75 ppm with 0.31 ppm limit of detection.

5. References

[1] Chen Y C, Lee I L, Sung Y M and Wu S P 2013 *Chemical* **188** 354-9
[2] Shadreck M and Mugadza T 2013 *African Journal of Pure and Applied Chemistry* **7** (9) 310-7
[3] Hirata T, Umezaki Y and Ikeda M 1986 *Analytical Chemistry* **58** (13), 2602-6
[4] Hirata S, Honda K, Shikino O, Maekawa N and Aihara M 2000 *Atomic Spectroscopy* **55** (7), 1089-99
[5] Kiran K, Kumar K S, Prasad B, Suvardhan K, Lekkala R B and Janardhanam K 2008 *Journal of Hazardous Materials* **150** pp 582-6
[6] Tang B, Yue T, Wu J, Dong Y, Ding Y and Wang H 2004 *Talanta* **64** pp 955-60
[7] Renedo O D, Lomillo M A A and Martinez M J A 2004 *Analytica chimica acta* **521** pp 215-21
[8] Buffle J and Tercier-Waeber M L 2005 *Trends in Analytical Chemistry* **24** pp 172-91
[9] Wyantuti S, Hartati Y W, Firdaus M L, Panatarani C and Tjokronegoro R 2015 *Int. J. Sci. Tech. Res.* **4** pp 135-9
[10] Wang J 2006 *Analytical electrochemistry* (John Wiley & Sons)

Process of Argumentation in High School Biology Class: A Qualitative Analysis

M Ramli*, E Rakhmawati, P Hendarto and Winarni

Department of Biology Teacher Education, Universitas Sebelas Maret, Jl. Ir. Sutami No. 36 A, Kentingan Surakarta 57126, Indonesia
Corresponding author: mramlim04@fkip.uns.ac.id

Abstract. Argumentation skill can be nurtured by designing a lesson in which students are provided with the opportunity to argue. This research aims to analyse argumentation process in biology class. The participants were students of three biology classes from different high schools in Surakarta Indonesia. One of the classroom was taught by a student teacher, and the rest were instructed by the assigned teachers. Through a classroom observation, oral activities were noted, audio-recorded and video-taped. Coding was done based on the existence of claiming-reasoning-evidence (CRE) process by McNeill and Krajcik. Data was analysed qualitatively focusing on the role of teachers to initiate questioning to support argumentation process. The lesson design of three were also analysed. The result shows that pedagogical skill of teachers to support argumentation process, such as skill to ask, answer, and respond to students' question and statements need to be trained intensively. Most of the argumentation found were only claiming, without reasoning and evidence. Teachers have to change the routine of mostly posing open-ended questions to students, and giving directly a correct answer to students' questions. Knowledge and skills to encourage student to follow inquiry-based learning have to be acquired by teachers.

1. Introduction

The 21st century is an era of globalization, marked by the advancement of science and technology. Advances in information and communications technology in this century freed everyone to access information around the world. The 21st century requires the global community to have a good scientific reasoning to dealing with every aspect of life that has always been associated with science. Scientific reasoning plays important role in the problems solving process [1], facilitating communities to understand and applying the science in daily life.

Scientific reasoning can be trained through the nurturing of the argumentation skill. According to Sadler & Fowler, developing students' scientific reasoning can be done with the verbalizing argumentation using social-scientific issues [2]. Scientific argumentation skill contributes to produce, evaluate and advance scientific knowledge [3]. Scientific arguments constitute formal evidence-based arguments [4] to generate a knowledge involving the coordination of data, claims and evidence [5]. The arguing is one of the most complex thinking skills in the learning process [6]. Learning by developing the argumentation skill aims to introduce scientific literacy to prepare students to face the future [7].

Argumentation skill is important to develop in the learning process, because it can change the students' concept understanding in biology. Biology learning process facilitates students to learn and to find a concept by applying scientific methods. The argumentation skills include aspects, such as claims (statements that address), evidence (scientific data to support a statement) and reasoning (justification related statements and evidence) [8]. The argumentation skills are used to repair and rebuild ideas based on scientific evidence for better understand the reality that occurs in nature as well as to provide a statement reinforced by reason and evidence to support the claim.

Factual condition in education practices in Indonesia indicates that the Indonesian students' argumentation skill needs to be improved. One research shows the profile of high school students in Indonesia tend to arguing improperly and without supporting evidence [9], and small number of students who can arguing correctly based on the information and the theory [10]; [11]. Based on the results of scientific literacy score of Indonesian students in PISA 2012, Indonesian students' argumentation skill in science also needs to be trained more.

The low argumentation skill is due to teaching and learning activities do not provide the opportunity for students to arguing. The learning activities in class are dominated by teachers that tend to control the class by explaining the learning material too much and ask questions that do not lead students to arguing. The majority of questions asked by teacher are questions that require short answers, or rhetorical questions. Therefore, it is less accommodating student to arguing. Implementation of teacher-oriented learning process makes the students are not able to master the concept independently. Weak concept mastery led to a lack of a student's arguing skill. Therefore, the learning process carried out by the teacher needs to be modified in order to engage students in arguing.

The solutions for that problem is by implementing the learning activities that guide students to argue through the scaffolding or the provision of assistance by teacher. A research by Zembal-Saul states that the evidence-based arguments can be enhanced through the argument scaffolding and argument frameworks [12]. Meanwhile, according to McNeill & Krajcik, teachers can provide support for students to develop the argument in the scientific explanation by the following ways: 1) create an explicit framework, 2) discuss the reasons for the scientific explanation, 3) modeling to construct scientific explanations, 4) discusses the similarities and differences with daily scientific explanations, 5) provide opportunities for students to develop a scientific explanation, 6) provides an opportunity for students to criticize the other students' explanations and 7) provide feedback to students [13]. Teacher is a key figure in learning to engage students in the process of argumentation to develop the ability to argue. Therefore, the ability of teachers to lead students to argue through the scaffolding should be good.

Scaffolding is a guidance or direction given by adults or people who are more competent to assist children to doing their activities [14]. Scaffolding is the approach proposed by Vygotsky stating that the scaffolding is done at an early stage to encourage students to study independently by providing direction, encouragement, warning, describing the problem in solving steps, providing examples, and other measures that enable learners learning independently [15]. Scaffolding is given in the early stages of learning and will be reduced gradually and suspended until the student can complete the task independently [15]. Scaffolding in argumentation provided by the teacher by sharing the example of ways to arguing with the statement delivered along with the reasons and evidence, and provide questions that lead to answers in the form of a statement with reasons and evidence. Scaffolding which engage students in arguing will be reduced gradually when student began to assert arguments independently with minimal guidance from the teacher, and then stopped when the students are able to argue independently. Scaffolding by teachers to students in the process of arguing is expected to guide students to arguing independently. Therefore, teachers should be able to provide the appropriate scaffolding to help students develop their argumentation skills.

2. Experimental Method

Three classroom action research (CAR) had been done simultaneously from February to April 2016, in three different classes of three different high schools in Surakarta and Karanganyar, two regencies in Central Java Province, Indonesia. The first CAR (A class) was an action research done by researcher (lecturer of university), and student teacher as the role teacher, who has actually learnt pedagogical content knowledge during their 3-years pre-service teacher training, and did three-months internship at the target school. The second CAR (B class) was collaboratively done by the researcher, student teacher with similar experience as the one in the first CAR, and teacher with more than 10 years work experiences. The third action research (C class) was conducted simultaneously by the teacher who have 4-years experiences, student teacher, and researcher. All lesson designs were prepared collaboratively by student teacher and teacher, guided by researcher. The first research was set on the topic of Plantae (Grade 10), the second CAR was Reproductive System (Grade 11), and Kingdom of Animalia was selected as the topic of the third CAR (Grade 10).

All CARs followed the cycles of action research introduced by Kemmis & McTaggart, which was started with the reflection of teacher and students of the previous condition and practices of learning, and continued with analysing the pre-action situation. Many problems had occurred and found during the previous learning process, and researcher, student teachers, and teachers finally decided to focus on how to improve students' argumentation process, since the occurred of dialogic argumentation was identified very rare during the process of learning.

The first CAR used an Assessment for Learning (AfL) as the primary action to improve argumentation of students. The action was consistently applied for teaching Plantae in one month or 3 hours per week (1 hour equals to 45 minutes). AfL is commonly known as the divergent assessment which will give students a challenge to do self-evaluation, and allow them to know and decide what the criteria of assessment that they have to accomplish. Teacher also gives feedback to students based on students' works, and furthermore advise students to find solutions in advanced learning on the topic that students do not understand well. AfL in this research was described in the first meeting as students directly know what the target they are expected to be acquired. At the end of each meeting, teacher gave students a couple of argumentative questions and asked students to answer it, and sometimes students brought it back as homework.

Meanwhile, the second CAR opted to develop a lesson design based on the syntax of lowest level of inquiry models, i.e. discovery learning, as the main interrupted action in each cycle to nurture argumentative skills of students. The design of lesson followed the syntax of discovery, and learning experiences in each cycle was flourished with the various of animal specimens, that student teacher provided. Since school does not have those specimens, student teacher borrowed them from the Laboratory of Department of Biology Teacher Education Sebelas Maret University. Invertebrate was taught prior to the vertebrate, and students were encouraged to discover the characteristic of classification in invertebrate, and discuss what the basis or determinants that scientist use to classify invertebrate. The same activities done when students learned about vertebrate.

Data of dialogic argumentation was noted, and audio-visual recorded. All dialogues in each cycle were analysed to know whether the students and role teacher have developed a sequence of argumentation in their oral activities during the learning process. The simple of McNeil and Krajick model of argumentation, which consists of three aspects of argumentation, i.e. claiming (C), reasoning (R), and evidence (E) was applied to identify the fact of argumentation.

3. Result and Discussion

Learning process before the action research was observed randomly, and all dialogues during a 45 minutes' class were noted. It shows that no sufficient argumentative dialog during the learning process. The dialogue on the beginning of the learning process is displayed in Table 1. The A class, which taught by student teacher on the topic Kingdom of Plantae, indicates the lack initiative of role teacher to encourage students to do argumentative dialogue. By asking student with a question: "What do you know about the Plantae?", it seemed that teacher expected students will response with more than single answers. One student answered briefly that Plantae is plant (in original language student used Indonesian language for plant as tumbuhan). There are two words in Indonesian language commonly used to translate the word plant, i.e. tumbuhan (from the basic word tumbuh or to grow), and tanaman (from the basic word tanam or to plant). In Indonesian terms, plants are the one which grows without human effort, and the one which is planted or cultivated by human. Role teacher in this case, tried to furtherly confirmed student's concept about the plant, however student thought the simple point by mention about the common colour of plant. At the end, another student suggested about the photosynthesis, which is a main determinant to differ plant from animal.

At the B class, which was learning about the Plantae too, teacher started the dialog by directing student to talk about spermatophyte or seed plants. Argumentative sequences are not available in this dialog, since teacher asked the open-ended question, and always confirmed and directly corrected students' answers. The same situation is also found at the C class, where teacher taught the digestive system and brought student to discuss on disturbances in digestive process. In this class, role teacher also seems to quickly moved to the main topic, without providing students with sufficient space and time to gradually think and understand the concepts.

Table 1. Comparison dialogue in the beginning of the learning process at pre action research.

A. Role Teacher: Student Teacher (Expository model with AfL)	B. Role Teacher: Teacher and Student Teacher (Discovery model)	C. Role Teacher: Teacher (Guided Inquiry Model)
T: What is in your mind when you hear the word Plantae? S: Plants, Ma'am (C) T: Plants, that like....? S: Plants that green colour, Ma'am (C) T: Is it just green? S: No, there are red, brown, and yellow (C) T: Does anyone else want to argue about plants? S: A living being, photosynthetic, a lot of type, Ma'am. C T: (show picture moss) Is this image also includes plants? S: Yes, that's the moss, lichens, including plants, Ma'am	T: Let's talk about habitat and life adaptation of spermatophyte. T: Where is the habitat of seed plants? S: On the ground, in the pots (C) T: Yeah right, on the ground, in the pots, in the water... T: And then.... T: Seed plants are usually found everywhere T: What the character of plant? S: Antagonist (C) T: No, isn't. Spermatophyte is photoautotroph, because they have a chlorophyll to photosynthesize T: for example.....	T: Is there anybody ever experienced diarrhea? S: Yes, we are (answer by some students) (C) T: Is diarrhea a disturbance in digestive system? S: Yes, Mam (C) T: All right, what the topic that we will learn today? S: Digestive system disorder (C) T: Exactly, we will learn about disturbances in digestive system T: Today we will immediately form the discussion groups. Please look for a variety of disorders of digestive system. Then find out the causes, prevention and treatment. Your time is 20 minutes, so please make the groups

Description: The symbol "C" indicates the claiming, "R" indicates *reasoning* and "E" indicate the *evidence*

After three cycles of action research (almost 4 weeks), the argumentative dialogue still did not improve significantly (Table 2). However, teachers had tried to use the question word, "why" or "what is the reason", which are the type of stimulus questions to encourage students to accomplish their claims with the reasoning behind. Student teacher at the first class action research (A case), try to always ended questions with a question tag, and students have been trained to make claiming followed by its reason. These types of dialogue are probably influenced by the AfL, which give students a sort of assessment, such as 2-tier assessment, in which students were trained weekly on doing some argumentative questions, and how to always strengthen their answers with a good reasoning and evidence. A 2-tier test might be consisted of multiple choice or essay test. In the multiple choice type, students have to choose a correct answer and follow it by selecting the correct reason. Meanwhile, in the essay test, student have to write the answer and also its reason. This treatment probably has a good effect on how students and teacher develop an argumentative dialog. Moreover, students also got the feedback of their works a week after, hence the argumentative skills of students are kept on the right track, as mentioned by McNeill & Krajcik [13].

The different pattern found in the second action research (B case), where students had not directly state the claiming followed by reasoning. It can be noticed that students always mentioned the reasoning soon after the teacher asked them to come with the reason. Interestingly to analyse the snapshot dialog among students in a group, when they discuss about Aves. It is absolutely great to start argumentative dialog when one student said, "Why did the chicken can not fly even though they have the wings while the birds can?" This question forced other students to think the characteristic of both birds and chicken, and make sure the distinguished point. Students are also try to always find an evidence of their opinion, and teacher had reduced her provisions to always directly answer students' questions.

However, the role teacher at the third action research (C case) still confused on how to pose questions to trigger the argument of students. Teacher also did not understand well what the different between reasoning and giving evidence, as she asked the following question, "There is relation between lungs and TBC. What the evidence is?" It seems students faced a bit puzzlement at the moment, as he answered, "The existence of bacteria". And the conversation became more unclear, when the teacher continued asking the reason of student's claim. Student did not have a chance to understand why bacteria can be a cause of TBC. One student had tried to define what TBC is, and how bacteria are able to infect human lungs, and caused TBC. It can be concluded that teacher could not keep the sequences of argumentation, as some scholars argued [16].

Table 2. Snapshots of the dialog at the last cycle

Role Teacher: Student Teacher Action: Expository model with AfL	B. Role Teacher: Teacher and Student Teacher Action: Discovery model	C. Role Teacher: Teacher Action: Guided Inquiry Model
T: The statement that the angiosperm is gymnosperm is true or false? S: Yes, angiosperm is gymnosperm (C) T: *Cyperus rotundus* and *Eulisine indica* are monocots that included in the group of Graminae. The statement is true or false? S: That's true, because the morphology between the two is almost the same as characteristic of the Graminae (C R) T: Dicotyledonous plant plays a role in human life, one of them in the health field, is this a true statement?? S: Yes, ma'am, because it can be used as medicine/remedies such as lime is used for cough (C R E) T: Today we will learn about the latest material on the Plantae, the gymnosperms. What do you know about by gymnosperms? S: Plant with open seed (C)	T: Have you visiting the zoo? T: Come on ... who wants to argue ... hands up! T: What do you think about vertebrates? S: Similar to Adit, Ma'am...(C) (nick name of *one student*) (C) T: Yeah, right. So is Adit also vertebrates? S: Yes. (C) T: What?? Why did you Adit include as vertebrates? S: Because he has skull. (R) T: Yes, he has cranium. T: Besides that, what do you know about vertebrates. What do you think distinguishes vertebrates to invertebrates?? S: The Vertebrates have endoskeleton, and invertebrate do not. (R) ------------------ S (36): Why did the chicken can not fly even though they have the wings while the birds can? S (16): They could... But only to a certain short distance only. (C) S (34): No, they can't. T: Can anyone answer her?	T: Depending on what the cause. Because of a bubbling in the alveoli. What was the causes of tuberculosis (TBC)? S: Genetic or heredity and also lifestyle (C) T: Lifestyle. Then, what else? S: Bacteria. (C) T: The bacteria also exist. Could be in the form of infections and non-infectious. That's the conclusion of the first and the second. The third conclusion. How about the TBC, the hypothesis is accepted or not? S (19): Yes, ma'am, there is a relationship between the lungs with tuberculosis. (C) T: What organs are attacked? S: Lungs. (C) T: Well, so there is a relationship between the lungs with tuberculosis.

T: Are these images including plants gymnosperms? S: Yes, Ma'am, (C) T: Then, what are the characteristics of the gymnosperms? S: They have Strobilus, ma'am (C) S: Yes, Strobilus, which is in this image? S: That browny cone shape, in pine (C)	S (33): They can't, but bird can, because they have an air bag. (E) S (34): air pocket. 8 S (9): Because the bird has an air pocket when it flew, different from the chicken, they not have the air pocket. (E) S (16): Because most of chickens is terrestrial. (R) S (20): The chicken did not have airbags. (R) S (3): So they can not fly. (R) S (31): Why did the chicken has the syrinx? S (16): No, they do not have. (C) S (33): Bird have that. (R) S (16): Syrinx is for singing. (R)	What the proof of that statement? S: The presence of bacteria. (C) T: What is your reason? S: According to the journal obtained earlier mentioned that tuberculosis is a disease of the respiratory system, caused by bacteria that get into the respiratory organs. Then, the bacteria that cause dysfunction in the lungs. So that the lungs do not function properly. (R E)

Description: The number in parentheses is the number of the students, while the symbol "C" indicates the claim, "R" indicates *reasoning* and "E" indicate the *evidence.*

The process of argumentation depends on many factors, such as what kind of the trigger questions to start the argumentative answers or statements that lead the conversation between teacher and student, or students and their peer. The skills and experiences of role teacher to facilitate the process of argumentation is also a prominent aspect. It is also influenced by the students' response which are actually as representative of knowledge that students have. The training to always claiming followed with the reasoning and evidence, is also important. And the various learning experiences that provided for students. All factors are involved together in the process of argumentation. Therefore, it is a must to apply all aspects in the process of learning simultaneously.

When teacher addresses to only adopt model or method of learning to improve argumentative skills of students, then it will come with the fact that students have experiences to deepen their concepts, but they cannot argue correctly. Thus, teacher has to thickened the model with assessment that will drill students to do routine argumentation. It can be also added the argumentative dialog, such as Socrates dialog to help students do correct argumentation, during the learning process.

The learning design that role teachers used in these action research were constructed based on the principles that students have to be provided with various experiences of learning. Role teachers, teachers, and researchers worked together to design appropriate lesson design which were concentrating on how students construct the knowledge. Therefore, the design was totally different with what have been routinely used by teachers. However, those lesson plans still follow the standard that determined in The Curriculum 2013 of Indonesian schools, and also met the time allocation of each topic of learning.

4. Conclusion

The process of argumentation of three action research which were conducted in different schools showed that a good action is the one that focus on training the students to do routine argumentative dialog by always posing argumentative questions and answer with clear claiming followed by reasoning and evidence. The process has absolutely influenced by the skill of teacher to ask trigger questions, and to keep the dialog sequences on the track of argumentation process. Formative Assessments which provides test with answer have to follow the sequence of argumentation will help students to be familiar with the argumentative skills.

The three action research had significantly changed the habit of teachers to directly answer the open-ended questions, and make corrections of students' answer. This changed practices have a good impact on how students learn how to argue. However, there is no significant impact of working experiences of role teacher in the process of argumentation.

5. References

[1] Moore J C and Rubbo L J 2012 *American Physical Soc.* **8** 1-8
[2] Sadler T D and Fowler S R 2006 *Science Educ.* **90** 986–1004
[3] Bricker L A and Bell P 2008 *Science Educ.* **92** p 473-98
[4] Schen M S 2007 *Scientific reasoning skills development in the introductory biology courses for undergraduates* (USA: The Ohio State University)
[5] Chen Y C 2011 *Examining the Integration of Talk and Writing for Student Knowledge Construction through Argumentation* (United States of America: The University of Iowa)
[6] Deane Paul and Song 2014 *Psicologia Educativa.* **20** 109-15
[7] Ryu S, Corrigan S, Knight A M and McNeill K L 2013 Assessing student's ability to argue across multiple modalities. Paper presented at the annual meeting of the *National Association of Research in Science Teaching*
[8] McNeill K L and Krajcik J 2006 *Supporting students construction of scientific explanation through generic versus context-specific written scaffolds* (Michigan: University of Michigan)
[9] Viyanti 2015 *J. Pendidikan IPA Indonesia.* **4** p 86-9
[10] Handayani P and Murniati S 2015 *J Inovasi Dan Pembelajaran Fisika* **2**
[11] Pritasari A C, Dwiastuti S, Probosari R M and Sajidan 2015 *J. Pendidikan IPA Indonesia.* **4** 158-63
[12] Zembal-Saul C 2009 *Learning to teach elementary school science as argument Science Education* **93** p 687-719
[13] McNeill K L and Krajcik J 2009 *Annual Intl Conf Grand Challenges and Great Opportunities in Science Education National Association for Research in Science Teaching* (California: Hyatt Regency Orange County) pp 1-17
[14] Ormrod J E 2008 *Psikologi Pendidikan Jilid 1* (Jakarta: Erlangga)
[15] Slavin R E 2011 *Psikologi Pendidikan Jilid Teori dan Praktik* (Jakarta: Indeks)
[16] Nielsen J A 2013 *Res Sci Educ.* **43** 371-93

IOP Conf. Series: Journal of Physics: Conf. Series **812** (2017) 012008 doi:10.1088/1742-6596/812/1/012008

Improvement of The Ability of Junior High SchoolStudentsThinkingThroughVisualLearningAssisted Geogbra Tutorial

M Elvi[1], Nurjanah[1]

[1] Department of Mathematics Education Universitas Pendidikan Indonesia, Jln. Dr. Setiabudhi No. 229 Bandung, Jawa Barat, Indonesia 40154
Email: *mariyantielvi@gmail.com*

Abstract. This research is distributed on the issue of the lack of visual thinking ability is a must-have basic ability of students in learning geometry. The purpose of this research is to investigate and elucide: 1) the enhancement of visual thinking ability of students to acquire learning assisted with geogebra tutorial learning: 2) the increase in visual thinking ability of students who obtained a model of learning assisted with geogebra and students who obtained a regular study of KAM (high, medium, and low). This research population is grade VII in Bandung Junior High School. The instruments used to collect data in this study consisted of instruments of the test and the observation sheet. The data obtained were analyzed using thetest average difference i.e. Test-t and ANOVA Test one line to two lines. The results showed that: 1) the attainment and enhancement of visual thinking ability of students to acquire learning assisted geogebra tutorial better than atudents who acquire learning; 2) there may be diffrences of visual upgrade thinking students who acquire the learning model assisted with geogebra tutorial earn regular learning of KAM (high, medium and low).

1. Introduction

Geometry is one of scope in mathematics. Geometry in mathematics is the science that studies the points, lines, areas and spaces and the relationships among them. As one of the material scope of mathematics, geometry was studied at various level of education divided into several subjects at every level. In geometry, students will learn about the shape, structure and analyze the characteristics and relationships of geometric objects [1].

According to the process of learning geometry Budiarto include (1) visualization, construction and draw a picture; (2) the study of aspects of space; (3) using the tool for presenting math concepts; (4) the presentation of formal mathematical systems as Suparyan [2]. The ability of visualization is the most important and basic ability in learning geometry that supports the ability of mathematical concepts. According to Hershkowitz [3] visualization is the ability of interpretation of drawings, diagrams, and thoughts by writing or tech tools with the aim of describing or communicate information and develop ideas and increase understanding Visual thinking or also called the visual thinking is ability to transform information into images or other forms that can help communicate information [4].

The survey results Trends International Mathematics Science Study (TIMSS) 2007 [5] in the domain of visual thinking ability of geometry content has not been high that is said to be only 19% Indonesia students answered correctly. In addition, studies conducted (2011), obtained the results that students can't think visually. In addition, the ability of the students in mengkontruksi geometryis still classified as low. This, due to students not used to visualize math problems.

One of the efforts to develop the capabilities ofthevisualthinking is a method of teaching or learning carried out by the teacher.The learning process should be supported by alearningmedia can developvisualthinking or visualizationcapabilitiesagainstmaterialgeometry. The application of appropriate learning model will encourage students increase and development of student's visual thinking ability.

The model of learning can be colaborate with the use of a customized learning media with the advancement of technology. The utilization of technology in learning is by applying a computer-assisted learning. One of the computer-assisted learning model is a model of learning tutorials. This learning model provides a complexstudy designs that contain material learning, exercises that accompanied feedback [6]. The tutorial program is a program using software in the form of computer programs. As for the software used to assist in the learning of mathematics such as software geogebra.

Geogebra is developed by Markus Hohenwarter in 2001. According to Hohenwarter [7], geogebra is of project computers (software) to learning math. Geogebra combines geometry, algebra, statistics and calculus. Several studies [8] revealed that geogebra very effectively applied in learning mathematics. Geogebra-assisted learning improve independent study students. Guven and Kosa[9] which suggests that as a software geogebra geometry can be used as an effective tool in learning for the visualization object as wellasimprove understanding.

Based on the descriptions above, outline of problems on the research is "Whether the application of the model of learning assisted geogebra tutorial can improve the ability of visual thinking of Junior High School students?". The outline of the research issues elaborated in the form of research questions as follows: 1) If increase visual thinking students who obtain a model of learning assisted geogebra tutorial better than students who obtain a regular learning?; 2) if there are differences of increase visual thinking students who acquire the learning model assisted with geogebra tutorial students earn regular learning of KAM (high, medium,and low)?

This research aims to examine and elucidate 1) Examines the increase in visual thinking ability of students who obtained a model of learning assisted geogebra tutorial students acquire learning; 2) Elucidate the difference of visual upgrade thinking students who acquire the learning model assisted geogebra tutorial students earn regular learning of KAM (high, medium,and low)?

2. Method

Thisresearchusesquasiexperimentaldesignresearch.Thisisbecauseresearcherscouldn'tpossiblyclassifysu bjectsrandomlyso that researchers receive a State subject. This study uses two research groups, namely the Group of experimental and control group. As for the design of the research plan are presented as follows:

Experimental group	:	*O*	*X*	*O*
The control group	:	*O*		*O*

The population in this research is the wholegradeVII in theSMPNegeri5Bandung.Sampling of the research done with thepurposivesamplingtechnique,namely the determination ofsampleswithspecificconsiderations[10].Consideration of samplingin this studyusing classes thathave the characteristics and the samerelative academic ability.Grade VII intheSMP consists of 10 classesfromClass VII-A to J.Ofthe 10 classes, selected two classes that will be used as a sample in this research.

The instruments used in the research are of two types i.e. instrument test and non-test instruments. Instruments in the form of the test consists of pretes and postes to measure visual thinking ability of students, while the instrument is in the form of a non-test, namely the observation sheet which contains indicators of activity of teachers and students in learning.

The visual thinking ability test instrument consists of a written question 5. Stage of preparation of the matter to the test, beginning with the preparation of the lattice problem followed by compile problem with alternative answer keys of each grain of matter. Beforethe question ofvisual thinking ability test used, first do tests to find out the degree of validity, reliability, power and difficulty distinguishing instrument.

Table 1. The Results of the Analysis of The Trial of Visual Thinking AbilityTest Per Grain Problem

No.	Validity	Diviculty level	Criterion
1	Valid	Easy	Enough
2	Valid	Medium	Enough
3	Valid	Medium	Good
4	Valid	Medium	Good
5	Valid	Difficult	Enough

Based on the trial results, obtained interprets that all matter is worthy. After the analysis of the tests obtained $r_{11} = 0.66$, thu inferred that tests that tested had levels of reliability are. Data analysis carried out quantitatively. Statistical tests used are the average difference test, and calculations performed using Microsoft Office Excel and IBM SPSS Statistics 20 with the following steps:

- To see a visual thinking ability of students between the before and after is calculated by using the formula learning gain score normalization
- $g = \frac{posttest\ score - pretest\ score}{maximum\ ideal\ score - pretest\ score}$
- To perform a test of the hypothesis, first set level of significance that is a = 0.05, as well as the test data of its homogeneity and normality.
- Hypothesis testing using t-test 1 and test 2 using hypothesis testing ANOVA two lines.

3. Result And Discussion

3.1 Result

Visual thinking abilityofdataobtainedthroughthepretestandpostest.Frompretestdataandvisualthinkingab ilityposttestcalculatedscoregainsnormalization (N-Gain)inbothclassroomlearning-assistedtutorialclassesnamelygeogebra(PTBG)andtheusuallearningclass(PB). The following are the data N-gain the ability of visual thinking.

Table 2. Descriptive Statistics of theData Visual Thinking Ability of Students

	Class	N	$\bar{x}$
N-gain	PTBG	30	0,6673
	PB	30	0,5080

Table 2 shows that increased ability of visual thinking in classroom PTBG after applied learning assisted geogebra tutorial better than classroom learning with PB. The data increased the ability of visual thinking or score N-gain is grouped by category KAM (high, medium and low). The following table presents the descriptive data N-gain visual thinking ability of the students of class PTBG and class PB already grouped based on requirements KAM.

Table 3. Descriptive Statistics N-Gain The Ability of Visual Thinking Based On Category KAM

N-gain	KAM Category	N	Mean
PTBG Class	High	6	0,71
	Medium	19	0,67
	Low	5	0,61
PB Class	High	4	0,64
	Medium	21	0,54
	Low	5	0,26

Table3 describesthescoreN-gaintheabilityofvisualthinkingclassPTBGandclassPB.TheaveragescoreofN-gain PTBG classhigherthantheclassofPBoneachcategoryKAM.

3.2 Analysis

3.2.1 N-gain Data analysis of Visual Thinking Ability

Based on table 3 that average N-gain visual thinking ability of the students of the PTBG class is higher than the average of N-gain PB. If you want to know the significance of the difference the average N-gain visual thinking ability of the students of PTBG class and PB class, then performed a test of the difference the average N-gain. Before that, first conducted the test of its homogeneity and normality test against N-score gain visual thinking ability of students in both classes. Test result data normality N-gain visual thinking ability of students both classes are presented in the following table.

Table 4. Test result Data Normality N-gain the ability of Visual thinking

	Class	Shapiro-wilk			Explanation
		Statistic	Df	Sig	
N-gain	PTBG	0,946	30	0,136	Normal
.	PB	0,965	30	0,414	Normal

It is known that the data table 4 N-gain the ability of visual thinking both of the class. The nexttest is done to find out whether its homogeneity of varianceclassPTBG and classPB are the same or different. As for its homogeneity test resultsdata N-gainvisualthinkingabilityof students both classes are presented in the following table.

Table 5. Test results of its homogeneity of Data N-gain the ability of Visual Thinking

	Levene Statistic	df1	df2	Sig
N-gain	0,134	1	58	0,716

In table 5 it appears that data N-gain the ability of visual thinking shows a value greater than the significance $\propto = 0.05$ i.e. of 0.716, so H_0 received. This means that the data N-gain visual thinking ability of the students of class PTBG and class PBhasavariancehomogeneity. Based on tabel 5 Note thatN-gain datavisualthinkingabilityhasvariancehomogeneity.Therefore, to know the differences of averagedataN-gainusedTestt. As for the workinghypothesisused is

$H_0 : \mu_1 = \mu_2$

$H_1 : \mu_1 > \mu_2$

Test result average difference data N-gain visual thinking ability of students are presented in the following table.

Table 6. The results of the test t Data N-gain the ability of Visual Thinking

N-Gain	t	df	Sig. (2-tailed)	Sig. (1-tailed)
	3,038	58	0,004	0,002

Table 6 showsthevalueofthesig(2-tailed)issmallerthanfrom $\propto = 0.05$ sothatH_0 isrejected.Itmeansanincreaseintheabilityofvisualthinking PTBGgradestudentsacquirelearningassisted geogebra tutorial better than grade PB which acquiredlearning.

3.2.2 N-gain Data analysis capabilities of Visual Thinking based on Category KAM

Based on table 4, note that score N-gain visual thinking ability of the students of the class PTBG is higher than the score of N-gain PB class on each category KAM. To find out the difference of score N-gain significantly based on ANOVA 2 Way. Before that first test conducted a test of its homogeneity and normality. Test result data normality N gain visual thinking ability of the students of the third category KAM served in the following table.

Table 7. Test result Data Normality N-gain the ability of Visual Thinking based KAM

Class	KAM Category	Shapiro-wilk			Explanation
		Statistic	Df	Sig	

PTBG	High	0,923	6	0,527	Normal
	Medium	0,943	19	0,294	Normal
	Low	0,927	5	0,574	Normal
PB	High	0,905	4	0,457	Normal
	Medium	0,960	21	0,523	Normal
	Low	0,896	5	0,388	Normal

Table 7 shows that all data N-gain the ability of visual thinking in both the class of each category KAM. Furthermore its homogeneity of variance test done. The following summary of test results of its homogeneity.

Table 8. Its homogeneity test Score N-gain the ability of Visual thinking and Learning based KAM

Levene Statistik	df1	df2	Sig	Keputusan
1,953	5	54	0,101	H_0Receive

Test results of its homogeneity and normality is known that N-gain data visual thinking ability based on the category of KAM distributed class on both normal and homogeneous. ANOVA test done the next two way. As for the working hypothesis used is

Colum Factor

$H_0 : \mu_{PTBG} = \mu_{PB}$

$H_1 : \mu_{PTBG} \neq \mu_{PB}$

Line Factor

$H_0 : \mu_1 = \mu_2 = \mu_3$

$H_1 : \mu_i \neq \mu_j$, *wherei, j= 1,2,3*

ANOVA test results two lines can be seen in the following table.

Table 9.ANOVA test results two lines of Data N-gain the ability of Visual Thinking

Factor	Sum of Squares	Df	Mean Square	F	Sig.
KAM Category	0,319	2	0,159	4,384	0,017

Table 9 shows that the category KAM give significant effects against visual thinking ability of students. It can be seen from the significance value less than $\propto = 0.05$, namely 0.017. This means that there is a difference of visual upgrade thinking students who obtain learning tutorials and regular learning geogebra assisted by category KAM. Next, to find out which category of KAM differs significantly to the enhancement of the capabilities of visual thinking done test Advanced ANOVA Tukey HSD Test I. Tukey HSD test results can be seen in the following table.

Table 10. Advanced Anova Tukey HSD Test

	Pair of KAM	Mean Difference	Sig	Explanation
HSD	High-Medium	0,0797	0,468	Receive H_0
Tukey	High-Low	0,2470	0,015	Reject H_0
	Medium-Low	0,1673	0,042	Reject H_0

Based on the table above it can be concluded that there was no significant difference in visual thinking ability of students improved on the category high with medium. While there is a significant difference of visual thinking ability enhancement in the categories high-low and medium-low category.

4. Discussion

Based on the analysis results data N-gain inferred that increased the ability of visual thinking, it was concluded that upgrades visual thinking PTBG grade students significantly better than PB. Preferably visual upgrade thinking of students in the class of PTBG of learning is done in the class. Learning-assisted tutorial geogebra (PTBG) gives students the opportunity to discover their own knowledge or concepts of transformation and develop visualization resources with students who are in the program geogebra.

The process of discovery and knowledge of the construction of a new look at the use of learning activities is categorized. Power visualization students developed through the course of geogebra. Geogebra program gives students the ease in exploring the ability of visualization by shifting, reflect, rotate and zoom in or out of the object. According to Saha [8] assisted learning geogebra is very effective in learning geometry. Gumanti [11] in his research mentioned that the program displays a simple visualization of geogebra some concepts geomeri. The existence of the visualization features in geogebra is very effective in helping students understand the concept of geometry of the particular transformation.

Geogebra also assisted tutorial learning dilengkap with exercise matter contained on the tutorials. Administering the practice question interactive learning is useful to check for understanding student learning to the materials. Good or not understanding the students towards the subject matter will be assessed by computer. This is in accordance with the opinion of Rusman [12] which stated that the practice of reserved in interactive media can maximize learning activities and subject matter, as well as helping students independently to perform reflection against the learning material. Assessment or feedback is very important as a basis to refine learning [13]. From the explanation above, it can be concluded that learning assistedgeogebra tutorial better than usual in developing learning and to improve the ability of visual thinking students.

Based on the analysis of test ANOVA 2 lines, note that there is an increase in the ability of the visual thinking of students based on the category of KAM. From the analysis of ANOVA Tukey HSD further noted that there was no difference in improvement of visual thinking ability of students of a high singnificant between categories category with KAM currently, but there is a difference of visual thinking ability of students improved significantly between categories with KAM low category. And there is the visual thinking ability of students improved significantly between categories are categories with KAM.

This happens because students are high category of KAM a lot more benefit from the learning that has been implemented. Students of higher and medium KAM had mathematical ability is stronger than students with KAM students low so high and medium able to absorb information properly. In line with the results of the research of Elda [14] that students with high and medium KAM can receive the material well and are able to teach it to students who are low-skilled.

5. Summary And Suggestions

Based on the results of data analysis and discussion can be drawn the conclusion that the increased ability of visual thinking students who obtain learning assisted geogebra tutorial better than students who obtain the usual learning, and there is an increased ability of visual thinking students based on category KAM.

From the summary above, then it is recommended to teachers to make learning assisted geogebra tutorial as an alternative learning that can enhance the visual thinking ability of students. Next learning assisted geogebra tutorial suitable for Junior High School students applied to the categories medium and low for develop the visual power of students.

6. References

[1] National Council of Teachers of Mathematics (NCTM) 2000 *Principles and Standars for School Mathematics*(Resto, VA: NCTM)

[2] Suparyan 2007 *Kajian Kemampuan Keruangan (Spasial Ability) dan Kemampuan Penguasaan Materi Geometri Ruang Mahasiswa Program Studi Pendidikan Matematika FMIPA Universitas Negeri Semarang* (Semarang: Tesis Universitas Negeri Semarang)

[3] Rosken B, and Rolka K 2006 *Proceedings 30th Conference of the International Group for The Psychology of Mathematics Education* vol **4**,ed J Novotná, H Moraová, M Krátká and N Stehlíková (Prague: PME) pp 457-464.

[4] Stoke S 2001 *Electronic Journal for The Integration of Technology in Education*, **1** pp 10-19
[5] Wardhani and Rumiati 2011 *Modul Matematika SMP Program Bermutu, Instrument Penilaian Hasil Belajar Matematika SMP: Belajar dari PISA dan TIMSS* (Jakarta: Kementerian Pendidikan Nasional Badan Pengembangan Sumber daya Manusia Pendidikan dan Penjaminan Mutu Pendidikan Pusat Pengembengan dan PPPPTK Matematika)
[6] Wena M 2012 *Strategi Pembelajaran Inovatif Kontemporer: Suatu Tinjauan Konseptual Operasional* (Jakarta: Bumi Aksara)
[7] Hohenwarter M, and Fuchs K 2004 *Combination of dynamic geometry, algebra and calculus in the software system Geogebra* (Austria: Computer Algebra Systems and Dynamic Geometry Systems in Mathematics Teaching Conference)
[8] Saha R A, Ayub A F, and Tarmizi RA2010 *Procedia Social and Behavioral Sciences***8** pp 686-693
[9] Guven B, and Kosa T 2008 *Turkish Online Journal of Educational Technology***7** pp 100-107
[10] Sugiyono 2013 *Metode Penelitian Kuantitatif, Kualitatif, dan R&D* (Bandung: Alfabeta)
[11] Gumanti S 2014 *Pengaruh Pembelajaran Geogebra Terhadap Peningkatan Kemampuan Pemahaman dan Visual Thinking Siswa SMP* (Bandung: Tesis Universitas Pendidikan Indonesia)
[12] Rusman 2014 *Model-Model Pembelajaran; Mengembangkan Profesionalisme Guru* (Bandung: Alfabeta)
[13] Suherman E 2003 *Evaluasi Pembelajaran Matematika* (Bandung: JICA FPMIPA UPI)
[14] Elda H 2015 *Peningkatan Kemampuan Advanced Mathematical Thinking dan Disposisi Berpikir Kreatif Matematis Mahasiswa Melalui Pendekatan M-Apos*(Bandung: DisertasiUniversitas Pendidikan Indonesia)

Construct Validity of Science Motivation and BeliefsInstrument (SLA-MB): A Case study in Sumedang, Indonesia

A Rachmatullah[1], R P Octavianda[2], M Ha[1], N Y Rustaman[3] and S Diana[3]

[1] Division of Science Education, College of Education, Kangwon National University, Chuncheon-si, South Korea

[2] Department of Bio-Management, School of Life Sciences and Technology, Institut Teknologi Bandung, Jl. Ganesha No.10, Bandung, Indonesia

[3] Department of Biology Education, Faculty of Mathematics and Science Education, Universitas Pendidikan Indonesia, Jl. Dr. Setiabudhi No. 229, Bandung 40154, Indonesia

Email: arifraach@gmail.com

Abstract. Along with numerous instruments developed and used in science education researches, some of those instruments have been translated to local language in the country where the instruments were used. Most of researchers that used those translated instruments did not report the quality of those translated instruments. One of the instruments is the Scientific Literacy Assessment (SLA) including the Science Motivation and Beliefs (SLA-MB) as part of the SLA. In this study, the SLA-MB has been translated into Indonesian Language (Bahasa). The purpose of this study is to investigate the SLA-MB instrument that has been translated to Indonesian language from the view of dimensionality, reliability, item quality and differential item functioning (DIF) based on IRT-Rasch analysis. We used Conquest and Winstep as the program for IRT-Rasch analysis. We employed quantitative research method with school-survey on this study. Research subjects are 223 Indonesian Middle school students (age 13-16), with 64 boys and 159 girls. IRT-Rasch analysis of the SLA-MB Indonesian version indicated that a three-dimensional model fit significantly better than one-dimension model, and the reliability of each dimensions are about 0.60 to 0.82. As well as those findings, fit values of all items are acceptable, moreover we found no DIF for all of the SLA-MB items. Overall, our study suggests that Indonesian version of SLA-MB is acceptable to be implemented as research instrument conducted in Indonesia.

1. Introduction

The rapid development of science and technology in the 21st century demands humans not only have the basic literacy such as reading, writing and counting but requires another ability that makes human can survive and follow this century which full of science and technological products. Basic capability that enables humans to survive in the rapid development of science and technology is called as scientific literacy. As described by the Organisation for Economic Cooperation and Development or OECD [1], scientific literacy is the ability to understand scientific knowledge, identify scientific issues, drawconclusions based on the evidences and make decisions on how the human impact the environment. Thus, scientifically literate society can make decisions and participate actively in the discussions about scientific issues which are currently being ardently discussed [2]. Unfortunately, based on the results from international survey, such as the Program for International Student Assessment (PISA) which is a popular triannual international survey, from 2000 until 2012 Indonesian students achievement on scientific literacy are always categorized in the low category, even in the 2012 survey from 65 countries that participated on PISA Indonesian students achievement on scientific literacy was placed on rank 64 [3]. The Indonesian Government is not standing still to follow up the results of this "long-standing" low achievement. It is proved in line with the publication of a new Curriculum 2013, which one of the goals of Indonesian education is to increasethe performance of Indonesian students in the International surveys and studies especially in PISA [4]. However, scientific literacy is not only deal with the cognitive aspects, but also affective aspects which is from several studies and some experts prove and mentioned that affective could significantly affect cognitive aspects [5],[6].One of the affective component that has been scrutinied and has a relationship with scienctific literacy is the *motivation to learn science* [7].

The *motivation to learn science* is one's internal state which aroses, directs and sustains in learning-science behaviours [7]. Motivation is essentially an elaboration of Bandura's social cognitive theory whichassumesthat students will learn efectively when it is self-regulated, which takes place when students can control their motivation against what they learn and it will eventually lead towards their learning outcomes [7],[8]. Making students become motivated self-learner is the goal of every teacher [9], because students, who have become a motivated self-learner in science classes, will have much curiosity to solve and discuss vigorously on science-related problems. This kind of capability will eventually lead students to implement it on their daily lives which are multi-discipline. It is aligned with one of the definitions and purposes of scientific literacy which stated that individuals can implement the methods, concepts and theories of science in their daily lives. Hence, the *motivation to learn science* has an important role in improving the student's scientific literacy, especially Indonesian students, which besidesthe results from PISA, the results from Rachmatullah *et al.* showed thatIndonesian student's motivation towards science is still low which is in the value of 60% [10].

In order to evaluate Indonesian students' attitude and motivation in learning science, we need a valid, reliable and qualify instrument which meets the psychometric standard of instruments, so that the results which are obtained from the instrument can be trusted. Moreover, recently Fives *et al.* developed a set of instruments to measure scientific literacy ofmiddle school students, and the instrument is not only measure stidents cognitive aspect, but alsoit measures student motivation towards science which is called Scientific Literacy Assessment-Science Motivation and Beliefs (SLA-MB) [6]. In previous studies, we have been using SLA-MB which has been translated into Indonesian (Bahasa), and we only reported the validation of its content, but did not report the validation of the construct [10]. In fact the construct validity is the important thing that should be reported because as Messick stated about construct validity, it provides evidence to support the trustworthiness of the score interpretation in the explanatory concept that describes the relationship between scores in the performance and test with other variables. In short, construct validity is evidence-based to interpret the score that obtained from the studnets or participants [11].

Based on Messick [11] in addition to the content validity that has mentioned above, there are five other aspects of the Construct Validity namely substantive, structural, generalizabity, external and consequential aspects. A substantive aspect of the construct validity is the empirical evidence that related to the consistency of the responses on the instrument [12]. Structural aspect of construct validity is an aspect that related to the consistency of responses that it could describe a particular domain or specific dimensions that lie in the instrument. Aspects generalizabity assumes that an instrument must be fair from a variety of respondents, no group of respondents who get disadvantage when doing the instrument. External aspects of contruct validity sees the correlation between the developed instrumentwithother similar instruments (e.g. same content), while consequences aspects of construct validity predicts how the consequences or effect on respondents after doing the instruments. For more information about the construct validity see Messick "Validity of Psychological Assessment"[11]. In order to determine the construct validity of an instrument, especially in the realm of social sciences namelypsychology and education including science education, the most well-known methods are through Classical Test Theory (CTT) and Item Response Theory (IRT).

CTT assumes that the individual's attitudesscores on the rating scale form a linear combination with total score, which in CTT each item is presumed to havethe same difficulty level and standard error [13], so that when analyzing rating scale data (e.g. Likert scale) is often considered as interval data, and sum it directly from the raw data. In fact the numbers from the Likert-scale are ordinal data, which the differences between one points to another is exatly not known. As an example of the unclear distinction "strongly agree", "agree", "disagree", whether it was true that "agree" to "strongly agree" to have one point different attitudes? Point or score is a number belonging to the categorical data or namely ordinal data, which can not be used when testing statistical parametric tests [13], because only the data in the form of interval and ratio one which can be used in the statisticstests [14]. Thus, with the CTT method most reasearches are not same, and the results of the instrument, which were analyzed using CTT, can only be used in the same research subjects and can not be used for wider subjects [15].

IRT is one of psychometric methods to examine an instrument to be more reliable, valid and widely used [16]. IRT such as Rasch analysis thatis coinducted in this study will convert ordinal data into ratioscale data and produceitem parameters and person's parameters which are ratio data too [16]. The data which have been analyzed through Rasch analysis will become equal with other types of measurements such as height measurement, and so forth that is continuous data and has specificunit. The data that have been analyzed using Rasch method will have a unit called the logit Rasch (logarithm of odds) which isin interval/ratio level [13],[17].Therefore, recently many science education researchers mainly those who related to educational assessment havestarted using IRT as a method of testing the reliability, validity and legality of the instrument and began to leave the CTT method [18]. For more details about Rasch models see Boone *et al* [19]. Generally speaking, based on the above explanation the purpose of this study is to uncover the construct validity (dimensionality, generality and reliability) of SLA-MB which has been translated into Indonesian by using IRT-Rasch model where the method was not used by the instruments'developers.

2. Research methods

Proviously this research had a purpose to uncover Indonesian middle school students' scientific literacy in one district called Sumedang, so the data collection procedureis referred to how PISA collects the data, in which the subject is the student with average age of 13-16 years. In Indonesia, students with that age-range are in the ninth grade of middle school. Therefore, the research subjects are 223 middle school students of grade ninth consisting of 159 girls and 64 boys. Students who used directly as the subject of our research are derived from eight middle schools in four sub-regions in Sumedang district. Four of eight those middle schools are A accreditated schools and the other fourschools are the school with B accreditation, so that each sub-region have representation in both A and Bacreditation. Because it is only used for the pilot study and the purpose of generalization, we only collected samples one class from each school.

As has already been mentioned and described in the previous section about research instrument that we used, we employed one part of scientific literacy instruments that have been developed by Fives *et al* [6] which is the Scientific Literacy Assessments-Science Motivation and Beliefs. At the time of developing the SLA-MB, Fives *et al* divided it into three indicators which are the *value of science*, *self-efficacy* and *personal epistemology*. SLA-MB is formatted rating scale with Likert- scale 1 to 5types. Before employing SLA-MB as the research instrument, it was translated first from English into Indonesian by experts in the field of English-Indonesian and then re-checked by an expert in the field of science education for its contextualization fit. Table 1 is listed each item on each indicator of SLA-MB both Original English version and Indonesian version.

The construct validity of an instrument or assessment consists of six aspects: content, substantive, structural, generalizabity, external and consequential. In this intensive research we did not examine all of these aspects, but we just examined three aspects that we believe are the crucial aspects and highly correlated with the definition of construct validity. The third aspects that we have done are substantive, structural and generalizabity. Before analyzing these aspects, at the time of coding data, we reversed the score or points items that have a negative attitude. Then after that, firstly we examined the structural aspects. Exploring structural aspects, we employed the dimensionality test based on IRT-Rasch analysis. We tested two types of Rasch models for SLA-MB instruments, whether such instruments isunidimension or multidimension (three-dimension) that is accordance with the validity of the content.The higher chi-square (X2) value and lower value on Final Deviance and the Akaike Information Criteria (AIC) than other models indicatethat the model fit well to the instrument. Examining substantive aspects, we conducted an analysis of reliability by using IRT-Rasch (through PV-reliability) and CTT (Cronbach's Alpha), in addition we also reporteditem fit indices of each dimension by reporting mean-square (MNSQ), which the benchmark for MNSQ rating scale is between 0.7-1.4 [19]. And lastly, in order to uncover the generalizabity aspects, we did Differential Item Functioning (DIF) gender analysis and examines whether there is gender bias for each item or not. According to Boone *et al.* items contained bias could be seen from the DIFcontrast > 0.64 [19]. We used software Conquest 4.5.0, SPSS 22 and Winstep 3.68.2 to analyze our data.

Table 1. Original version and indonesian version of SLA-MB.

Indicator	Original Version [6]	Indonesian Version[10]	Code
Value of Science	In general, I find working on science assignments	Secara umum, saya melakukan suatu kerja dalam tugas sains.	VS1
	Compared to most of your other activities, how useful is what you learn in science?	Dibandingkan dengan banyaknya aktivitas kamu, seberapa bergunakah kamu belajar sains?	VS2
	For me, being good in science is	Bagi saya, menjadi pintar dalam sains adalah	VS3
	Compared to most of your other activities, how important is it for you to be good at science?	Dibandingkan dengan aktivitas lain, seberapa pentingkah untuk pintar dalam sains?	VS4
	How much do you like doing science?	Seberapa suka-kah kamu dalam mengerjakan sains?	VS5
	Some things that you learn in school help you do things better outside of class, that is, they are useful. For example, learning about plants might help you grow a garden. In general, how useful is what you learn in science?	Beberapa hal yang kamu pelajari di sekolah membantu kamu untuk melakukan hal yang lebih baik di luar kelas, berarti ini sangat berguna. Sebagai contoh, mempelajari tumbuhan yang kamu tanam di kebun. Secara umum, seberapa berguna-kah apa yang kamu pelajari di dalam sains?	VS6
Self-Efficacy	I know when to use science to answer questions.	Saya tahu kapan saatnya menggunakan sains untuk menjawab suatu masalah	SE1
	I can use science to make decisions about my daily life.	Saya dapat menggunakan sains untuk membuat suatu keputusan mengenai kehidupan sehari-hari saya.	SE2
	I know how to use the scientific method to solve problems.	Saya tahu bagaimana menggunakan metode ilmiah untuk memecahkan suatu masalah.	SE3
	It is easy for me to tell the difference between scientific findings and advertisements.	Sangat mudah bagi saya untuk membedakan antara temuan ilmiah dan yang bukan.	SE4
	When I do my work in science class, I am able to find the important ideas.	Ketika saya melakukan pekerjaan saya di kelas sains, saya dapat menemukan ide-ide penting.	SE5
	I can use math to answer scientific questions.	Saya dapat menggunakan matematika untuk menjawab pertanyaan ilmiah.	SE6
	I can tell the difference between observations and conclusions in a story.	Saya dapat membedakan antara observasi dan kesimpulan pada sebuah cerita.	SE7
	It is easy for me to make a graph of my data.	Sangat mudah untuk membuat grafik dari data yang saya dapat.	SE8
Personal Epistemo-logy	Everybody has to believe what scientists say.	Semua orang harus percaya apa yang dikatakan ilmuwan.	EB1
	All questions in science have one right answer.	Semua pertanyaan dalam sains hanya mempunyai satu jawaban benar.	EB2
	Scientific knowledge is always true.	Pengetahuan ilmiah selalu benar.	EB3
	In science, you have to believe what the science books say about stuff.	Dalam sains, kamu harus mempercayai sesuatu yang dikatakan dalam buku sains.	EB4
	The most important part of doing science is coming up with the right answer.	Bagian terpenting dalam mengerjakan sains yaitu dengan mendapatkan jawaban yang benar.	EB5
	Whatever the teacher says in science class is true.	Apapun yang guru katakan dalam kelas sains adalah benar.	EB6
	Scientists pretty much know everything about science; there is not much more to know.	Ilmuwan sangat mengetahui mengenai segala sesuatu tentang sains, sehingga tidak ada lagi keraguan.	EB7
	If you read something in a science book, you can be sure it's true	Jika kamu membaca sesuatu di buku sains, kamu dapat meyakini itu benar.	EB8
	Once scientists have a result from an experiment that is the only answer	Begitu para ilmuwan mendapatkan hasil dari sebuah penelitian, itu menjadi satu-satunya jawaban.	EB9
	Scientists always agree about what is true in science	Para ilmuwan selalu setuju tentang apa yang benar dalam sains	EB10
	Only scientists know for sure what is true in science.	Hanya para ilmuwan yang tahu pasti apa yang benar dalam sains.	EB11

3. Findings

As can be seen on Table 2, a dimensionality comparison test between unidimension and three-dimension of the SLA-MB is showed that SLA-MB fit the Rasch model with three dimensions. This is supported by the acquisition of three-dimensional has higher X^2 value than unidimension model, which are 976.76 and 904.62, respectively. In addition three-dimensional model also has lower FinalDeviance and lower AIC (13432.49 and 13500.49 respectively) compared to SLA-MB unidimension model which has 13712.81 for Final Deviance and 13770.81 for AIC.

Table 2. Item dimensionality test results using a Rasch analysis.

Model	X^2	df	Sig.	Final Deviance	Akaike Information Criterion (AIC)
One Dimension	904.62	24	0.00	13712.81	13770.81
Three Dimensions	976.76	22	0.00	13432.49	13500.49

Substantive aspect of SLA-MB was examined through reliability test, which the results are shown on Table 3. based on Table 3 the reliability values of each dimensions through CTT methods (Cronbach's alpha) and IRT-Rasch (PV-reliability) are above 0.6, which means that thethe results obtained from instrument can be accepted.

Indicator Value of Science has a reliability value on CTT and IRT-Rasch 0.691 and 0.735 (respectively), for self-efficacy has value of reliability 0.603 and 0.662, and for the Personal Epistemology dimension has reliability value 0.785 and 0.817. In addition to value the reliability, we examined the substantive aspects to explore the fitof each item with the Rasch model, the results are shown on the value weighted and unweighted MNSQ on Table 4. For dimensions Value of Science, the MNSQ value of each item ranged from 0.87 – 1.24 logit, for the dimensions of self-efficacy MNSQ values ranged from 0.79 – 1.21 logit and for personal epistemology dimension ranged from 0.65 – 1.31 logit. According to the MNSQ value rangedbetween 0.7-1.4 logit for almost all of the items in each dimension of SLA-MB, it means that the items fit Rasch model.

Table 3. PV-Reliability and Cronbach's Alpha.

Model	Indicators	Reliability	
		Cronbach's Alpha	PV Reliability
Three-Dimension	Value of Science	0.691	0.735
	Self-Efficacy	0.603	0.662
	Personal Epistemology	0.785	0.817

As shownon Table 4, we did not find any DIF gender or gender bias from all of SLA-MB items. The DIF contrast values of each item are still in the value of lower than 0.64. DIF contrast that is rangedon the values of science dimension is from 0.03 to 0.27, on the dimensions of self-efficacy is ranged from 0.03 to 0.54 and Personal Epistemology from 0.09 to 0.35.

Table 4. Item fits for three dimensions and differential item functioning (DIF) gender of SLA-MB Items.

Dimensions	Item Code	Estimates	Unweighted MNSQ	Weighted MNSQ	DIF Gender Contrast
Value of Science	VS1	0.75	0.87	0.87	0.12
	VS2	-0.21	1.17	1.16	0.27
	VS3	-1.08	0.94	0.97	0.03
	VS4	0.24	0.99	1.01	0.24
	VS5	0.94	1.06	1.07	0.07
	VS6	-0.64	1.16	1.24	0.12
Self-Efficacy	SE1	-0.18	0.79	0.80	0.54
	SE2	-0.16	0.96	0.96	0.09
	SE3	0.22	0.80	0.80	0.21
	SE4	0.43	0.85	0.85	0.22
	SE5	-0.47	1.02	1.01	0.03
	SE6	0.19	1.21	1.20	0.35
	SE7	0.23	1.01	1.02	0.08
	SE8	-0.26	1.08	1.09	0.09
Personal Epistemology	EB1	-0.04	1.05	1.03	0.15
	EB2	0.06	1.23	1.23	0.35
	EB3	0.10	0.84	0.84	0.21
	EB4	0.16	0.66	0.65	0.09
	EB5	-0.84	1.19	1.21	0.33
	EB6	0.37	0.85	0.85	0.12
	EB7	-0.03	0.93	0.92	0.12
	EB8	0.07	0.79	0.78	0.31
	EB9	-0.11	1.28	1.28	0.17
	EB10	-0.24	0.93	0.93	0.17
	EB11	0.50	1.31	1.29	0.23

4. Discussions

The validity of an instrument is an overall evaluative assessment of the empirical and theoretical evidence supporting the feasibility and suitability of interpretation and feed-back based on the test scores or other parts of the assessment [11]. Therefore, the validity of an instrument is not only as a supporting instrument or assessment but also as a reflection of what we explore through the instruments. A valid instrument will produce a finding, data and other thingsthat we searched through itby reflecting what it should be reflected in and obtained from the instument. Based on our findings on the construct validaty of SLA-MB that have been described in the previous section, in this section we will discuss the extent to which aspects of construct validity are traced from this study might be explained.

First, the structural aspect of construct validity is related to the consistency of response so that from participants that can describe a particular domain or specific dimensions. Based on Table 2, it is shown that the numbers of domains that exist on the SLA-MB based IRT Rasch analysis arethree domains, which indeed it is in accordance with the content validity that has been validated by the original developers of SLA-MB. Those three domains or dimensions are the value of science, self-efficacy and personal epistemology. The importance of disclosure the domains or dimension besides indeed reveal the structural form of an assessment or research instruments, with the disclosure of an instrumentdimensionsionality will provide clarity regarding what are the variables that traced from research instruments that we used.Basically an instrument or assessment can uncover one or more variables whether it will be knowledge, performance or attitude of the respondents [18]. Why we should use a statistical analysis, in this case IRT-Rasch analysis, and why do not it jut divided based on its content? As explained previously that the construct validity not only deal with the theoretical aspect (in this case the content), but it should be dealt with empirical aspect which isreflected from data sample. Thus, the validity of the content and structural in this case are complemented to each other. The implications of the acquisition of an instrument dimensionality is to uncover the boundaries of an item in instruments that measure the properties that are supposed to be measured by the instrument which is in the domain or the same dimensions with other items that measure the same variable.

The results of the dimensionality investigation will impact on the quality of every item that exists in every dimension. Based on the results that shown in Table 4 especially on the fit indices or MNSQ of each item, we found that almost all items has MNSQ values are met the benchmark. It shows that every item is fit with Rasch models, in which case each item fit with every dimension they are occupied. But, we found one of the items on the Personal Epistemology dimension that overfit Rasch model, the item is EB4 with MNSQ 0.66. We assume that the item is still fit the Rasch model because the relative value of 0.66 is still being near the cutoff.In addition to the items fit, substantive aspects of an instrument can also be explained by the reliability of the instrument or the internal consistency of the instrument. With sufficient reliability, then a predictable instrument can provide consistent measurement results [18]. Based on the results of reliability test through CTT with Cronbachs' alpha and reliability based on the Rasch model with the PV-Value, every dimension of SLA-MB possessed reliability values which belong to the instrument which has good internal consistency [20].

The last finding ofour study is Differential item Functioning (DIF) gender of SLA-MB. The purpose of analyzing DIFgender is to examine whether one gender group benefited at the time doing on each item in the SLA-MB and othergender groups disadvantaged. Based on the results that shown in TABLE 4, we found that there is no big gender bias from all of the items (DIF contrast <0.64). It reveals that the SLA-MB instrument is not an instrument that is biased or in favor of one gender group, it also supports the generality of SLA-MB. The absence of DIF or bias of the instrument SLA-MB implies that the obtained survey results can perform a comparative study of gender because no significant gender bias in each item. In contrast, if we found gender bias, for example, some items more easily answered by girls than boys, then the findings could not do a comparative analysis of gender, because it is quite predictable that girls might have higher results.

5. Conclusions and limitations

Based on our findings, the exploration of contruct validity on SLA-MB instrument using IRT-Rasch, it was found that the SLA-MB fit with the three dimensions Rasch model which consist of the Value of Science dimension, Self-efficacy dimension and the Personal Epistemology dimension, which corresponds to the validity of the content that have been analyzed by SLA-MB developers. Based on substantive aspects, SLA-MB has acceptable value of internal consistency and has MNSQ values that indicateeach item of SLA-MB fit with Rasch models. In addition, in the aspect of generalizabity, there is no gender bias of each item ofthe SLA-MB, so that each item has the same difficulty level if done well by female students and male students. With such results, we conclude that the SLA-MB Indonesian version can be used as an instrument to explore motivation, values and beliefs of Indonesian students towards science.

Although this study has done to produce and reveal the construct validity of the SLA-MB using IRT-Rasch, but not all aspects of construct validity are revealed. Therefore, for further study more aspects of the disclosure of the construct validity are strongly advised to obtain the better validity and legitimacy of instruments that fulfill the pre-requisites of psychometric and psychology.

6. References

[1] OECD 2006 *Assesing Scientific, Reading and Mathematical Literacy: A Framework for PISA* (OECD Publishing) pp 1-187
[2] Glynn S M, Taasoobshirazi G and Brickman P 2007 *Journal of Research in Science Teaching* **44** 1088-107
[3] OECD 2014 *PISA 2012 Results in Focus: What 15-Year-Olds Know and What They can do with What They Know* (OECD Publishing) pp 1-39
[4] Indonesian Ministry of Education 2013 *Dokumen Kurikulum 2013*
[5] Holbrook J and Rannikmae M 2007 *International Journal of Science Education* **29** 1347-62

[6] Fives H, Hubnuer W, Bimbaum A S and Nicolich M 2014 *Science Education* **98** 549–80
[7] Glynn S M, Brickman P, Norris A and Taasoobshirazi G 2011 *Journal of Research in Science Teaching* **48** 1159-76
[8] Bandura A 2006 *Applied psychology: New Frontiers and Rewarding Careers* (London: Psychology Press) pp 53-79
[9] Druger 2006 Experiential learning in a large introductory biology course *Handbook of College Science Teaching* (Arlington: National Science Teacher Association Press) pp 37-43
[10] Rachmatullah A, Diana S and Rustaman N Y 2016 *International Seminar on Mathematics, Science and Computer Science Education 2015 (Bandung)* (New York: AIP Conference Proceedings) ed Hidayat *et al* pp 080008-1– 080008-5
[11] Messick S 1995 *American Psychologist* **50** 741
[12] Loevinger J 1957 *Psychological Report*
[13] Sondergeld T A and Johnson C C 2014 *Science Education* **98** 581-613
[14] Bode R K and Wright B D 1999 Rasch measurement in higher education *Higher Education: Handbook of Theory and Research* (Netherlands: Springer Netherlands) pp 287-316
[15] Hambleton R K 2000 *Medical Care* **38** 60-5
[16] Bond T and Fox C M 2015 *Applying the Rasch model: Fundamental Measurement in The Human Sciences* (New York: Routledge) pp 1-382
[17] Boone W J, Townsend J S and Staver J 2011 *Science Education* **95** 258-80
[18] Neumann I, Neumann K and Nehm R 2010 *Int. J. Science Education* **33** 1373-405
[19] Boone W J, Staver J R and Yale M S 2005 *Rasch Analysis in The Human Sciences* (Dordrecht: Springer) pp 69-91
[20] Field A 2013 *Discovering Statistics using IBM SPSS Statistics* (London: Sage Publication)

IOP Conf. Series: Journal of Physics: Conf. Series **812** (2017) 012010 doi:10.1088/1742-6596/812/1/012010

The Effect of Concept Attainment Model on Mathematically Critical Thinking Ability of The University Students

L MAngraini[1,2], B Kartasasmita [1,3],and D Dasari[1]

[1]Departement of Mathematics Education, Universitas Pendidikan Indonesia, Jl. Dr. Setiabudhi No. 229,Bandung 40154, Indonesia.
[2]Departement of Mathematics Education, Universitas Islam Riau, Jl. Kaharuddin Nasution No. 113,Pekanbaru 28284, Indonesia.
[3]Departement of Mathematics Education, Universitas Pasundan, Jl. Dr. Setiabudhi No. 193, Bandung 40154, Indonesia.
Email: lilismarina@student.upi.edu

Abstract. This study examined the university students' mathematically critical thinking ability through Concept Attainment Model learning. The Kolmogorov-Smirnov test, Levene test, t test, ANOVA one and two ways were used to analyse the data. The results of this study showed that (1) there is no difference grade on the student's mathematical critical thinking ability between experimental group and conventional group as a whole, (2) there is no difference on the students' mathematical critical thinking ability of experimental classes based on their mathematical early ability (3) there is no interaction between the learning that is used with the students' mathematical early ability on the students' mathematical critical thinking ability.

1. Introduction

Committee on the Undergraduate Program in mathematics (CUPM) [1] provided six basic recommendations for the department, programs, and courses in mathematics. One of recommendation stated that each course in mathematics should be an activity that will assist students in the development of analytical, critical reasoning, problem solving, and communication skills. The CUPM Recommendation above explained, the task of the institution in charge of educating prospective teachers who will teach mathematics, including prepare students to have mathematically critical thinking ability. Institutions of Higher Education Personnel (LPTK), Elementary School Teacher (PGSD) which is in charge of childbirth prospective teachers who will teach math partially take responsibility to prepare their students to have mathematically critical thinking ability.Mathematically critical thinking ability is the ability that could be cultivated, so that the lecturer had a role in the development efforts of students' mathematical critical thinking ability.

Inch, *et a*l [2] say*"Critical thinking is a process in which a person tries to answer rationally those questions that cannot be easily answered and for which all the relevant information is not available"*. Critical thinking is more focused on doing something, the students are able to think critically not immediately believe the explanation from the lecturer, but students were trying to consider the reasoning and find other information to obtain the truth.

According [3] critical thinking is characterized by rational activities such as: interpreting information based on the framework of a particular theory, linking theory to practice, file a claim and justify it, utilizing the data in support of the argument, making the relationship between ideas, ask questions, evaluate knowledge, predict, describe something, analyzing, synthesizing, categorizing, comparing or contrasting, identify problems and solve them.

According to Bruner, learning process will go well and creatively, if lecturer give students an opportunity to find a rule (including concepts, theories, definitions, and so on) through examples that describe / represent rules that became the source [4], in other words, students are guided inductively to understand a common truth. Study process that presented by Bruner is in line with the theory of concept attainment model according to Bruce [5] concept attainment model is more focused on ways to strengthen the internal human impulses in understanding science, by digging and organizing, as well as developing the language to express it. Bruner, Goodnow, and Austin reveal concept attainment model deliberately designed to help students learn the concepts that can be used to organize information, so it makes easy for students to learn concepts in more effective way [6]

Concept attainment model has several stages of learning, the stages in concept attainment models help train students in critical thinking mathematically. The first learning stage, namely the presentation of the data and the identification of the concept, in this stage, students are asked to compare the characteristic features in the example and non-example, students were asked to create and test the hypothesis, then the student making the definition of the concept on the essential characteristic features, here the students are trained to think critically because students asked to compare the characteristic features in the examples and non-examples, so that they can find the definition of the concept on the characteristic of concept essential features. The second stage is the stage of testing the achievement of the concept, at this stage the students were asked to identify examples of concepts and create additional instances, here the students are trained to think critically for students to think about what characteristics are represented on a concept and what kind of example that meets the criteria of the concept. The third stage is the stage of analytical thinking strategy, at this stage the students are trained to communicate math, because students are asked to express the concept with their own words, reveals the reasons relating to create additional instances, and write down the steps to resolve the task of the concept being studied, and formulated the mathematical concept, so that this learning model proposed by Bruce, *et al* , theoretically believed that students' mathematical critical thinking ability become more better.

Kauchak and Eggen suggested that concept attainment model is an inductive learning model, designed by lecturer to help students learn the concepts and train the students in practicing high-level thinking skills [7]. Concept attainment model is very relevant in the teaching of mathematics [8], because the concept attainment models can foster understanding and appreciation of students to the concepts, principles so grows the power of reason, think logically, critically, systematically and others.

Concept attainment model is a learning model that aims to help students understand a particular concept, this model is more appropriate when the emphasize of learning is more focused on the introduction of a new concept, so as to train high-level thinking skills [4].Formulation of the problems are:Is there any significant difference between students' mathematically critical thinking who taught by concept attainment model with the students who taught by conventional teaching as a whole?Is there any significant difference on the students' mathematically critical thinking ability who taught by concept attainment model based on their mathematical early ability (KAM)? Is there an interaction between the learning that is used and the mathematical early ability (KAM) on the students' mathematical critical thinking ability?

2. Experimental Method

The research plan design for this experiment is depicted in Figure 1 below.

XO

O

Figure 1.Design a plan[9]

IOP Conf. Series: Journal of Physics: Conf. Series **812** (2017) 012010 doi:10.1088/1742-6596/812/1/012010

Information :
O: Give the test (mathematically critical thinking ability)
X: Learning concept attainment model

This research was conducted in one of the Islamic State University. The population of this research was all students of Elementary school Teacher Education (PGMI) 5th Semester. All the population selected as sample of the study, the control class and experiment class.

The materials tested are two-dimensional figure that contained in the subject of Mathematics Education II. The test instrument of mathematically critical thinking consists of three questions that formed as description. Time Allocation for execution of the test is 3 x 50 minutes.

3. Result and Discussion

The data of Mathematically Early Ability (KAM) is collected and analyzed to determine the students' Mathematically Early Ability before this research was conducted. Mathematically Early Ability was obtained from the grades of mathematics education subject, which they acquired in the fourth semester. Then the grade is grouped by early capability categories of high, medium and low. The following Table 1 is the test results KAM Data equality both groups learning:

Table 1. Data Equality Test of KAM

Test-t	Data	Criteria
N	82	H_0 Accepted
Sig. (2-tailed)	0,23	

The table above shows there is no difference of KAM average between students who taught byconcept attainment model learning and students who taught byconventional model learning. The test results of data equivalence of Students' mathematically critical thinking ability based learning model that used is presented in the following Table 2:

Table 2. Data Equality Test of Matematically Critical Thinking Ability

Test-t	*Data*	Criteria
N	82	H_0Accepted
Sig	0,122	

The table above shows there is no significant difference between students' mathematically critical thinking ability who taught by concept attainment model learning and students who taught byconventional model learning. The results of test calculations differences of students' mathematically critical thinking ability who taught by concept attainment model presented in the following Table 3.

Table 3.ANOVA Test One Way

	The sum of squares Squares	Df	Average	F	Sig.
Inter-group	112,93	2	61,46	1,29	0,28
In the group	1949,61	41	47,55		
Total	2072,54	43			

The table above shows there is no difference of students' mathematically critical thinking ability who taught by concept attainment model based on KAM. Furthermore, The results of calculation of the effect of interaction between the learning that used and KAM on the students' mathematically critical thinking ability will be tested using ANOVA two ways are presented in the following Table 4:

Table 4. ANOVA Test Two Ways

	The sum of squares Squares	Dk	Average	F	Sig.	H_0
Learning	134,59	1	134,59	2,88	0,09	Accepted
KAM	302,16	2	151,08	3,23	0,04	Denied
Interaction	74,45	2	37,22	0,79	0,45	Accepted
Error	3546,24	76	46,66			
Total	12127,00	82				

The table above shows that the learning factor used in each study group did not have a significant impact on the students' mathematically critical thinking ability. Afterward, KAM factor that owned by each study group had a significant influence on the students' mathematically critical thinking ability. Furthermore, there is no significant effect caused by the interaction between the learning factor that used and each KAM group learning on the students' mathematically critical thinking ability.

4. Conclusion

The conclusions found in this study are as follows. First, the time of the study is only six meetings and one meeting to the test, causing limitations of students to develop mathematically critical thinking ability. Especially in mathematically critical thinking ability that takes practice continuously so that students become increasingly critical in learning. The implementation time which is lasted for two months there not too make students used to learning by using the concept attainment model, so that the results are as expected, more research time should be extended again.

Second, there are many experimental class students who do not understand the basic concepts of algebra operations, so during the learning process the researcher more preoccupied with fixing concepts that are still wrong, as a result of learning with concept attainment model less effective due to the ability of students that are not balanced. Learning by concept attainment model is more stresses on the discovery of concept by each individual. Because many students still wrong with the basic mathematical concepts, the learning much more constrained, lecturers should improve the concepts that are still wrong, and then pass the learning materials.

Third, a large number of students in one class (experiment class), were 44 students, so the researchers somewhat overwhelmed in providing stimuli so they can achieve the concept. Of the 44 students, only about 15 students who have basic mathematical concepts were good. During the teaching process the lecturer come to the students (which basic mathematical concepts was not yet good enough) one by one giving the stimuli of learning materials to keep students achieve its own concept.

Fourth, 5th semester PGMI Students, overall not have interest in learning mathematics. Furthermore, in the 7th semester they will be grouped according to their specialization. From 44 students of Experimental class only about 15 people who wish to join math majors, others wish to choose Indonesian majors and Natural Science. Those who will choose math majors, seemed enthusiastic during the course in following the subject matter. But those who chose Indonesian Language (Bahasa) and Natural Science major less enthusiastic for learning, in addition of their basic concepts of mathematics are still wrong, they also felt that learning mathematic is caused dizziness, and later they also do not want to be a teacher Madrasah Ibtida'iyah (MI) who teaches mathematics, It seems most of their interest in learning math is not good.

Fifth, students used to learn to accept the subject matter. Students usually pay attention, and lecturers explain and use power point. This kind of habit learning is unfamiliar and difficulties for students with learning concept attainment model, because learning by using this model requires students to discover concepts on their own without direct lecturer explained, the role of lecturer only provide stimuli until they can find their own concept. Because learning by new concept attainment model is new experienced by students, therefore students need habituation so that students are not confused following study.

Sixth, the Mathematically critical thinking ability is a mathematical ability that have a high grade. So that the effect of concept attainment model learning model looks better, it takes a habituation process of learning that focuses on students' mathematically critical thinking ability.

5. References

[1] Mathematics Association of America 2004 *Undergraduate Programs and Courses in the Mathematical Sciences: CUPM Curriculum Guide 2004* (Washington DC: The Mathematics Association of America Published) p 2
[2] Inch,*et al* 2006 *Critical Thinking and Communication* (Boston: Pearson Education, Inc) p 5
[3] Sihotang,*et a*l 2012 *Critical Thinking* (Jakarta: Pustaka Sinar Harapan) p 2
[4] Uno B H 2008 *Model Pembelajaran* (Jakarta: BumiAksara) p 10-12
[5] BruceJ,*et a*l 1992 *Models of Teaching* (Boston: Allyn and Bacon) p 125
[6] Aunurrahman 2009 *Belajar dan Pembelajaran* (Bandung: Alfabeta) p 158
[7] Silitonga M 2006 *Jurnal Penelitian Bidang Pendidikan*. **12** 172
[8] Mustamin A 2005 *Pembelajaran Matematika Menggunakan Model Pembelajaran Pencapaian Konsep* (Kendari: WAKAPENDIK) p 72
[9] Ruseffendi E T 2005 *Dasar-dasar Penelitian Pendidikan dan Bidang Non-eksakta Lainnya* (Bandung: Tarsito) p 49

Acknowledgments
We thank Siti Mukhsinah who has edited this paper.

Collembola Diversity between Chemical Pesticide and Bioinsecticide in Broccoli Farm (*Brassica oleraceae* var. *italica*)

Y Sanjaya and Suhara

Biology Education Program, Universitas Pendidikan Indonesia, Jl. Dr. Setiabudi No. 229, Bandung 40154, Indonesia
Corresponding author: yayan@upi.edu

Abstract. The existance of Collembola diversity was determined by how land system work. Farming systems with excessive pesticide application can reduce number of Collembola. On the other hand nowaday people aware of environment by using bioinsecticide. The Method were comparing two land system which use Chemical pestisice and Bioinsecticide. Procedure were using Trapping wells (PMS) in three plots; T0: control without treatment, T1: Chemical Insecticide, T2 : Bioinsecticide for 24 hours. The factors that measure are abiotic factors by taking 10 grams of soil planting Broccoli (*Brassica oleraceae* var Italica), after 24 hours of taking separates it with other land animals, then identifying Collembola species were using Microcam based on identification book of Collembola. The result showed that density and Biodiversity of land system bioinsecticide was the highest value and indic. It was found also that in Broccoli farm dicovered 3 Familia and 8 species of Collembola both litter and soil. Species found that *Isotomurus* sp, *Seira* sp, *Lepidosira* sp, *Coecobrya* sp, *Callyntura* sp, *Homidia* sp, *Sallina* sp and *Ascocytrus* sp, three Family is derived from Isotomidae, Entomobryidae and Paronellid.

1. Introduction

Nowadays organic farm system is developed to a friendly environment since chemical-based agriculture not used anymore. The role of soil organisms in soil, decomposition processes of is more critical in organic farming than in chemical-based agriculture. According to Soehardjono et *al.*, (2012) there are some advantages of organic system that connected in health food, clean environment, well organsed land structure, mineral cycle, land fertility restore.

One that play role of decomposition is Collembola. Collembolan together with mites, are among the most abundant soil microarthropods that play an important role in decomposing soil food webs [13]. Mites are dominant in forest soils and undisturbed habitats, whereas Collembola are increasingly important in managed grasslands and especially in arable land. Because of their large population sizes (10^4 to 10^5 m^{-2}), rapid reproduction rates, role in fragmenting detritus, and grazing upon microbes, Collembola have a significant impact on microbial dynamics [10].. Although, Collembola and other mesofauna have a function similar to macrofauna in attacking and fragmenting plant residues, they are considered to be more important in regulating microbial populations rather than macrofauna, e.g., amphipods, isopods, centipedes and millipedes, earthworms and mollusks like snails and slugs [17]. Collembola closely interact with all elements of the decomposer food web [9] and are active under most environmental conditions – unlike nematodes, bacteria or earthworms.

The supply and availability of plant nutrients in the soil is mainly regulated by rate of release of plant nutrients from soil organic matter. The soil fauna is likely to play an important role in this process. The effect of the use of organic fertiliter on decomposition and mineralization of organic matter by soil fauna is a question of interest. Hence, a better understanding of these processes is needed to find out whether the decomposition and mineralization of nutrients from organic matter could be enhanced by improving soil conditions for soil fauna. This will have significant implications for sustainable agroecosystem management in the future. Thus the aim of this study is to investigate diversity of Collembola as decomposer with pesticide and biopesticide treatment in Brocolli farm.

2. Experimental Method

This Research was conducted in a vegetable farm that located in North Bandung West java Inonesia. Plastic bags filled with 3 kg soil were prepared for each treatment using Randomized Complete Block Design with five treatments and one control, with three replications . Each treatment comprised of a raised-bed plot measuring 1 m x 7 m with a distance of 30 cm between them. The field experiment had the same treatments as the greenhouse experiment. The treatments in this study were formulated as follows:

T0: control without treatment
T1: Chemical Insecticide
T2 : Bioinsecticide (*Tephrosia vogelii*)

2.1. Soil Sampling, Extraction and Collembola Identification

Soil samples for the study of Collembola were taken three times during the growth season (three replicate plots per treatment). Samples were collected at three sampling points per plot using steel cylinders (0-10 cm depth, 100 cm2). Soil microarthropods were extracted from the soil samples using an extractor based on the method by Tullgren (1918) for 24 hours. The extraction used standard 25 W electric light bulbs. Microarthropods were collected in small plastic bottles filled with 70% isopropanol. The extraction was terminated after 5 days. Collembola were identified at the Research Center for Biology, Indonesian Institute of Science, Cibinong, West Java.

2.2. Application of chemical insectiside and Tephrosia vegolii extract

Two weeks later in the three plots of field in the planting of broccoli (Brassica oleraceae var Italica) apply chemical pesticides at T0 and T1 while on T1 added insecticide treatment plant of Tephrosia vegolii. Chemical pesticides by mixing 25 pesticide in into 7 liters of tap water. Giving pesticide and insecticide plant by spraying the surface of the leaves of Cauliflower (Brassica oleraceae var Italica). Two weeks later in the three plots of land in the planting of Broccoli (Brassica oleraceae var Italica) is re-treated by adding chemical pesticides at T0 and T1 while on T1 added insecticide treatment plant of leaves bean pig (Tephrosia vegolii). Chemical pesticides by mixing 5 ml Antracol, 10 ml and 10 ml Cucrakon Dursban into 7 liters of tap water. Giving pesticide and insecticide plant by spraying the surface of the leaves of broccoli (Brassica oleraceae var Italica).

3. Collection data

Soil samples taken were taking after treatment to get Collembola. Samples were collected per plot using plastic cylinders (0-10 cm depth, 250 cm2). Samples were identified in animal lab. and confirmed at the Research Center for Biology, Indonesian Institute of Science, Cibinong, West Java. Collembola was identified up to the genus level According to [6] .

Trapping wells. Thre traps were design int three plots and were observed for 24 hours. Take 10 grams of ground planting Cauliflower (Brassica oleraceae var Italica), after 24 hours of taking the Trapping wells and separating Collembola with other land animals, then identify Collembola are already caught using Microcam based identification book Collembola.

4. Result and Discussion

4.1. Relative Abundance of Each Genus of Collembola
A higher number of genera of Collembola were *Isotomurus, Ascocytrus sp and Homidia sp. Ascocytrus sp* , was the most dominant. It is notable that some species from four genus were most abundant under organic crop management [2].

Tabel 1. Collembola abundance before treatment on Broccoli plantation

Genus	Control		Pesticide		Biopesticide	
	Before treatment	After treatment	Before treatment	After treatment	Before treaatment	After treatmeent
Isotomurus sp	8	19	7	12	8	16
Ascocytrus sp	26	28	22	29	22	27
Lepidosira sp	-	-	-	-	-	4
Seira sp	-	-	-	-	-	7
Homidia sp	9	8	13	13	10	18
Coecobrya sp	-	-	-	-	-	
Callyntura sp	-	2	-	2	-	
Sallina sp	-	-	-	-	-	
Total	**43**	**57**	**42**	**56**	**40**	**80**

In dry periods, Collembola normally would seek a more favorable habitat. They will go into the deeper layers in order to survive until the rain again [10]. As a result, Collembola who move on the surface will go into the litter and soil. This is consistent with the above results, which led to high or low number of individuals using traps "pit fall trap" due to the migration of Collembola ground. Soil temperature can affect the migration of Collembola, due Collembola will find a suitable place for its survival [14] . Organic matter, temperature, and humidity of the air and soil in the forest nature reserve is low. Organic matter, temperature, and humidity can affect the abundance of Collembola. Because there are several species of Collembola are sensitive to soil moisture so that the variation in species composition and different populations [7].

On the other hand, under chemical fertilizer + insecticide conditions lower density of Collembola was observed. Obviously, under organic condition, Collembola preferred organic manure compared to control as their diet or substrate. It is well known that Collembola food preference depends on the substrate on which the fungus is grown [8]. Field condition is likely to provide better condition for microorganisms, including organic fertilizer and pesticide which contains bacteria, actinomycetes and fungi, to grow on green manure. While in the field condition, this association between microorganisms, Collembola and green manure were greatly influenced by heavy rain.

[10] mentions that Collembola have a high response to differences in environmental and ecological factors such as changes in the chemical content of the soil, the configuration microhabitat and agricultural activities contained therein. So that the reconstruction in this Broccoli plantation can change the situation in the surrounding microhabitat that directly impact the presence of Collembola in it. Some studies suggest that Collembola have their own preferences in choosing its habitat. The existence of Collembola more on natural forest than in agricultural areas. Additionally, [12] states that the density and diversity of Collembola more numerous in areas not disturbed than disturbed areas.

4.2. Abiotic factors
The results of measurements of abiotic parameters in Broccoli plantation include air temparatute, light density soil moisture and soil pH. Can be seen on Picture 1.This is according to the results of research [16] which states that the presence of soil fauna is strongly influenced by environmental factors, such as soil moisture and soil pH. Based on Table 4.1 it can be seen that the temperature of the soil in the Broccoli plantation of 25,3 oC-26,6oC. This is consistent with the results of the study of [1] which states that Collembola that live in the area have a tendency survive better in winter than

summer. Overall the optimum temperature required by Collembola to live including a lowly situated between 5-15°C, but there was also active at temperatures -2°C or 28°C (Ganjari, 2012). Resistance to high and low temperature varies depending on the type and age (Amir and Andrew, 2008). Humidity in the botanical garden that is 76.7%, the moisture in the Botanical garden UPI were high. This is consistent with the proposed [5] soil moisture suitable for the survival of the Collembola high humidity. If the humidity is low Collembola will migrate to deeper soil layers. Based on Table 4.1 it can be noted that in the botanical garden soil pH is acidic which is equal to 6.5. This happens because the Collembola belong to the land mesofauna whose population is most prominent in the land with soil pH that is acidic ([18]; [4])

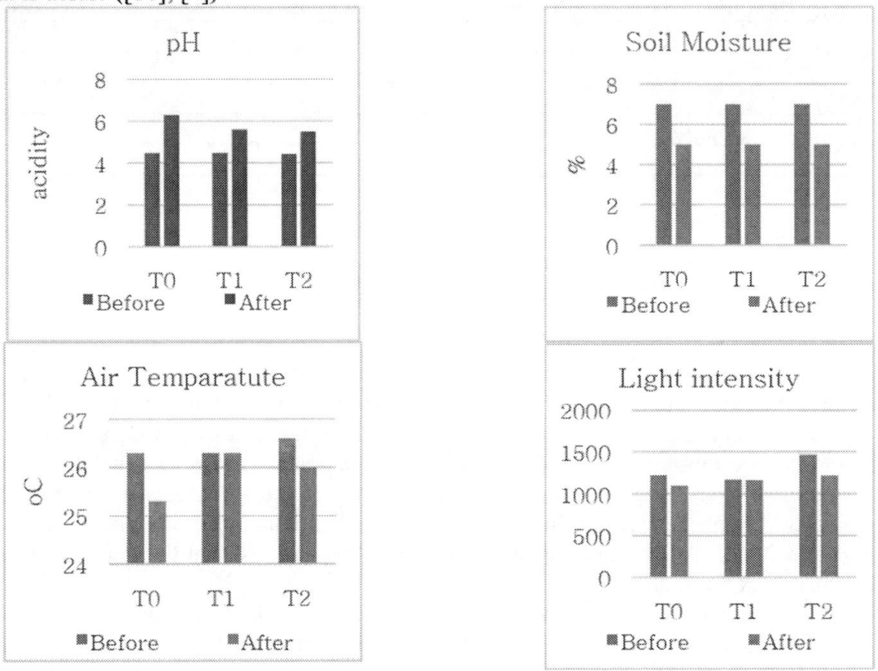

Figure 1. Abiotic factors of Callombolan existance in Broccoli plantation

4.3. Callombolan Diversity

Based on the research that has been conducted in the area of Broccoli plantations, there are seventh genera, where the four genera is included into three families, namely Entomobryidae, Isotomidae, and Paronellidae. The number of individuals at most that the Entomobrydae Familia, precisely genus Acrocyrtus by the number of 241 individuals with an abundance of 0.267 while the smallest number of individuals in the Family Isotomidae, precisely genus Isotomurus by the number of individual 4 with an abundance of 0.0004. This is caused by the dryness of soil and temperature of the surface layer [5].

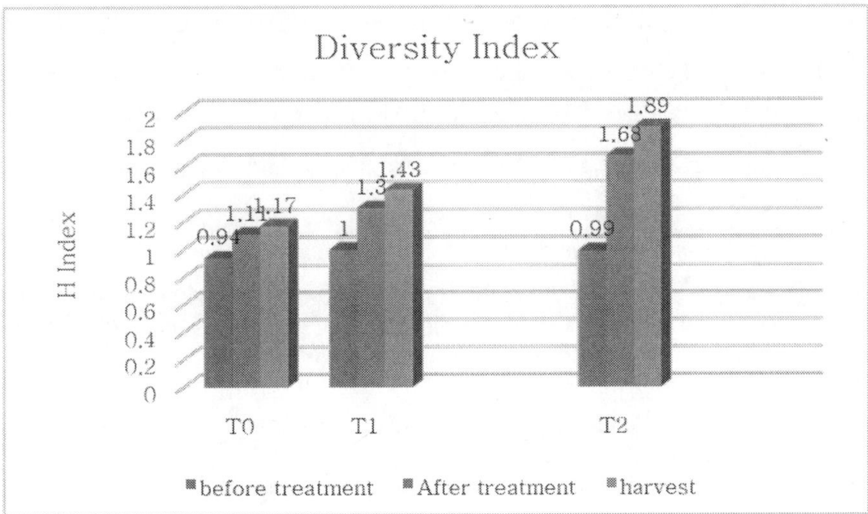

Figure 2. Diversity index of Callombolan with non synthetic treatment

Collembola low diversity can be seen also from biotic and abiotic factors in Broccoli plantation. Vegetation is one of the biotic factors that indirectly affect the diversity of Collembola [11]. Dense plantation will form a stable microclimate that supports the activities and development of soil arthropods. The density of the vegetation in an area can lead to fall-out, so that the thickness of the litter can be increased. The existence of this litter thickness may increase the availability of food resources better to ground arthropods [15]. Vegetation in Broccoli plantation is dense. So that the nutrients contained in it is only capable of supporting life some Collembola only. [11] states that the more varied vegetation will become more varied litter also described as a food source. The existence of diverse sources of feed is likely to impact the increase in animals that consume them. Vegetation in early growth plan relatively homogeneous, so that Collembola contained in the vicinity is relatively not much different. The existence of a less dense vegetation in the botanical gardens influenced the abiotic factors in the vicinity. This happens because the sunlight can be directly on the ground without any obstructions. So it can improve soil moisture and temperature in it. At temperatures and low soil moisture, Collembola will move towards the deeper layers. As a result, the Collembola sampling will be more difficult.

5. References

[1] Abbas M 2012 *Indian J. of Fundamental and Applied Life Sciences.* **2** 18 -25

[2] Amir and Andrew M 2008 *The role of insect tail Pegas (Collembola) in order to Improve Soil fertility*. Agency for Agricultural Research and Development, Research and Development Center of Plantations Warta

[3] Ganjari L 2012 Abundance of Collembola on Habitat Vermi komposting type *Widya News* 1 131-44

[4] Coleman DC, Crossley DA and Hendrix PF 2004 *Fundamentals of Soil Ecology* (USA: Elseiver Academic Pr)

[5] Detsis V 2000 Vertical distribution of Collebola in decidous forest under Mediterranean climatic conditions. *Belg Zool* **130** 55-9

[6] Greenslade P, Deharveng L, Suhardjono YR 2000 *Handbook to Collembola of Indonesia.* Advisor Willem N. Ellis. (Bogor: Museum Zoologicum Bogoriense)

[7] Imler U 2004 *Pedobilogia* **48 (4)** 349-63

[8] Kaneko N, Lean M A, Parkinson D 1995 *Pedobiologia* **39** p 538-46

[9] Lee Q and Widden P 1996 *Biochem* **28** 689-90

[10] Hopkin S P 1997 *Biology of the springtails Insecta: Collembola* (Oxford: Oxford University Press)
[11] Materna J 2004 *Pedobiologia* **48** 339-34
[12] Muturi J J, Mbugi JP, Mueke JM, Lagerlöf J, Mungátu J K, Nyamasy G, and Gikungu 2009 *Tropical and Subtropical Agroecosystems* **11** 415- 22
[13] Petersen H 2002 General aspects of Collembolan ecology at the turn of the millennium *Pedobiologia* **46** p 246-60
[14] Tiana W 2008 *Collembola community in Lima Type Land Surface Area Habitat at Telaga Warna Bogor and Cianjur*. Bogor
[15] Sebayang D, Suryati, T and Adianto 2000 *Arthropod diversity and abundance of soil in natural forests, pine forests, vegetable gardens and open land in Mount Tangkuban* Perahu. p 75-79
[16] Suhardjono R Y, Deharveng L A and Bedos 2012 *Collembola (spring tail)*. (Bogor: PT Vega Briantama Vandanesia)
[17] Swift M J, Heal O W and Anderson JM 1979 *Decomposition in terrestrial ecosystems* (Berkeley: University of California Press)
[18] Wallwork JA 1970 *Ecology of Soil Animal* (London: Mc Graw-Hill)

Acknowledgments
I would like to thanks to the management of "Litle Farm" that facilitate us counducting the biopesticide experiment.

IOP Conf. Series: Journal of Physics: Conf. Series **812** (2017) 012012 doi:10.1088/1742-6596/812/1/012012

Implementation of Socioscientific Issues Instruction to Fostering Students' Decision Making Based Gender on Environmental Pollution

H P Rizal[1], P Siahaan[2] and G Yuliani[3]

[1] Science Education, Postgraduate Student, Universitas Pendidikan Indonesia, Jl. Dr. Setiabudi No. 229, Bandung 40154, Indonesia
[2] Department of Physics Education, Universitas Pendidikan Indonesia, Jl. Dr. Setiabudi No. 229, Bandung 40154, Indonesia
[3] Department of Chemistry Education, Universitas Pendidikan Indonesia, Jl. Dr. Setiabudi No. 229, Bandung 40154, Indonesia
Email: parsaoransiahaan@upi.edu

Abstract. The purpose of this study was to investigate the implementation socioscientific issues (SSI) instruction to examine students' decision making based gender. Selection of environmental problem based on the consideration this issue has potential to evaluate from various scientific disciplines and may initiate to make decision. The participant was students grade VII in South Sulawesi. Pre experiment method was utilized in study with one group pre-test and post-test design. The instrument used in this study comprised of open-ended question, observation sheets for group discussion, sheet for interview and observation sheet of implementation instruction in class. The result indicated that instruction has been implemented successfully. It was found that pre-test and post-test was fostered with N-gain for male and female in medium category. Based on independent t test was found there was no difference improvement of decision making between male and female students. The group showed there was no difference between male group and female group, but a significant difference was noted in mix group. Students and teacher give positive response to the implementation of instruction. It is concluded that science instruction based SSI can be implemented to foster male and female decision making by forming group discussion.

1. Introduction

Scientific literacy is worthy of note in education because become an indicator of quality education in a country. In addition, prepare students who literate science is the main purpose of learning curriculum of science in some western countries such as Hong Kong Curriculum Development Council, National Research Council (NRC), Qualifications and Curriculum Authority (QCA) and the American Association for the Advancement of Science (AAAS) [1].

According to Jho, Gyoung and Kim scientific literacy includes positive attitude towards science, understanding of scientific knowledge in everyday life such as science knowledge associated with the ideas contained in paper, media, internet and make informed decisions [2]. In other words, scientific literacy summarizes the relationship between the understanding of science, attitudes toward science and decision making in the context of everyday life. Thus, provided students' decision making ability related phenomena and social science become one of the specific objectives of science and important to developed not only to improve scientific literacy of students, but also to improve students' critical thinking skills to participate as citizens who provide solution of various issues encountered.

Tawil & Liliasari suggested that decision-making is the process of thinking to identify and decide choice of various options [3]. The ability to make decisions leading to the ability to systematically evaluate the possibility of factual and ethical action on complex situation of the phenomenon as well as existing issues and systematically make the final decision. One effort to fostering decision making ability of students is by integrate a variety of issues related to the phenomena that occur in community into science lessons. Issues that integrated science, technology and social dimensions. The issue is known as Socioscientific Issues (SSI). SSI term is mostly used for designing real world scenarios in which not only focuses on the facts related to science, but also normative values (Sadler in Botcher and Meisert) [4]. One of SSI that can trigger student to use their ability to make decisions is issue about environment. Environmental issues often contain values, ill structure and controversial [5]. This issue can lead student to making decision.

Lewis and Leach explained that students with better science knowledge shows better reasoning and use the resources that are supported by the evidence in making decisions [6]. Contrary, Hong and Hong and Chang reveals that students' decision-making is not related to its scientific knowledge [7]. While Fleming explains that there are other factors that are very strong compared to the science knowledge that greatly affects decision making such as emotions and personal experiences. These factors need to be considered because it's varying for each individual primarily between male and female. In SSI discussion differences of individuals or group can build a different perspective. Therefore, in SSI discussion gender differences may become factor that distinguishes students' ability to make decisions beside science knowledge. In addition, with regard to gender stereotypes which are developed in community about the differences between male and female in decision making. Female are considered as a highly emotional, full of feeling and male are not affected by emotional feelings and tend to be more logical. This stereotype is very strong and pervasive impression on our culture [8] The stereotype judged female less able to make decision than male, however this stereotype still need to prove because research about gender still unable to show consistent result. According to Dagun male known to be more able and responsive in recognizing a problem when starts entering adulthood than female [9]. Differences in the ability possessed by male and female influenced by the thinking male and female are different, that male are more analytical and flexible than female. Contrary, Halpern explains that in general there is no difference between males and female in the overall IQ, learning, memory, problem solving, and concept formation [10].

Based on the description, this study examine junior high school students and divide student and group discussion not only based on the knowledge level or characteristic, but also gender of student to find out how the difference of improvement decision making ability between female and male and how the difference of decision making ability between group based gender by implementation of socioscientific issues instruction.

2. Research methodology

Research design used in this study is a one group pretest-posttest design [11]. With these designs, the research subject is an experimental class without comparison. In the design of one group pretest-posttest, single-subject groups were given a pretest (T1), treatment (X), and posttest (O). Instruments on the pretest and posttest at same time, but given at different times. This study was conducted in Junior High School, Sidenreng Rappang province of South Sulawesi with a population students grade seventh. Study throughout to students with second semester who have five classes with the composition of 25-30 students in a class. The sample in this research is student class VIIa Academic Year 2015-2016. The research start by given pretest to student and learning process for 4 days to discuss environmental issues with worksheet as a guide for student and in the last meeting with post test.

2.1. Instrument test
The instrument used in this study is pre test and post test instrument consist of five indicators according to Tawil & Liliasari [3]: a) identify the problem, b) collect information, c) Determine alternative solution, d) Analyze pros and cons, e) Make Decision. Instrument used in the form of essay question with three issues (fish bombing in south Sulawesi, pollution of motor oil, increasing of vehicles in Makassar, south Sulawesi), each issue consists of a brief article that sparked the curiosity of students and contains six questions that lead students to make decision with the total questions for all issues contains 18 number. The instrument asked student to identify problem based on the article and collect information to more understand about the issue and main problem, determine alternative solution of the main problem and then weigh pros and cons, advantage and disadvantage of the alternatives for the environment, society, human health. Based on that, students asked to make decision with the reasons. Beside pre test and post test of decision making instrument, student worksheet in classroom based on decision making indicators was prepared to assist students in group to discuss and make decision related environment issues. Group discussion divided into 3 groups; male group, female group, mix group (consist of female and male student) with formation six students each group. The second indicator, collect information assessed during the learning process by using observation sheets, sheet for teacher and student interview to know the response about implementation of instruction in class.

2.2. Data analyze
Decision making instrument was tested to determine the validity and reliability by used anates application before used in this study. Validity is the measure that indicates the level of validity of an instrument [12]. Reliability test used in this study is the product moment correlation while to interpreting the value of reliability each item problems, the value was consulted to table of r product moment critical value, with the provision that if r calculate > r table at $N = 34$ and the significant level $\alpha = 0.05$ means the item problem was reliable [12]. The improvement of decision-making ability of students each indicator, each item and in generally was calculated used Normalized Gain (%). In this study, the first hypothesis about how the difference improvement of decision making ability between female and male was tested using Independent Sample t – test for small sample and the second hypothesis about the differences improvement of ability of decision-making among male group, female group and mix group were tested using ANOVA. If the result of ANOVA reject the null hypothesis means that there are differences in average of decision making ability between group of male, female and mix, then carried across the Tukey test to compare the average male group, female group and mix group and determine which groups differed significantly. If each group obtained p value < 0.05 means that there is a statistically significant difference among the three groups, if p values > 0.05 means that there is no statistically significant difference among the three groups [13].

3. Result and discussion

3.1. Decision making between male and female students
Decision-making ability of students was measured by using instrument that represented of 5 indicator of the ability of decision-making by Tawil & Liliasari [3]. At the First meeting, students are given pretest to represent the ability of students in decision making. In the implementation of SSI instruction students are given issue or phenomena then discussed and in the end students are given posttest. The recapitulation result showed in Table 1.

Table 1. The comparison of pretest, posttest and n-gain of decision making male and female students.

Indicator	Number of problem	Pretest Average (%)		Posttest Average (%)		N-gain	
		Male	Female	Male	Female	Male	Female
1[a]	1,7,13	44.44	41.97	64.19	74.07	0.33	0.57
	2,8,14	43.21	40.74	65.43	74.07		
	3,9,15	38.27	58.02	54.32	82.72		
		41.97	46.91	61.31	76.95		
3[b]	4,10,16	41.97	60.49	74.07	83.95	0.55	0.59
4[c]	5,11,17	25.92	40.74	49.38	65.43	0.32	0.42
5[d]	6,12,18	32.09	40.74	64.20	75.31	0.47	0.58
Highest score		41.97	46.91	74.07	83.95		
Lowest score		25.92	40.74	49.38	65.43		
Average		35.49	47.22	62.24	75.41	0.42	0.54

[a] Identify problem
[b] Determine alternatives solution
[c] Analyze pros and cons
[d] make decision

Table 1 shows level of decision making ability of male and female differ by 11%. It is seen from the results of pretest two groups for male gain 35.49% from 100 ideal value and female 47.22% from 100 ideal values. Decision making ability was improved in general and each indicator after implementation of SSI instruction. The average of N-gain was increased with 0.42 for male (medium category) and 0.54 for female (medium category). This result in line with Klosterman and Sadler opinion that learning with socioscientific issues requires involvement of student, presented challenge to students to explore the controversial issue, integrate social aspects based on their investigation that can improve decision-making ability of students [14].

The first indicator identify problems requires student's ability to create questions and answers with regard to the issues with the aim to identify the main problem of issue. The ability of students to identify problems obtained through the average in make question, make a temporary answer and determine the problem. Determines an alternative solution is an indicator of the highest achievements in both groups of students. This is because the issue given close to student. In addition, during the learning process students are also trained to make decisions by providing socioscientific issues, so that students can specify a range of alternative solutions to make decisions on a given issue. The last indicator makes decision students are assigned to select the most appropriate alternatives to solve problem of environmental pollution and enclosing the reasons for selecting the alternatives. Based on the results in this study, it can be seen that students are able to write good problem solving alternatives, but still difficult to analyze pros and cons of issue. The second indicator that collects information assessed during the learning process by using observation sheet.

Based on the test results of the difference in average improvement of decision making ability showed that there is no difference average of improvement decision making ability between male and female to the implementation of socioscientific issues instruction. These mean that generally the abilities of male and female equally in decision-making. In line with the opinion of Halpern & Lamay that on the average, male and female showed the same results on general intelligence tests [15]. According to Halpern in general there is no difference between male and female in the overall IQ, learning, memory, problem solving, and concept formation [10]. Fraine, Damme and Onghena explained a weak correlation between gender and academic achievement, gender factor is unable to become the only aspect of the benchmarking of student achievement [16]. The difference between male and female is due to social and cultural factors.

3.2. Decision making between male group, female group and mix group

Sadler suggested that characterizes the SSI is a discussion group to discuss the issue or phenomenon presented [17]. Discussion group consisting of members of heterogeneous enables a variety of different perspectives on the issues. In addition, a heterogeneous group with different abilities of students can help low-ability students to master the concepts included in the issue because students with high ability guide students who have low capacity. Recapitulation of the assessment of the group discussions illustrated in Table 2.

Table 2. The comparison average difference of group discussion.

Indicators	No.problem	Pretest Average			Postest Average			N-gain		
		Male	Female	Mix	Male	Female	Mix	Male	Female	Mix
1[a]	1,7,13	48.15	68.52	50.00	62.96	90.74	66.67	0.37	0.65	0.27
	2,8,14	38.89	61.11	35.18	59.26	85.19	42.60			
	3,9,15	25.93	33.33	12.96	59.26	75.93	42.59			
		37.65	54.32	32.71	60.49	83.92	50.62			
3[b]	4,10,16	53.70	72.22	55.56	83.33	88.89	70.37	0.64	0.60	0.33
4[c]	5,11,17	44.44	61.11	25.93	77.78	88.89	57.41	0.60	0.71	0.43
5[d]	6,12,18	38.89	33.33	18.52	53.70	77.78	42.59	0.24	0.67	0.30
Highest		53.70	72.22	55.56	83.33	88.89	70.37			
Lowest		37.65	33.33	18.52	53.70	77.78	42.59			
Average		43.67	55.24	33.18	68.82	84.87	55.25	0.45	0.66	0.33

[a] Identify problem
[b] Determine alternatives solution
[c] Analyze pros and cons
[d] make decision

Table 2 shows the mix group obtains lower results than male and female group, but N-gain category for each group equal in medium category. The highest result of each group showed in determine alternative solution indicator. The second indicator collect information assessed during the learning process by using observation sheet with the average of collect information for female group gain 76.39, male group gain 70.83 and mix group gain 65.28. Calculation with ANOVA showed difference average in ability of decision-making between groups on the implementation of SSI. Further testing to compare the entire pair-group average between male, female and mix group that differed significantly by using Tukey test. The result showed in Table 3

Table 3. The comparison tukey test of group discussion.

Group		Sig	α	Interpretation
1[a]	2	0.352	0.05	Not Significantly difference
	3	0.038		Significantly difference
2[b]	1	0.352		Not Significantly difference
	3	0.002		Significantly difference
3[c]	1	0.038		Significantly difference
	2	0.002		Significantly difference

[a] Male Group
[b] Female Group
[c] Mix Group

Table 3 shows female and male group not significantly difference based on the sig number that higher than α, but mix group with female and male group significantly difference (sig < α). In group discussion, differences in environmental characteristics of students, student background, values in society also influence the interaction of students in group. According Ormrod the distribution group that is too heterogeneous make differences in ability among students more clearly [15]. High ability students may dominate the discussion and make low-ability students less participates. Low-ability students may be embarrassed to ask for help when they do not understand and give other members do most or all of the work. Further Ormrod explains that the group would be more effective if all group members are female or all male because female students are more likely to express their opinions in small groups or all of its members are female [15]. In line with these opinions, this study dividing groups based on gender differences and cognitive abilities caused the mix group too heterogeneous, so that cooperation within the group was not optimal when compared to group of all male or female and when given test to make independent decision, the result vary between members of group.

4. Conclusion

Based on the results of research and discussion that has been described can be concluded that students' decision making ability was fostered with the implementation socioscientific issues instruction. The result showed no difference in improving of decision making ability between male and female in generally although in some indicators female student get higher score than male, but generally no significant difference. In socioscientific issues group discussion showed that female and male group no significant difference, but mix group was noted. The male and female group more active and make informed decision than mix group. This result may occur because the formation of group too heterogeneous and students difficult to cooperate each other. According to result in this study, in making decision individually, gender does not really give a significant influence, but when asked to make decision as a group, gender different affect decision making student. Suggestion for other researchers who are interested to doing similar research, important to take into account about the allocation of time because to discuss about controversial issues need more time to express the opinion. The other factor that can influence students' decision making such as ethnic, cultures and the environment condition should be more attention.

5. References

[1] Lee C Y and Grace M 2012 *Science Education Journal* **96** pp 787-807
[2] Jho H, Gyoung H and Kim M 2013 *Science & Education Journal* **23** 1131-51
[3] Tawil M and Liliasari 2013 *Berpikir Kompleks dan Implementasinya dalam Pembelajaran IPA* (Makassar: Badan Penerbit UNM) pp 125-40
[4] Bottcher F and Meisert 2011 *Research in Science Education* **43** 479-506
[5] Weinberger Y and Dreyfus A 2013 *International Journal of Environment & Science Education* **8** 627-643
[6] Lewis J and Leach J 2006 *International Journal of Science Education* **28** pp 1267- 87
[7] Hong J L and Chang N 2004 *Research is Science Education* **34** 97-111
[8] Santrock J W 2003 *Adolescence Perkembangan Remaja* (Jakarta: Erlangga) p 376
[9] Dagun S 1992 *Maskulin dan Feminim: Perbedaan Pria dan Wanita dalam Fisiologi* (Jakarta: Rineka Cipta) p 68
[10] Feldman R 2011 *Pengantar Psikologi Understanding Psychology* (Jakarta: Salemba Humanika) p 56
[11] Fraenkel J R *et al* 1993 *How to Design and Evaluate Research in Education* (New York: McGraw-Hill)
[12] Arikunto S 2012 *Prosedur Penelitian Suatu Pendekatan Praktek* (Jakarta: Rineka Cipta)
[13] Sarwono J 2006 *Analisis Data Penelitian Menggunakan SPSS* (Yogyakarta: Andi) p 88
[14] Klosterman M L and Sadler T D 2010 *International Journal of Science Education* **32** pp 1017-043
[15] Ormrod J E 2008 *Educational Psychology* (United States of America: Pearson Education) pp 118-88
[16] Fraine B D, Damme J V and Onghena P 2007 *Contemporary Educational Psychology* **32** 132-50
[17] Sadler T D, Barab S A and Scott B 2007 *Research in Science Education* **37** 371-91

IOP Conf. Series: Journal of Physics: Conf. Series **812** (2017) 012013 doi:10.1088/1742-6596/812/1/012013

The Application of Problem Based Learning Model to Improve Mathematical Literacy Skill and The Independent Learning of Student

F A Sari[1] ,I A V Yandari[2]and Fakhrudin[2]

[1]Departementof Mathematics Education, UniversitasPendidikan Indonesia, Jl. Dr. Setiabudhi No. 229, Bandung 40154, Indonesia.
[2]Departementof Mathematics Education, Universitas Sultan AgengTirtayasa, Banten 42124, Indonesia
E-mail: fitriaidasari@student.upi.edu

Abstract. The background of the research was due to the lack skill of student mathematical literacy. The objective of the research is to know the ability and the improvement of the students in term of the mathematical literacy skill and the independent learning of the students by using problem based learning modelbetter than students who received conventional learning. This research conducted in one of Junior High School in Serang in academic year of 2013/2014. The samples of the research were two classes. The test instrument consists of fouressay items. The results showed that ability and the improvement of the students in term of the mathematical literacy skill and the independent learning of the students by using problem based learning model better than students who received conventional learning. Based on the research results, as well as the improvement of the research, we suggest that the learning model of problem based learning can be taken into consideration by the teacher as an alternative learning at school. Additionally, we recommend further research that examines the effects of problem based learning models on different aspects for a more diverse material.

1. Introduction

Mathematics is one of humanity's great achievements. By enchancing the capabilities of the human mind, mathematics has facilitated the development of science, technology, engineering, business, and government. For people to participate fully in society, they must know basic mathematics. Citizens who cannot reason mathematically are cut off from whole realmsof human endeavor. Innumeracy deprives them not of opportunity but also of competence in everyday tasks [1].In thus is the reason for the importance of mathematics to be learned. There are five competency in mathematics, namely mathematical problem solving, mathematical communication, mathematical reasoning, mathematical connections, and the mathematical representation [2]. The fifth of these competencies is neededby students to be able to use mathematics in daily activity. Capabilities that include five of these competencies is mathematical literacy. Mathematical literacy helps a person to understand the role or usefulness of mathematics in daily life as well use it to make the right decisionsas a citizen [3].However, the mathematical literacy skill of students in Indonesia arestill quite low. PISA as the programs carried out by the OECD in 2012 has been conducting research to see mathematical literacy skills of students aged 15 years in 65 countries. From the results of these studies indicate that the mathematicalliteracy of children in Indonesia is ranked 64 of 65 countries with a score of 375 [4]. Indonesian's students are only able to answer questions PISA level 1, 2 and 3, and a few students can solve level 4 [5].Indonesian student success in solving problems of PISA is largely determined by the evaluation system and the ability of teacher to develop student's matematical literacy. So, teachers need to develop about a matter which is equivalent to PISA assessment [6].

On the other hand, to improve mathematical literacy skills required independent learning of studentin mathematics. Independent learning is learning in which the learning goals, methods or direction to achieve the learning goals, and regulation of learning process is decided, guided and managed by the learner [7].Independent learning can be seen quite broadly to encompass a variety of situations and contexts where students are interpreting and scaffolding new knowledge and skills independently from those around them [8].The aim of independent learning is to teach students to learn for themselves and in turn empower them in their learning process regardless of their learning context or environment [9].Being an independent learner does not mean 'learning on your own' or in an isolated way [10]. We can encourage engaged form of independent learning by facilitating peer study groups, who may meet up for weekly revision or for exam preparation [11].

One model that can be used to improve mathematical literacy skills and independent learning of students in the learning process is problem based learning model (PBL).This model began to be appointed because in general terms, problem based learning model presenting an authentic and meaningful situation to provide convenience for students to conduct an investigation [12]. Through these authentic experiences, preservice teachers develop a better understanding of constructivist teaching practices, became more successful at implementing constructivist methods [13].Learners work collaboratively with others as they analyze complex and ill-defined problems [14].Learners also work independently to collect information they then bring back to the group as they resume their collectiveproblemsolving and subsequent reflection on both the issue at hand and the group's functioning [15].They use a priro and post priori knowledge to reason intectually and make learners collaboration with others in small groups [16].

2. Experimental Method

The method used in this study is a quasi-experimentalmethod. While the design of the study is the non-equivalent control group design. The design is depicted inFigure 1.

0 $X_1$0

----

0 $X_2$0

Figure 1. Non-equivalent control group design [17]

Information:

0	= Pretest / posttest experimental class / controlclass
X_1	= The treatment on experimental class that is learning by problembased learning model.
X_2	= The treatment on control class that is learning by conventional learning.
---	= Subject is not chosen at random

The population in this study were students of class VIII Junior High School 4, Serang academic year 2013/2014. The sample in this study was taken using cluster random sampling method of determining the sample if the population consists of classes of individuals or clusters [18].In this study, samples were taken two classes, namely the control and experimental classes. Selection of the control and experimental classes chosen randomly.Students in the experimental class will learning with problem based learning model, while the control class will be given conventional learning which will then be used as a comparison of the experimental class.

The instrument that was used was test and nontest instruments. In this research, the test instrument to be used is essay question that includes indicators of matematical literacy. The given problems were tested first to know about the validity, reliability, power differentiator, and the level of the six questions. From the test results that have been calculated, the entire matter is qualified to serve as a test of mathematics literacy ability at the beginning and end of the test execution. The non test instrumen in this research use scaling subject. Scaling subject method is a method that oriented to the subject that has purpose to put student on a continum rating so the relative position of individual can be obtained [19].Analysis of the data in this research is the analysis of quantitative data. In the analysis of quantitative data, to beanalyzed is pretest of mathematical literacy, scale of initial learning independence, posttest of mathematical literacy and scale of finale learning independence, and also gain of mathematical literacy skill and independent learning of student.

3. Result and Discussion

In this research, the ability that measured was mathematical literacyof experimental class and control class. The research starts from the initial test in class VIII in Junior High School 4, Serang to determine the ability of mathematical literacy. From the initial tests conducted obtained information that mathematics literacy ability of students still unsatisfactory (see Figure 2).

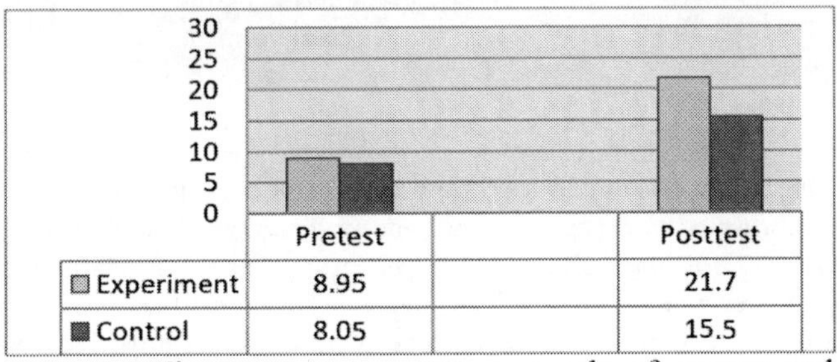

Figure 2. Histogram average score result of pretest and posttest mathematical literacy ability

Increased mathematics literacy is significant from occuring in the experimental class that uses the learning model of problem based learning (See Figure 2). Experimental class has higher average than control class that uses conventional learning. It's supported also by the result increase in the experimental class is 0,62 while the control class is 0,35 (See Figure 3).

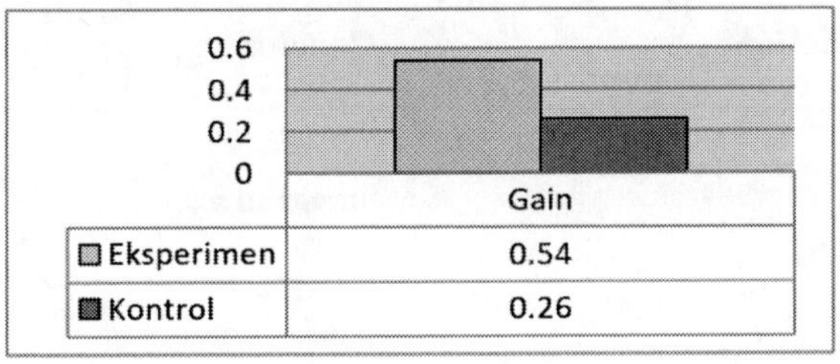

Figure 3. Histogram result for the normalized gain

Table 1. Recapitulation of the difference of two average test

Pretest			Posttest			Gain		
Result Count	**Value of Table**	**Conclusion**	**Result Count**	**Value of Table**	**Conclusion**	**Result Count**	**Value of Table**	**Conclusion**
1,11	1,675	There is no difference in ability between the experimental class and control class	7,65	1,675	Mathematical literacy skills in experimental class is better than control class	9,00	1,675	Increase student's mathematical literacy skill in experimental class is better than control class

Parametric tests, one part t-test, used for know about the conclusion of posttest and gain. This is because the data were normally distributed and homogeneous. The result of the posttest among experimental class and control class showed difference(see Table 1). Mathematical literacy skills in experimental class is better than control class. This experiment also indicates that learning using problem based learning model more involved than conventional learning. The result of the gain among experimental class and control class also showed difference(see Table 1). Increase student's mathematical literacy skill in experimental class is better than control class. This suggests that learning by usingproblembased learning model can improve students' mathematical literacy.

Table2. Recapitulation of the difference of two average test

Initial Scale Score			Last Scale Score			Gain		
Result Count	**Value of Table**	**Conclusion**	**Result Count**	**Value of Table**	**Conclusion**	**Result Count**	**Value of Table**	**Conclusion**
0,78	1,675	There is no difference	2,30	1,675	The independent learning of student in experimental class is better than control class	3,86	1,675	Increase of the independent learning of student's in experimental class is better than control class

In this research also measured independent learning of student in experimen and control class using the difference of two average test. This is because the data were normally distributed and homogeneous. From the initial test, show that there are no differences in the ability of the experimental class and control class (see Table 2), it is because the research were conducted in the same chapterand the same treatment between these two classes.Parametric tests, one part t-test, used for know about the conclusion of posttest and gain. The result showed independent learning of student in experimental class is better than control class (See Table 2). This is evidenced from the acquisition the average score of the experimental class 132 (82.5%) was better than the average score of independent learning control class 121 (75.63%).

When viewed from the gain among the last scale score and the initial scale score, gain among pretest and posttest, it show that the gain of the experimental class is higher than the gain of control class (See Table 2). This suggests that problembased learning model can improve mathematical literacy skills and learning independence of junior high school students. The results of this study show that problem based learning model is better than the conventional learning in terms of improving the literacy skills of mathematical and the independent learning of student.

It is influenced by several factors, using problem based learning modelmake mathematics more interesting, fun and meaningful so that the atmosphere in the learning process is not monotonous that make students bored and not interested at all to the material presented by the teacher.American preservice elementary teachers believed that "making mathematics fun" was central to the learning process [20].However they also found that students expected that enjoyable lessons succeed in teaching mathematics regardless of content or context of the lessons.

Another factor were, with PBL model the students can build their own knowledge, so they were not easy to forget the things they have acquired. This is consistent with the constructive theory which states that students must find their own and transform complex information, check the new information with the old rules and revise them if the rules were no longer appropriate [12].High levels of student engagemnt in PBL classrooms because it places students in s real world, problem-solving context [21].Studies have revealed that PBL has positive effect on the development of higher-order thinking skills in specific groups of students [22].An experimental study of 76 teachers who utilized PBL in their classrooms revealed that, compared to the control group of students in traditional classes, their students scored higher on standardized exams, as well as ability tests that measured problem-solving skilss and content application to real-world problems [23].

In addition, using PBL model make students possible to think and exchange opinions with their friends. Learning theory by Vygotsky believe that social interaction with friends can stimulate the formation of new ideas and enrich the intellectual development of students [24].The exchange of information between students can build new knowledge so that students can better understand the learning materials studied. Students also reported higher enjoyment of their PBL work because it gave them opportunities to interact with their friends and to make new friends through cooperative projects [25].

4. Conclusions

Based on the results and discussion of research, conclusions can be drawn that Problem Based Learning model can improve the mathematics literacy ability class VIII Junior High School 4, Serang. It can be summarized as follows: (1) The ability of mathematical literacy of students who get the learning by problem based learning model is better than the students who received conventional learning; (2) Increasing the matematical literacy skills of students who get learning by problem based learning model is better than the students who received conventional learning; (3) Independence learning of students who get learning by problem based learning model is better than the students who received conventional learning; (4) Improved learning independence of students who had learning byproblembased learning model is better than the students who received conventional learning.

Based on the above conclusion, the researchers recommend that (1) The problem based learning model can be taken into consideration by the teacher as an alternative learning at school; (2) Problem based learning model consists of five stages. In order for all stages of the learning process implemented to the fullest, it is necessary allocation of time as well as possible; (3) Further research that examines the effects of problem based learning models on different aspects for more diverse material as well.

5. References

[1] Kilpatrick J *et al* 2001 *Adding it Up Helping Children Learn Mathematics* (Washington DC: National Academy Press)

[2] National Council of Teacher of Mathemtics 2000 *Principles and Standards for School Mathematics* (USA: NCTM)

[3] Kamaliyah,*et al* 2013*Jurnal IndoMSJME***4** 9-28

[4] OECD2013 *PISA 2012 Results in Focus What 15-year-olds Know and What They Can Do with What They Know*(USA: OECD)

[5] EdoSI,*et al*2013 *JIMS Journal***4**41-58

[6] JoharR 2012*Jurnal Peluang***1** 30-41

[7] Balapumi R and Aitken A 2012 *Proc. of 23rd Australian Conf. on Information System 2012*(Victoria: Geelong) p 1-10

[8] NagpalK2013*Journal of Social Science & Interdisciplinary Research***2** 27-35
[9] Broad J 2006 *Journal of Further and Higher Education***30** 119-143
[10] Meyer B,*et al* 2008 *Research Brief:Independent Learning-Literature Review*. (London:Department for Children Schools and Families)
[11] Damon W & Phelps E 1989*International Journal of Educational Research***13** 9-19
[12] Trianto2013*Mendesain Model Pembelajaran Inovatif Progresif* (Jakarta: Kencana Prenada Media Group)
[13] Park S H and Ertmer P A 2007 *JRTE Journal***40** 247-267
[14] Hmelo S C 2004 *Educational Psychology Review***16** 235-266
[15] Simone C D 2014 *Teacher Education: Trajectories of Change***4** 17-29
[16] Carroll K 2009 *Proc. Problem-Based Learning, Pedagogy & Practice*vol 2 (Singapore) p 494-501
[17] Ruseffendi2005*Dasar-dasar Penelitian Pendidikan & Bidang Non-Eksata Lainnya* (Bandung: Tarsito)
[18] Sudaryono2011*Metode Penelitian Pendidikan* (Banten: Dinas Pendidikan Provinsi Banten)
[19] Azwar S 2013 *Penyusunan Skala Psikologi* (Yogyakarta: Pustaka Pelajar)
[20] Foss D H and Kleinsasser R 1996 *Teaching and Leacher Education***12** 429-442
[21] Brush T and Saye J 2008 *Journal of Problem-Based Learning***2** 21-56
[22] Mergendoller J,*et a*l 2006 *Journal of Problem-Based Learning***1** 49-69
[23] Finkelstein N,*et al* 2010 *Effect of Problem Based Economics on High School Economics Instruction NCEE*(Washington DC: National Center for Education Evaluation and Regional Assistance)
[24] Rusman2012*Model-Model Pembelajaran* (Jakarta: Raja Grafindo Persada)
[25] Belland B,*et al* 2006 *Journal of Problem-Based Learning***1** 1-18

IOP Conf. Series: Journal of Physics: Conf. Series **812** (2017) 012014 doi:10.1088/1742-6596/812/1/012014

The Effect of Integrated Learning Model and Critical Thinking Skill of Science Learning Outcomes

N Fazriyah[1], Y Supriyati[2] and W Rahayu[2]

[1] Department of Primary School Education, Universitas Pasundan,
Jln. Tamansari No. 6-8 Kota Bandung, Jawa Barat Bandung 40154, Indonesia
[2] Department of Assessment and Evaluation Education, Universitas Negeri Jakarta,
Jakarta 13220, Indonesia
E-mail: nurulfazriyah@unpas.ac.id

Abstract. This study aimed to determine the effect of integrated learning model and critical thinking skill toward science learning outcomes. The study was conducted in SDN Kemiri Muka 1 Depok in fifth grade school year 2014/2015 using cluster random sampling was done to 80 students. Retrieval of data obtained through tests and analysis by Variance (ANOVA) and two lines with the design treatment by level 2x2. The results showed that: (1) science learning outcomes students that given thematic integrated learning model is higher than in the group of students given fragmented learning model, (2) there is an interaction effect between critical thinking skills with integrated learning model, (3) for students who have high critical thinking skills, science learning outcomes students who given by thematic integrated learning model higher than fragmented learning model and (4) for students who have the ability to think critically low yield higher learning science fragmented model. The results of this study indicate that thematic learning model with critical thinking skills can improve science learning outcomes of students.

1. Introduction

Learning outcomes can be interpreted as what is expected in the students, ability, or perceived value can be completed in a part of learning [1]. The purpose of learning science in elementary school which is that students can have the ability to learn, responsible, understanding and interest in learning that will enable them to survive in people's social environment [2]. The problem that occurs is the result of learning science has not shown satisfactory results [3, 4]. The learning process that takes place is still oriented on a cognitive ability, which is currently learning 21st century learning skills demanded a lot of bias held by the students [5]. Hence the need for improvement in the learning process and learning outcomes achieved so science learning objectives can be obtained.

One of 21st century skill needs to be owned by the students is the ability to think critically [5]. Critical thinking skill is one of the internal factors that affect learning outcomes [6]. This capability can be developed and studied [5,7]. Gagne developed a category of learning outcomes based on the characteristics of the content, or content must be controlled by student [8]. Bloom and colleagues divides into several domains of learning outcomes or aspects of capabilities with a variety of characteristics, namely: (1) *cognitive* behaviour, that is the thought process or behavior including the work of the brain, (2) *affective* behaviour, that raised a person as a sign of his tendency to make choices or decisions to act in a particular environment, (3) *psychomotor* behaviour, raised by the human work [9]. Then body functions of Bloom's taxonomy of cognitive that revised by Anderson and

Krathwohl define into two dimensions: the dimension of cognitive processes and dimensions of knowledge [10].

Learning model that can help to achieve the learning outcomes effectively is thematic integrated learning model. Thematic learning model is a model moved from integrated learning model. Thematic integrated model is a model by connecting multiple fields of study which is pedestrianized with a theme (webbed) [11]

There's a tendency for teachers to package student learning experience fragmented firmly among the other subjects, learning that separates the subjects expressly only will make learning difficult for students because such separation provide a learning experience that is artificial[12]. Lack of understanding of a topic has not been able to realize in the form of the ability to ask properly. Therefore packing learning experience will greatly affect the meaningfulness of the experience for them. The importance of the concept of self-sustainability and the environment is like a note in learning science in elementary school, encourage thematic integrative teaching model needs to be held in the elementary school.

One of the internal factors that affect learning outcomes is the ability to think critically. Critical thinking skills need to be developed. One reason is to prepare learners for life maturity. Thinking skills can be learned and improved through the process of learning science in school [13], because learning science develop curiosity and critical attitude towards the phenomena of nature.

Learning science should develop the realm of learning to a higher level so that students can develop skills. Since learning is only studying the realm of thinking low level (lower order thinking skills), the students' thinking skills just to remember and know. Learning science at school is based on the observation of researchers, yet maximize the realm of thought advanced (higher order thinking skills) in the form of analysis, synthesis and evaluation. Expected learning science paying attention to the realm of thought is primarily on critical thinking skills. Therefore, it is necessary to study the factors that affect student learning outcomes. In particular, study the effect of thematic integrative model of learning and critical thinking skills with the results of elementary school students to learn science.

This study aims to determine: 1) The difference in learning outcomes among students learning science by using thematic integrative learning model and student learning using learning model fragmented. 2) The effect of the interaction between the learning model and the critical thinking ability of students to learn science results. 3) The difference between the results of learning science students who have high critical thinking skills that are taught using an integrative model of thematic learning and using model fragmented. 4) The difference between the results of learning science students who have a lower critical thinking skills that are taught using an integrative model of thematic learning and using model fragmented.

2. Research methodology

The research was conducted on elementary school fifth grade students at SDN Kemiri Muka 1 Depok. The research conducted in the first half (odd) in the academic year 2014/2015, which took place in September -October 2014. This research used experimental methods, and variables in this study consists of: (1). The independent variables were integrated learning model, and (2). The dependent variable is the result of science learning outcome and variable attributes is critical thinking skill.

The population in this study were all students of class V The design of this study using experimental design by level 2 x 2. Sampling technique done by cluster random sampling. Choosing sample that this research did it by raffle the school that have minimum B accreditation which is define good school. Then raffle the homogen class that will be used as experimental class and control class. Each class selected by test into two group that high critical student and low critical thinking students.

Experimental class are given thematic learning model and class control by fragmented learning for 1 month that researcher already prepare the lesson plan before. After treatment finished then student do the test by instrument. The instrument used are: (1) The test instrument science student learning outcomes in the form of a multiple choice test, (2) critical thinking skills instrument is test description. Instrument reliability coefficient of 0.868 Science learning outcomes while reliability students' critical thinking skills by 0.732. Data were analyzed using ANOVA two lanes and testing simple effect with Tuckey test. Testing was performed using Liliefors normality and homogenity test with Bartlett's test.

3. Result and discussion

This section describe data that related to the description of the variables were: the dependent variable science learning outcomes, independent variables which consists of integrated learning model and critical thinking skills. Description of research data these three variables would otherwise form the size of centralization of data, among others: (1) mean, (2) median, (3) modus and (4) standard deviation and (5) varians.

Table 1. Result research.

Data	A_1	A_2	A_1B_1	A_1B_2	A_2B_1	A_2B_2
Mean	74, 1	68,6	84,6	63,7	63,5	73,8
Standar Deviation	12,47	8,97	7,34	5,85	5,37	9,00
Median	64,4	67,5	85,7	64,1	61,6	76,7
Modus	68,5	67,8	80,9	66,5	60,2	75,1
Varians	155,46	80,43	53,87	34,22	28,87	80,96

Testing the hypothesis of this study was performed using two way Analysis of Variance (ANOVA 2x2). The results of the ANOVA was briefly can be seen from the following table:

Table 2. Calculation results with anova two paths.

Sources	df	SS	MS	F	F table $\alpha = 0{,}05$
Factor A	1	332,75	332,75	6,72	4,08
Factor B	1	300,57	300,57	6,07	
Interaction AB	1	2673,84	2673,84	54,04	4,07
Within Group (Err)	40	1979,27	49,48		
Total (Corrected)	43				

Based on the results of testing the first hypothesis, it is known that the application of thematic learning model more effective in improving student learning outcomes. This is in line with that expressed by the results of data analysis using ANOVA two ways at significance level $\alpha = 0.05$, above, gives a value of $F = 6.72$ is greater than the F table = 4.08. This means that H_0 is rejected, means that the average student science learning outcomes by thematic integrative learning model is higher than the average student by learning model fragmented. Since the average score science learning outcomes of student by thematic learning model integrative 74,14 higher than the average score of science learning outcomes of students who use the model of fragmented 68.64 it was concluded students science learning outcomes that used thematic integrative model is higher than using models fragmented.

Evidence of this strengthens the research hypothesis and reject H_0, could be reinforced by the results of relevant research [14]. Thematic integrative model creating students active, creative, effective, and critical thinking with extensive knowledge and insight, which in turn puts the student as well as the subject and object of study, because it involves students in learning [15].

The test results show that the second hypothesis is an integrated learning model gives different effect on the learning outcomes of science when applied to students who have the ability to think critically different. Model of integrated learning and critical thinking ability of students to the science learning outcomes effectiveness is at a different level. The students who have high critical thinking ability thematic integrative models are more effective than the fragmented models, contrast to the students who have low ability to think critically, thematic integrative models are less effective than fragmented model. Data results analysis using ANOVA two way at significance level $\alpha = 0.05$, mentioned above, gives a value of $F = 54.04 >$ Ftable (Ft) = 4.07, this means that H_0 is rejected. Which means that there is a very significant interaction effect between integrated learning model with critical thinking ability of students to the sciencelearning outcomes. Interraction is proved by the relevant research [16], based on the positive there is a direct influence on the variables critical thinking on science learning outcomes.

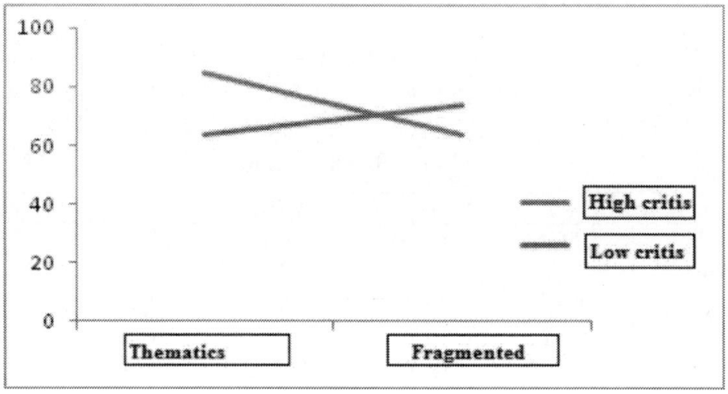

Figure 1. Grafic interaction.

The graphic images provide information that students who have high critical thinking skills when given thematic learning model of integrative science learning outcomes higher than using model fragmented. While different from the critical thinking skills of students is low. Lower learning results when given a thematic learning model compared using model fragmented. Integrative model of thematic learning stimulate students to think more highly of having to connect the various concepts and knowledge which is pedestrianized with an interesting theme. Higher thinking skills is what is said Costa [6] that needs to be honed and developed further in learning.

Results of testing the third hypothesis showed that the group of students who have critical thinking skill height of more effectively using the Model Thematic Integrative than Model Fragmented in his research found that influence the ability of critical thinking of students in the subject of Physics of the results studying Physics. Children who have high critical thinking skills when stimulated by using integrative thematic learning will increase learning outcomes in accordance with the opinion of [10].

Results of the fourth hypothesis suggest that students who have low critical thinking skills, scores of learning outcomes by fragmented groups higher. The test results Tuckey on a group of students who have the ability to think critically low values obtained $q = 4.76$ is greater than qtabel (0.05) (4; 11) = 4.26. This means that there are differences in learning outcomes science group of students who were given a thematic model of integrative learning and student groups by learning models fragmented for group of students who have the ability to think critically low. It can be concluded that the group of students who have the ability to think critically low, the result of groups of students learn science by learning fragmented models than science learning outcomes groups of students by thematic integrative learning models.

The acceptance of the research hypothesis that the fourth line with the opinion [11] that characteristic children who have critical capabilities should be able to make critical decisions and creative, however, students ability to think critical low cannot make decisions complicated, let alone connect the various concepts and themes presented by the teacher through learning.

Then the students ability to think critically low belajarannya results would be lower if using an integrative model of thematic learning than those who studied with fragmented learning model. Because they are not skilled connecting the various concepts, and themes.

4. Conclusion

Based on the results, it can be resumed that integrated learning model can effectively critical thinking skill and both effect to science learning outcomes specially at elementary learner. This result can answer the question how to increase high-level thinking skills through a learning process especially science. Also raise important question for current assessment in process learning.

5. References

[1] Nitko A J 2001 *Educational Assesment of Students* (Colombia Ohio: Merril Prentice Hill) p 22
[2] Close B 1973 *Teaching Science in Elementary Schools* (Sydney: McGraw-hill) p 5
[3] OECD PISA 2013 *Snapshot of Student Performance in Mathematics Reading and Science* (OECD Publishing)
[4] Kemendikbud 2011 *Survei International PISA (Programme for International Student Assessment)* (Jakarta: Kemendikbud)
[5] Trilling B and Fades C 2009 *21st Century Skills Learning for Life in Our Times* (San Fransisco: Jossey-Bass) pp 7-11
[6] Kowiyah 2012 *Jurnal Pendidikan Dasar* **3** 178
[7] Paul W R 1995 *Critical Thinking How to Prepared Student for Arapidly Changing World* (Santa Rosa: Foundation for Critical Thinking) p 503
[8] Hannum W 2015 Robert Gagne's theory *Learning Theory Fundamentals*
[9] Siregar E and Hartati N 2010 *Teori Belajar dan Pembelajaran* (Bogor: Ghalia) p 8
[10] Anderson R W and Krathwohl D 2001 *A Taxonomy for Learning, Teaching and Assessing: A Revision of Bloom's Taxonomy of Educational Objectives* (New York: Addison Wesley Longman, Inc) p 28
[11] Fogarty R 1991 *The Mindful School How to Integrate the Curricula* (Illinois: Skylight Training and Publishing Inc.) p xlv
[12] Semiawan C 2007 *Landasan Pembelajaran dalam Perkembangan Manusia* (Jakarta: Pusat Pengembangan Kemampuan Manusia) p 88
[13] Potts B 1994 *Practical Assessment, Research & Evaluation Journal* **4** 1-3
[13] Sekaringtyas T 2013 *The Effect of Motivation Learning and Critical Thinking Skills to Science Learning Outcomes at Sukatani IV Elementary School* (Jakarta: Thesis University Nation of Jakarta)
[14] Trianto 2010 *Model Pembelajaran Terpadu* (Jakarta: Bumi Aksara) p 41
[15] May L Q and Kam H W 1998 *Thinking Process Going Beyond the Surface Curriculum* (Singapore: Sprint Print Prentice Hall) p 91

Development of Teaching Materials Based Interactive Scientific Approach towards the Concept of Social Arithmetic For Junior High School Student

M K Abadi[1], H Pujiastuti[2] and L D Assaat[2]

[1]Departemen of Mathematics Education, Post Graduate School, Universitas Pendidikan Indonesia, Jl. Dr. Setiabudhi No. 229, Bandung 40154, Indonesia.
[2]Departemen of Mathematics Education, Mathematics Education Faculty, Universitas Sultan Ageng Tirtayasa, Jl. Raya Jakarta KM.04 Pakupatan, Serang 42124, Indonesia
Email: kharisabadi@student.upi.edu

Abstract. The scientific approach is the characteristic of the curriculum 2013. In learning to use a scientific approach, learning process consists of five stages: observe, ask, try, reasoning and convey. In the curriculum 2013 the source of learning is a book, print media, electronic and about nature or relevant learning resources. Most of the print instructional materials on the market does not appropriate in the curriculum 2013. Teaching materials with a scientific approach, beside that to the teaching materials should motivate students to not be lazy, do not get bored, and more eager to learn mathematics. So the development of scientific-based interactive teaching materials that if this approach to answer the challenge. The purpose of this research is to create teaching materials appropriate to the curriculum 2013 that is based on scientific approach and interactive. This study used research and developed methodology. The results of this study are scientific based interactive teaching materials can be used by learners. That can be used by learners are then expected to study teaching materials can be used in android smartphone and be used portable.

1. Introduction

Mathematics is a discipline that was very important in their daily lives. This is evident in our daily activities can not be separated from activities related to mathematics. But awareness of the importance of mathematics is still not evenly distributed. This can be seen in students who still feel scared and lazy to learn mathematics. Many attempts by the government and teachers to improve students mathematical abilities. Efforts are being made among other teachers in the learning process in the classroom so that students are more interested and diligent in studying mathematics. In addition to the efforts of teachers is also important government role in improving the quality of students' mathematical abilities in Indonesia. Efforts have been made by the government among others to enhance the curriculum.

Curriculum is a set of plans and arrangements regarding the objectives, content and learning materials as well as the means used to guide the implementation of learning activities to achieve specific educational objectives. The new curriculum is currently at 2013 [1]. Curriculum 2013 has distinctive features namely the application of scientific approach. Efforts application of scientific or scholarly approaches in the learning process is often touted as a characteristic of the curriculum 2013, which is certainly interesting to learn and further elaborated [2]. Scientific approach in learning it includes components: observe, ask, try, processing, presenting, concluded, and create [1][2][3].

Implicitly, the statement becomes a challenge to learn more in-depth curriculum 2013 so that the educators are better equipped to develop the ability of learners.

Curriculum 2013 requires students to use learning resources to acquire knowledge about the subject matter. Definition of learning resources based on the Standard Process Primary and Secondary Education, learning resources can be in the form of books, print and electronic media, the natural surroundings or other relevant learning resources [4]. It can be concluded, books as teaching materials, including the learning resources that are used in the curriculum of students in 2013, so this is an important thing that needs to be developed.

Books or instructional materials used in teaching the curriculum in 2013 should correspond to the five stages of learning: to observe, ask, try, reasoning, and presentation. In the course curriculum in 2013 is still a lack of book that refers to the scientific approach to the curriculum in 2013, it can be known by comparing the book provided by the government with the books sold in public places. In addition to the lack of teaching materials based scientific approach we have challenges in curriculum in 2013 itself, which students use technology/electronic media as learning support. In developing teaching materials used in the curriculum in 2013 should be approached scientific and students use technology as a learning support. Based on the above preliminary researchers chose a title based interactive development of teaching materials scientific approach on the subject of social arithmetic SMP.

The purpose of this research is to produce interactive teaching materials based scientific approach on the subject of Social Arithmetic SMP. Limitation of research and development on these materials are as follows, first the teaching materials that developed is teaching material in the form of soft copy files to Portable Document Format. Second product development of teaching materials is limited in mathematic subjects the material social arithmetic for student class VII. Third measures undertaken in development research, seven of these, namely the potential and problems, data collection, product design, design validation, design revisions, product testing, and product revision.

2. Literature Review

2.1. Interactive Teaching Materials

To define interactive teaching materials we have to understand the meaning of the teaching materials and interactive. Instructional materials are all kinds of materials that are used to assist educators in implementing the teaching and learning activities in the classroom [5]. We understand anything related to classroom learning and provide an acceleration in understand the material being taught by educators to students called teaching materials. To increase students' interest in learning teaching materials should be interesting. Because of human nature that can be bored. Basically all the people do not want their boredom in his life [6]. This makes the challenge to create a more pleasant teaching material. To enhance the learning material to make it more attractive, effective when used in a predetermined learning purposes [7]. Besides the sense of teaching materials include products of science, technology, art, or sports, among others, in the form of articles, designs, patents, or teaching materials[8].

Through the learning model based interactive multimedia intended to make learning more interesting and the material is abstract can be visualized in animation media in accordance with the actual conditions on the ground to follow the development of science and technology continues to evolve so that student motivation to learn and build knowledge becomes easier to do [9].

The purpose of the interactive teaching material is to visualize things that are abstract. Classification within the scope of multimedia interactive learning system lies not in hardware, but rather refers to the learning characteristics of students in response to the stimulus shown a computer screen [9]. In making the instructional materials interactive emphasis on making learning software that has characteristic students who can respond teaching materials. It can be concluded that interactive teaching material is a form of learning in this book-based software that provides stimulus in the form of animations to users [10].

IOP Conf. Series: Journal of Physics: Conf. Series **812** (2017) 012015 doi:10.1088/1742-6596/812/1/012015

2.2. Scientific Approach
Provides skills in the conception of the curriculum in 2013 that uses a scientific approach is to observe, ask, try, reasoning, Present, create [3][4]. In learning this should be an important basis in each learning approach. The components of the scientific approach as follows:

2.2.1. Viewing. Viewing is an activity to find out is to focus on what is being searched out. As activities seek out a merchant to sell his wares, selling way this is viewed and searched out by us. Event to find out is what is called observing.

2.2.2. Ask. Propose an activity to find out something by way of exchange of ideas. In contrast to observe, in a more focused activity observed observers do see something observed activities. Propose activities are activities to find out something by digging information to someone either directly or through the media. The cause of the students asked the educators, among others because the students do not understand the relevant subject matter, in addition to the students want to associate the knowledge that has been gained with the new knowledge gained.

2.2.3. Attempt. Trying to figure out something is activity results by giving the new treatment. Try activities in the field of mathematics is usually done when solving problems. Solving problems is done with tries to solve the problem with trying to work on a paper or doing new things based on observing first. This is the result of trying to know anything what happens after the treatment was given.

2.2.4. Reasoning. Reasoning is the process by which new knowledge is processed in mind. These cognitive skills are closely related with the speed to understand and analyse the material presented. Reason is used in questions and high rate ability.

2.2.5. Present. Present are activities to provide information to the audience with material that has been determined. Examples of seminars are wont to do, there are people as informants who provide specific information in a seminar attended by many audiences. Present can be concluded that the activities carried out by the presenter of material to the audience with material that has been determined.

2.2.6. Creating. Creating an activity to make it to have a new distinction. In mathematics learning activities to create a professional manner is to make the method, manner, or a new theory. But the simple, while answering students answer in their own way is a creative activity. Meaning of creating itself is made so that in the field of education students can create knowledge that exist in the real world into his mind.

2.3. *Arithmetic social.* Buying and selling activities are carried out in the supermarket or on the market, is one example of social arithmetic in economic activity [11]. We know that the economic activity we did was matter of discussion of social arithmetic. Social arithmetic studied in class VII SMP. Social arithmetic in economic activity, a rebate (discount), gross, tare, net, interest and tax savings [12]. The value of all, the value per unit, the value in part, the purchase price, purchase price, profit, loss, discount, gross, net, tare, tax and single flower [11].

3. Experimental Method
Educational R n D is an industry-based development model in which the findings of research are used to design new products and procedures[13]. This research aims to create products based interactive teaching materials scientific approach on the subject of social arithmetic junior high, so that the research will be carried out is a research and development or research and development.

Research and development is a research method that is used to produce a particular product, and test the effectiveness of these products [14]. Research and development is a process or steps to develop a new product or improve existing products that can be justified [15]. So that it can be concluded that the development of the research is to improve or make things so much better.

3.1. Model Development

The products resulting from this research is based interactive teaching materials scientific approach on the subject of social arithmetic SMP. Research departing development of their potentials and problems of data collection, product design, design validation, test user, product revision, test products, design revisions, revision of the product, product mass [14]. For more details, step-by-step development of interactive teaching materials based scientific approach on the subject of social arithmetic SMP as shown in the figure below.

However, because given the time and budget constraints, this study only involves these steps until the seventh step. The seventh step is the potential and problems, data collection, product design, design validation, design revisions, product testing, and product revision. This is because the products are produced only in small scale.

3.2. Development Procedure

Interactive teaching materials based scientific approach on the subject of social arithmetic was developed using seven stages of development. Stages include the potential and problems, data collection, product design, design validation, design revisions, test products and product revision.

3.3. Testing Products

After the making of interactive teaching materials based scientific approach on the subject of social arithmetic is completed the next stage of product trials. Product testing is done twice: expert testing and limited test. Test validation is performed before performing the test because the experts to determine the extent to which the goals and objectives of product development experts on the development. For this test was carried out by expert mathematician and expert education.

Test on the field at 1 to 3 schools with 6 to 12 test subjects (teachers) [15]. In this limited test will involve six junior high students of class VII. In these trials will be used questionnaire assessment of teaching materials, filling a questionnaire carried out by six students after the test is done.

Scale model of computation that are used are Likert scale. Six students had been asked to fill out a questionnaire with the option strongly agree (SS), agree (S), disagree (TS), and strongly disagree (STS). For the positive statements it will be to the value of SS = 4, C = 3, TS = 2, STS = 1 and vice versa for negative statement then SS = 1, C = 2, TS = 3, STS = 4. For the assessment based interactive teaching materials scientific approach used by students open questionnaire.

3.4. Instruments

In this study, data collection using research instruments such as questionnaires. Technique of data collection is done by giving a set of questions or a written statement to the respondent to answer [14]. This questionnaire will be given to student who have been using interactive teaching materials and is completed by the student. The questionnaire used in the form of closed and open questionnaire questionnaires so that researchers will get more feedback from users of this interactive teaching material.

3.5. Data Analysis

These calculations are used to determine the results achieved in the trials conducted to students. This description uses the following formula[16]:

$$\textit{Final value} = \frac{\textit{raw score or total score of responden}}{\textit{scores ideal or highest total score answers}} \times 100\% \qquad (1)$$

3.6. Indicators of Success
In this study, the indicator is the assessment of test trials mathematician and education experts. Indicators of success of research and development-based interactive teaching materials scientific this approach is assessed by greater than 70% of the tester mathematicians and education experts.

4. Result and Discussion

4.1. Test Expert
Mathematicians have come from the environment department of Mathematics Education University of Sultan Ageng Tirtayasa. In case this is a lecturer of mathematics education so that scientific mathematics is the field of science. Assessment is done objectively about interactive teaching materials scientific-based approach that was developed from the viewpoint of mathematical concepts or materials. The result are mathematic expert 1 is 82.14% and mathematic expert 2 is 92.86%.

Education experts whose role in the assessment of this questionnaire derived from one tester supervisor and an examiner of SMPN 1 Cinangka mathematics teachers who have experience in educating students and understand the curriculum, 2013. Following the results of the assessment tester education expert 1 93.75% and education experts 2 is 73.96%.

Media experts who play a role in these two testers assessment questionnaire derived from SMAN 1 Cinangka computer teacher. Following the assessment of examiner media expert 1 is 83.75% and media expert 2 is 77.5%.

4.2. Test Results Limited
After testing expert and product improvement of teaching materials, and then do the test product against a small group (limited test). Test a small group consisted of 13 students of VII class of SMP Satu Atap Gelam 2 Kota Serang. The result of Limited test is 85.77%.

4.3. Final Product
Storage media using data storage device in the form of soft copy (flash dish, hard disk, CD, DVD). Teaching material size is 21 cm x 29.7 cm (A4). The file size is 116 Mb. The number of pages is 33 pages of teaching materials. The format of teaching materials is a Portable Document Format (.PDF). Typeface on teaching materials is Times New Roman 12 and 12. The material presented Cambria is a social arithmetic.

4.4. Linkage Teaching Material
Learning curriculum emphasizes activity observed in 2013 contains, ask, digging up information / tries, reasoning, inform. The meaning of the five thing above are educators can not provide material information directly to learners. In this case the teacher becomes a facilitator of learners in constructing their own knowledge. In this teaching material teaching material users are required to construction own knowledge. Teaching material automatic constructknowledgeby giving problems from simple to complexes. This problem is the usefulness of the material being studied, so that users can easily imagine for example refers to real-world problems. In this teaching material at each end of the material is given the correct definition, so that after the next phase of his knowledge to constructstraighten information if there is some mistake. This is the basis of the manufacturing-based interactive teaching materials scientific approach.

5. Conclusion
After the various processes of testing by several experts as well as a limited test of the number of students obtained percentage of the assessment results. Results of votes mathematician first test of 82.14% and test mathematician second at 93.86%, so by testing mathematic expert that materials based interactive scientific approach include category is very strong. Results of the assessment of the educational expert test is 93.75% so get conclusion these materials include the category is very strong

and the two education experts test is 73.96% so that the resulting conclusion of these materials include strong category. The results of expert assessment of the media is 83.75% and the two media expert assessment is thus 77.5% based interactive teaching materials scientific approach is categorized as strong. The limited test results were excellent which got 85.77%, so very strong in the category.

Based on the results of expert testing and limited testing has been done addressing that interactive teaching material based scientific approach is fit for use in the learning process. Limited test results obtained show students' response to interactive teaching material based scientific approach is good.

Based on the conclusion, some suggestions that could be addressed are: (1) The only teaching materials developed in social arithmetic material so expect further researchers develop interactive teaching material based on a scientific approach to the subject of another; (2) for researchers who want to develop further teaching materials, it is advisable to increase the population in a limited test; (3) Researchers who wish to further develop this materialis advisable to increase the exercises more challenging and should be tailored sequence material level of difficulty so that students feel more challenged; (4) Preparation of teaching materials also requires the ability of researchers to operate the computer to support the manufacture so that further research is expected to be ready in the use of computer technology; (5) In making interactive teaching materials adapted to the design of a computer and mobile phone screen; and (6)It is expected that the first secondary school teachers try to use products based interactive teaching materials of this scientific approach.

6. References

[1] Undang-undang no. 20 Tahun 2003 tentang Sistem Pendidikan Nasional (Jakarta: Depdikbud)
[2] Kurniasih and Sani B 2014 *Implementasi Kurikulum 2013: Konsep & Penerapan* (Surabaya: Kata Pena) pp 40–141
[3] Kusmaryono I and Suyitno H 2015 The effect of constructivist learning using scientific approach on mathematical power and conceptual understanding of students grade IV *IOP PublishingJournal of Physics: Conference Series* **693** (2016) 012019
[4] Peraturan Menteri Pendidikan dan Kebudayaan Republik Indonesia Nomor 65 Tahun 2013 (Jakarta: Depdikbud)
[5] Ahmadi 2011 *Strategi Pembelajaran Sekolah Terpadu* (Jakarta:Prestasi Pustakarya) p 208
[6] Djamarah S B and Zain A 2010 *Strategi Belajar Mengajar* (Jakarta:Rineka Cipta) p 160
[7] Uno H B 2011 *Perencanaan Pembelajaran* (Jakarta:Bumi Aksara) p 32
[8] Permendikbud RI No 65 2013 Standar Proses Pendidikan Dasar dan Menengah (Jakarta: Permendikbud RI)
[9] Ali M 2009 *Jurnal Edukasi @Elekto* pp 12–14
[10] Kusmaryono I and Suyitno H 2015 *The Effect of Constructivist Learning Using Scientific Approach on Mathematical Power and Conceptual Understanding of Students Grade* IV *IOP PublishingJournal of Physics: Conference Series* **693** (2016) 012019
[11] Manik D R 2009 *Penunjang Belajar: Matematika : Untuk SMP dan MTs Kelas 7* (Jakarta: Pusat Perbukuan Departemen Pendidikan Nasional) pp 117–118
[12] Nuharini D and Wahyuni T 2008 *Matematika 1: Konsep dan Aplikasinya: untuk Kelas VII SMP/MTs* (Jakarta: Pusat Perbukuan Departemen Pendidikan Nasional) p 136
[13] Borg and Gall 2003 *Educational Research: An Introduction, 7th ed* (New York: Longman) p 569
[14] Sugiyono 2011 *Metode Penelitian Pendidikan: Pendekatan Kuantitatif, Kualitatif, dan R & D*, (Bandung:Alfabeta) pp 199–409
[15] Sukmadinata N S 2009 *Metode Penelitian Pendidikan* (Bandung:Rosda Karya) pp 164–170
[16] Riduwan 2010 *Dasar-Dasar Statistika* (Bandung:Alfabeta) p 40

Students' Misconceptions on Titration

H R Widarti[1], A Permanasari[2,]*, and S Mulyani[2]

[1]Chemistry Department, Universitas Negeri Malang (UM), Jl. Semarang 5, Malang 65145, Indonesia
[2] School Science Program Study of Postgraduate School, Universitas Pendidikan Indonesia, Jl. Dr. Setiabudhi 229, Bandung 40154, Indonesia
*Corresponding author: E-mail:anna.permasari@upi.edu

Abstract. Misconceptionscanlead to the problems of students' understanding, if it is not resolved immediately. Through this research, the potency of misconceptions has been investigated on analytical chemistry subject, especially on titration concepts. Descriptive method was used in this study, involving 66students who were takingAnalytical Chemistry course. The seven items of multiple choice tests with reasons were used as an instrumentand the result was then analyzed on using modified CertaintyofResponseIndex technique. The results show that 51.6% of students understand the concepts of titration well; 13.3% of students do not understand the concept, and 35.1% of students have misconceptions. The misconceptions mainly occurred in macroscopic and symbolic levelofrepresentation on choosing measuring equipment for titration, using titration equipment, and calculating titration.

1. Introduction

Titration concept has been taught to students since at senior high school. Students' understanding of the titration process can vary depend on their own knowledge and experience. Some students' do not understand, but some others face on misconceptions. Misconception may occur to students if the understanding of concepts does not correspond to the actual concepts[1,2].Louga et al.[3]also stated that misconception is inconsistency between students' views and experts' views.Misconceptions will lead to problems affecting students' understanding level if not resolved immediately. Misconceptions may be occurred in chemistry. Many researches have been done to anticipate the potential misconceptions, because misconceptions always arise in lectures.

Pinarbasi[4]reported misconception on colligative properties subject, and advised that the strategies of learning should be applied on using substantial review. Moreover, Yarroch (in Nakleh[5]) has carried out the research on misconceptions associated with chemical reaction.He reported that all students managed to equalize the equation, but most students could not draw molecular diagrams correctly to explain the similarities sub-microscopically. Regarding the size of atoms, Eymur et al.[6] reported that students who are prospective teachers have misconceptions which are almost similarly experienced by high school students. Misconceptions are highly dependent on many factors, such as experience, creativity, perception, and textbooks.

Pinarbasi[7] reported that Turkish students have a number of common misconceptions on the topic of acids and bases. The same research of misconceptions on acidsand bases has performed by Demircioglu et al.[8]Rahayu et al. [9], Damanhuri et al. [10], and Tumay [11]. Pan &Henriques [12]at least the six types of misconceptions on acids and bases concepts were proposed.The type of misconceptions that mostly occurred was the notion that the endpoint or equivalence point of a titration is always at pH 7.

Misconceptions on titration concepts need to be studiedbecause it is a basic concept for learning analytical chemistry.Research on acid-base titrationlearning of students in junior hight school has been reported by Sheppard [13]. The results indicated that students have major difficulty on acid-base concept as well as they were not able to accurately describe the concept of acid-base, such as pH, neutralization, acid-base strength, and acid-base theory. Several factors contribute to these difficulties;basics of chemistry on acid-base are too visceral, with more emphasis on mathematical calculations for learning and textbooks-dependent. The density of acid-base concept, the confusing acid-base terminology and lack of consensus on what material should be included in the curriculum of chemistry were also identified as problems. Widarti et al. [14] reported that they found misconceptions on redox titration, where misconceptions occurred on concepts involving concentration measurements with chemical equation, species existing in solution from titration process, and the characteristics of redox titration involving potential calculation.

A research on titration was carried out involving 38 students who were taking basic analytical chemistry (BAC) course. The study showed that despite the students' ability in BAC was considered as fair in category, but three times tests indicated significant deterioration (the score respectively 72.79; 66.00; and 59.00). Based on the research on students' answers, they are generally weak in mathematics literacy and ability to provide relevant analysis of how to determine the choice of indicators in a titration (Widarti et al. [15]). It is highly associated with a students' low ability in explaining phenomenon in a submiscroscopic and symbolic way. Therefore, a research is done to investigate students' misconceptions on the concept of titration.

2. Experimental Method

Descriptive method was used in this research. The subjects of research were 66 students in the 3rd semester of Chemistry Education from 2014-2015 batch in a university in East Java. Misconceptions in the study arerevealed by the merger of two instruments that have been developed, namely the multiple-choice test of 7 questions on the concept of titrationwith open reason and Certainty of Response Index (CRI) technique by Hasan, et al. [16] which has been modified. The technique can distinguish among knowing the concept well, not knowing the concept and having misconceptions.The reasonings is needed as a reflection of their thinking and understanding of the proposed concept. From the answers and the reasonings, the congruence of students' understanding with the scientific concept can be discovered.The characteristics of the misconceptions are durable, firmly entrenched in ones' mind (Louga, 2013). Relates to personal belief, so the used of used CRI techniques is relevant to reveal student misconceptions.

The modified CRI, as proposed by Hakim et al. [17], was used to accommodatebiasfrom Indonesian student culture. Mostly, students understand concept but not confident whit their answers, show the adding reasoning in every students' answer will help to analyze data. According to Kurbanoglu et al. [18], students who master the concept well also have high confidence.The modification of CRI scaling was also performed as recommended by Potgieter et al. [19], by using four scales (1-4), 1 = guessing, 2 = not certain, 3 = certain, 4 = very certain.

Based on the answers, reasonings and CRI, there are some possibilities happening. If students' answer is correct with correct reasoning and with CRI>2, then students are classified as understanding the concept well. If students' answer is correct with correct reasoning but with CRI<2, students are still classified as understanding the concept well. If students' answer is wrong with correct or wrong reasoning and with CRI>2, then students are classified as having misconception. If students' answer is wrong with correct or wrong reasoning and with CRI<2, then students are classified as not understanding the concept. If students' answer is correct with wrong reasoning and with CRI<2, then students are classified as not understanding the concept. Possibilities in utilizing modified CRI for each of students' answer are as seen on Table 1.

IOP Conf. Series: Journal of Physics: Conf. Series **812** (2017) 012016 doi:10.1088/1742-6596/812/1/012016

Table 1. Classification of modified CRI on students' answers (Hakim et al. [17], Potgieter et al. [19])

Answer	Reasoning	CRI	Description
True	True	> 2	Understand the concept well
True	True	<2	Understand the concept but not certain of the answer given
True	False	> 2	Misconception
True	False	<2	Not understand the concept
False	True	> 2	Misconception
False	True	<2	Not understand the concept
False	False	> 2	Misconception
False	False	<2	Not understand the concept

3. Results and Discussion

Data test on volumetric analysis in BAC course reflected the students' skills level. Each student has different level of comprehension and understanding of concepts although they study the same material with the same lecturer. The different ability may cause the different comprehension and misconceptionsto the concepts learned.As it is stated before, titration concepts has already taught since senior high school. In reality, the BAC course could not reduce all of the misconceptions. The total number and percentage of the students' answers on titration are as seen on Table 2.

Table 2. Totalnumber and percentage of studentsmisconception on titration

No	Concept	Total and Percentage (%) of Students Having Misconception					
		Misconception (M)		Not Understand (N)		Understand (U)	
		Total	Percentage	Total	Percentage	Total	Percentage
1	Selection of equipment	29	44	4	6	33	50
2	Titration processing	10	15	3	5	53	80
3	Washing and rinsing burette	5	8	2	3	59	89
4	Washing and rinsing Erlenmeyer	42	64	3	4	21	32
5	Rinsing technique along titration	41	62	3	4	22	34
6	Selection of titration method	1	1	31	47	34	52
7	Stoichiometric titration	34	52	16	24	16	24
	Average		35,1		13,3		51,6

Number of student: 66

As it is shown in Table 1, between the 7 testswere given, the students'answer number 4, 5, and 7 (Washing and rinsing erlenmeyer, Rinsing technique along titration, and Stoichiometric titration) lead to misconceptions (more than 50% of students). Meanwhile, almost of all student were not misconceptions on the concepts in test number 2 (Titration processing) and 3 (Washing and rinsing burette). From the answer test number 1 (Selection of equipment) and 6 (Selection of titration method), number of students who have misconceptionsis proportional with the number of students who have understand the concepts.This is encouraging that only a few students who not understand the concepts (13,3% of all students). The portrait of overall misconceptioncompare to the understand concepts is shown in Figure 1.

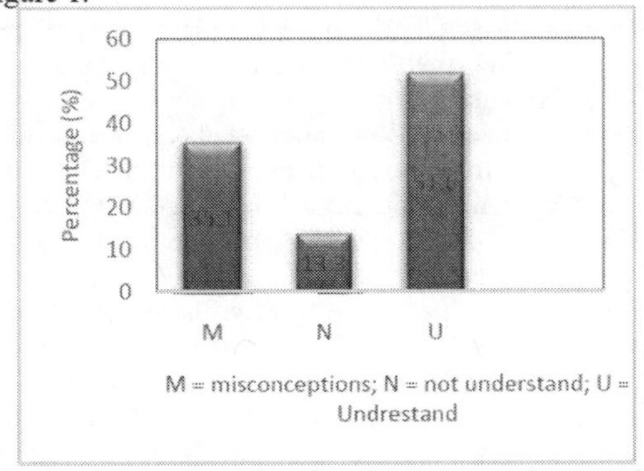

Figure1. Percentage of students' misconceptions on titration

Misconception most commonly occurred in the answer to question number 4 (64%). The questionasked:"what has to be done when provided erlenmeyer flasks has already rinsed off but not dried and it will be used immediately for titrating HCl with NaOH solution?". Most of students answer with minsconceptions potential.Students explained erlenmeyer has to be rinsed by HCl solution instead of aquadest. Some of them said that the erlenmeyer should be dried, and some other else answered to use erlenmeyer immediately. Erlenmeyer as a container for sample or titrant (HCl) has to be clean and dry.If not, it will affect the quantitative result. Wet apparatuscan be used but it must be rinsed off with aquadest. However, if the apparatus are already dry and clean, it is not need to rinse them off. When dried with tissue, erlenmeyer will become dirty because the wet tissue will stick to the erlenmeyer.

The same case also occurred on the question number 5 regarding to the process of titration. The students were asked to explain whether it is allowed to rinse erlenmeyer wall with aquadestalong titration.As many as 62% of student showed misconceptions by answering that it is not allowed because the number of substance will change. However, rinsing or adding aquadest to erlenmeyer flask do not affect the amount of substance, because the number of moles of solute (HCl) will not change regardless the addition of solvent volume. The students must understand that the number of dissolved substance or the number of moles of a solute(HCl) do not change by solvent addition. The change will only be occurred in concentration and the molarity of a solution. There is a relationship between concentration with the amount of solute.Concentration is the number of moles of solute in a volume of solution.

Moreover, Misconceptions also occur in the answer of the question number 7 (the calculation of moles of NaOH needed for titration with 10 mL 0,1 M oxalic acid solution). In student's perception, themmoles of oxalic acid is equal to mmoles of NaOH at the end point of titration, without considering the chemical equation. It is also predicted that students possibly do not know the chemical formula of oxalic acid. Actually, they have to take account into equivalence of H^+ and OH^- ions in chemical equation. Oxalic acid produces two H^+ion.The reaction between oxalic acid and NaOH is as follows:

$$2H^+ + C_2O_4^{2-} + 2Na^+ + 2OH^- \longrightarrow 2H_2O(l) + C_2O_4^{2-} + 2Na^+$$

One mole oxalic acid (produce two moles H^+) is equivalent with two moles of NaOH (produces two moles).

Test number 1 is about the equippment needed to measure the volume of sample solution for titration. As many as44% of students answer measuring pipette, not volumetric pipette to measure the volume of solution sample. The students still assume that measuring pipette has the highest accuracy in measuring the volume of a solution, instead of choosing volumetric pipette.

Most of students did not understand the concepts that should be answered in the question number 6. The concept is about titration application, especially in selection of appropriate method that can be used to determine the content of a sample containing caustic soda. Commonly, they chose the other techniques instead of acid-base titration. They proposed some reasons. Some of them thought that caustic soda can be precipitated on using $AgNO_3$ to produce silver hydroxide. Some others argued that hydroxide ion can be oxidized into oxygen gas.

Most of students have the good understand to the concepts on what to do to the washed but not dried burette when it will soon be used to titrate HCl sample with NaOH solution. Most of the students gave argument they have to rinse the burette with NaOH solution that will be used as titrant.

Based on research, there were still found some potential misconceptions on titration subject. It should not be the case because actually this concept has been studied both during senior high school as well as in basic chemistry course in graduate level. Mostly, the misconceptions occur on macro and symbolic representation. Students confused on choosing the appropriate apparatus for measuring volume of sample. Students were also ambiguous in titration calculations. This phenomena occurs in less meaningful learning as stated by Berg, Cross et al., and Pinarbasi[5,6,7]. Some researches recommend learning has to be plan with the more students active and partisipatory as well as accommodate the multiple representative especially in learning chemistry[3, 4,8, 9].The important thing is student has to have a high motivation on learning. Some strategies are recommended by some researchers, such as collaborative[20], problem possed[21], or cognitive dissonance[22]. the last strategy has not used in chemistry learning yet, so it is interest to combine the multiple representation and cognitive dissonance strategies, as a challenge interactive learning.

4. Coclusion

Misconceptions will lead to the problems on the level of students' understanding, if it is not resolved immediately. By the descriptive research involving 66 students of basic analytical chemistry course, it is shows that 51.6% of students understand the concepts of titration well; 13.3% of students do not understand the concept, and 35.1% of students have misconceptions. The misconceptions mainly occurred in macro and symbolic skills on choosing measuring equipment for titration, using titration equipment, and calculating titration.

5. Acknowledgments

The writers are grateful for the BPPS-doctor scholarship from the Directorate of Higher Education, Republic of Indonesia. The writers also would like to convey gratitude for my college (Universitas negeri Malang) who have been extremely helpful along the study.

6. References

[1] Suparno, P 2005 *Miskonsepsi & Perubahan Konsep Pendidikan Fisika* (Jakarta: Grasindo)
[2] Van den Berg, E 1991 *Miskonsepsi Fisika dan Remediasinya* (Salatiga: Universitas Kristen SatyaWacana)
[3] Luoga, N E., Ndunguru, A P, and Mkoma, L S 2013 *Tanzania Journal of Natural & Applied Sciences* **4** pp 575–81
[4] Pinarbasi, T, Sozbilir, M, and Canpolat N 2009 *Chemistry Education Research and Practice* **10** pp 273–80.
[5] Nakhleh, B M 1992 *Journal of Chemical Education* **69**(3) pp 191–6
[6] Eymur, G, Cetin, P, and Geban O 2013 *Journal of Chemical Education* **90** pp 976–80.
[7] Pinarbasi, T 2007 *Journal of Baltic Science Education*, **16**(1) pp 23–34
[8] Demircioglu, G, Ayas, A. and .Demircioglu,H 2005 *Chemistry Education Research and Practice,* **6**(1) pp 36-51
[9] Rahayu, S, Chandrasegaran, L A, Treagust, F D, Kita, M, and Ibnu, S 2011 *International Journal of Science and Mathematics Education*, **9** pp 1439-58.

[10] Damanhuri, M I B, Treagust, F D, Won, M, and Chandrasegaran, L A 2016 *International Journal of Environmental & Science Education* **11** pp 9-27
[11] Tumay, H 2016 *Science and Education* **25** pp 21–46
[12] Pan, H, and Henriques, L 2015 *School Science and Mathematics* **115** pp 237-44
[13] Sheppard, K. 2006 *Chemistry Education Research and Practice* 7 pp 32-45
[14] Widarti, R H, Permanasari, A, and Mulyani, S 2016 *Jurnal Pendidikan IPA Indonesia* **5** pp 56-62
[15] Widarti, R H, Permanasari, A and Mulyani S 2014 *Analisis Kesulitan Mahasiswa Pada Perkuliahan Dasar-Dasar Kimia Analitik di Universitas Negeri Malang* (Malang: Seminar SNKP)
[16] Hasan, S, Bagayoko, D, and Kelley, E L 1999 *Physics Education* **34** pp 294–9.
[17] Hakim, A, Liliasari, and Kadarohman, A 2012 *International Online Journal of Educational Sciences*, **4**(3) pp 544–53
[18] Kurbanoglu, N I, and Akin, A 2010 *Australian Journal of Teacher Education* **35** pp 48–59
[19] Potgieter, M, Rogan, M J and Howie, S 2005 *African journal of research in SMT Education* **9** pp 121-34
[20] Bell, T, Urhahne, D, Schanze S and Rolf 2010 *International Journal of Science Education* **32** pp 349-77
[21] Rachmawati, N and Rusmini 2012 *Unesa journal of chemical education* **1** pp 34-39
[22] Linenberger, J K and Bretz, L S 2012 *Chemical Education Research &Practice* **1** pp 172-8

Improving Students' Science Process Skills through Simple Computer Simulations on Linear Motion Conceptions

P Siahaan, A Suryani, I Kaniawati, E Suhendi and A Samsudin*

Universitas Pendidikan Indonesia, Departemen Pendidikan Fisika.
Jl. Setiabudhi No. 229, Isola, Sukasari, Isola, Sukasari, Kota Bandung, Jawa Barat, 40154, Indonesia

Email: achmadsamsudin@upi.edu

Abstract. The purpose of this research is to identify the development of students' science process skills (SPS) on linear motion concept by utilizing simple computer simulation. In order to simplify the learning process, the concept is able to be divided into three sub-concepts: 1) the definition of motion, 2) the uniform linear motion and 3) the uniformly accelerated motion. This research was administered via pre-experimental method with one group pretest-posttest design. The respondents which were involved in this research were 23 students of seventh grade in one of junior high schools in Bandung City. The improving process of students' science process skill is examined based on normalized gain analysis from pretest and posttest scores for all sub-concepts. The result of this research shows that students' science process skills are dramatically improved by 47% (moderate) on observation skill; 43% (moderate) on summarizing skill, 70% (high) on prediction skill, 44% (moderate) on communication skill and 49% (moderate) on classification skill. These results clarify that the utilizing simple computer simulations in physics learning is be able to improve overall science skills at moderate level.

1. Introduction

Nowadays, science and technology develops rapidly. Technology development has an impact on the development of learning methods and media, particularly in the use of computer and the internet. The use of information and communication technology (ICT) will make the learning process more interesting and challenging, facilitatingthe teachers and students in teaching-learning processes [1, 2,3]. Because of that ICT is able to deliver information simply and quickly. In broad-spectrum, the use of computer media by teachers in schools is restricted on Microsoft Power Point program to enhance students' understanding of the concept. Therefore, the skills obtained by the students are merely memorizing skill. It is in line with Prince [4] and Bobadilla [5] who stated that at the moment, learning tends to be based on memorizing the theories without utilizing students' experiences. In contrast, the most important point in learning is improving wide range of skills that can be used to solve problems in the real life situation, such as science process skills (SPS).

Learning model applied by teachers in secondary schools, both junior and senior high school are 70% lecturing models (transfer of knowledge), and the rest are discussion, demonstration and experiment models with the percentage of 10% for each model [6]. Based on the questionnaire distributed to high school students in Bandung City, the score obtained by physics students in the test or quiz is still low because of some reasons such as unfavorable class conditions, students' poor math skills, and monotonous learning process (lecturing), occasional experiments, and demonstrations without showing a video or animation that support the explanation of the concept. In fact, sufficient amount of computer media is available in the laboratory and projector is able to be used in the classrooms.

On the contrary, based on interviews to one of physics teachers, lecturing is generally used as the learning because the time is limited compared with the amount of discussion topics demanded by the curriculum. The elaboration above shows that the learning is conducted with the teacher as a source of knowledge, and students rely on the memorizing skill rather than discovery skill, and learning with multimedia is occasionally done.

Science process skills are skills that focus on the learning process to develop students' skills in understanding the knowledge or concepts, independently discovering and developing necessary facts, concepts, and values [4, 7]. Science process skills are a learning approach that integrates science process skills into the system of integrated material presentation [8, 9]. The learning approach is not only transferring knowledge, but also emphasizing on the process of scientific inquiry. In this learning approach, the teacher acts as a facilitator who guides and manages students' learning activities so that students are able to construct necessary facts, concepts, and new values in their lives independently.

The skills in the science process skills according to Gultepe [10] and Ramig, Bailey & Ramsey [11] are observing, summarizing, identifying and manipulating variables, predicting, hypothesizing, organizing and interpreting data, investigating, experiments and surveys, and reproducible for independent investigation. Meanwhile, according to Rauf et al. [12] aspects of the science process skills are skills to observe, to conclude, measure, predict, communicate and classify. In this research, science process skills indicators used are five skills proposed by Rauf et al. [13] that are, observation skill, summarizing skill, predicting skill, communication skill and classifying skill.

Based on these considerations, researchers did research to determine the development of students' science process skills in learning linear motion topic using computer multimedia. Several previous studies suggest that the use of interactive multimedia on the uniformly accelerated motion topic appeals to be very interesting [13], the motivation of students after studying with computer multimedia increases by 6,016% [7, 8], and the increase of science process skills when learning by utilizing interactive multimedia is about 29% [8]. It shows that the use of computer multimedia is able to develop students' science process skills.

2. Methods

The method utilized in this study is a pre-experimental research. The research design is one group pre-test-post-test. In this study, students were given an early test in the form of questions that had been adjusted based sub-concepts and science process skills indicators. After that, students acquire learning by using multimedia computer as a moderate of learning and the last activity is the students obtain the final test. The research design is shown in the Figure 1 as follows.

O_1	X	O_2
Pre-test	*Treatment*	*Post-test*

Figure 1. One Group Pretest-Posttest Research Design.

The data obtained in this study is a quantitative data of pre-test and post-test results is the form of students' multiple choice test score. The population in this study is the seventh grade students of Junior High School (SMP) in Bandung, which consists of eleven classes with the average number of students in each class are 35 students, while the sample is one of grade VII in the school with 36 students. But the test results that can be analyzed are only from 23 students.

The data processing in this research began by calculating pre-test and post-test score. Subsequently, the effect of multimedia computers learning on students' science process skills was determined by using normalized gain analysis. The development before and after the computer multimedia learning was calculated by computing normalized gain equation <g> as follows.

$$\langle g \rangle = \frac{score\ of\ posttest - score\ of\ pretest}{score\ of\ ideal - score\ of\ pretest} \qquad (1)$$

Normalized gain criteria which proposed by Hake [14] can be seen in Table 1.

Table 1. Normalized gain criteria.

<g>	Criteria
(<g>) ≥ 0.7	High
0.3 < (<g>) < 0.7	Moderate
(<g>) < 0.3	Low

3. Results and Discussions

Learning linear motion by using computer multimedia was implemented in three sub-concepts that should be discussed: the definition of motion; uniformly linear motion and uniformly accelerated motion. The results of the normalized gain analysis of each topic are presented in Table 2.

Table 2. Normalized gain recapitulation in each sub-concept.

	Average Score			
Sub-concept	Pre-test	Post-test	<g>	Criteria
1	1.57	2.74	0.34	Moderate
2	1.65	3.83	0.65	Moderate
3	1.30	3.52	0.60	Moderate
Overall	4.52	10.09	0.53	Moderate

Based on Table 2, there is a gradually increase in students' science process skills at each sub-concept with similar normalized gain scores and they can be included in the moderate category. As found in previous studies [12, 13], those studies also showed the identical results that learning with computer multimedia is able to improve students' science process skills. The reason is the use of multimedia computers in the learning process will stimulate students to conduct their own investigations in order to understand knowledge, so that students will be more active during the learning process. The significant rise occurred in a second sub-concept in which uniformly linear motion was discussed with N-Gain score is at 0.65. While the lowest increase occurred in the first sub-concept with N-Gain score is only 0.34. However, it is still in the moderate category [14].

At the first sub-concept, a relatively lower increase is reasonable because in this first lesson students were not familiar with the learning method. During the learning process, students were required to be more active in comprehending knowledge independently. Meanwhile, the nature of the students was still accustomed to learning with the traditional way by making the teacher as the only source of learning. In the other words learning process was centered on the teacher (teacher-centered) and it focused on rote learning by memorizing the knowledge from textbooks. In addition, students were not familiar with the types of questions given to measure the knowledge they had understanding from the learning process. Another point showed that students were totally preoccupied to complete LKS (Students' Worksheet) even though they had not been accustomed to such a way of learning.

Table 3. Normalized gain recapitulation for each science process skills indicator.

Indicators	Average score		. <g>	2. Criteria
	Pre-test	Post-test		
Observing Skill	0.87	1.87	0.47	Moderate
SummarizingSkill	0.57	1.61	0.43	Moderate
PredictingSkill	1.13	2.43	0.70	High
CommunicationSkill	0.91	1.83	0.44	Moderate
ClassifyingSkill	1.04	2.00	0.49	Moderate

In the next sub-concept, students' science process skills arose at its peak because the students got used to the method applied in learning and the questions given. On the other side, at the third sub-concept students' science process skills was slightly decreased due to more difficult sub-concept and longer duration for group discussion so the time for class discussion and confirmation was limited. Aspects of science process skills observed in this study include five indicators, namely observing skill, summarizing skill, predicting skill, communication skill, and classifying skill. The development at each indicator of the science process skills is presented in Table 3.

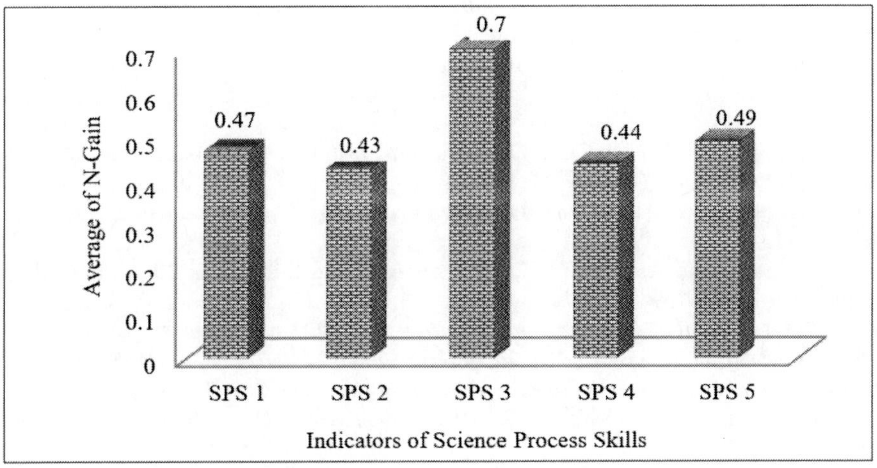

Descriptions:
SPS 1 : Observing Skill
SPS 2 : Summarizing Skill
SPS 3 : Predicting Skill
SPS 4 : Communication Skill
SPS 5 : Classifying Skill

Figure 2. The development of each science process skills indicators

The developments on students' science process skills are similar for each indicator except for predicting skill which scored the most significant rise with N-Gain (<g>) score at 0.70. On the other hand, the lowest increase occurred in summarizing skill (0.43) and communication skill (0.44) even though all questions developed for these two aspects (respectively 3 questions) are in easy and moderate difficulty level. In addition, communication skill is a skill frequently trained by students in the learning process, as well as predicting skills compared with other skills. However, according to interviews conducted randomly to some students and the observation done during learning process, students did not fully understand how to process and interpret data, either from table to charts, or from chart to chart which is one of the capabilities that exist on the communication skill. Unlike this skill, students mastered the skill to solve problems in predicting skill well. So, naturally the development result of predicting skill is higher than other skills. In line with Rauf et al. [12], the research result shows that the utilizing of simple computer simulation is able to improve students' science process skills.

4. Conclusions

Based on the research that has been conducted, it was concluded that the use of multimedia computer in linear motion concept for junior high school students is able to increase science process skills with moderate normalized gain score in each sub-concept learnt. The development of science process skills is also found in each indicator with the highest increase in predicting skill indicator, while the lowest increase is on summarizing skill and communication skill indicator.

In practical terms, the use of computer multimedia can be used as an alternative to improve students' science process skills. However, it has not been able to improve summarizing and communication skills well. It is supported by the results of this research which shows that the normalized gain scores are still low. So, to improve these skills, computer multimedia can only be used as the media that supports learning process. It is necessary to get direct experience of learning by doing experiments using real tools even if it is only a simple experiment. For further research, it would be better if the tests are conducted using an essay test in order to access more detail information of students' thinking process in answering questions particularly in interfering the scientific representation, one indicator of communication skill. Besides, performance assessment should be done so that those skills can be observed directly and more accurate.

5. References

[1] Samsudin A, Suhandi A, Rusdiana, D, Kaniawati, I and CoştuB 2016 *Asia-Pacific Forum on Science Learning and Teaching***17(1)**1
[2] Samsudin A, Suhandi A, Rusdiana, D and Kaniawati, I 2016 *Journal of Physics: Conference Series* **739**
[3] Kaniawati I, Samsudin A, Hasopa Y, Sutrisno A D and Suhendi E 2016 *Journal of Physics: Conference Series* **739**
[4] Prince M 2004 *Journal of Engineering Education* **93(3)** 223
[5] Bobadilla C, Marina, Lorza L, Rubén, González V, Eliseo P, Gómez S and Fátima 2016 *International Journal on Advances in Education Research* **3(2)** 38
[6] Crook S J, Sharma M D and Wilson R 2015*Contemporary Issues in Technology and Teacher Education* **15(2)** 126
[7] Yang K andHeh J 2007 *Journal of Science Education and Technology* **16(5)** 451
[8] Park J C and Slykhuis D A 2006 *Contemporary Issues in Technology and Teacher Education* **6(2)** 218
[9] Beyer B K 1991 *Teaching thinking skill: a handbook for elementary school teachers* (New York: Allyn & Bacon)
[10] Gultepe N 2013 *International Journal of Environmental & Science Education* **11(5)** 779
[11] Ramig J E, Bailer J and Ramsey M J 2012*Teaching science process skills* (California: Good Apple Inc)
[12] Rauf R A A, Rasul M S, Mansor A N, Othman Z and Lyndon, N 2013 *Asian Social Science* **9(8)** 47
[13] Samsudin A, Liliawati W, Sutrisno A D, Suhendi E and Kaniawati I 2015 *Advances in Social Science, Education and Humanities Research* (doi:10.2991/icaet-14.2014.16)
[14] Hake R R 1999 *Analyzing Change/Gain Scores* (Indiana: Indiana University)

Acknowledgment

This work was financially supported by "Hibah Fundamental" Kementerian RISTEK DIKTI, Republik Indonesia's Research Grants in the fiscal year 2016.

IOP Conf. Series: Journal of Physics: Conf. Series **812** (2017) 012018 doi:10.1088/1742-6596/812/1/012018

Development of Assessment Instrument of Critical Thinking in Physics at Senior High School

T Sugiarti*, I Kaniawati and L Aviyanti

Post-graduate of physics education, Universitas Pendidikan Indonedia
Jl. Dr. Setiabudhi No. 229, Bandung 40154 Indonesia

*Corresponding author's email : tiarsugiarti@student.upi.edu

Abstract. The result of preliminary study shows that the assessment of physics in school did not train students' critical thinking skill. The assessment instrument just measured low cognitive aspects. Supposedly, critical thinking skill is trained in the assessment activity. The study aims to determine the characteristics and the quality of critical thinking skill instrument. It employs descriptive-qualitative method with research and development as the research design. The research participants are 35 students involved in the limited trial and 188 students in the wider trial from three public senior high school in Ciamis which in high level school. The data was collected through expert validation, tests and interviews. The results indicate that the characteristics of the assessment instrument of critical thinking skill is open-ended. The instrument fulfills some indicators namely analyzing argument, deduction, induction, and display information in the form of scenario, text, graphic and table. In addition, the data processing through V4 Anates program shows that the instrument reliability achieves 0.67 with high interpretation of 0.67 and the validity is 0.47 with enough interpretation. Thus, the assessment instrument of critical thinking skill in the form of open-ended essay meets the criteria of quality test,so it can use as instrument of assessment critical thinking skill.

1. Introduction

A preliminary study suggested that assessment instrument in Physics at school only measured the students' basic thinking skill. Critical thinking skill, which includes to one of high level thinking skills, has not been examined by the teachers. It is due to the fact that they refer to the National Examination's item sheets. The analysis of the National Examination's item sheets in last five years, showed that 7% of them examines memory, 28% measures understanding, and 62,4% of the items evaluates application. Those three aspects include to basic thinking skills (Hawker: 2011).

Teachers said that they rarely analyzed each item given to the students because they do not have enough time to do it, so the teachers do not recognize the quality of each item they have made. According to Arikunto (2008), a test is considered good if it has two test requirements, namely validity and reliability. So, teachers have to analyze each item they employed.

Generally, teachers trained critical thinking skills during teaching by asking orally about physical phenomena. Students were expected to gain information and respond by stating opinion about the questions asked.

Critical thinking skills have not been examined in the form of item sheets because there are small amount of items measuring physics learning performance in terms of critical thinking skills. Whereas, according to Tuncay (2006), assessment tools employed by teachers played important roles in assessing students' achievement and development of students' critical thinking skills. Therefore, assessment tools not only useful for examining the students' mastery, but also to improve their thinking ability.

One of the alternatives that can be used to overcome the problems mentioned above is by developing assessment instrument of critical thinking skills in physics for senior high school. Thus, that development process showed characteristics and quality of instrument of critical thinking skills.

Critical thinking is defined as ability to give reasons *(reasonable)* and think reflectively focused on belief and behavior (Ennis: 2011). Ennis (2011) says that critical thinking skills include several aspects, namely basic explanation, foundation of making decision, conclusion, further explaination, supposition and combination, and other additional skills. In examining critical thinking skills, Ennis (2001) stated that *open-ended* problems are suitable to assess critical thinking because it can be adapted easily and more comprehensive. *Open- ended problems* are questions or problems that have more than one possible answer and more than one strategies to answer (Pelfrey, as cited in Azizah: 2013). Based on information complexity displayed on the problems, Ennis (2001) divided open-ended problems into three structures, namely high structure, medium structure, and minimal structure.

Stobaugh (2013) explains that in constructing an assessment instrument of critical thinking skills can be done by giving information on the problems. The information given are either in the form of scenario, concrete examples, or aunthentical tasks. In addition, they can be delivered by using illustrations (pictures, cartoons, advertisements), tables, graphics, and maps. The information can also be communicated by using quotations or texts.

2. Method

The present study employed a descriptive qualitative research method with research and development design.This research and development were conducted until the level of instrument development covering limited test and broader test. The participants of limited test were from one of the classes in a public senior high school in Kabupaten Ciamis which has 35 students. They were selected using purposive sampling. Whereas for broader test, 188 from those three public senior high school were involved.

The developed assessment instrument of critical thinking skills was examined on instrument validity sheets. Validity sheets contains assessments of instrument contents and constructions, and validated by three expert lecturers. Those assessment instruments were in the form of essays on *open-ended* problems, consist of six items, examined to the students to obtain empirical validity, so that the quality of developed instrument was ascertainable. Moreover, an interview was conducted to the participants in order to gain information about clarity of the developed problems. Consequently, those informations are helpful to improve the developing instrument of critical thinking skills.

Analysing items, including validity analysis, reliability, and discrimination power, were used to identify quality of instrument. Reliability on limited test was determined by a test-retestmethod, namely giving similar test to the same object in certain periods of time (Aikunto: 2008). Reliability was counted by employing Pearson's corelation coefficient

$$r_{xy} = \frac{N\sum XY - (\sum X)(\sum Y)}{\sqrt{(N\sum X^2 - (\sum X)^2)(N\sum Y^2 - (\sum Y)^2)}} \quad (1)$$

Software Anantes V4 was employed to analyze the results of validity, reliability, discrimination power, and level of difficulty.

3. Finding and Dissucsion

There were nine items of assessment instrument which were validated by expert lecturers. They covered an aspect of basic explanation with indicator of argument analysis, and aspect of conclusion, including indicator of deduction and induction. The results of expert lecturer's validation were shown in the table below:

Tabel 1. Recapitulation of the Results of Content Validation of Critical Thinking Skills Instrument in Physics at Senior High School

Problem No.	CTS Indicators	Problem Types	Lecturer 1	Lecturer 2	Lecturer 3
1	Argument analysis	HS	VWI	VWI	VWI
2	Deduction	MS	VWI	NV	VWI
3	Induction	MS	VWI	VWI	VWI
4	Deduction	HS	VWI	NV	VWI
5	Argument analysis	MS	VWI	VWI	VWI
6	Induction	MS	VWI	VWI	VWI
7	Argument analysis	MS	VWI	NV	VWI
8	Deduction	MS	VWI	V	VWI
9	Induction	HS	VWI	VWI	VWI

Note:
MS: Medium Structure
VWI: Valid with improvement
V: Valid
NV : Not Valid
HS : High Structure

Table 1 shows that six out of nine developed problems of critical thinking skills in physics at senior high school number 1, 3, 5, 6, 8, and 9 were considered valid with improvement, both in terms of constructing interrogative sentences, spelling, picture information, and scoring column. Meanwhile, the other 3 items, namely number 2, 4, and 7 were considered not valid because they did not suit to indicators of critical thinking skills. Those three items were not discarded but they were revised by adding clearer information, changing interrogative sentence into more appropriate form according to the lecturer's suggestions, so that the problems could be used to measure the students' critical thinking skills.

There were six problems used in instrument test namely number 1, 3, 4, 5, 8 dan 9 because those six problems have fulfilled the indicators of critical thinking skills, namely argument analysis, deduction and induction. Those six problems also represented two types of problems of critical thinking skills. Problems number 1, 5 and 9 represented high structure, while numbers 3, 4, 8 corresponds to medium structure.

3.1. The Results of Limited Test of Assessment Instrument of Critical Thinking Skills

The analysis of the limited test problems was conducted by using Anates V4, with validity score of 0,47, interpreted adequate. Meanwhile, reliability was determined by a *test-retest* method and counted by employing product moment correlation(Arikunto: 81), with realiabillity score of 0,82, considered very high. Table 2 shows the results of problems analysis of limited tes

Tabel 2. The Result of Problems Analysis of Limited Test

Problems No	CTS Aspects	Item Validity		Discrimination Power		Level of Difficulty		Conclusion
		Score	**Category**	**Sco re**	**Category**	**Score**	**Category**	
1	Argument analysis	0,48	Adequate	0,22	Adequate	0,40	Medium	Accepted
2	Deduction	0,65	High	0,51	Fine	0,54	Medium	Accepted
3	Induction	0,70	High	0,34	Adequate	0,42	Medium	Accepted
4	Argument analysis	0,58	Adequate	0,15	Poor	0,39	Medium	Accepted with revision
5	Deduction	0,75	High	0,30	Adequate	0,29	Medium	Accepted with revision
6	Induction	0,53	Adequate	0,29	Adequate	0,44	Medium	Accepted

After analyzing, the problems considered unfavorable or less than adequate were changed or repaired. Decision making towards those problems need repairing was done with several considerations about the result of analysis or item validity (r), level of difficulty (P), and discrimination power (D). According to Mulyatiningsih (2009), if two out of three criteria of good problems have been fulfilled or consistent, the problems could be employed. On the contrary, if two out of three criteria of good problems have not been fulfilled, the problems must be revised. In addition to problem analysis, revision was also done by considering the problems readability by the students through an interview, asking the clarity of information and questions on the problems.

3.2. The Result of Broader Test of Assessment Instrument of Critical Thinking Skills

The result of broader test problems analysis processed with Anates V4 was 0.47 for validity and 0.67 for reliability test as a whole, interpreted adequate and high. Table 3 shows the results of problems analysis with broader test.

Table 3. The Results of Problems Analysis with Broader Test

Problems No	CTS Aspects	Item validity		Discrimination power		Level of Difficulty		Conclusion
		Score	Category	Score	Category	Score	Category	
1	Argument analysis	0,55	Adequate	0,34	Adequate	0,34	Medium	Accepted
2	Deduction	0,65	High	0,47	Fine	0,47	Medium	Accepted
3	Induction	0,66	High	0,40	Adequate	0,36	Medium	Accepted
4	Argument analysis	0,29	Low	0,07	Poor	0,38	Medium	Accepted with revision
5	Deduction	0,82	Very high	0,55	Fine	0,39	Medium	Accepted
6	Induction	0,44	Adequate	0,16	Poor	0,36	Medium	Accepted

After analyzing problems, decision making towards those problems need repairing was done with several considerations about the result of analysis or item validity (r), level of difficulty (P), and discrimination power (D). In addition to problem analysis, revision was also done by considering the problems readability by the students through an interview, description of students' answer, and analysis of problems compatibility to physics for senior high school.The followings an examples of problems analysis of assessment instrument

1) Problems number 3

Problems number 3 covers the aspects of conclusion, indicating induction of relationship between rising of temperature and heat based on the results of experiment. This problem is a kind of medium structureand outlined in the form of table of experiment data containing information about heat, initial temperature, final temperature, and temperature changing. In this problem the students were asked to induce by generalizing based on the existing data and completing their answer with graphic.

An analysis to students' answer showed that they were able to conclude the relationship between heat and rising of temperature. They could make a conclusion based on systematic data. The students were also able to conclude that temperature changing was influenced by number of heat given. They could also convert the information into the form of graphic.

Based on the interview, the clarity of problems number 3 was clear enough, both information given and the questions asked. Problems number three has suited the materials given in the classroom. The students have learned about some factors influencing amount of heat. This problem has also suited the quality of good problems based on an analysis conducted and displayed on table 3. Thus, problems number 3 could be used as assessment instrument of critical thinking skills without any revision.

3.3. Characteristics of Assessment Instrument of Critical Thinking Skills in Physics at Senior High School

The problems analysis with broader test showed that those problems have validity score of 0,47, interpreted "adequate" and reability score of 0,67, interpreted "high". It resembled Amalia (2012) that the characteristics of problems to examine critical thinking skills are those which have validity and realibility scores of more than adequate.

The characteristics of problems to examine critical thinking skills are those which written in the form of open-endedproblems divided based on complexity of information,namely high structure, medium structure, and minimal structure(Ennis: 2001). The problems which were developed in this research were high structure and medium open-ended essay problems. High structureproblems give information in form of text to the students. The high structure problems to examine critical thinking skills was provided with three indicators, namely argument analysis (number 1), deducting (number 5), and inducting (number 6). Problems with medium structure were examined by giving simpler information than that of high structure.

Such problems were examined through problems number 2 (indicator of deduction), number 3 (indicator of induction), and number 4 (argument analysis).

Moreover, this assessment instrument of critical thinking skills were also ordered based on two aspects of critical thinking skills, namely involving basic explanation and concluding. Those two aspects covers three indicators, namely argument analysis, deducting, and inducting. The developed aspects have suited to aspects of critical thinking skills stated by Ennis (2011), namely involving basic clarification, foundation of decision making, infering, explaining further, approximating and combining, and additional skills. Aspects and indicators of critical thinking skills measured in this study were explained below.

1) Involving basic clarification

The aspects of Involve basice clarificationwere examined through problems number 1 and 4. This aspect presented an ability to analyze arguments, which could be tested by identifying irrelevancy and considering possible answer as well as searching for similarities and differences between something.

2) Infering

The aspect of infering presented someone's ability to infer both in inductive and deductive way. Indicator of deducting was examined to question number two and five. Those problems were presented by giving some statements which are assumed correct, alternatives that consists of likely correct conclusion and asking students to check those conclusions.

Indicator of induction was examined through problems number 2 and 6. Those problems were given by presenting data. The students were asked to make a generalization based on information given on the problems.Critical thinking skills in the present study adopted those according to Ennis (2011). One of the characterictics of this type of problems is that there are some relevant information so that the students could think reasonably and give reasons to make decision of what should be believed and done. In this case, information were important things to construct the problems of critical thinking skills.

Based on the description of problems that have been explained before, table 4 following displayed the characteristics of assessment instrument of critical thinking according to information attached to the problems.

Table 4. Characteristics of Assessment Instrument of CTS Based on Information Attached to the Problems

Problems No	CTS Aspects	CTS Indicators	Description of Information attached to the problems
1	Involving basic clarification	Argument analysis	Text about phase transition Graphic about phase transition of CO2 Graphic about phase transition of water
2	Concluding	Deducting	Picture showing percentage of sun radiation Conclusions made based on the picture
3	Concluding	Inducting	Table showing results of experiment to determine relationship between heat and change of temperature
4	Involving basic clarification	Argument analysis	Scenario containing opinions of two experts about AC installation
5	Concluding	Deduction	Text containing information about ice field in Antarctic Statements showing conclusion of the text
6	Concluding	Inducting	Scenario of daily life Table showing amount of metal's conductivity Table showing elements diversity in the universe

Table 4 showed that information attached to the problems are in the form of graphic, text, table, and scenario. Such information suited to those used to test critical thinking skills employed by Stobaugh (2013), namely scenario, real examples, and authentic assignments. Visual displays were in the form of illustration including advertisement, painting, picture and cartoon, diagram and table, map and graphic. Other information were in the form of quotations which consists of short quotation or text. Those informations were combined in a problem, for instance scenario and picture or scenario and table. These studies found a characteristic that information displayed in the instrument were scenario, visual information in the form of table and graphic as well as text.

An experiment in the form of results of lecturer's judgementand interviews to the students that have been elaborated before, showed a characteristic of assessment instrument of critical thinking skills: the problems should be written in effective sentence, suited to the standard Bahasa Indonesia. The information should be clear. For example, the picture should be colored and the paragraph should be written simply. Information content should be related to materials learned by the students. If it was a new context, it would better to asked questions related to materials that have been learned by the students. It was to avoid who could not answer the questions. Assessment was aimed at measuring and examining purpose of learning (Surapranata: 2007).

4. Conclusions

Present study concluded that the characteristics of assessment instrument of critical thinking skills in physics for senior high school covering two aspects, namely involving basic clarification indicating argument analysis, and the aspects of inferring indicating deduction and induction. As a result, further studies, discussing instrument development based on other aspects and indicators of critical thinking skills, are needed either refers to the same or different expert.

The problems developed in this study are high structure open-ended essays which have complex enough information, and medium structure with simpler information. Therefore, supplementary studies should be conducted to explain about assessment instrument of critical thinking skills in different form of problems, like multiple choice.

Assessment instrument of critical thinking skills is a kind of problems that contains various informations. The information to use is scenario, table, graphic, and text. Thus, further studies about the development of assessment instrument of critical thinking skills are required, based on the types of given information, such as authentic assignment, comic, and quotation.

Based on problems analysis, assessment instruments developed in this study have fulfilled the quality of a good test by validity and reability scores of 0,47, interpreted adequate, and 0,67, interpreted high. However, according to thwe result of interview, one out of six problems were revised by changing the content of the text, so that the problems suited the materials learned by the students. Two problems were changed in visual display, and another one was revised by reducing given information.

Problems development in this study only covers heat. Therefore, future studies were required to develop assessment instrument of critical thinking skills in other topics learned in senior high school. In addition, the problems should be written by considering materials that have been learned by the students.

5. References

[1] Amalia S N 2012 *Pengembangan Soal Esai Berpikir Kritis dan Profil Pencapaianya di SMA Negeri Kota Bandung Pada Tema Sistem Organ.* A Thesis (School of Postgraduate Studies UPI Bandung: Unpublished)

[2] Arikunto S 2008 *Dasar Dasar Evaluasi Pendidikan* (Jakarta: Bumi Aksara)

[3] Azizah Y N 2013 *Pengembangan dan Analisis Soal Open-Ended Problem Dalam Mengukur Keterampilan Berpikir Kreatif pada Materi Minyak Bumi* (An Undergraduate Thesis UPI Bandung: Unpublished)

[4] Ennis R H 2001 *Critical Thinking Assessment* [Online]
[5] Ennis R H 2011 *The Nature of Critical Thinking: An Outline Of Critical Thinking Disposition and Abilities* [Online]
[6] Hawker B 2011 *How To Assess High-Order Thinking In Your Classroom* [Online]
[7] Mulyatiningsih E 2012 *Metode Penelitian Terapan Bidang Pendidikan* (Bandung: Alfa Beta)

IOP Conf. Series: Journal of Physics: Conf. Series **812** (2017) 012019 doi:10.1088/1742-6596/812/1/012019

A Review of Gamification in Technological Pedagogical Content Knowledge

H W Prabawa

Departement of Computer Science Education, Universitas Pendidikan Indonesia, Jl. Dr. Setiabudi No. 229, Bandung 40154, Indonesia

Email: harsawara@upi.edu

Abstract. This paper review 10 papers that relating to gamification adoption in developing technological pedagogical content knowledge (TPACK) framework. Technological developments lately led to the trend of increased use of ICT in the learning process, one of which is gamification. Gamification is the concept of applying game mechanics and game design techniques to engage and motivate people to achieve their goals. Gamification in education as an intersection of learning and fun. The problem is that not all game's attributes suitable for use in presents a teaching material. TPACK is a framework for the teacher that described a complex interaction among three bodies of knowledge : content, pedagogy and technology. TPACK engagement has an impact on the teacher mastery in dimension of teaching material content, in addition to improve teachers skill in developing technology in classroom learning.

1. Introduction

The rapid development of technology clearly had a significant impact in the world of education. One of which is with commonly Information and Communication Technology (ICT) used as a learning resource (in the form of instructional media, or just a repository of material) and even the management of learning activities (eg e-learning, distance learning, etc). Student interaction with ICT was certainly increasing, due to the demands of the learning process (other than the reasons related the devices ownership, need for self-actualization, etc). Therefore, ICT becomes important to be a concern for educators [1].

This condition raises the interesting issues about the importance of mastering ICT for teachers. Integrating ICT into classroom has always been a challenging task for many teachers [2]. Teacher feel not quite ready prearing learning using ICT, partly caused by the mastery of the theoretical framework that relate to the development of ICT for learning is lacking [3]. Research conducted by the Son, Robb and Charismiaji [4] regarding the Computer Literacy and Competency revealed that teachers in Indonesia generally have basic computing skills are quite high, but the frequency of utilization of these capabilities are limited only to a few applications only (one quite dominant is a word processing application). It's obvious impact on the ability of computing and computer skills of students.

The gap between the potential of technology and its implementation in learning gave rise to various related questions about the reason involvement of technology in classroom learning. In many cases the use of technology in education, teacher impressed only play the role of *gatekeepers* [5] [6], which makes decisions on the selection of appropriate technology to be used in learning and make decisions on exactly when the students allowed to use these technologies [7].

Gamification is using elements of game mechanics to provide practical solutions by way of building interest (engagement) certain groups [8]. Gamification not only use the game elements and game design techniques in non-game contexts [9], but also to empower and engage learners with the ability of motivation on learning approach and maintains a relaxed atmosphere

In understanding technology as an integral partner for education, a conceptual framework called Technological Pedagogical Content Knowledge (TPACK) deserves to become the reference. The TPACK was first proposed by Mishra and Koehler [10] to describe an integrated connection between content knowledge, pedagogical knowledge, and technological knowledge. The framework illustrates essential knowledge of how teachers could integrate technological tools into their teaching of specific content in their school practice [11]. Figure 1 shows a complex interaction among three bodies of knowledge : content, pedagogy and technology.

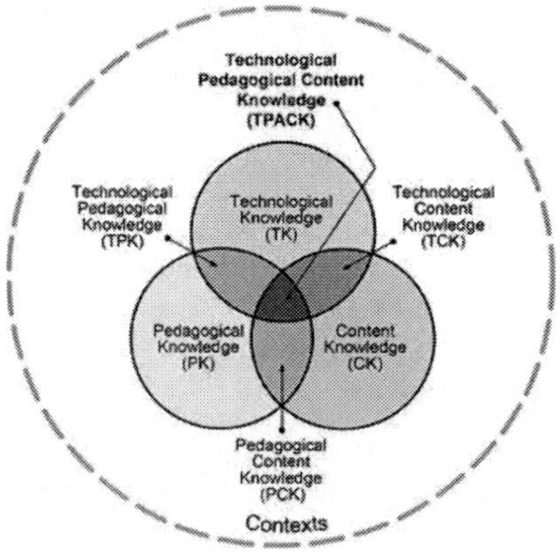

Figure 1. Technological Pedagogical Content Knowledge (TPACK) framework.

2. Method

The literature were identified by exploring the google search engine with keyword "gamification". As a result, total 10 articles were located. This paper review using three main categories were employed the reseachers to make sense of the articles, they are : basic data (authors, year of publication, journal/conference); content analysis (pedagogy, content area), and gamification element. Table 1 shown the taxonomy which will be used in analyzing the elements of gamification that are used in developing instructional learning object base on the TPACK.

Table 1. Taxonomy of gamification element.

Element	Type
Game based	rules, interactivity and feedback to produce measurable results
Game mechanics	level, earnings badge, point system, score and time challenge
Game thinking	competition, cooperation, exploration and storytelling

Games based elements are intended to create a system in which learners or players involved in the abstract challenge, defined by rules, interactivity and feedback to produce measurable results ideally bring out the emotional reactions; *Game mechanics* can be interpreted as level, earnings badge, point system, score and time challenge is the element used in the gamification; *Game thinking* is the idea of thinking about everyday experiences such as jogging turn it into an activity that has elements of competition, cooperation, exploration and storytelling [12].

3. Result and discussion

Table 2. Gamification in technological pedagogical content knowledge.

Reference	Pedogogical Approach	Content Area	Gamification Element
Cook, Smith, Maglaras, Janicke (2016) – The 4th International Symposium for ICS & SCADA Cyber Security Research 2016 [13]	Role Playing	Network Security	*Game mechanics* : points system, level, leaderboards, badges, challanges *Game thinking* : Collaborative to commpetitive, instrinsic to extrinsic, multilayer to solitary,campaign to endless, emergent to scripted
Prambayun, Farozi (2015) – Proceeding Seminar Nasional Teknologi Informasi dan Multimedia [14]	Communicative Language Teaching (CLT), including Task Based Language Teaching (TBLT) and Content Based Instruction	Nature Science for Promary School	*Game mechanics* : points system, level, leaderboards, badges, challanges
Figueroa (2015) - Digital Education Review, Number 27, June 2015 [15]		Second Language Learning	*Game based* : progress bar, reward system, *Game mechanics* : point system, level, leaderboards, badges, *Game thinking* : social element
Lister, West, Cannon, Sax, Brodegard (2014) – Journal of Medical Internet Reasearch (JMIR) Serious Games, Vol.2, Issue 2 [16]		Health and Fitness	*Game based* : digital reward, real world prizes, rank of achievement, *Game mechanics* : level, leaderboards, badges, time pressure *Game thinking* : social or peer pressure, competitions
Brewer, Anthony, Brown, Irwin, Nias, Tate (2013) – Proceedings of the 12th International Conference on Interaction Design and Children [17]		Touch and Gesture interactions	*Game based* : reward

4. Conclusion

It should be understood that the gamification not mean to make a game, make a special application to apply the concept of gamification would be better, but if resources owned is not possible to make a special application for gamification then gamification can use simple tools to implement gamification in the learning process in class. The most important thing is the right concept, clear goals and be able to build for student engagement in learning. TPACK engagement has an impact on the teacher mastery in dimension of teaching material content, in addition to improve teachers skill in using and developing technology in classroom learning.

5. References

[1] Polly D, Mims C E, Shepherd and Inan F 2010 *An International Journal of Research and Studies* **26** 863-70
[2] So H J and Kim B 2009 *Australasian Journal of Educational Technology* **25** 101-16
[3] Kramarski B and Michalsky T 2010 *Learning and Instruction* **20** 434-47
[4] Son J B, Robb T and Charismiadji I 2011 *Computer-Assisted Language Learning - Electronic Journal* **12** 26-42
[5] Ertmer P A 2005 *Educational Technology Research & Development* **53** 25-39
[6] Zhao Y, Pugh K, Sheldon S and Byers J L 2002 *Teachers College Record* **104** 482-515
[7] Forssell K S 2011 *Technological Pedagogical Content Knowledge : Relationship to Learning Ecologies and Social Learning Networks* (Standford: A Dissertation Submitted to The School of Education and The Committee on Graduate Studies of Standford University)
[8] Vianna Y, Vianna M and Medina B 2014 *Recreating Companies Through Games* 1ª edition vol 3 (Rio de Janeiro: MPV Press)
[9] Werbach K and Hunter D 2012 *For the Win: How Game Thinking can Revolutionize Your Business* (Philadelphia: Wharton Digital Press)
[10] Koehler M J and Mishra P 2005 *Journal Educational Computing Research* **32** 131-52
[11] Srisawasdi N 2012 *Procedia - Social and Behavioral Sciences* **46** 4031–38
[12] Kapp K M 2012 *The Gamification of Learning and Instruction: Game-based Methods and Strategies for Training and Education* 1 edition (San Francisco: Pfeiffer & Company)
[13] Cook A, Smith R, Maglaras L and Janicke H 2016 *The 4th International Symposium for ICS & SCADA Cyber Security Research 2016 (Belfast)* (Swindon: BCS Learning & Development Ltd.)
[14] Prambayun and Farozi M 2015 Pola perancangan gamifikasi untuk membangun engangement siswa dalam belajar *Seminar Nasional Teknologi Informasi dan Multimedia 2015* Yogyakarta
[15] Figueroa J F 2015 *Digital Education Review* **27** 32 – 54
[16] Lister, West J H, Cannon B, Sax T and Brodegard D 2014 Just a fad? gamification in health and fitness apps *JMIR Serious Games* 2(2014)2
[17] Brewer R, Anthony L, Brown Q, Irwin G, Nias J and Tate B 2013 *The 12th International Conference on Interaction Design and Children (New York)* (New York: ACM Digital Library, The Association for Computering Machinery)

Acknowledgments

Grateful to the Department of Computer Science Education, Universitas Pendidikan Indonesia who have given support to the writing of this manuscript.

Synthesis and Morphology Properties of Mn-doped Zinc Aluminate ($ZnAl_{2-x}O_4$: xMn^{4+}, x=1%) Using Co-Precipitation Methods with Microwave Thermal Assisted

B W Nuryadin[1, 2*], Asti Sawitri[1, 2], E C S Mahen[1, 3], and A Y Nuryantini[1, 3]

[1]Laboratorium Nanomaterial Fisika, Universitas Islam Negeri Sunan Gunung Djati Jalan A.H. Nasution 105, Cibiru – Bandung 40614, Indonesia
[2]Departemen Fisika, Universitas Islam Negeri Sunan Gunung Djati Bandung Jalan A.H. Nasution 105, Cibiru – Bandung 40614, Indonesia
[3]Departemen Pendidikan Fisika, Universitas Islam Negeri Sunan Gunung Djati Jalan A.H. Nasution 105, Cibiru – Bandung 40614, Indonesia

Email: bebehwahid102@uinsgd.ac.id

Abstract. ZnAl2O4: Mn4+ particle were synthesize using co-precipitation using NaOH and PVA as precipitating agent and surfactants, followed by microwave-assisted thermal treatment. The morphology properties and absorption spectral of particles were investigate using SEM and UV-Vis spectrometer. The results showed that the $ZnAl_2O_4$ particles has a spinel structure with various particle size. Absorption spectral exhibits that particle has a potential photo catalytic properties at visible light spectra, especially for the pollutant degradation.

1. Introduction

Zinc aluminate (ZnAl2O4), related to the electronic structure Fd3m [1], is a semiconductor material with a wide energy band gap (3.8-3.9 eV). ZnAl2O4 material has superior semiconductor properties, such as low surface acidity levels, high thermal stability, electronic and optical properties, and hydrophobic. Therefore, these materials could be develop as electroluminescence material, sensor, or a heterogeneous catalyst [2, 3].

In recent years, these various physical and chemical processes have been used to synthesize ZnAl2O4, such as sol-gel method [3], solid state reaction [4], hydrothermal [5], deposition [6] etc. Compared with other methods, the deposition method is faster, high levels of chemical homogeneity, easy in setting the composition and structure, and setting the particle size and morphology. However, this method requires a sintering process in order to produce the proper oxide particles.

In this paper, $ZnAl_{2-x}O_4$:xMn^{4+} (x=1%) have successfully synthesized using the co-precipitation method.

Moreover, the process of microwave heating were use as part of the process of sintering material ZnAl2O4:Mn. Microscopic structure, morphology and optical properties of the sample of interest to be investigate and discussed closely.

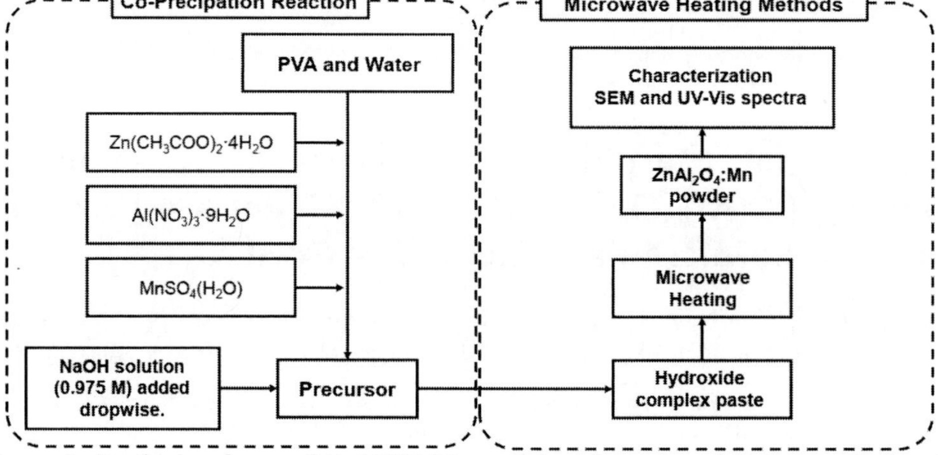

Figure 1. Synthesis scheme of ZnAl2O4:Mn particle using co-precipitation methods with microwave thermal assisted.

2. Experiment

Manganese doped Zinc Aluminate (ZnAl2O4:Mn) synthesized using 0.5 g zinc acetate tetrahidrate (Zn(CH3COO)2•4H2O, BM= g/mol), 1.708 g aluminum nitrate nonahidrate (Al(NO3)3•9H2O, BM= g/mol) dan 0.0025 g mangan sulfat hidrate (MnSO4(H2O), BM= g/mol) as zinc, aluminum and mangan source, respectively. All of these chemicals will produce a material with the following composition ZnAl1,99O4: 1% Mn4+. First, 0.01 g of PVA (polyvinyl alcohol, BM = g / mol) as a surfactant dissolved in 100 ml of pure water. After that, all basic materials are dissolve to obtain a clear precursor solution. Then, a solution of NaOH (0.78 g NaOH in 20 ml of water solvent) is slowly dripped into a precursor solution (drop wise). From this process,coloured solution became white and a precipitate formed. Subsequently, the precipitate is separated from the solvent water using filter paper until obtain white paste. The white paste is a complex particles hydroxide ($Zn(OH)_2$, $Al(OH)_3$, and $Mn(OH)_2$). To determine the effect of microwave heating on particle morphology $ZnAl_2O_4$:Mn, 1g complex hydroxide paste sample were heated using the microwave (800 watts) for 0 - 2 minutes, until white ceramic powder produced. The morphology and absorbance spectra of $ZnAl_2O_4$:Mn particle characterized using Scanning Electron Microscope (JEOL JCM-6000 NeoScope Benchtop SEM, Japan) and UV-Visible Spectroscopy (Home-Made Spectroscopy from LD-Didactic).

3. Results and Discussion

Figure 2 is a SEM image of the sample ZnAl2O4: Mn for some time variations microwave heating. Results precipitation reaction in the form of a white paste that we call complex hydroxide particles has a size in the range of 400 nm. Figure 2 (a). SEM observation showed that the hydroxide particles covered by the polymer layer of PVA. These causes the particles are not mutually agglomeration, and produce a particle size that is small enough. Microwave heating causes the particles undergo localized heating process. This is due to the microwave energy is absorbed by the particle turns into heat which is sufficient for the oxidation processes. Therefore, microwave heating transform complex particles into particles ZnAl2O4: Mn. In addition, the length of time the heating causes agglomeration caused by high heat and energy loss due to the polymer layer decomposes during microwave heating process, see Figure 2 (b) and 2 (c). It is characterize by an increase in particle size $ZnAl_2O_4$:Mn as the length of time the microwave heating, from 1.6 μm to 2.3 μm (Figure 2 (d)). This observation indicates that microwave heating is very effective in the process of sintering and particle oxidation.

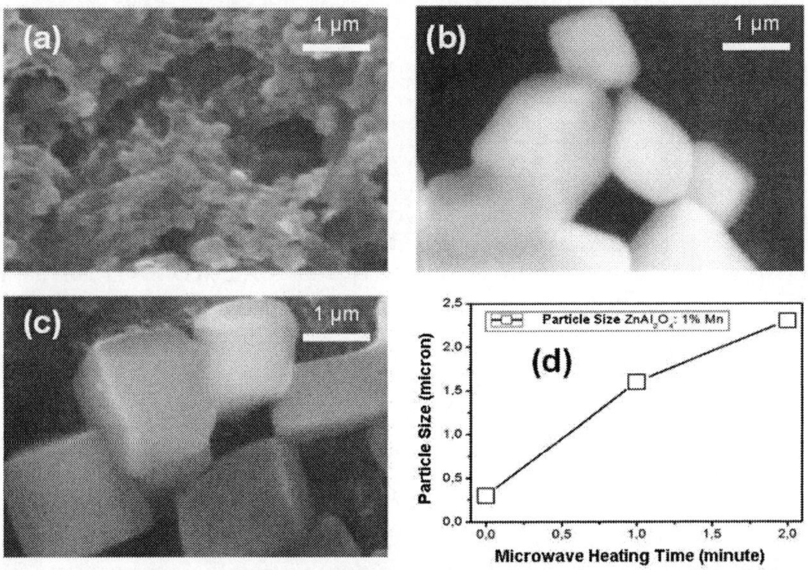

Figure 2. SEM image of ZnAl2O4:Mn paricle for various of microwave heating time, (a). 0 min, (b). 1 min, and(c) 2 min. While, (d).realtion of ZnAl2O4:Mn particle size distribution with heating time.

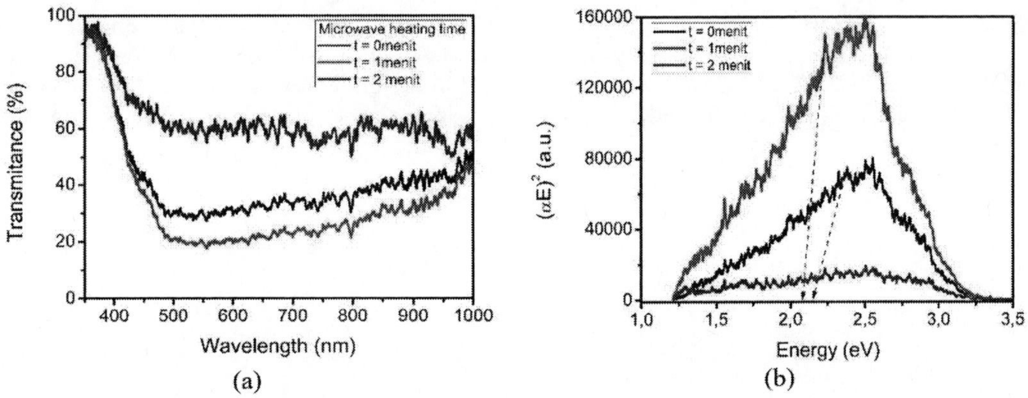

Figure 3. (a) Absorbance spectra, and (b) Tauc plot analysis of ZnAl2O4:Mn particle with various of heating time, all sample prepared with 0,01 g of ZnAl2O4:Mn particle at 5 ml of pure water.

Figure 3 is an absorbance spectrum of particle observations ZnAl2O4: Mn in the synthesized at some time variations synthesis. The results of these observations have managed to show that the sample absorbance spectrum of wide area, hingg 900 nm to 500 nm, with a peak at around 650 nm. Observations showed differences absorbance spectrum intensity for each sample time microwave heating. This probably caused by the quality of crystallization and particle size ZnAl2O4: Mn. The intensity of the absorbance spectrum in the wavelength range of blue-yellow (400-500 nm) is very low. This probably caused by the emission properties (luminescence) particle ZnAl2O4: Mn, this part needs to be studied further.

The optical band gap calculated using the Tauc plots relationship. Expressed by,

$$(\alpha h v) = C(hv - E_g)^n \quad (1)$$

where C is a constant, α is absorbance coefficient, E_g is the optical band gap of material, and n depends on the type of transition (n = 2 or 1/2 for direct band gap and indirect band gap) [????]. The calculated optical band gap of $ZnAl_2O_4$:Mn particle from Tauc plots was found to be 2.7 eV as shown in Fig. 3(b). The calculation results showed that all samples have the similar value of optical band gap.

4. Conclusions

Particle ZnAl2O4: Mn was synthesized using microwave heating assisted deposition. Observations show that the particle morphology occurs localized heating process that causes agglomeration or formation of large particles. Moreover, observations with absorbance, particle ZnAl2O4: Mn is estimate to have photocatalytic properties. Thus the particles ZnAl2O4: Mn has the potential to be develope as a photocatalytic material especially for liquid waste degradation.

5. References

[1] Hill R J, Craig J R and Gibbs G V1979 *Phys. Chem. Miner.* **4**317–39
[2] Kapse S D,Raghuwanshi F C, Kapse V D, and Patil D R 2012 *Curr.Appl. Phys.* pp 307–12
[3] Xiaojuan W, Zhiqiang W, Xiujuan C, Xuan W, Hua Y and Jinlong J 2015 *Sci.: Mater. Electro.* **26** 6606-11
[4] Pin S, Suardelli M, D'Acapito F, Spinolo G, Zema M, Tarantino S C and Ghigna P J 2013 *Phys. Chem.* **C117** 6105–112
[5] Zawadzki M and Wrzyszcz J2000 *Mater. Res. Bull.* **35** 109–14
[6] Fernandez-Osorio A, Pineda-Villanueva E, Chavez-Fernandez J2002 *Mater. Res. Bull.* **47** 445–52.

Acknowledgment

This work was supported by a research grant (B-200/B2-63/V.2/PP.00.9/06/2016) from UIN Sunan Gunung Djati Bandung, the Ministry of Religious Affairs, Republic of Indonesia. Author contributions: B.W.N, E.C.S.M, and A.Y.N contributed in design, experimental, and data processing. A.Y.N, A.S and E.C.S.M contributed partly in writing the manuscript draft and B.W.N. completely wrote the manuscript.

Ethnomathematics study: uncovering units of length, area, and volume in Kampung Naga Society

T Septianawati[1,2], Turmudi [1,2], and E Puspita[2]

[1]Departemen of Mathematics Education, Post Graduate School, Universitas Pendidikan Indonesia, Jl. Dr.Setiabudhi no.229, Bandung 40154, Indonesia
[2]Departemen of Mathematics Education, Faculty of Mathematics and Science Education, Universitas Pendidikan Indonesia, Jl. Dr.Setiabudhi no.229, Bandung 40154, Indonesia
Email: tiaseptianawati@student.upi.edu

Abstract. During this time, mathematics is considered as something neutral and not associated with culture. It can be seen from mathematics learning in the school which adopt many of foreign mathematics learning are considered more advanced (western). In fact, Indonesia is a rich country in cultural diversity. In the cultural activities, there are mathematical ideas that were considered a important thing in the mathematics learning. A study that examines the idea or mathematical practices in a variety of cultural activities are known as ethnomathematics. In Indonesia, there are some ethnic maintain their ancestral traditions, one of them is Kampung Naga. Therefore, this study was conducted in Kampung Naga. This study aims to uncover units of length, area, and volume used by Kampung Naga society. This study used a qualitative approach and ethnography methods. In this research, data collection is done through the principles of ethnography such as observation, interviews, documentation, and field notes. The results of this study are units of length, area, and volume used by Kampung Naga society and its conversion into standard units. This research is expected to give information to the public that mathematics has a relationship with culture and become recommendation to mathematics curriculum in Indonesia.

1. Introduction

During this time, mathematics is regarded as a perfect science with absolute truth and not related to everyday life.Mathematics always taught at schools as a culturally free subject involved learning that supposedly universally accepted facts, concepts, and contents [1]. Sumardyono [2] states that the objects of mathematics are social-cultural-historical, meaning that mathematics and learning the common property of all the people. Therefore, mathematics has always been part of human culture even in the simplest form. This means that culture and mathematics related each other.

A field of study which is used to show relevancy between culture and mathematics is ethnomatematics. Ethnomathematics defined as a cultural anthropology of mathematics and mathematical education[3,4]. Ethnomathematics consists of two words, "ethno" and "mathematics". The mean of ethno is a cultural group and mathematics considered as a set of activities such as measuring, classifying, sorting, summing, and modelling [5]. Definition of ethnomathematics also expressed by William D Barton. In his dissertation, William D Barton [6] argued that ethnomathematics is a field of study which examines the way people from other cultures understand, articulate and use concepts and practices which are from their culture and which the researcher describes as mathematical.

According to Alan J Bishop [7] mathematical ideas develop anywhere because people living in different cultures may be doing the same thing. Some of the activities done by humans is very important in developing mathematical ideas. Such activity includes counting, locating, measuring, playing games, designing, and explaining. The activity classification is helpful in designing learning to introduce students to mathematical knowledge contained in a culture. With these design students are expected to understand that learning math is fun and useful for real life.

Ethnomatematics research is suitable to do in Indonesia which has a lot of culture. Despite the progress of the times happening, many ethnic in Indonesia who survived by clinging to the doctrine of his ancestors. One of themis Kampung Naga society located in Neglasari Village, Salawu subdisctrict, Tasikmalaya district, West Java Province.

Kampung Naga is a village which adamant to his ancestral traditions, namely Sundanese. Kampung Naga society usesSundanese language as everyday language. They have unique shape of the building, which is shaped house on stilts. Kampung Naga has an area 1.5 hectares. Kampung Naga area may not be increased or reduced. Therefore, no additional buildings or houses in Kampung Naga.

Kampung Naga is beautiful place. To reach Kampung Naga, visitors have to pass through 439 stairs. Rice fields, forests, and rivers can you see when walking down the stairs. Kampung Naga is restricted by a small ditch in the north and south, restricted by a small hill in the west, and restricted by Ciwulanriver in the east. So that, Kampung Naga is surrounded by the natural environment.

The life principle of Kampung Naga society is a live with nature. This is evident from Kampung Naga society efforts to keep preserve nature. They limit the development of technology to get into Kampung Naga. One of them is that there are not an electricity. They did not forbid the technology, but because there is no electricity then it would automatically filter incoming technology. Moreover, the reasons why there are no electricity is because theirhouses are made of materials that flammable and that does not cause social gap between Kampung Naga society.

Based on the results of preliminary studies, one of the cultural activities done by the Kampung Naga society is measurement concept. In measuring activities, Kampung Naga society using auniqueunit of length, area, and volume. Therefore, the aims of this study is to uncover units of length, area, and volume used by Kampung Naga society.Accordingly, the research problems are: (1) How the units of length, area, and volume used by Kampung Naga society? (2) Canthe units are converted into standard units?

2. Experimental Method

The qualitative approach is used in this research because a problems studied related to social life or complex society, holistic and meaningful. Besides that, qualitative approach also allows the researchers to uncover ethnomathematics [6,8]. Ethnographic methods used for this research because researchers will uncover mathematical ideas contained in measuring activity done by Kampung Naga society so that we can uncover the units of length, area, and volume in Kampung Naga society.

In this study, data collection is done by using the ethnography principles that is observation, interviews, documentation, and field notes. Once the data is collected, then analyzed with the measures, namely reducing the data, present data in a short description and a table, and draw conclusions. Test validity of the data is done by using credibility test, transferability test, dependability test, and confirmability test.

3. Result and Discussion

Measuring is an activity that is often done by people in daily activities, including by Kampung Naga society. When measuring length, area, and volume of an object,one used a comparison called unit. Units of length, area, and volume used by a tribe is different depending on the language and measuring instruments used. Because civilization and human culture are growing and expanding with their relationships between communities around the world, then to make easier people to interact with each other made standard units of measurement or what is often referred to International Units.

3.1. Unit of Lenght

In the past times, before using standard measuring tools, Kampung Naga society measured the length of an object by using objects in the surroundings, for example a rope, a stick or a part of the human body. To measure the length and width of the land, Kampung Naga society use a rope. Then the rope was measured by *jeungkal*and marked every one meter by making a knot. The number of knot indicated how many meters length or width of the land.

Units of length which until now used by Kampung Naga society, namely *jeungkal* and *deupa*. There are two opinions in Kampung Naga describing size of *jeungkal*. The first opinion says that one *jeungkal* is a distance from the tip of the thumb to the tip of the middle finger when stretched. Meanwhile, another opinion says that one *jeungkal* is a distance from the tip of the thumb to the tip of the little finger when stretched. Second opinions believe that *sajeungkal*size is about 20 cm.Second opinion of Kampung Naga society the same asElis Suryaniopinion [9]which states that size of *jeungkal*is from the tip of the thumb to the little finger, approximately 20 cm. Figure 1 shows us about how Kampung Naga society use *jeungkal* to measure.

Deupa is a distance from the right hand fingertip and left hand fingertip when stretched. Kampung Naga society believes that if it is converted to meters, then one *deupa* equal to 1.5 meters. According Danadibrata [10]"*deupa nyaéta antara tungtung ramo leungeun katuhu jeung tungtung ramo leungeun kenca lamun leungeun duanana dipanjangkeun, sadeupa kira-kira 1.7 meter*". Based on these opinions, *deupa* is a distance between right hand fingertips and left hand fingertips when arms are both stretched, one *deupa* equals to 1.7 meters. This is different from Kampung Naga society that state*s a deupa* equal to 1.5 meters. These differences may occur because the body size of each person is different.

Beside *jeungkal* and *deupa*, Kampung Naga society also measured object by using the distance between the right hand tip with the left shoulder like Figure 2. The distance between the tip of the right hand with the tip of the left shoulder of approximately 1 meter. Therefore, Kampung Naga society call this measure with *sameter*.

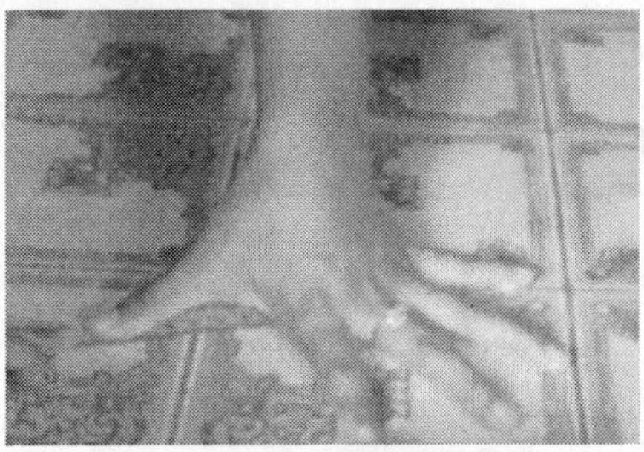

Figure 1. Measurementof *Jeungkal*

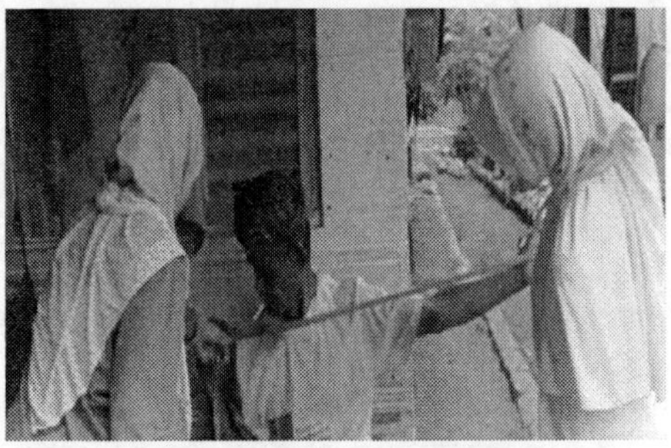

Figure 2. Measurementof *Sameter*

Measurement of length with parts of human body have been used by many countries in the world. Each state has units of length almost equal to other regions but with a different name because of differences in the language used. For example, Swahili society, East Africa used the term*shibiri* to the span (*jeungkal*), that is the distance from the tip of the thumb to the tip of your little finger (about 9 inches), *mkono* = 2 *shibiri* to pinch (about a yard), and *pima* = 4*mkono* to fathom (about 2 yards) [11]. Mexico society using the units of length as *cemacolli* (arm), *cemmolicplitl* (bone), *cenmaitl* (hand) or *cenxocpalli* (footprints)[12].

3.2. Unit of Area

Nowadays, land measurementin Kampung Naga society is done by using a meter. Length and width are calculated in meters, but then the result is converted into a unit, namely *bata* or *tumbak*. *Bata* was being used from ancient times to the present. Kampung Naga society believe that 1 *bata* equals to 14 m^2. If the land is rectangular, then one *bata* of that land area is the area of land which is have length 3.75 m.

Besides*bata*, unit of area known by Kampung Naga society are *areu* and *bau*. If *areu* is converted to square meter, then 1 *areu* = 100 m^2. Meanwhile, if converted into *bata*, then 1 *areu* = 7 *bata*. So, 1 *areu* = 100 m^2 = 7 *bata*. Thus obtained 1 m^2 = 0.07 *bata*. Size 100 m^2 = 7 *bata* is used by the government to convert square meters into *bata*. For *bau* unit,Kampung Naga society states that one *bau* equal to 100 *bata*. Thus, *sabau* = 100 *bata* = 1.400 m^2. Whereas, if use the government system, 1 *bau* = 100 *bata* = 1.428, 57 m^2.

Danadibrata[10] states that *"bata nyaéta ukuran legana sawah atawa tanah, sok disebut ogé tumbak. Sabata = satumbak= 14.193 m^2"*. Based on these opinions *bata*is a size of farm or land area, also called *tumbak*. In this case, the size of the batain Kampung Naga society equal to Danadibrata's opinion. This means, the size of the batain Kampung Naga society is equal to the size of *bata* in Sundanese people in general, just the size 1 *bata*in Kampung Naga society is rounded to 14 m^2.

According Danadibrata[10], *"bau nyaéta ukuran legana sawah atanapi tanah. Sabau = 500 tumbak/bata = 7,096.5 m^2"*. Based on these opinions *bau*is a size of land area. It can be concluded that *bau* used by Kampung Naga society and Sundanese people in general wasdifferent. If normally in Sundasociety 1 *bau* was equal to 500 *bata* (*tumbak*), in Kampung Naga society 1 *bau* equal to 100 *bata*.

According Claudia Zaslavsky[11], the Ethiopian ancient used the unit *acre* for area of land. *Acres* represents the number of land plowed by a pair of bulls in one day. In Indonesia in the 9th century widely known unit of measure, such as *barih, latir, tu, tampah, tampah haji, suku, hamat, blah, jong, lirih, kunci,* and*pecal*. These terms werefound on inscription in the early 9th century, found in the Temanggung. During this period, the size of the land area is usually calculated based on the number of seeds that can be planted on the land. Therefore, the size of the unit area is always preceded by the number of seeds[13].

3.3. Unit of Volume

The use of unit volume saw in the measurement of wood. Wood was one of the materials used in the construction of houses in Kampung Naga society. The wood was taken from the garden, located about 500 m from Kampung Naga. Construction of houses in Kampung Naga is almost nothing else because of the limited area. However, the calculation of the wood is still used in a house renovation. In the construction of the house, wood used as a pillar, rafters, *palang dada*, and the board. A form of pillar, rafters, *palang dada*, and the boardcan be observed as at fFigure 3 and Figure 4. The calculation volume of wood is done by multiplying the length, width, and thickness of the wood. Mathematically, the calculation volume of wood is written: "Volume = length × width × thickness". Then, the units expressed in *elo, dim, strip*, and *kibik*.

Figure 3. Pillar, *Palang Dada*, and Board

Figure 4. Rafters

3.3.1. Pillar

Pillar used by Kampung Naga society has widths and thicknesses of 10 cm wood. Whereas, for its length adjusted to the needs. Calculation of pillar volume in KampungNaga was done without converting the units into same unit. But at the end of calculation, a result divide by a hundred. Then, the results are expressed in units called *élo*. From the calculation of pillar volume in Kampung Naga, this can be developed a mathematical model: Vt = (p)*élo,*with Vt ispillar volume and p is pillar length (in meters).

In this section, researchers also tried to calculate a pillar volume which has two meters of length. The first, researcher convert meter unit into centimeter. Here is the process of counting

Pillar Volume = 2 m × 10 cm ×10 cm = 200 cm × 10 cm × 10 cm= 20,000 cm^3

Based on the mathematical model above, volume of the pillar which has length 2 meters is 2 *élo*. Thus, the pillar volume = 2 *élo* = 20,000 cm^3. Then, we conclude that 1 *élo* = 10.000 cm^3 = 0,01m^3

Kampung Naga society call 1 m^3 with *sakibik,* so that 1 *kibik*= 1 m^3 = 100*élo*. Kampung Naga society has learned that 1 *kibik* = 100 *élo*. This shows that since old time Kampung Naga society has the ability to convert the unit well. But, a term of unitsbeing used in Kampung Naga have different with the unit that has been standardized.

3.3.2. Rafters

Kampung Naga society has been using rafters which has a size of 6 cm for a thick and 6 cm for a wide. For a size of a length adjusted to a needs. Calculation of volume rafters is slightly different to the pillar volume. If at the end of pillar volume calculations are performed division with a hundred number and the results are expressed in élo units, at the calculation of rafting volume divide with ten number and expressed in a dim units. Calculation of rafters volume in Kampung Naga society can be developed a mathematical model: Vk = (3,6× p) dim, with Vk is rafters volume and p is rafters length (in meters).

Researchers attempt to calculate rafters volume with previously convert meters unit to centimeters.

Volume kaso-kaso = 2 m × 6 cm × 6 cm = 200 cm × 6 cm × 6 cm = 7.200 cm^3= 0,0072 m^3

From a calculation of Kampung Naga people, rafters volume is 7 dim 2 strip. So:

7 dim 2 *strip* = 7.2 *dim*

2 *strip* = 0.2 *dim*

1 *strip* = 0.1 *dim* atau 1 *dim* = 10 *strip*

7 dim 2 *strip* = 7.2 *dim* = 0.0072 m^3

1 *dim* = 0.001 m^3 = 1 dm^3

1.000 *dim* = 1 m^3 = 1,000 dm^3

From the previous calculations, be obtained that 1 *kibik* = 1 m^3 = 100 *élo*. Because 1 m^3 = 1,000 dim, then 1 *kibik* = 1 m^3 = 100 *élo* = 1,000 *dim* = 1,000 dm^3.

Based on the calculations above, it can be concluded that 1 *kibik* = 100 élo = 1,000 dim, 1 *élo* = 10 dim, and 1 *dim* = 1 dm^3 = 1 liter.

Thus, the mathematical model for rafters volume can be written as follows.

Vk = (3.6 × p) dim = (3.6 × p) dm^3orVk = (3.6 × p) *élo*, with Vk is rafters volume and p is rafters length (in meters).

3.3.3. Palang Dada

Size of *palang dada*have been using by Kampung Naga society is 10 cm for wide and 5 cm for thick, with a length adapted to the needs. The process of calculating *palang dada*volumeis equal to pillar volume. Here is a form of mathematical models of the *palang dada*volume :Vpd = (0.5 × p) *élo.*With Vpd is*palang dada* volume and p is*palang dada* length (in meters).

3.3.4. Board

Wooden boardwas used for flooring of buildings in Kampung Naga. The floor boards are used in almost every room of the house, except in the kitchen that uses Palupuh floor made from bamboo. The size of the floor board used by Kampung Naga society is 20 cm of wide and 3 cm of thick, with a length adapted to the needs. Calculation of volume board is the same as the calculation of pillar volume and *palang dada*. Here is a form of mathematical models of the board volume: $Vp = (0.6 \times p)$ *élo*.With Vp is board volume and p is board length (in meters).

In olden times, humans measure the volume and weight by using the objects so that the unit was usually still associated with the objects used. For example, if the measure of sugar with a spoon, then the units used are spoons. According Claudia Zaslavsky[10], Swahili, East Africa know *kibaba* unit, *kisaga*, and *pishi* to unit of volume. One *kibaba* is one cask (pint), one *kisaga* = 2 *kibaba* or about a quarter gallon, and one *pishi* = 2 *kisaga* or half-gallon size. Meanwhile, In Mexico known units of volume are *acalli, cemixcolli, cemacuahuitl, cenxumatli, centcuauhxumatli, cempopolli, centlachipinilli,* and*centlamapictli*[12].

From explanation above, it was clear that each region has unique units in the measurement of length, area, and volume. It shows the cultural richness of each region in the world. Therefore, researchers hope that mathematical knowledge incorporated into the school curriculum not only mathematical knowledge that comes from the West, but also mathematical knowledge comes from every region in the world.

4. Conclusion

Ethnomathematics study in the domain of education can be used to uncover the ideas in a cultural activities or social groups to develop mathematics curriculum for, with, and by the group. Thus, mathematics can have different shapes and develop suitable with each culture. The result of this study is a form of units of length, area, and volume in Kampung Naga society. Until now, Kampung Naga society uses typical units obtained hereditary. Unit of length used by Kampung Naga society, namely *jeungkal*, *deupa*, and *sameter*. Unit of area used by Kampung Naga society, namely *bata*, *areu*, and *bau*. Unit of volume used by Kampung Naga society, namely *élo, dim, strip,* and *kibik*. This study revealed the conversion of units of length, area, and volume in Kampung Naga society into standard units. In addition, in the context of the wood volume measurements revealed models of mathematics that can facilitate cultural actors in calculating the volume of wood (pillar, rafters, *palang dada*, and board). The results of this study can be used as a reference to develop learning materials of contextual mathematicsbased on local culture, which is expected to reduce the public perception that mathematics does not have relevance to daily life.

5. References

[1] Milton R and Daniel C *Revista Latinoamericana de Etnomatemática* **4** 32-54
[2] Sumardyono 2004 *Karakteristik Matematika dan Implikasinya Terhadap Pembelajaran Matematika* (Yogyakarta: Depdiknas)
[3] Paulus G 1996 Ethnomathematics and mathematics education *International Handbook of Mathematical Education* (Dordrecht: Kluwer Academic Publiser) pp 909-943
[4] Ubiratan D 1997Ethnomathematics and its place in the history and pedagogy of mathematics *Ethnomathematics Challenging Eurocentrism in Mathematics Education* (Albany: State University of New York) pp 13-24
[5] Marcelo C1997 Ethnomathematics and education *Ethnomathematics Challenging Eurocentrism in Mathematics Education* ed Arthur B. Powell and Marilyn Frankenstein (State University of New York Press, Albany) pp 261-2
[6] William D 1996 *Ethnomathematics: exploring cultural diversity in mathematics* (Auckland: Dissertation University of Auckland)
[7] Alan J 1997. *Papua New Guinea Journal of Teacher Education* **4** pp 17-20
[8] Wilfredo V 2010 Stone *walls and water flows: interrogating cultural practice and mathematics* (Auckland: Dissertation University of Auckland)
[9] Elis S 2011 *Ragam Pesona Budaya Sunda* (Bogor: Penerbit Ghalia Indonesia)
[10] Danadibrata 2006 *Kamus Basa Sunda* (Bandung: Kiblat Buku Utama)

[11] Claudia Z 1973 *Africa Counts* (Boston, Massachusetts: Prindle, Weber and Schmidt)
[12] Vicente Tand Geiselda S 2002 *Science Scope* **25** 12-17
[13] Mashadi 2010 *Proc. of the Int. Seminar on Mathematics and Its Usage in Other Areas* ed Mashadi, *et al* (Riau) pp 177-184

An Implementation Of *Icare* Approach (Introduction, Connection, Application, Reflection, Extension) to Improve The Creative Thinking Skills

Carni[1*], J Maknun[2] and P Siahaan[2]

[1]Post-Graduate School, Universitas Pendidikan Indonesia
[2]Department of Physics Education, Universitas Pendidikan Indonesia, Jl. Dr. Setiabudhi No. 229, Bandung 40154 Indonesia

*Corresponding author's email: carniyamin@gmail.com

Abstract. This study is aimed to get an overview about the increase of creative thinking skills in ten grades high school students as the impact of the implementation of the ICARE approach to the dynamic electrical material. This study is using pre-experimental method. And, the research design is one-group-pretest-posttest. In this case, the participants of this study are students in ten grades in one senior high school in West Java which is randomly selected. The data is collected from the students by doing pretest and posttest in order to measure the increase of students' creative thinking skills. In the final analysis, the results of this study presents that the implementation of the ICARE approach generally increase the students' creative thinking skills. The result of the N-Gain showed that the students' creative thinking skills increased by the average score of 0.52, categorized as medium. This is caused by the implementation of ICARE approach to the application stage.

1. Introduction

The era of globalization and technology not only facilitate people in the matter of using but also confronts human with a challenge and some complex problems. The development of technology was increasing the changes of human life rapidly. Consequently, every individual should be able to solving all the problems in daily life. Creative thinking skills are necessary in order to solve the complex problems in the future [1], especially in the 21st century.

Assuming that students have a potential to be creative, then there are many ways to awaken, stimulate, develop, and access the students to the creative potential. In fact, an educator suggested to prioritizing the suitable establishment and implementation approach in order to enhance the students' creativity. Supposedly, creative thinking skills should be practiced at all levels of education from primary schools to universities to achieve the goals in accordance with the demands of the educational curriculum [2]. Hopefully, with good thinking skills students can stimulate their creative thinking skills to solve problems in daily life by using their knowledge.

In contrast, according to a preliminary study conducted in one Senior High School in Bandung shows the result of the creative thinking test s that are given is in the amount of 32,5, is still very low. The result is triggered by five main factors: (1) learning process is still informative; (2) learning is still dependent on the teacher; (3) learning is a memorizing activity; (4) teachers rarely practicing the comprehension ability; (5) teachers rarely practicing some experiment activities, and learning of physics is still the focus on the count. For this reason, many students assume that physics is a difficult and complex subject.

Therefore, as the solution to the problem, we recommend for the use of ICARE approach to enhance the students' creative thinking skills. When the students have been interested in some subjects that taught in the school, especially Physics, the students are expected to be more comprehensive on the subject. Moreover, when the students have been able to find the relation between the concepts they learned in class and its application in everyday life.

2. Theory

2.1. Creative thinking

Creative thinking skills is the skills to think of many possibilities; to use various methods; to use different point of view; to think of something new and innovative in order to guide the teachers in generating and selecting alternatives, Isaksen [2], According to Jawar [3] creative thinking known as mental activity that aimed to build a strong desire in order to find a solution or solutions previously unknown.

The source of the creativity is in someone who realizes that there is something wrong, less, and mysterious. One of the prominent features of the creative person is that he or she is be able to discover a problem while the others do not. On the contrary, the person that is unable to discover the problem behind the issues can be categorized as not creative person. Thus, think creatively is one element in the problem solving process.

One of the necessary capabilities of the 21st century is the problem solving abilities. Generally, problem solving is usually perceived as an activity of making a solution based on the knowledge of the person has been learned. Problem solving is an effort to find answers based on knowledge, understanding, and ability that the person has in order to provide demands from the uncommon situation [4]. Guilord [1] states the close relationship between creative thinking and problem solving. He said that creative thinking is a part of the solution. Furthermore, think creatively produce new products, and problem solving involves producing a new response to the new situation of the new result, Guilord [1]. It can be concluded that any solution has creative aspects.

Equally important, Torrance [5] suggested that there are four main indicators of creative thinking skills, those are fluency, flexibility, originality, and elaboration. Firstly, fluency is used to create ideas, answers, problem solving and questions, provide some suggestions to do various things also always prepare for more than one answer. Secondly, flexibility is the ability to understand the questions or topics from different perspectives, and be able to change the approach or the way of thinking, the ability to express some solutions or approaches into the problems. Examples to this indicator such as, generating ideas and alternative solutions to problems. After that, originality is the ability to create new and unique expression, thinking the unusual way to express yourself and be able to make combinations that unusual from the elements. Lastly, elaboration is the ability to add a situation or problem that it becomes complete and detail, in which there are tables, charts, models, or ideas, including the ability to develop new product ideas in order to add details to an object. This study focuses on creative thinking skills in problem solving.

3. ICARE Learning Approach

ICARE (Introduction, Connection, Application, Reflection, and Extension) is a model of learning that is focus on understanding the concepts and applying learners' knowledge [6]. Historically, the model was introduced in 1997 at San Diego State University by Bob Hoffman. This model is a framework pedagogic with online learning system at San Diego State University. But it is possible to be used at United Stated Agency International Development (USAID) of Indonesia through the Decentralized Basic Education (DBE) which is in 2006 was introduced and using ICARE learning model as a framework for the research about teachers and learning process in schools [7].

The most interesting thing from the ICARE learning model is its flexibility. ICARE learning model can provide an opportunity for teachers to change students' learning experience through an emphasis on every stage. For example, teachers put more emphasis on the Connection stage, and then the teacher should use the methods or approaches that could improve the ability to understand concepts to the students. Meanwhile, on the Application and Reflection stage, teachers are required to use an approach that reflects the constructivism learning and teachers play role as the learning facilitators for students. ICARE models also provide an opportunity for the teachers to create a special module depends on the consideration of the students' ability (Ress; 2011). Refer to the standard of educational process; there must be an increase and balance between hard skills and soft skills in a learning process. Fortunately, at the Connection and Application stage, the students can get the increase and balance their hard skill and soft skill.

3.1. Definition of ICARE model

Model ICARE is a learning model that was developed in 1997 initiated by Bob Hoffman. Initially, ICARE Learning model is designed for learning online system. As time goes by, in 2006 ICARE models began to be used as a model of learning in the classroom. The stages of ICARE models are:

a. Introduction : at this stage, the teacher establishes the content of the lesson to the students. This must include an explanation of the objectives and goals that are expected to be achieved during the lesson. At this stage the teacher introduce the students about the phenomenon that have been designed for contextual learning. Then, the students observe the phenomenon, and they have the opportunity to asking question(s) about phenomenon that is being displayed. Other than that, motivation should also be given at this stage in order to make the students interested in learning material that will be given in the classroom.
b. Connection :at this stage, the teacher tried to relate the new knowledge with something already known by the students from learning or experience in daily life. In the classroom, teacher demonstrates the materials, and there is a question and answer session that makes students tell what they remember from previous learning experience. The most important thing at this stage is an investment of the concept in a way to invite the students to plan and conduct some activities independently or group that performs the example in the context of real-world based on the inquiry.
c. Application :the application stage is the most important stage of learning. After the students acquire new knowledge or skills from the connection stage, they should be given the opportunities to implement these knowledge and skills in their daily activities at school. The application stage of learning takes the longest period in the learning process because students are required to conduct experiments or to implement their knowledge in a real world context. And this is different than a sample application that has been done at the previous stage, Connection stage.
d. Reflection : this stage of learning is a resume of the whole learning activity. In this stage, the students have the opportunity to reflect on what they have learned in the classroom. By all means, the teachers' task is to assess the extent of learning. Reflection or resume activities may involve a group discussion where the instructor asks students to make presentations or explain what they have learned. As an alternative, they can make an individual writing activity in which the students write a summary of the learning outcomes. Additionally, the resume activity can also be a quick quiz where the teacher provides some questions based on the content of the object. An important point to remember with reflection is that teachers need to give opportunities for students to express what they have learned in the classroom.
e. Extension :Although the learning activity has finished, it does not mean that all students are able to use what they have learned. In that case, the teachers have to do the continuing action that supports the continued learning of the students that is the extension model of learning. The extension stage is an activity where the teacher presents activities that can do following the lesson to reinforce and extend the learning. In other words, the extension activities is the teachers'

activity to give homework to the students. Extension activities may include the provision of supplementary reading materials, resuming the next material or some questions for exercise.

4. Research Methodology

This study is using pre-experimental method. The research design of this study is using one-group pretest-posttest [8]. With these designs, the participants are the students of an experimental class without comparison. In the design of one-group pretest-posttest single group of subjects were given a pretest (initial test), treatment (X), and posttest (final test). The participant of the pretest and posttest is same but given in a different time. The form design as shown in Table 1.

Table 1. Research design one-group-pretest-posttest

Pretest	Treatment	Posttest
O_1	X	O_1

Information :
O_1 = Pretest and posttest creative thinking skills
X = Treatment approaches ICARE Approach

The participants of this study are students in ten grades in one senior high school in West Java the second semester of the 2015/2016 academic year. And, the sample of this study is class X.6 consisting of 32 students. In the process of school selection is using the random sampling technique.

5. Findings And Discussion

The findings in the study include the initial capability test, final capability test, and the average value of N-gain normalized creative thinking skills, as presented in Figure 1.

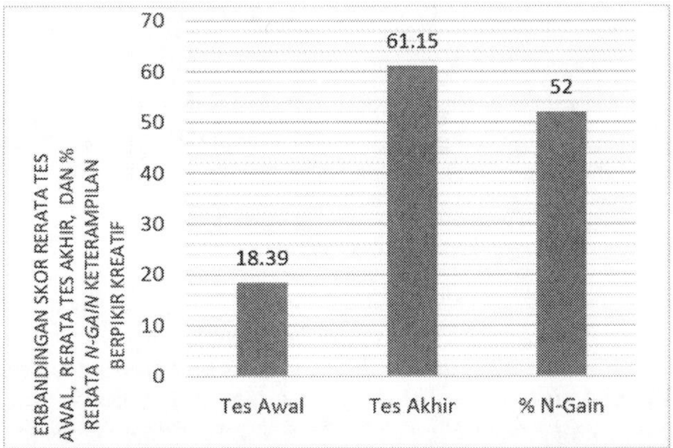

Figure 1. Diagram of average initial test, final test average, and average of normalized Gain scores <g> on creative thinking skills tests

This study was conducted to discover an increase in creative thinking skills after ICARE learning approach implemented in the learning process. In general, students' creative thinking skills increased with middle category. It is shown in the results of the data analysis that was illustrated in Figure 1, the average gain scores $\langle g \rangle$were normalized. As can be seen on the diagram, it showed the creative thinking skills is in amount of 0.52. The result of the data analysis when it confirmed to the category of Hake [9], the average scores of the normalized gain is classified into medium category. In the initial tests of students' creative thinking skills is in the amount of lower average. On the other hand, once ICARE approach learning has been implemented, the average score increased. This indicates that the ICARE approach can increase creative thinking skills.

Based on the analysis of data showed that the average score of normalized gain $\langle g \rangle$creative thinking skills on dynamic electrical material for each indicator as in Figure 2.

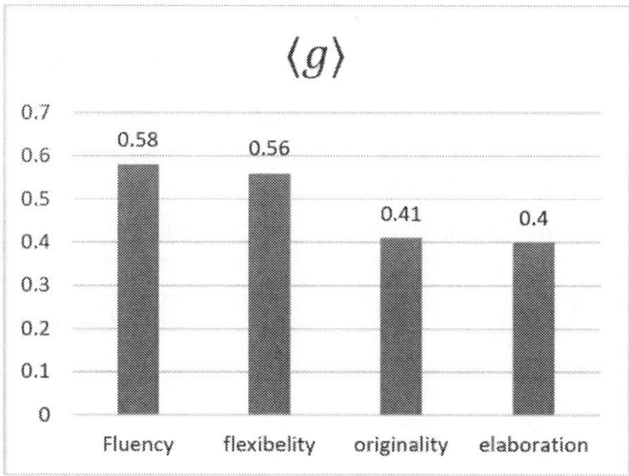

Figure 2. Comparison chart$\langle g \rangle$Creative thinking skills indicator

The increase of students' creative thinking skills, in all the indicators based on the criteria of normalized gain [9], is in the medium category. This shows that the students' creative thinking skills in all four main indicators generally increased with medium category after the teachers implemented the ICARE approach in the learning process, especially in learning physics. It is in line with the results of a research conducted by Maskur [10] which states that learning Math with ICARE approach can actually enhance the creative thinking skills of students in the three-dimensional material.

6. Conclusion

Based on the results of the study and data analysis has been done, then it can conclude the that :

1. Implementation of ICARE approach can further enhance creative thinking skills high school students of class X for the dynamic electricity material.
2. The average students' creative thinking skills is in medium category.

7. References

[1] Treffinger D J 1995*Educational Psychology Review***7**(3) pp 301-12
[2] Hajiyakhchali A 2013 *Social and Behavioral Sciences***84** pp 549 – 52
[3] Bilal and Al-Khatib 2012 *American Int. Journal of Contemporary Research* **2**(10)
[4] Krulik S and Rudnick J A 1996 *The New Sourcebook for Teaching Reasoning and Problem Solving in Junior and Senior High School* (Boston: Allyn and Bacon)
[5] Torrance E P and Safter H T 1999 *Making the Creative Leap and Beyond* (Buffalo, NY: Creative Education Foundation Press)
[6] Hoffman B and Ritchie D 1998*Society for Information Technology & Teacher Education International Conf.* pp 119–23
[7] DBE3 2006 *Matematika Untuk Kehidupan, Pembelajaran dan Pekerjaan* (Jakarta: USAID DBE3 Life Skills for youth)
[8] Millan Mc J 2001 *Research in Education a Conseptual Introduction* (New York & London: Longman
[9] Hake R R 1999 *Analyzing Change/Gain Score.* [Online]
[10] MaskuR A dkk 2012 *Journal of Primary Educational* **1**(2) pp 85-90

Edutourism Taka Bonerate National Park through Scientific Approach to Improve Student Learning Outcomes

R S Hayati[1,a)]
[1]Biology Education Major FKIP Muhammadiyah Makassar University, Makasar, Sulawesi Selatan, Indonesia
[a)]rizasativanihayati@gmail.com.

Abstract. This research aim is develop the potential of Taka Bonerate National Park as learning resources through edutourism with scientific approach to improve student learning outcomes. Focus of student learning outcomes are students psychomotor abilities and comprehension on Biodiversity of Marine Biota, Corals Ecosystem, and Conservation topics. The edutourism development products are teacher manual, edutourism worksheet, material booklet, *guide's* manual, and Taka Bonerate National Park governor manual. The method to develop edutourism products is ADDIE research and development model that consist of analysis, design, development and production, implementation, and evaluation step. The subjects in the implementation step were given a pretest and posttest and observation sheet to see the effect of edutourism Taka Bonerate National Park through scientific approach to student learning outcomes on Biodiversity of Marine Biota, Corals Ecosystem, and Conservation topics. The data were analyzed qualitative descriptively. The research result is edutourism Taka Bonerate National Park through scientific approach can improve students learning outcomes on Biodiversity of Marine Biota, Corals Ecosystem, and Conservation topics. Edutourism Taka Bonerate National Park can be an alternative of learning method on Biodiversity of Marine Biota, Corals Ecosystem, and Conservation topics.

1. INTRODUCTION

Local potential of tourism can be used as a learning resource through edutourism activities. Tourism should have function to fulfill the needs of the physical, spiritual, and intellectual every tourist (ConstitutionNumber 10 Year 2009 about Tourism). Facts show that the tourist areas have not been optimally utilized as an educational media, many tourists who visit just for refreshing only. Local potential attractions have the potential to be developed as a source of biology learning. This is possible because of the availability of the biology study object in tourism location. Each region requires education in accordance with the characteristics of the region, so the curriculum should contain that diversity to produce graduates that are relevant to the region development needs (BSNP, 2006: 7). Learning through the student'senvironmentis better to develop a balance hard skill and soft skills of students, because students intersect and interact directly with the environment as an object of study (Darsiharjo, 2010). Therefore, in learning biology teachers should include local potential as a learning resource.

In the education and tourism field, we know the term edutourism can be a solution in using local potential of tourism as a source of learning. Edutourism in the 1st Annual International Conference on

IOP Conf. Series: Journal of Physics: Conf. Series **812** (2017) 012023 doi:10.1088/1742-6596/812/1/012023

Sustainable Edutourism interpreted as a journey that aims to learn formally or informally, in a tourism environment that has a local unique. Edutourism can be used as one solution the use of local potential as a learning resource for biology learning.

Taka Bonerate National Park is one of the attractions located in Selayar, South Sulawesi. Taka Bonerateis a nature conservation area which is geographically located in the Flores Sea, an area is 220,000 ha, and this area is the third largest coral atoll in the world, after Atoll Kwajifein in the Marshall Islands and Atoll Suvadiva in Maldive (Effendi, 2011: 15). Taka Bonerate has high diversity of marine biota and habitat for many species of marine animals that are endangered and protected. Therefore, it's very appropriate if these attractions packed into a learning resource on Diversity of Marine Biota, Coral Reef Ecosystem, and Marine Biota Conservation topics for student tourists through edutourism.

During this time Taka Bonerate formally untapped as a source of learning for student tourist. There are no media to supports biology learning activities to student tourist. Edutourism development research is aimed to increase students learning outcomes. It is need attractions mapping of the local potential as biology learning resource and make edutourism products in form of teacher manual, edutourism worksheet, material booklet, guide manual, and Taka Bonerate National Park governor manual. After the implementation stage of edutourism with scientific approach using the edutourism product, edutourism can improve psychomotor and cognitive of SMA N Taka Bonerate students on Biodiversity of Marine Biota, Reef Ecosystem, and Conservation topics.

2. METHOD

This is a research and development (R & D) with ADDIE development model. ADDIE consist of five steps, there are Analysis, Design, Development and Production, Implementation, Evaluation. It was adapted from Dick and Carey model (Dick, Carey, Carey, 2005: 6-8). Analysis stage consist of learning objects identification, student'sneed analysis, curriculum analysis, and instructionalanalysis. Design stage consist of develop learning objectives, framework (outline) of edutourism development product, and design evaluation tools. Development and productionstageconsist of developedinstrument to assess edutourism, develop learning strategy, materials selection, writing draft of edutourism development product, draft review, draft assess, and firstrevision. Implementation stage is trial for edutourism development product.Implementation subject are students from SMA Taka Bonerate who school location adjacent to Taka Bonerate National Park and previously they never visit to Taka Bonerate National Park. Implementation of development products are quasi use Pretest Posttest Group Design, where students were tested before and after edutourism implementation.

Table 1. Research Design, Pretest Posttest Group Design

Table 1. Research Design, Pretest Posttest Group Design

Group	Pre-Test	Treatment	*Post-Test*
Treatment Class	O	X	O

Note:

X=*Edutourism* Taka Bonerate National Park

O=Student ability before and after edutourism implementation

Measuring gain score conducted to understand whether edutourismthrough scientific approach can improvestudent's cognitive learning outcomes in Marine Biota Diversity, Coral Reef Ecosystem, and Marine Biota Conservation topics:

Table 2.Criteria Improving Learning Outcomes of Gain Score

Category	Gain Score (g)
High	$g \geq 0,7$
Medium	$0,7 > g \geq 0,3$
Low	$g < 0,3$

(Hake. 1998: 64-74)

After edutourism product implemented, the inplementation result is used to do second revision. The las step is evaluation stage wheresummative evaluation conducted to determine the usefulness of the product and make recommendations on further product development.

3. RESULT AND DISCUSSION

3.1. Taka Bonerate National Park Potential as Learning Resources

Taka Bonerate National Park has many potential as learning resources, it can be developed edutourism concept. Tourist, especially students tourist can study in Taka Bonerate National Park. Potential analysis conducted to find learning object and decide edutourism activity in Taka Bonerate National Park.

Table 3. Learning Objects and Edutourism Activity in Taka Bonerate National Park

Learning Objects	Edutourism Activity	Learning Post
Marine Biota Diversity	1st Learning Activity: Observe and Identify Marine Biota Diversity by diving or snorkeling	Tinabo Island: Baby Shark Spot, Ibel Orange Spot, Softcoral Points, Acropora Points, Corina Corner, Seagrass Spot, Kima Spot
Corals Diversity	2nd Learning Activity: Observe and Identify Corals Diversity by diving or snorkeling	Tinabo Island: Ibel Orange Spot, Softcoral Points, Acropora Points, Corina Corner
Coral Reef Ecosystem	3rd Learning Activity: Identify components of coral reef ecosystems and their interactions	Tinabo Island: Ibel Orange Spot, Softcoral Points, Acropora Points, Corina Corner
Conservation Efforts in Taka Bonerate National Park	4th Learning Activity: Case Study of Marine Biota Conservation	Tinabo Island: Coral Transplantation Spot, Acropora Points, Corina Corner, Kima Spot

Edutourism is an integration of education into tourism, tourist refresh but do not forget to study about what they observe in tourist area. Here are edutourism activities that can be offered as a package holiday for students:

Table 4.EdutourismPackage in Taka Bonerate Island

Edutourism package "Marine Biota"		
Learning Post	**Learning Aspects (Education)**	**Refreshing Aspects (Tourism)**
a. Baby Shark Spot b. Corina Corner c. Seagrass Spot d. Kima Spot	a. Observediversity of marine biota b. Studydiversity of marine biota c. Study the importance of marine biota diversity d. Study many cases of marine biota conservation in Taka Bonerate National Park e. Study how to conserve the diversity of marine biota	a. Diving b. Snorkeling c. Photography below sea level d. Canoeing e. Boating f. Swimming with baby shark
Edutourism package "Coral Reef"		
Learning Post	**Learning Aspects (Education)**	**Refreshing Aspects (Tourism)**

a. Ibel Orange Spot b. Softcoral Points c. Acropora Points d. Other Corals Spot	a. Observe the diversity of corals b. Study the components of coral reef ecosystem and their interactions c. Study the importance of coral reef ecosystem d. Study how to conserve coral reef ecosystem e. Practice Coral Transplantation	a. Diving b. Snorkeling c. Photography below sea level d. Boating

3.2. Curriculum Analysis Result

Curriculum analysis conducted to make competencies mapping that will be delivered by edutourism activities in Taka Bonerate National Park. Basic competencies (KD) was developed from core competencies (KI)National Curriculum Year 2013. Below are the basic competencies:

1.1 Admire the order and complexity of God's creation on biodiversity, ecosystems and the environment
1.2 Be aware and admire the scientific thinking in observing capabilities bioprocess
1.3 Be sensitive and concerned about environmental issues, maintaining and caring for the environment as a manifestation of the practice of their religion
2.1 Conduct scientifically: conscientious, diligent, honest to the data and facts, discipline, responsibility, and care in observation and experimentation, daring and polite in asking questions and arguing, environmental care, mutual aid, cooperation, love peace, argues the scientific and critical, responsive and proactive in every action and in conducting observations and experiments in the classroom/laboratory and outside the classroom/ laboratory
2.2 Concerned for the safety of themselves and the environment by applying the principle of safety when conducting observations and experiments in the laboratory and in the neighborhood
3.5 Applying the principles of classification to classify protist based on common characteristics of the class and its role in life through thorough and systematic observation.
3.8 Applying the principles of classification to classify animalia into phyla based on observation of anatomy and morphology and associatestheir role in life.
3.9 Analyze information/data from various sources about the ecosystem and all the interactions that take place therein.
3.10 Analyzing the data changes in the environment and impact on the lives
4.5 Plan and carry out observations about the characteristics and role of protist in life and present the results of observations in the form of model/chart/ image.
4.8 Presenting data on the comparison of network complexity constituent bodies of animals and its role in various aspects of life in the form of a written report.
4.9 Designing a chart of the interaction between the components of the ecosystem and food networking that takes place in the ecosystem and presenting the results in various forms of media.
4.10 Solving environmental problems by making environmental conservation efforts.

3.3 Edutourism Development Product

Learning objects that potential to be studied by students tourist in Taka Bonerate then developed into instructional organization that consist of teacher manual, edutourism worksheet, Booklet Material, *guide's* manual, and Taka Bonerate National Park governor manual.

1. Teacher Manual
   Teacher manual consist of three parts, Part I. Introduction of EdutourismTakaBonerate National Park, Part II. Edutourism Activities with Scientific Approach, and Part III.Edutourism Learning Equipment consistsof syllabus and lesson plan
2. Edutourism Worksheet
   Worksheet consists of four learning activities, 1st Learning Activity: Marine Biota Diversity, 2nd Learning Activity: Understand Coral Reef, 3rd Learning Activity: Coral Reef Ecosystem, 4thLearning Activity: Marine

Biota Conservation. Before learning activities, students introduced work safety to have activities in the sea. After learning activities, students evaluated with assess instruments that included in edutourism worksheet.

3. Booklet Material
Booklet material consists of three topics, namely: Diversity of Marine Biota, Coral Reef Ecosystem and Marine Biota Conservation. Each topic begins with an explanation of learning objectives and ends with "Do You Know?".In the end of the booklet therea are given conclusion and glossary
4. Guide's Manual
Guide's manual consists of two parts. First part, guide are introduced about the potential of Taka Bonerate National Park for edutourism development. Second part, guide can study how to guide edutourism activity in Taka Bonerate National Park.

f. Taka Bonerate National Park Governor Manual
Governor manual consist of two parts. First parts, governor of Taka Bonerate National Park are introduced about the potential of Taka Bonerate National Park for edutourism development. Second part, governor can study edutourism activity in Taka Bonerate National Park.

3.3 Edutourism Implementation Result

Edutourism products that have been developed were implemented on students of SMAN Taka Bonerate. Each student does activities based on worksheet and learns the material from bookletmaterial. Edutourism activities conducted with scientific approach and cover three topics; there are Marine Biota Diversity, Coral Reef Ecosystems, and Marine Biota Conservation. Stages of edutourismwith scientific approach are observing, questioning, associating, experimenting, and networking. Each stage of learning in edutourism still included fun or refreshing activity. Students learning outcomes focus on cognitive and psychomotor. Cognitive learning outcomes measured by providing pretest posttest to students, while psychomotor learning outcomes observed with the observation sheet. Edutourism implementation results prove that there is an increase in students' cognitive learning outcomes. The results of the pretest posttest can be seen from Figure 1. Calculation of gain score was also conducted to determine whether there is an increase in the ability of students between before to after learning through edutourism. The average gain score of students 0.71. Students who have a gain score as high category is 75%, while the other medium. So the 75% of students increase cognitive learning outcomes after Edutourism Taka Bonerate National Park with scientific approach. The percentage gain score of students can be seen from Figure 2.

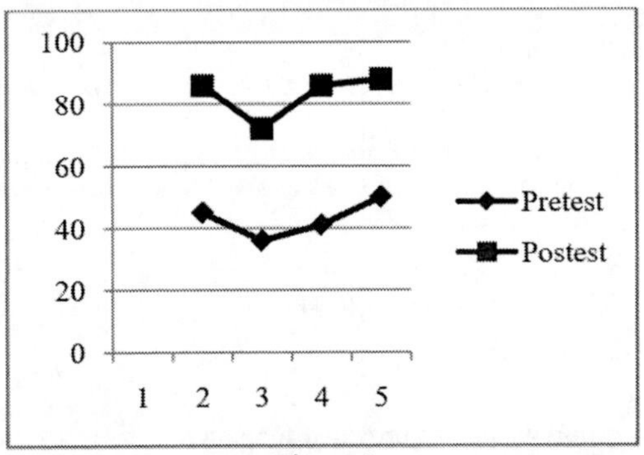

Figure 1

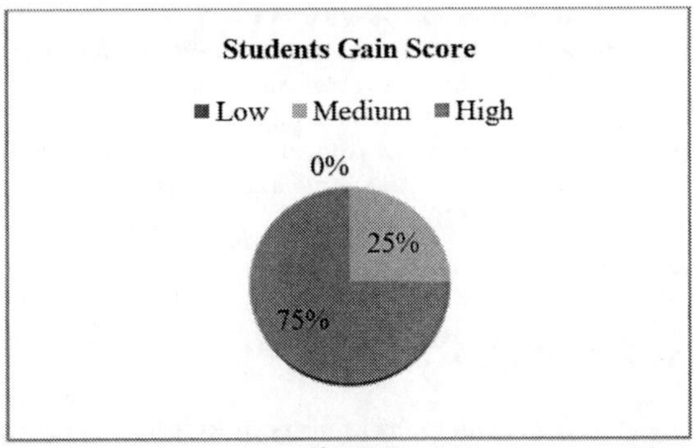

Figure 2

The analysis results of observation sheets explain that learning through Edutourism can encourage students psychomotor. Psychomotor learning outcomes associated with achievement through manipulation skills that involve muscle and physical strength. Psychomotor learning outcomes focus was on moving, manipulating, communicating, and creating ability. Psychomotor learning outcomes are a continuation of the cognitive and affective learning outcomes. The successful development of the cognitive aspects will positively affect the development of psychomotor aspects.Psychomotor learning outcomes that appear when students do EdutourismTaka Bonerate National Park:

1. The ability of moving, students always maximizes the use of muscles to move in a tourist sites to fullfill worksheet by observation or data collection. Students do snorkeling in almost Edutourism activity.
2. The ability of manipulating, students needs the coordination of body movements such as the eyes, ears, hands, and fingers to conduct data collection at tourist sites.
3. The ability of communicating, students always interact and communicate effectively with the guide, teacher, and surrounding communities in the tourist sites for data collection and problem solving.
4. The ability of creating, edutourism activities also forces students to think to make an idea of conservation, complete the edutourism worksheet, and make conservation poster. So from this edutourism activity students generate products.

4. CONCLUSION
Edutourism of Taka Bonerate National Park can improve the cognitive aspects of learning outcomes and encourage the emergence of psychomotor aspects. Therefore, Edutourism Taka Bonerate National Park can be used by teachers as an alternative method of learning on diversity of marine biota, coral reefs ecosystems, and marine biota conservation topics.

5. REFERENCES

[1] BSNP 2006 *PanduanPenyusunan KTSP Jenjang Pendidikan Dasar dan Menengah* (Jakarta: BSNP)

[2] Canadian Edutourism Associates Cooperative Limited (CETA). *What is Edutourism?* http://www.*edutourism*conference.org/whatis*edutourism*. html, 2010
[3] Constitution Number 10 Year 2009 about Tourism
[4] Dick W, Carey L and Carey J O 2005 *The Systematic Design of Instruction 6th Edition* (Addison-Wesley Educational Publisher Inc, Glenview) pp 6-8
[5] Darsiharjo 2010 *Pembangunan Kepariwisataan yang Ilmiah, Edukatif, danReligius* in Professor Appointment Speech (UPI, Bandung)
[6] Ahmad E D *Enjoy The Beautiful Underwater Diversity in Takabonerate National Park Kima Buletin* Volume 7 3rd Edition No.2086-8979 pp.15
[7] Hake R 1998 *American Association of Physics Teachers Journal* pp. 64-74
[8] Hendriyan, "*AnalisisKemampuanPsikomotorSiswapadaPembelajaran Hands On Teknik Challenge Exploration Activity,*" S.Pd. Thesis, UIN Syarif Hidayatullah
[9] Kumalasari, Hartoko, Anggoro, "*Toward a New Paradigm of Ecosystem and Endemic Organism Based Spatial Zonation for Taka Bonerate Marine Protected Area,*"Paper submitted for J. Of Coastal and Marine Resources Management (2013)

ACKNOWLEDGMENTS

Authors are grateful to the Ministry of Research and Higher Education which has provided funding for this research and the governor ofTaka Bonerate National Park who has provided an opportunity for researchers to develop EdutourismTaka BonerateNational Park in Tinabo Island.

Effect of Fruit Lemon Juice Addition to The Content of Protein, Fat, Lactose and Probiotic on Soy Yogurt

F M T Supriyanti[1,a], Zackiyah[1] and N Azizah[1]

[1]Chemistry Education Department, UniversitasPendidikan Indonesia, Jln. Dr.Setiabudhi.No. 299 Bandung Jawa Barat Indonesia 40154
[a]florentinasupriyanti@yahoo.co.id

Abstract. This research aimed to determine the effect of lemon juice to the content of protein, fat, lactose and probiotics, in beverages soy yogurt. Soy yogurt which produced was a multifunction yogurt drink high levels of antioxidants, contains probiotics and can be used by people with lactose intolerance. The research method includes the production of fortified soy yogurt with lemon juice, were made with the ratio between the lemon juice and soy yogurt were 0:10 (L0) ; 1 : 9 (L1) ; 2 : 8 (L2) ; and 3 : 7 (L3) . Analysis of the results include the content of protein by Kjeldahl method , the content of fat by Soxhletasi method, lactose test by Luff Schoorls method and content of probiotics with total plate count enumeration techniques. The results showed fortified yogurt had a protein content greater than before fortification (L3 > L2 > L1 > L0); The fat content L0 > L1 > L2 > L3. Fortified yogurt lactose content is smaller than before fortification (L0 > L1 > L2 > L3). The content of probiotic yogurt fortified L1 > L3 > L2. From this research can be concluded that the yoghurt fortified (L3) is the best, with the highest protein content, low fat, low lactose than L1 and L2, and had probiotics content. It is advised to conduct further research on the expired time of fortified soy yogurt products

1. Introduction

Yogurt in the market usually comes from cow's milk. Related Research yogurt production uses cow's milk has been done by some previous researchers. The addition of star fruit extracts in the manufacture of yogurt drinks can be used as a flavor enhancer and to assist the growth of lactic acid bacteria[1].Fortified soy yogurt lemon juice is useful as a source of protein drinks, containing probiotics, low-lactose, and high levels of antioxidants. Drinks can be used as a functional beverage, the nutritional content needs to be known. According to Jannah[1], some additional research on the fruit in yogurt have been done like Strawberry[2], oranges and grapes[3], cerry, peach, raspberry, blueberry[4] coconut water [1]. The fortification is done using yogurt were made from cow's milk. While the manufacture of yogurt using soy milk has not been done. Soy milk has a protein content of 4.4% is higher than cow milk is 2.9%[5]. Lee et al.[6], do the research by adding mango extract soy milk to streng the content of vitamins A, C and minerals. The addition can provide sweetness and reduce the unpleasant taste of soy milk. Soy milk and cow's milk have similar amino acid composition, so that soy milk can be used as a substitute for cow's milk for those who are allergic to animal proteins. Soy yogurt was fermented using lactic acid bacteria[7]. Vij et al,[8] states fermentation can increase the presence of isoflavones, help cutting peptide bonds of proteins, increases the solubility of calcium, improve health and help the immune system. Fermented milk requires bacteria, one of which is *lactic acid* bacteria. *Lactic acid* bacteria are added into the production of yogurt can act as probiotics. Havenar,[9] states that probiotics are single or mixed cultures of microbial life is consumed by humans or animals, beneficial to the host to maintain the balance of microflora in the digestive tract. Donkor et al,[10] states that the proteins in fermented soy milk can promote the growth of various strains of probiotics such as, *Lactobacillus casei* and *Streptococcus thermophillus*.

Fortification has been done on other sources of protein foods such as milk, soy milk and yogurt[11]. According to Santoso[12], in general, have soy milk contains vitamin B1, B2, niacin, pyridoxine and vitamin B group is high. Other vitamins that are contained in sufficient quantities of vitamin E and K, while vitamin C is contained in small amounts. Micronutrients such as vitamins in addition to functioning in helping work the various types of enzymes are also some of which act as antioxidants[13].Vitamin C may act as an antioxidant. According to the USDA National Nutrient data base in 100 grams Lemon (Citrus limon) vitamin C contained as many as 53 mg, or about 88%. In this study produced soy yogurt fortified with vitamin C from lemon juice.

Soy yogurt can be used by people with lactose intolerance, which in his body does not have the enzyme lactase. In yogurt contains bacteria that can produce the lactase enzyme. This enzyme can convert sugar contained in milk is lactose into glucose and galactose, which causes lactose content is reduced, so the yogurt can be consumed by people with lactose intolerance. Research on the effect of the addition of lemon juice on the content of protein, fat, lactose and the amount of probiotics in yogurt fortified soy vitamin C has not been done before.

This study aims to determine the content of protein, fat, lactose and the amount of soy yogurt probiotic before and after fortification using lemon juice . This research made some yogurt with a variety of ratios between lemon juice and soy yogurt , ie 0:10 (L0) ; 1 : 9 (L1) ; 2 : 8 (L2) ; and 3 : 7 (L3) , and then analyzed the content of protein , fat , lactose and probiotics . Products of yogurt fortified vitamin C each have a data content of protein, fat, lactose and probiotics. So , the yogurt can to be used as a multi functional drinks, which contain probiotics, with a low lactose content and high antioxidant.

2. Methods

2.1. Material

The materials was used in the research were Soybean (Bandung , Indonesia) , lemon fruit (Bandung, Indonesia), *Lactobacillus bulgaricus* and *Streptococcus thermopillus* (Biofarma, Bandung, Indonesia), $CuSO_4.5\ H_2O$, K_2SO_4, sulfuric acid (H_2SO_4,98%,Bratachem,Indonesia), sodium hydroxide pellets pure (NaOH) (s), Boric acid GR (H_3BO_3) (s), indicator Tashiro, Hydrochloric acid (HCl) 25%, Sodium thiosulfate 5-hydrate ($Na_2S_20_3.5\ H_2O$), Potasium Iodate GR (KIO_3), and Aquades.

2.2. Method

2.2.1. The production of fortified yogurt lemon juice

Production of fortified yogurt lemon juice preceded by the production of soy milk and soy yogurt with appropriate procedures performed in previous studies[14].Yogurt fortification with vitamin C was resulted from lemon juice obtained three variants: L1; L2 and L3. As for the variation of the ratio between lemon juice and soy yogurt is 1: 9 (L1); 2: 8 (L2) and 3: 7 (L3). As a control use soy yogurt without the addition of lemon juice 0:10 (L0). Lemon juice sterilized using a 0.2 µm Millipore size.

2.2.2. Analysis of the results of fortified soy yogurt

2.2.2.1.Test of the protein content by Kjeldahl method

In the determination of protein content, we use the Kjeldahl method (material was used consisted of $CuSO_4.5H_2O$ and K_2SO_4 mass ratio of 1: 3, added 10 mL of concentrated H_2SO_4, NaOH 30%, 10 mL H_3BO_3 3%, HCl 0.1 N and indicators Tashiro)

2.2.2.2. Test of the fat content by Soxhlet method

Determination of the fat content we use Soxhletation methods (materials used include 25% HCl and petroleum ether).

2.2.2.3. Test of lactose content using Luff Schoorls methods
Lactose content analysis used Luff Schorll method, (the material used solution of KI 20%, 26.5% H_2SO_4, sodium tio sulfate 0.1 N and 1% starch indicator.

2.2.2.4. Probiotic content test using total plate count enumeration techniques.
Probiotics test performed using total plate count enumeration techniques with materials *Lactobacillus bulgaricus and Streptococcus thermopillus* and Luria Berthani media (LB).

3. Findings and Discussion

3.1. Production and caracteritation of Fortified Yoghurt
Fortified soy yogurt lemon juice has been successfully made in three variants. In Figure 1. we analyzed the protein content (blue line) , fat (red line) , and lactose (green line) yogurt before and after fortification . We analyze the difference between the ratio of lemon extract and soy yogurt : 0:10 (L0) ; 1 : 9 (L1) ; 2 : 8 (L2) ; and 3 : 7 (L3) . In comparison to the raw material used we also tested against soy milk (LK).

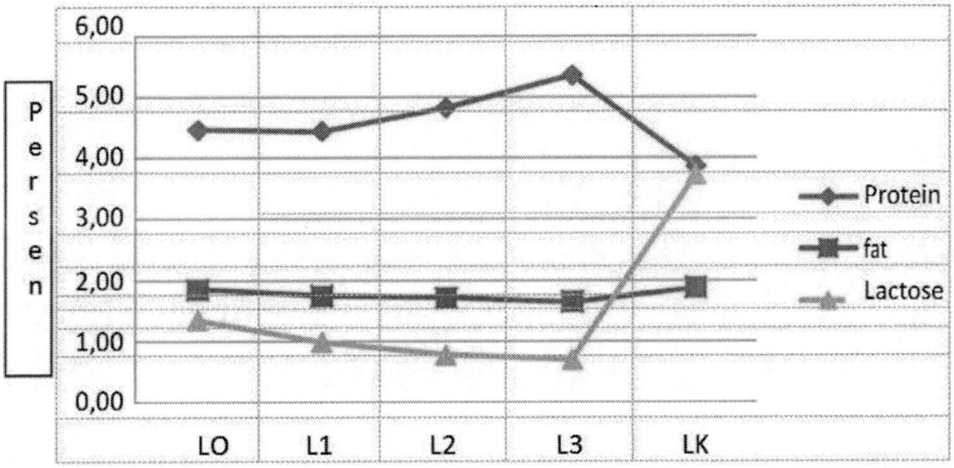

Figure 1.Results of the analysis of protein, fat and lactose yogurt before and after fortification

From the experimental data obtained that the protein content of soy milk is 3.87%, while the protein content of soy milk theoretically is 3.6 to 4.5% [7]. It shows that based on the protein content of soy milk used meets the criteria as soy milk. While the protein content in soy yogurt without lemon juice extract is 4.47%. It shows that without the addition of lemon juice protein content of yogurt (fermented milk) which is produced in accordance theoretical data is 4.44%[5]. From the experimental data obtained that more lemon juice are added, the protein content of yogurt fortified products is higher, it is supported by the opinion starting theoretically lemon juice contains 1.1% protein[5]. So the protein content of yogurt fortified sortable L1 (4.45%) <L2 (4.82%) <L3 (5.36%). Fortified yogurt product was selected with the highest protein content is L3.

From the experimental data obtained that the fat content in soy milk 1.87% while the fat content in soy milk theoretically is 2.5%, it is influenced by the quality of soymilk in the production, which is of course influenced also by the quality of soybean seed. If we look at the fat content of yogurt before fortification was 1.85%, and the fat content decreased with the increase of lemon juice, then the decline will affect the smell of rotten on soy yogurt, because the taste and the smell was caused by work lipoxygenase enzyme present in soybean seeds. Enzymes that react with fat in soybean grinding time, particularly if used in cold water. The results of the reaction at least in the form of volatile compounds is ethyl-phenyl-ketone. While the lemon juice has a low fat content. In lemon juice contains omega-3 fatty acids a total of 26 mg / 100gr, omega-6 fatty acid total of 63 mg / 100g lemon juice. From the research results when sorted the fat content, the fortified yogurt lemon juice L1 (1.74%)> L2 (1.71%)> L3 (1.64%).

Figure 1, shows that more of lemon juice added, lactose content decreases. Known as milk sugar lactose, in the presence of the enzyme lactase, lactose is converted into glucose and galactose. Lactase enzyme production decreases with age a person, then the consumption of milk can cause diarrhea. L0 ie without fortification yogurt with lactose content of 1.35%; L1 with a lactose content of 0.99%; while the L2 with a lactose content of 0.77%; and L3 is 0.67%. With the addition of lemon juice can be concluded that the more lemon juice added lactose content decreases, L3 has a low lactose content.

3.2. Analyzed the number of Probiotics

Figure 2, we analyze the number of probiotics contained in yogurt before and after fortification. Probiotics are live microorganisms, when obtained in sufficient quantities would be beneficial to the health of the host[15]. In the production of soy yogurt's probiotics used were *Lactobacillus bulgaricus and Streptococcus thermopillus* both types of bacteria are used to accelerate the process of acid production, increase the amount of lactic acid and improve aroma[7].

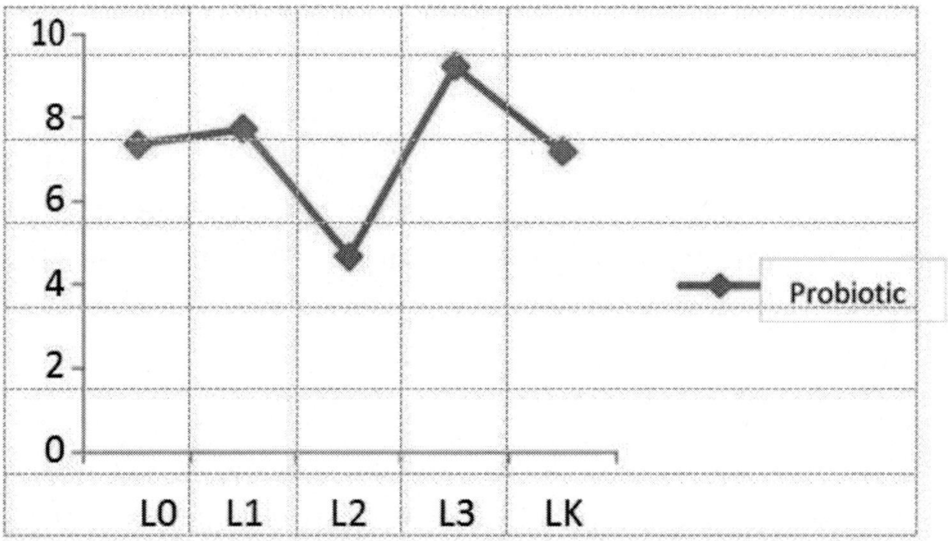

Figure 2. Content of probiotics in yogurt before and after fortification

Figure 2, the data content of probiotics in yogurt fortified lemon juice obtained all products containing probiotic. The highest content of probiotic products obtained at L3, followed by L1 and L2, or if sorted L3> L1> L2. In this study used two types of probiotics namely *Steptococcus thermophilus* and *Lactobacillus bulgaricus*. Evangelista, et al,[16] stated that among the various lactic acid bacteria potential, *Streptococcus thermophilus* and *Lactobacillus bulgaricus* traditionally used, the bacteria can be produced quickly in acid medium, through theformation of organic acids, especially lactic acid, and compounds such as acetic acid, ethanol, aromatic hydrocarbons, exopolysaccharide and various enzymes that are very important.

3.3. Antioxidant Activity of fortified soy yogurt

From previous studies[14], obtained antioxidant activity to control 40.04% of soy yogurt and yogurt products fortified with lemon juice, soy yogurt comparison: lemon juice 9: 1 (1) was 77.13%; ratio 8: 2 (2) was 93.88% and ratio of 7: 3 (3) was 88.97%.

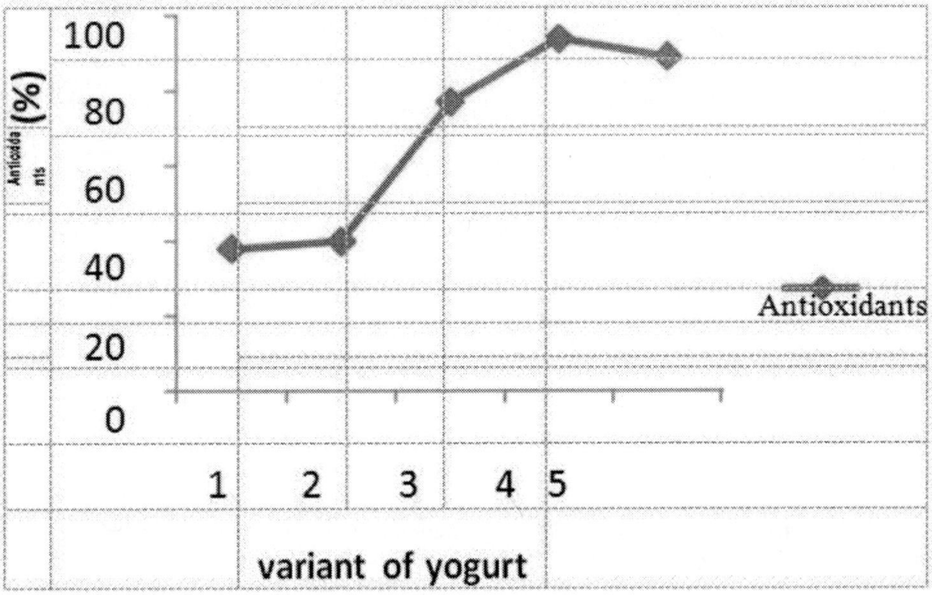

Figure 3.The content of antioxidants in yogurt before and after fortification

The addition of lemon juice into the yogurt can increase the antioxidant content, because the lemon juice contains ascorbic acid (vitamin C) that can function as an antioxidant . In fortified yogurt product number antioxidant content of 40.04 %. Increased antioxidant occurred after fortification lemon juice. The highest amount of antioxidants obtained in yogurt fortified L2, ie 93.88 %. From the research it can be seen that the addition of lemon juice can produce multifunctional soy yogurt, that contain probiotics, high levels of antioxidants, and has a lactose content and lower fat and higher protein content than yogurt without fortification .

4. CONCLUSION

This study succeeded in making yogurt fortified fruit juice lemon as a multifunctional beverage. The drink containing the probiotic , with a high antioxidant , can be used by people with lactose intolerance. Yogurt fortified fruit juice are known high protein and low fat. We Obtained 4 variations of yogurt products with the ratio between lemon juice and soy yogurt are 0:10 (L0) ; 1 : 9(L1) ; 2 : 8 (L2) ; and 3 : 7 (L3) . Analysis of the results includes protein, fat, lactose, and probiotics. The results showed fortified yogurt had a protein content greater than before fortification (L3 > L2 > L1 > L0) ; The fat content of L3 < L2 < L1. The content of lactose yogurt fortified smaller than before fortification (L3 < L2 < L1 < L0) . The content of probiotic yogurt fortified L3 > L1 > L2.

References

[1] Jannah A M, Legowo A M, Pramono Y B, Al-Baarri A N, Abduh S B M 2014 *J. Aplikasi Teknologi Pangan Vol. 3(2)*
[2] Thomson J L, Lopetcharat K, and Drake M A 2007 *J. Dairy Sci.* Volume **85** (4) : 721-9
[3] Hossain M N, Fakrudin M and Islam M N 2012 *American J. of Food Tech.* Vol. **7**(9) : 562-70

IOP Conf. Series: Journal of Physics: Conf. Series **812** (2017) 012024 doi:10.1088/1742-6596/812/1/012024

[4] Con A H, Cakmak-ci, Caglar A and Gokalp H Y 1996 *J. of Food Protect.* Vol. **59** (4) : 402-6
[5] Cahyadi W 2007 *Kedelai: Khasiat dan teknologi* (Jakarta: Bumi Aksara)
[6] Lee S Y, Morr C V and Seo A 1990 *J. of Food Sci.* Volume **55**: 532-6
[7] Syamsuddin Y, Hesti M, Friday S, Rahmat D 2013 *Internasional journal on Advanced Science Engineering Information Technology* Vol.**3** no. 4
[8] Vij S , Hati S, Yaday D 2011 *Food and Nutrition Sciences Volume* (2) pp 502-9
[9] Havenaar R, and Jos H, Huis in't Veld J 1992 *Probiotics: A General View in the Lactic Acid Bacteria in Healthand Disease* Wood, B.J.B. (ed) (Blackie Academic and Profesional, New York, 1992).
[10] Donkor O N, Henrikson A, Vasiljevic T & Shah N P 2005 *Journal of Food Science* Volume **70** 375-81
[11] Panomai N, Sanchaisuriya P and Rakoin L 2010 *Journal of Science* Volume **51** : 325-32
[12] Santoso S P 2009 *Susu dan Yoghurt Kedelai* (Laboratorium Kimia Pangan Faperta Universitas Widya Gama, Malang)
[13] Maria M B 2014 *Fortifikasi vitamin C sebagai antioksidan dalam minuman sari buah.*(Online) Tersedia:http://chemistranger.blogspot.com/2013/01/fortifikasi-vitamin C.html (Diakses, 19 April 2014)
[14] Supriyanti F M T, Azizah Z N 2015 *Multifunction yogurt based on soy milk and lemon juice* , Proceding Seminar MSCEIS. CH-63(2015).
[15] Araya M, Morelli L, Reid G, Sanders M E, Stanton C, Pineiro M, & Embarek P B 2002 *Giedline for theevaluation of probiotics in food.* Joint FAO/WHO Working Group Report Drafting Guiedline Eval. ProbioticFood, London (Ontario Canada). 30,1-11
[16] Evangelista S R, Ghiselli G, Filho F M 2012 *Food and Nutrition Sciences Volume* (3) pp 1128-35

IOP Conf. Series: Journal of Physics: Conf. Series **812** (2017) 012025 doi:10.1088/1742-6596/812/1/012025

A Development of Mathematical Connecting Ability of Students in Junior High School through a Problem-Based Learning with Course Review Horay Method

P N Malasari[1], H Nindiasari[2], and Jaenudin[2]

[1]Departement of Mathematics Education,Universitas Pendidikan Indonesia, Jl Dr. Setiabudhi 229,Bandung 40154, Indonesia

[2] Departement of Mathematics Education, Universitas Sultan AgengTirtayasa, Jl. Raya Jakarta KM 04, Serang 42124, Indonesia

Email: putrinurmalasari@student.upi.edu

Abstract. The aim of this research is to develop the lower ability of mathematical connection. The population of this research was the entire students of grade VII at one of state junior high school in Tangerang. The instrument used on the research is math connecting ability test. The results of the research showed that: 1) the average of post-test of mathematical connection in experimenting group is 74.78 and 66.61 for controlling group; 2) the average of gain of mathematical connection in experimenting group is 0.39 and 0.22 for controlling group. The conclusions of the research are: 1) the final result of mathematical connection in experimenting group with adequate criterion is getting way better than controlling group with poor criterion; 2) a developing mathematical connection in experimenting group with sufficient criterion is getting way better than controlling group with sufficient criterion. According to this study results, the author presented some suggestions that: 1) problem-based learning with course review horay method can be used as an alternative to improve mathematical connecting ability; 2) mathematics learning alternatives for implementing curriculum 2013; 3) continue with other aspect of research on a broader study.

1. Introduction

Learning in school includes a wide range of science presented in the form of subjects. Each subject has its own role to develop students' skills. Mathematics is one of the important subjects taught in school. Many problems in daily life can be solved with mathematics. Mathematics also has a role in supporting the advancement of other sciences, such as education, physics, astronomy and others. Mathematics courses consist of various topics and concepts related to each other. This is consistent with the argument of attribution Bruner that in mathematics there is a close relationship between one concept with another, not only in terms of content, but also in terms of formulas are delivered [1].

Educational Unit Level Curriculum (KTSP) explains that one of the purposes of mathematics courses at the level of secondary school is that students have the ability to explain the relationship between concepts. This objective can be accomplished if the student has the ability mathematical connections. Mathematical connection capability is one of the five standard capabilities that must be possessed students in learning mathematics [2] Importance of mathematical connections, that if the students were able to see the connection of topics in mathematics, the students are able to associate mathematical ideas, so the understanding of math will be deeper and last longer [2].

Based on the results of preliminary studies Malasari[3] on mathematical connection capability using instruments Listyotami[4] states that the average score of students' mathematical proficiency test connections in one of state junior high school Tangerang amounted to 33.8 while the maximum score of the test is 81, with the lowest indicators that write object relationships and mathematical concepts. Based on the results of the preliminary study, it appears that the ability of students' mathematical connections in one of state junior high school Tangerang is low.

In connection with the above problems, it can be affirmed that the performance improvement of mathematical connections through a learning process as well as innovative in one of state junior high school is an important need to be implemented. A lesson that can be used to enhance these capabilities is a problem-based learning with course review horay method. The process of problem-based learning with problem scenarios and their order can help students develop cognitive connections [5]. Problem-based learning with course review horay method presents a problem at the beginning of learning to support the achievement of the concept, while the conventional learning presents a problem at the end of the learning and serves as an exercise of the material that has been studied.

Based on these problems, this research relates to the development of mathematical connection, entitled a development of mathematical connecting ability of students in junior high school through a problem-based learning with course review horay method.These research questionsare whether the final result and developing in mathematical connection ability of students who received problem-based learning with course review horay method better than students who received the conventional learning.

2. Theoretical Background

2.1. Mathematical Connection

Mathematical connection derived from the two words, mathematics and connection. The connection is a relationship that can facilitate all activities, while the mathematical is concerned with mathematics [6]. Mathematical connection is the ability to make a meaningful relationship between the concepts of mathematics, between concepts with other areas, and with the daily life [7]. Based on the explanations that have been described above, it can be concluded that the mathematical connections is the ability to connect internally mathematics, mathematics with other sciences and mathematics to daily life. There are two general types of mathematical connections, modeling connection and mathematical connection. Modeling connection is the relationship between problem situation that arise in the real world or in other disciplines with a mathematical representation, while the mathematical connection is the relationship between the two representations are equivalent and the completion of each representation [2]. According to the statement, the mathematical connection can be classified into three types, namely, the connection between math topics, connections with other disciplines and real-world connections with students or connection with everyday life. Mathematics is not a collection of separate topics, although in reality it is often partitioned math and taught in several branches [2]. There are topics in mathematics prerequisites as a basis for understanding the next topic [8]. Like building a multi-story building, the second floor and the next will not be realized if the foundation and the floor before being prerequisites have not really mastered. Mathematics as a discipline of science can be useful for other disciplines and mathematics has a role as the queen of science at the same attendant [8].

2.2. Problem-Based Learning with Course Review Horay Method

Problem-based learning is an approach to learning where students work on authentic problems with a view to preparing students' own knowledge, develop discoveries and thinking skills, develop independence and confidence with problem-solving activity [9]. Course review horay method based is a method of learning with comprehension testing using a box that is loaded with a number to write down the answer, which first gained true sign immediately shouted horay [10]. Problem-based learning with course review horay is problem-based learning that in its implementation using the course review horay with steps which course, the formation of the group, team work, review, horay and evaluation [11].

3. Experimental Method

This research was a quasi-experimental research designs with nonequivalent control group design [12]. The design of the research show as follows:

O X O
- -
O O

Description:

O : Pretest and posttest mathematical connection
X : Problem-based learning with course review horay
--- : The subject is not grouped randomly

The population of this research was the entire students of grade VII at one of state junior high school in Tangerang. The sampling technique used was a cluster sampling technique. There were 76 students taken as sample, and they were divided into two groups, namely the experiment group and the control group. The students in experiment group learned rectangular topic used problem-based learning with course review horay method and the control group learned the same topic used conventional learning. The teaching and learning process was done within five weeks.

The indicators used in this study, as follows: (a) Recognize equivalent representations of the same concept; (b) Recognizing the relationship mathematical procedure a procedural representation equivalent representation; (c) Use and assess the linkages between different mathematical topics and linkages outside mathematics; (d) Use of mathematics in daily life [13].

The data was collected using a test of mathematical connection ability, before the instrument test is used to collect data, the instrument tested for validity, reliability, distinguishing features, and level of difficulty, and also the test instrument was validated by expert. In the pre-test, normality test using chi square test and F test for homogeneity test. With $\propto = 0.05$, sample come from normal distributed population and homogeneous. Method of hypothesis testing used is t-test (right tailed-test).

Post-test is used to determine the final result of mathematical connection. The modify criterion of the final result of mathematical connection as follows: (1) score 90 – 100 (excellent criterion); (2) score 80 – 89 (good criterion); score 70 – 79 (adequate criterion), and score < 70 (poor criterion) [1]. Normalized gain is used in this study to determine the improvement in mathematical connection ability. The formula of gain level [14]:

$$\text{gain}=\frac{\text{posttest - pretest}}{\text{maximum score - pretest}} \quad (1)$$

4. Result and Discussion

Data processing mathematical connection tests such as pretest, posttest and gain using descriptive statistics to determine the average value, standard deviation and variance. According to Table 1, the average score of post-test both groups are presented in the diagram below and shown in Figure 1 that the final result of mathematical connection of the experimental group was better than the control group. It is supported by t-test (right tailed-test). This means, the final result of students' mathematical connection in experimental group who obtained problem-based learning course review horay method better than control group students who get conventional learning.The data processing results can be seen in the following Table 1:

Table 1. Descriptive Data

Data	Group	N	Max Score	Highest Score	Lowest Score	$\bar{x}$	S	S^2
Pretest	Experiment	38	100	55	10	36,95	10,99	120,86
	Control	38		60	15	38,82	11,33	128,42
Posttest	Experiment	38	100	85	60	74,78	6,22	38,78
	Control	38		80	55	66,61	7,01	49,16
Gain	Experiment	38	1	0,67	0,38	0,57	0,08	0,0061
	Control	38		0,55	0,18	0,42	0,10	0,0099

Based on the results of data analysis, gain experimental group is better than the gain of the control group. This is supported by the results of t-test analysis one side (right side). This means that an increase in students' mathematical connection in experimental group who obtained problem-based learning course review horay method better than control group students who get conventional learning. For more details about comparing final result and gain of mathematical connection can be seen in the Figure 1below.

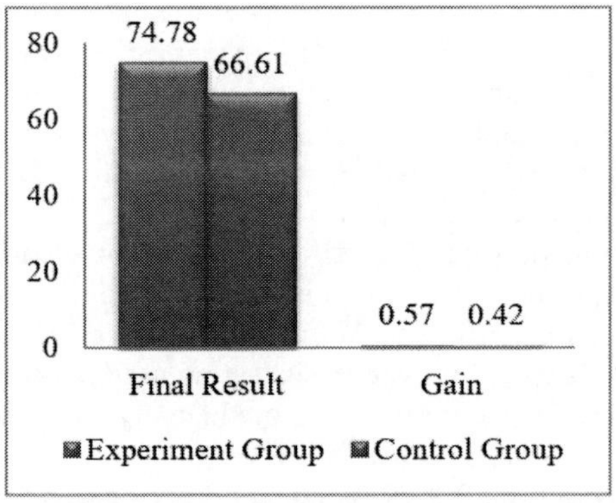

Figure 1. Average Score of Final Result and Gain of Mathematical Connection

The average of final result experimental group is 74.78 and 66.61 forcontrollinggroup. Based on these results, it appears that the final result of the experimental group included in adequate criterion and the final result of the control group were in poor criterion. Data from the gain obtained by the average value of the gain in the experimental group of 0.57, while the average gain in the control group of 0.42. Increased ability to connect mathematical experimental and control groups included into the sufficient criterion. Indicators connection with the mathematical average of the highest value is the use of mathematics in daily life, while the indicator with the lowest average value is used and assessing the linkages between different mathematical topics and linkages outside mathematics.

Students still have difficulties when work on the problems of mathematical connections. The experimental group experienced relatively fewer difficulties than the control group. Difficulties students during work on the problems that require mathematical connection capabilities are presented in Table 2 form below.

Table 2. Difficulty of Student

Number	Indicator	Difficulty	
		Experiment	Control
1	Recognize equivalent representations of the same concept	Some students have difficulty in drawing conclusions kite concept.	Students difficulty in writing the steps equivalent representations of the same concepts and the conclusion is still not quite right.
2	Recognizing the relationship mathematical procedure an equivalent representation of procedural representation.	There are some students who are not taking the equivalent mathematical procedures.	Most students write answers directly regardless of mathematical procedures are equivalent.
3	Use and assess the linkages between different mathematical topics and linkages outside mathematics.	Drawing wake through the process of GLBB.	Determining the time on a moving object, resulting in errors in drawing graphs.
4	Use mathematics in daily life	Some students are still less appropriate to write down the steps answers, so the connection is less visible.	Mathematical modeling and write down the steps to answer the many framed photos that will be drawn.

The achievement of the research objectives as mentioned above, is influenced by several factors including: (1) problem-based learning with course review horay method apply active learning by students as the center of learning and the teacher acts as a facilitator, mediator and motivator, (2) activity team work, including group discussions, completing worksheets in groups and exercises while playing, (3) review their activities, the teacher explains the return of the materials studied as well as reviewing the material that supports student understanding of the material being studied, (4) activity evaluation do the students by communicating orally, subject that has been gained in learning and evaluating the course of the discussion process.So, problem-based learning course review horay method can be applied to improve the mathematical connections in junior high school students. This is supported by Rusman [5] that one alternative learning approach that enables the development of thinking skills of students (connection) is problem-based learning.

Relating to the implementation of curriculum 2013 through a scientific approach to learning steps such as observing, ask, experiment (explore), association and communication, the relation with this research is problem-based learning course review horay method in practice there some stages of learning according to curriculum 2013. These stages are, according to the team work and the experiment to question (explore), where students can talk to each other and to ask the teacher about the nature, circumference and area of quadrilateraland wake phases of team work as well in accordance with the stages of the association, the students can analyze and link the properties wake quadrilateral, as well as the stages of evaluation in accordance with the communication, in this case the student can present orally the things they have learned and do a resume with the teacher about the learning materials have been obtained.Similar tostudents felt that they had improved their personal abilities, such as collaboration, overcoming shyness, listening and communication ability with problem-solving group work [15]. Core competencies in the curriculum in 2013, namely to live responsible behavior,

IOP Conf. Series: Journal of Physics: Conf. Series **812** (2017) 012025 doi:10.1088/1742-6596/812/1/012025

care (tolerance) and interact with the social environment can also be realized with problem-based learning course review horay method, because this learning is a group learning that fosters responsibility for the student's own and a responsibility to share knowledge through discussion in order to understand the group of their friends.

5. Conclusion

Based on the formulation of the problem and the result of research and discussion of the result of the study as described in the previous chapter, the conclusions of the research are (1) The final result of mathematical connection in experimenting group with adequate criterion is getting way better than controlling group with poor criterion, (2) A developing mathematical connection in experimenting group with sufficient criterion is getting way better than controlling group with sufficient criterion.

According to this study results, the author presented some suggestions that (1) Problem-based learning with course review horay method can be used as an alternative to improve mathematical connecting ability, (2) Mathematics learning alternatives for implementing curriculum 2013, (3) Continue with other aspects of research on a broader study.

6. References

[1] Suherman E 2001*Evaluasi Proses dan Hasil Belajar Matematika* (Jakarta: Universitas Terbuka)p 48
[2] National Council of Teacher of Mathematics 2000 *Principle and Standar for School Mathematics* (United State of America: NCTM) pp 7-64
[3] Malasari P N 2012 *Studi Pendahuluan Kemampuan Koneksi Matematis Siswa SMP* (Tangerang)
[4] Listyotami M K 2011*Upaya Meningkatkan Kemampuan Koneksi Matematika Siswa Kelas VIII SMPN 15 Yogyakarta Melalui Model Pembelajaran Learning Cycle (5E)* (Yogyakarta: UNY)
[5] Rusman 2010 *Model-model Pembelajaran Mengembangkan Profesionalisme Guru* (Jakarta: Rajawali Pers) pp 229-236
[6] Departemen Pendidikan Nasional 2008 *Kamus Besar Bahasa Indonesia (KBBI)*(Jakarta: Balai Pustaka).
[7] Noto M S, Hartono W and Sundawan D 2016 *Infinity Journal* **5** 99-108
[8] Tim MKPBM 2003*Strategi Pembelajaran Matematika Kontemporer*(Bandung:JICA UPI) pp 23-25
[9] Tawfik A A and Kolodner J L 2016 *InterdisciplinaryJournal of Problem-Based Learning***106**
[10] Huda M 2013 *Cooperative Learning* (Yogyakarta: Pustaka Pelajar) p 229
[11] Suprijono A 2010 *Cooperative Lerning* (Yogyakarta: Pustaka Pelajar)
[12] Ruseffendi E T 2010*Dasar-dasar Penelitian Pendidikan dan Bidang Non-Eksakta Lainnya* (Bandung: Transito) p 53
[13] Sumarmo U 2014 *Penilaian Pembelajaran Matematika* (Bandung: Reflika Aditama) p 27
[14] Hake R 1999 *Analyzing Change/Gain Scores* (Bloomington: Indiana University)
[15] Tan D K Koppi A and Field D J 2016*International Journal of Innovation in Science and Mathematics Education***2** 54-66

The Analysis of Learning Obstacleand Students Learning Motivation of Prospective Math Teachers in Basic Physics Class

D T Kurniawan[1*], A Suhandi[2], I Kaniawati[2] and D Rusdiana[2]

[1]Graduate School of Universitas Pendidikan Indonesia
[2]Departemen Pendidikan Fisika, Universitas Pendidikan Indonesia
Jl. Dr. Setiabudhi No. 229, Bandung 40154 Indonesia

*Corresponding author's email : dedetriekurniawan@fkip-unswagati.ac.id

Abstract. Learning motivation revealed as a whole intrinsic factor that created, maintained and supported students to achieve the goal of learning. As the bigger motivation came with bigger success, motivation was considered as the main key to reach what students have planned. There were intrinsic and extrinsic factors that influence both the students and lecturers' motivation. The factors in one hand, were essential to be defined by the lecturers in order to maintain and enhance the students' enthusiasm. On the other hand, they also encouraged and thrilled the students to learn. The study aimed to expose and describe the motivational tendency and to knowledge and analyze learning obstacles faced by the students in basic physics class on students of prospective math teachers in FKIP Unswagati Cirebon. In addition, the study focused on the description of the six motivational components stated by Glyn and Koballa. The six were intrinsic motivation, extrinsic motivation, the relevance of studying physics for subjective purposes, willpower, self assessment and anxiety. Class responses were determined through questionnaire with four main indicators; the causes of being less popular subject, the cause of being disfavored subject, the description of the way the students draw the examination on basic physics subject and the academic background of the students. The results showed that 54% students stated that physics was disfavored because the subject was difficult to understand, 49% stated that the cause of being disfavored of the subject was because physics required complicated mathematics. Most of the students preferred to have game based activities that boosted thinking skill. According to the analysis of the students' motivation, the findings revealed that the students' had high level of anxiety in learning the subject. They mostly expressed their anxiety appeared from the material density and text book based assignments.

1. Introduction

Low mastery of the material and the ability of prospective teachers certainly related to the difficulties and problems faced by students. Mathematical competence of students will grow more optimal if the teachers are competent in managing the learning process, both from learning material as well as boost the activities students in learning mathematics. The teacher's competence can not be separated from his ability when he obtained the learning of mathematics in college (as student teachers).

The lecture in the classroom is one of the activities in learning activities. There are several factors that determine the teaching and learning activities that faculty, students, curriculum, learning environment, Media, and psychological factors. The psychological factor is a factor of the which is the main thing that determines the intensity of study that includes: interest, intelligence, aptitude, motivation and cognitive ability (Djamarah, 2002). Interests are subject steady tendency to feel attracted to study or subject-specific and were delighted to learn the material (Winkel, 2004). Each learning activity will be found lazy students who participated and actively participates to follow the lessons. Interest has a great influence on learning activities. The learning process will run smoothly when accompanied by an interest because the interest is the tool main motivation that can encourage student learning within a certain timeframe Motivation to learn is overall the driving force of psychic within the students who lead learning activities, ensure the continuity of learning activities, and provide direction on learning activities to achieve objectives (Winkel, 2004). Motivation can determine whether or not achieve the goal, so the greater the motivation will be even greater success achieved. Student motivation in following the practice learning can arise intrinsically and extrinsically (Winkel, 2004). A set curriculum in mathematics education courses FKIP Unswagati Cirebon become one of extrinsic motivation for students in the following study because it follows these activities is compulsory for students in order to apply the theories and concepts acquired during the learning in the classroom. So like it or not the student should follow. Intrinsic motivation is more important than extrinsic motivation for intrinsic motivation comes from within the individual itself without any external stimuli intrinsic motivation is more genuine and lasting, and does not rely on impulse or the influence of others.

Motivation plays an important role in the learning process for both lecturers and students. For professors find the motivation to learn from the students was very necessary to maintain and enhance the students' learning spirit. For students learning motivation can foster the spirit of learning so that students are motivated to do the act of learning. student learning activities with pleasure because it is driven motivation.

Based on the exposure of the background mentioned before, it is necessary to do a research on the analysis of learning difficulties and learning motivation student teachers of mathematics in general physics coursework.

2. Research Methodology

2.1. Participants

The Research was conducted on mathematics education courses FKIP Unswagati Cirebon. This study was conducted to find the motivation to learn and difficulty learning mathematics student teachers in the following general physics lectures. Student teachers of mathematics in question are the first semester students who signed the general physics course in mathematics education courses in FKIP Unswagati Cirebon in the first semester of 2014 to 2015 and the first semester of 2015-2016.

2.2. Instrument

*2.2.1. Observation.*Conducting observations directly to the college mathematics education courses FKIP Unswagati Cirebon. Observations made during the learning takes place on the campus. These observations were made to dig up information all about the difficulties and motivation to learn mathematics student teachers in the following general physics lectures.

2.2.2. *Questionnaire.* Questionnaires were distributed to the students was a questionnaire to solicit a response / feedback regarding the difficulties and motivation to learn mathematics student teachers in the following general physics lectures. Motivation Questionnaire This study focused on six components description Disclosure motivation to learn. Six component motivation to learn by Glyn and Koballa is intrinsic motivation, extrinsic motivation, the relevance of studying physics for personal purpose, determination, self-assessment ability to general physics lecture and the last is Anxiety

2.2.3. Interview. Interviews were conducted with faculty that aims to find teaching methods that are used, the barriers faced by lecturers in providing general physics lectures. In addition, interviews were conducted with some student teachers of mathematics to know the difficulties and motivation to learn other less revealed through questionnaires.

2.2.4. Analysis of document. After seeing the results of questionnaires student teachers of mathematics and interviews with parties related to the research, to more convincing results of the report, the documents related to the study analyzed. These documents include the value obtained by the student in a particular subject, such as the Irish Lectuer GBPP and SAP as well as an evaluation tool used by lecturers like Quis, UTS and UAS.In general, these research activities can be seen in this figure :

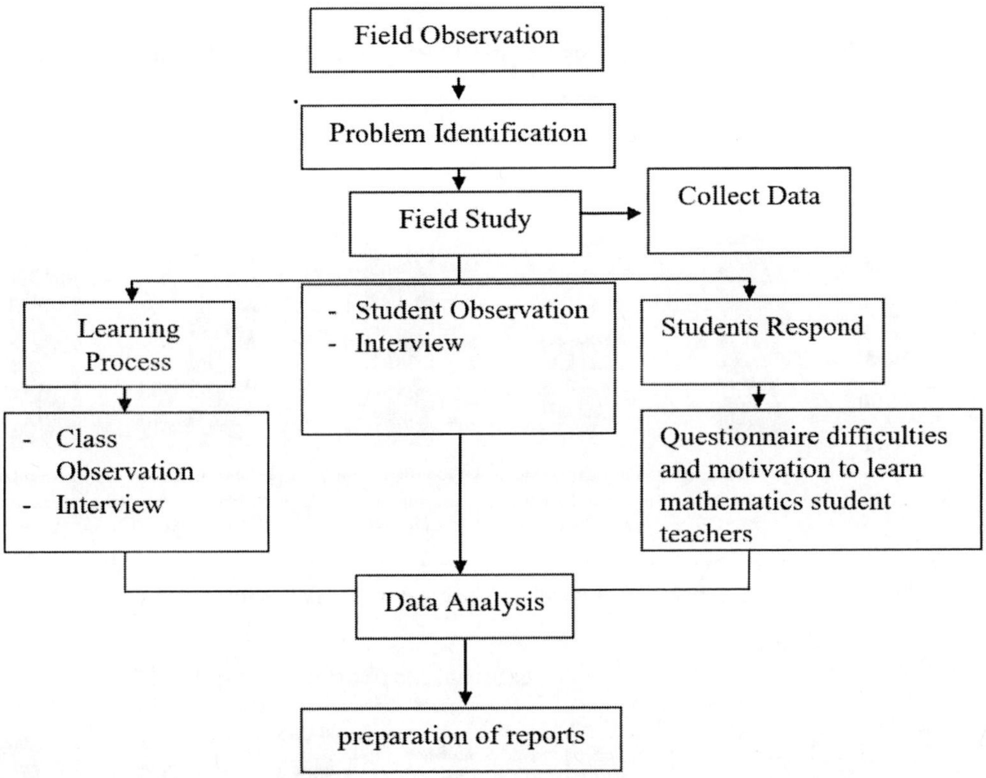

Figure 1. Research Step

3. Result And Analysis

Data research consisted of motivation and difficulty student teachers of mathematics in elementary physics coursework. The data is then in the recap, analyzed and presented then used to answer research questions. Data for motivation to learn decomposes in the form of a diagram in figure 2 Until figure 8. For learning obstacle data presented infigure 9 up to figure 12.

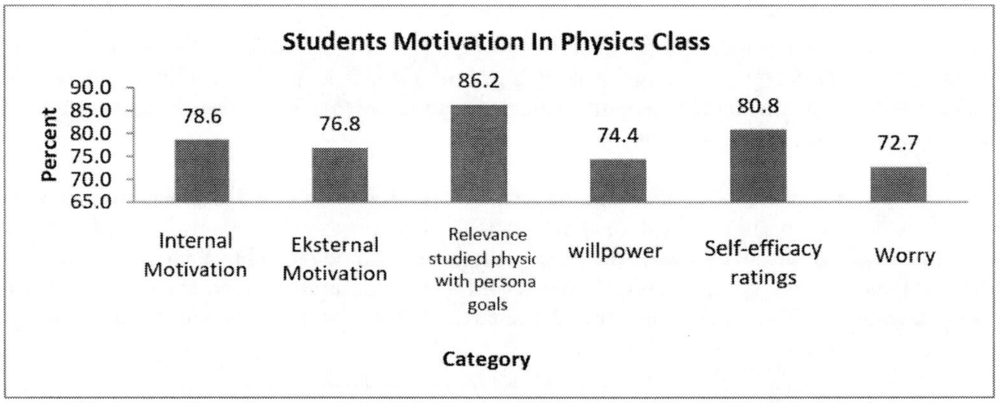

Figure 2. Motivation prospective student learning math teacher in the lecture physics

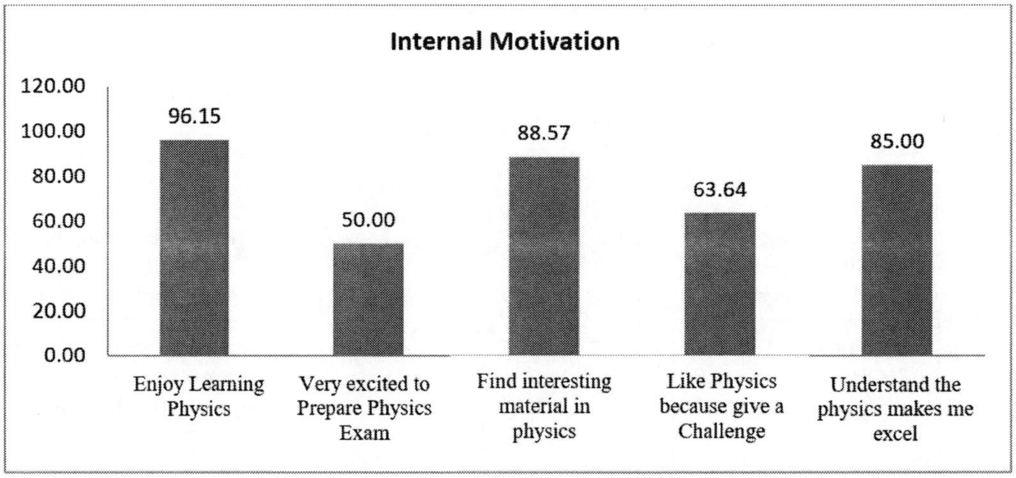

Figure 3. Internal motivation students prospective mathematics teachers in the course physics

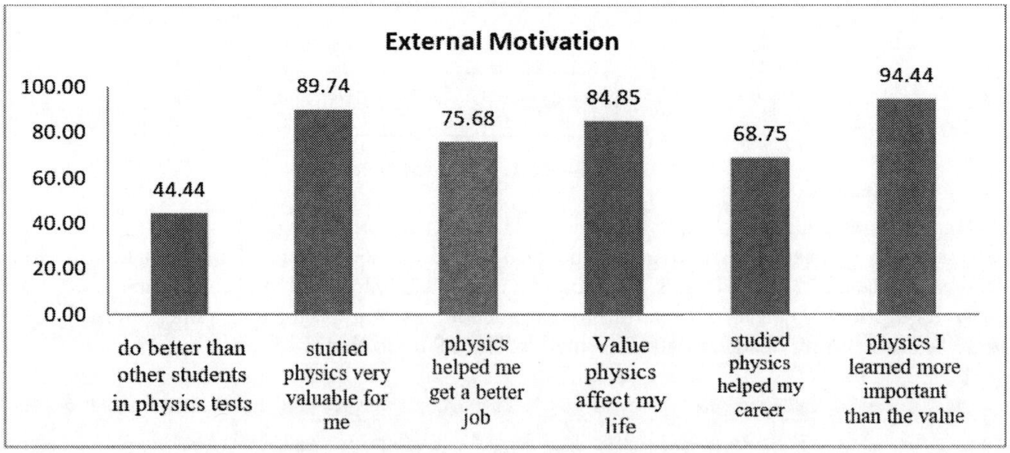

Figure 4. External motivation students prospective mathematics teachers in the course physics

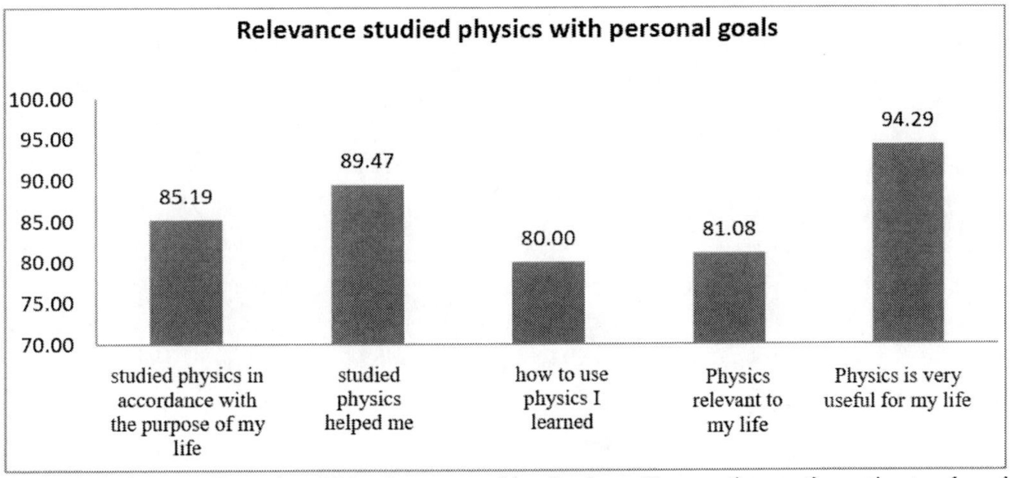

Figure 5. Relevance Learning Objectives owned by Students Prospective mathematics teachers in the course Physics

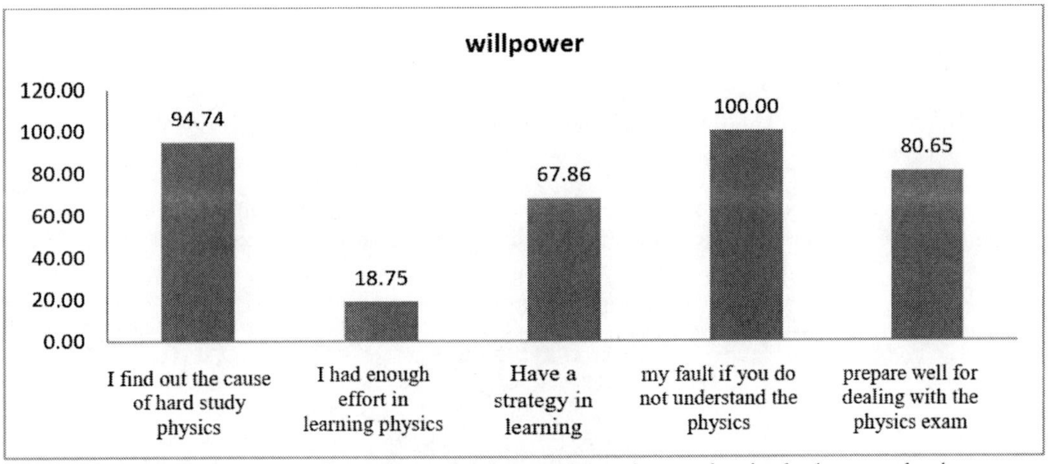

Figure 6. Prospective students determination mathematics teacher in the lecture physics

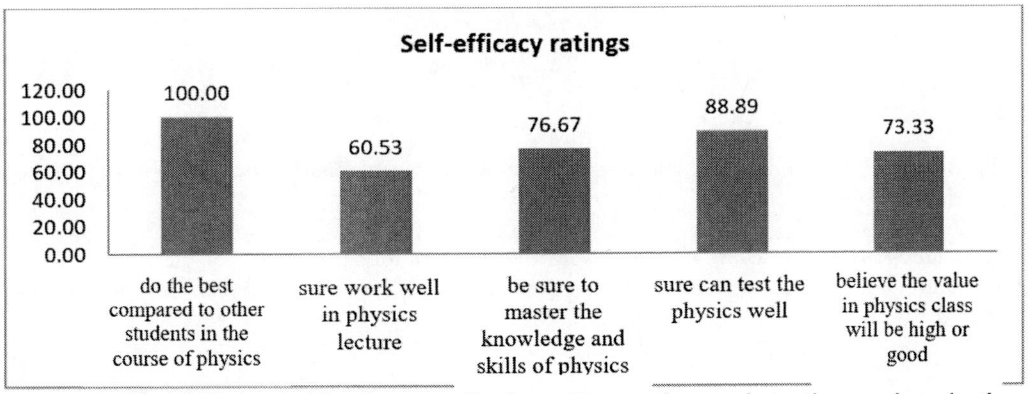

Figure 7. The ability to self-assessment Students Prospective mathematics teachers in the course Physics

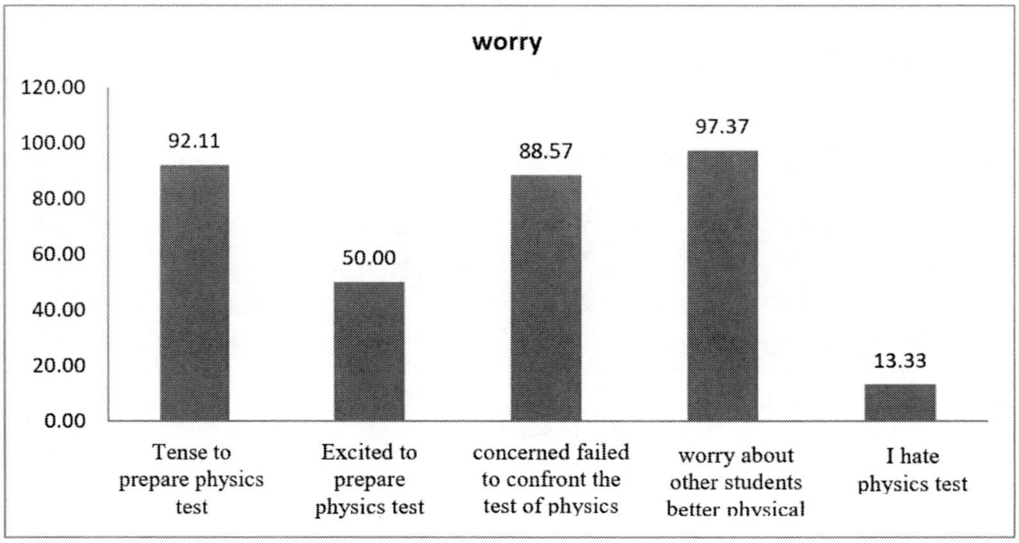

Figure 8. Prospective students anxiety level mathematics teacher in the lecture physics

3.1. Learning Obstacle Analysis Class Basic Physics Candidate for Math Teachers in Unswagati

3.1.1. Background Education Students Prospective math teachers. Questionnaire results of the student teachers about math education background can be seen in figure 9 below.

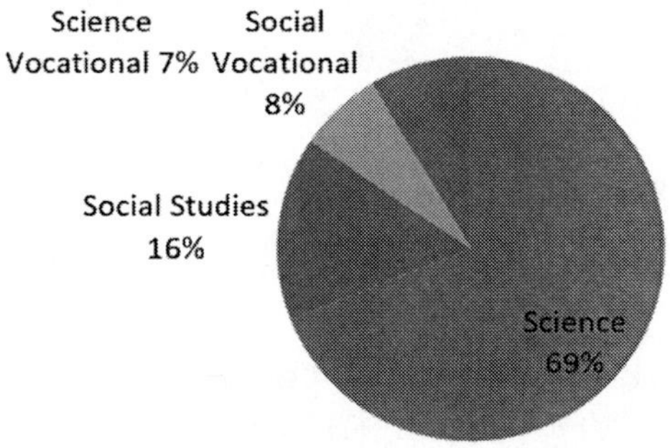

Figure 9. Profile education background mathematics student teachers in FKIP Unswagati Cirebon

3.1.2. Opinions on for Un-fun basic physics class. Questionnaire results of the student teachers of mathematics into the causes unpleasant subject of basic physics lectures can be seen in Figure 10 below.

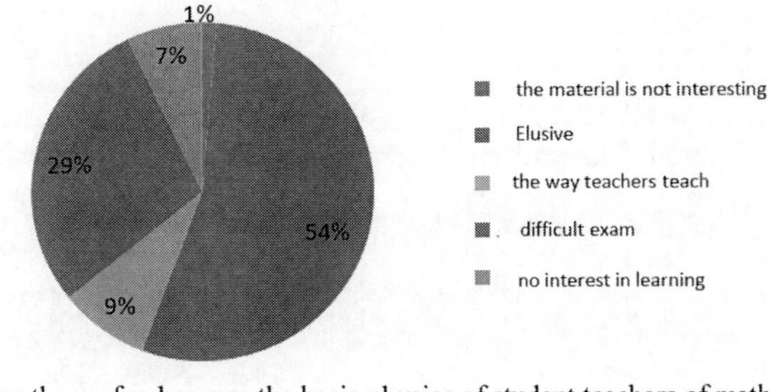

Figure 10. Opinions on the un-fun because the basic physics of student teachers of mathematics

3.1.3. Opinions on because he did not like the basic physics Class. Questionnaire results of the student teachers disliked math as to the cause of basic physics lectures can be seen in figure 11 below.

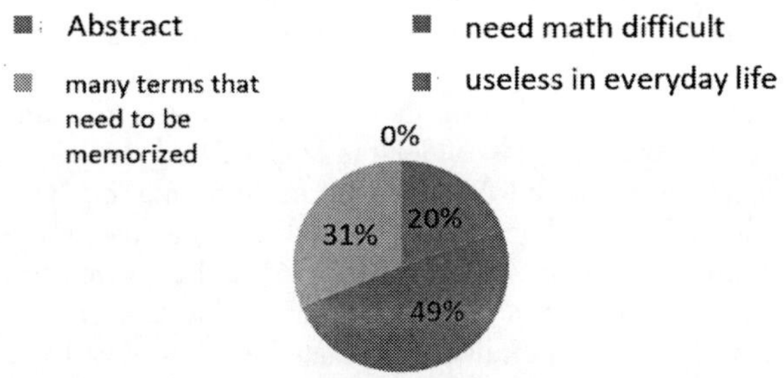

Figure11. Opinions on because he did not like the basic physics of student teachers of mathematics in FKIP Unswagati Cirebon

3.1.4. Assessment Lecturer of Physics Students on the basis of the results of Questionnaire. Rating student teachers of mathematics to physics lecturer base can be seen in Figure 12 below

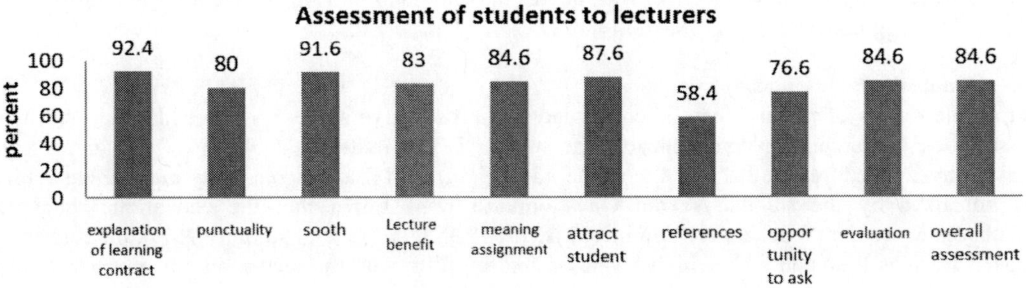

Figure 12. Assessment of student teachers of mathematics to the basic physics lecturer at FKIP Unswagati Cirebon

Figure 13. Research Documentation

4. Discussion

From the analysis of the motivation to study physics for student teachers of mathematics known to turn out students have high levels of anxiety following the lecture in physics. Most students expressed concern that high is because the material is quite dense and system administration tasks to students were pretty much out of textbooks.

Data answer the questionnaire results showed that student teachers of mathematics do not like basic physics lectures because the subject is less understandable (54%). in addition to the basic physics requires mathematical considered difficult (49%) are increasingly making basic physics is not liked by the students. From Questionnaire was known that the educational background mathematics student teachers are dominated by the IPA (69%), social studies (16%), the rest are from SMK.

It is considered to be pleasant from basic physics lecture is a lecture is very applicable for exposing the problem and the phenomenon of natural events that day - the day we can find (39%). Physics lectures an important basis for the provision of prospective teachers of mathematics, and they shall be aware of it. how the processes, products and attitude are supplied to them through this course will be an advanced provisions that are very useful for their future. Osborne and Freyberg (1985: 91) reveals that one of the roles that science teachers as a motivator (motivator). A teacher should provide motivation and instill the perception to students that learning is something that is needed to achieve success in the future not compulsion. Therefore, the learning activities in the classroom must be made so that students are motivated to learn, for example by the method of teaching a fun, exciting games, because it is known that students generally like a game that uses the process of thinking (66.7%). Performance Assessment of lecturers by students appears that the value of the lowest faculty currently on the teaching materials used (58.4%).

5. Conclusion

From the results of research can be concluded that prospective students do not like the math teacher lecture the fundamental physics because the subject is less understandable (54%). in addition to the basic physics requires mathematical considered difficult (49%) are increasingly making basic physics is not liked by the students. From Questionnaire was known that the educational background mathematics student teachers are dominated by the IPA (69%), social studies (16%), the rest are from SMK. Besides It students have high levels of anxiety following the lecture in physics. Most students expressed concern that high is because the material is quite dense and system administration tasks to students were pretty much out of textbooks.

6. References

[1] Abin S M 1998 *Psikologi Kependidikan* (Bandung : PT. Remaja Rosda Karya)
[2] Thomas A *1996 Multiple Intelligences in The Classroom* (Virginia: Association for Supervision and Curriculum Development)

IOP Conf. Series: Journal of Physics: Conf. Series **812** (2017) 012026 doi:10.1088/1742-6596/812/1/012026

[3] Ali I 2010 *Potrait prestasi Belajar fisika SMA pada jurusan teknik gambar bangunan, teknik komputer jaringan, teknik otomotif dan teknik pemanfaatan energi di SMKN X Garut.* (Bandung : Sekolah Pascasarjana UPI)

[4] Dede T K 2013 *Jurnal LOGIKA Unswagati Cirebon* **1**(2) pp 93-104

[5] Dede T K 2015 *Analisis Gaya Belajar (Learning Styles) dan Profil Kecerdasan Majemuk (Multiple Intelligence) Mahasiswa Calon Guru Matematika FKIP Unswagati Cirebon.* Penelitian LEMLIT Unswagati

[6] Dede T K 2015 *Upaya Meningkatkan Penguasan Konsep Dan Keterampilan Proses Sains Mahasiswa Calon Guru Matematika Fkip Unswagati Cirebon Pada Konsep Fluida Statis* Penelitian Lemlit Unswagati 2014

[7] Lazear D 2004 *High order Thunking the multiple intelligence way* (Chicago, USA : Zephyr Press)

[8] Makmun A S 2003 *Psikologi Pendidikan* (Bandung: Rosdakarya)

[9] Ratna W D 1991 *Teori-teori Belajar* (Jakarta : Erlangga)

[10] Sardiman 2005 *Interaksi dan Motivasi Siswa*(Jakarta: Raja Grafindo Persada

[11] Robert E S 2009 *Educational Psychology : Active Learning Edition 12th Edition* (Ohio : Pearson International)

[12] Sumadi S 2004 *Psikologi Pendidikan* (Jakarta: Rajawali Pers)

Biology Students' Initial Mental Model about Microorganism

Y Hamdiyati[1], F Sudargo[2], S Redjeki[2], and A Fitriani[2]
[1]*Biology Education Department, Universitas Pendidikan Indonesia (UPI), Indonesia*
[2]*Graduate School of Science Education, Universitas Pendidikan Indonesia (UPI), Indonesia*

Email: yhamdiyati@upi.edu

Abstract: The purpose of this study was to identify biology students' initial mental model about microorganism. This research used descriptive method with 32 sixth semester biology students at Biology Education Departement-Universitas Pendidikan Indonesia as its respondents. Data was taken at the beginning of the 6th semester before respondents endure microbiology course. Instrument used to assess mental model was drawing-writing test in which it contains concepts such as structure of bacteria, archaea, virus, and fungi. Students were asked to describe their imagination about the structure of microorganisms and subsequently asked to explain the structure of microorganisms in writing through open-ended questions. Students' response was then compared to scientists or experts' mental models as the targeted mental model. Student mental models were categorized into five levels (levels 1-5), namely "there is no drawing/ writing," "wrong or irrelevant drawing/ writing of question," "partially correct drawing/ writing," "the drawing/ writing that has some deficiencies," and "completely correct and complete drawing/ writing." Results showed that the level of mental models through drawing or writing about the four concepts were varied. The highest level of mental models through drawing (D5) was found in the concept of bacteria, while the highest level of mental models through writing (W3) was found in the concept of bacteria, virus, and fungi. Mental model levels most commonly found in each concept through drawing-writing tests (D/ W) were bacteria (D2/ W2), Archaea (D1/ W1 and D2/ W2), virus (D3/ W3), and fungi (D2/ W1). From these results it is advisable to improve lectures and assessment strategy to enhance or complement students' mental models about microorganisms.

Keywords: Mental model, biology students, microorganism, drawing-writing test.

1 Introduction

Microorganism plays a pivotal role for life on earth. Physically and chemically, microorganisms as one of the subjects in microbiology have a central role in biosphere [1]. Two important themes in microbiology are understand microorganism characteristics and function as well as applying these understanding for the benefit of human and earth [2]. Increasing use of microorganisms in health, food, and other industry, as well as in environment and certain aspects of biotechnology make students' knowledge about this science aspect is essential. Genetic engineering and cell culture are closely associated with microbiology because microorganism (especially plasmid) is a tool to insert genes from one organism to another. Therefore, good understanding of microbiology concepts is vital for thoroughly understands these phenomenons.

Research on learners' understanding and knowledge about scientific phenomenons has been done previously and some research has also delved on how learners conceptualize a biological phenomenon, but learners' ideas about microbiology concepts is still understated in the literature [3]. Results from several studies showed that all participants at each education level have different misconceptions about microbes [3,4]. The participants have alternative views about microbes' size and morphology, associate microorganisms with horrible animals and assume that the shape of microorganisms is like that of small animals. The formation of misconceptions or alternative conceptions is undesirable in learning process, because students will draw the wrong conclusions, and it will negatively affect the process of constructing new knowledge. Learners' conception about microbiology and interconnection between concepts can be fully detected by exploring their mental models.

Mental model is a depiction, personal idea, or internal representation of an individual about a phenomenon, a collection of ideas or concepts. Mental model provides an indication of the ideas believed by someone at a particular time, although it is not considered as a static entity. Through the interview and drawing-writing techniques, non-expert groups (students and teachers) already have a complex mental model, but it is not scientifically correct yet. Meanwhile, group of experts discussed more topics [5]. Through free word association test, drawing-writing techniques, and semantic differences attitude scale, Biology education students already have knowledge about some concepts of microorganism but there are concepts that they have not yet mastered, and misconceptions are still occurred[4]. Through free word association test and drawing-writing techniques, biology education students mainly wrote about the definition of a virus and there are still misconceptions about virus [6].

Mental model as a dynamic representation that is never complete, continuously grow and develop along with the increase of information [3]. Mental models are generative because it can lead the student to new information, and they can subsequently use it to predict and explain [7]. Therefore, a teacher or lecturers need to know the initial level of students' mental models before learning process begins in order to create more effective and meaningful learning.

The purpose of this study was to identify students' mental models about the concept of microorganism's (bacteria, archaea, viruses, and fungi) structure. The main goal of exploring mental model is to improve the quality of learning based on the description of students' conception and the internal representation. This study was based on constructivism framework in which learners' ideas about a concept, phenomenon, or event are usually obtained from previous experience both in everyday life as well as in formal occasions.

2 Research Methods

This research used descriptive method. Respondents for this study were 32 sixth semester students in Biology Education Departement at Universitas Pendidikan Indonesia (UPI). Data was taken at the beginning of 6th semester before students endure microbiology course. Students' mental model was probed by using drawing-writing test in which it contains several microorganism concepts such as structure of bacteria, archaea, virus, and fungi.

Table 1. Mental model category for drawing-writing test results.

Level	Statement	Drawing	Writing
Level 1	There is no drawing/writing	D1	W1
Level 2	Wrong or irrelevant drawing/writing of question	D2	W2
Level 3	Partially correct drawing/writing	D3	W3
Level 4	The drawing/writing that has some deficiencies	D4	W4
Level 5	Completely correct and complete drawing/writing	D5	W5

Adapted from Yayla and Eyceyurt [8].

Students were asked to describe their imagination about the structure of microorganisms, and they subsequently asked to explain the structure of microorganisms in writing through open-ended questions. Students' answer was then compared with scientists' mental model as the targeted mental model. Student mental models are categorized into five levels [8]. Five levels of mental models for each concept (structure of bacteria, archaea, virus, and fungi) are illustrated in Table 1.

3 Results and Discussion

Student mental model on microorganism structure (bacteria, archaea, virus, and fungi) through drawing and writing tests are depicted in Table 2 and 3. Students' mental models through drawing test (D) show a higher level than writing test (W) except in structure of a virus concept. Student learning experiences during junior high school as well as in prerequisite courses for microbiology always used pictures to explain about microorganism. It appears that through an image, a mental model about microorganisms is easily represented. Image could represent a model expressed by a student about a particular phenomenon that exist in the public domain [3]. Moreover, in terms of time, drawing is much easier to use. Drawing techniques are also very good tools in collecting data from various perspectives. Furthermore, data in a form of image is also easier to internalized [3]. Figure 1-4 shows students' mental model about microorganisms found during this study. Figure 1 illustrates students' mental models about bacteria were at D2 to D5 level. Figure 2 depicts student mental models about Archaea were at D2 and D3 level. Figure 3 illustrates student mental models about virus at D2 and D3 level. Figure 4 depicts students' mental models about fungi at D2, D3, and D4 level. Students' mental models about microorganisms (writing test) are illustrated in Table 4.

Table 2. Students Mental Model about Bacteria and Archaea structure.

	Bacteria					Archae				
	W1	W2	W3	W4	W5	W1	W2	W3	W4	W5
D1	-	1	-	-	-	11	1	-	-	-
D2	2	11	-	-	-	4	11	-	-	-
D3	4	7	4	-	-	4	1	-	-	-
D4	-	-	2	-	-	-	-	-	-	-
D5	-	-	1	-	-	-	-	-	-	-

D = Drawing, W = Writing

Table 3. Students Mental Model about Virus and Fungi structure.

	Virus					Fungi				
	W1	W2	W3	W4	W5	W1	W2	W3	W4	W5
D1	1	-	1	-	-	2	-	-	-	-
D2	-	1	2	-	-	9	7	1	-	-
D3	3	4	20	-	-	1	8	3	-	-
D4	-	-	-	-	-	-	1	-	-	-
D5	-	-	-	-	-	-	-	-	-	-

D = Drawing, W = Writing

Based on Table 2 and Table 3, one student reached D5 mental model for structure of bacteria concept, one student achieved D4 mental model for fungi concept, although from four concepts about microorganism structure, mental models were mostly at D3 (partially correct drawing) level. It means that for drawing test, only one student achieved the targeted mental model (in line with experts or scientist's mental model) while mental models for writing were still in D3 level. Compatibility between drawing and writing level were found in structure of a virus mental model (D3/ W3, 20 respondents). This result suggested that good image representation does not reflect equally good verbal representations.

The highest mental model level from writing test (W) is found at W3 level for structure of bacteria, virus, and fungi concept, while mental models for structure of Archaea were still at W2 level. These results indicated that verbal representation is harder to store in long-term memory compared when it is in a form of image.

Based on Table 2 and Table 3, the levels of student mental models for structure of bacteria and fungi concepts were higher. This was because students have been studying these concepts since junior high school. In high school, those concepts were discussed in an even more comprehensive manner as a form of compliance to the curriculum implemented in Indonesia. In college, students are getting the information about the structure of bacteria in general biology and cell biology courses. The structure of fungi has been studied in more detail in Botany Cryptogame course, while archaea and virus were studied in 10th grade [9]. None of the students could describe or write characteristics of archaea cell walls, because they did not obtain this information in high school. In prerequisite courses (general biology and cell biology), they also did not get information about archaea. Similar case were also found for virus concept.

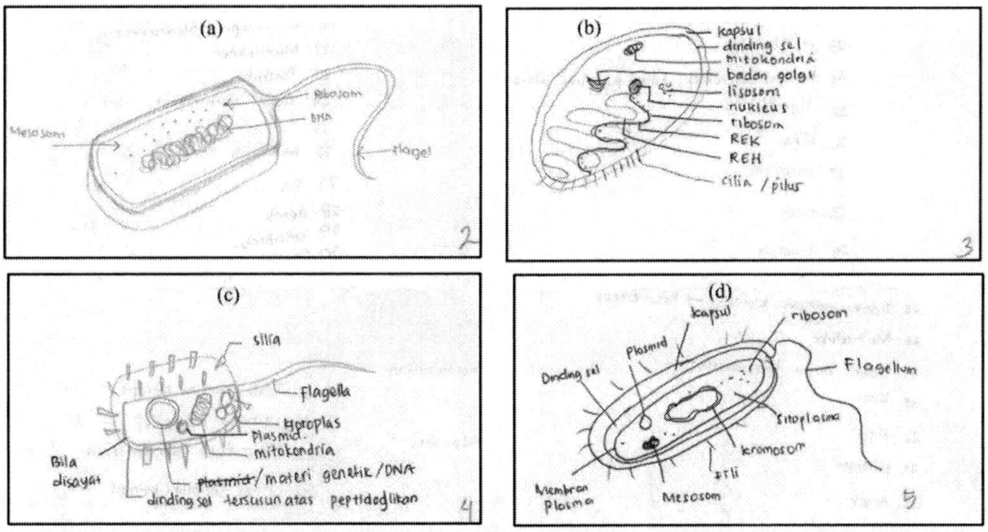

Figure 1. Students mental model about bacteria structure, (a) irrelevant drawing/D2, (b) partially correct drawing/D3, (c) drawing that has some deficiencies/D4, and (d) completely correct and complete drawing/D5.

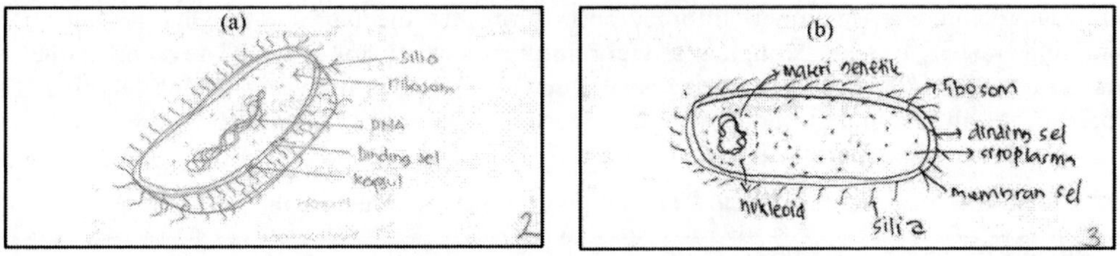

Figure 2. Students mental model about archaea structure, (a) irrelevant drawing/D2 and (b) partially correct drawing/D3.

Figure 3. Students mental model about virus structure, (a) irrelevant drawing/D2, and (b) partially correct drawing/D3.

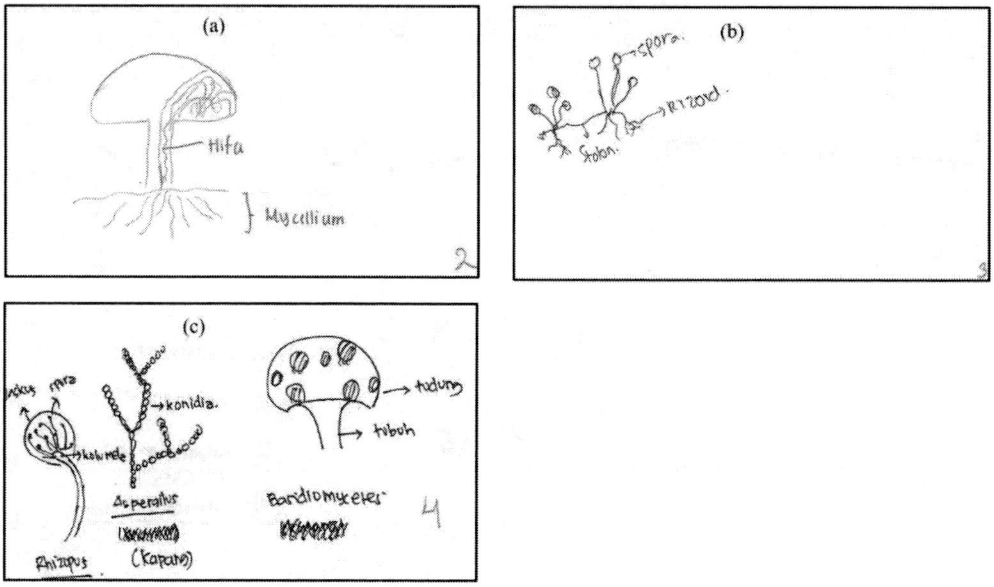

Figure 4. Students mental model about fungi structure, (a) irrelevant drawing/D2, (b) partially correct drawing/D3, and (c) drawing that has some deficiencies/D4.

All of the students choose to describe bacteriophage virus structure. Similarly, most students choose to depict Basidiomycetes fungi (56%) while the remaining students choose to illustrate molds. There was not even one student decided to describe yeast. Students' initial mental model about microorganisms is strongly influenced by the process of inserting new information into existing schemas. Initial mental model was resulted from mental model evolution in which prior knowledge converges with new information to form new mental models. This mental model-building process can be through a model-reinforcement, elaboration and revision model, or model-rejection [10]. Drawing-writing is an efficient technique to reveal student's learning. This technique is an important technique because it enables researchers to collect data about learners' mind, understanding, perspectives and attitudes [4].

Table 4. Students mental model about microorganism structure through writing test.

Microorganism	Writing
Bacteria	Bacteria is a unicellular organism with peptidoglycan cell wall (W2). Bacteria is microorganism with peptidoglycan cell wall, do not have

	core membrane, and has a circular DNA (W3).
Archaea	Archaea has similar characteristics with bacteria (W2).
Virus	Virus consisted of capsid (W2). Virus has DNA/RNA protected by capsid (W3).
Fungi	Fungi has a cap forming-hypha structure and do not have chloroplast (W2). Fungi is a eukaryotic organism, do not have chloroplast, consisted of hypha, and this hypha formed a structure called mycelium (W3).

New ideas are assimilated or accommodated into the existing scheme by creating a new scheme. This may cause cognitive dissonance if new ideas are scientifically accurate but are not consistent with the scheme that learners have, in which rejection of new ideas may also occur. When new ideas are accepted and connected with existing ideas, learning occurs by increasing and improving learners' mental models, although this is of course will take time [3]. Based on research results, it is advisable to develop a strategy to correct or improve student mental models about microorganism so that learners' mental models will be in compliance with targeted mental model. Course's strategy development also needs to be supplemented with appropriate assessment to encompass students' mental model.

4 Conclusion

Results showed that the level of mental models through drawing or writing about four microorganism concepts were varied. The highest level of mental models through drawing test (D5) was found in the concept of bacteria, while the highest level of mental models through writing test (W3) was found in the concept of bacteria, viruses, and fungi. Mental model level most commonly found in each concept through drawing-writing test (D/W) were bacteria (D2/W2), Archaea (D1/W1 and D2/ W2), virus (D3/ W3), and fungi (D2/ W1). From these results it is advisable to improve lectures and assessment strategy to enhance or complement students' mental models about microorganisms.

5 References

[1] Schaechter M, Kolter R and Buckley M 2004 *Microbiology in the 21st Century: Where Are We and Where Are We Going?* (South Carolina: American Academy of Microbiology Charleston)

[2] Madigan MT, Mantinko JM, Bender KS, Buckley DH and Stahl DA 2015 *Brock Biology of Microorganisms* Fourteenth ed. (Prentice-Hall International Inc.) p 2

[3] Byrne J 2011 Models of Micro-Organisms: Children's knowledge and understanding of micro-organisms from 7 to 14 years old *International Journal of Science Education*

[4] Kurt H 2013 Turkish student biology teachers'conceptual stucture and semantic attitudes towards microbes *Journal of Baltic Sience Education* **12** 5

[5] Jee BJ, Uttal DH, Spiegel A, and Diamond J 2013 Expert–novice differences in mental models of viruses, vaccines, and the causes of infectious disease. *Public Understanding of Science* (PUS)

[6] Kurt H and Ekici G 2013 What is a virus? Prospective biology teacher'cognitive structure on the concept of virus *Internasional Online Journal of Educational Sciences* **5** 3

[7] Wiji 2014 Pengembangan Desain Perkuliahan Kimia Sekolah Berbasis Model Mental untuk Meningkatkan Pemahaman Materi Subyek Mahasiswa Calon Guru Kimia *Doctoral Thesis* Sekolah Pasca Sarjana UPI Bandung Unpublished

[8] Yayla RG and Eyceyurt G 2011 Mental Models of Pre-service Science Teachers about Basic Concepts in Chemistry *Western Anatolia Journal of Educational Science (WAJES).* Selected Papers Presented at WCNTSE Dokuz Eylul University Institute, Izmir, Turkey ISSN 1308-8971

[9] Minister of Education Regulation No. 64 Year 2013 about *Content Standard* (Permendikbud Nomor 64 tahun 2013 tentang Standar Isi)

[10] Buckley BC and Boulter CJ 2000 Investigating the role of representations and expressed model in building mental models In Gilbert, J.K. & Boulter C.J. (Ed.) *Developing models in science education* (Dordrecht:Kluwer Academic Publishers)

IOP Conf. Series: Journal of Physics: Conf. Series **812** (2017) 012028 doi:10.1088/1742-6596/812/1/012028

Comparing Science Virtual and Paper-Based Test to Measure Students' Critical Thinking based on VAK Learning Style Model

T H Rosyidah[1], H Firman[2] and L Rusyati[1]

[1] Department of international Program on Science Education, Universitas Pendidikan Indonesia, Jl. Dr.Setiabudi No. 229, Bandung 40154, Indonesia
[2] Department of Chemistry Education, Universitas Pendidikan Indonesia, Jl. Dr.Setiabudi No. 229, Bandung 40154, Indonesia
Email : tiyas.rosyidah@gmail.com

Abstract. This research was comparing virtual and paper-based test to measure students' critical thinking based on VAK (Visual-Auditory-Kynesthetic) learning style model. Quasi experiment method with one group post-test only design is applied in this research in order to analyze the data. There was 40 eight grade students at one of public junior high school in Bandung becoming the sample in this research. The quantitative data was obtained through 26 questions about living thing and environment sustainability which is constructed based on the eight elements of critical thinking and be provided in the form of virtual and paper-based test. Based on analysis of the result, it is shown that within visual, auditory, and kinesthetic were not significantly difference in virtual and paper-based test. Besides, all result was supported by quistionnaire about students' respond on virtual test which shows 3.47 in the scale of 4. Means that student showed positive respond in all aspet measured, which are interest, impression, and expectation.

1. Introduction

Nowadays, educational world has been try to emphasize on digital multimedia known as Technology-based assessment (TBA).. Soon TBA will replace paper-based testing, extend business and substance of assessment in education[1]. Parallel to these tendencies, paper-based assessment reached its limits[2]. Further development such as reduction of costs, logistic and feedback time is unexecutable with paper-based test tools[3]. Supporting the information about computer based test, the creation of a technology-enriched classroom appears to have a positive effect on student acquisition of higher-order thinking skills[4]. Critical thinking as one of higher order thinking skill can be simply put in contrast to illogical or irrational ways of thinking[5]. However, it cannot be equated with argumentative types of thinking or making criticisms[6]. Critical thinking further involves reflective types of thinking, that is thinking about activities[7]. Refers to that, critical thinking was necessary to be brought as a skill that possible to be measured in implementation od computer based test. Other thing which also interesting to discuss is a finding which stated that male and female students have significantly different learning styles. It was the responsibility of the instructor to address the diversity of learning styles and develop appropriate learning approaches[8].

2. Research method

The method used in this research was experiment method which is pre-experiment methods specified into one group postest only design. Firstly, the data about students' learning style was obtained through VAK (Visual-Auditory-Kynesthetic) learning style inventory by Victoria Chislett. The learning style inventory was a set of multiple choice questions with three options where each option indicated one kind of learning style. Further, one group of student was given set of question in form of paper-based test. In the next following weeks, the students were given set of questions provided in science virtual test. This method was apropriate with the objective of the research which is comparing vitual test and paper-based test to measure student critical thinking on living thing and environmental sustainability.

3. Result and discussion

3.1. Comparison between science virtual test and paper-based test

The data about general comparison between science virtual test and paper-based test is obtained by implementing those two tests on the population. As it is analyzed using statistical analysis, both science virtual test and paper-based test are shown that the data is not homogen as well it doesn't distribute normally. It can be indicated from *Asymp. Sig* which is less than 0.05. Refers to that, there should be nonparametric statistical analysis used. This research is using *Mann-Whitney (U)* in order to analyze the data.

Table 1. Comparison between science virtual and paper-based test.

Element of Critical Thinking	Science virtual Test		Paper-based Test		Mean diff.	*U*	*Asymp.* Sig
	Mean	SD	Mean	SD			
Element 1	82.05	20.64	60.00	25.82	22.05	412.0	0.000*
Element 2	49.36	21.06	45.00	24.81	4.36	743.0	0.563
Element 3	82.05	26.87	28.75	31.79	**53.30**	2085	0.000*
Element 4	54.70	22.28	47.50	32.80	7.20	751.0	0.606
Element 5	87.82	18.02	63.75	26.52	**14.07**	385.0	0.000*
Element 6	73.68	26.22	55.83	24.33	17.85	512.0	0.003*
Element 7	76.92	27.74	49.16	30.18	27.76	403.5	0.000*
Element 8	78.63	24.76	46.63	33.59	32.00	409.0	0.000*
Average	73.15	23.45	49.58	28.73	23.57	6.000	**0.006***

* significant differences
Green mark : most different
Orange mark : less different

From Table 1, it can be infered that the mean difference between science virtual test and paper-based test is all show positive value. Means that, science virtual test has higher result compare to paper based test. Specifically, the most different is showed in element three which is about "assumption". While the less difference is showed in element five which is about "information". From eight elements of critical thinking, there are six elements which shows that science virtual test is significantly different compare to paper-based test where it is signed by *Asymp Sig.* which is less than 0.05. Meanwhile, the other two elements showed that there is no significant difference in sciene virtual test compare to paper-based test where it is signed by Asymp. Sig which is greather than 0.05.

One potential limitation for realizing the benefits of computer-based assessment in both instructional assessment and large scale testing comes in designing questions and tasks with which computers can effectively interfacee. It was also paralel with the implementation of computer based test in Indonesia. Science virtual test is available to be implemented in order to improve the technology based assessment. It is available to make the assessment process becoming more effective as it was paperless, costless, and drive the student to enjoy the test more compare to conventional testing known as paper and pencil test. Besides, the creation of a technology-enriched classroom appears to have a positive effect on student acquisition of higher-order thinking skills[4]. As one of higher order thinking skill, students' critical thinking which is measured in this research by using science virtual test is significantly higher than what is resulted on paper-based test.

On the other hand, the implementation of science virtual test is dependently require such kind of supported technology included laptop, personal computer, LCD Projector, and suitable audio speaker or earphone. The absence of technology absolutly influence the implementation of science virtual test especially in most of Indonesian public school. But anyway, the clarity of test item is still fulfilled by the direct visualization of certain concept regarding living thing and environmental sustainability.

Never the last, there are two elements of critical thinking which showed that there is no significant difference between science virtual and paper-based test. Those are about "question at issue and concept" critical thinking elements. Refers to the result, students' skill regarding questioning was not dependent on technology. In addition, King on the study about effects of self-questioning training on college students' comprehension of lectures stated that teacher should also engage the students to provide a learning environment which enable the student to raise their questioning skill thus their critical thinking especially question at issue element will be improved[10].

The result also showed that there is no significant difference between science virtual test and paper-based test regarding the element of concept. Concept has strong relation to cognitive perspective. Cognitive perspectives drives the students to set certain concept which is required to be understood and criticized through questionning the exists concept. In addition, student willingness to gain their concept knowledge has also expected to be increased as the pre requisite on critisize certain concept[6]. Paralel to that, the psychologist Albert Bandura discovered the importance of behavioral models when he was working with patients with snake phobias. Observational learning could not occur unless cognitive processes were at work[11]. Therefore, improvement of observational experience is required in order to increas students' critical thnking especially in the element of concept.

Other than that, education must be started to act on a new strategies as 21st century educational demand, one of them is through innovation on leaning. It drives the student to improve their learning behavior. Learning behavior proceed could change the assumption, belief, and habit[12].

3.2. Comparison between science virtual and paper-based test based on vak lerning style

The data about the effect of learning style in science virtual test and paper-based test is analyzed by mean of investigating the difference within visual, auditory, and kinesthetics in each element. The statistical analysis in this part has been conducted through anova since the variable which is compared are more than two. Besides, the result obtained for the whole element is also investigated in order to know whether there is a diffenrence within visual, auditory, and kinesthetics for the total element as cn be seen in Table 2. The same analysis is also implemented for paper-based test in order to find out the comparison between both test.

Table 2. Comparison between virtual and paper-based test based on vak learning style.

Test	Element of critical thinking	Visual		Auditory		Kinesthetic		F	Asymp. Sig
		Mean	SD	Mean	SD	Mean	SD		
Science Virtual Test	Element 1	80.88	18.81	90.62	12.93	80.00	25.36	0.799	0.466
	Element 2	48.53	20.67	59.38	22.90	46.67	20.85	1.011	0.374
	Element 3	85.29	23.48	93.75	17.68	73.33	31.99	1.756	0.187
	Element 4	60.79	21.20	45.83	17.25	51.11	24.77	1.491	0.238
	Element 5	86.75	19.99	90.63	12.93	86.67	18.58	0.146	0.865
	Element 6	70.59	26.04	63.33	31.27	80.00	24.56	1.116	0.338
	Element 7	78.43	26.19	87.50	17.25	71.11	33.02	0.932	0.403
	Element 8	82.35	23.91	75.00	29.55	71.11	24.77	0.676	0.515
	Average	74.20	22.54	75.75	20.22	70.00	25.49	0.305	**0.740**
Paper-Based Test	Element 1	55.82	26.43	59.38	22.90	61.67	28.14	0.490	0.952
	Element 2	39.71	23.48	37.50	23.14	55.00	25.35	2.081	0.780
	Element 3	32.35	35.09	25.00	37.29	26.67	25.82	0.189	0.829
	Element 4	50.98	35.59	45.88	35.36	44.45	29.59	0.164	0.850
	Element 5	67.65	26.70	59.38	22.90	61.67	29.68	0.327	0.723
	Element 6	64.70	21.96	45.83	30.54	51.11	21.33	2.220	0.123
	Element 7	49.02	35.59	49.99	25.19	48.88	27.79	0.004	0.996
	Element 8	45.09	40.78	54.16	24.80	44.44	29.99	0.241	0.787
	Averge	50.665	30.70	47.14	27.77	49.24	27.21	0.186	**0.832**
OVERALL COMPARISON								14.0	**0.059**

* significant differences
Green mark : most different
Orange mark : less different

From Table 2, it can be inferred that within visual, auditory, and kinesthetic there is no element which showed that those three group has significant difference in each element as well as what are showed in the result showe in each element on paper-based test. The same as in the total comparison where it is showed that there is no significant difference within visual, auditory, and kinesthetic group. Specifically, the most different showed in science virtual test is in element three which is about "assumption", while the less difference is showed in element five which is about "information". Hence, the most different showed in paper-based test is in the element one which is about "purpose", while the less difference is showed in element six which is about "concept".

Learning style is a component of the wider concept of personality. McAdams and Pals (2006) offer a five-principle model of the whole person that encompasses evolutionary design for human nature, dispositional traits, characteristic adaptations, self-defining life narratives, and socio-culture contexts[13]. Learning style falls into the categories of dispositional traits and characteristic adaptations where there are differences across individual humans but there are groupings of humans who have common or similar learning style[14]. Other than that, Gender, competitiveness, and the familiarity were not related to the students' performance in both science virtual test and paper-based test[15]. Science virtual test could still implemented for all students regardless those factors. Specifically, the research showed that there is no difference between students' critical thinking between science virtual test and paper- based test measured based on VAK Learning style. Means that science virtual test can be imlemented for student with any kind of learning style. It might be because science virtual test provides such kind of media which was available for visual, auditory and kinesthetic group[16]. It was in line with research done by Myers and Dyer in the research entitled "The Influence of Students' Learning Style on Critical Thinking Skill" showed that Males and females in this study possess similar levels of critical thinking skills[16].

4. Conclusion

The research about comparing science virtual test and paper-based test to measure students' critical thinking based on living thing and environmental sustainability has been conducted systematically based on the result and discussion. Science virtual test shows higher score compare to paper-based test score in most elements of critical thinking, there are only two elements shows insignificant result of comparison. Hence, the total comparison shows that science virtual test is significantly difference which is higher than paper-based test. Besides, comparison between science virtual test and paper-based test do not influenced by the effect of learning style, and students' experience. It is available to be implemented in all students regardless the classification of those three aspect.

5. References

[1] Bandura A 2007 *Journal of Social Clinical Psychology* **26** 641-58
[2] Bennett R E 2002 *Journal of Technology, Learning and Assessment* **1**
[3] Clariana R and Wallace P 2002 *British Journal of Educational Technology* **33** 593-602
[4] Chislett V and Chapman A 2005 *Journal of Research on Technology in Education* **34** 107-19
[5] Dantas W M 2002 *SYSTEM ELSEVIER Journal* **30** 543-555
[6] DuFour R 2010 The role of professional learning communities in advancing 21st century skills *21st Century Skills: Rethinking How Students Learn,* ed Bellanca J and Brandt R (Bloomington: Solution Tree Press) pp 77-95
[7] Facione P A 2011 *Critical Thinking: What It Is And Why It Counts* (San Jose: Insight Assessment - A division of California Academic Press) pp 1-30
[7] Hawk T F and Shah A J 2007 *Decision Sciences Journal of Innovative Education* **5** 1-19
[8] Hopson M H, Simms R L and Knezek G A 2004 *Journal of Research on Technology in Education* **34** 109-19
[9] King A 2002 *Contemporary Educational Psychology Elsevier* **14** 366-381
[9] Lau J 2009 *A Mini Guide to Critical Thinking* (Department of Philosophy, The University of Hong Kong) pp 1-23
[10] Letourneau D K et al 2011 *Ecological Applications* **21** 9-21
[11] McAdam and Pals E 2006 *Decision Sciences Journal of Innovative Education* **5** 1-19
[12] Molnar G, Toth K R and Csapo B 2011 *Comparing Paper-Based and Computer-Based Testing in The First Grade* (American Educational Research Association)
[13] Myers B E and Dyer J E 2006 *Journal of Agricultural Education* **47** 43-52
[14] Scalise K and Gifford B 2006 *The Journal of Technology, Learning and Assessment* **4** 5-44

Development of Interactive Learning Media on Kinetic Gas Theory at SMAN 2 Takalar

M Yanti*, N Ihsan, and Subaer

State University of Makassar, Department of Physics.
Jl. Dg. Tata Raya. Kec. Tamalate, Makassar Sulawesi Selatan. 90224.

Email: meiliiyantii@gmail.com

Abstract. Learning media is the one of the most factor in supporting successfully in the learning process. The purpose of this interactive media is preparing students to improve skills in laboratory practice without need for assistance and are not bound by time and place. The subject of this study was 30 students grade XI IPA SMAN 2 Takalar. This paper discuss about the development of learning media based in theory of gas kinetic. This media designed to assist students in learning independently. This media made using four software, they are Microsoft word, Snagit Editor, Macromedia Flash Player and Lectora. This media are interactive, dynamic and could support the users desires to learn and understand course of gas theory. The development produce followed the four D models. Consisted of definition phase, design phase, development phase and disseminate phase. The results showed 1) the media were valid and reliable, 2) learning tools as well as hardcopy and softcopy which links to website 3) activity learners above 80% and 4) according to the test results, the concept of comprehension of student was improved than before given interactive media.

1. Introduction

Improvement quality of education is an important thing in this country. Specially for human resources which can affect in this case. In addition, facilities and infrastructure for helping learning and teaching process already developed by government. Which aims to make students easy to understand the explanation of their teacher. There are many way to achieve that goals, they are utilization of technology which existence can't be denied. In development of technology, there is no doubt that internet has important role to share the information. Technology can have a reciprocal relationship with teaching [3]. So that they can be a good solution in several aspects of life. So the increase Human Resources can be supported by the ability to use technology wisely, especially in education.

Education world always change in create learning innovation. Especially in develop method, teaching material and the media. Therefore as people who play a role in education have to able in utilize technology to develop learning media system that is interactive. Through this media students can do an active learning and increase student's activity.

Physics is an esoteric knowledge which systematic and unified. Called systematic because one product be related with the other and unified because the product can support each other. Sometimes character of Physics is abstract and concrete. Physics with abstract character is hard to visualized and make student hard in study concept of physics.

This case makes students have assumption that physics is hard and boring. Therefore we need a good approach to solve this problem and good media can help students to develop their ability in physics. Based on previous research, it can be seen that high percentage of students have highly motivated to follow leaning process using media because the media help students to understand concept of physics[1]. The Presence of technology is growing on the classroom. The new generations of student come ready to work with these new technology[2].

For make a good utilization of technology, this study will make a interactive media learning that uses by student and teacher. There are 3 variable measured, validity, learning outcomes and student activity in class during the learning process.

2. Experimental Method

2.1 Research Method

This Study includes Research and Development (R&D)[8]. This study has aims to develop an interactive learning media in Kinetic Gas Theory. The development follow the four D style. They are Define, Design, Development and Dissiminate phase. In Define phase, the data is collected from observe the student, school and teacher. Obtained the characterisics from students that they already be able to use computer and internet technology. Internet have the potential to innovate, accelerate, enrich, and deepen skills, to motivate and engage students, to help relate school experience to work practices [6]. Further in design phase, obtained Study plan, students worksheet and draft media. Study plan and worksheet are be planned from define phase. These pictures below shows the Lesson plan and worksheet that has been developed.

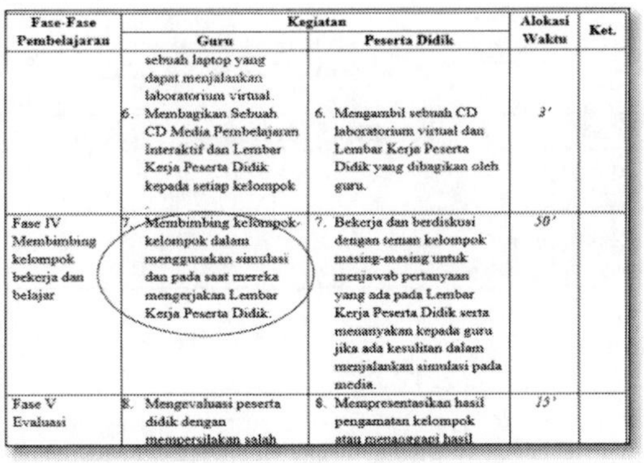

Fase-Fase Pembelajaran	Kegiatan		Alokasi Waktu	Ket.
	Guru	Peserta Didik		
	sebuah laptop yang dapat menjalankan laboratorium virtual. 6. Membagikan Sebuah CD Media Pembelajaran Interaktif dan Lembar Kerja Peserta Didik kepada setiap kelompok	6. Mengambil sebuah CD laboratorium virtual dan Lembar Kerja Peserta Didik yang dibagikan oleh guru.	3'	
Fase IV Membimbing kelompok bekerja dan belajar	7. Membimbing kelompok-kelompok dalam menggunakan simulasi dan pada saat mereka mengerjakan Lembar Kerja Peserta Didik.	7. Bekerja dan berdiskusi dengan teman kelompok masing-masing untuk menjawab pertanyaan yang ada pada Lembar Kerja Peserta Didik serta menanyakan kepada guru jika ada kesulitan dalam menjalankan simulasi pada media.	50'	
Fase V Evaluasi	8. Mengevaluasi peserta didik dengan mempersilakan salah	8. Mempresentasikan hasil pengamatan kelompok atau menanggapi hasil	15'	

Figure 1. Display of Lesson Plan

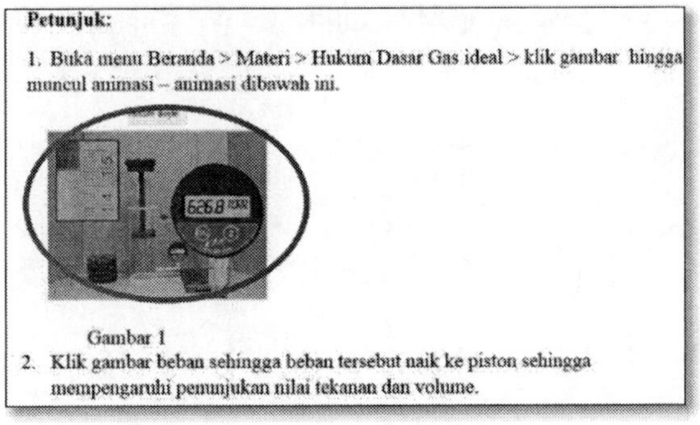

Petunjuk:

1. Buka menu Beranda > Materi > Hukum Dasar Gas ideal > klik gambar hingga muncul animasi – animasi dibawah ini.

Gambar 1

2. Klik gambar beban sehingga beban tersebut naik ke piston sehingga mempengaruhi penunjukan nilai tekanan dan volume.

Figure 2. Display of Worksheet

One step of the development phase is manufacture digital pocket book as a product of this research. Making an application is adapted to the design plan. Firstly be collecting the material that will be used to fill the content in applications, such as materials, images, animations and simulations. From the define phase, materials prepared by the main material. Then the material is divided into several sub-materials, ie characteristics of the ideal gas law - the law of ideal gas , ideal gas equation and the kinetic energy of an ideal gas. After that, media through the validation phase . At this stage the media validated by two people who are members of the Association of Indonesian Education Evaluation.

2.2 Research Instrument

Instrument is evaluating of Learning Media Sheet. This instruments are used to obtain data on the opinions of experts (validator) on the validity of instructional media that have been arranged as the reference/guidance in revising media. Data validation experts reviewed and analyzed the results of experts to study the media. The results of the study were used as input to revise media enhance learning developed. Second Instrument is validation of learning outcomes test. Learning outcomes test instrument used to measure cognitive abilities of students after being taught by using interactive learning media. These tests were conducted at the end of the learning process. And the last instruments is observation sheet of student activities. This instruments are used to obtain data on the activity of students.

2.3 Analysis Data

Learning outcomes data analysis results done simply by using descriptive analysis. After that Categorize the learning outcomes of students based on the interval of values obtained. Observations of Students activities done by giving a score based on the following conditions :

Table 3. Coding Responden

Category	Positive Stat. Score	Negative Statement Score
Very Agree	4	1
Agree	3	2
Less Agree	2	3
Disagree	1	4

3. Result and Discussion

3.1 Digital Pocket Book

Digital pocket book contains several items, such as pictures, animations and interactive simulations that are relevant. Addition of picture and the animation is expected to help students understand the concept of kinetic theory of gases properly. Applications developed starting with the home display and has several items to choose. For examples material, summary evaluation etc. The page is shown in the figure below

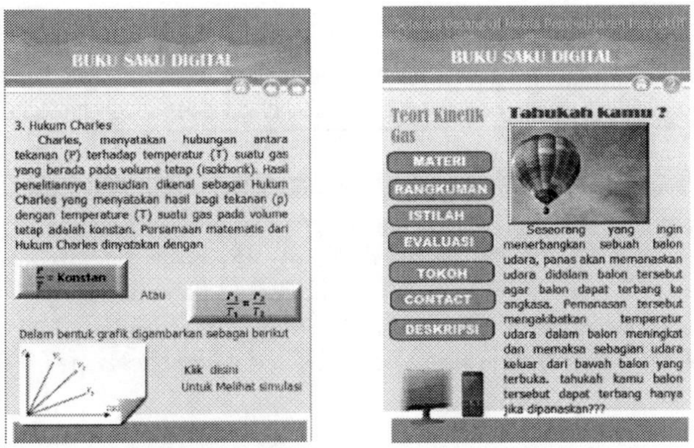

Figure 3. Display of Home menu and material menu

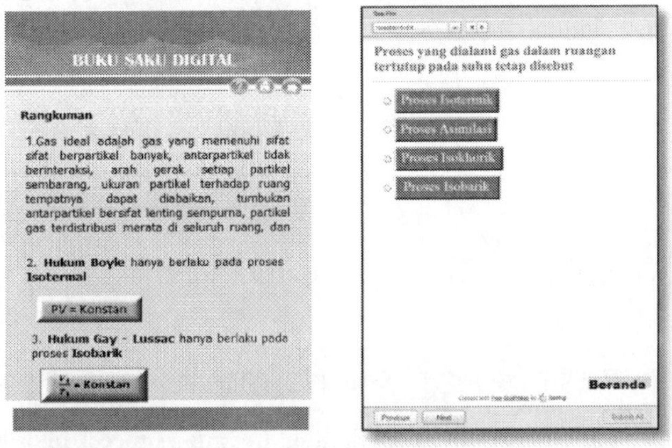

Figure 4. Display of Summary menu and Quiz menu

On the menu the material contains a brief description of the material along with equations and graphs . In addition each end submateri equipped with animation and simulation features that facilitate learners to understand the explanation before. The features shown in the picture (b). On summary and a glossary contains a brief description and a list of terms that are used during the learning takes place. The features shown in the picture (d). Menu evaluation is given exercises in the form of multiple choice questions with the number of questions are 10 items. Before learners were asked to do the exercises, there are some guidance in media. So that student could do their task easily.

3.2 Learning Outcomes

Physics learning outcomes of students taught using interactive learning media can be seen in the table and diagram below

Table 4. Descriptive Analysis of learning outcomes

Statistic	Learning outcomes
Average	10,75
Highest Score	13
Lowest score	5
Ideal score	15
Number of students	40
Data range	8

Number of interval class	5
Length of class	1.6
Standard deviation	2.5
Variants	6.24

Based on analysis result performed by students tend to show that learning outcomes are in the good category. The table above shows that the highest scores were achieved is 13, the lowest score is 5 while the average scores is 10,75. The application of educational technology enhances skills and cognitive characteristics [6]. According to the data conducted there are students who have a high score of learning outcomes and accompanied by high percentage of activities. Based on graph below it can be seen regression analysis between learning outcomes and sudent activities.

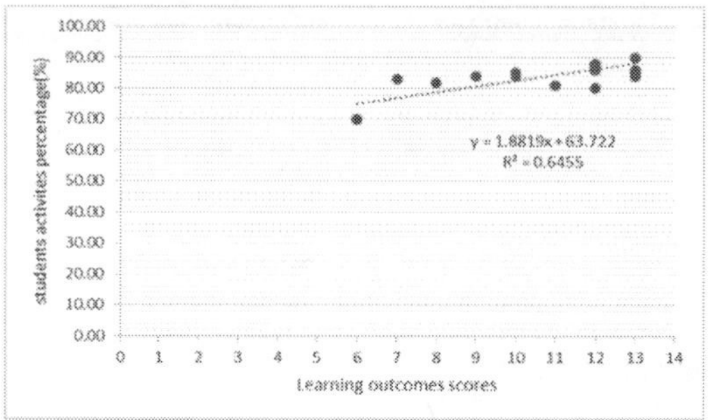

Figure 5. Relationship between Student activities and learning outcomes

The value of R2 is 0,645 it shown that learning outcomes and student activities have relationship. The greater percentage of activities can make greater score in leaning outcomes. The principle, the form of learning activities based on ICT help student in establishing the concepts, procedures, knowledge and express opinion of students in learning. Learning process in oriented by activities as an approach that emphasises the activity of student are optimal to obtain a balanced learning outcomes [7].

3.3 Students Activities

Activities of students observed at each meeting consists of several aspects of the observations. Rating activity of students is done by giving a score to each individual who works in each group. The result of analysis is shown in figures below

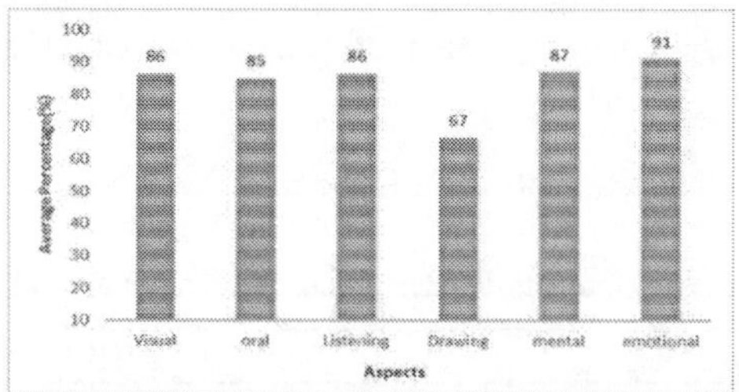

Figure 6. Percentage of Students activities at first meeting

Based on the analysis of the activity of students, found that on any criteria, the majority of learners categorized in very well criteria. All students pay attention to the teacher's explanation at the time of introduction until the end of the lesson. On the visual activities criteria with sub-indicator reading instructional materials and use media, students were active reading instructional materials. On the oral criteria with sub-indicators answered questions and discussion, students were also seen very active in this criteria. They always work together and sometimes forward ideas. At the listening criteria most learners hear an explanation from the teacher and from friends who are expressing their opinions. Often the teacher gives feedback on this activity by giving a question on the side-lines of learning about what has been described before. So it is apparent that learners really listen to what is described by the teacher.

Drawing activities has a lower percentage than the others. This is due to lack of skills of learners in placing the correct scale to create a graph. The second lowest percentage after drawing a graph is doing the calculations. Most learners perform calculations correctly. Including a pair magnitudes asked by the formula. But in kinetic energy matery learners do not understand of using constants. Because the kinetic energy has two constants used, they are R and K. In addition little things like includes units at each end of the answers are rarely done by students.

Based on the analysis, this is in line with the opinion which says that learning with interactive media is able to change principle of learning from teacher center to the student center dynamically [5]. Learning which use student center principle will make concept assignment be better. This type of learning can make students more autonomous and efficient in time, it means that when they need remedial or enrichment their time no depend to the schedule in school[3]. Although the students were active following study, students still need to be familiarized with the learning patterns using media. The aim is constructing knowledge with minimal teacher assistance [9].

4. Conclusion

Based on the data analysis we can conclude that interactive learning media valid and reliable to use as a learning media. The media can be used well in school due to its portable and can be used anytime and anywhere. When the media applied in school, there are two variable have been measured. First, for learning outcomes conducted average score is 10,75 and it means the score is in high category. Second is activity of students with average percentage is 80%. The high percentage is emotional activity. It means that the students interest with the learning process using interactive media.

5. References

[1] Andi M Y 2016 *Jurnal Sains dan Pendidikan Fisika. Jilid* ***12*** (2) 102-06. *ISSN : 1858-330*

[2] Davis N E and Tearle P 1999 *A core curriculum for telematics in teacher training.* Available: *www.ex.ac.uk/telematics.T3/corecurr/tteach98.htm*

[3] Eric K, Scot O, Jennifer G and Jason H 2009 *The Instructional Power of Digital Games Social Networking Simulations*

[4] Hermawanto dkk. *Jurnal Pendidikan Fisika Indonesia* 1693-246

[5] Izuddin S 2012 *Jurnal Pendidikan Vokasi* **2** (2)

[6] Lazar S 2015 *The Importance Of Educational Technology In Teaching (IJCRSEE) International Journal of Cognitive Research in Science, Engineering and Education* **3** (1) 111-4

[7] Sanjaya, W 2008 *Perencanaan dan Desain Sistem Pembelajaran* (Jakarta: Kencana)

[8] Sugiyono 2013 *Metode Penelitian Kuantitatif Kualitatif dan R & D* (Bandung: Alfabeta)

[9] Sukmawati et all. 2016 *Jurnal Sains dan Pendidikan Fisika. Jilid* ***12*** (2) 102-06

Acknowledgments

Praise and great gratitude to Almighty God submitted by the author for blessings and help to complete this paper on time. On this occasion with great humility, I would like to thank you to Headmaster and physics teacher at SMAN 2 Takalar for given me a facility to applied the media. Next for HEPI, validation organization for evaluating the media and test instrument. Then for academic adviser to all and guidance so that paper can be finished

IOP Conf. Series: Journal of Physics: Conf. Series **812** (2017) 012030 doi:10.1088/1742-6596/812/1/012030

A Study of Oil Viscosity Mental Model

Albaiti[1], Liliasari[2]*, Omay Sumarna[2] and Muhamad Abdulkadir Martoprawiro[3]

[1] Graduate School, Universitas Pendidikan Indonesia, Jl. Dr.Setiabudi No. 229, Bandung 40154, Indonesia

[2] Department of Chemistry Education, UniversitasPendidikan Indonesia, Jl. Dr.Setiabudi No. 229, Bandung 40154, Indonesia

[3]Department of Chemistry, Institut Teknologi Bandung, Ganesa Street No. 10, Bandung 40132, Indonesia

*Corresponding author: e-mail : liliasari@upi.edu

Abstract. There is no study regarding on how to learn viscosity of the liquid (e.g. oil) by interconnecting macroscopic, sub-microscopic and symbolic levels. Therefore, the purpose of this research was to study the mental model of the oil viscosity. Intermolecular attractive force of oil constituent on the sub-microscopic level is depicted in the form of mental models. In this research, the viscosity data for some types of oil was measured by using Hoppler method. Viscosity of mineral oil SAE 20W-50, mineral oil SAE 15W-40 and synthetic oil SAE 10W-40 were 1.75, 1.31, and 1.03 Pa s, and the densities of these oils were 908.64, 885.04, and 877.02 kg/m^3, respectively. The results showed that the greater density of the mineral oil that is assumed to be composed of linear chains of hydrocarbons, the longer the chain of hydrocarbon linear. Consequently, thereare stronger the London force and greater the oil viscosity. The density and viscosity of synthetic oil are lower than that of both mineral oils. Synthetic oil structurally forms polymers with large branching. This structure affects a lower synthetic oil viscosity. This study contributes to construct a mental model of pre-service chemistry teachers.

1. Introduction

The property that characterizes a fluid's resistance to flow is its viscosity, η, eta [1].Viscous fluid tends to adhere to solid surfaces that contact with fluid. The fluid that flows easily -such as water or kerosene- has less viscous than thick fluid, such as honey or lubricating oil. One of viscosity concept application is oil lubricating as engine lubricant. The viscosity of oil is considered in reducing the frictional forces generated by the machine moving and contacted each other, to prevent attrition, especially in piston area.According to the oil viscosity, it is known oil identifier code, SAE (Society of Automotive Engineers). The numbers followed that code, showing the level of the oil viscosity. For example lubricating oil with SAE 10W-30, it means that oil has a viscosity SAE 10W in low temperature and SAE 30 in high temperature [2].The greater number that follows to the oil code indicates more viscous the oil. SAE 30 motor oil is more viscous than SAE 10 motor oil [3].

Application of viscosity concept in oil as engine lubricant is very interesting to be used as a context in studying liquid viscosity. Viscosity is one of the topics studied by pre-service chemistry teachers in physical chemistry lecture. So far, there is no study regarding on how to learn viscosity of the liquid (such as oil) by interconnecting the three levels of representation: macroscopic, sub-microscopic and symbolic levels. The ability of students to interconnect the three levels of chemical representation reflects their mental model [4]. An internal representation of cognitive constructed by someone, including a visual-pictorial component and propositional components to understand or to give a rational explanation of a real-world phenomenon, imaginary situations, events, or processes that their structures reflect perceived structure of the situations, events or process is called mental model [5, 6].

Pre-service chemistry teachers are expected to interconnect the three levels of chemical representation in order to have a holistic understanding of the chemistry. In learning chemistry, pre-service chemistry teachers also study physical chemistry. However, the results of the study that conducted by Nyachwaya& Wood showed that most physical chemistry textbooks omitted discussion on how to interconnect mathematical representations to the macroscopic level or sub-microscopic level [7]. The use of multiple representations to interconnect macroscopic, symbolic and sub-microscopic level was very minimal, not more than 1% of the overall picture in the text. It was also found in student's activity in the laboratory in physical chemistry lecture. The lack in using multiple representations is certainly an impact on the construction of student mental models.

Based on the above explanation, it should be an effort how to design a lesson on this topic by interconnecting the three levels of chemical representation. As an initial step, it has been studied viscosity of the liquid mental models, such as oil. Therefore, the purpose of this research was to study the mental model of the oil viscosity. Mental model describes intermolecular attractive force of oil constituent on the sub-microscopic level. This study contributes to construct a mental model of pre service chemistry teachers.

2. Methods

Viscosity data of various types of oil was obtained by using Hoppler method (falling ball method). Materials were mineral motor oil SAE 20W-50, mineral motor oil SAE 15W-40 and synthetic motor oil SAE 10W-40. Apparatus were a 100 ml cylinder, a solid glass ball, a ruler, a stopwatch, a clipper, an analytical balance, a digital thermometer and a pycnometer.

Firstly, the third types of oil measuring in the same temperature. Secondly, the density of solid glass balls and oil was determined. Thirdly, cylinder was marked at a distance of 5 cm from the oil surface. Furthermore, oil was poured into a cylinder. Solid glass ball was dropped into the oil and the transmission time of solid glass ball in a distance of 5 cm from the surface of oil will be recorded. These steps were carried out to a distance of 10 cm and 15 cm from the oil surface.Data obtained from the optimization procedure laboratory were studied descriptively with explanation at the submicroscopic and symbolic levels.

3. Results and Discussion

Table 1 shows the results data in measuring oil viscosity. It was time measured for a solid glass balls in the oils at a certain distance along with its viscosity. The radius of solid glass balls used was 0.78 cm.

Table 1. Time for a solid glass balls in the oils at a certain distance along with their viscosity

Oil SAE	The density of oil (kg m^{-3})	The density of glass ball (kg m^{-3})	The time for ball (seconds)			Oil Viscosity (Pa. s)			The Average of oil viscosity (Pa. s)
			at 5 cm	at 10 cm	at 15 cm	at 5 cm	at 10 cm	at 15 cm	
20W-50 (oil 1)	908.64	2250.629	0.55	0.85	1.5	1.96	1.51	1.78	1.75
15W-40 (oil 2)	885.04	2250.629	0.37	0.70	1.1	1.34	1.27	1.32	1.31
10W-40 (oil 3)	877.02	2250.629	0.30	0.50	0.9	1.09	0.91	1.09	1.03

3.1. Mental Model of Oil Viscosity

Oil can be grouped into two majors, mineral oil and synthetic oil. Mineral oils are the chemical components obtained from crude oil (petroleum crude oil). Crude oil is classified as paraffinic or

Naphthenic. Paraffinic crude oil contains lots of linear and branched hydrocarbons. Naphthenic crude oil contains few of the long-chain "wax-like" hydrocarbon components which carbon atoms linked in rings. These rings are known as cyclo-paraffins. Paraffinic and Naphthenic structure are presented in Figure 1.

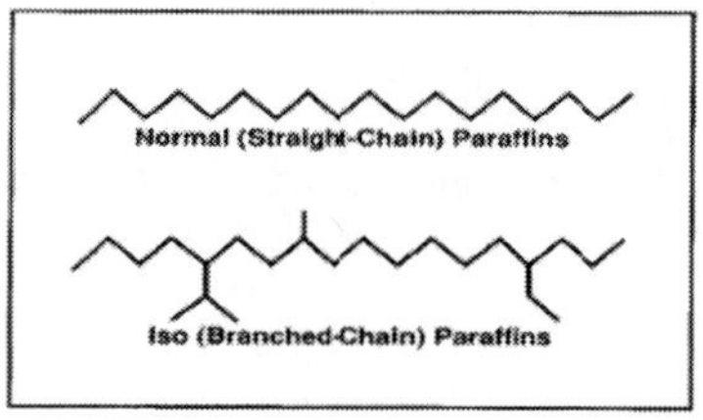

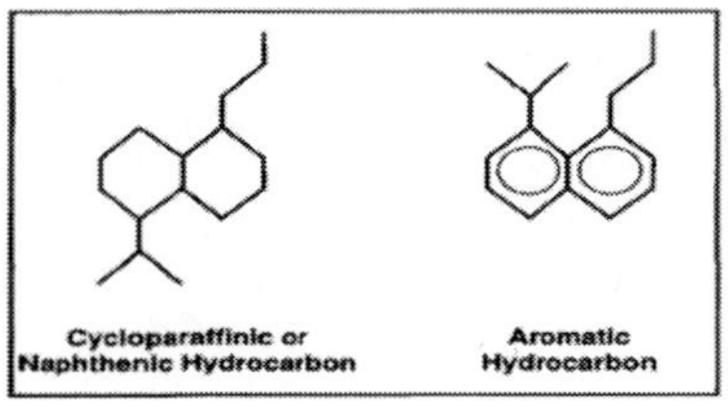

(a) (b)

Figure 1. Structure of paraffinic and naphthenic [8]

According to Figure 1, the structure of oil constituents affects to the stronger of intermolecular forces among oil constituents. The closer distance between molecules so that intermolecular forces of hydrocarbon even greater. Cyclo-paraffinic has stronger intermolecular forces than paraffinic because of its structure. It forms rigid structure. Aromatic compound has more stable structure than others because of electron delocalization.

3.1.1. Mineral Oil SAE 20W-50 and SAE 15W-40

In macroscopic models mineral oil SAE 20W-50 (oil motor 1) more viscous than oil mineral SAE 15W-40 (oil motor 2). Table 1 shows that oil motor 1 has density (908. 64 kg m-3) greater than oil motor 2 (density was 885 kg m^{-3}). Based on this phenomenon, hydrocarbon constituents in both mineral oil are assumed in the form of linear hydrocarbons. Hydrocarbon constituent mineral oil 1 is longer than mineral oil 2.

Oil lubricating is composed by hydrocarbon C_{20}-C_{70} [9]. The mineral oil SAE 20W-50 is assumed to be composed by linear hydrocarbon dotriacontane, $C_{32}H_{66}$. In sub-microscopic, intermolecular forces occurred between dotriacontane constituents of mineral oil 1 through London forces. The hydrogen atoms of an alkyl group at dotriacontene molecule will interact with hydrogen atoms from other molecules dotriacontane. Visualization intermolecular forces on the sub-microscopic level between dotriacontane molecules through London forces is shown in figure 2.

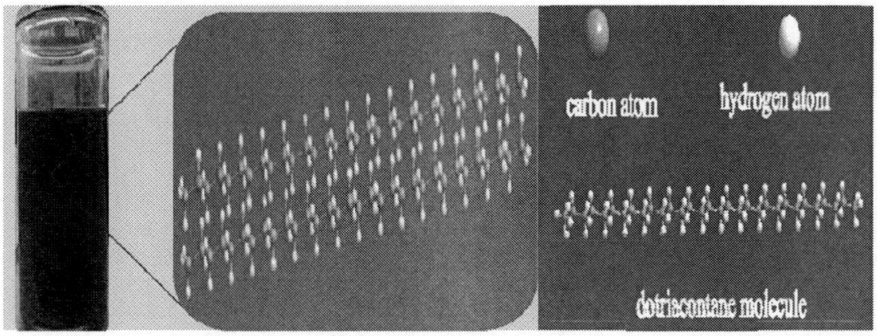

Figure 2. Mental model of mineral oil SAE 20W-50

In the meantime for mineral oil SAE 15W-40, it is composed by linear hydrocarbon hentriacontane, $C_{31}H_{64}$. There are intermolecular forces between hentriacontane molecules through London forces too. The hydrogen atoms of an alkyl group at hentriacontane molecule will interact with hydrogen atoms from other molecules of hentriacontane. Figure 3 shows a visualization of intermolecular forces between hentriacontane molecules through London forces at the sub-microscopic level.

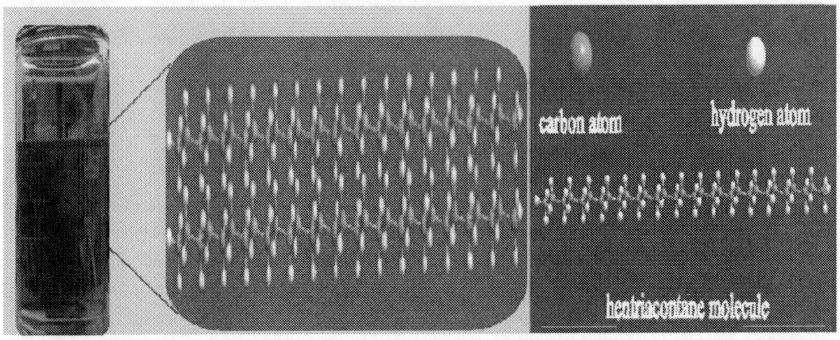

Figure 3. Mental model of mineral oil SAE 15W-40

The longer chain hydrocarbon constituent mineral oil and the closer the distance of the molecules affect intermolecular forces (London forces) even bigger. As a result, the increasing friction occurs between the ball and mineral oil and its viscosity also grow. Viscosity of the both oil is increasing with the increase in the value of SAE. It was supported by the average viscosity data obtained in this experiment. Viscosity of mineral oil SAE 20W-50 was 1.75 Pa s and 1.31 Pa s for mineral oil SAE 15W-40.

3.1.2. Synthetic Oil SAE 10W-40

Poly-Alpha-Olefins (PAO) are constituent components of synthetic oil that more widely used by industrial synthetic lubricants (oil) in the United States and Europe [8]. Motor-oils today can be designed and manufactured therefore the performance closely matches PAO in most commercially significant in applications [10]. PAO has a double bond and forms polymer. PAO synthesized through two stages (sequence) of linear Alpha-Olefins derived from ethylene. The first stage is the synthesis of oligomers with the aid of a catalyst, for example BF_3. The second stage is the hydrogenation of the double bond in the oligomers in order to produce a more stable structure (not easily oxidized). The polymer formed is estimated to have a relative molecular mass of ≤ 10.000 [11]. Figure 4 shows the structure of PAO [8].

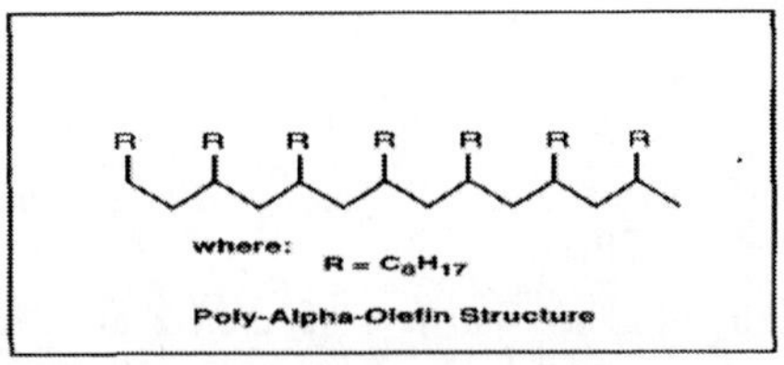

Figure4. PAO Structure

Synthetic oil SAE 10W-40 is less viscous than others. The result of this experiment showed that density of synthetic oil (877. 02 kg m^{-3}) was lower than others. Synthetic oil was assumed contains PAO, Poly-Alpha-Decene. It forms polymer $(C_{10}H_{20})n$. Synthetic oil structurally forms polymers with large branching. This structure affects the decrease of molecule orientation and weaker of intermolecular forces. Finally, the decrease friction occurs between the ball and this synthetic oil. Furthermore, visualization of this molecule structure in the trimmers-form $(C_{10}H_{20})_3$ is presented in Figure 5.

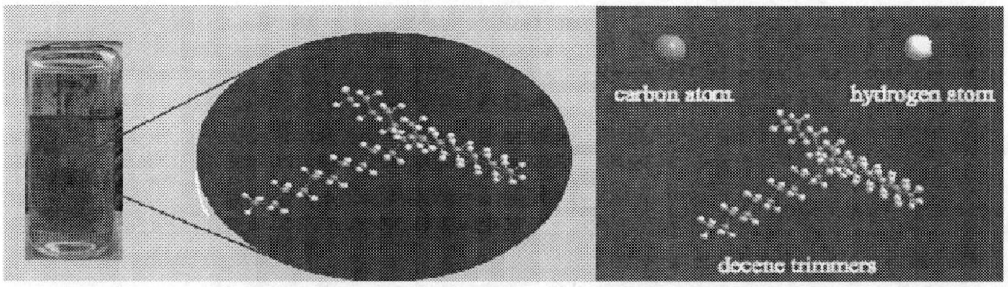

Figure 5. Mental model of synthetic oil SAE 10W-40

The lowest SAE of this synthetic oil than others was supported by the average of viscosity data that obtained in this experiment. It is shown in the table 1. The mental models of various types of oil with different SAE is an internal representation of researchers cognitive in the form of visualization components intermolecular forces that occurs in the oil and propositional components as well as the definition of mental models synthesized from several experts [5, 6]. The components contained in each type of oil are an analogy to explain the observed phenomena. This is agreement with those of Coll's [12] that mental models used to produce a simpler concept, to provide stimulation and support for the visualization and provide explanations for scientific phenomena.

3.2. Different Types Oil Viscosity to the Time Solid Glass Ball at a Certain Distance

Furthermore, the result data was obtained in measuring oil viscosity and the explanation in sub-microscopic level (intermolecular forces oil constituent) can be expressed in mathematical modelling. This graph shows the relationship of the viscosity of various types of oil to the time solid glass ball at a certain distance from the surface of the oil.

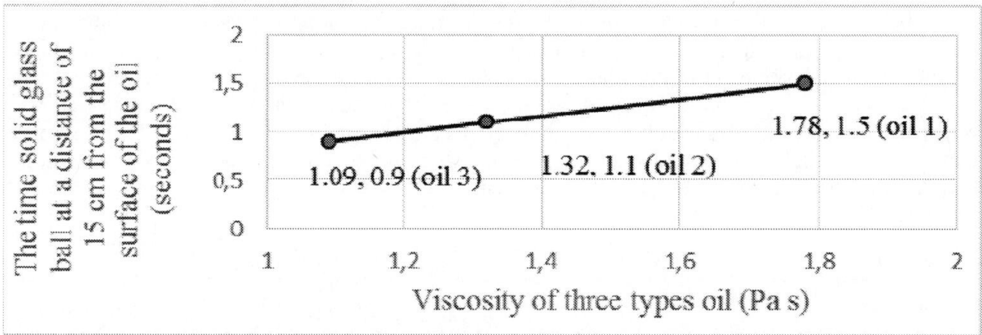

Figure 6. Graph relationship viscosity of various types of oil to the time dropping solid glass ball at a distance of 15 cm from the surface of the oil

Figure 6 shows that viscosity of mineral oil 1, mineral oil 2 and synthetic oil were 1.78 Pa s, 1.32 Pa s and 1.09 Pa s. The time for solid glass ball at a distance of 15 cm from the surface of the oil at the oil temperature of 26.6°C for all three kinds of oil are 1.5 seconds (mineral oil 1), 1.1 seconds (mineral oil 2) and 0.9 seconds (synthetic oil). The biggerviscous of fluids(the fluid more viscous)the greater time of a ball dropping in the fluids. However, the value of the various types of motor oil viscosity in this research is not the same because of differences in the value of the density of motor oil. This suggests that oil constituent and its molecular structure can vary. The density of motor oil SAE 15W-40 by this research at oil temperature 26.6°C is 885.04 kg / m^3 with a viscosity of 1.31 Pa s. The density of motor oil SAE 15W-40 by Viscopedia data at temperature 26.6°C is between 878.7 kg / m^3 and 872.5 kg / m^3withviscosity between 0.15531 and 0.28723 Pa s [13].Meanwhile, motor oil viscosity SAE 40 according to V & P Scientific Inc. is 0.65 to 0.9 Pa s at temperature 21.11°C [14].

The oil temperature in this research was controlled to be the same. The increasing oil temperature affects a number of energy absorbed by the oil so that it can break the interactions that occur in the oil.

Movement of oil particles is more quickly when the temperature is increased. As a result, intermolecular attractive force and friction occurs between the oil and the ball will decrease. Then, oil viscosity will decrease too. Anyway the higher the temperature the smaller liquid viscosity. If the viscosity is smaller, then the time of the glass ball at a certain distance will be smaller. Therefore, it can be concluded that the temperature is inversely proportional to viscosity.

Based on this study, the thickness phenomenon of oil which indicated by the viscosity SAE grades can be explained by using the concept of intermolecular forces and molecular structure. Pre-service chemistry teachers understanding of these concepts may affect to the construction of their mental models about oil viscosity. This opinion is the same as Tumay's [5] that the ability to construct and use mental models can impact the conceptualization learners about chemistry concepts. Therefore, the study of oil viscosity mental models is very useful to the construction of mental models and increased understanding of pre-service chemistry teachers on this matter. In order to conduct knowledge into long-term memory, students should be encouraged to use mental models in order to see how to connect all three levels of representation [15]. The efforts to develop a basic understanding of chemistry concepts, learners should be able to interconnect their understanding of the symbolic level to the macroscopic level and the sub-microscopic level [16, 17]. It certainly needs to be supported by the pattern of teaching facilities are useful for generating discussion connectedness between levels of representation [18].

4. Conclusion

Study of oil viscosity mental models was expressed in visualization of intermolecular forces that occur between oil constituents. Oil that has biggerSAEmacroscopically is more viscous, indicating a strong intermolecular attractive force. The longerhydrocarbon chain constituent of mineral oil, the greater London forces,therefore the viscosity of the oil increases. Synthetic oil structurally forms polymers with large branching. This structure affects a lower synthetic oil viscosity. The results of this study can be followed by designing a fluid viscosity by interconnecting of the three levels of representation as improving mental models and increasing concept understanding of pre-services chemistry teachers.

5. Acknowledgments

The authors would like to thank the Chairman of Chemistry Education Department, Head and Laboratory assistant of the Physical and Inorganic Chemistry Laboratory UniversitasPendidikan Indonesia Bandung who have facilitated to this research.

References

[1] Levine I R 2009 *Physical Chemistry* (New York : Mc-Graw Hill) pp 479-87
[2] Thibault R 2001 *Machinery Lubrication1* (Tulsa : Noria Corporation) pp 3–6
[3] George N J, Obianwu V I, Akpan A E and Obot I B 2010 *Arch. Phys. Res.* **1** pp 103-11
[4] Chittleborough G D 2004 *Ph. D. Thesis* (Australia : Curtin University of Technology).
[5] Tumay H 2014 *Chem. Educ. Res. Pract.* **15** pp 366-79.
[6] Wang C Y and Barrow L H 2010 *Res. Sci. Educ.***41** pp 561-86
[7] Nyachwaya J M and Wood N B 2014 *Chem. Educ. Res. Pract.* **15** pp 720-8
[8] Culpon D H and Mead T C1992 *Lubrication*ed Oberender F G (New York : Texaco Inc) p1
[9] Ahmed N S and Nassar A M 2011 *Tribology Fundamentals and Advancements* (In Tech) p 250
[10] Balabin R M and Safieva R Z 2008 *Fuel* **87** pp 2745-52
[11] Song W and Heilma W J *U.S. Patent* No. 7129197
[12] Coll R K 2009 *Educ. Quim* pp 18-31
[13] Viscopedia *Engine Oil* see http://www.viscopedia.com/viscosity-tables/substances/engine-oil/
[14] V&P Scientific, inc. *Viscosity Tables* see http://www.vp-scientific.com?Viscosity_Tables.htm .
[15] Jansoon N,Coll R K and Somsook E 2009 *Int. J. Env. Sci. Educ.***4(2)** pp 147-68
[16] Bain K, Moon A, Mack M R and Towns M H 2014 *Chem. Educ. Res. Pract.* **15** pp 320–35
[17] Hernandez G E, Criswell B A, Kirk N J, Sauder D G and Rushton G T 2014 *Chem. Educ. Res.*

Pract. **15** pp 354-65
[18] Becker N, Stanford C, Towns M and Cole R 2015 *Chem. Educ. Res. Pract.* **16** pp 769-85

IOP Conf. Series: Journal of Physics: Conf. Series **812** (2017) 012031 doi:10.1088/1742-6596/812/1/012031

A Method for Evaluating Information Security Governance (ISG) Components in Banking Environment

M Ula, M Ula and W Fuadi

Informatics Engineering, Malikussaleh University, Jl. Cot Tengku Nie Reuleut, Aceh Utara, Indonesia

Nanggroe@gmail.com

Abstract. As modern banking increasingly relies on the internet and computer technologies to operate their businesses and market interactions, the threats and security breaches have highly increased in recent years. Insider and outsider attacks have caused global businesses lost trillions of Dollars a year. Therefore, that is a need for a proper framework to govern the information security in the banking system. The aim of this research is to propose and design an enhanced method to evaluate information security governance (ISG) implementation in banking environment. This research examines and compares the elements from the commonly used information security governance frameworks, standards and best practices. Their strength and weakness are considered in its approaches. The initial framework for governing the information security in banking system was constructed from document review. The framework was categorized into three levels which are Governance level, Managerial level, and technical level. The study further conducts an online survey for banking security professionals to get their professional judgment about the ISG most critical components and the importance for each ISG component that should be implemented in banking environment. Data from the survey was used to construct a mathematical model for ISG evaluation, component importance data used as weighting coefficient for the related component in the mathematical model. The research further develops a method for evaluating ISG implementation in banking based on the mathematical model. The proposed method was tested through real bank case study in an Indonesian local bank. The study evidently proves that the proposed method has sufficient coverage of ISG in banking environment and effectively evaluates the ISG implementation in banking environment.

1. Introduction

The growth of information technology has been so explosive in the recent decade. Computer has been widely applied in every aspect of our life from business, government, education, finance, healthcare, and aerospace to defense system. With society's increasing dependency on information technology (IT), the consequences of computer *crime* can be extremely grave [1]. Security breach and computer viruses cost global businesses $1.6 trillion a year and 39,363 human years of productivity. In 2009, Symantec has detected 59,526 phishing hosts around the globe, that number is increased by 7% compared to phishing hosts detected in 2008. The percentage of threats to confidential information is increased to 98% in 2009 compared to 83% in 2008, 89% of the threats have the ability to export user data and 86% of them have keystroke-logging component [2].

The information system has become the heart of modern banking in our world today, and information has become the most valuable asset to protect from insiders, outsiders and competitors.

Customers are very concerned about privacy and identity theft. Business partners, suppliers, and vendors are seeing security as the top requirement, particularly when providing mutual network and information access. Banks' ability to take advantage of new opportunities often depends on its ability to provide open, accessible, available, and secure network services. Having a good reputation for safeguarding information will increase market share and profit. Therefore, banks have to be responsible for fraudulent activity perpetrated via the internet channel. Banks have to reimburse most customers for losses, although the customer clearly compromised their account credentials. An empirical research in Information Security Governance in banking is noted to be lacking [3]. There are some ISG frameworks, have been developed and widely practiced in types of businesses, but each of them has its own advantages and weaknesses. Commonly, it must be customized to fit with organizational structure and environment. ISG frameworks employ in banking system is adopted from general business ISG governance framework. Different business type has different characteristic so that bank need customized framework that's more suitable for bank environment. Hence, this research seeks to fill this research gap. The research aims to identify the degree of importance of each ISG component to be implemented in banking systems. This paper is organized into five sections. This section introduces the background of the study and research concern. Section two portrays the literature review of information security governance in banking. Section three discusses research methodology. The section four discusses the degree of importance of each ISG component. Finally, the paper ends with conclusions.

2. Information security governance evaluation

The ability to evaluate governance processes provides the best measure of governance performance [4]. Practical evaluation of security processes, such as malware statistics, is considered to be the most useful of all security tests. Having metrics provides the necessary quantitative information required to manage information risks and threats [5]. Literature study identifies many papers which proposed a wide range of approaches for evaluating corporate governance, ICT governance and information security governance. There are two ISG evaluation approaches found in the literature which are proposed Corporate Governance Task Force [6] and EDUCAUSE [7]. The security assessment approach developed by EDUCAUSE is a modification from the CGTF (2004) Information Security Governance Framework. EDUCAUSE (2004) developed a scorecard approach to effectively implement information security governance within institutions of higher education such as colleges and universities. However, both of the methods use direct summation method for grading and scoring the ISG implementation. In this research, more realistic ISG evaluation method is developed for banking industry with focus on component importancy and critical components that mapped to bank corporate governance [8].

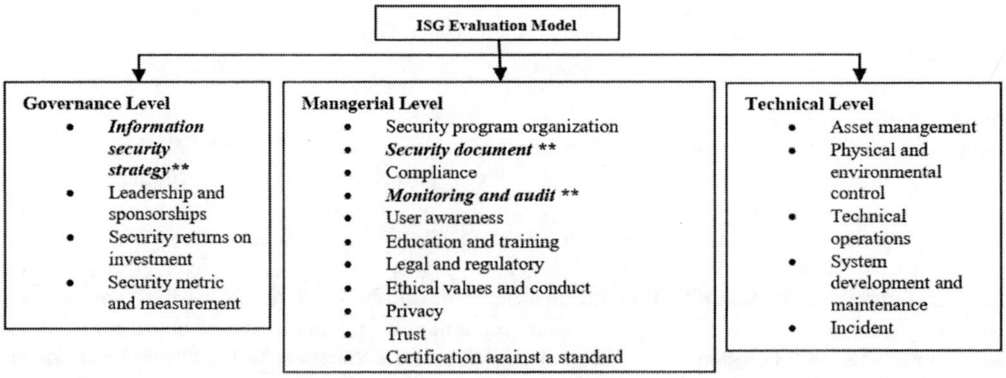

Figure 1. The structure of ISG evaluation model.

The structure of the ISG evaluation model is given in Figure 1. For this problem structure, a modified equation from Simple multi-attribute rating technique (SMART), Edwards and Barron proposed a variant named SMARTS (SMART using Swings) that in the course of the comparison of the importance of the criteria also considers the amplitude of the utility values, in examples; the changes from the worst utility value level to the best level among the alternatives [9].

3. Research methodology

The research started by documents and standard review. Almost all security standards available were examined to identify security components suggested by the documents. The information obtained in this preliminary study guides in developing the instrument for the main survey. The population of this study refers to all Bank Leaders, IT/IS professionals and Vendors related to banking industry all over the world which are Bank IT Staff, Security Manager, Security Consultant, IS Auditor, IT Practitioner. The IT/IS professional should have recognized certificates (CCNA, CCP, CEH, CISA, CISM, CISSP, Comptia Security +, ISO2001, ISP, ITIL, RHCT). Sampling come from the bank security professional signed up for linkedin.com social network and some security professional Group in Yahoo group and Google group. Data collection strategy used in this research is a web-based survey using Google document application. A strong reason for choosing web based survey instead of other traditional methods is because of the demography and location of the respondents are scattered around the world [10]. A self-designed and administered survey using Google Document application is developed. This method involves posting a questionnaire on Googledocs web site with respondents typically replying from remote computers. An introductory and acknowledgement page accompanies the questionnaire. 930 personal massage invitations were sent to security professionals at Linkedin.com inviting them to participate in the study. As a result of the effort, it received 189 responses. Within the 189 responses, 7 lacked complete answers for every question; the total number of effective responses is 182. The questionnaire was structured into three parts which are Respondent's Profile, Critical Components, and Importance level.

The survey questionnaire is designed to get feedback and opinion of the respondents. Part I of the questionnaire examines the demographic information about the respondents and their bank, such as respondent's gender, professional level, type of bank and size of bank. Part II was designed to determine the critical components of information security governance to be implemented in banking system. The ISG components used in this questionnaire are derived from information security documents and standards reviews done prior to the survey development. If one of these components is missing, there is no security for the whole system even though all other components are implemented. The respondents were requested to choose several ISG components that he thought as the critical components to be implemented in banking system. The questionnaire questions were structured as follows; "*Please check the items which you consider as the critical components to be implemented at ... level in banking system*". Part III of the questionnaire is designed to gather the opinion from respondents about the degree of importance of ISG to be implemented in banking system. The questionnaire questions were structured as follows; "*How important is it that bank ...?*". The response option is a Likert type five points scale response categories, ranging from "unrelated" to "extremely important". Each respondent is given a numerical score to describe respondent degree of importance to the preserved each component. The questionnaire is collapsing and mapping for logical groupings (technical level, managerial level, and governance level) to avoid confusing the respondents. By structuring the questions in this manner, would minimize the respondent's burden because of short time consuming to complete the survey and therefore the quality of the data could be maximized. List of component was as structured in Figure 1. The data result from ISG critical components and importance level serve in the mathematical model to develop ISG evaluation Method.

4. Result and analysis

This section describes the survey result of the most critical components should be implemented in banking ISG and the importance of each components. Table 1 shows the the survey result of the critical component and the importance of each component.

Table 1. ISG critical components and importance.

	Components	Critical Percentage (%)	Importance
Strategic Level	Information Security Strategy*	83.5	4.47
	Leaderships and Sponsorships	48.4	4.37
	Security Return on Investment	24.7	3.74
	Security Metric and Measurement	49.5	3.92
	Auditor Security Program	42.3	3.92
Managerial Level	Security program organization	43.7	4.22
	Security policies, procedure and guidelines*	88.0	4.47
	Compliance	52.5	4.21
	Monitoring and audit*	71.0	4.25
	User awareness	60.1	4.35
	Education and training	51.4	4.2
	Legal and regulatory	41.0	4.15
	Ethical values and conduct	30.1	3.99
	Privacy	35.5	4.46
	Trust	26.8	4.04
	Certification against a standard	20.8	3.63
	Risk management	67.2	4.21
	Risk assessment process	48.6	4.14
	Best practice and baseline consideration	29.5	4.26
Technical Level	Asset management	43.7	4.07
	Physical and environmental control	55.7	4.16
	Technical operations	54.6	3.88
	System acquisition, development and	49.2	3.85
	Incident management	63.4	4.2
	Business continuity plan*	77.6	4.35
	User management	47.5	4.08

4.1 ISG critical components in banking environment

In this section, the survey result of identifying the most critical components of ISG that should be implemented in banking will be discussed. The frequency statistical analysis result for Governance, Managerial and Technical level is provided in Table 1. The survey result in critical component table shows that, for Stategic Lavel , 83.5 % respondents chose Information Security Strategy as a critical security component in banking system. For Managerial Level, 88.0 % chose security policies, procedure and guidelines and 71.0% chose monitoring and audit as the critical security components in banking system. For technical level, 63.4% chose incident management, 77.8% chose business continuity plan, as a critical security components in banking system. This critical components would include in developing the mathematical formula for evaluating ISG implementation in banking system.

4.2 Importance of the ISG components

Regarding the importance, the result in Table 1 shows that all ISG components have means values higher than 3. It shows that all components discussed are important for banks at Governance Managerial and Technical level. At Governance level, the result shows that the Information Security Strategy and Leaderships and Sponsorships are perceived as the most important components that should be implemented by banks. The security guideline, procedure and policy, and privacy are perceived as the most important components that should be implemented by banks at Managerial level. At Technical level, Business continuity plan is perceived as the most importance component in that should be implemented in banking system.

5. Developing ISG evaluation method

The objective of the ISG evaluation is to identify how much amount of ISG components had been implemented in a bank as a countermeasure to latest security threats. From the structure shows in Figure 1, the mathematical model for ISG Evaluation proposed as in equation 1.

$$ISGLevel = \frac{\left(\sum_{j=1}^{n} W_j M_j\right)\left(\prod_{u=1}^{v} C_u\right)}{\left(\sum_{j=1}^{n} W_j\right)} \tag{1}$$

Where M_j is the level of the jth ISG component implementation in a bank. The implementation level of each ISG component in a bank can be determined by using the proposed evaluation document given in Appendix. The document derived from ISO 17799 [11], FISMA and some other security checklists. This proposed tools use six Likert scales scoring to evaluate the implementation level of each ISG component. The scoring is given in Table 2 below.

Table 2. Implementation level scoring.

Score	Stand for
0	Component is not exist
1	Component exists but is not implemented
2	Component that is occasionally implemented
3	Component is implemented with much vulnerability.
4	Component that is implemented with little vulnerability,
5	Component fully implemented.

W_j is the weight of the each ISG component in banking system. The component weighting values are derived from the importance degree resulted from the survey data as shown in Table 1. C_u is critical component of uth security component and v is the number of total critical components. he critical components used in this mathematical model also derived from the survey result. The value of $\prod_{u=1}^{v} C_u$ is 1 if the entire critical components are well implemented, otherwise it would be 0, if the critical components not well implemented.

6. Testing the developed method in real bank

X Bank, a local provincial governance bank in Indonesia, selected to use the proposed method. X Bank is an actual bank; however, its name has been changed in this document to comply with its policy. The bank has divided into two business mainstreams which are conventional banking and Islamic (sharia') banking. The testing was done by five evaluators, in main office Banda Aceh, and branches in Sigli, Bireuen and Lhoksumawe. They were asked to evaluate their information security program implementation based on the method provided in the developed method tool. Each evaluator was given a softcopy of the developed method (scorecard) document in Microsoft Excel file with computed using the ISGlevel equation. The summary of the testing result is provided in Table 3.

Table 3. Summary of X bank ISG implementation evaluation result.

Level	S.B. Bireuen	S.B. Sigli	C.B. Bireuen	C.B. Lhoksumawe	Main Office	Average
Governance	2.90	3.13	2.87	2.53	3.19	2.92
Managerial	3.93	3.64	3.48	3.8	3.96	3.79
Technical	4.27	4.06	4.19	4.1	4.28	4.16

Table 3 shows the summary of the testing result done by the evaluators to their branch. This testing done by the evaluator using developed method that given to them as a tool to evaluates ISG components implementation in their bank. The results in Table 3 clearly indicate the weakness areas of X Bank Information Security Program. At governance level, the most of the evaluation score are lower than 3. Refer to the values in Table 2, It can be conclude that, at bank X governance level, the ISG components are occasionally implemented. However, at Managerial level, the most of the evaluation score are higher than 3. It can be conclude that, at bank X Manajerial level, the ISG components are implemented but with much vulnerability. Futhermore, at Technical level, the most of the evaluation score are higher than 4. It can be conclude that, at bank X Technical level, the ISG components are implemented with little vulnerability.

Therefore, it show that the developed method can highlight the weakness of bank X information security program and which level grant the responsibility. Hence, the developed method proves that its can achieve its objectives as an alternative of ISG evaluation method in banking envireonment.

7. Conclusion

To established a proper information security program, a bank need to define the critical component to be implemented in their ISG framework. This study conduct a web-based survey to gathering professional judgment about the critical components must be implemented in bank ISG program. From the survey result, it is found that the critical components for ISG implementation in banking system are: Information Security Strategy, Security policies, procedure and guidelines, Monitoring and audit, Business continuity plan, Disaster recovery plan. The ISG components implemented in banking have different degree of importance. Generally, "Information Security Strategy" and Leaderships and Sponsorships" perceived as very important components should be implemented by a bank at Governance level, "Guideline" and Privacy" perceived as very important components should be implemented by a bank at managerial level, and "Business Continuity Plan" is a very important component at technical level. So that current method on security metric and audit is not realistic because it still use direct summation method and treat all component as the same important. It is need more realistic method which has a weighting coefficient for each ISG component measured in banking environment. In this research, a method to evaluate the ISG implementation was developed and tested in a real bank environment. This developed method used the importance level as a weighting coefficient and the existence of the critical components also taken in to consideration when calculation the level of ISG implementation in a bank. The developed method had highlight the weakness of bank X information security program and which level grant the responsibility, therefore, it proves that its can achieve its objectives as an alternative of ISG evaluation method in banking envireonment.

8. References

[1] Mahncke R J, McDermid D C and Williams P A 2009 *Proceedings of the 7th Australian Information Security Management Conference (Melbourne)* (New York: Springer)

[2] Symantec Enterprise Security 2010 *Symantec Global Internet Security Threat Report Trends for 2009* vol XV (Mountain View: Symantec Corporation)

[3] Kurt B and Tentra G M 2004 *Corporate Information Security Governance in Swiss Private Banking* (Zurich: Master's Thesis University of Zurich)

[4] Fitzgerald T 2008 *Information Security Governance: What Is It and How can We Accomplish It?* (Milwaukee, WI: ISACA Kettle-Moraine Chapter Meeting)

[5] KabayM E 2009 Security metrics research *Network World*

[6] Corporate Governance Task Force 2004 Information security governance: A call to action *Corporate Governance Task Force Report* (USA: National Cyber Security Summit)
[7] EDUCAUSE 2004 *Information Security Governance Assessment Tool for Higher Education*
[8] Ula M, Ismail Z, Zidek S M 2011 *Journal of Information Assurance & Cybersecurity*
[9] Edwards W and Barron F H 1994 *Elsevier* **60** 306-25
[10] Cooper D R and Schindle P S 2008 *Business Research Method* 10th Edition (Singapore: McGraw-Hill)
[11] Standards Australia 2005 *Information Technology Security Techniques Code of Practice for Information Security Management* (Sydney: Standards Australia Limited)
[12] Brotby W K 2006 *Information Security Governance, Guidance for Boards of Directors and Executive Management* 2nd edision (Rolling Meadows: IT Governance Institute, Rolling Meadows)

Undergraduate Students' Difficulties in Reading and Constructing Phylogenetic Tree

S Sa'adah[1,2*], F S Tapilouw[1] and T Hidayat[1]

[1] Biology Department, Universitas Pendidikan Indonesia, Jl. Dr.Setiabudi No. 229, Bandung 40154, Indonesia
[2] Jalan A.H. Nasution No. 105, Cipadung, Cibiru, Kota Bandung, Jawa Barat 40614 UIN Sunan Gunung Djati Bandung, Indonesia
Corresponding author : topikhidayat@upi.edu

Abstract. Representation is a very important communication tool to communicate scientific concepts. Biologists produce phylogenetic representation to express their understanding of evolutionary relationships. The phylogenetic tree is visual representation depict a hypothesis about the evolutionary relationship and widely used in the biological sciences. Phylogenetic tree currently growing for many disciplines in biology. Consequently, learning about phylogenetic tree become an important part of biological education and an interesting area for biology education research. However, research showed many students often struggle with interpreting the information that phylogenetic trees depict. The purpose of this study was to investigate undergraduate students' difficulties in reading and constructing a phylogenetic tree. The method of this study is a descriptive method. In this study, we used questionnaires, interviews, multiple choice and open-ended questions, reflective journals and observations. The findings showed students experiencing difficulties, especially in constructing a phylogenetic tree. The students' responds indicated that main reasons for difficulties in constructing a phylogenetic tree are difficult to placing taxa in a phylogenetic tree based on the data provided so that the phylogenetic tree constructed does not describe the actual evolutionary relationship (incorrect relatedness). Students also have difficulties in determining the sister group, character synapomorphy, autapomorphy from data provided (character table) and comparing among phylogenetic tree. According to them building the phylogenetic tree is more difficult than reading the phylogenetic tree. Finding this studies provide information to undergraduate instructor and students to overcome learning difficulties of reading and constructing phylogenetic tree.

1. Introduction

The main purpose of science education is to develop scientific literacy [1]. One component of science literacy is the ability to represent the phenomenon, the object of events, abstract concepts, ideas, and the mechanism of the system [2], [3]. Representation is an entity in which all thinking is considered to take place, therefore important in the learning process [4]. To really understand the science, students need to know how to interpret, represent and assess scientific claims, all of which implies a fundamental role for representational work [5]. Biologists produce phylogenetic representation to express their understanding of evolutionary relationships.

The Phylogenetic tree is visual representation depict a hypothesis about the evolutionary relationship among taxa and widely used in the biological sciences [6], [7], [8]. As visualizations, phylogenetic trees are a type of schematic diagram that illustrates abstract concepts rather than appearances of objects (iconic diagrams) or quantitative relationships [9]. The ability to read and understand a phylogenetic tree is called tree-thinking [10]. Tree thinking is an important skill for biology students [11]. However, numerous studies indicate that phylogenetic trees are difficult to comprehend for college-level students [12] and students often struggle with interpreting the information that phylogenetic trees depict [13], [14]. The purpose of this study was to investigate undergraduate students' difficulties in reading and constructing a phylogenetic tree. Information obtained from this study can give an idea to educators in developing undergraduate students' skill to reading and constructing phylogenetic trees.

2. Method

The Method of this study is descriptive research. The descriptive research provides information about conditions, situations, and events that occur in the present [15]. The research has been conducted on Program Studi Pendidikan Biologi UIN SGD in Bandung West Java. The subjects in this study were 40 undergraduate students who take vertebrate zoology course in even semester 2016. The instruments used to collect data were questionnaires, interviews, multiple choice and open-ended questions, reflective journals, and observations. All the instruments conduct to reveal undergraduate students' difficulties in reading and constructing a phylogenetic tree. Each question that given to the students are adjusted with tree thinking skill (modification from Novic and Catley [16]), namely describe the character of a taxon from the phylogenetic tree, determine the character of evolution, comparing the type of the phylogenetic tree, applying the concept of clade, Evaluate relative evolutionary relatedness, determine the common ancestor between two or more species (MRCA / Most Recent Common Ancestor) and built a phylogenetic tree. Sample question can be seen in Table 1.

Table 1. Sample Question

No	Indicator	Question
1	Determine the most recent common ancestor (MRCA) between two or more species	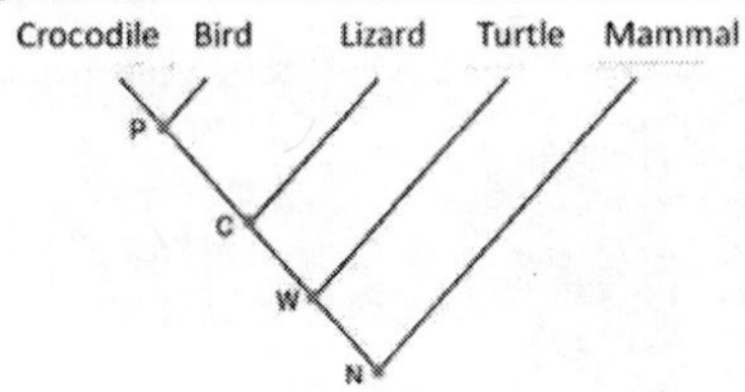 Based on the phylogenetic tree above, the most recent common ancestor (MRCA) of birds and lizards are P, justify that statement? True, because the P position is closest to birds and lizards True, because P developed into C and C developing towards lizard Wrong, because the common ancestor of birds and lizards are C Wrong, because the bird is the common ancestor of crocodiles and lizards Wrong, because the common ancestor does not show in the phylogenetic tree
2	Evaluate relative evolutionary relatedness,	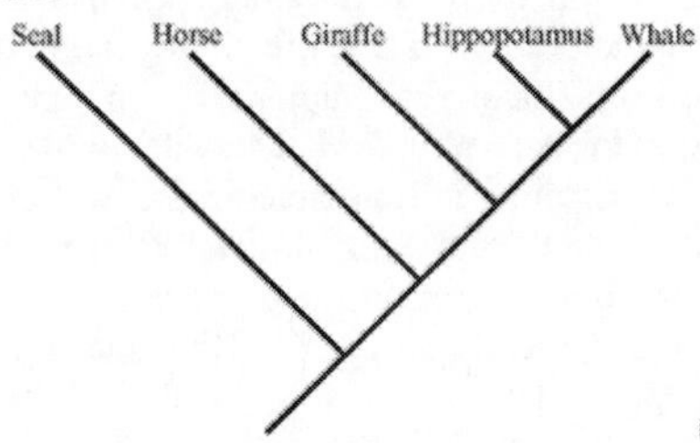 (Baum, 2005) Observe the phylogenetic tree above, based on the tree Which of these

	statements is true about a giraffe?
	Giraffe more closely related to the horse than with hippopotamus
	Giraffe more closely related to the whale compared with hippopotamus
	Giraffe is closely related to the horse but was not related to hippopotamus
	Giraffe equally related to a hippopotamus and whale
	Giraffe more closely related to the whale compared with hippopotamus

3. Result and Discussion

Results of questionnaire to the students revealed can be seen in Table 2. and Figure 1.

Table 2. Students' Responses to the Difficulties in Read and Construct a Phylogenetic Tree

Question: Which do you think is the most difficult of the following matters relating to the phylogenetic tree?	
Indicator	Students' Responses
Describe the character of a taxon from the phylogenetic tree	10%
Determining the character of evolution	43%
Comparing the type of the phylogenetic tree	38%
Applying the concept of clade	10%
Evaluate relative evolutionary relatedness	29%
Determining the Most Recent Common Ancestor between two or more species	5%
Construct a phylogenetic tree	86%

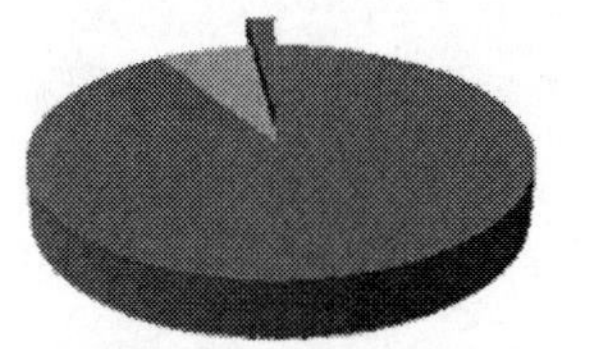

Figure 1. The Percentage Students' Responses to The Question "According to You, Which One is Easier to Read or Construct Phylogenetic Tree?"

Based on Table 1 and Figure 1, an interview and reflective journal, the students stated that constructing the phylogenetic tree was more difficult than reading the phylogenetic tree. The students' responses indicated that main reasons for difficulties to constructing phylogenetic tree are difficult to placing taxa in a phylogenetic tree based on the data provided because they are still confused to determining the sister group from the character table. Thus, the phylogenetic tree that is made does not describe the actual evolutionary relationship (incorrect relatedness). Students also have difficulties in determininig character synapomorphy, autapomorphy from character table. According to them, the more taxa and characters presented in a table of characters, the more difficult to construct a phylogenetic tree. The result of student responses, not much different from previous studies that tree building is more conceptually difficult and builds upon tree reading skills [14].

Percentage of students who answered correctly in each of the questions in the tests related to reading the phylogenetic tree shown in Figure 2.

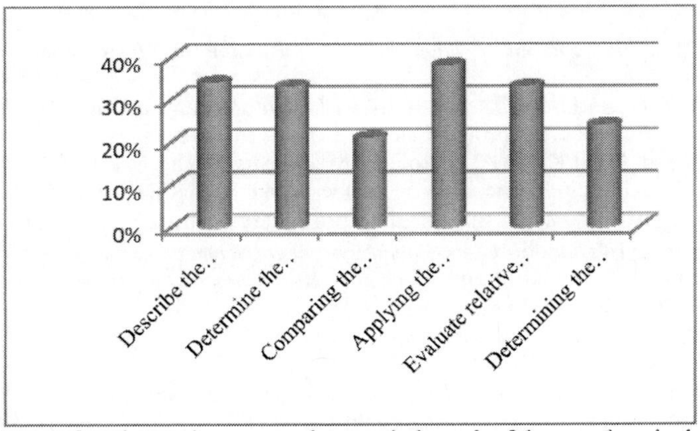

Figure 2. Percentage of students who answered correctly in each of the questions in the tests related to read phylogenetic tree

Based on Figure 2, it appears that the students' ability to read the phylogenetic tree is low, as well as the students' ability to construct a phylogenetic tree. Some reasons have given student-related difficulties in reading phylogenetic tree included (1) If taxa and characters presented in the phylogenetic tree in large number; (2) if the phylogenetic tree is given in ladder format, because students had much greater difficulty understanding the ladder than the tree format (Figure 3.); (3) they forgot about synapomorphy, autapomorphy, so often confused.

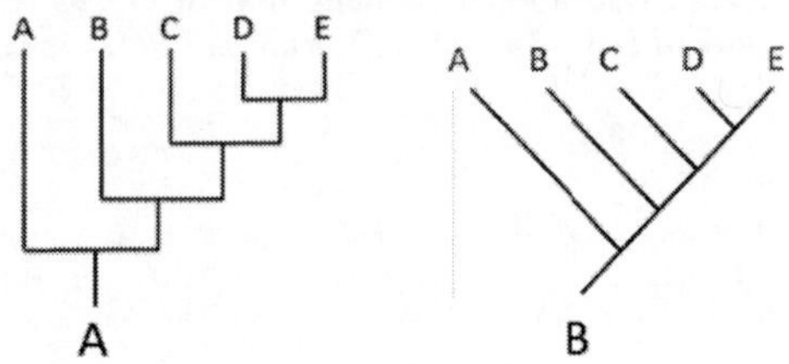

Figure 3. Tree Format (A) Ladder Format (B)

Understanding a phylogenetic tree is not an easy thing. The researcher has an opinion that understanding phylogenetic tree is a complex cognitive task and without proper scaffolding, many students can't transfer the empirical data into visual structure [17]. So that students can read and construct a phylogenetic tree such as an expert, the student must understand the structure of the phylogenetic tree, the mechanism of evolution, inheritance, and genomics [2]. Students also need to continue the practice in reading and building the phylogenetic tree.

4. Conclussion

Based on this study concluded that undergraduate students having difficulties in reading and constructing a phylogenetic tree, such as determining the sister group, character synapomorphy, autapomorphy from data provided. According to them, building the phylogenetic tree is more difficult than reading the phylogenetic tree. Thus, it would appear that there is a strong necessity to find a suitable strategy to cover most of these difficulties for improving students in reading and constructing a phylogenetic tree. The result of this study will help educator and researchers to design learning techniques to overcome the mentioned difficulties.

5. References

[1] National R C 1996 *National Science Educational Standard* (Washington: National Academy Press) pp 1-3

[2] Halverson K L and Friedrichsen P F 2013 *Multiple Representations In Biological Education* Treagust DF, Tsui C Y Springer Science Business Media pp 185-201

[3] Chiu M H and Wu H K 2009 The Role Of Multimedia In The Teaching And Learning Of The Triplet Relationship In Chemistry J K Gilbert and D Treagust (Dordrecht: Springer) p 1

[4] Gilbert J K 2010 The Role of Visual Representations in the Learning and Teaching of Science: an Introduction *Asia-Pacific Forum On Science Learning And Teaching* **11** 1

[5] Tytler R, Prain V, Hubber P and Waldrip B 2013 *Constructing Representations To Learn in Sciene.* (Rotterdam, Netherland: Sense Publishers) p 1

[6] Baum D A 2008 Groups *Nature* Education **1(1)** 190

[7] Baum D A and Offner S 2008 *Am Biol Teach.* **70 (4)** 22–9

[8] Baum D A and Smith S D 2012 *Tree thinking an Introduction to Phylogenetic Biology* (Colorado: Roberts And Company Publishers) p 35

[9] Dees J, Momsen J L and Niemi J and Montplaisir L 2014 *Life Sciences Education* **13** 666–76.

[10] Gregory T R 2008 *Evol Educ Outreach* **1** 121–37

[11] Brenda C, Phillips, Laura R, Novick, Kefyn M, Catley, Daniel J and Funk 2012 *Evo Edu Outreach* **5** 595–602

[12] Meir E, Perry J, Herron J C and Kingsolver J 2007 *The American Biology Teacher* **71** 6

[13] Omland K E, Cook L G and Crisp M D 2008 *Bio Essays* **30(9)** 854–67

[14] Halverson K L 2011 *Evo Edu Outreach* **4** 95–106

[15] Neville T 2005 Postlethwaite Educational research: some basic concepts and terminology, International Institute for Educational Planning/UNESCO p 9

[16] Novick L R and Catley K M 2013 *American Educational Research Journal* **50(1)** 138–77

Analysis of Student Errors on Division of Fractions

E Maelasari[1] and AJupri[2]

[1]Departement of Primary Education,Postgraduate Student, Universitas Pendidikan Indonesia, Jl. Dr. Setiabudhi 229, Bandung 40154, Indonesia
[2]Mathematics Education Department, Mathematics and Science Education Faculty, Universitas Pendidikan Indonesia, Jl. Dr. Setiabudhi 229, Bandung 40154, Indonesia
Email: maygunawan@student.upi.edu

Abstract. This study aims todescribe the type of student errors that typically occurs at the completion of the division arithmetic operations on fractions, and to describe the causes of students' mistakes. This research used a descriptive qualitative method, and involved 22 fifth grade studentsat one particular elementary school in Kuningan, Indonesia. The results of this study showed that students' error answers caused by students changing their way of thinking to solve multiplication and division operations on the same procedures, the changing of mix fractions to common fraction have made students confused, and students are careless in doing calculation. From student written work, in solving the fraction problems, we found that there is influence between the uses of learning methods and student response, and some of student responses beyond researchers' prediction. We conclude that the teaching method is not onlythe important thing that must be prepared, but the teacher should also prepare about predictions of students' answers to the problems that will be given in the learning process. This could be a reflection for teachers to be better and to achieve the expected learning goals.

1. Introduction

Fraction is one of the core mathematics topics that should be learned by students in elementary school. This topic includes arithmetical operations, addition, subtraction, multiplication, and division, on proper fractions, and decimals. By studying this topic, students are expected to understand the material form of algebraic fractions operations in the next level of education.

In a current situation, students are more likely to make errors in operationsbecause the learning material of fractions focused only on memorizing formulas and procedures, without sufficient attention to the meaning of processes. In the division of fractions for example, this is always accompanied by a divisions algorithm used by multiplying the number of students divided by the form of the inverse of the divisor.The students' errors in solving division of fractions often caused by the error use of this algorithm.

The errorsappearbecause the algorithm is only seen as a set of procedure that is not meaningful for students. Students do not know why they should reverse the form of fractions when they do division of fractions.

There are some different students' mistakes in using the division algorithm. For example, students simply divide the numerator of the fraction by the fraction divisor, and also split both the denominator, and multiplying fractions divided by the fractional divisor without changing the fractional divisor in the form of its *inverse*[1].

Some misconceptions, one of the errors found is using the concept of multiplication in the division of fractions. For example, in the division of fractions by integers, students directly multiply the integerwith the fractions. This error occurs from an accidental students and error based on formal knowledge of the student, such as the inability of student to interpert $2 : \frac{1}{2}$ as " how many $\frac{1}{2}$ in 2" ([2], [3]).

From the above description, this study focuses toinvestigate types of student errors that typically occur when doing division of fractions, and to explain the causes that students make. Classification errors made by participants when dividing fractions is the following [1]:

1.1. The algorithm-based error
Different ways to calculate falls into this category. This procedure includes reverse common shares in lieu of divisor or reversal before multiplying the numerator and denominator. This typically describes the results of memorization algorithms. When the algorithm displays a measure that does not mean, allow students to forget that step or change the way that it could be a mistake.

1.2. Errors are not intentional
Research on how to distribute the operation showed that students in balancing operations with fractions and integers on to explain the distribution of primer using a long way in its entire distribution model. In this distribution model of an object divided into separate figures or the smallest group

1.3. An error based on formal knowledge
Errors in the limited thinking of alleged breakdowns and lack of knowledge in connecting operations are included in this category. Lack of knowledge is probably the source of the bad response to a variety of tasks including the division of fractions.

The errors of students in solving mathematical problems can be used byteachers to know the difficulties being faced by students and to improve the next learning and teaching process. One way to reduce errors is to analyze students' mistakes. The error analysis is an attempt to observe, discover and clarify mistakes with certain rules. Classification according to the specific rules in question are classifies errors by type of concept error, algorithm error, operation error, and random errors [4].

Cognitive state about students with intellectual upon ability to digest the subject matter at hand. Nature, weights, media and others in pursuit transfer students are all factors students' mistakes in answering problems[5].

Fractions is a part of the rational numbers that can be written in the form $\frac{a}{b}$where*a, b*are integers, *b* is not equal to zero, and *b* is not a factor of *a*. In this case "*a*" is called the numenator and "*b*" is called the denominator.

2. Experimental Method
This research used a qualitative descriptive method[6], and was conducted in the second semester of the 2015/2016 teachings of knowledge. Subjects in this study were all students of class V at one particular elementary school in Kuningan. The subjectsof this study are 22 fifth grade students of the subject who will have some students representing the error on each items to be used as a research subject respondents or interviews that match the criteria. A criterion for selection of interview subjects is determined based on the variation of the error and the number of mistakes made by students in solving each item.

The instrument of this study included ten questions and an interview guideline. Problem test was used to locate the fault of students in solving problems. Interviews were carried out, to students in order to determine causes of error factor in solving problems, after the result of the students' written works have been analyzed.

3. Results and Discussion
The research was conducted in grade five, in one of elementary schools inKuningan, Indonesia, in the second semester of the 2015/2016 academic year. Based on the research results, the language is

relaxed, clear and understandable , the teacher explained the purpose of the learning to be achieved at the time of learning. This is in accordance with aspects of learning in Piaget's theory, namely, using language that is appropriate to the child's way of thinking. Because of the language and the way children think differently from adults. Elementary school children are in the concrete operational stage according to Piaget's theory. Where children are invited to think logically start with the help of concrete objects. The teacher started the lesson by asking students to recall simple fraction multiplication, so that children do not feel confused[5]. The results of analysis of student errors are shown in Table 1.

Table 1. The category of student errors

Categories	Pattern of Error	The students answer
The algorithm-based error	Students assume that the completion of the operation of division of fractions is equal to solve addition operations on fractions, i.e., by equating the denominator.	$6 : \frac{2}{3} = \frac{6}{1} \times \frac{2}{3} = \frac{18}{3} \times \frac{2}{3} = \frac{36}{3} = 12,$
	Students complete the operationof division of fractionsand integers by dividing those numbers directly.	$\frac{5}{9} : 3\frac{1}{3} = 3\,(\frac{5}{9} : \frac{1}{3}) = 3\frac{5}{3}$
An error based on formal knowledge	Students assume that the division of fractions and integers, wherever the location of the breakdown of the fractions that have to be reversed.	$5\frac{1}{2} : 3 = \frac{6}{2} : 3 = \frac{2}{6} \times 3 = \frac{6}{6} = 1$
	Students think that the changingof fractions into equivalent fractions, can be solved by multiplying the numerator and the denominator.	$5\frac{1}{2} : 3\frac{2}{3} = \frac{6}{2} : \frac{5}{3} = \frac{6}{2} \times \frac{3}{5} = \frac{18}{10}$

Type of errors of the fifth grade students to solve problems in the division of fractions.On the basis of an understanding of division of fractions algorithm, we found the following patterns of errors:

- Students assume that the division of fractions and integers, wherever the location of the breakdown, the fractions have to be reversed. This error occurs because of a lack of understanding of the basic concepts of fractions and division of integers.
- Students assume that the completion of the division of fractions is equal to solve addition operations on fractions, by equating the denominator. This error occurs because of the application of the law and relevant strategies that students who do not use the concept of fractional sum in solving division operations on fractions.
- Students complete the operations division of integer fractions by dividing direct those numbers. This error occurs because of a lack of understanding of the basic concepts of

fractions and division of integers or vice versa, as well as a lack of understanding of the basic concepts of integer division.

- Student think the alternation fractions into equivalent fractions can be solved by simplifying the numerator and the denominator. Students made a mistake in understanding the breakdown of the fractions, including the numerator and the denominator; do not adjust the numerator of the fraction, and cannot change a whole number into a fraction [8].

Based on the above analysis, one of the causes of these errors is to be short of mastery of prerequisite skills, such as students do not know that integers can be expressed as a fraction. In addition, other factors such as a lack of understanding of the concept of multiplying fractions with integers.

4. Conclusion

Based on the resultsand discussion as described in the previous section, we conclude that themistakes made by students in solving fractions, can be classified into four types of errors are: (1) Students assume that the division of fractions and integers, wherever the location of the breakdown of the fractions that have to be reversed; (2) Students assume that the completion of the operation division of fractions equal to solve addition operations on fractions, by equating the denominator; (3) Students completing the operations division of the integer fractions by dividing direct those numbers; and (4) Students think that the change fractions into equivalent fractions, can be solved by simplifying the numerator and the denominator remains.

The causes of student mistakes are the following: (1) Implementation of laws and strategies that are not relevant, one example of the application of the law and the evidence irrelevant strategy that students use the concept of the sum fractions in solving division operations on fractions; (2) Be short of understanding of the basic concepts of multiplication and division of whole numbers with fractions; (3) Be short of mastery of prerequisite skills on fractions. For example, students do not know that integers can be expressed as a fraction.

Based on this results, we suggest that (1) Teachers should give emphasis to the material preconditions on the material fractions so that students are not making a mistake in applying the arithmetic operation algorithm shards;(2) Teachers should often give a lot of exercises fractional arithmetic operation so that students are more skilled in solving them;(3) For researchers who want to do research on a material matter fractions should be used when the test is more focused research materials (e.g.specialized on addition and subtraction of fractions or multiplication and division fractions only) in order to be in a study in order to obtain more accurate results.

6. References

[1] Tirosh D2000Journal for Research in Mathematics Education **31** 5-25

[2] BergesonT 2000*Teaching and Learning Mathematics, Using Research to Shift From the "Yesterday" Mind to the "Tommorow" Mind*(Wahington: State Superintendent of Public Instruction)

[3] Newstead K. & Murray H 1998 *Proc. of the Twenty-second Int. Conf. for the Psychology of Mathematics Education* (Stellenbosch) vol 3 (South Africa: Western Cape) pp.295-302

[4] Septy S Y,*et al* 2012Menyelesaikan pembagian pecahan tanpa algoritma*Forum Pendidikan Matematika* 1(2012)2

[5] Gagne RM1992*The Condition of Learning and Theory of Instruction* (NewYork: RinehartAndWinston)

[6] Gall MD 2003*Educational Research an Introduction*(United States:Pearson Education, Inc)

[7] Suherman E,*et al* 2003*Evaluasi pembelajaran Matematika* (Bandung: FPMIPA UPI)

[8] Zakaria E 2010*InternationalEducationStudies*3 105-10

[9] Wintarti,*et al* 2008*Contextual Teaching and Learning Matematika SMP/MT Kelas VII Edisi4*(Jakarta:PusatPerbukuanDepartemenPendidikan Nasional)

Acknowledgments
We thank students and teacher for their active participation in this research

IOP Conf. Series: Journal of Physics: Conf. Series **812** (2017) 012034 doi:10.1088/1742-6596/812/1/012034

Non-Spatial Analysis of Relative Risk of Dengue Disease in Bandung Using Poisson-gamma and Log-normal Models: A Case Study of Dengue Data from Santo Borromeus Hospital in 2013

R Irawan, B Yong and F Kristiani

Department of Mathematics, Universitas Katholik Parahiyangan, Jl.Ciumbuleuit 94, Bandung 40141, Indonesia
Email: robynirawan0209@gmail.com

Abstract. Bandung, one of the cities in Indonesia, is vulnerable to dengue disease for both early-stage (Dengue Fever) and severe-stage (Dengue Haemorrhagic Fever and Dengue Shock Syndrome). In 2013, there were 5,749 patients in Bandung and 2,032 of the patients were hospitalized in Santo Borromeus Hospital. In this paper, there are two models, Poisson-gamma and Log-normal models, that use Bayesian inference to estimate the value of the relative risk. The calculation is done by Markov Chain Monte Carlo method which is the simulation using Gibbs Sampling algorithm in WinBUGS 1.4.3 software. The analysis results for dengue disease of 30 sub-districts in Bandung in 2013 based on Santo Borromeus Hospital's data are Coblong and Bandung Wetan sub-districts had the highest relative risk using both models for the early-stage, severe-stage, and all stages. Meanwhile, Cinambo sub-district had the lowest relative risk using both models for the severe-stage and all stages and BojongloaKaler sub-district had the lowest relative risk using both models for the early-stage. For the model comparison using DIC (Deviance Information Criterion) method, the Log-normal model is a better model for the early-stage and severe-stage, but for the all stages, the Poisson-gamma model is a better model which fits the data.

1. Introduction

Dengue disease is one of the most dangerous disease in tropical region and it may kill its patients. This disease is caused by virus DEN 1-4 which is transmitted by Aedes aegypti's bites [8]. This disease is divided into two stages, which are the early-stage and severe-stage [3]. The early-stage of dengue disease is called DF (Dengue Fever), meanwhile the severe-stage is divided into two types, which are the DHF (Dengue Haemorrhagic Fever) and DSS (Dengue Shock Syndrome). DSS is the next stage from the DHF with the highest possibility which can cause death.

The number of dengue cases in the worldwide that had been reported annually are kept increasing every year. For nearly 75% out of the number cases, the disease burden due to dengue were reported from Asia region. The epidemic of the dengue is a major public health problem in some countries in the southern part of Asia such as Indonesia, Myanmar, Sri Lanka, Thailand, and Timor-Leste. In Indonesia, there are about 20% of the total number of the cases in the whole country that reported from both Jakarta and West Java province [8].

Bandung is one of the most populous cities in Indonesia and geographically located on the highland part of West Java province. Due to the location of Bandung, this city has higher level of humidity than the other cities in Indonesia. Bandung is located 791 meters above sea level with the temperature of

18.5℃ until 30.1℃, meanwhile Aedes aegypti can live in a place with the temperature above 16℃ and under 1,000 meters above sea level. Therefore, Bandung is one of the cities that suitable for the Aedes Aegypti to live at and breed. Based on data from the Health Department of Bandung, the number of dengue cases increases every year. The number of the cases that occurred from 2011 until 2013 respectively are 3,911 cases, 5,107 cases, and 5,749 cases. Based on those data, then Bandung is classified as the critical area of dengue outbreak. Moreover, from Santo Borromeus hospital in year 2013, about one-third out of the total cases in 2013 were hospitalized in this hospital. From this fact, it is reasonable to use dengue data from this reputable hospital.

Consider that Bandung has 30 sub-districts, which means there must be different possibility for each sub-district to be vulnerable by dengue disease. Based on the number of patients and the number of the population in each sub-district, we can construct a model to estimate the relative risk level for each sub-district.

In this paper, the spatial effect would not be considered because the data that are used here are only from one hospital which Santo Borromeus hospital is located in the northern part of Bandung that will caused the other sub-dictricts that located far from this hospital will not have significant influences from one sub-district to the other. There are many methods to estimate the relative risk estimation without considering the spatial effect. For several paper using Poisson model by [3], the model that used was the frequentist statistics inference model. In this paper, the relative risk will be estimated by the numerical simulation of Bayesian statistics inference using Poisson-gamma model and Log-normal model that will be processed by WinBUGS 1.4.3 software. The models are used due to the supposition that the number of the patients and the number of the population are correlated and each of them are statistically distributed. In this paper, the relative risk estimation for each sub-district will be estimated using both models and will be compared using DIC (Deviance Information Criterion) method to determine the best model which fits the data.

2. Relative Risk

Relative risk (θ) is the ratio between the probabilities of an event in a certain region to the probability of an event in a whole population [6]. There are 3 possibilities of relative risk estimation. If θ=1, it means that the risk level in a certain region is as same as the risk level in a whole population. If θ<1, it means that the risk level in a certain region is lower than the risk level in a whole population or less vulnerable. The last possibility, θ>1 means that the risk level in a certain region is higher than the risk level in a whole population or more vulnerable.

In this paper, with the notation of θ_i which means the relative risk level in i^{th} region, the relative risk levels are divided into 5 categories. Those are VL (Very Low) if $\theta_i<0.5$, L (Low) if $0.5\leq \theta_i<1$, M (Medium) if $1\leq \theta_i<1.5$, H (High) if $1.5\leq \theta_i<2$, and VH (Very High) if $\theta_i \geq 2$.

In frequentist statistics inference, the formula to estimate the relative risk estimationis defined as [2]:

$$\theta_i = \frac{y_i}{e_i} \tag{1}$$

where y_i is the number of the patients in i^{th} region based on observation and e_i is the expected number of the patients in the ith region. The parameter e_i is defined as $e_i = N_i.\frac{\sum_{i=1}^{n} y_i}{\sum_{i=1}^{n} N_i}, i = 1,2,...,n$, where N_i is the number of the civilization in the ith region and n is the number of the region.

3. Non-Spatial Models

There are several models that can be used to estimate the relative risk θ_i for each region. Beside the frequentist statistics inference to estimate the relative risk explained above, there is also other statistics inference method which is also implemented. In Bayesian statistics inference, the parameters of each random variables are considered to be statistically distributed too. It is considered by the fact that the parameters in some statistics distributions are not always have a constant values, it may sometimes change according to the situation. In this paper, there will be two models that used to estimate θ_i. Both of the models are using MCMC (Markov Chain Monte Carlo) method with Gibbs Sampling algorithm and simulated using WinBUGS 1.4.3. The models are Poisson-gamma model and Log-normal model which used the Bayesian statistics inference.

Both of these models are required three elements, which are likelihood distribution, prior distribution, and posterior distribution. Likelihood distribution is a function of the observed data with the given parameters that remains the data are assumed to be conditionally independent and would in fact be correlated and informs the parameter of the data. The prior distribution is a distribution assigned to the parameter before seeing the data as it is informing the prior beliefs or assumptions. Posterior distribution is the product of the likelihood and prior distributions that describes the behavior of the parameters after the data are observed and prior assumptions are made.

For instance, the observed data of a certain event in m areas is notated with $y = \{y_i\}, i = 1,...,m$ and the relative risk is notated with $\theta = \{\theta_i\}, i = 1,...,m$, then the likelihood distribution is defined as [1]:

$$L(y \mid \theta) = \prod_{i=1}^{m} f(y_i \mid \theta) \tag{2}$$

which $L(y \mid \theta)$ is the likelihood distribution function and $f(y_i \mid \theta)$ defines the probability function if the data $y = \{y_i\}$ with given relative risk estimation θ.

The prior distribution is notated by $g(\theta)$ and posterior distribution, notated with $p(\theta \mid y)$ defines the probability of the θ estimation after the data are observed and given. So, the posterior distribution is defined as [1]:

$$p(\theta \mid y) = \frac{L(y \mid \theta) g(\theta)}{p(y)}, p(y) = \int L(y \mid \theta) g(\theta) d\theta \tag{3}$$

with $p(y)$ defined as the probability of the number of the patients and this function can be written as a proportionality function of $p(\theta \mid y) \propto L(y \mid \theta) g(\theta)$.

4. Poisson-gamma Model

In Bayesian inference statistics, every parameters that involved in the model may have specific distribution. In Poisson-gamma model, the observed data $y = \{y_i\}, i = 1,...,m$ are Poisson distributed as well as the likelihood function, where $y \sim Poisson(e_i \theta_i)$. Meanwhile, the prior distribution is assumed to be Gamma distributed, so the posterior distribution is Gamma distributed as well. The likelihood function and prior distribution for this model are defined as:

$$L(y \mid \theta) = \prod_{i=1}^{m} \frac{(e_i \theta)^{y_i} e^{-e_i \theta}}{y_i!} \tag{4}$$

$$g(\theta) = \frac{\lambda e^{-\lambda\theta} (\lambda\theta)^{\alpha-1}}{\Gamma(\alpha)}, \Gamma(\alpha) = (\alpha - 1)!, \alpha \geq 0 \tag{5}$$

As for the posterior distribution function, it is used the proportionality function of $p(\theta \mid y) \propto L(y \mid \theta)g(\theta)$, so that the posterior distribution is defined as:

$$p(\theta \mid y) \propto \left(\prod_{i=1}^{m} \frac{(e_i\theta)^{y_i} e^{-e_i\theta}}{y_i!} \right) \left(\frac{\lambda e^{-\lambda\theta} (\lambda\theta)^{\alpha-1}}{\Gamma(\alpha)} \right) \tag{6}$$

$$p(\theta \mid y) \propto \theta^{\sum y_i + \alpha - 1} e^{-\theta(\sum e_i + \lambda)} \tag{7}$$

$$p(\theta \mid y) = \frac{\left(\sum e_i + \lambda\right)^{\sum y_i + \alpha} e^{-\theta(\sum e_i + \lambda)} \theta^{\sum y_i + \alpha - 1}}{\Gamma\left(\sum y_i + \alpha\right)} \tag{8}$$

To estimate the estimation of the relative risk θ_i, it is tended to use the numerical simulation of MCMC using Gibbs Sampling algorithm for j times iterations. The first procedures in this algorithm for the jth iterations are choosing the estimation of $\theta_i = \theta_i^{(j-1)}$, where $\theta_i^{(j-1)}$ is the estimation of θ_i in the $(j-1)^{\text{th}}$ iteration and then generate the estimation of $\alpha \sim Exp(0,1)$ and $\lambda \sim Exp(0,1)$, where α and λ are the hyperparameters with the certain statistical distributions that called as hyperprior distribution. Next, calculate $\sum y_i + \alpha$ and $\sum e_i + \lambda$ and then generate $\theta_i \mid \theta_{(-i)}, y \sim Gamma\left(\sum y_i + \alpha, \sum e_i + \lambda\right)$, where $\theta_{(-i)}$ are the all of the estimation of θ_i in each region except for the one in i^{th} region. Finally, choose $\theta_i^{(j)} = \theta_i$ as the θ_i is the estimated value of relative risk after j times iteration.

5. Log-normal Model

In the real life, most of the events are considered to follow the pattern Log-normal distribution [5]. Therefore, the value of the relative risk θ_i may as well follow the pattern of Log-normal distribution. In Log-normal model, the observed data $y = \{y_i\}, i = 1,...,m$ are Poisson distributed as same as the one in Poisson-gamma model. In this model the relative risk estimation θ_i is Log-normal distributed and estimated by $\theta_i = e^{v_i}$ where $v_i \sim N(\mu, \vartheta_v^2)$ and for simplification, choose $\mu = 0$, so that $v_i \sim N(0, \vartheta_v^2)$. Review that in the Poisson model is not defined if θ_i is zero, so there is another vague parameter α [7], that caused the estimation of the θ_i becomes $\theta_i = e^{v_i + \alpha}$ where $\log \theta_i = v_i + \alpha$ [2]. The parameter α later will be defined as the mean relative risk in study region. This model is considered to be more flexible than the Poisson-gamma model because the θ_i can be estimated by calculating $\theta_i = e^{v_i + \alpha}$ [2].

After the defining above, the likelihood of this model is defined as $L(y \mid \theta) = \prod_{i=1}^{m} \frac{(e_i\theta)^{y_i} e^{-e_i\theta}}{y_i!}$ and the prior distribution where $v_i \sim N\left(0, \tau_v^2\right)$ is defined as $g(v_i) = \frac{1}{\sqrt{2\pi}\vartheta_v} e^{\frac{-v_i^2}{2\vartheta_v^2}}$.

The first procedure to estimate the relative risk θ_i for the jth iterations is choosing the estimation of $\theta_i = \theta_i^{(j-1)}$, where $\theta_i^{(j-1)}$ is the estimation of θ_i in the $(j-1)^{\text{th}}$ iteration. Next, generate the estimation of $\vartheta_v \sim Gamma(0.5, 0.0005)$, where ϑ_v is the hyperparameter and then generate $v_i \sim N(0, \vartheta_v^2)$ and

$\alpha \sim N(0,10^{-5})$. Finally, the relative risk can be estimated by calculating $\theta_i = e^{\alpha+v_i}$ and choose $\theta_i^{(j)} = \theta_i$.

6. DIC (Deviance Information Criterion)

Both of those models will be compared using DIC (Deviance Information Criterion) that can be calculated in WinBUGS 1.4.3 software. The DIC is the combination between the measuring of fit $\overline{D}$ and the measuring of complexity p_D. The DIC is defined as [4]:

$$DIC = \overline{D} + p_D = D(\overline{\theta}) + 2p_D \tag{9}$$

where p_D is the effective number of parameters and $\overline{D}$ is the expected estimation of deviance where $D(\overline{\theta}) = -2\log p(y \mid \overline{\theta}), p_D = \overline{D} - D(\overline{\theta})$. The estimation of $p(y \mid \overline{\theta})$ is the estimation of the posterior distribution probability in the model that have been generated. The smallest DIC value of a model shows that the model is the best one to fit the data.

7. Data Analysis

In this paper, the calculation of the relative risk for both Poisson-gamma and Log-normal models are done by using WinBUGS 1.4.3 software. The data that being used are the number of Bandung civilization in 2013 from Bandung Statistics Center and the number of the observed cases data are used from Santo Borromeus Hospital in 2013. The results for each stage of dengue disease (early-stage, severe stage, and all stages) for both models are given in one table in Table 1.

In the WinBUGS 1.4.3 software, it provides the user to see how does the simulation been done in the figures and the density graph of the θ_i that have been simulated for 10,000 times. In the Figure 1, it is shown the simulation of Poisson-gamma models and the density for relative risk estimation of dengue disease all stages in Coblong sub-district after 10,000 times iterations. From the Figure 1(a), simulation of calculation of the relative risk of dengue disease all stages that done for 10,000 iterations in Coblong sub-district as it is also shown the range of the θ_i estimation as it is always bigger than 3.5. As for Figure 1(b), it is shown the density of the simulation for the relative risk estimations using Poisson-gamma model in Coblong sub-district, as we can see the density is tending to resemble to the Gamma distribution density graph. Meanwhile, the Log-normal model simulation results are given in Figure 1(c) and (d), in this figure, the simulation and density result for relative risk of dengue disease all stages in Cinambo sub-districts. In Figure 1(c), we can see the simulation of calculation of the relative risk of dengue disease all stages that done for 10,000 iterations in Cinambo sub-district and the range of the θ_i estimation which is always not bigger than 0.5 and the density of the relative risk estimation in Figure 1(d) is resemble to the Log-normal distribution density graph.

Figure 1. Simulation of relative risk estimation using Poisson-gamma model in Coblong sub-district (a) and the density of the relative risk estimation from the simulation that resemble to Gamma distribution density graph (b), Simulation of relative risk estimation using Log-normal model in Cinambo sub-district (c) and the density of the relative risk estimation from the simulation that resemble to Log-normal distribution density graph (d)

Table 1. Number of Civilizations (N_i), Number of Observed Patients for Early-stage (y_i e), Number of Observed Patients for Severe-stage (y_i s), Number of Observed Patients for all stages (y_i a), Relative Risk of Dengue Disease Early-stage for Poisson-gamma&Log-normal Models (θ_i eP and θ_i eL), Relative Risk of Dengue Disease Severe-stage for Poisson-gamma &Log-normal Models (θ_i sP and θ_i sL), Relative Risk of Dengue Disease All Stages for Poisson-gamma&Log-normal Models (θ_i aP and θ_i aL), and The Relative Risk Categories (C)

i	Sub-districts	N_i	y_i e	y_i s	y_i a	θ_i eP	C	θ_i eP	C	θ_i aP	C	θ_i eL	C	θ_i sL	C	θ_i aL	C
1	Andir	97,553	12	20	32	0.8706	L	0.3218	VL	0.4134	VL	0.8444	L	0.3246	VL	0.4132	VL
2	Antapani	74,461	10	41	51	0.9455	L	0.8217	L	0.8411	L	0.9098	L	0.8157	L	0.8348	L
3	Arcamanik	69,313	15	41	56	1.4320	M	0.8840	L	0.9857	L	1.4120	M	0.8725	L	0.9815	L
4	Astana Anyar	68,830	5	38	43	0.5770	L	0.8227	L	0.7700	L	0.5551	L	0.8163	L	0.7629	L
5	Babakan Ciparay	147,096	4	39	43	0.2520	VL	0.4038	VL	0.3670	VL	0.2643	VL	0.4030	VL	0.3671	VL
6	Bandung Kidul	58,957	5	41	46	0.6615	L	1.0270	M	0.9575	L	0.6285	L	1.0220	M	0.9482	L
7	Bandung Kulon	142,412	8	45	53	0.4396	VL	0.4772	VL	0.4627	VL	0.4306	VL	0.4763	VL	0.4618	VL
8	Bandung Wetan	31,124	19	83	102	3.3500	VH	3.7340	VH	3.8080	VH	3.7130	VH	3.8410	VH	3.9070	VH
9	Batununggal	120,928	9	39	48	0.5603	L	0.4876	VL	0.4957	VL	0.5441	L	0.4852	VL	0.4925	VL
10	Bojongloa Kaler	120,406	3	18	21	0.2503	VL	0.2369	VL	0.2278	VL	0.2640	VL	0.2434	VL	0.2319	VL
11	Bojongloa Kidul	85,668	3	31	34	0.3363	VL	0.5504	L	0.4969	VL	0.3439	VL	0.5463	L	0.4949	VL
12	Buah Batu	95,108	14	55	69	1.0210	M	0.8634	L	0.8912	L	0.9976	L	0.8555	L	0.8842	L
13	Cibeunying Kaler	70,924	25	95	120	2.2430	VH	1.9540	H	2.0380	VH	2.2890	VH	1.9560	H	2.0440	VH
14	Cibeunying Kidul	107,806	14	36	50	0.9200	L	0.5076	L	0.5759	L	0.8917	L	0.5053	L	0.5732	L
15	Cibiru	72,016	2	10	12	0.3112	VL	0.2341	VL	0.2264	VL	0.3228	VL	0.2413	VL	0.2336	VL
16	Cicendo	99,752	10	84	94	0.7299	L	1.2440	M	1.1490	M	0.7054	L	1.2370	M	1.1450	M
17	Cidadap	58,672	15	68	83	1.6510	H	1.6880	H	1.7040	H	1.6400	H	1.6900	H	1.7030	H
18	Cinambo	25,231	1	1	2	0.5080	L	0.1439	VL	0.1684	VL	0.4930	VL	0.1896	VL	0.1990	VL
19	Coblong	131,530	81	334	415	4.0030	VH	3.7130	VH	3.8090	VH	4.1510	VH	3.7380	VH	3.8310	VH
20	Gedebage	37,082	3	10	13	0.6721	L	0.4378	VL	0.4600	VL	0.6309	L	0.4365	VL	0.4563	VL
21	Kiara Condong	131,972	4	80	84	0.2781	VL	0.9010	L	0.7805	L	0.2877	VL	0.8960	L	0.7769	L
22	Lengkong	71,187	21	82	103	1.9080	H	1.6830	H	1.7460	H	1.9160	H	1.6860	H	1.7470	H
23	Mandalajati	63,578	4	14	18	0.5267	L	0.3529	VL	0.3675	VL	0.5082	L	0.3550	VL	0.3692	VL
24	Panyileukan	40,248	1	6	7	0.3672	VL	0.2691	VL	0.2526	VL	0.3705	VL	0.2782	VL	0.2620	VL
25	Rancasari	76,896	9	41	50	0.8398	L	0.7961	L	0.8000	L	0.8049	L	0.7898	L	0.7947	L
26	Regol	81,467	15	72	87	1.2440	M	1.3010	M	1.2990	M	1.2200	M	1.2960	M	1.2920	M
27	Sukajadi	108,375	16	80	96	1.0290	M	1.0920	M	1.0810	M	1.0010	M	1.0870	M	1.0770	M
28	Sukasari	81,908	21	120	141	1.6800	H	2.1370	VH	2.0760	VH	1.6780	H	2.1450	VH	2.0790	VH
29	Sumur Bandung	36,579	4	32	36	0.8268	L	1.2820	M	1.1930	M	0.7693	L	1.2630	M	1.1810	M
30	Ujung Berung	76,902	4	19	23	0.4466	VL	0.3883	VL	0.3830	VL	0.4419	VL	0.3871	VL	0.3820	VL

IOP Conf. Series: Journal of Physics: Conf. Series **812** (2017) 012034 doi:10.1088/1742-6596/812/1/012034

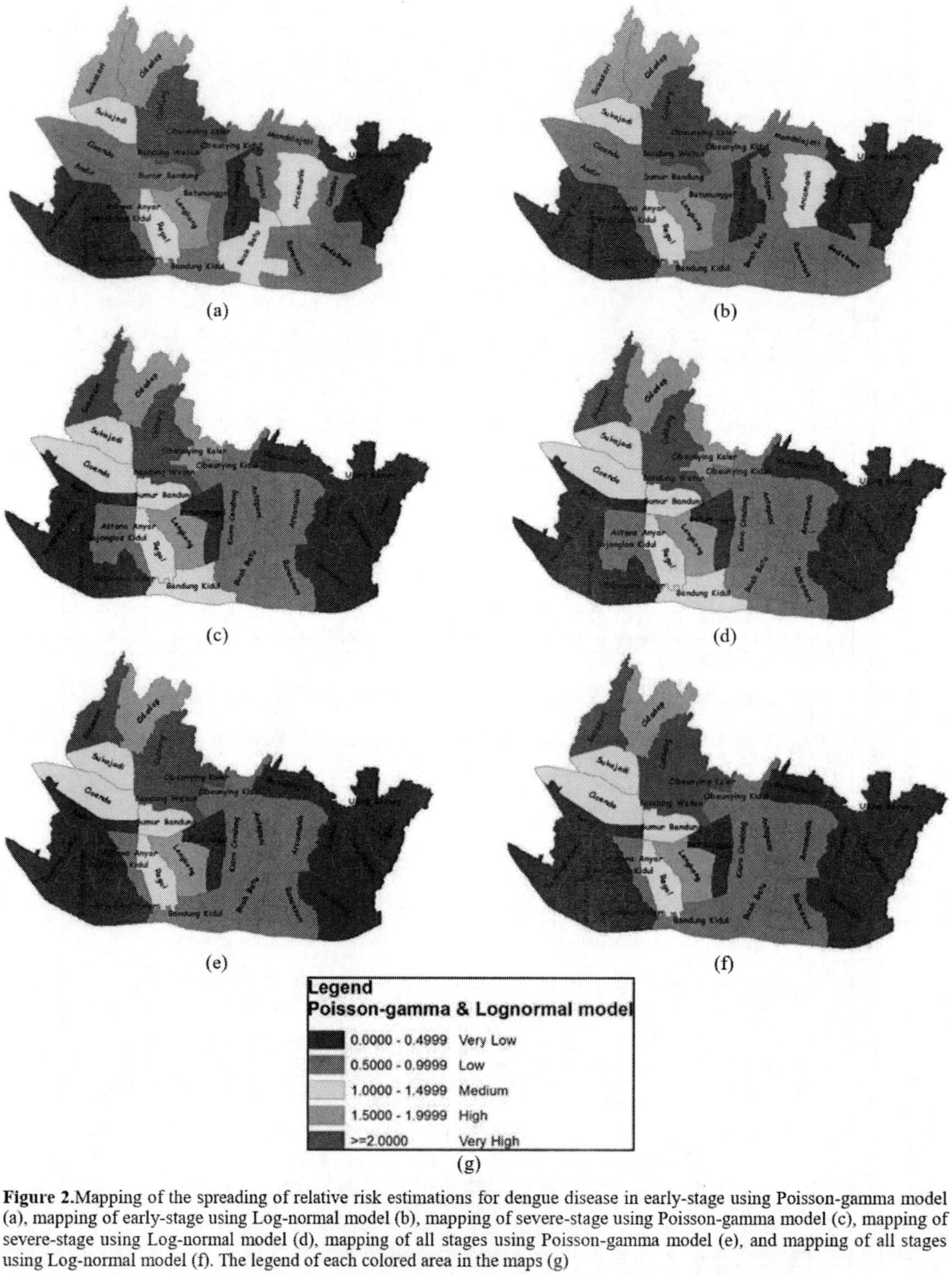

Figure 2. Mapping of the spreading of relative risk estimations for dengue disease in early-stage using Poisson-gamma model (a), mapping of early-stage using Log-normal model (b), mapping of severe-stage using Poisson-gamma model (c), mapping of severe-stage using Log-normal model (d), mapping of all stages using Poisson-gamma model (e), and mapping of all stages using Log-normal model (f). The legend of each colored area in the maps (g)

The mapping for the spreading of the relative risk level for 3 stages in both models are given in Figure 2. The mapping for the early-stages of dengue disease for Poisson-gamma and Log-normal models are respectively given by Figure 2(a) and Figure 2(b), the severe-stages are given by Figure 2(c) and Figure 2(d), and the all stages mapping are given by Figure 2(e) and Figure 2(f), the legend for the map is given in Figure 2(g). The relative risk estimation in both models are the same for the dengue disease in severe-stage and all stages are the same. In the early-stage, most of the sub-districts have the same level of relative risk estimation, there is one sub-district that different, which is BuahBatu sub-district. From the mapping result, it is easier to see that both models are giving a close estimation of the relative risk.

Both models are then be compared using DIC from Equation (1) which can be calculated in WinBUGS 1.4.3 and the results are given in Table 2. The calculation involved the simulations and the results are compared by each severity category. The result is then determined by taking the smaller estimation of the DIC which is the deviance for the data.

Table 2. The value of $\overline{D}$, $D(\overline{\theta})$, p_D , and DIC for each severity stages and models.

Severity and Models	$\overline{D}$	$D(\overline{\theta})$	p_D	DIC
Early-stage in Poisson-gamma	146.727	122.634	24.094	170.821
Early-stage in Log-normal	145.584	120.789	24.795	170.380
Severe-stage in Poisson-gamma	193.637	165.677	27.960	221.597
Severe-stage in Log-normal	194.048	165.674	28.374	222.423
All stages in Poisson-gamma	199.654	171.294	28.360	228.013
All stages in Log-normal	199.780	171.155	28.625	228.405

8. Conclusion

Coblong and Bandung Wetan sub-districts had the highest relative risk using both models for the early-stage, severe-stage, and all stages. Coblong sub-district had the highest relative risk estimation in early-stage of dengue disease in both models and all stages in Poisson-gamma models. Meanwhile, Bandung Wetan sub-district had the highest relative risk estimation for severe-stage dengue disease in both models and all stages in Log-normal model. Cinambo sub-district had the lowest relative risk using both models for the severe-stage and all stages and BojongloaKaler sub-district had the lowest relative risk using both models for the early-stage. The models that are used in this paper are both almost have the same results, but after comparing both models, Log-normal model is the best model to fit the early-stage and severe-stage data, meanwhile the Poisson-gamma model is the best model to fit the all stages data.

9. References

[1] LawsonA B 2013 *Bayesian Disease Mapping Hierarchical Modeling in Spatial Epidemiology 2nd edition (*Boca Raton: CRC Press)

[2] Lawson A B,Browne W J and Rodeiro C L V 2003 *Disease Mapping with WinBUGS and MlwiN* (Chichester: John Wiley & Sons Ltd., Chichester)

[3] Yong B, KristianiF and Irawan R2016*CR Journal***2** 39-53

[4] Lunn D, Jackson C, Best N, Thomas Aand Spiegelhalter D 2012*The BUGS Book - A Practical Introduction to Bayesian Analysis* (Boca Raton: CRC Press, Boca Raton)

[5] Limpert E, Stahel W Aand AbbtM2001*BioScience***51** 341-52

[6] Samat N A andMa'arof S H M I 2013 *International Journal of Mathematical, Computational, Physical, Electrical and Computer Engineering*71299-303

[7] S. R. and V. P. 2014 *Annals of Biological Research* **5**23-31

[8] World Health Organization 2009 *Dengue Guidelines for Diagnosis, Treatment, Prevention And Control*(Geneva: World Health Organization)

Acknowledgments
The authors appreciate the valuable insights and comments provided by anonymous referees which improve this paper. The authors also acknowledge support from the Health Department of Bandung and also Santo Borromeus Hospital for their data.

IOP Conf. Series: Journal of Physics: Conf. Series **812** (2017) 012035 doi:10.1088/1742-6596/812/1/012035

Virtual Reality Website of Indonesia National Monument and Its Environment

B A Wardijono, F Hendajani and S A Sudiro

Computer Engineering, STMIK Jakarta STI&K, Jl. BRI Radio Dalam, Jakarta Selatan, Daerah Khusus Ibu Kota Jakarta 12140, Indonesia

E-mail: bheta@jak-stik.ac.id

Abstract. National Monument (Monumen Nasional) is an Indonesia National Monument building where located in Jakarta. This monument is a symbol of Jakarta and it is a pride monument of the people in Jakarta and Indonesia country. This National Monument also has a museum about the history of the Indonesian country. To provide information to the general public, in this research we created and developed models of 3D graphics from the National Monument and the surrounding environment. Virtual Reality technology was used to display the visualization of the National Monument and the surrounding environment in 3D graphics form. Latest programming technology makes it possible to display 3D objects via the internet browser. This research used Unity3D and WebGL to make virtual reality models that can be implemented and showed on a Website. The result from this research is the development of 3-dimensional Website of the National Monument and its objects surrounding the environment that can be displayed through the Web browser. The virtual reality of whole objects was divided into a number of scenes, so that it can be displayed in real time visualization.

1. Introduction

National Monument (Monumen Nasional) or popularly abbreviated with the "Tugu Monas" is a 132-meter-high memorial monument as the memory of the resistance and the Indonesian people's struggle for independence from Dutch colonial. National Monument was built during the reign of president Sukarno on August 17th, 1961. Tugu Monas was opened to the public on July 12, 1975. Monas has gilded flame symbolizing the spirit of Indonesian struggle blazing. The National Monument is located in the centre of Merdeka Square, Central Jakarta.

The development of Monas commemorate and preserve the Indonesian struggle for independence during the revolution of 1945, in order to continue to inspire patriotism and spirit of present and future generations. Figure 1 shows the real National Monument in a picture.

Submission and dissemination of information about the monument to the public becomes important because every region in Indonesia has experience a form of colonialism and fighting to repel the invaders. Dissemination of this information is intended to make the public and the citizens of Indonesia always remember and not to forget the history of the struggle for Indonesian independence patriot.

One way of disseminating information today that effective is to utilize computer technology, such as web-based application programs and the Internet. Computers today can not only be used to process data and information in the form of text, but also in other formats such as images, moving images (animation), 3-dimensional models, simulations, sound and video. Through this diverse format that allows information from a source can be packaged into a form more attractive, easier, and effective in their delivery.

Figure 1. National monument statue.

The issues raised in this research is how to design and develope a three-dimensional graphic models of the National Monument and inform it the people of Indonesia as well as foreign tourists through 3D computer graphics technology and Web technology. By modeling the National Monument in the form of a 3D graphic model in visual way, the important information from the National Monument and it surrounding objects can be conveyed to the public.

The purpose of this study include: disseminating information to the public about the National Monument and the its objects in surrounding environment, providing information about the struggle for independence patriot in dioramas room, and apply the knowledge in the field of information technology and computer graphics by developing the virtual reality website.

A number of previous studies have used virtual reality technology to display 3D graphics along with the surrounding environment. An example that shows a 3D model are displayed in VRML format in which the model can be interactive and onshore users navigate the 3D virtual environment, through the control of the keyboard or other control device [1]. Another VRML model was Heidelberg city as shown in Figure 2.

Figure 2. Visualization of virtual reality with VRML technology [2].

2. Literature review

2.1. 3D Model and virtual reality

3D modelling is the process of developing a mathematical representation of any three-dimensional surface of an object (either inanimate or living) via specialized software. Modelling results is referred to as 3D models which can be created automatically or manually.

Currently, there are many software as 3D modeller. There are commercial software 3D Studio MaxTM and there is also freeware like BlenderTM. Such software is the 3D computer graphics program has feature to create animations, 3D models, and image by having the capabilities and features to create three-dimensional models, particle systems, dynamic simulation, shader, illumination, as well as animation.

Virtual Reality (VR) is a technology that allows one to simulate on a real object using a computer that is able to generate a 3D atmosphere that makes the user as physically involved. This technology can be used for learning in visual form so as to facilitate the learning process. VR technology itself has actually been around since the 1960s and is known in some form, implementation of VR itself with regard to the depiction of a scope that is presented in the use of computers, VR technology itself is actually a method used to simulate an environment and make the user as being in situations and places that become the object of VR, both in taking control and interact directly with the environment.

2.2. WebGL

WebGL is part of HTML[5] technology, as bundled in most browsers that support HTML5 technology. WebGL can work on multiple browser-based desktop and mobile. WebGL has developed and makes searching the web more beautiful in appearance because it displays a visualization of moving, from a game to data visualization. WebGL was developed by the Khronos Group, which is an agency that also regulates OpenGL, and is a free cross-platform API that brings OpenGLES 2.0 to the web as a 3D drawing context in HTML.

This study used Unity3D, which is software used for several kinds of things, such as for the visualization of the model, making 3D games and animated objects. Unity3D WebGL allows developers produce 3 dimensional VR models that can later be displayed through a web browser.

A number of previous studies that utilize WebGL to display 3-dimensional objects in a large size, for example, to produce a virtual museum [3], Cultural Heritage [4], Architectural Heritage Leaning [5], Cultural Artifacts [6], Virtual Educational [7] and Huge City Model [8].

3. Research method

In this study, basically there are two main things those are 3-dimensional modeling and visualizing in the form of virtual reality. It was needed to be prepared number of steps so that the end product of the 3D modeling of the National Monument and its objects surrounding area can be displayed visually on the Web. Users can explore the 3D models and its environments by controlling buttons on the keyboard or by using other control devices (mouse). The research method that was used in this research is presented in Figure 3.

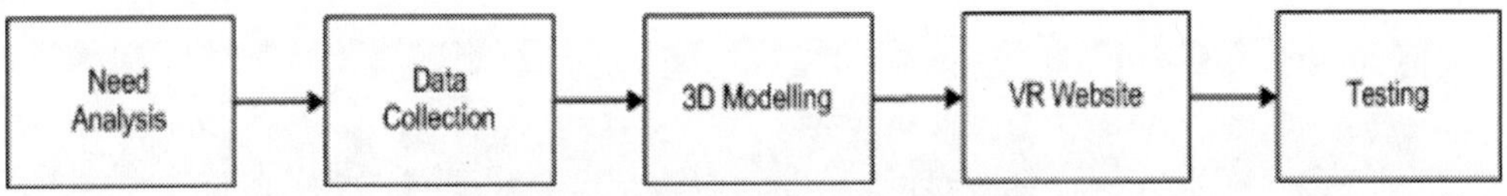

Figure 3. Step by step of research.

The research was begun by need analysis. This is analyzing hardware and software that needed to generate a web-based application. Data Collection step was done by visiting the location of research i.e the National Monument. This step research was collected information of shape, dimension and texture of objects. 3-Dimensional modeling step was creating and modeling real National Monument and the objects that exist in the form of three-dimensional graphic models. The research continued with creating VR website, converting the 3-dimensional model that has been created in the form of virtual reality that can be showed with a web browser and developing a website. Testing step was testing of the models with examining the models to display whether the model correct or error to display.

4. Designing

The website was designed in 5 main menus. The menus are: home, gallery, history, about and virtual reality. Home menu shows the National Monument description, Gallery consists of the slide show many pictures of National Monument in real condition, History consists of the history of National Monument about menu shows the author of website and Virtual Reality menu consists of some scene virtual reality model of National Monument and Its environment. This navigation for the website is presented in figure 4.

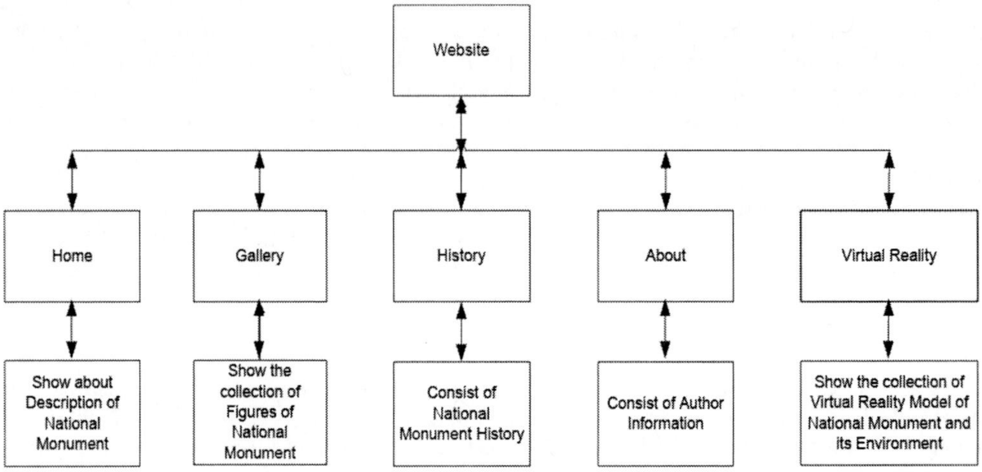

Figure 4. Navigation structure of website.

5. Result

Result of the study was a three-dimensional model of the National Monument and the objects in surrounding environment, which can be displayed in a web browser and the website of the National Monument. The research data have been taken from the real National Monument. Texture data taken from real National Monument and then added to each object as original object. To generate the texture according to the dimensions of the original model, adding some texture on large objects must be adapted to the actual dimensions. Figure 5 shows the Website of National Monument with VR technology inside.

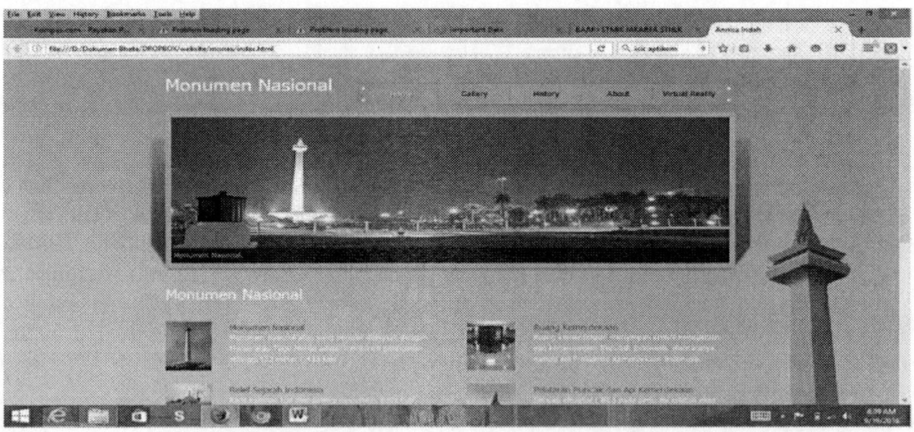

Figure 5. The national monument website.

IOP Conf. Series: Journal of Physics: Conf. Series **812** (2017) 012035 doi:10.1088/1742-6596/812/1/012035

Virtual Reality shown through the website is placed in Virtual Reality and Gallery menu. Virtual reality modelling is done by dividing into a number of scenes, where in every scene featuring a specific object, either inside or outside the monument. This is done due consideration of the size of large models that will be hard for a computer that is used when rendering.

In the Gallery menu, there are 9 pieces VR which can be accessed by the user, i.e.: the National Monument, the Statue of Kartini, Diponegoro statue, Statue of MH Thamrin, the Statue of Five Hero, Reflection Garden, Deer Garden, Independence room, Top of Monas, and Museum. At any existing VR models, users can perform a walkthrough through via the keyboard by pressing the up arrow, down left and right. Image Virtual Reality of National Monument looks beyond is as shown in Figure 6.

Figure 6. Virtual reality model of National Monument.

Furthermore, in Figure 7 (a), (b) dan (c) shows the virtual reality models for other scenes

(a) (b) (c)

Figure 7. Scene of (a) Diponegoro Statue, (b) Deer Garden, and (c) Kartini Statue.

Another scene are from Top of National Monument and scene inside of the Monument i.e. the History museum of Indonesia, as seen in figure 8.

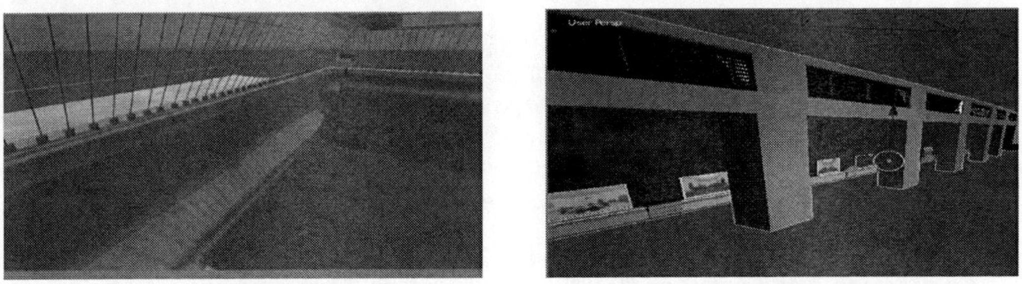

(a) (b)

Figure 8. Scene (a) from top of National Monument, (b) inside National Monument.

The Website has been tested on localhost, and will be published after the end of project on December 2016 on http://www.vr-monas.com. The testing was done by examining every existing menu: Home, Gallery, History, About and Virtual Reality. Especially for Virtual Reality menu of website, it was tested using control via the keyboard with arrow keys up, down, left and right. Besides, it was also tested by using mouse to navigate the virtual reality models.

The testing results of navigation in virtual reality models showed that the models can be controlled by the user, where the user can do a walkthrough the virtual reality models through keyboard and mouse. Testing was done by using two web browsers i.e. Mozilla Firefox and Google Chrome. The result of testing by a web browser Mozilla Firefox and Google Chrome obtained good results. Both browsers can display the virtual reality models in real time and users can explore virtual reality models with easy. Table 1 shows the result of testing to load virtual reality model of some scene of National Monument web. The testing was done using computer specification: Processor intel Core i7 5500U CPU @2,4GHz , RAM 4GB, NVIDIA Geforce 920M 4GB, and Operating System Win 8.1.

Table 1. Testing result to load VR-Model of scenes.

Scene	Loading Time (second)
National Monument Statue	12.07
Moh. Thamrin Statue	26.16
Diponegoro Statue	29.68
Deer Garden	24.98
Five Hero Statue	25.72
Kartini Statue	23.70
Freedom Room	23.69
Reflection Garden	26.14
Top of Monas	23.62

6. Conclusion

This research has successfully generated three-dimensional graphics modelling of Indonesia National Monument and objects surrounding the environment. Result of this 3d modelling and virtual reality models was successfully displayed in a web. The test results showed the model can be displayed in the web and the user can roam (walkthrough) in real time on the virtual reality models. Users can control the virtual reality models with a keyboard and mouse.

Suggestions for next development of this research is to add models incorporate more dynamic or animated elements in it, such as a human object, animal, or a moving vehicle. Thus the generated virtual reality can resemble the real conditions.

7. References

[1] Jeginovic S, Rizvic S and Chalmers 2004 *Interactive 3D Models–From 3ds Max to VRML* (Electrical Engineering Faculty, University of Sarajevo)

[2] Schilling A and Zipf A 2003 *Proc. of the Eighth International Conference on 3D Web Technology (Saint-Malo)* pp 39-47

[3] Aderhold A, Jung Y, Wilkosinska K and Fellner D W 2013 *Distributed 3d Model Optimization for The Web with The Common Implementation Framework for Online Virtual Museums* (Digital Heritage International Congress, IEEE Xplore Digital Library) vol 2 pp 719-26

[4] Auer M, Agugiarob G, Billena N, Loos L and Zipf A 2014 *Web-Based Visualization and Query of Semantically Segmented Multiresolution 3D Models in The Field of Cultural Heritage* (ISPRS Annals of the Photogrammetry, Remote Sensing and Spatial Information Sciences) vol II-5 pp 33-9

[5] Bakar J A A, Jahnkassim P S and Mahmud M 2013 *Int. J. Interactive Digital Media* **1** 37-45

[6] Chanhoom W 2014 *International Journal of Advances in Computer Science & Its Applications – IJCSIA* **4** 10-4

[7] Seok K H and Kim Y S 2014 *Int. J. Multimedia and Ubiquitous Engineering* **9** 273-84

[8] Prandi F, Devigili F, Soave M, Di Staso U and De Amicis R 2015 *3D Web Visualization of Huge CityGML Models* (The International Archives of the Photogrammetry, Remote Sensing and Spatial Information Sciences, ISPRS Geospatial Week) vol XL-3/W3 pp 601-5

Acknowledgments
This research was funded and sponsored by Dirjen Dikti on Competitive Research Grant scheme in 2016 with a Letter of Assignment No: 007/PHB-DIKTI/BP-LP2M/STI&K/II/2016.

Spatial and Activities Models of Airport Based on GIS and Dynamic Model

R M Masri and I M Purwaamijaya

Department of Civil Engineering Education, Universitas Pendidikan Indonesia, Jl. Dr Setiabudhi No. 229, Bandung 40154, Indonesia
Email: rinamasri@upi.edu

Abstract. The purpose of research were (1) a conceptual, functional model designed and implementation for spatial airports, (2) a causal, flow diagrams and mathematical equations made for airport activity, (3) obtained information on the conditions of space and activities at airports assessment, (4) the space and activities evaluation at airports based on national and international airport services standards, (5) options provided to improve the spatial and airport activities performance become the international standards airport. Descriptive method is used for the research. Husein Sastranegara Airport in Bandung, West Java, Indonesia was study location. The research was conducted on September 2015 to April 2016. A spatial analysis is used to obtain runway, taxiway and building airport geometric information. A system analysis is used to obtain the relationship between components in airports, dynamic simulation activity at airports and information on the results tables and graphs of dynamic model. Airport national and international standard could not be fulfilled by spatial and activity existing condition of Husein Sastranegara. Idea of re-location program is proposed as problem solving for constructing new airport which could be serving international air transportation.

1. Introduction

Geographical Information Systems (GIS) is used for transport modeling because can significantly contribute to a more efficient modeling process and more reliable model results, namely, geospatial data, disaggregated transport models and the role of geo-vizualitation. This paper is presented to stengthen the spatial perspective in transport modeling and to call for a further integration of GIS in the domain of transport modeling [1].

Many elaborate dynamic model are explored effectively in transport system modeling. We various decision rules that embed distance and direction, density thresholds and transition or mutation probabilities into the model's dynamics, and we then outline the software designed to generate effective transport simulation consistent with GIS data inputs, outputs and related functionality [2].

The modeling of spatial processes is involved as an emerging branch of geo-computing. The most important being traditional regionalized system dynamics approaches were used at large spatial scales [3].

Airport modeling is of highest diversity combining presentation of navigation airspace, sophisticated constructions, buildings, different transportation subsystems, etc. Airports technological and organizational complexity forces the airport planners to collect highly accurate spatial data, develop better and more precise navigation and layout plans, manmade construction models, and employ state-of-the-art geographic information systems for running their business. Several national aviation administrations and the International Civil Aviation Organization (ICAO) have issued documents stating two important objectives. First objective instructs the airport planners how to capture and store spatial data and develop an airport database, particularly an electronic airport layout plan (eALP) that will serve for the development an enterprise airport geographic information system (eAGIS). Second objective requires the deployment of avionic equipment that will convert airspace traffic management from being ground-based to satellite-based, which requires highly accurate three-dimensional (3D) airport models. With GIS, Computer Aided Drafting and Design (CADD), and Building Information Modeling (BIM) technologies currently in place, sophisticated airspace and manmade modeling has to be fulfilled by airport planning professionals. As airports move toward the implementation of new traffic management, the airspace and navigation obstruction analyses are of highest priority as these analyses are based on different spatial data models that are simultaneously involved. The most important models are of: navigation data and imaginary protective surfaces; ground and runway surfaces; manmade structures [4].

Geographic information science and technologies are revolutionizing basic and applied science by allowing integrated holistic approaches to the analysis of geographic locations and their attributes. However, the increasing mobility and connectivity of many people in the world means that the relationships between people and place are becoming more subtle and complex, rendering a place-based perspective incomplete [5].

The economic behavior of airports changed dramatically in the last quarter-century, with consequences on the internal management structure, the relationship with other competitive airports, and the way airports interact with their nearby city. Airport commercialization is the transformation from a public transportation utility towards a profit-orientated enterprise. Parking lots, shops and shopping malls, restaurants, entertainment centers, hotels and congress facilities can be found at every bigger airport these days and belong to the non-aviation sector. Revenues created by non-aviation businesses sum up to 50 % and more of the airports income. The new spatial quality in this is that non-aviation activities are also placed outside the airport fence [6].

The Geography of Transport Systems, concerned with movements of freight, people and information, tries to link spatial constraints and attributes with the origin, the destination, the extent, the nature and the purpose of movements. It is divided into nine topics, each covering a specific conceptual dimension, including: Networks, modes and terminals, international transportation, urban transportation dan environmental impacts [7].

Transportation researchers and airport engineers joined hands to explore practical aspects of airport simulation and their utilization to enhance airport planning (both airside and landside) and their implementation for complex airport and airspace capacity studies. The primary objective here was to solve the lingering capacity shortages plaguing busy airports, the NAS airspace, and the airport system at large [8].

2. Research method

The descriptive method was used for spatial analysis. The descriptive method was applied to the analysis of spatial form of presentation of thematic maps that go through stages, namely: (1) user needs identification, (2) data collection (maps and statistics), (3) data groupped by types and levels, (4) input, processing and output of data conceptual modeling, (5) design data structures and database management functional modeling, (6) spatial development system implementation, (7) analog data converted into digital, (8) a model programmed of the interaction between the database management systems with the users, (9) spatial system tested to obtain feedback, (10) the spatial system revision, (11) the use of a spatial system, (12) spatial system maintenance.

The method used is mechanistic explanatory for airport activities model. Methods mechanistic explanatory through stages, namely: (1) a literature study and analysis of real-world system, (2) the components identification that make up the real world system, (3) a causal loop diagram design, (4) flow chart modeling, (5) primary and secondary data collection, (6) data entry into the flow chart model, (7) the flow chart model tested, (8) feedback, (9) the flow chart, (10) the flow chart model revised implementation, (11) tables and charts printed out, (12) mathematical model printed out, (13) the model validity tested, (14) the model sensitivity tested, (15) conclusions and recommendations.

3. Result and discussion

3.1.Spatial information and model

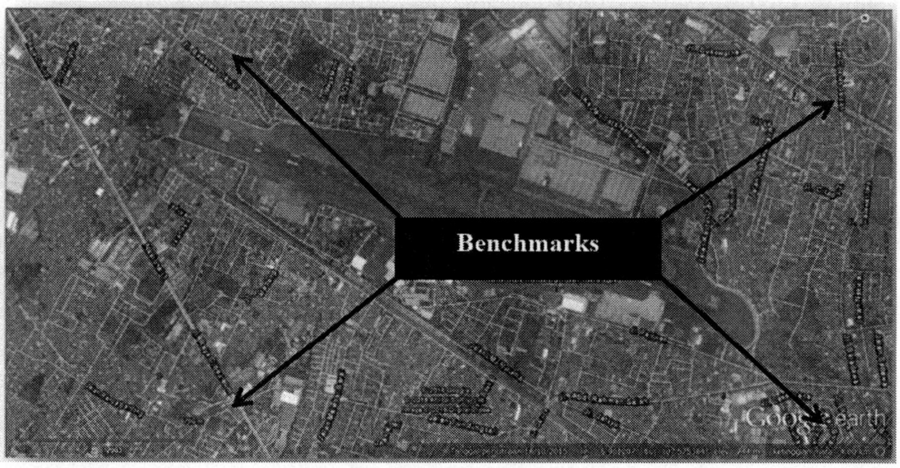

Figure 1. Satellite image benchmarks of Husein Sastranegara Airport from 4 kms height from earth surface (Source : Google Earth, 2016).

Benchmarks was used as reference points for registering and transforming satellite image coordinates to zone 48 UTM (universal transverse mercator) coordinates. Raster data were converted become vector data through on screen digitizing. Affine method was used as registration and transformation process.

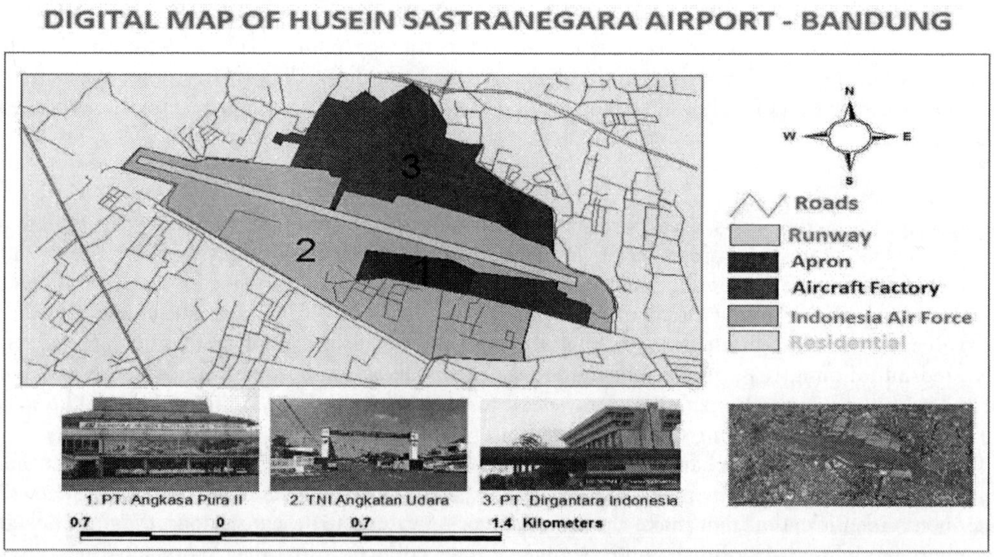

Figure 2. Digital map of Husein Sastranegara Airport (Satellite map digitalization, 2016).

Digital map was produced from digitizing on screen which was integrating graphical (point, line, polygon) and attributes (numeric, string, boolean, date) data. Spatial analysis could be done on digital map resulted.

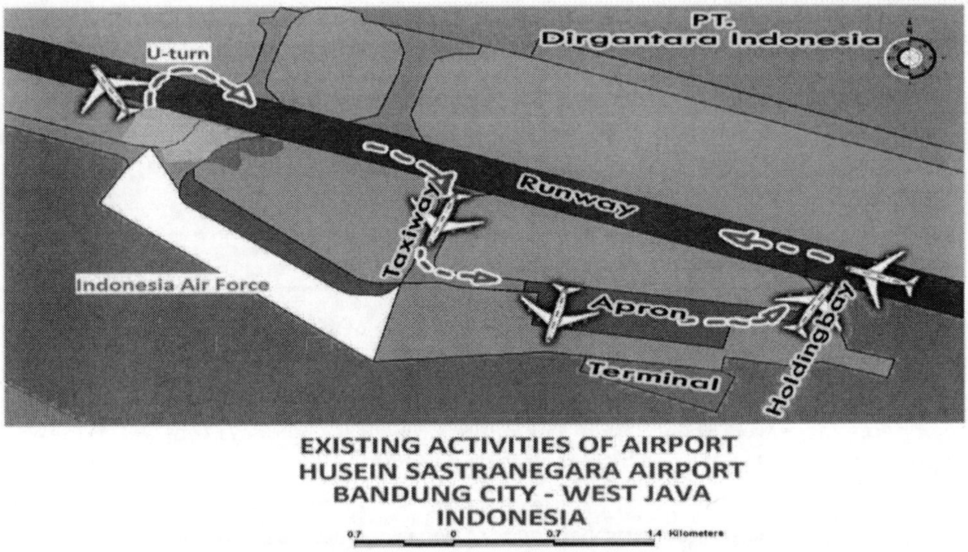

Figure 3. Existing activities map of Husein Sastranegara Airport (spatial analysis, 2016).

Airplane was landed on runway from east to west. Airplane was going to change its direction to taxiway then apron from west to east and stopped near of terminal. Only one runway was constructed in Husein Sastranegara Airport.

Table 1. Area calculation of Husein Sastranegara Airport.

No	Owner	Area
1.	Indonesia Air Force	358,000 m^2
2.	Angkasa Pura Company	5,000 m^2
3.	Dirgantara Indonesia Aircraft Factory	150,000 m^2
4.	Runway	99,000 m^2

Area of Indonesia Air Force, Angkasa Pura Company, Dirgantara Indonesia Aircraft Factory and Runway were produced from spatial analysis using GIS software. The largest area was founded on Indonesia Air Force parcel.

3.2. *Dynamic simulation model*

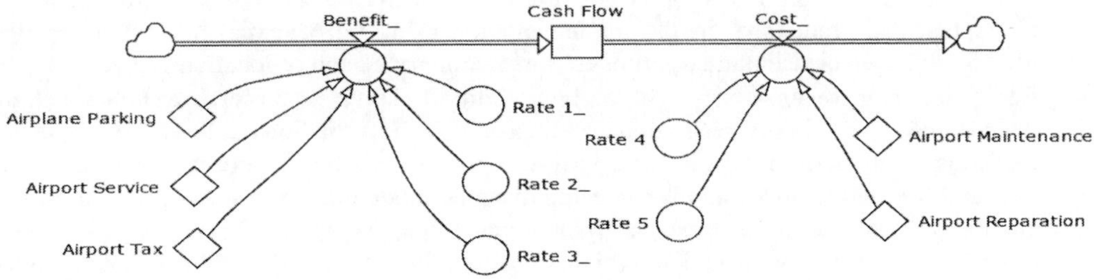

Figure 4. Flow chart diagram of spatial and activities models of airport based on gis and dynamic model.

IOP Conf. Series: Journal of Physics: Conf. Series **812** (2017) 012036 doi:10.1088/1742-6596/812/1/012036

Airplane parking, airport service, airport tax, airport maintenance and airport reparation were grouped into constant symbols. Rate 1, 2, 3, 4, 5 were grouped into auxiliary symbols. Inflow (benefit) or outflow (cost) is constant time auxiliary. Reservoir (cash flow) is inflow (benefit) minus outflow (cost).

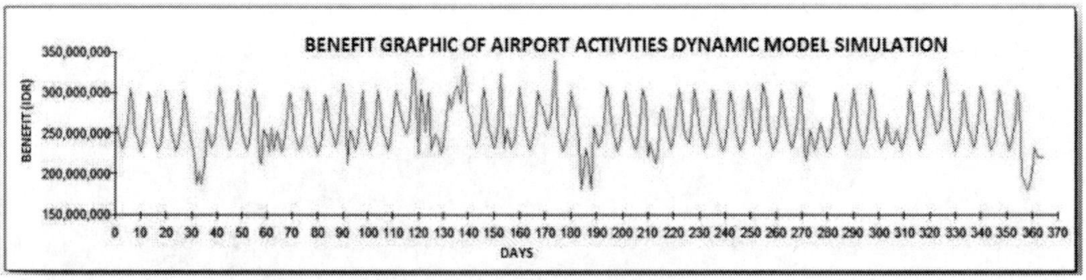

Figure 5. Benefit graphic of spatial and activities models of airport based on gis and dynamic model.

Graph fluctuation for beneficial components were shown from dynamic model simulation. Beneficial prices range were about 150 to 350 millions rupiah for 365 days (one year). Lowest price was happened on day 180 to 190 and highest price on day 170 to 175.

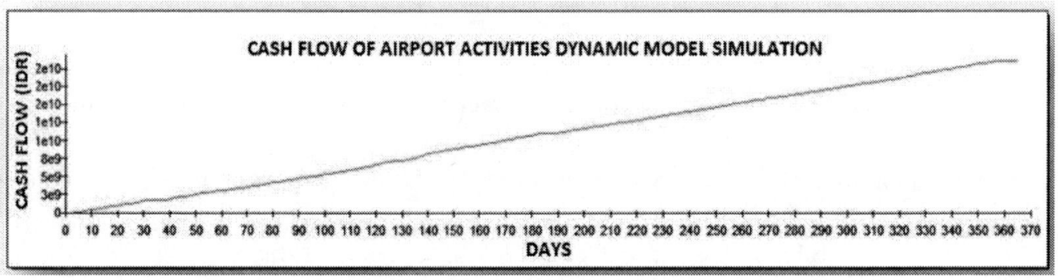

Figure 6. Cash flow graphic of spatial and activities models of airport based on gis and dynamic model.

Graph was continuing up for cash flow were shown from dynamic model simulation. Cash flow price was started from 0 rupiah on day 0 become 2 billion rupiah on day 365. Cash flow was beneficial minus cost components.

4. Conclusion

- Airport parcel area information, take off and landing aircraft movement and properties neighbourhood based on GIS topology (closure, connectivity, conjuguity) were resulted from conceptual and functional model implementation using GIS. Spatial information could be used for accurate decision in airport operation, maintenance and re-location project.
- Beneficial components (airplane parking, airport service, airport tax), cost (airport maintenance and airport reparation) and cash flow informations referenced on dynamic modeling symbols (reservoir, inflow, outflow, constant, auxiliary) were produced from causal loop and flow chart model simulation using dynamic modeling. Tables and graphs information could be contributed to performance evaluation and financial audit.
- Spatial and activity information which were resulted from GIS and dynamic modeling simulation were describing existing spatial condition not feasible for wide body aircraft international flight. Only one runway for airplane landing and take off. New runway construction could not be realized because of land acquisition.

- Airport National Standard still could be fulfilled by Husein Sastranegara but International Standard could not. Passenger increased in the future (5 to 10 years later) could be caused Husein Sastranegara airport performance decreased.
- Existing performance condition of Husein Sastranegara airport was very difficult for being increased in spatial and activity perspective. Airport re-location project is one option for being choiced.

5. References

[1] Loidl M, Wallentin G, Cyganski R, Graser A, Scholz J and Haslauer E 2016 *International Journal of Geo-information* 1-23
[2] Batty M, Xie Y and Sun Z 1999 *Elsevier* **23** 205-33
[3] White R and Engelen G 2000 *Elsevier* **24** 383-400
[4] Panayotov A and Georgiev 2012 *International Conference on Computer Systems and Technologies - CompSysTech'12(Ruse)* (USA: Researchgate) pp 1-9
[5] Miller H 2007 *Geography Compass Journal Compilation* **1** 503-35
[6] Zepezauer F 2012 *A case study about gainesville regional airport and its effects on the economic development of Gainesville, Florida* 2012 Paper for NEURUS Program Spring
[7] Rodrigue J P, Comtois C and Slack B 2006 *The Geography of Transport Systems* (Abingdon: Routledge) pp 1-297
[8] Griforyev I 2016 *Airport Modeling and Simulation* (Surabaya: Universitas Narotama) pp 602-58

Acknowledgement

The authors are grateful to Directorate of Higher Education, Rector of Indonesia University of Education, Director of Husein Sastranegara Airport, Civil Engineering Department – Indonesia University of Education for providing data, recommendation, logistic and financial support for Spatial and Activities Models of Airport Based on GIS and Dynamic Model.

A Facile Synthesis and Photoluminescence Properties of Boron Carbon Oxynitride (BCNO) Phosphor Materials for Security Ink Application

A Y Nuryantini[1, 2*], B W Nuryadin[1, 3], E C S Mahen[1, 2], and A Sawitri[1, 3]

[1]Laboratorium Nanomaterial Fisika, Universitas Islam Negeri Sunan Gunung Djati Jalan A.H. Nasution 105, Cibiru – Bandung, 40614, Indonesia
[2]Departemen Pendidikan Fisika, Universitas Islam Negeri Sunan Gunung Djati Jalan A.H. Nasution 105, Cibiru – Bandung, 40614, Indonesia
[3]Departemen Fisika, Universitas Islam Negeri Sunan Gunung Djati. Jalan A.H. Nasution 105, Cibiru – Bandung, 40614, Indonesia

Email: ade.yeti@uinsgd.ac.id

Abstract. A facile synthesis of rare-earth free using boron carbon oxynitride (BCNO) phosphor material for security ink has been investigate. BCNO were synthesize by low temperature microwave heating methods, with H_3BO_3, citric acid and urea to be used as boron, carbon and nitrogen source, respectively. Then, the BCNO nanocrystals were disperse in water-polymer based solution until they became evenly spreading and turn into security ink without rare earth metals. The photoluminescence (PL) and UV-Visible spectroscopy were use to characterize the optical properties of BCNO and the security ink. The characterization results showed that BCNO and the security ink had similar PL properties (PL Peak and PL Peak Intensity). In addition, the UV-Vis spectra proved that the security ink had electronic properties such as being semiconductors based phosphor materials. The results indicate that BCNO phosphor material can be potentially be developed as rare-earth free security inks, and other applications such as optoelectronic, white LED, lighting, etc.

1. Introduction

In recent years, there are many researches about the compounds of oxynitride and nitride as phosphorus materials due to their superior characteristics, such as being non-toxic, having good thermal and chemical stability, having wide excitation and emission wavelength, and also possessing high luminance efficiency when being activated by rare earth ions such as Ce3 +, Eu 2+, and Li + [1]. However, the aforementioned elements of rare earth ions, which act as the activator and the centres of luminescence, are expensive, scarce, and also becoming the sources of environmental pollution [2].

In 2008, Ogi et al did manage to synthesize boron carbon oxynitride (BCNO) phosphor materials without the use of rare earth metal ion doping at a relatively low heating temperature between 700-900 0C and at an atmospheric pressure condition. BCNO has a broad spectrum of excitation wavelength (from UV to blue ray) which causes luminescence in the visible light spectrum (400 nm - 585 nm). It also has half-life of luminescence (emission lifetime) from nano to milliseconds. The range of emission/luminescence colour of BCNO is from purple to near red and it can be adjust easily by varying the ratio of carbon source using an optimum heating temperature and reaction time [3]. The chemicals used to produce phosphorus material BCNO are economical and non-toxic. The chemicals used include H3BO3, (NH2)2CO, and polyethylene glycol (PEG) as the source of boron, nitrogen, and carbon, respectively. A continuous development of the manufacture of phosphor from BCNO have managed to generate new findings which include the knowledge on the luminescence mechanism of BCNO, a better synthesis method to adjust the amount of emissions [4], the selection of carbon sources [4, 5] addition of nanoparticles SiO2 [6], BCNO with red luminescence [7]. As well as possible applications in the field of optoelectronics [8].

In recent years, many study about phosphorescent have been published in regards of the development of non-colour ink [9, 10]. Non-color ink applied for security ink, which is an effective and low cost (efficient) method, for the protection of valuable documents and products against fraud, forgery, alteration, and unauthorized trading. Non-colour ink applied on the field of optoelectronics such as light-emitting diodes (LED). Non-colour inks can be customize with inkjet printing technology to print confidential documents. Inkjet printing process has a number of advantages such as the exact deposition of ink on the paper or substrate with a good layout, and low ink consumption hence it will therefore reduce the cost. Non-colour inks are generally develop from colloidal made from a mixture of polymers and phosphorescent materials.

To the best of our knowledge, there is still no existing study about the development of a non-colour security ink based on BCNO. This is due to the difficulty in dispersing the material of BCNO in a suitable solvent and the difficulty of producing BCNO in nanoparticle size. In this paper, we report on the research about the BCNO nanophosphor using microwave method. It were expected that the solution of BCNO nanoparticle can be applied as a non-colour security ink that can replace the use of phosphorescent materials, which based on rare earth metal ions so that the application non-colour security ink becomes more economical, applicable, and pollution-free to the environment.

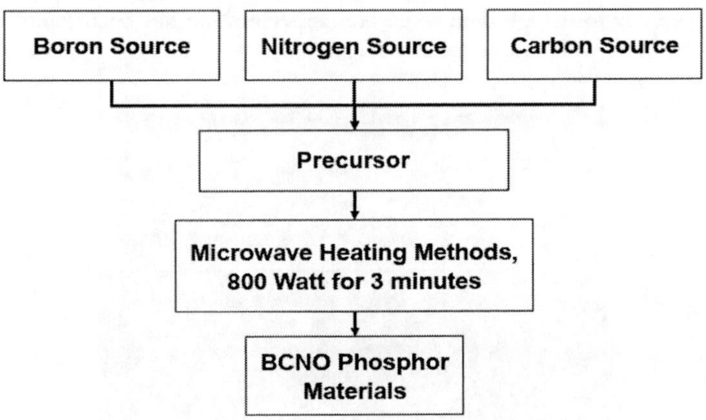

Figure 1. Flow diagram of the synthesis of BCNO nanoparticle using microwave-heating methods.

2. Experiment

The BCNO nanophosphors for security ink application were synthesized using boric acid ($B(OH)_3$), urea ((NH2)2CO) and citric acid (C6H8O7) as the boron, nitrogen and carbon sources, respectively. All chemicals were purchase from Merck Co., Ltd., Germany without further purification. Precursor solutions were prepared by mixing boric acid, urea, and citric acid in distilled water, followed by stirring with high rotational speed and at room temperature for 10 min to obtain homogeneous solutions. The citric acid/boric acid (C/B) mass ratio was varied at 0.2 (Sample B), 0.4 (Sample C), 0.8 (Sample D), 1.2 (Sample E), 1.6 (Sample F) and 2 (Sample G), and the urea/boric acid (N/B) mass ratio for all samples was prepared at 4. The precursors were heated using microwave at 800 watt for 3 min under ambient atmospheric pressure. Then, the BCNO nanocrystals dispersed in water-polymer based solution until they evenly spread and disperse. The flow diagram to describe the synthesis process of BCNO phosphor materials using microwave-heating method shown in Figure 1. The photoluminescence (PL), photoluminescence excitation (PLE) and UV-Visible spectra of each sample measured at room temperature using a spectrofluorophotometer (Cary Eclipse Spectrophotometer, Agilent, Australia) equipped with a xenon laser source and Homemade UV-Visible spectroscopy (LD Didactic, Leybold, Germany). All PL, PLE and UV-Vis spectra analyses were performs at room temperature with 365 nm excitation.

3. Results and Discussion

Figure 2 shows the samples of BCNO phosphor solution synthesized using a microwave with variation in carbon concentration, without being exposed to UV light (a) and during irradiation process by UV light (b). From Figure 2 (a) it appears that the increase of carbon concentration caused the phosphoric solution of BCNO became brownish. The color of the solution turned darker when the concentration of the carbon increased which was possibly due to the formation of free carbon in the sample. The phosphoric solution of BCNO that was irradiated by UV light produced blue luminescence as can be seen in figure 2 (b). The amount of carbon content was very influential to the photoluminescence of BCNO in which the blue color became brighter as the content of carbon increased.

The influence of the amount of carbon content on the photoluminescence of BCNO can be seen from the characterization results of Photoluminescence (PL) spectrofluoro-photometer in Figure 3 (a). The luminescence emission of BCNO was within the range of wavelength from 461 to 475 nm which showed that the luminescence of the sample was in blue region. The wavelength shifted toward larger number as the concentration of carbon in the solution increased. The shift of peak wavelength of the BCNO luminescence due to the addition of carbon concentration has been reported by Ogi, et al., in the manufacture of BCNO with simple heating method [3].

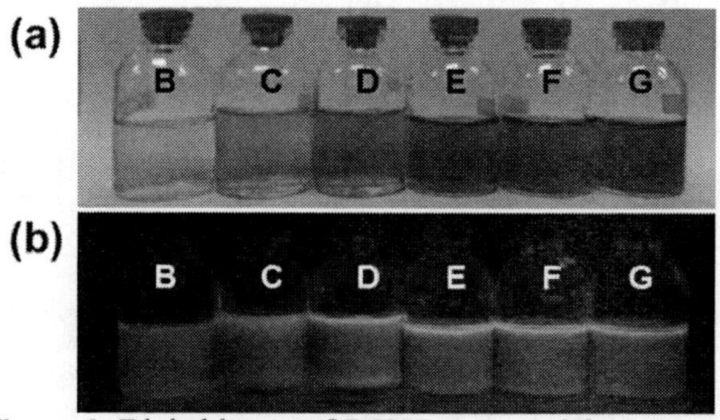

Figure 2. Digital image of BCNO nanoparticles on water for various carbon concentrations, under (a) visible light, and (b) UV light.

IOP Conf. Series: Journal of Physics: Conf. Series **812** (2017) 012037 doi:10.1088/1742-6596/812/1/012037

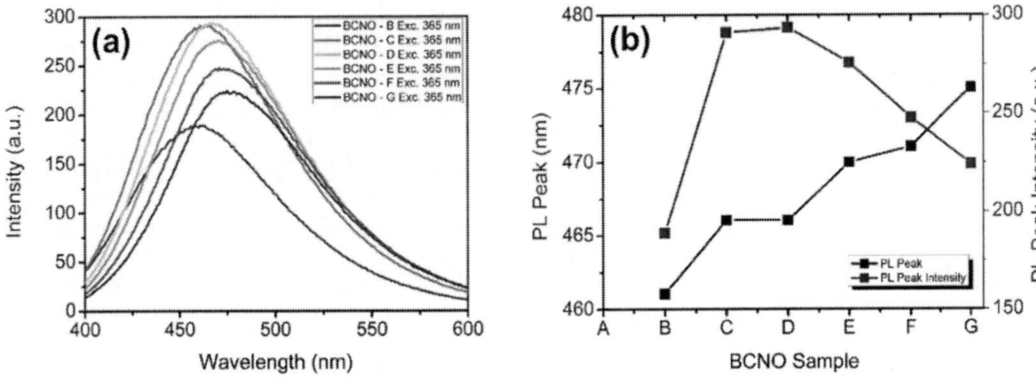

Figure 3. Effect of carbon concentration on the Photoluminescence properties of BCNO nanoparticle: (a) PL Spectra and (b) PL Properties.

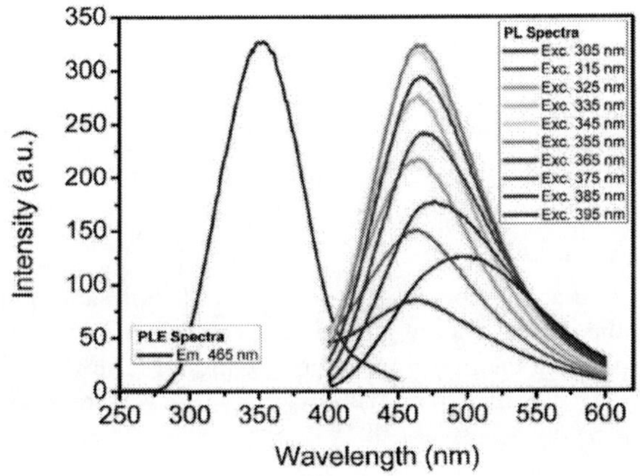

Figure 4. PLE spectra of BCNO phosphor material (Sample D) under UV light excitation.

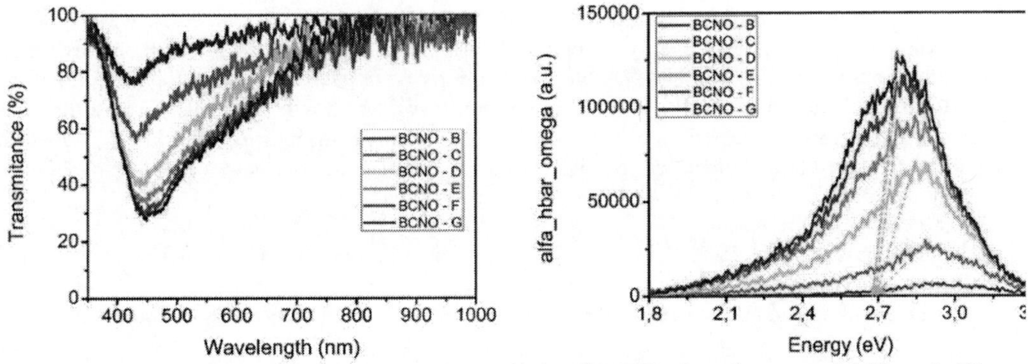

Figure 5. Transmitance spectra and tauc plot analysis of BCNO phosphor material (Sample D).

IOP Conf. Series: Journal of Physics: Conf. Series **812** (2017) 012037 doi:10.1088/1742-6596/812/1/012037

The intensity of luminescence produced was also highly dependent on the concentration of carbon as shown in Figure 3 (a) and (b). The addition of carbon in samples B, C and D caused the luminescence intensity to increase. However, the luminescence intensity in samples E, F, G decreased which was possibly due to the excessive carbon content in the solutions so that the carbon did absorb the excitation light and dim the light emission of BCNO phosphor. The lowest intensity was generated by sample B, which was allegedly happen to occur due to low carbon reaction so BCNO had not been formed. Meanwhile, the optimum intensity was obtained by sample D.

To complete the PL data, sample D was characterized using Photoluminescence Exciation (PLE) to determine the effect of the excitation wavelength on the emission color intensity of the photoluminescence material of BCNO at the wavelength of 465 nm. The excitation wavelength used was varied from 305 to 395 nm. The highest emission color intensity was produced at the excitation wavelength of 365 nm (Fig. 4).

To determine the energy gap of BCNO which had been successfully created, a characterization by UVVIS spectrometer was conducted. The measurement results are shown in Figure 4 which shows the transmittance spectrum of BCNO phosphor material synthesized by a microwave. From the graph, it was proven that the large amount of carbon concentration in the solution resulted in larger absorbance at the wavelength of 450 nm.

The optical band gap in a semiconductor can be evaluated by observing the absorption coefficient as a function of the photon energy using Tauc's law [11]:

$$\alpha = \frac{B(E-E_g)^P}{E}$$

where B is a constant, Eg is the optical energy gap and p is a constant which determines the type of the optical transition (p = 2 and 1/2 for direct allowed transition and indirect allowed transition, respectively). Therefore, the optical energy gap of the prepared BCNO was determined by assuming direct transition and interpolating the linear portion of the plot of $(\alpha E)^2$ *vs. E*. The value of the optical energy gap was about 2.7 eV which is well within 2.25-3.19 eV as reported by Wang *et al.* [12] for BCNO phosphors.

4. Conclusions

We have successfully synthesized rare-earth free security ink using carbon boron oxynitride (BCNO) phosphor material by microwave heating method. The BCNO was synthesized by low temperature microwave heating methods, with H3BO3, citric acid and urea to be used as boron, carbon and nitrogen source, respectively. Then, the BCNO nanocrystals were dispersed in water-polymer based solution until they became evenly spread and turn into security ink without rare earth metals. The photoluminescence (PL) and UV-Visible spectroscopy were used to characterize the optical properties of BCNO and the security ink. The characterization results showed that BCNO and the security ink had similar PL properties (PL Peak and PL Peak Intensity). In addition, the UV-Vis spectra proved that the security ink had electronic properties such as being semiconductors based phosphor materials.

5. References

[1] Hirosaki N, Xie R J, Sekiguchi T, Yamamoto Y, Suehiro T and Mitomo M 2005 *Applied Physics Letters* **86** 211905-1 - 211905-3

[2] D B MA and L K, 2013 *Scarcity of rare earth elements ChemSusChem* **6** (11) 2045-55

[3] Ogi T, Kaihatsu Y, Iskandar F, Wang W N and Okuyama K 2008 *Advanced Material* **20** 3235-38

[4] Kaihatsu Y, Iskandar F, Widiyandari H, Wang W N and Okuyama K 2009 *Fabrication and Characterization of Yellow Emitting Diodes Electrochemical and Solid State Letters* **12** (3) J33-J36

[5] Nuryadin B W, Pratiwi T P, Abdullah M, Khairurrijal K, Ogi T and Okuyama K 2010 *Journal Electrochemisty Society* **157** J329-J333

[6] Faryuni I D, Nuryadin B W, Iskandar F, Abdullah M, Khairurrijal K, Ogi T and Okuyama K 2014 *Journal of Luminescence* **148** 165-8.

[7] Nuryadin B W, Mahen E C S, Iskandar F, Ogi T, Okuyama K, Abdullah M and Khairurrijal K 2014 *Advance Material Research* **896** 464-7

[8] Mahen E C S, Nuryadin B W, Iskandar F, Abdullah M and Khairurrijal K *2013 Pajajaran International Physics Symp.2013* Bandung

[9] Akihiro K, Yamashita T, Yamaguchi K and Aida T 2005 *Nature Materials* **4** (7) 546-9

[10] Mutai T, Satou H and Araki K 2005 *Nature Material* **4** (9) 685-7

[11] Pimpabute N, Burinprakson T and Somkhunthot W 2011 *Optica Applicata* XLI (**1**) 257-68

[12] Wang N W, Ogi T, Kaihatsu Y, Iskandar F and Okuyama F 2011 *Journal of Material Chemistry* **21** 5183-9

Acknowledgment

This work was supported by a research grant (DIKTIS PUIK/37/2016) from the Ministry of Religious Affairs, Republic of Indonesia. The authors would like to thank the Physics of Nanomaterial Laboratory member espesialy to Ririn Nurjanah and Yuli Yuliani to conduct experimental and characterization at UIN Sunan Gunung Djati Bandung.

IOP Conf. Series: Journal of Physics: Conf. Series **812** (2017) 012038 doi:10.1088/1742-6596/812/1/012038

Fabrication and Characterization Of Volatile Organic Compound Gas Sensor Based GaN Thin Film

D Rusdiana* and A Aminudin

Departement of Physics Education, Universitas Pendidikan Indonesia, Jl. Dr. Setiabudhi No. 229, Bandung, Jawa Barat, 40154, Indonesia

Email: dadirusdiana@upi.edu

Abstract. Gas sensor using GaN film has been proposed on silicon substrate by sol gel spin coating technique with parameters growth temperature of 850^{0}C, spinner rate of 1000 rpm, Ga_2O_3 molarity of 1.33 M and N sources derived from nitrogen gas at a flow rate of 100 sccm. From the results of testing the electrical properties of the gas environment type volatile organic compound, especially hydrogen gas turns electric resistance GaN semiconductor thin films declined sharply from 1.5 x 10^{-2} to 7 x 10^{-3} Ohm.cm. The GaN thin film not only exhibited good sensitivity to hydrogen gas but also showed good linearity in the characteristic of the sensitivity to hydrogen gas concentration.

1. Introduction

Historically, research and development of semiconductor alloy material was closely related to the development of optoelectronic devices and electronic devices. One device electronics is quite important is the gas sensor to monitor the gases that are harmful such as hydrogen (H_2), methane (CH_4), carbon monoxide (CO), acetylene (C_2H_2), and nitrous oxide (NO_2) at relatively high temperature. Semiconductor materials used for gas sensor applications are being studied extensively by researchers is a semiconductor alloy metal oxides such as SnO_2 (Yamazoe et.al, 1992, Morrison, 1982), but because this material has an energy band gap is not so wide, thermochemical relatively low stability of the gas sensor made of SnO_2 have less good quality. Gas sensor made of SnO_2 are known to have poor sensitivity for many gas and has a response time that is long on the sensor signal (Gopel and Schierbaum, 1995). To increase the sensitivity of the gas sensor is required of semiconductor materials that have wide energy band gap and thermochemical stable at high temperatures (Shur et.al, 1999, Sardar et.al, 2003, Kung et. Al, 1995, Striteet. al, 1992, Akasaki et.al, 1994, Fertita et.al, 1994). GaN material is a material most appropriate alternative, because in addition to these materials have a wide energy band gap (3.4 eV) also has the energy band gap structure with direct transition (direct bandgap). Another advantage of this material is to have the chemical stability, mechanical, and thermal high, so that the stable used in extreme environmental conditions. Therefore in this study examined the potential of GaN material for gas sensor applications.

This research is directed to the study of the manufacture of gas sensors to produce devices gas sensor has high sensitivity for monitoring gases that are harmful such as hydrogen (H_2), methane (CH_4), carbon monoxide (CO), acetylene (C_2H_2), and nitric oxide (NO_2) at relatively high temperatures. Thus the problem in this research are: "How sensitivity gas sensors are made of a semiconductor material GaN? " . To answer this research has been carried out a study of growing GaN thin films with the techniques of sol gel spin coating by optimizing the parameters of growth like molarity gel, the rpm of the spinner, and the growth temperature thus found parameter of growth that is optimum to produce GaN thin films which have physical characteristics are good for gas sensor applications. Furthermore, the next stage will be made of a gas sensor GaN thin films to be tested the sensitivity of some kind of gas, in terms of the initial phase of test gas used is nitrogen gas and hydrogen gas.

2. Experimental

The gas sensor made of a GaN material that growth with sol gel spin coating technique with the parameters growth temperature of 850°C , the spinner rate of 1000 rpm , Ga_2O_3 molarity of 1.33 M and N sources derived from nitrogen gas 100 sccm. The structure of the gas sensor made in this study as shown in Figure 1.

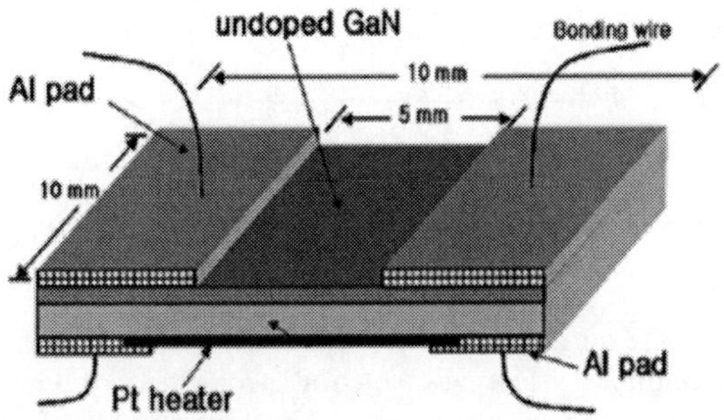

Figure 1. Structure of gas sensor

The undoped GaN thin films were grown by sol gel spin coating on a Silicon (111) substrate. For ohmic contact, Al electrodes was deposited on the front surface of the substrate. To investigate the sensitivity of sensor shown in Figure 2.

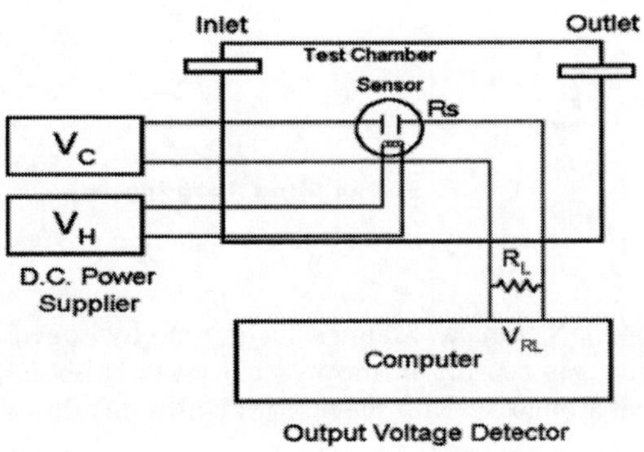

Figure 2. Schematic diagram for gas sensor measurement

Where Rs is the resistance of GaN, Vc is the circuit voltage, VRL is the output voltage, and RL is the load resistance. The sensitivity was defined as $((R_{air} - R_{gas}) / R_{air}) \times 100\ \%)$ (Dae, 2003), where R_{gas} and R_{air} are the electrical resistance of GaN sample in gases to be detected and the in air, respectively.

3. Results and Discussions

To determine the potensial for using GaN as gas sensor the resistance of GaN films was measured on hot plate 300^0 C. Figure 3 shown profile of I-V characterization of GaN gas sensor in nitrogen and hydrogen gases.

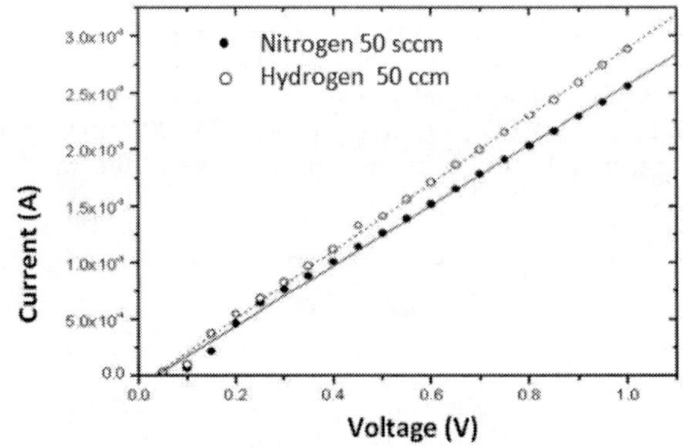

Figure 3. Current variation of GaN film as function of voltage

As shown in Figure 3, the current increased with increasing voltage, which is general property of ohmic contact of semiconductor. The gas sensing properties of GaN based sensor was tested in chamber containing hydrogen and nitrogen gases. The sensitivities of the sensor, which correspond to the resistance variation of GaN when it was exposedto these gases, are shown in Figure 4.

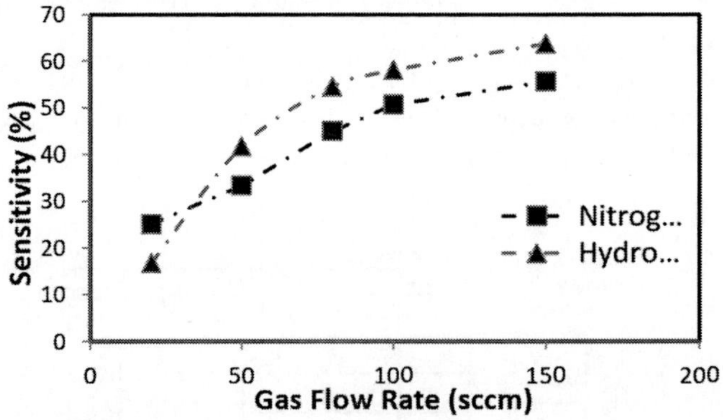

Figure 4. The sensitivity of GaN film as function of Gas flow rate at 300^0 C

As shown in Figure 4,GaN film have high selectivity to hydrogen gas compared to nitrogen gases. Environmental sensitivity gas sensors in the hydrogen gas is about 60 % more sensitive than in a nitrogen gas environment around 50% for the test gas with a gas flow rate of 100 sccm.

4. Conclusion

From the results of the characterization of the sample gas sensor made of GaN semiconductors with spin coating techniques with the parameters of growth: growth temperature of 850^0 C, the spinner rate of 1000 rpm, Ga_2O_3 molarity of 1.33 M and N sources derived from nitrogen gas 100 sccm, it shows the level of sensitivity which is good especially for the testing of nitrogen gas and hydrogen gas .

Environmental sensitivity gas sensors in the hydrogen gas is about 60 % more sensitive than in a nitrogen gas environment around 50% for the test gas with a gas flow rate of 100 sccm. From the test results on samples of the gas sensor, it can be developed a prototype gas sensor for gas leak detection systems are very useful for the many security laboratory use or exploit gas.

5. References

[1] YamazoeN and Miura N 1992*Chemical Sensor Technology* Vol 4 (NewYork: Elsevier)
[2] MorrisonSR 1982 *Sensor Actuators***B2**329
[3] GopelW andScierbaumK D 1995 *Sensor Actuators***B26**1
[4] ShurMS, Bykhovski R A D,Gasak andKhan MA 1999 *MIJ-NSR***41** G1.6
[5] SardarK, Raju ARand Subbanna GN 2003 *Solid State Comm.***125** 355
[6] Kung P, Saxler A, Zhang X, Walker D, Wang TC, Ferguson I and Razeghi 1995*Appl. Phys.Lett.*,**66** (22) 2958
[7] StriteS and Morkoc H 1992 *J. Vac. Sci. Technol.***B10**(4) 1237
[8] AkasakiIand Amano H 1994 *J. Electrochem. Soc.***141** (8) 2266
[9] FertittaKG, HolmesAL, NeffJG, CiubaFJand Dupuis RD 1994*Appl.Phys.Lett.* **65** (14) 1823
[10] DaeSL, Jung HL, YongHLand Duk DL 2003*Sensors and Actuators***B89**305

IOP Conf. Series: Journal of Physics: Conf. Series **812** (2017) 012039 doi:10.1088/1742-6596/812/1/012039

The Different Patterns of Gesture between Genders in Mathematical Problem Solving of Geometry

Y Harisman[1,2], M S Noto[1,3], M T Bakar[1,4] and A Amam[1,5]

[1]Departement of Mathematics Education, UniversitasPendidikan Indonesia, Jl. Dr. Setiabudi No. 229, Bandung 40154, Indonesia
[2]Departement of Mathematics Education, STKIP PGRI Sumatera Barat, Jl. GunungPangilun Padang Utara Kota Padang Sumatera Barat 25000, Indonesia
[3]Departement of Mathematics Education, UniversitasSwadayaGunungDjati, Jl. Perjuangan No.1 Cirebon, Indonesia
[4]Departement of Mathematics Education, UniversitasKhairun, Jl. Batu Angus Dufa-dufa Kota Ternate Maluku Utara, Indonesia
[5]Departement of Mathematics Education, UniversitasGaluh, Jl. R.E Martadinata No. 150 Ciamis 46274 Jawa Barat, Indonesia
E-mail: yulyantiharisman_mtk@student.upi.edu

Abstract. This article discusses about students' gesture between genders in answering problems of geometry. Gesture aims to check students' understanding which is undefined from their writings. This study is a qualitative research, there were seven questions given to two students of eight grade Junior High School who had the equal ability. The data of this study were collected from mathematical problem solving test, videoing students' presentation, and interviewing students by asking questions to check their understandings in geometry problems, in this case the researchers would observe the students' gesture. The result of this study revealed that there were patterns of gesture through students' conversation and prosodic cues, such as tones, intonation, speech rate and pause. Female students tended to give indecisive gestures, for instance bowing, hesitating, embarrassing, nodding many times in shifting cognitive comprehension, forwarding their body and asking questions to the interviewer when they found tough questions. However, male students acted some gestures such as playing their fingers, focusing on questions, taking longer time to answer hard questions, staying calm in shifting cognitive comprehension. We suggest to observe more sample and focus on students' gesture consistency in showing their understanding to solve the given problems.

1. Introduction

According to Mustafa [1], gesture widely can be defined as hands' or body's movement to state or help in expressing thought or feeling. In the last decade, gesture improvedexponentially from literature work [3]. One of them was a study done by Roth [1]. In this study, some cases were given to students, then the function of gesture was observed in supporting students' conceptual comprehensive. The findings showed that gesture was a bridge between experience of body language and conceptual abstraction. This study has significant impact toward teachers to diagnose students's understanding about geometry. Furthermore, Kelly [5] said that gesture has specific correlation with what the students say in facing a problem.

In addition, Alibali [6] has opinion that mathematical cognition can be created by two main keys, perception and action. This claim is strongly supported by fact where teachers and students do the gesture while they are showing their ideas and concepts about mathematics. The previous study was done by Edwards [7], he did it to 14 college students who took mathematics course. He used David McNeill's work as a basic foundation to analyze spontaneity movement that he observed from the students. The findings showed that gesture could be divided into four kinds: iconicphysical, icon-symbolic, metaphoric, and deictic. He also proved that the hypothesis that is showed by McNeill where students have specific gesture in solving mathematical questions.

These studies generally see the correlation between gesture and students' comprehension in answering mathematical questions. It is rarely found that a study sees the consistency of students' gesture in solving a problem. This study described the difference patterns of gesture between male and female whether they are consistent or unique individually. But, this study has not done repeatedly to different subjects but based on different case.

The previous study about gender was a study which explored male and female communication way [8]. This study referenced to Reynold's research [4]. The study reported the function of gesture generally in dyadic communication where the main concept was hard to describe. A study was done to two students by giving mathematical questions about graphic and route of a bus. The students were set to collaborate to make sense in solving about speed and time. The report showed by Reynold [4] was completed with prosodic signs such as tone of voice, intonation, speech rate and pause so the signs help the observer in understanding the two students. These studies include a few functions of gesture to solve collaborative problem in interpreting the given problems, they are: 1) gesture helps to reach and keep on the same cases, 2) gesture takes part as a cognitive media, if there is students' shifting comprehensive, it indicates that there is uncertain cognitive, 3) gesture is an index from cognitive understanding [4]. Reynold's study left a question for us, Is gesture unique for each individual? or Are the differences used to identify student's gesture patterns in different age, gender, and so forth? The difference patterns of gesture as a diagnostic tool teachers to see the male and female students' understanding when the material is delivered.

This study answered about the questions: 1) what gestures does the female student show in solving mathematical problems? 2) What gestures does the male student show in solving mathematical problems? The problems given to the students were about geometry. It was chosen based on Junior High School teacher's report where that topic was a hard topic to be understood by the students.

2. Experimental Method

This study reports one aspect to see gesture that is showed by the students in solving mathematical problems about geometry. Two students from Junior High School with different gender were chosen to answer seven questions about mathematical problems. They have same ability in mathematics. It was done based on their mathematical achievements. The reason for choosing Junior High School for this study was they could give expression and movement (gesture) to indicate understanding because they are in in way to change from kids to teenagers. The things observed were how male and female student shows his or her gesture while he or she is solving the mathematical problems.

Here the three examples of seven given problems were chosen in geometry topic can be seen in Table 1 below:

IOP Conf. Series: Journal of Physics: Conf. Series **812** (2017) 012039 doi:10.1088/1742-6596/812/1/012039

Table 1. Description of the problems

No	Problems in geometry material
1.	Mr. Hok Guan, the owner of pipes shop wants to tie his pipes with rope, each tie is 6 pipes. He is confused to tie them (r= 10 cm). Should it be tied in triangle or square model? Because he has a limited rope to tie the pipes. He expects that he will save the rope to tie. Please help him to solve it. Which tie should he follow to save the rope? 
2.	Mr. Ahmad wants to put several books into a box which size 30cm x 25cm x 25 cm. The books are in the same sizes: length is 20 cm, width is 15 cm, and thick is 2 cm. How many books can be put in the box maximumly?
3.	It is known that there is a swimming pool with the length is 60 meters, width is 15 meters, depth on the left is 1 meter and it goes to slope slightly until 5 meters. How much water is needed to fill it until full?

The problems given were set and designed by mentioning all problem solving strategies like guessing, and checking, drawing picture, making list, making table, revising, looking at the patterns, using a logic reason, solving a simple problem, and making questions. Both of two students were asked to answer the seven questions. Then, they were interviewed by using semi structural approach. They were also asked to explain orally about what they had done.

The data collection of students' presentation and interview recording among researcher and students were taken after the students finished answering the questions. The recordings are videos while presentation and interview process where the researcher observed and found the information through several questions to see students' comprehension about the given problems in order to see the gesture of the students. The result of interview is explained clearly in this paper. Based on the findings, it was found that the patterns of both two students'gesturs in answering the questions. It was showed not only the interview with the students but also prosodic signs such as tone of voice, intonation, speech rate, and pause that might help in understanding the both two students' gestures. The interview questions aimed to see the consistency of both two students' gesture.

3. Result and Discussion

Here is the description of the result video recording from male and female student and the prosodic signs such as voice pressure, intonation, speed of talking, a pause that might help in understanding the both two students' gestures completely.

3.1. Episode 1: Oral presentation on question one for female student

Annisa answered the question number 1 with less correct answer. Annisa only tried to answer the length of the rope to tie the pipes squarely than for pipes which are tied in triangle, she could not answer it. For the last answer, she said that using the triangle pattern to tie the pipes is better because it saves more rope. It is a wrong answer because the two patterns of tie need the same length rope. Figure 1 is the description of her answer on the question in writing.

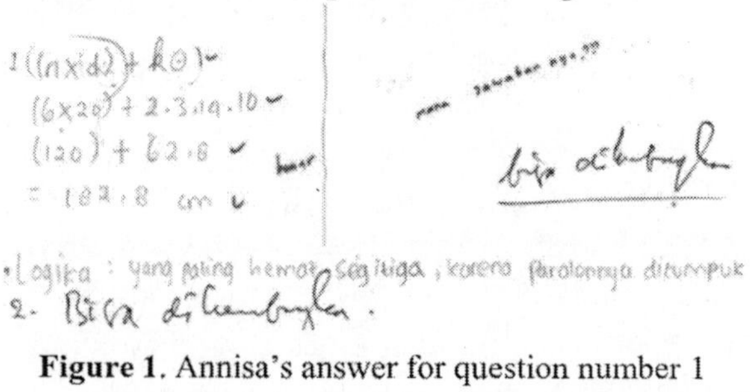

Figure 1. Annisa's answer for question number 1

Related to the answer, Annisa was interviewed with semi structural questions to see her gesture in showing her comprehension about the question. The Table 2 below will show the recording of interviewee and Annisa.

Table 2.Protocol from conversation and gesture between Annisa and interviewee

Turn	Conversation	Gesture
1.1	Interviewee: triangle pattern is better cause it will save more rope and the pipes are piled, [1.02 sec] ... the square pattern is also piled, isn't it? (Reading Annisa's answer).	She pointed to the last answer by using a pen and moved the pen to the piled square pattern.
1.2	Anissa: hehehe, [1.53 sec.] But the pile is up not to the sideline.	She played her tip veil, put her head down then let her left hand on the table, moved her right hand to make a triangle, and touched her eyes.
1.3	Interviewee: So to know a better rope to save is based on the pile, right? Should we find what we have to look for first, should not we?	The interviewee's hands moved from the question to Annisa's answer then the interviewee looked at Annisa.
1,4	Anissa: hehehehe, No, we should not [2.38 sec] but I usually cannot answer this question, I mean only for this one.	She leaned her body to the interviewee and pointed to the counted answer. She pointed to the square picture and matched it with her answer sheet.
1,5	Interviewee: Why don't you use the same formula to find the square pile?	The interviewee moved his or her body tall and looked at Annisa seriously.
1,6	Anissa: Mmm [2.01], If I use the same formula, the answer will be same.	She leaned her body to the interviewee seriously and tried to set her veil and put her head down.
1,7	Interviewee: Oh, why will the answer be same? This is used for all. By the way, what formula is this?	The interviewee pointed to formula Annisa wrote for the length of the rope and the pile of the pipes.
1,9	Anissa: [104 sec]... around formula	She noded her head and played her hands like a circle.
1,10	Interviewee: around formula, right? This formula can be used, can't it? Whatever the patterns of the pile pipes, *n* is the number of the pipes, isn't it?	The interviewee pointed to formula *nxd* that was written by Annisa.
1,8	Anissa: [2.34 sec] oh yeah. This formula can be used, isn't it? And will the result be same?	She leaded her body more to the researcher by playing her pen to the picture of the question.
1,9	Interviewee: What do you think, is the result same? How to count it? What is *n*?	The interviewee pointed to the two pictures tied pipes and moved to formula that is done by Annisa on her answer sheet.
1.10	Anissa: mmmmm... [1.24 sec]. The number of pipes...[2.03 sec].	She node her head then counted the number of the pipes on the square pile pattern of the question.
1.11	Interviewee: By the way, what rope does show the length 6x20?	The interviewee turned around the question sheet to Annisa and asked her to point out which rope on the question.
1.12	Anissa: [2.34 sec]..What does it mean?	She pointed the question and leaned her body forward to the interviewee.
1.13	Interviewee: Pay attention on 6 ropes which ones.	The interviewee pointed to all rope on the picture.
1.14	Anissa: it means all...[2.43]	She turned her pen around and followed the whole ropes.
1.15	Interviewee: Why all? The formula has to sides, doesn't it? *nxd* + kll of the circle, which ones are the *nxd*?	The interviewee tried to illustrate a half of picture by trying one.
1.16	Anissa: I see... don't continue it...this one...this one...this one..one sideline, one diameter, and next.. they are six.	She was spirit by following the length of the rope and underlined what she wanted.
1.17	Interviewee: yes, that is right, then what twice around circles do you mean?	The interviewee asked Annisa again to look at the picture.
1.18	Annisa: I know...I know...this part...here it is...here...and here...it becomes a round of circle.	She followed the left around circle by pen.
1.19	Interviewee...that's really correct.	The interviewee gave thump to Annisa.

We found five dominant patterns of gesture on Annisa. She often put her head down. It can be seen on turn 1.2 and 1.6. This happened when she felt that she was blamed or what she did was wrong. In addition, she showed her lessconfidence and hesitating by playing her tip veil. It also can be seen on turn 1.2 and 1.6. Hesitating also can be seen when Annisa often laugheda bit longer. We can see it on turn 1.3, 1.6, 1.8, 1.10. She often noded her head when the shifting cognitive comprehension. This can be seen on turn 1.10 and 1.9. When the interviewee tried to make her understand about the concept, she directly showed a gesture by nodding her head and when her own cognitive comprehension moved to a new one. She suddenly would move her body forward to the intervieweeand asked a question if there was something she did not understand. This is the dominant gesture done by Annisa when she found a new knowledge. The turn that showed it is turn 1.6, 1.4, and 1.8.

3.2. Episode 2: oral presentation on question 1 for male student

Alvaro also answered the question number 1with less correct answer but he counted the length of the rope for both square and triangle pipes. For the last answer, he said that the square pattern is better that triangle to save the rope. It is a wrong answer cause both of the patterns need a same length rope. Figure 2 is the description of his answer in writing.

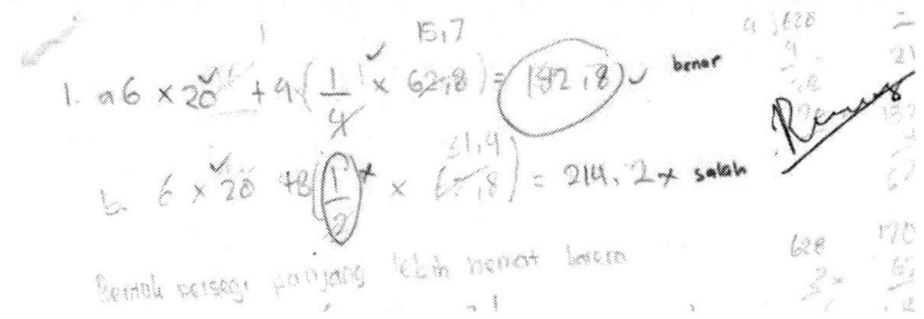

Figure 2. Alvaro's answers for question number 1

Based on the answers, Alvaro was asked with several semi structural questions to see his gesture in showing his understanding of the questions. The Table 3 below will show the recording of interviewee and Alvaro.

Table 3. Protocol from conversation and gesture between Alvaro and interviewee

Turn	Conversation	Gesture
1.1	Interviewee: Where does the formula come from?	The interviewee pointed to the formula that Alvaro wrote.
1.2	Alvaro: 6 is the number of the circles, 20 is diameter.	He pointed to the pile pipes picture.
1.3	Interviewee: So, where did you get 4?	The Interviewee pointed to 4 on the picture.
1.4	Alvaro: 4 is from...[2.45 sec].. What is that…ohhh number arch circles.	He kept calm and pointed to each arch circles.
1.5	Interviewee: How big is the arch?	The interviewee pointed to the arch.
1.6	Alvaro: 90 degrees.	He pointed to angle arch of the circle.
1.7	Interviewee: And how big is angle arch on triangle pile pipes?	The Interviewee pointed to the triangle
1.8	Alvaro: 180 degrees.	He put his pen up and pointed to angle arch of triangle.
1.9	Interviewee: what, Why 180?	Interviewee underlined the triangle pile.
1.10	Alvaro: [2.40 sec]...That's right 180.	He followed to draw the arch circle on a half triangle pile without considering left edges of circle.
1.11	Interviewee: why is it straight? If there is a line that touches the circle, what is it? You drew it like this, how could it be like this?	The Interviewee pointed to the line which touches the circle and tried to draw a correct one.
1.12	Alvaro: [3.63 sec] I don't know. I have never learned it before…	He kept looking at the question and didn't ask to the interviewee.
1.13	Interviewee: What is the relationship if a line touches circle.	The Interviewee followed to draw to the arch circle.
1.14	Alvaro: As vertical, yeah… oh yes I forgot..so from here the corner becomes 120.	He described a corner on triangle pile arch circles.
1.15	Interviewee: that's really right… so, how long is the rope for this one and this one?	The Interviewee pointed to the circle.
1.16	Alvaro: [3.45 sec]... one around circle	He remarked all the arch circles.

In general, there are four gestures that can be seen from male student. Male student is always active to move his hands while transferred his knowledge to the interviewee. It can be seen on turn 1.8, 1.10, and 1.14. On these turns, Alvaro always moves his hands on picture of the questions. If he cannot answer the questions, he will not look wonder by moving his body down like the female student does but he keeps looking at the questions. It can bee seen on turn 1.12, think a bit longer when he cannot answer a question. It can be seen of pause on turn 1.14 and 1.16, Alvaro thought first before showing a question while the female does. He keeps calm when the shiftingcognitive comprehension. It can be seen on turn 1.14 and 1.16. Alvaro changes his comprehension from 180 to 120 degrees. He does not laugh, nod his head, and hold something that shows his worry while the female student does.

4. Conclusion

Some gestures showed by female student were she often put her head down, looked lessconfidence, hesitated, noded her head when the shifting cognitive comprehension, directly moved her body to the interviewer, and showed question when there something was not understood. Furthermore, some gestures showed by male student were he moved his hand, kept to see the questions, thought a bit longer on question he could not answer, and stayed calm when the shiftingcognitive comprehension.In conclusion, there was a significant difference of gesture between female and male student. This study was just for one sample and in description step. For further study, it is suggested to see the consistency of gesture between male and female student. Does it usually happen to each individual or each individual has different and unique gesture in showing his or her comprehension in facing problems. Thesepatterns were found when the students were asked repeatedly by using the questions. The weakness of this study was it only used one subject of the research. To further study, it is suggested to do in some subjects of the research. The findings of this study can help teachers to diagnose how male and female students' face the problems, shift the cognitive comprehension, and when they do not understand about the problems.

5. References

[1] Mustafa S 2013 *Gesture dalamPembelajaranMatematikaKNPM V Proc.*(Malang: Malang State University)pp.521-4.
[2] Roth W M 2001 *Review of Educational Research***71** 365
[3] Reynolds F J &Reeve R A2001 *Journal of Mathematical Behaviour***20** 447
[4] Roth W M& Lawless D2002*Journal Learning and Instructions***12** 285.
[5] Kelly S D 2008 *Language and Linguistics Compass***2**20
[6] Alibali M W and Nathan M J 2012*The Journal of The Learning Sciences***21** 247
[7] EdwardsL D 2007*American Educational Research Association Annual Conf.Proc.*(Chicago)pp. 1-21.
[8] Mohindra V andAzhar S 2012 *Journal of Humanities and Social Science***2** 27

Acknowledgments
We thank Dr. Kusnadi who helped and guided us in developing geometry problems and in interviewing students. Also, we thank Joko and Anny Sovia who edited this manuscript paper.

The Perceptions of Dayak Society of Losarang Indramayu to The Conservation of Natural Resources as Subsistence

Dian Permana Putri

Faculty of Education and Science Education, Swadaya Gunung Jati University
Jl.Perjuangan no.1, Cirebon West Java Indonesia
Corresponding author : dianpp@lecturer.unswagati.ac.id

Abstract. This study aims to investigate the perception of local society, especially in the Dayak society in Losarang Indramayu to the importance of the natural resources conservation as subsistence. The research method of this study is qualitative approach to investigate the perception of Dayak society. The result of the research shows that the majority of Dayak local society in Losarang Indramayu is farming rice and vegetables through intercropping. Dayak society of Losarang Indramayu really appreciates to the preservation of natural resources. They believe that if they keep nature, then nature will also be friendly with them. One of the way to save the nature is by doing Pepe ritual, becoming vegetarian, and do not destroy animals around them. Dayak society of Losarang believes that each animal would also like to have the desire to live, taste, and imagination just like humans. Furthermore, they also build a special building that is used to respect all kinds of animal. The building uses bamboo as walls and fibers as roof. In that building, they raise many kinds of farm animal such as chickens, cows, goats, and pigs. By preserving the fauna, they believe that the crops will be abundant every year. By making the society to be aware about the importance of the natural resource and express their aspiration to the policymakers is a step to establish the sustainability of natural resources. Furthermore, in the management of natural resources by the stakeholders, the local society perception needs to be appreciated, understood, and considered.

1. Introduction

Biodiversity is a term often heard today. The term biodiversity is used to define both the abundance of animals and plants found on earth. Biological resources have great benefits to support human life. The necessity of food, clothing, housing, cosmetics, medicines, and recreations can be fulfilled by biological resources. Every day we are dependent on the availability of vegetables, fruits, livestock, and medicines.

Indonesia has the potential beneficial biological resources of ecologically vital and valuable in the world. The variety of natural resources in Indonesia contribute to the biological riches of the world about 13% (Anonymous, 1989). Indonesia has an abundance of species richness of plants, animals and microbes. It is estimated that as many as 28,000 species of plants, 350,000 species of animals and 10,000 thousand microbes living naturally in Indonesia. Although, the area of Indonesia is only 1.32% of lands in the earth, 10% turned out to be the habitat of species of flowering plants, 12% of mammals, 16% of reptiles and amphibians, 17% of birds, 25% of fish, and 15% insects in the world. 36% of 515 species of mammals of the world are endemic in Indonesia, 18% of 33 species of primates are endemic, 40% of 78 types of parrots are endemic, and 44% of 121 species of butterflies of the world are endemic in Indonesia (Mc Neely et al., 1990). Indonesia is known as one of the countries with greatest biodiversity in the world and is the center of the world's biodiversity (Mac Kinnon, 1992).

There are many problems to preserve our biodiversity such as in 2007, Intergovernmental Panel on Climate Change at US announced that the temperature of earth has been increased 0,74 0C within 10 years from 1996 to 2005 (Saijo et al., 2010). In the last 500 years, human activity is identified as one of the causes that affects to species extinction. Since 1996 (IUCN, 2016) the number of animal species that is threatened increased from 5.205 to 8.462. The existence of biological resources need to be maintained to keep the balance of ecosystems and biochemical cycles. According to Ramadoss & PolyaMolli (Putri, 2014) Biodiversity is one of the main assets that must be protected to ensure human welfare of current and future.

It is important to do some efforts to preserve the diversity of biological resources. It needs a way that conservation efforts provide benefits to the community and not making obstacles. It needs understanding of the basic needs of local communities and conservation in order to influence decision-making by policy makers. Sheil et al (2006) reported the results of conservation not only from the extent of protected areas, conservationists or just for professionals only, but biodiversity should be maintained in the region other uses. It requires the cooperation of local communities and other stakeholders, which also reflects a wide perception as well as their choice.

It needs to gather information about public perception related to maintaining biological resources mainly as a source of subsistence. Local people's perceptions about the importance of biodiversity will build mutual understanding, enrich the knowledge regarding land use, improve biodiversity conservation efforts, protect the interests of local as well as considerations in decision making by policy makers.

There hasn't been a lot of information about community perceptions Dayak Losarang Indramayu on natural resources as a source of subsistence, the society of Dayak Indramayu Losarang is different from the Dayak community living in Borneo. Administratively, this Dayak Segandu tribe lives in the Krimun village Losarang district, Indramayu. This community does not live and isolate in the jungle, but they blend with the local community. They have beliefs, habits, and different appearance to the public in general. Besides, their life philosophy is nature, nature is the center of all their activities, so the perception of Dayak Losarang Indramayu regarding biological resources as a source of subsistence is interesting to be investigated.

2. Research Method

The approach used in this study is a qualitative approach. This study was conducted on May and July 2016. This study is located in the village of Dayak Segandu, Krimun village, Losarang Indramayu. In general, data collection techniques performed in this study refers to observation, interview and documentation. Methods and data analysis technique used qualitative descriptive analysis. This analysis emphasizes on deductive and inductive inference and scientific logic. There are two kinds of descriptive analysis used in this study that isdescriptive explorative analysis. This method emphasizes on exploring / investigating information deeply and focused on the goal of analysis results to be achieved. This method is used to identify and describe the characteristics of the environment, social conditions, perceptions and aspirations of the communities about biodiversity and its conservation.

3. Result and Discussion

3.1. The History of Losarang Dayak Indramayu

Segandu Dayak village located in the village of Krimun Losarang Indramayu. This community has beliefs and lifestyle different from the communit /villager as usual. There is no relation between this community with Dayak community in Kalimantan. They call themselves as Suku Dayak Hindu – Budha Segandu Indramayu. But it doesn't mean Hindu or Budha religion. They have beliefs to one God but not wrapped in the performance of religion.The founder of this tribe is EranTa'mad Diningrat Gusti Alam in 1970. In the former time, this community named as Dayak tribe SS Multipurpose instituation. It because of the founder feel bored with the government rules. Because of seeing the condition of the environment,Ta'mad start to self-reflection and realize that it is the best way for humans. In 1972 the instituation of SS Dayak Multipurpose Sciences changed its name into Dayak JakaUtama, 1994 turned into Dayak Siswa and in 2000 turned back into a Dayak Hindu-Budha Bumi Segandu Indramayu. Based on the interviews with the warlorlds of the Dayak Hindu-Budha Bumi Segandu Indramayu, Tarxim bin Kalsim, *Suku* means the foot. The point is that every human being standing on their own feet. *Dayak* is derived from the word meaning sifter or sieve that means can sieve or filter out which ones are good andwhich one is correct. Hindu means uterus or womb, Buddha means *Udha* or naked. The philosophy is that every human being is born naked (nature) does not bring anything. *Bumi* means thing, which *Segandu* means the whole body, Indramayu "In" means the core of the deepest, meaningful *Darma* means the elderly and*Ayu* means women. Overall from the literal meaning, the tribe Dayak Hindu-Budha BumiSeganduIndramayu, means foot stepped on beliefs that has been taken from the womb to devote to nature, the elderly and women. Their philosophy is natural. It is the best way to get closer to nature. They believe that the core beliefs in life is nature.

In 2016, Dayak Hindu-Budha Bumi Segandu Indramayu renamed Dayak Adat Jawa Petani Bumi Segandu Indramayu. *Adat* means habit, anywhere and in any way, this community will not eliminate habits or traits that have been attached, both in appearance and attitude. *Jawa* means Losarang native Dayak people is Javanese people not Kalimantan. *Petani* means this community come from the farmers. The origin of Dayak community is farmers that are humble, patient, work hard, and hones. So that philosophy means the foot stepped on faith, habits, patience, work hard and honest who had been taken from the womb to devote nature, the elderly and women.

The beliefs to preserve the nature always accurs in society. Orang Rimba Jambi has traditional rule not destroy the forest (Takiddin, 2014). Suku Baduy assume that they are part of nature that must protect and preserve natural reasources in this world (Erwinantu, 2012). Every society has a unique habit and tradition but they have similarities in gaining the purpose of honoring and preserving the nature.

Based on the way they dressed and habits of the Dayak community Losarang Indramayu divided into three groups such as:

3.1.1. Dayak Preman. This class is the majority of the people living in the village and blend with other society in the village as usual. Forms of housing, clothing , food , habits , lifestyle, like society in general. They have an identity as a citizen.

3.1.2. Dayak Seragam. This group knowing by the color of cloth. They are wearing black shirt and black trouser. They don't have a religion and no identity card as a citizen.

3.1.3. Dayak Asli. This Dayak is only wearing a pair of black and white pants without a shirt. This outfit symbolizes the sky and the earth. Or it could be said to represent the state of the world, that everything is in pairs. There is good and bad.They don't have a religion and no identity card as a citizen. They doing some rituals and being vegetarian.

Costume is as the salience that Baduy society has. Baduy society can be distinguished by their habit of life such as Baduy Luar and Baduy Dalam. The salience of Baduy Dalam is that they are wearing white shirt and black pants (Erwinantu, 2012). Different from Dayak society that they are only wearing short black and white pants without shirt. The similarity of the way they dressed shows that there is equality, there is no one is higher social statutes than other.

3.2. The Perception of Dayak Community to The Sustainability of Natural Resources as a Subsistence. This community lived in the area which has approximately 100m^2. There are a gate and a main hall, there are also a building that serves as an animal cage. They live with other communities that are not a community of Krimun village. Most of members of this community are farmers (80%) and others are entrepreneur. Based on the way they dressed and habits of life, Dayak Losarang Indramayu has three categories namely, Dayak Preman Dayak Seragam, and Dayak Asli. This Dayak are the ones who are considered as the original Dayak that is not dressed and wearing only short pants with black and white color. The member of this Dayak are 100 followers. Beside the use of clothing that became his trademark they also have different habit with two other Dayak category. This community opened to the development of technology. In farming they also use tractors plowing their fields. But they do not know about formal education, because this community refused to fill out the form ID card (KTP). The reason is because they have animism beliefs, while in making KTP, they should fill the name of religion and take a picture that wearing cloth. They are not associated with political parties, social organizations and elections. This condition is the same with Baduy society, although they respect to the government as the main authority in a country but they don't follow the politics party. Time to time the phenomenon and life serve some choices, although for the strict community. The society of Baduy Dalam supported the election by praying to make the election goes honestly and safely, they try to rely their aspiration on Baduy luar society that is more familiar with the culture of other society (Erwinantu, 2012).

Dayak Asli community has some customs and rituals. They have a philosophy of nature. They believe that if they safe the nature, so the nature will also be friendly with them. One of their habits is to become vegetarian. This Dayak did not consume animal foods, such as meat and eggs. They believe that every animal also would like to have the desire to live and taste just like humans. To Respect the welfare and conservation of animals, they build a special building. The building is a cage with bamboo walls and fibers roof. In that place lives some various type of livestock such as chickens, cows, goats, and pigs (figure1a). By preserving the fauna, they believe that the agricultural products that they generate will remain abundant annually.

(a)

(b)

Figure 1. The conditions on Dayak community, Losarang Indramayu. (a) the special building for various livestock, (b) Place for kungkum ritual

The Dayaks have a ritual sunbathing or known as pepe tradition. They do this tradition in the morning while enjoying the morning air and sun. This tradition is one of the teachings of Ta'mad Diningrat Gusti Alam or Ki Ta'mad who is the leader and founder of the Dayak community Losarang. Pepe is one of the rituals that they do in their daily live. Then continued with the kungkum ritual or soak in the afternoon when the sun is scorching (figure1b). It Begans at 9 until noon, kungkum or soaking in water to the extent of the neck. Chanting and compliment the natural and mender or telling story about wayang (traditional drama performance of Java that play with leather puppets) . Pepe ritual and kungkum carried for four months in a year. The aim of this activity is to feel someone's feeling. This ritual also aims to appreciate the preservation of natural resources, reassurance and control anger. Tradition ritual that has not only spiritual value but also becoming salience of Baduy society. They do the Ngawalu ritual for 3 months that usually start from the end of December to March. Kawalu month marked when the rice start efflorescing to harvesting. In this ritual the society of Baduy do one day fasting in every Kawalu month and Kawalu ritual closed by Ngalaksa (grinding the rice together, then making the grinding rice becomes Laksa - like noodle made from rice flour) then it will be distributed to all Baduy society (Erwinantu, 2012). But different from Dayak society that is always welcome to the visitor in the moth ritual, Baduy society don't welcome to the visitor in ritual month.

Being a vegetarian and carry out the rituals is one way in which the Suku Dayak Losarang Indramayu for the preservation of nature, especially to meet their needs in food. Besides the needs of the construction, accessories and future needs. Accessories worn by the community in the form of bracelets, necklaces and belts pants. These accessories have several different meanings based on its constituent materials. The accessories made of pring ireng or black bamboo means historical significance of Java. Then the yellow bamboo or ivory means farmers, while those made from barley means pure cultural history. The use of this material based on abundance of natural resources around them. What is interesting in their neck emblem Pancasila always used. Based on the interview with one of the members of the Dayak Tarxim bin Kalsim this case means "Besides symbolizing the country, Pancasila also symbolizes the unity of us all. Despite differences, but we are united with the symbol of Pancasila ". Dayak community to be open, in daily they also mingle with local residents, yet limited. Although impressed exclusive, but they are known are friendly and helpful. Like in Baduy society, Erwinantu (2012) states that Baduy society is welcome and gracious to other society but always watchful, and one of the effort to maintain the rule that the visitors don't allow to stay in more than one night. If it is needed, the visitors can stay outside the village like in shelter in the field. The visitors should follow the traditional rule of Baduy society if they come to their society.

Based on interviews and questionnaires which are represented by the 10 respondents (table 1) showed that the Dayak community Losarang Indramayu , the majority argued that maintaining the preservation of animals and plants is important as a form of inheritance to descendants. As well as one effort to maintain the availability of food and guarantee harvests abundant. Furthermore, other respondents argued that maintaining natural resources as valuable to the local community and as an auxiliary seed dispersal. This opinion in line with the traditional society of Kuta and Baduy (Efendi, 2014;Erwinantu, 2012) that community states that knowledge, attitude, and behavior affect the existence of their life and environment. Baduy has the proverb "Gunung teu meunang dilebur, lebak teu meunang dirusak, Arey teu meunang diteuteuk, cai teu meunang ditua' (mountain cannot be changed, valley cannot be destroyed, shrub cannot be cutted, water springs and river cannot be contaminated. Kuta traditional society has proverbs Leuweung Ruksak, cai Beak, Manusa Balangsak" (the destroyed forest, the water is empty, human will suffer). That proverbs is part of knowledge to the environment of life that is maintained and taught from generation to generation.

Table 1. Public opinion on the factors to preserve natural resources

Statement	Respondents who chose the answer
1. Worth to local communities	2
2. Animals as auxiliary seed dispersal	1
3. Animals and plants as a legacy for descendants	6
4. Do not know	1

While the public advice on tackling environmental damage at the surrounding , which is represented by the 10 respondents by multiplying the planted crops (70 %), reforestation (20 %) and to the protection of plants considered rare (10 %). Assuring the public to realize the importance of the natural resources that they have and express its aspirations to policymakers is a step that determines the sustainability of natural resources. One of the way that the government and stakeholder can do to support the Dayak community is Education for Sustainable Development (ESD). Segara (2014) states that ESD is a conscious way to protect and preserve the nature environment, social, and culture as the existence of human life. Furthermore, in the management of natural resources by the stakeholders, the local perception needs to be appreciated, understood and consider. In this case supported by Barr et al (2001) that the development of local democracy opens the opportunity to manage the field and resources. The paradigm may not follow the change of local democracy, for example in planning land used is still not effective and consultative. Not only the politicion and government employee. All parts that involved in conservation policy need to be encouraged to have more consultation with the local people.

4. Conclusions and Suggestions

Dayak Losarang community has three categories, namely Dayak Preman, Dayak Seragam, and Dayak. Among the three classes of Dayak majority of farmers are the ones who are very consistent in maintaining the sustainability of natural resources. Their efforts, among others, to become a vegetarian, do some rituals like *pepe* and *kungkum* and build a special building to respect of livestock. They believe that these measures will keep the nature and maintain an abundance of crops. Several factors are the reason for the community to sustain natural resources, among others because of natural resources is a legacy for the descendants, valuable for local people and animals as helpers seed dispersal. While the public advice on tackling environmental damage that occurred around that is by multiplying planting crops, reforestation and perform protection for plants considered rare.

This research can be used as a baseline for future studies. Further research is needed on how to identify the species by the Dayak community, who need it, how much it needed and the way to maintain the value of natural resources in the future.

6. References

[1] Anonymous 1989 *Strategi Nasional Pengelolaan Keanekaragaman Hayati* (Jakarta: Kantor Menteri Negara KLH RI) p 3

[2] Barr C, Wollenberg E, Limberg G, Anau N, Iwan, R, Sudana I M, Moeliono, M and Djogo T 2001 *The Impacts of Decentralization on Forests and Forest-Dependent Communities in Malinau District, East Kalimantan: Case studies on Decentralisation and Forests in Indonesia* (Bogor: CIFOR) p 25

[3] Efendi A 2014 *Social Science Education Journal* **1(2) 2014** 211-8

[4] Erwinantu 2012 *Saba Baduy: Sebuah Perjalanan Wisata Budaya Inspiratif* (Jakarta: PT. Gramedia Pustaka Utama) pp 41-86

[4] IUCN 2016 *Annual Report 2015* (Gland Switzerland) p 17

[5] Mac K K 1992 *Nature's Treasurehouse-The Wildlife of Indonesia* (Jakarta: PT Gramedia Pustaka Utama) p 292

[6] Mc Neely J A, Miller K R, Reid V W, Mittermeier A R and Werner T B 1990 *Conserving The World's Biological Diversity* (Gland: Switzerland) p 25

[7] Putri P D 2014 *J. Ilmiah Lemlit Unswagati Cirebon* March 2014 **10(7)** pp 42-49.

[8] Saijo T and Hamasaki H 2010 *Chapter 6: Designing Post-Kyoto Institutions: From the Reduction Rate to the Emissions Amount. at Adaptation and Mitigation Strategies in Climate Change.* (Tokyo: Springer) p 85

[9] Segara N B 2015 *Social Science Education Journal 2*(**1) 2015** 22-30

[10] Sheil D, Puri R K, Wan M, Basuki I, van Heist M, Liswanti N, Rukmiyati, Rachmatika I and Samsoedin I 2006 *Ambio* **35** 17–24

[11] Tarkiddin 2014 *Social Science Education Journal* **1(2)** 161-9

Acknowledgments

Thanks to Mr. Tarxim bin Kalsim who have helped provide information about the history , customs and practices of the Dayak community LosarangIndramayu. Ms. Hesti Muliawati and Lala Lestiana for their help so that this research took place without any major obstacles.

IOP Conf. Series: Journal of Physics: Conf. Series **812** (2017) 012041 doi:10.1088/1742-6596/812/1/012041

Analysis of Students' Incorrect Answer on Two-Dimensional Shape Lesson Unit of the Third-Grade of a Primary School

Ulfah[1] and A Jupri[2]

[1]Departement of Primary Education, Postgraduate Student, Universitas Pendidikan Indonesia, Jl Dr. Setiabudi 229, Bandung 40154, Indonesia
[2] Departement of Mathematics Education, Faculty of Mathematics and Science Education, Universitas Pendidikan Indonesia, Jl Dr. Setiabudi 229, Bandung 40154, Indonesia
E-mail: ulfah.pgsd@gmail.com

Abstract. This research aims to analyze a comparison between predictionsmade by researchers and the actual facts occurred in the real classroom situation. This study useda qualitative method, and involved 37 third-grade students from a primary school in Bandung. We analyzed students' responses to questions that are given by the teacher on the topic of two-dimensional shape. The results of this research showed that the learning method used by the teacher influenced students' written responses. It can be another approach for the teacher to be used according to the lesson unitsthat are being presented, therefore the learning processes wouldrun optimally and the learning goals can be achieved. Based on the results, we conclude that: (1) Students' incorrect responsesare not only influenced by the teaching method used by the teachers,but also influenced by otherintricate factors; (2) The teachers need to comprehend that every student has some different ways of thinking in answering the given problems.

1. Introduction

Mathematics is a science which plays an important role in human life and will not be separated from everyday life. Various activities that we do is always associated with mathematics, such as buying and selling activities in the market, the use of money for transportation, payment of bills, etc. In addition, mathematics is also a means for a person to think in order to develop logical, critical and systematic thinking. It helps people to be able to anticipate, plan, decide, and properly solve each problem in daily life [1]. Therefore, mathematics is used as a subject that has to be studied by students ofprimary education [2].The mathematics concepts that should be learned by primary students are including the aspects of numbers, geometries and measurements, and data processing. Square and rectangular perimeters are thelesson units that belong to the aspects of measurements and geometries. These lesson units are studied in the second semester of the third grade. In general, the learning conducted by the teachers use a classical model so the students are directly told the formula of square or rectangular perimeters. Instead, the teachers need to invite students to find the definition of perimeter and the formula of rectangleby themselves. The learning activities that get the students to discover the learning purpose of the lesson unit. As Bruner [3] states in his principle, he assumes that learning to discover what is appropriate to the preferred knowledge actively by one'sown self will give the best results. Besides that, trying independently to resolve problems and the knowledge that is attached with it produces a meaningful knowledge.

When a learning process is meaningful, students will be able to easily remember the definition and the formula they have been applied in learning. To calculate square and rectangularperimeters, the knowledge about the length and the width of a figure and the formula which should be used is definitely required. Moreover, the precision in using the formula and the calculation of thelesson units are also needed. If the students are not precise in using the formula and the calculation, the final results would be incorrect. Therefore, it is also important for the students to be able to comprehend basic addition, subtraction, multiplication and division. The lesson units at mathematics would be all connected.If those basic lesson units could not be comprehended, the students would find difficulties in studying the further lesson units. In overcoming the students' problems, the teachers need to use a certain model, method approach, technique, and learning strategy to improve the students' ability in understanding the lesson units.

Learning model is a system that is used to conduct a learning process by applying a particular procedure. While an approach is a way that is used by teachers in elaborating the learning process to achieve the learning goals. Furthermore, a method is a way to convey the subject's materials to the students that can be applied for every subject. While a technique is a method that requires a specificway for certain subjects. Meanwhile a strategy is a specific way to teach a main substance of a certain material. Despite there are many required procedures, according to Nisbet [4], it is said that there are no (single) most correct learning ways and no best teaching ways. Every person has a different intellectual ability and different characteristic and personality, so they tend to adopt different characteristic approaches for learning. Therefore, in using a learning method, teachers needto adjust to the materials that would be conveyed and to the students' different characteristics.

On learning activities, students are taught how to solve the problem. The teacher always gives examples of how to solve a mathematics problem without giving an opportunity to students to try to discover a way by themselves. It made the students to be likely less creative in solving problems. Therefore, it is necessary to conduct a research on the influence of the learning process which is carried out with the students' ability to solve problems or to answer questions. The study was conducted to predict the models, methods, techniques, learning strategies used by teachers and the predictions of how students solve problems. The difficulties faced by students could be analysed, so we could give solutions to overcome those problems. The results of the analysis could be used to assess teachers' enhancement performances, learning processes and improvement of exercises that are given to the students.

Based on this problems, the formulation of the problem for this research are (1) Is the learning process that teachers use in the classroom similar to the predictions made by researchers?; (2) What are the factors that cause students having difficulty in answering the questions?

2. Experimental Method

To address the research questions, we carried out a qualitative study through administering an individual written test on the topic of geometry. The data was collected through explorative abilities of primary students at two-dimensional shape lesson unit. The participants of the research were 37 third-grade students of a primary school in the city of Bandung. The instruments included a written test consisted of ten questions about the perimeter of rectangle and were taken from the school's textbooks.

The procedure of this qualitative research was the following. Firstly, researchers made predictions towards the learning sessions that would be conducted by the teacher and made predictions about the exercise's answers that would be emerged from students. Secondly, we collected data of the students' answers. Data analysis was done by comparing our predictions with the reality that occur in the classroom, that is the learning performed by the teacher and students' exercises answers. Finally, we made conclusions and recommendations.

3. Results and Discussions

3.1. Learning Process

When the learning was being conducted, teachers began the class by making perceptions about square and rectangularperimeters on the next meetings. Then, the teachers invited students to remember its formula. The teachers wrote the formula of squares and rectangles on the class' board and gave the students the exercise examples. The students answered the given exercise questions. After the students understood, the teachers give exercise questions for about ten questions. The students answered the questions individually. However, there were still many students going to the teachers to ask about the way to solve the math problems. The students were given approximately 30 minutes to answer the questions. When there were students who have finished answered the questions, the teachers allowed them to take a break outside the class. So in the last session of the learning activity, teachers do not dismiss the activity or conclude it.

After the observation has been conducted, the researcher tried to compare the predictions of learning methods with the actual data that is discovered in the classroom. The elaboration about predictions and the actual data are explained in the Table 1 below.

Table 1. Comparison of Predictions of the Usage of Learning Model with the Actual Data

Usage of Learning Model	Prediction(s)	Actual Data
Model	Classical	Classical
Approach	Contextual	Expository
Method	Lectures, Questions and Answers, Exercises	Lectures, Questions and Answers, Exercises
Technique	Expository	Expository
Strategy	Papers, books, rulers	X

According to the observation results that have been conducted, learning model used is the classical models. This is similar to the predictions that have been made previously. The classical model which is used seem to be traditional, for example a teacher explained mathematical concepts with corresponding examples and given exercise (mainly bare mathematics tasks), while students paid attention, took notes and did the exercises. This teaching approach is quite common in Indonesia [5]. However, to be considered in the classical model of a teacher should be a good model in accordance with the theory expressed by Baruda [6]. If a teacher has a good hand writing and speaks with a good manner, explains clearly and systematically, then students will emulate them. Hence, the teacher shouldtake those things into consideration while teaching. When the teacher explained that in answering the questions,the first thing that need to be done is to write the length and width which are known in the matter. After that, the teacher also wrote the formula and how it is calculated. The students imitate the teacher's command, as seen from the students' answers which are almost entirely similar to the examples of the answers written by the teacher.

The learning approach used by teacher is expository approach. This is not the same with the predictions that have been made based on the contextual approach. At the time of learning, the teacher dominated and became an information center for the students.

The learning method used by teacher were lectures, discussion and exercises. This is consistent with the predictions that have been made previously. The teacher dominated teaching and learning activities, definitions and formulas about two-dimensional shape lesson units. The teacher explained to the students what should they do, gave the example problems and told how to solved the problems.

Learning techniques used by teacher is expository. This is similar with the predictions that have been made previously. This technique is chosen because there are a lot of number of students and the time is relatively short, so that the use of expository technique is suitable.

The teachers did not use learning strategies. This is not consistent with the predictions that have been made. The reason is because two-dimensional shape lesson units have been discussed for four times previously. Therefore, the teachers did not use instructional media. The possibility of using instructional media is done in the beginning of the discussion of the lesson material.

3.2. Student' Answer Analysis at Two-Dimensional Shape Problems
After making predictions about students' answers, the next step is the analysis of student' answers at two-dimensional shape problems. This part will explain about the possible reasons why the students' answer the question, how many students answered correctly and incorrectly, and how the comparison between predictions with actual data. Here is a Table 2 on the analysis of students' answers to questions provided by the teacher. The questions whichare given by the teachers are ten questions. But this article will only discuss one question that is Question 1. Question discussed representative of the matter as a whole.

Table 2. Students' Answer Analysis

Problems	Prediction(s)	Students' actual strategies	n/N
1. 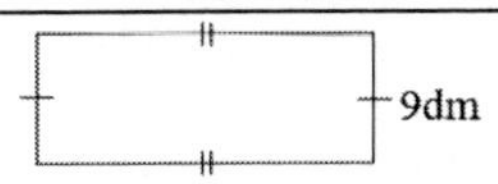 Perimeter = . . .	a. l= 17 dm w = 9 dm p = 2 x (l + w) = 2 x (17 + 9) = 2 x 26 = 52 dm	- a answer appear - b answer appear - c answer appears - There are four students who answer no prediction - d, e, and f answers not appear	25/37 7/37 1/37 4/37
	b. l= 17 w = 9 p = 2 x (l + w) = 2 x (17 + 9) = 2 x 26 = 52		
	c. p= 2 x (l + w) = 2 x (17 + 9) = 2 x 26 = 52 dm		
	d. l= 17 dm w = 9 dm p = 2 + (l + w) = 2 + (17 + 9) = 2 + 26 = 28 dm		
	e. l = 17 dm w = 9 dm p = 2 x (l - w) = 2 x (17 - 9) = 2 x 8 = 16 dm		
	f. l = 17 dm w = 9 dm p = 2 + (l - w) = 2 + (17 - 9) = 2 + 8 = 10 dm		

Note: n = number of students used a kind of strategies; N = number of all students

Based on observations of the students' answer, overall at this number, many students answered correctly. The answers above are similar with the researchers' predictions. In the actual data, there are 25 students who answered correctly in its process of writing formulas up to the final result. The students also wrote the length and width which are known in the matter, such as l = 17 and w = 9 dm. In addition, the students also wrote a calculation formula appropriately, i.e. p = 2 x (l + w) with the

IOP Conf. Series: Journal of Physics: Conf. Series **812** (2017) 012041 doi:10.1088/1742-6596/812/1/012041

final result 26 dm (see Figure 1). The students also did not forget to write the length unit in accordance with a matter that is decimetre (dm).

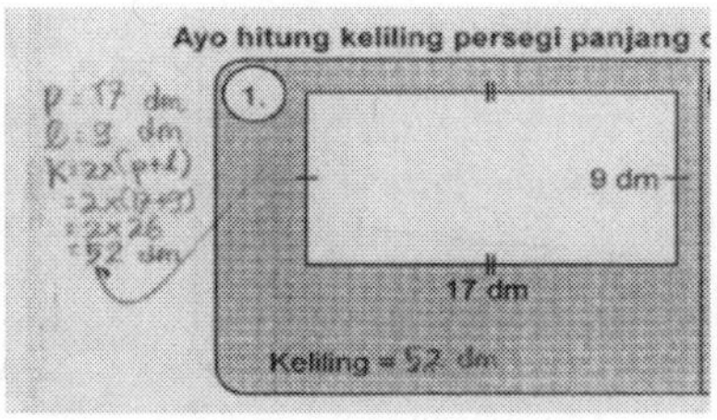

Figure 1. An example of a answer

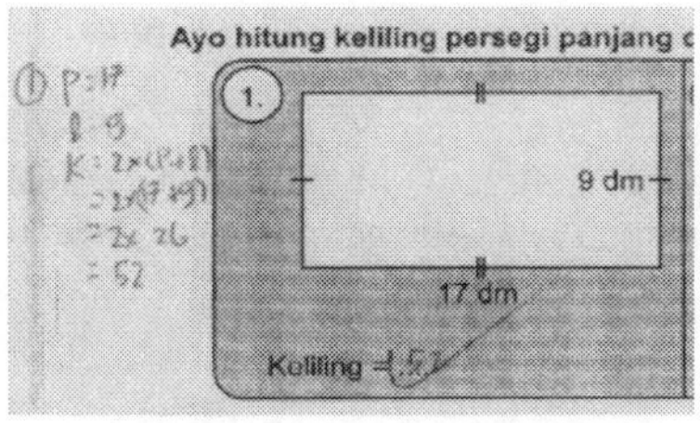

Figure 2. An example of b answer

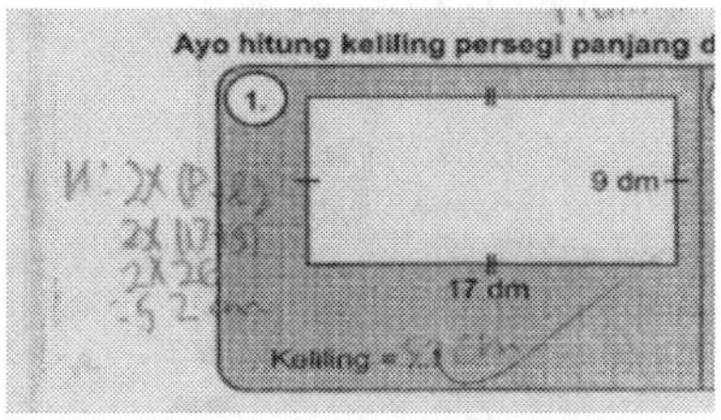

Figure 3. An example of c answer

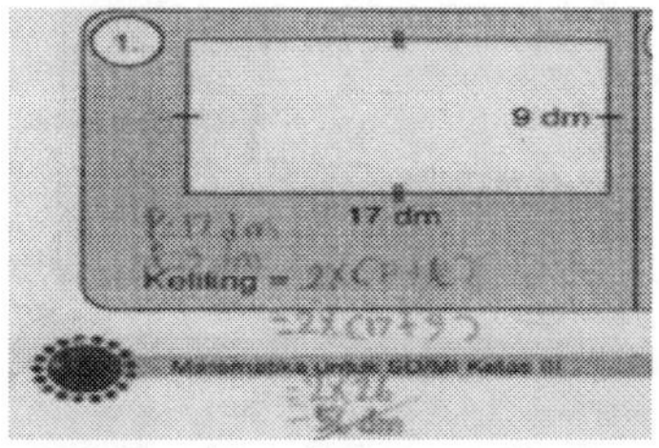

Figure 4. An example of student answer that are not in the prediction

Note: p = panjang (length), l = lebar (width), k = keliling (perimeter)

The students' answer that has been predicted to likely appear is b answer. It was found in seven students. Both students wrote an answer without using the length unit, like just wrote l = 17 and w = 9 (see Figure 2). Meanwhile, five other students hadsome errors in the writing of length unit, there was incomplete answer and some students who wrote cm instead of dm. For example there was a student who wrote l = 17 cm and w = 9 cm. The students wrote the formulas and calculations correctly. Meanwhile they wrote the length unit incorrectly.

One student chose c choice, yet he did not write the length and the width. Writing formulas and the final result were correct, but the writing of length unit was incorrect. The student wrote 52 cm, while the correct answer was supposed to be 52 dm (see Figure 3).

There were four students who did not answer as whathas been predicted. One of the students wrote the numbers incorrectly. Meanwhile, three other students made incorrect answers in the calculation (see Figure 4). The students did not do comprehensively in completing multiplication of arithmetic operation task. Difficulties can occur when students do calculation. In this case students calculated with pencil and paper methods.

Based on the result of predictions that have been made, there are three predictions which do not appear such as d, e, and f answers. The student did not answer similar with the prediction probably because the students were already familiar with the perimeterof rectangular formula. That is why no student who wrote the formula incorrectly. Therefore, only three of the six responses that appeared on the students' answers.In general, the students be able toanswer the question, but students look like just understand as 'instrumental understanding'.Skemp [7, 8] describes two kinds of understanding as instrumental understanding and relational understanding. Instrumental understanding is manifested when students know rules and formula and have the ability to use them without reason, not knowing where those rules and formula come from. For example, many students know that the formula to calculate the perimeter of a rectangle is 'two multiplied by length and width', but they do not know why this is so. He argues that students should develop a relational understanding of perimeter. In other words, students should know both 'what' to do and 'why' when dealing with problems that involve perimeter [7, 8]. This also implies that learners should be able to associate or relate the concepts of perimeter with other mathematical concepts and their everyday life experiences. Teacher must teach wisely, helping students deepen their understanding of rectangular perimeter from simple recognition to analysing and justifying geometric statements and to solving problem involving geometry. Geometry is a topic that is often neglected in primary school, yet it has many benefits for children. Geometry is used to help us represent the space in which we live and to describe the movements and the relationships between objects in space [9].

4. Conclusion

There is no learning method is the best method. All the methods used by the teacher is good because the use of models, approaches, methods, techniques, and strategies cannot be separated from the relation to the lesson unit that will be taught. But it is quite good that the teacher used lesson plans for each lesson, so that the teaching and learning are well planned. In addition, the teachers should useparticular instructional strategies to improve student learning outcomes. The teacher can also use instructional media that are in the surrounding so that learning will be more meaningful for students.

Based on the results of comparative analysis between the predictions and the actual data that is found in the field on students' answers on the topic of two-dimensional shape problems, we found that the result of the conducted research is similar with the predictions that have been made. But there are also answers that are not in the prediction appear on students' worksheets. The unexpected answers mostly arise because students are not conscientious in performing arithmetic operations that led to the error final result. In general, the students already know the formula of rectangular perimeter, an error occurred in miscalculations and the writing oflength unit. But the students who made incorrect answer in writing the length unit are justified by their teacher. However, the teachers should thoroughly correct the students' answers and remind them to always see the length unit in every answer of question. In addition, to improve student learning, the teachers should always analyze the student' answers for questions that have been given. Therefore the future problems in student learning barriers can be overcome.

Based on these results, we consider that (1) The error responses of the students are not only influenced by the teaching methods used by the teachers, but also influenced by other complex factors. So, we recommend further research can assess students' answer errorscaused by other factors; (2) The teachers need to understand that every student has different ways in answering the question, so that the teachers can be more prudent in assessing the students' answers.

5. References

[1] Phonapichat P, Wongwanich S, and Sujiva S 2013 *Procedia – Social and Behavioral Sciences***116**3169-3174
[2] Undang-undang no. 20 Tahun 2003 tentang Sistem Pendidikan Nasional (Jakarta: Depdikbud)
[3] Dahar R W 2011 *Teori-Teori Belajar dan Pembelajaran*(Jakarta: Erlangga) p 79
[4] Erman S,*et al* 2003 *Strategi Pembelajaran Matematika Kontemporer*(Bandung: Jurusan Pendidikan Matematika FPMIPA UPI) p 74
[5] Jupri A,Drijvers Pand Van den Heuvel-Panhuizen M 2014 *Mathematics Education Research Journal***26**683-710
[6] Nasution M 2015*Jurnal Ilmu-ilmu Kependidikan dan Kependidikan Sains* **3** 109-121
[7] Skemp R 1976 *Mathematics Teaching* **77** 20-26
[8] Machaba F M 2016 The concepts of area and perimeter: insight and misconceptions of grade 10 learners *Pythagoras 37* (1) a304
[9] Reys R, Linquist M M, Lambdin D V, and Smith N L 2004*Helping Children LearnMathematics*(United States of America: John Wiley & Sons, Inc) p 353

Acknowledgements
We thank students and teacher who actively participated in this study.

IOP Conf. Series: Journal of Physics: Conf. Series **812** (2017) 012042 doi:10.1088/1742-6596/812/1/012042

The Effect of Generate Argument' Instruction Model to Increase Reasoning Ability of Seventh Grade Students on Interactions of Living Thing with their Environment

Y Darmawanti[1], P Siahaan[2] and A Widodo[3]

[1] Department of Science Education, Universitas Pendidikan Indonesia, Jl. Dr.Setiabudi No. 229, Bandung 40154, Indonesia
[2] Department of Physics Education, Universitas Pendidikan Indonesia, Jl. Dr.Setiabudi No. 229, Bandung 40154, Indonesia
[3] Department of Biology Education, Universitas Pendidikan Indonesia, Jl. Dr.Setiabudi No. 229, Bandung 40154, Indonesia
Email: darmawanti.yuan@gmail.com

Abstract. This study aim to examine the effect of generates an argument instruction model to increase students' thinking skills, especially reasoning ability in lesson material of interactions of living thing with their environment. The study use weak experimental method with and the design is One-group pretest-posttest design. Sample in this study consists of 34 junior high school students of Seventh Grade in one of the junior high school in Ciamis. The instrument used to collect data is the essay questions of reasoning ability test according to reasoning Marzano's framework which consist of the eight indicators that are comparing, classifying, induction, deduction, constructing support, analyzing perspectives, analyzing errors, and abstraction. In generally, the results show there is an increase in the students' reasoning ability is significantly (Sig = 0.000). In addition, an increase in the ability of reasoning also viewed based on gender, and the result show there is not significantly (Sig = 0.168) the difference of reasoning ability between male student and female student. Increasing the ability of reasoning divided into two categories that is middle and low category.

1. Introduction

One of the thinking skills that need to be developed for students in studying science is reasoning abilities that also associated with the argument. Both of these abilities are related to one another, this is in line with that proposed by the Basel et al. in his research on the analysis of students' arguments on the concept of evolutionary theory which suggests that the analysis of the arguments gives us the opportunity to concentrate on the ability when the practice in terms of reasoning and arguing [4]. Beside it, a low complexity argument of students seem obvious but on the other hand, the use of various schemes of argument suggests that this argument comes from everyday experience. This shows that all students seem to have basic skills in argumentation derived from everyday life or other scientific contexts. Venville and Dawson argumentation skills play an important role in the formal reasoning [24]. In addition, any intervention based on the argument also positively impact on the quality of formal reasoning about a problem mainly related to sociosainstific.

Konstantinidou and Macagno explained that there are strategy and methods that can encourage the active participation of students in developing the ability of reasoning through argumentation, the results of research suggests that the latest research shows the increase in the application of the arguments used as a strategy of education and teaching methods [7]. Based on these problems, we need a teaching strategies or teaching methods that can be used to develop reasoning skills. One of the way to do is through the implementation of Generate An Argument Instructional Model. This model is a learning model that encourages students to be active in the group argued. Sampson and Schleigh this learning model is designed to provide opportunities for small groups of students to develop the claim that answer research questions based on the amount of data available [17].

In the process of learning in the classroom, sometimes not all students are actively involved in their group, so the contribution of students in learning as well as a discussion group becomes uneven. Active participation of students during the learning process is influenced by gender. Lowrie and Diezmann has conducted research on gender differences in secondary school students in a test of problem solving related to the graph, the results showed gender differences favoring boys among 9 to 12 year old students' graphical languages in mathematics [10]. Other studies also showed (Lorenzo, Crouch & Mazur, 2006; Pollock, Finkelstein & Kost, 2007 in Nieminen, et. Al 2012) male student is better than female student in understanding the concepts of force, motion, electricity, and the concept of magnetism in materials physics.

Sampson and Schleigh argues that Generate An Argument Instructional Model was designed to provide opportunities for small groups of students to develop claims answered questions according the problem is based on the data available [17]. As part of this process, the group made a tentative argument then give their claims and evidence that supports the claim by using media that can be viewed by others. Each group then had the opportunity to share their ideas during the session argument. Application of this model shows that it should provide benefits that are useful for teachers because it can be used as a template or guide to design learning aimed at describing the content in the existing curriculum [16]. Here are the stages Generate An Argument Instructional Model that consists of four stages identification of problem, question, and task, generation of a tentative argument, the argumentation session and group sense-making [16], [13], [20].

Stiggins suggests that the reasoning is the application of knowledge in the context of problem solving [21]. Septiana argues that the reasoning of students is the result of the receipt and processing of information received during the learning process [18]. Kusumawati and Woro [8] argues that the reasoning requires one to describe the specific results obtained from the process of observation, facts and allegations. Reasoning ability is one aspect of cognitive intelligence possessed by each individual. Basically, the individual has the ability to create a reasoning are different from each other. Reasoning ability is required of students when faced with tests such as the completion of the problems and remedies during the learning process.

Student reasoning ability develop naturally and continuously in line with increasing levels of education. Based on Inhelder & Piaget Basic reasoning ability have evolved since someone was 4 years old, but growing more rapidly again when it enters adolescence [5]. The development of reasoning ability differ at all ages, in adolescence reasoning ability can be developed through simple problem-solving strategies, through this strategy one can integrate knowledge and problem solving ability as part of the learning process [19].

Although everyone basically has the reasoning ability from an early age, but this ability should continue to be trained in order to develop properly. The development of reasoning ability are also a result of the teachings and practices of successive [15]. Students reasoning ability need to be developed in adolescence because reasoning ability was essential to learn the concepts of science [9] In order to develop the ability of reasoning can be done through a series of instructional design that facilitates students use reasoning ability, for example by learning science and mathematics-based inquiry can help develop students' reasoning ability [3].

Reasoning leads to the complex thought [12]. Reasoning ability can be measured through a series of tests were prepared based on any particular framework. The conceptual framework offered by Marzano consist of cognitive and affective components [21]. The reasoning process consists of five dimensions, there are attitude and a good perception of learning acquiring and combining knowledge, extending and refining knowledge, using meaningful knowledge, and the habits of mind [11] [12].

Based on Dugari cognitive dimensions of reasoning Marzano can be applied as a learning strategy that can improve achievement in science [2]. In addition, according Salamat that through learning based on cognitive dimension Marzano can improve critical thinking skills in understanding the concepts of physics [2]. The results of another study (Baz 2001)stating that the chemistry learning through learning strategies adopted from the cognitive dimension Marzano shows the results that can improve student achievement in high-level thinking skills such as decision making, critical thinking, and creative thinking [2]. In this study, assessing student reasoning ability focus to the second cognitive dimension which consists of eight aspects of reasoning that is comparing, classifying, induction, deduction, error analysis, constructing support, abstraction, and analyze perspectives.

2. Research methodology

This research use Weak Experimental method with the research design is the One-group pretest-posttest design. The number of samples consists of 34 students of 7th grade at one junior high school in Ciamis. Reasoning skills of students are assessing through reasoning ability test which consists of 13 essay by 8 indicators reasoning abilities. The test is done twice: before give treatment (pretest) and after give treatment (posttest).

3. Results and discussion

Based on the results of data processing, it can be seen an increase in the average score of posttest when compared with the score pretest reasoning abilities of students. Results of significance tests through paired sample T test show the significance results with obtained Sig. (0.000 < 0.05) and shown in detail in Table 1 and 2.

Table 1. Recapitulation of reasoning ability test.

N	Ideal Score	Average of Pre test	Average of Post test	Average of Gain (G)	Average of N-Gain (<g>)	Sig.
34	100	55,28	64,93	9,65	0,222	0,000

Indicators of reasoning ability assessd in this study consisted of eight indicators include the ability to comparing, classifying, induction, deduction, constructing support, analyzing perspectives, analyzing errors, and abstraction. The score of each is shown in Table 2.

Table 2. Recapitulation indicators value of reasoning.

Indicators of Reasoning	Average Pre test	Average Post test	N-Gain <g>	Category
Comparing	54,41	60,78	0,140	Low
Classifying	36,27	46,08	0,154	Low
Induction	58,33	62,25	0,094	Low
Deduction	58,22	73,53	0,366	Middle
Constructing Support	52,45	63,73	0,237	Low
Analyzing Perspectives	64,71	73,53	0,250	Low
Analyzing Errors	64,71	78,43	0,389	Middle
Abstraction	49,51	62,75	0,262	Low

In addition to seeing the reasoning ability of students, this research also viewed ability of students by gender, which distinguished group of male students and female students. Based on the results of data processing, increasing the average score of posttest compared with pretest is better in both of group that is male students and female students. The results show in Table 3.

Table 3. Recapitulation of reasoning ability evaluated by gender.

Student Group	N	Ideal Score	Average of Pretest	Average of Posttest	Average of Gain (G)	Average of N-Gain (<g>)
Male	16	100	52,90	64,58	11,68	0,248
Female	18	100	56,84	65,24	8,40	0,208

The stages that exist in the generate argument instructional model facilitate students develop reasoning relating to the third dimension according to Marzano's framework of reasoning that extending and refining their knowledge. Through the first stage to the final stage students are required to actively develop the knowledge, the stimulus is given at the beginning of learning in the form of presentation of the problem. In addition, students are given the opportunity to select the knowledge through sessions arguments, which gives students the opportunity to express opinions not only on his native groups but also other groups who visit. So that students not only acquire knowledge through a learning process, but it can develop and select the knowledge gained.

The learning experience is very important in the achievement and student learning outcomes. Through various phenomena and problems presented during the learning process, students gain knowledge and experience that is more real because they are directly involved and experience it yourself. Learning through the generate an argument instruction model applied in this study pasted direct observation method to develop an argument while being prepared by the students. Generate an argument instruction model more emphasis on the study conducted outside the laboratory, so it is not required to obtain the data directly, but from a variety of literature and data from previous observations, Generate An Argument Instructional Model designed to develop arguments based on data obtained through the literature because of limitations in observation and data collection of objects, if done directly [16]. The ability of the student's thinking in this regard reasoning abilities of students are a significant improvement after the study is completed. Giving the problems in early learning, developing a tentative argument, and the argumentation sessions have helped develop the students to think systematically and logically and to be able to develop answers based on observation, combined with theory or concept based on the literature.

Although an increase in students' reasoning abilities significantly seen from the average value post test results, basically still occur obstacles when learning takes place. The low category of improvement based on the value of N-Gain is likely due to the obstacle when learning takes place. Responses were fairly well demonstrated by the students was also evident from the enthusiasm of the students increased from one learning process to the next learning process. Although the first learning process, most students are still confused about the learning patterns applied by teachers who are different from the usual pattern they use every day. Students' ability to communicate is still relatively low, it looks at the first learning process where the students still do not brave to ask questions, express or refute opinions. This is because the learning process is commonly used in everyday is not much or not familiar high-level thinking skills exercises.

If we look at the score of N-Gain, an average increase students' reasoning abilities of male and female student included in the low category. However, the value of N-Gain group of male students is higher than the score of N-Gain group of female students. Researchers initially suspected increase in student groups of male student will differ significantly from the group of female students. But the concept of gender itself is not an absolute, but rather is influenced also by other factors in the surroundings. The concept of gender is the inherent nature of men and women that is socially and culturally, but nature itself can be interchanged with one another, may change over time and differ from one place to another [6]. However, there is research that shows that there are differences in science achievement more positive on male students compared to female students [23].

No significant differences between the groups of male students and female students assumption group for learning model used is the Generate An Argument Instructional Model equally well for both groups of students, so the achievements of the two group students did not differ significantly. The same thing also expressed by the results of other studies of gender differences in the understanding of concepts and the ability of scientific reasoning in style concept which states that learning the concept of style does not show a significant difference between male students and female students, although the results of pretest and posttest male student better than female students, which indicates that the learning methods are used effectively for the two groups of students [14]. The different results show that the reasoning abilities of male students is lower than female students [22]. In addition the results of other studies also demonstrate reasoning abilities of male students are generally lower than on girls, except in drawing conclusions logically indicators [8].

All stages of learning that exist in the Generate An Argument Instructional Model followed by a group of male students and group of female students, so that all students have an equal opportunity to develop thinking patterns. In addition, the materials studied are concepts that are close to the students' everyday life, namely the interaction of living thing with the environment. Therefore male students and female students are much different reasoning ability at the end of learning. It shows the characteristics of the subject matter also influence student performance achievements of men and women students, so that the characteristics of such academic subjects also affect the level of significance of gender differences.

Seeing the results of observation on student learning activity showed each stage generally follow well. But if you look at the enthusiasm to learn, male students look more enthusiastic when learning of group and observation activities. This is presumably because male student are more like a learning process with student-centered and relevant to problem solving, but female student are more like learning to write activity [1]. So, it can also influence is no difference between the reasoning ability of male students and female students due to screen test reasoning ability in writing. So, in this case, the gender factor is not the major determinant in the successful completion of a test-based reasoning because there are external factors that influence such differences in learning style preferred by male students and female students also indirectly affect the achievement of reasoning abilities.

4. Conclusion

Based on the analysis, it can be concluded that there are significant differences between the results of the pretest and posttest of students' reasoning abilities after learning to use the Generate An Argument Instructional Model with significant score of T test results of paired samples T test is 0,000 < α (0.05). It also means the Generate An Argument Instructional Model sufficient to give effect to an increase in the ability of reasoning even though the score of N-Gain is still in the low category. Increase the ability of reasoning viewed based on gender in general there is no difference, this is indicated by the score of N-gain average are in the same category that is low.

5. References

[1] Acar Ö, Türkmen L and Bilgin A 2015 *Eurasia Journal of Mathematics, Science & Technology Education* **11** 1027-040
[2] Alkhateeb O 2015 *Journal of Education and Practice* **6** 146-59
[3] Bao L *et al* 2009 *Science* **32** 586-87
[4] Basel N, Harms U and Prechtl H 2013 *Journal of Biological Education* **47** 1-8
[5] Fabby C and Kathleen K 2015 *Journal of STEM Education* **16** 20-6
[6] Fakih M 2013 *Analysis of Gender and Social Transformation* (Yogyakarta: Pustaka Pelajar)
[7] Konstantinidou A and Macagno F 2013 *Science & Education* **22** 1069-087
[8] Kusumawati D and Woro S 2014 *Jurnal Inovasi Pendidikan Fisika (JIPF)* **3** 95-9
[9] Lawson A E, Banks D L and Logvin M 2007 *Journal of Research in Science Teaching* **44** 706–24
[10] Lowrie T and Diezmann C M 2011 *Learning and Instruction Elsevier* **21** 109-25
[11] Marzano R J 1992 *A Different Kind of Classroom: Teaching with Dimensions of Learning* (Aexandria: Association for Supervision and Curriculum Development)
[12] Marzano R J, Pickering D and McTighe J 1994 *Assessing Students Outcomes, Performance Assessment Using the Dimensions of Learning Model* (Alexandria: Association for Supervision and Curriculum Development)
[13] Muslim 2014 *Program Development of Argumentative Skills Oriented School Physics Course for Prospective Physics Teacher* (Bandung: Disertation UPI)
[14] Nieminen P, Savinainen A and Viiri J 2012 *International Journal of Science and Mathematics Education* **11** 1137-56

[15] Rich J D, William, F Jr and Willis D 2011 *Hispanic Journal of Behavioral Sciences* **33** 261-77
[16] Sampson V and Gerbino F 2010 Two instructional models that teachers can use to promote & support scientific argumentation in the biology classroom *Research and Learning* vol 72 (University of California Press) pp 427-31
[17] Sampson V and Schleigh S 2013 *Scientific Argumentation in Biology : 30 Classroom Activities* (Virginia: National Science Teachers Association)
[18] Septiana R 2015 *Reducing Cognitive Load Students Senior High School Using Framing on Plant Classification Practice Activity Based Learning Styles* (Bandung: Magister Thesis UPI)
[19] Shrager J and Siegler R S 1998 *SAGE Journals* **9** 405-10
[20] Siswanto 2014 *The Implementation Of "Generate Argument" Instructional Model Using the Scientific Methode to Increase Cognitive Abilities and Argumentation Skills Of Senior High School Students* (Bandung: Magister Thesis UPI)
[21] Stiggins R J 1994 *Student-Centered Classroom Assessment* (New York: Macmillan College Publishing Company)
[22] Suharyani E D 2012 *The Implementation of Classification Approach to Improve the Mastery Concept of World Plants and Reasoning Students Senior High School Based on Gender* (Bandung: Magister Thesis UPI)
[23] Temitayo O and Abayomi 2015 *Journal of Education and Practice* **6** 87-92
[24] Venville G J and Dawson V M 2010 *Journal of Research in Science Teaching* **47** 952-77

Detecting and Reducing Science Teacher Candidate's (STC) Misconception About Motion and Force By Using Force Concept Inventory (FCI) and problem Based Learning (PBL)

Y Oktarisa*, I S Utami and Y R Denny

University of Sultan Ageng Tirtayasa, Dept. of Chemical Engineering, Engineering Faculty, Jakarta Street KM. 4, Serang – Banten, Indonesia

*Corresponding author's email : yuvita.oktarisa@untirta.ac.id

Abstract. This study has been done to 34 science teacher candidates of Teachers' Training and Education Faculty of Sultan Ageng Tirtayasa University at their first year of study during 2015-2016 school years. This research focused on student's misconception about motion and force and how Problem Based Learning (PBL) reducing it. Diagnostic test of misconception about motion and force has been detected by using Force Concept Inventory (FCI). FCI had been used in pretest and posttest, and to find the reducing of students' misconception N-Gain pretest and posttest of each student had been calculated. Quasi experiment one group pretest and posttest had been used as the research method, and Problem Based Learning (PBL) used as the treatment of manipulation. After two weeks learning motion and force with PBL approach, N-gain which obtained prove that misconception about motion and force had been reducing.

1. Introduction

In science education, research on conceptual change has been a central topic in science education literature for more than two decades. Student's native intelligence mostly has effected by student's experience in daily life. Mainly student's conceptual about motion and force come from their common sense belief, which influenced by their daily experience of motion and force it self. Students tend to explain every case of their environment from their own learning perspective. Misconception arises when the intuitions and impression of students are being used to process the information from their environment.

Mostly, students common sense belief about motion and force inappropriate with Newtonian concepts in most respect (Hestenes et al., 1992; Muller and Sharma, 2012). When scientific information or concepts clarified improper way, this could lead to wrong understanding and misconception (Kartal et al., 2011). Based on observation in some of high schools, most of the students learned the most basic Newtonian concepts in wrong way, it making students failed to comprehend the next material in physics course. In the world of science teaching, some research has been done to detect and reduce misconception. In 1992, David Hestenes et al, design an instrument to probe students belief on Newtonian concept, this instrument called Forced Coencept inventory (FCI). Although this instrument was made more than two decades ago, it is still cab be used to detect students' misconception about motion and force. FCI had been given to more than 1500 high-school students and more than 500 universities in United States (Hestenes et al., 1992). Based on experienced on what had been done in United Stated in using FCI for detecting misconception about motion and force, it will be interesting if FCI also used to detect misconception abut motion and force to science teacher candiates in Indonesia. Banten province as a part of Indonesia had been selected to be a subject for this study because the test to identifying misconception about motion and force never had been done yet.

2. Material and Method

The data of the Misconception about motion and force had been detected by using multiple choices in Force Concept Inventory. Force Concept Inventory (FCI) can be used in three purposes, firstly as diagnostic tool; secondly, to evaluate instruction, for evaluating instruction and finally for final exam. In the first year of FCI had been used 1500 high schools and more than 500 university students in America to identify their misconception about motion and force (Hestenes et al., 1992). In general FCI consist of several topics, which are force, kinematics, Newton's first law, Newton's second law, Newton's third law, principle of action-reaction and kind of force(Poutot and Blandin, 2015). FCI almost had been used in detecting student's misconception more over two decades and it is still can be used until now because the validity and reliability of FCI had been at very good level (Hestenes et al., 1992).

Method of this research is quasi experiment one group pretest posttest. This research method often described by giving pretest, implementing treatment and giving posttest (Cohen and Ledford, 1994). In this study, the test had been taken before learning process and after the learning process, so that every science teacher candidate followed two test, pretest and posttests. Quasi experiment had been chosen because the aim of the study was to see the reducing of misconception a group of science teacher candidate before and after implementing problem-based learning. 34 Science teacher candidates from Science Education Department school year 2015-2016 had been answering 29 questions from FCI which are given by the lecturing who taught Basic Physics. After two weeks learning process by implementing Problem Based Learning as the teaching method the FCI had been used again to see the reducing of misconception of science teacher candidate.

The performance of science teacher candidate in reducing their misconception in pretest and posttest the normalized gain had been used(Bao, 2006; Wiyono et al., 2012). Since its introduction, the normalized gain has been widely used in assessing students' performance in pretest and posttest (Bao, 2006). How far the STC's misconception decreasing after implementing Problem Based learning will be described at result discussion.

$$n - gain = \frac{Posttest\ Score - Pretest\ Score}{Maximum\ Score - Pretest\ Score} \quad (1)$$

3. Result and Discussion

The misconception about motion and force of Science Teacher Candidate (STC) in pretest and posttest had been grading by calculating correct and incorrect answer. Incorrect answers gave information about misconception regarding motion and force of science teacher candidate. In pretest only one question of FCI got the score higher than 50%, the rest of the questions got the score lower than 40%. The description about the number of student answering correct and incorrect answer along with STC's most common misconception is on table below.

Table 1. STC's most common misconception along with correct and incorrect answer.

Item No	Correct Answer	Incorrect Answer	STC's Most Common Misconception
1	0	34	Heavier objects fall faster
2	1	33	Greater mass implies greater force
3	0	34	Heavier object fall faster
4	4	30	The heavier object exerts greater force
5	0	34	Gravity increases as objects fall
6	1	33	Last force to act determines motion
7	6	28	Last force to act determines motion
8	1	33	Gradual/delayed Impetus build up
9	1	33	Impetus supplied by "hit"
10	12	22	Circular impetus
11	1	33	Most active agent produces greatest force
12	8	26	Only active agents exerts force
13	2	32	Most active agent produces greatest force
14	10	24	Most active agent produces greatest force
15	9	25	Only active agents exerts force
16	5	29	Impetus dissipation
17	2	32	Acceleration implies increasing force, gravity increase as objects fall
18	1	33	Largest force determines motion
19	17	27	Force compromise determines motion
20	5	29	Position-velocity un discriminated
21	4	30	Velocity-acceleration un discriminated
22	20	14	Impetus supplied by "hit"
23	2	32	Force compromise determines motion
24	0	34	Last force to act determines motion
25	6	28	Force causes acceleration to terminal velocity
26	0	34	Loss/recovery of original impetus
27	4	30	Impetus dissipation
28	3	31	Motion when force overcomes resistance
29	6	28	Mass makes thing stop, motion implies active force

Almost all of STC have misconception in every topic, which asked by FCI. In topic Gravity which had been appear in number 1, 3, 5, 9, 12,16,17 and 22 all of these questions had score below 30%. All of the STC candidate taught that heavier object will fast faster than light object. In kinematics topic which had been appeared in 5 questions which are number 7, 20, 21 23, 24 and number 25 the score gotten also below 30%. Most of STC couldn't discriminate position, velocity and acceleration. In first law of motion, which is in number 4, 6, 8, 10, 18, 26, 27 and 28, the science teacher candidate taught that largest force determines motion. In second law of Newton, which is appeared in number 6, 7, 24 and 25, the entire STC taught that the last force to act determines motion. In third law of Newton, which is appeared in number 2, 14, 11 and 13, STC taught that greater mass implies greater force.

After detecting misconception of STC about motion and force the learning process continued by doing problem based learning. Science teacher candidate had been given a problem and they did the discussion to solve the problem given. Problem based learning had been chosen due to its successfully implementation in some studies(Poutot and Blandin, 2015). In two weeks learning process, problem based learning successfully in reducing misconception of STC.

One way, which had been done by STC in investigating which one goes faster, heavy object or light object. By dropping the heavy object and the light object at the same amount of height and recorded the time consuming by both objects with video tracker STC proved that both of the objects consuming same amount of time. STC used two objects whit different mass but same shape. Video tracker had been used to show the time used for both objects is the same.

Figure 1. Problem Based Learning done by STC

After doing Problem Based Learning in two weeks, STC had been done posttest to see the reducing of misconception about motion and force. Figure 2 below shows the difference amount of the correct answer in answering FCI.

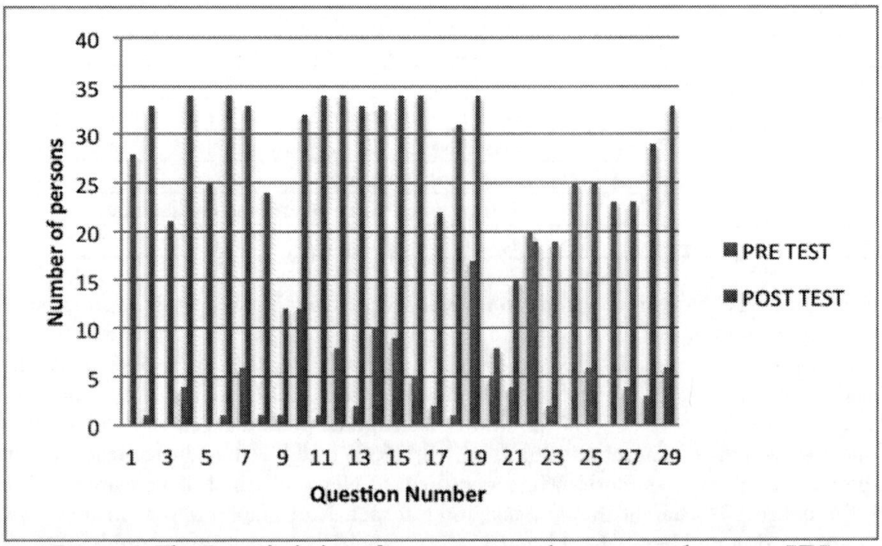

Figure 2. The Recapitulation of correct answer in pretest and posttest STC

Figure above shows the reducing in misconception about force and motion of STC. In posttest most of the questions in FCI had gotten score above 70 %. The decreasing of misconception STC had been reducing after implementing Problem Based Learning as the method. After getting pretest of STC, the next step, which had been done to calculate the reducing misconception of STC is N-gain. The figure above shows the n-gain for 34 Science Teacher Candidate.

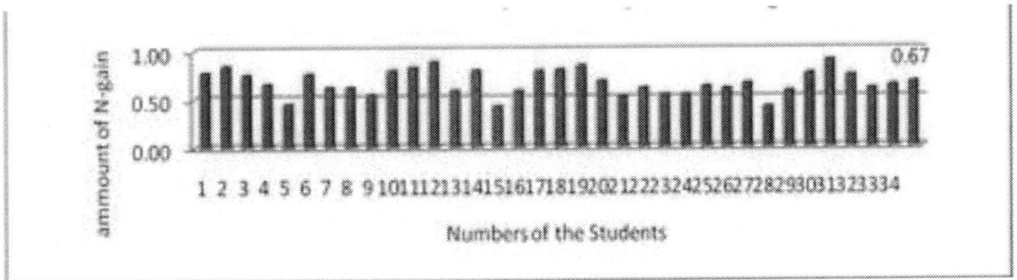

Figure 3. N-Gain 34 Science Teacher Candidates

Each student had positive N-gain, misconception of every Science teacher candidate had been decreasing. The average of N-gain is 0.67, it means problem based learning which had been implemented successfully to reduce misconception. Most of the problem in FCI had been investigated by STC with doing experiment and prove the hypothesis which mad by them. Some of the question hard for them to investigate by doing the experiment since the case given is hard to visualization. Beside that the lack of time also become the main factor why STC couldn't maximize the implementation of problem based learning.

4. Conclusion

In this study Force Concept Inventory had detected misconception of science teacher candidate (STC). Most of SCT had misconception in all of the topic cover by FCI. STC still taught that heavier object fall faster than light object. Couldn't discriminate position, velocity and acceleration also describe misconception of STC. Not only in gravity but also in Law of Newton STC shows misconception, which can be seen from the result in pretest.

Problem based learning method can be used to reduce misconception of STC, because by this method STC can prove their hypotheses, which had been predicted as the solution of the problem. But because most of STC solved the problem given by doing the experiment, not all of the problem given could solved by the student with problem based learning.

5. References

[1] Bao L 2006 *Am. J. Phys.* **74** pp 917–22.
[2] Cohen S G, Ledford G E 1994 *Sage Journals Hum. Relat.* **47**(1) pp 13–43.
[3] Hestenes D, Wells M, Swackhamer G 1992 Phys. Teach. **30** 141–58.
[4] Kartal T, Öztürk N, Yalvaç H G, 2011 *Procedia-Soc. Behav. Sci.* **15** pp 2758–63.
[5] Muller D A and Sharma M D 2012 *Proceedings of the Australian Conference on Science and Mathematics Education (formerly UniServe Science Conference)* pp 58-63
[6] Poutot G, Blandin B 2015 *J. Educ. Res. 3* pp 116–20
[7] Wiyono K, Setiawan A and Paulus C T 2012 *J. Pendidik. Fis. Indonesia J. Phys. Educ.* 8(1) pp 74-82
[8] Bao, L 2006 *Am. J. Phys.* **74** pp 917–22
[9] Cohen S G, Ledford G E 1994 *Hum. Relat. Sage Journals* 47 pp 13–43
[10] Hestenes D, Wells, M, Swackhamer G, others, 1992 Phys. Teach. Journals **30** pp 141–58
[11] Kartal T, Öztürk N, Yalvaç H G, 2011 *Procedia-Soc. Behav. Sci.* **15** 2758–63
[12] Muller D A, Sharma M D 2012 *Proc. of the Australian Conference on Science and Mathematics Education (formerly UniServe Science Conference)* pp 58-63
[13] Poutot G, Blandin B 2015 Am. J. Educ. Res. **3** pp 116–20

[14] Wiyono K, Setiawan A, Paulus C T, others 2012 *J. Pendidik. Fis. Indones J. Phys*. Educ. **8** pp 74-82

The development of CERDAS learning strategy guide for science education students of distance education

U Rahayu[1,2*], T Darmayanti [1] A Widodo[2] and S Redjeki[2]

[1]Departmen of science and mathematics education program, Universitas Terbuka, Jl.Cabe Raya, Pondok Cabe, Tangerang Selatan, Indonesia
[2]Sekolah Pasca Sarjana, Universitas Pendidikan Indonesia, Jl. Setiabudi 226, Bandung, Indonesia
*Corresponding author: urahayu@ecampus.ut.ac.id

Abstract. Self-regulated learning (SRL) is a part of students' skills in which they manage, regulate, and monitor their learning process so they can reach their study goal. Students of distance education should comprise this skill. The aim of this research is to describe the development of distance students learning guide, namely "CEDAS strategy" designed for science students. The students' guidance consists of seven principles, they are; selecting and applying learning strategy appropriately, managing time effectively, planning of learning realistically and accurately, achieving study goal, and doing self-evaluation continuously. The research method was qualitative descriptive. The research involved the students of Universitas Terbuka' Biology education who participated in Animal Embryology course. The data were collected using a questionnaire and interview. Furthermore, it was analyzed by descriptive analyses. Research finding showed that during try out, most of the students stated that the learning guide was easy to understand, concise, interesting and encouraging for students to continue reading and learning. In the implementation stage, most students commented that the guide is easy to understand, long enough, and helpful so it can be used as a reference to study independently and to apply it in the daily basis.

1. Introduction

Self-regulated learning (SRL) refers to the thoughts, feelings and actions planned and systematically adjusted according to the needs that affect learning and motivation of a person [1]. SRL is a process involving students in regulating, managing, and monitoring complex learning process so they can achieve their academic goal [2]. SRL consists of processes such as setting goals for learning, attending and concentrating on instruction, using effective strategy to organize, making code and rehearsing information to remember, establishing a productive work environment, using resource effectively, monitoring performance, managing time effectively, seeking help when needed, holding positive beliefs about one's capabilities, holding the value of learning, including the factors affecting learning, anticipating outcomes of action, and resulting satisfaction and pride of the effort that has been done [1]. The self-regulatory process is a cyclical form which consists of three phases; they are, forethought phase, performance or volitional control phase, and self-reflection phase. Forethought phase precedes an actual performance and refers to a process which sets the stage for action. Performance control phase includes the processes that occur during learning and affect attention and action. Self-reflection phase, which occurs after the performance, is used when people respond to their efforts [3].

During forethought phase, high independence learners are going to achieve their learning goal and strategy planning. This phase consists of task analysis and self-motivation. Task analysis includes goal setting and strategy planning. Goal setting is an activity to determine the learning objectives and if necessary to modify them. In reaching his goal, a student determines an appropriate plan of learning strategies. Strategic planning is a strategy that will be used to achieve academic goals such as determining the use of cognitive strategies to achieve academic goals. A student can modify his learning goal or change his learning strategy if he knows that his goal cannot be achieved or his strategy is not accurate. During the performance phase, students use self-control strategies [3]. Self-reflection is the third phase. In this phase, students do self-judgment and self-evaluation. Self-judgments mean that students compare the current level of performance with their purpose. During the self-evaluation, students compare the performance with specific standards [4].

The ability of students' SRL can be improved through systematic interventions [1]. Students' SRL can be developed through the improving of time management skills, understanding and summarizing the text, noting, anticipating and writing the test [5]; through the development of cognitive strategies such as learning strategies that are specific to the content; and through the development metacognitive strategies such as monitoring and self-reflection.□

SRL Skill of Open Distance Learning' (ODL) students are a must because of the character of distance education; a separation between teachers and learners. The level of SRL skill of ODL students should be well built due to a lack of interaction between teachers and students. However, SRL of ODL students is not as high as supposed to be.□

ODL is a learning form of andragogy, in which learning process is designed to address the particular needs of adults [6]. Therefore, the following matters should be considered in ODL such as adult limitation, learning styles, previous learning experience, linkage of what is being studied and its application in life, linkages of previous knowledge with new experience, involving actively in the learning process, a need to provide a process of self-reflection in the learning process and support to make improvements in a variety of situations. So, ODL institution should appreciate to this situation and provide a learning guide for its students, especially for science students. Learning guide is needed for ODL science students to handle learning problems they are facing, so they can successful in their study. Furthermore, in order to improve SRL skills of ODL science students, in this paper, I d' like to describe how to develop a distance student learning strategy guide, namely "CERDAS" designed for science students. It will describe techniques of learning strategy that can be used to improve SRL skills of ODL students especially Science ODL students.

2. Research methodology

The research method was qualitative descriptive. The research consisted of some steps. The first step was the development of the draft guidelines. The second step was trying out the readability and the usefulness of CERDAS learning strategy (CLS) for Biology Education students - Regional Offices of Bogor and Surabaya (n=10) and then it was validated by experts (n=5). The third step was its revision based on experts' and students' comments. The fourth step was used by students who joined online tutorial in Animal Embryology (n=30) course. Data was collected using a questionnaire and interview. Then all of the data were analyzed in descriptive analyses.

3. Result and discussion

3.1. The development of the CERDAS learning strategy (CLS)

SRL of ODL science students guide in this research is a modification of CLS developed by Darmayanti [7] and Puspitasari [8]. The Darmayanti's CLS [7] was chosen to be the basic draft of learning guide for science ODL of this research because her intervention had a positive impact on student self-directed learning at Universitas Terbuka, especially on the component of students' learning needs [7]. The Darmayanti's CLS version describes tips or strategies on how to select and use learning strategies, manage and motivate learning appropriately (C); to use study time effectively

(E); to plan learning realistically (R); to set goals that can be achieved (D); to plan learning accurately (A); and to plan specific learning objectives (S). Initially, the material of Darmayanti' CLS version [7] is a modification of SMART learning strategy developed by Andersen [9]. SMART learning strategy refers to the ability to determine the purpose of studying specifically (Specific), measurably (Measurable), can be achieved (Attainable), realistically (Realistic), and time management properly (Timely). In this study, Darmayanti' CLS version modified researcher by adding and replacing the old material with the new and relevant content or materials, images, and illustration as presented in Table 1.

Table 1. The modification of Darmayanti' CLS Version

Section of Darmayanti CLS	CERDAS for science ODL students
What do you know?	Modifying illustration with the relevant illustration.
What is cerdas strategy?	Addition of self-evaluation concept because this concept is a part of self-reflection that supposed to be there
The letter C for Cerdik (Smart)	Changing the title and content on being smart in selecting and using learning strategy. Addition of scientific learning techniques: The technique of Preview-Question-Read-Reflect-Recall [10] Tips for reading scientific materials [11] How to arrange mind mapping [12]
Smart in learning management	Changing and adjusting illustration and content with the relevant illustration and content.
Smart in recognizing surrounding environment	Changing the title and content to a recent situation.
The letter E for Effective	Changing and adjusting illustration with the interesting illustration.
The letter R for realistic	Changing and adjusting illustration with the interesting illustration.
The letter D for goal attainable	Changing and adjusting illustration with the relevant illustration.
The letter A for Accuracy	Changing and adjusting illustration with the relevant illustration.
The letter S for specific	Changing and adjusting illustration with the relevant illustration
Example implementation of CLS	Simplifying examples; adjusting example of learning through online tutorials because the examples did not connect with the online tutorial; Addition of sample form used in selecting and determining learning techniques; Addition of mind-mapping, writing questions and answer, monitoring of learning targets and mastery of concepts in student sheet, because there was not practiced how to implement learning techniques /strategy
Closing notes	Modifying closing notes to adjust to a recent situation

Furthermore, the CLS for science ODL students was also modified from the CLS of Puspitasari [8] especially at the part of the example implementation of CLS. In this research, we modifying the table of target and learning schedule weekly by adding media column.

The strategies explained in CLS guide actually referred to aspects or dimension of SRL [1]. To be a self-regulated learner, a student has to have skills in setting goals of learning, using effective strategy, establishing a productive work environment, using resource effectively, monitoring performance, managing time effectively, seeking help when needed, holding high motivation, and evaluating success or failure regularly. Therefore, these aspects are described in accordance with the condition as ODL Science students.

3.2. Step 2: Try out

Concerning the readability of the draft CLS guide for science ODL, through a questionnaire, 81.7 % students agreed that the draft CLS for Science ODL students was easy to understand. It was emphasized by the student during the interview. "The learning strategy is easy to understand, and its words are not so hard to digest." On the other hand, there was a student who commented differently. "The sentence was a bit difficult to understand. However, the illustrations can help me to understand the material". This comment shows that not all students understand the content of CLS guide for science ODL students. Therefore, CLS for science ODL students needed elaboration.

Besides that, 83.3 % students stated that the sentences that were used in the draft learning strategy guide were concise. It was supported by the result of the interview. "The learning guide is concise enough." However, there was a student commented in contrast, that the sentences used in the part of C for cerdik were so long, not concise. This comment shows that not all students agreed that the sentences that use in the CLS guide were concise. Therefore, it needed a little revision.□

Relating to the illustration that presented at CLS guide for science ODL students, 95.8% students stated that the draft of learning guide had interesting illustration. Even, the students said that they could understand the material by looking at the illustrations presented in the guide. This comment shows that most of the students agreed that the draft of CLS for science ODL students had good and interesting illustrations. Therefore, it needed a very little revision. □ In addition to the usefulness of the draft of CLS guide for science ODL, 100 % students agreed that the draft of learning guide encourages students to learn other section of the CLS guide. Then, 100% students agreed that the draft CLS guide useful for studying independently. One of them said that CLS helped students, particularly for freshmen students who did not fully understand about independent learning or ODL. Puspitasari [8] also found that CLS that she used in her research are readable and useful, even though it needed revision and elaboration.

3.3. Step 3: Revision

In accordance with students, experts' comments and inputs, researcher then made revisions. After revisions, CERDAS Strategy guide was printed and stored in the form of word files. Then two forms of CERDAS strategy guide were printed in digital forms and delivered to students who were actively participating in the online tutorial course of Animal Embryology.

3.4. Step 4: Implementation

The sections of CERDAS learning strategy that integrated with the online tutorial as an exercise or training for improving students' regulating learning were some tasks that had to be done at week 1, 2, 4, 6, and 8 through the online tutorial. The tasks included arranging learning planning and its monitoring. Through this task, students had to arrange the schedule of learning for a week and to monitor if it was either achieved or not; Developing mind mapping of a relevant topic. In this task, for example, students had to develop mind mapping about the mechanism of division and blastula stage on Amphioxus and Amphibia; Compiling questions and answers. Through this task, for example, students had to arrange a set of essay test about the mechanism of division and blastula stage on Amphioxus and Amphibia; Doing self-reflection related to mastery of concepts. Through this task, students had to do self-reflection, asked themselves whether the topics had already been mastered or not, and how to handle the problems. During that time the students were completing the tasks. They should refer to the module as the main resource of the course and CLS as guidance for learning independently.

At the last session of the online tutorial, the students filled the questionnaire and some of them were interviewed about the implementation. This paper only described and discussed things that were related to CLS guide for science ODL students and the results are below.

Out of 30 students followed the online tutorial that integrated CLS guide, 96% students said that the CLS guide was easy to understand. This was supported by the results of students' interview that the CLS guide presented systematically and straightforwardly, so made it was easy to read. In addition, 89% students said that the CLS guide was to the point, concise, simple and understandable. The usefulness of CLS guide was felt by the students. There were 96% students agreed that the CLS guide could be used as a reference for learning independently. Besides those, 93% of the students stated that the CSL guide encouraged them to apply it in a process of learning. Some of them wanted to implement this strategy when they taught the subject in the class. Moreover, there were students who said that the CLS motivated and inspired them to implement in a learning process. It was relevant to the study of Puspitasari [8] that the CLS that she used in her research are useful, motivate her students to be more discipline.

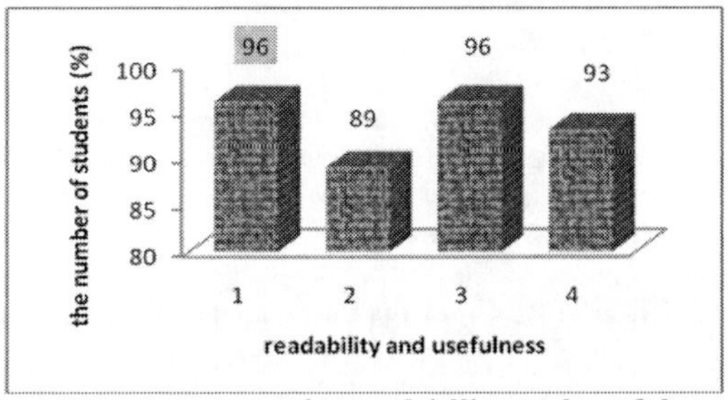

Figure 1. the average students' comments on the readability and usefulness of CLS guide for science ODL students

Note: 1 = easy to understand; 2= conciseness; 3= usefulness as SRL reference; 4= usefulness as encouragement of students' learning

Based on students' comments, it seems that the CLS guide for science ODL students can be used as a reference for independent learning. It is really helpful for ODL students who are demanded to have a high level of self-regulated learning skills. By reading and learning the guide, students are encouraged to plan, monitor, and regulate their learning better. □

Regarding the implementation of the CSL for science ODL, some of students faced some difficulties. First, some of students found difficulty in managing time effectively. The students could not manage time effectively. They lack skills in managing time for being busy in teaching and at the house. So, they had no time to make a summary, to read and learn the module as the schedule that has been made. Second, the students found the difficulty in establishing a realistic short-term goal. The students had already been registering many subjects. So finally, when they made a priority of doing a task, they chose easy tasks. The course of Animal embryology was a subject that was quite difficult to learn and it was not relevant to the subject that they taught to their students at a school. Third, some of students found difficulty in doing monitoring and evaluation of the learning target and schedule. It was hard to do because many schedules could not be obeyed, so the targets could not be achieved. Fourth, there was student found the difficulty in arranging mind mapping. It was hard to do because they did not really understand what they read.

Concerning these obstacles, it was concluded that the students' problems were related to the characteristics of ODL students. Learning for ODL should consider a need of the students. Most of the ODL students continue to study to a get relevant degree and better career. They need linkage between what is being studied with application in life [6] as a science teacher. The students preferred other course material that was related to their life and their need.

4. Conclusion

The Cerdas Learning Strategy Guide for science ODL students is readable and useful. It can be used as a reference. It assists students in independent learning and encourages students to adopt independent learning strategies. However, during eight-week practices, some of students experienced difficulties in applying it, specifically when managing time effectively, setting realistic goals for short-term learning, monitoring, and evaluation, and making a mind map. It needed more time to train for students to use this guide, so the obstacles could be handled.

5. References

[1] Schunk D H and Ertmer P A 2005 *Handbook of self-regulation* (San Diego: Academic Press) pp 631-46
[2] Zimmerman B J and Schunk D H 2001 *Self-regulated learning and academic achievement: theoretical perspectives* (Hisdale: Erbaum)
[3] Schunk D H 2012 *Learning Theories An educational Perspective* (Yogyakarta: Pustaka Pelajar)
[4] Zimmerman B J 2008 *Am. Ed. Res. J.* **45** (1)166-83
[5] Zimmermann B J, Bonner S and Kovach R 1996 *Developing self-regulated learners: beyond achievement to self-efficacy* (Washington D.C: American Psychological Association) p 147
[6] K Cercone 2008 *AACE J.* **16** (2),137 -59
[7] T Darmayanti 2008 *J. PJJ.* **9** (2) 68-82
[8] K Puspitasari 2012 Dissertation (The Florida State: University College Education retrieved)
[9] Andersen R 1995 *Getting ahead: Career skills that work for everyone* (New York: McGraw-Hill)
[10] Slavin R E 2011 *Psikologi pendidikan: teori dan praktik.* (Jakarta: PT Indeks) pp. 257.
[11] Thomas E L and Robinson H A 1972 *Improving reading in every class: A sourcebook for teachers* (Boston: Allyn & Bacon)
[12] Buzan T 2007 *Mind Map* (Jakarta: PT Gramedia Pustaka Utama)

Acknowledgment

The authors would like to thank Universitas Terbuka that has funded this research.

IOP Conf. Series: Journal of Physics: Conf. Series **812** (2017) 012045 doi:10.1088/1742-6596/812/1/012045

Crossed Products Related to Cyclically Ordered Semi Groups

R Rosjanuardi

Department of Mathematics Education, Indonesia University of Education, Jl. Dr. Setiabudhi 229, Bandung 40154, Indonesia

Email: rizky@upi.edu

Abstract. Suppose (G, R) is a cyclically ordered abelian group. Thru unwounding method of Rieger, we construct a semi group crossed product related with (G, R).

1. Introduction

The theory of crossed product by semigroup of endomorphisms is one of important and interesting area in modern theory of operator algebras. This theory has a plenty of applications such as in Cuntz algebras [1], Toeplitz algebras (Adji,*et al.*[2], Laca and Raeburn [3], Adji [4], Adji and Raeburn [5], Rosjanuardi [6], Adji and Rosjanuardi [7], Rosjanuardi and Albania [8] and Hecke algebras arising in number theory (Laca and Raeburn [9], Larsen and Raeburn [10]). This theory is a generalisation of the theory of crossed product by group of automorphism, which is a well developed area of the theory of operator algebras, and has rich of examples.

In some cases, the theory of crossed product by semigroup of endomorphisms heavily relies on theory of totally ordered group, for examples Adji *et. al* [2], Adji [4], Adji and Raeburn [5], Rosjanuardi [6], Adji and Rosjanuardi [7], Rosjanuardi and Albania [8], Rosjanuardi [11] are among others. The concept of order is essentials as the semigroup crossed product here is a generalisation of Stacey's [12] crossed products of single endomorphism which is related with the action of semigroup of natural numbers $\mathbb{N}$.

A generalisation of the notion of totally ordered group is cyclically ordered group. Cyclically ordered group which was originally introduced by Rieger in 1947 has attracted great deal of attention, for example Swierczkowski [13], Fuchs [14], Harminc [15], Jakubik [16, 17], Cernak [18], Giraudet, Leloup, Lucas [19] and Leloup, Lucas [20] are among others. The concept of this order is different with regular linear order, as it is a ternary relation instead of binary relation.

Given a cyclically ordered group (G, R), we will make use the theorem of Rieger to get its unwound totally ordered group $\mathrm{uw}(G)$ to obtain a dynamical system and its semigroup crossed product.

2. Preliminaries on Cyclically Ordered Group

Definition 1*A cyclically ordered group is a pair* $(\boldsymbol{G}, \boldsymbol{R})$ *consists of a group* $\boldsymbol{G}$ *and a ternary relation* $\boldsymbol{R}$ *on* $\boldsymbol{G}$ *satisfying the following axiom:*

R1: $\forall\, x, y, z, R(x, y, z) \Rightarrow x \neq y \neq z \neq x$ (R is strict),
R2: $\forall\, x, y, z, x \neq y \neq z \neq x \;\Rightarrow\; R(x, y, z)$or $R(x, z, y)$ (R is total),
R3: $\forall\, x, y, z, R(x, y, z) \Rightarrow R(y, z, x)$ (R is cyclic),
R4: $\forall\, x, y, z, u, R(x, y, z)$and $R(y, u, z) \Rightarrow R(x, u, z)$ (R is transitive),
R5: $\forall\, x, y, z, u, v,\; R(x, y, z) \;\Rightarrow R(uxv, uyv, uzv)$ (R is compatible with the operation in G)

Riegerin Leloup and Lucas [20] states that every cyclically ordered group can be obtained from a totally ordered group. Let $(G, \leq)$ be a totally ordered group and g_0 be a positive, central and cofinal element of G. The quotient group $G/\langle g_o \rangle$ can be cyclically ordered by setting

$$R(a\langle g_0\rangle, b\langle g_0\rangle, c\langle g_0\rangle)$$

$$\Leftrightarrow$$

$$\exists a' \in a\langle g_0\rangle,\ b' \in b\langle g_0\rangle,\ c' \in c\langle g_0\rangle \text{such that } 1_G \leq a' < g_0, 1_G \leq b' < g_0, 1_G \leq c' < g_0$$

and either

$$a' < b' < c' \text{or } b' < c' < a' \text{ or } c' < a' < b'.$$

The cyclically ordered group $G/\langle g_o \rangle$ is called the *wound round* associated to G and g_0.

Moreover theorem of Rieger says that for every cyclically ordered group (G, R), there is a totally ordered group $\mathrm{uw}(G)$, and an element z_G which is positive, central and cofinal in $\mathrm{uw}(G)$ such that

$$(G, R)$$

is isomorphic to the cyclically ordered

$$\mathrm{uw}(G)/\langle z_G \rangle.$$

The group $\mathrm{uw}(G)$ is called the *unwound* of (G, R). More precisely, the unwound group $\mathrm{uw}(G)$ of cyclically ordered group (G, R) is given by

$$\mathbb{Z} \times G$$

together with the linear order

$(n, c) < (n', c') \Leftrightarrow$ either $n < n'$, or $n = n'$and $R(1_G, c, c')$, or $n = n'$ and $c = 1_G$.

3. Semigroup Crossed Products

Definition 2*Let Γ^+ be the positive cone of a totally ordered abelian group. A dynamical system consists of a unital C^*-algebra A, and a semigroup homomorphism $\alpha : \Gamma^+ \to Endo(A)$. The homomorphism α is called an action of Γ^+ on A by endomorphisms, or just action for short when the context is clear . We denote the dynamical system by (A, Γ^+, α).*

Definition 3*A **covariant representation** of (A, Γ^+, α) is a pair (π, V) in which $\pi : A \to B(H)$ is a unital representation of A and $V : \Gamma^+ \to Isom(B(H))$ is an isometric representation of Γ^+, that is, $V_s V_t = V_{s+t}$ and the covariant condition $\pi(\alpha_t(a)) = V_t \pi(a) V_t^*$ for all $a \in A$, $t \in \Gamma^+$ is satisfied.*

When we consider the semigroup □ , we have the dynamical system considered by Stacey [12]. Focusing on the dynamical systems considered by Stacey [12] allows us to find an example of a dynamical system that has no covariant representation. Stacey pointed out a dynamical system of no covariant representation, as we recited below.

Definition 3*A **covariant representation** of (A, Γ^+, α) is a pair (π, V) in which $\pi : A \to B(H)$ is a unital representation of A and $V : \Gamma^+ \to Isom(B(H))$ is an isometric representation of Γ^+, that is, $V_s V_t = V_{s+t}$ and the covariant condition $\pi(\alpha_t(a)) = V_t \pi(a) V_t^*$ for all $a \in A$, $t \in \Gamma^+$ is satisfied.*

When we consider the semigroup □ , we have the dynamical system considered by Stacey[12]. Focusing on the dynamical systems considered by Stacey [12] allows us to find an example of a dynamical system that has no covariant representation. Stacey pointed out a dynamical system of no covariant representation, as we recited below.

Example 1 (Stacey [12]) Suppose $\alpha : c_0 \to c_0$ is the unilateral shift, defined by

$$\alpha((x_1, x_2, ...)) = (x_2, x_3, ...) \tag{1}$$

Then for each $x \in c_{00}$, the dense subalgebra of sequences of finite support, $\alpha^n(x) = 0$ for some n. Hence, if $(\pi, \{T_i\})$ is a covariant representation of (c_0, α) on H and β is the corresponding *-endomorphism of $B(H)$ then $\beta^n(\pi(x)) = 0$ implies $\pi(x) = 0$. Hence, by the continuity, π is the trivial (zero) representation. Therefore (c_0, α, N) has no covariant representation.

The concept of **semigroup crossed product** of the dynamical system (A, Γ^+, α) was introduced by Adji *et. al* [2]. It is a development of the crossed product by groups action of Raeburn [21]. This version of crossed product can also be viewed as the untwisted version of semigroup twisted crossed product of Rosjanuardi [6, 11], Adji and Rosjanuardi [7].

Definition 4 *A crossed product of the dynamical system (A, Γ^+, α) is a triple (B, i_A, i_{Γ^+}) of a unital C^*-algebra B together with a unital homomorphism $i_A : A \to B$ and embedding of Γ^+ as isometries $i_{\Gamma^+} : \Gamma^+ \to B$ such that :*

1. *(i_A, i_{Γ^+}) is a covariant pair for (A, Γ^+, α),*
2. *for any covariant representation (π, V) of (A, Γ^+, α) there is a unital representation $\pi \times V$ of B such that $(\pi \times V) \circ i_A = \pi$ and $(\pi \times V) \circ i_{\Gamma^+} = V$,*
3. *B is generated by $i_A(A)$ and $i_{\Gamma^+}(\Gamma^+)$.*

Existence of semigroup crossed product of a dynamical system heavily rely on the existence of non zero covariant representation. Adji,*et al.* [2] gave an example of dynamical system in which the crossed product exists, and is universal for isometric representation of the semi group. The algebra in the dynamical system is the closed subspace of $\ell^\infty(\Gamma)$ spanned by $\{1_x : x \in \Gamma^+\}$, where

$$1_x(y) = \begin{cases} 1 \text{ if } y \geq x, \\ 0 \text{ otherwise.} \end{cases}$$

This algebra is denoted by B_{Γ^+}. For each $x \in \Gamma^+$, the automorphism τ_x of $\ell^\infty(\Gamma)$ which is defined by

$$\tau_x(f)(y) = f(y - x) \text{ satisfies } \tau_x(1_y) = 1_{x+y}, \tag{2}$$

which implies that τ restricts to an action α of Γ^+ by endomorphism of B_{Γ^+}.

Suppose $T: \Gamma^+ \to B(\ell^2(\Gamma^+)$ defined by $T_x(\delta_y) = \delta_{y+x}$, then T is an isometric representation of Γ^+ . Proposition 2.2 of Adji, Laca, Nilsen, Raeburn [2] implies there is a representation π_V of B_{Γ^+} such that (π_V, V) is a covariant representation of the dynamical system $(B_{\Gamma^+}, \Gamma^+, \alpha)$. Therefore the crossed product $B_{\Gamma^+} \times_\alpha \Gamma^+$ exists.

4. Results and Discussion

Given a C^* −algebra A, a cyclically ordered abelian group (G, R) and an action α of the positive cone P of G . Since the class of the order is not a linear order, we can not directly make use the Definition 4 to obtain the semigroup crossed product of the dynamical system (A, P, α). But we can make use the method of Rieger to unwound the cyclic order to get the linear order related with R .

If $\text{uw}(G)$ is the unwound of (G, R), then $\text{uw}(G)$ is linearly ordered. *We* now consider the positive cone $(\text{uw}(G))^+$. If α' is an action of $(\text{uw}(G))^+$ by endomorphisms of A, and there is a nontrivial covariant representation (π, V) of the dynamical system $(A, (\text{uw}(G))^+, \alpha')$, then the semi group crossed $A \times_{\alpha'} (\text{uw}(G))^+$ exists. We call $A \times_{\alpha'} (\text{uw}(G))^+$ the semigroup crossed product related with the cyclically ordered abelian group (G, R).

Example:

Given a cyclically ordered abelian group (G,R). Let $\mathrm{uw}(G)$ be the unwound group of (G,R). We will make use the method of Adji, Laca, Nilsen and Raeburn [2] in constructing the semi group crossed product which is universal for the isometric representation of the semi group $(\mathrm{uw}(G))^+$, i.e the positive cone of $\mathrm{uw}(G)$.

The algebra in the dynamical system is the closed subspace of $\ell^\infty(\mathrm{uw}(G))$ spanned by

$$\{1_x : x \in (\mathrm{uw}(G))^+\},$$

where

$$1_x(y) = \begin{cases} 1 \text{ if } y \geq x, \\ 0 \text{ otherwise.} \end{cases}$$

This algebra is denoted by $B_{(\mathrm{uw}(G))^+}$. For each $x \in (\mathrm{uw}(G))^+$, the automorphism τ_x of $\ell^\infty(\mathrm{uw}(G))$ which is defined by

$$\tau_x(f)(y) = f(y-x) \text{ satisfies } \tau_x(1_y) = 1_{x+y},$$

which implies that τ restricts to an action α of $(\mathrm{uw}(G))^+$ by endomorphism of $B_{(\mathrm{uw}(G))^+}$.

Suppose $T: \Gamma^+ \to B(\ell^2((\mathrm{uw}(G))^+)$ defined by $T_x(\delta_y) = \delta_{y+x}$, then T is an isometric representation of $(\mathrm{uw}(G))^+$. Proposition 2.2 of Adji, Laca, Nilsen, Raeburn (1994) implies there is a representation π_V of $B_{\mathrm{uw}(G)^+}$ such that (π_V, V) is a covariant representation of the dynamical system $(B_{\mathrm{uw}(G)^+}, (\mathrm{uw}(G))^+, \alpha)$. Therefore the crossed product $B_{(\mathrm{uw}(G))^+} \times_\alpha (\mathrm{uw}(G))^+$ exists.

In particular, since the unwound group of (G,R) is the linearly ordered group $\mathbb{Z} \times G$, then the semi group crossed product $B_{(\mathbb{Z}\oplus\mathbb{Z})^+} \times_\tau (\mathbb{Z} \oplus \mathbb{Z})^+$ considered in Rosjanuardi and Albania [8] and Rosjanuardi [11] can be viewed as a semi group crossed product related with the cyclically ordered group $\mathbb{Z}$.

5. Conclusion

Given a cyclically ordered abelian group $(\boldsymbol{G},\boldsymbol{R})$. Thru unwounding method of Rieger, we can aply the method Adji,*et al* [2] to obtain a semi group crossed product related with $(\boldsymbol{G},\boldsymbol{R})$.

6. References

[1] Cuntz J 1977*Comm. Math. Phys***57** 173-85
[2] Adji S, Laca M, Nilsen M and I. Raeburn 1994*Proc. Amer. Math. Soc.* **122** 1133-41
[3] Laca M and Raeburn I 1999*J. London Math. Soc.***59** 330-44
[4] Adji S 2000*J. Op. Theory***44** 139–50
[5] Adji S and Raeburn I 2004*Integral Equation Operation Theory***48** 281-93
[6] Rosjanuardi R 2005 *Structure theory of twisted Toeplitz algebras and ideal structure of the (untwisted) Toeplitz algebras*(Bandung: Thesis Institut Teknologi Bandung)
[7] Adji S and Rosjanuardi R 2007*Acta Mathematica Sinica***23** 1639-48
[8] Rosjanuardi R and Albania I N 2012 *Far East Journal of Mathematical Sciences***67** 99-110
[9] Laca M and Raeburn I 1999 *J. London Math. Soc***139** 330-44
[10] Larsen N S and RaeburnI 2002*J. London. Math. Soc.***2**198-212
[11] Rosjanuardi R 2013*Far East Journal of Mathematical Sciences* **72** 69-78
[12] Stacey P J *1993 J. Austral. Math. Soc. (series A)***54** 204-212
[13] Swierczkowski S 1959*Fund. Math***47** 167-172
[14] Fuchs L 1963 *Partially ordered algebraic systems*(Oxford: Pergamon Press)
[15] Harminc M 1988 *Math. Slovaca***38** 249-253
[16] Jakubik J 1988*Casopis Pro pestovani matematiky***113** 184-196
[17] Jakubik J 1990 *Czech. Math. Journal***40** 534-538
[18] Cernak S 1995*Math. Slovaca***45** 29-38
[19] Giraudet M,Leloup G and Lucas F2013First order theory of cyclically ordered groups*Cornell University Libary* arXiv:1311.0499v1 [math.LO]

[20] Leloup G and Lucas F 2013c-Regular cyclically ordered groups*Cornell University Library*arXiv: 1312.5269v1 [math: LO]
[21] Raeburn I 1988*Proc. Edinburgh math. Soc.***108** 321-30

Acknowledgements
We acknowledged Direktorat Jendral Pendidikan Tinggi (Grant: Hibah Penelitian Fundamental, 2016).

IOP Conf. Series: Journal of Physics: Conf. Series **812** (2017) 012046 doi:10.1088/1742-6596/812/1/012046

Secondary Teachers' Mathematics-related Beliefs and Knowledge about Mathematical Problem-solving

T Y E Siswono[1], A W Kohar[1], S Hartono[1]

[1] Faculty of Mathematics and Natural Science, Universitas Negeri Surabaya, Surabaya 60231, Indonesia
Email: tatagsiswono@unesa.ac.id

Abstract. This study investigates secondary teachers' belief about the three mathematics-related beliefs, i.e. nature of mathematics, teaching mathematics, learning mathematics, and knowledge about mathematical problem solving. Data were gathered through a set of task-based semi-structured interviews of three selected teachers with different philosophical views of teaching mathematics, i.e. instrumental, platonist, and problem solving. Those teachers were selected from an interview using a belief-related task from purposively selected teachers in Surabaya and Sidoarjo. While the interviews about knowledge examine teachers' problem solving content and pedagogical knowledge, the interviews about beliefs examine their views on several cases extracted from each of such mathematics-related beliefs. Analysis included the categorization and comparison on each of beliefs and knowledge as well as their interaction. Results indicate that all the teachers did not show a high consistency in responding views of their mathematics-related beliefs, while they showed weaknesses primarily on problem solving content knowledge. Findings also point out that teachers' beliefs have a strong relationship with teachers' knowledge about problem solving. In particular, the instrumental teacher's beliefs were consistent with his insufficient knowledge about problem-solving, while both platonist and problem-solving teacher's beliefs were consistent with their sufficient knowledge of either content or pedagogical problem solving.

1. Introduction

Research on discussing relationship between teacher's belief and practice within mathematics instruction have been addressed by many scholars with a variety of purposes. Some tried to conceptualise models on such relationship [1,2,3,4], while others reported how such relationship was confirmed into practical interest [5,6].Either the models or the reports were identified to have several associated key variables which emphasize on explaining to what extend one variable influence another variable. Such key variables such as past school experience and immediate classroom situation, for instance, respectively have strong influence on teachers' mathematics beliefs and teacher's teaching practice [1].Another important variable, teacher's knowledge, is scrutinized to have strong interaction with beliefs in shaping teachers' teaching practice with varying degree being given to particular types of knowledge or beliefs in different situations [5].Fennema et al [3] added that beside students' behaviours, such interaction will directly affect teacher's decisions in planning classrooom instructional activity. Hence, more recent studies reported how such interaction influence teachers' teaching practice. Bray's study results [6] argued that teacher's beliefs seemed most likely influence on how teacher structure class discussion whereas teacher's knowledge appeared to drive quality of teachers' responses within classroom discussion. On the other hand, Anderson et al [1] summarised the knowledge component includes mathematical content and plans, professional development, teachers' knowledge and decisions, important mathematics and assessment procedures, use of mathematics texts, and teacher education program.

While such scholars above discussed general knowledge needed by teachers regarding mathematics instruction, in particular on problem solving, what knowledge does teachers need to hold? Some describe in pedagogical context, while others revealed in the context of content. Pedagogically, Franke, Kazemi, & Battey [7] explained that teachers need to orchestrate class discussion so that students share multiple problem-solving strategies, analyze relations among strategies, and explore contradictions in students' ideas to provide greater insight into the mathematical focus. In brief, Chapman [8]summarised three types of knowledge for teaching problem solving: problem solving content knowledge consisting of knowledge of mathematical problems, problem-solving proficiency, and problem posing, as well as pedagogical problem solving knowledge consisting of knowledge of students as problem-solvers, and instructional practices for problem-solving, and affective factors and beliefs.

Regarding types of mathematics-related beliefs which might be held by teachers in teaching problem solving, several authors proposed some categories, such as Ernest [9] with his prominent categories: instrumental, platonist, and problem solving. Table 1 summarizes several authors' view on how such three beliefs are characterized [10].

Table 1.Summary of beliefs about mathematics, mathematics teaching, and mathematics learning

Beliefs about the nature of mathematics	Beliefs about mathematics teaching	Beliefs about mathematics learning
Instrumentalist	Content focussed with an emphasis on performance	Skill mastery, passive reception of knowledge
Platonist	Content focussed with an emphasis on understanding	Active construction of understanding
Problem solving	Learner focussed	Autonomous exploration of own interests

Our study aims to describe teachers' mathematics-related belief and knowledge about mathematical problem solving as well as search for possible interaction between such two variables emerging in our teacher participants.

2. Experimental Method

For the purposes of our research, we describe qualitatively three Indonesian lower secondary mathematics teachers' belief and knowledge regarding problem-solving with different philosophical views of teaching mathematics: instrumental, platonist, and problem-solving. Those teachers were selected from the result of interviewing 12 teachers purposively chosen from Surabaya dan Sidoarjo city using a belief-related task discussing what best instructional sequences should be implemented into a mathematical problem within classroom activity. Of all the teachers, we selected teachers with the most significant feature describing their view about a particular type of belief. We selected Toro (male, *pseudo name*) as one of our further interest of participants since he seemed to portray his beliefs in an instrumental view. Two other selected teachers characterizing platonist and problem solving were represented by Yati (female, *pseudo name*) and Ana (female, *pseudo name*), each of whom agrees that students should be engaged actively in class activities as well as agree that lesson should be started by the contextual problem as shown by the task. The difference is that Yati tends to provide general steps on how to solve the problem, whereas Ana prefers to give feedback regarding her students' idea on solving the problem rather than recommend specific idea in solving the problem.

All the three teachers selected were individually interviewed about their beliefs and knowledge about mathematical problem solving. The beliefs include three components, i.e. nature of mathematics, how to teach mathematics, and how students learn mathematics; while the knowledge includes problem-solving content knowledge and pedagogical problem-solving knowledge. To collect the data of teachers' belief we used a set of task-based interview. Such tasks were arranged in the form of incomplete statements. We provided three options for each incomplete statement illustrating the description of each of beliefs. Each of the teachers was asked to choose one of the options which mostly represent his/her view and explain why he/she chose that option. Furthermore, we also asked them to compare his/her argumentation contrasted with the other two remaining options. As illustration, when asked about the what people should learn from mathematics, the teachers were confronted with options: (1) to have skills in calculating and applying mathematical formulae or procedures when solving daily life problem, (2) to be proficient in understanding topics in mathematics, such as algebra, statistics, probability, geometry, and the interrelationships among those topics entirely, and (3) to have thinking skills such as understanding regularities of phenomena, being critical and creative in solving any problems. Meanwhile, to collect data about teachers' knowledge, our questions were primarily inspired by Chapman's category of problem-solving knowledge for teaching, such as when a mathematics question becomes problem [the nature of problem], what processess needed to solve mathematical problem, how to implement problem solving process and strategies within classroom activity, etc.

Analysis of data was carried out by firstly reducing data, displaying data, and finally drawing conclusion and verification[11]. The conclusion was sought to understand the characteristics of belief and knowledge espoused by teachers and how such two variables interact each other shaping their idea of mathematical problem solving.

3. Results and Discussion

3.1. *Toro's beliefs and knowledge about mathematical problem-solving*

Toro's view of nature of mathematics was mostly aligned with Instrumental, especially when he argued about what one should learn from mathematics and when he defined mathematics. He prefers to have proficiency in calculating as well as applying formula and procedures because he said,"I think one needs to master mathematical concepts to understand how to apply a mathematical formula,...most of the mathematics teach us to calculate. "This is in line with his option emphasizing that mathematics is a discipline discussing calculation, numbers, and formula. He asserted his view by arguing that strategies of solving a problem will only be successful if one knows the similar problem he/she ever encountered. However, when he was asked to link between real life problem and mathematics, he prefers to choose the statement that there are parts of topics in mathematics used to solve real life problems and there are also parts of mathematics used to develop the topics of mathematics itself. In this case, however, it indicates his platonist view.

His instrumental view also appeared in most cases examined in the group of 'teaching mathematics' statements, i.e. viewing the role of teacher, the precise time to teach contextual problem, ways to minimize students' misunderstanding, and strategy to motivate students on particular topic. In viewing the latter two cases, for example, he told, "I choose to clarify their misunderstanding by giving a more detail explanation particularly on the parts in which they find difficulties instead of letting them discuss because if I pose questions, they are often be confused, no feed back.", and "to motivate students learn something difficult to understand its application, I used to explain more clearly on the prerequisite knowledge for learning, in this case, central and circular angle." Ultimately, when he responded on the best way to teach mathematics, he viewed, "teaching math should start with giving concepts, exercise from easy to difficult".

In responding how students should learn mathematics, Toro's beliefs also indicates an instrumental view. He argued that students need to practice many times to master concepts or procedures inherent in the problem, need to only focus on particular strategy which can be applied to many types of problem by repeating such strategy many times rather than learning various strategy that may not applicable in the task students are solving, and must not use calculator whatever the reason is. Nevertheless, in viewing formula, he told, "they [students] may use the formula given in the book or learning how such formula is constructed, but more importantly, they should develop their own 'formula' to solve any problem.". He seemed to have the problem-solving view in this case.

Regarding knowledge, Toro seemed to have partial understanding, primarily on the meaning of mathematical problem, types of problem, problem-solving process, and strategies. When asked about whether particular mathematical questions are a problem or not for students, he knows that a problem should be challenging, interesting, and has no immediate solution, but he asserted that a question which demands insufficient prerequisite knowledge is a problem. He said, “question ‘find the area bounded by curves $y=x^2+1$, $x=-4$, $x=5$, and x-axis’ is a problem for my students because they require certain formula which is not learned yet’. Besides, Toro was seemingly unable to select an appropriate problem for her lesson as indicated when he incorrectly explained about for what purpose a particular problem is designed. He said, "this problem [a problem about finding the concept of circle area] is designed for mastering students' skill in applying the formula of circle area. I don't think it works on my students if I give it in early learning of circle area". Data also indicates that Toro incompletely mentioned problem-solving process. He noticed, "when solving problem, I need to firstly understand, then choose suitable strategies, and apply such strategies". No indication appeared that solving problem needs also process of looking back as suggested by scholars. When explaining about implementation of problem solving process within classroom, Toro was not able to explain an ideal process of guiding students to solve problem as indicated by his statement, “...students are asked to read the problem and listened to what I explained to understand the problem...let them choose methods and apply such methods..”.Accordingly, his knowledge about guiding process is rather teacher-centered in the beginning of solving problem.

3.2. Yati's beliefs and knowledge about mathematical problem-solving

Yati expressed a deep belief that her nature of mathematics regards with platonist view as indicated by her responses to all the incomplete statements. She concluded that mathematics is a discipline of logic which is permanent over time. She argued, ‘something that can change from it is only the methods which may be more various in the future, but something such as numbers, geometrical shape or other objects in mathematics will be remaining same.” Her idea is supported by her statement that the strategy of solving problem will find its success depends on whether one ever works on it or not, regardless of whether his concepts and procedures is correct or not. She also agrees that mathematics teach us to understand topics in mathematics, such as algebra, statistics, probability, geometry, and the interrelationships among those topics entirely. Such interrelationships, she said, are important as basic skill to solve a more complex problem. Thus, her view on defining mathematics influences her view on the scope of what to learn from mathematics and her successful experiences when solving a mathematical problem.

While Yati’s belief about nature of mathematics is likely significant, her belief in teaching mathematic is equally mixed between instrumental, platonist and problem solving. Her instrumental view is indicated by her responses on the role of teacher when real life problem should be taught, and what to place reward, while her problem-solving view is described by her responses on how to minimize students' misunderstanding during the lesson. Meanwhile, platonist view is pointed out by her statements on responding cases on motivating students to learn and the best way to teach the particular topic and how to help students when finding difficulties in solving the problem. Her statements such as "I prefer to choose to provide a figural display to represent problem of this system of linear equation and let students explore such display to find further strategy rather than let them find their own form representation, it makes students not confident because of their insufficient knowledge of strategies" and supports such view.

Yati’s view on learning mathematics, however, is aligned with problem-solving view. All the cases examined, except the response on using calculator, are indicated to support an idea of autonomous exploration of students' interests. In viewing which strategies should be learned in solving a particular problem, for instance, she suggested students figure out all the strategies provided because she maintained, "although some strategies are not effective enough to solve a certain problem, but such strategies will be likely more effective for other problems."

Regarding Yati's problem-solving knowledge, same with Toro, she seemed to have a partial understanding, primarily on problem solving content knowledge. The difference is that Yati has a better understanding of the problem-solving process and strategies as pointed out by her explanation that solving problem, as she told, start from understanding problem, selecting appropriate strategies, trying to execute all possible the selected strategies, and ending with checking solution back to the problem. In responding the nature of problem, his idea that question about the application of integral is a problem for her students shows her incomplete understanding of what actually a mathematics problem is. However, her knowledge of pedagogical problem solving is likely better than that of content problem solving. Her idea of guiding students solve mathematics problem such as bringing a familiar situation into the problem, asking students to read the problem and pointed some students to reveal what the problem ask to do, asking students to identify what unknown that should be sought, until asking students share their strategies and solution within a presentation session to compare students' result illustrates this view.

3.3. *Ana's beliefs and knowledge about mathematical problem-solving*

The view of nature of mathematics espoused by Ana refers mostly to problem-solving, except on the way she defined mathematics. Her idea of defining mathematics includes his view that mathematics is dynamic and changeable on methods to invent such objects. She said, " you will surely find any changes in mathematics, even it is very easy to change, however, I'm not really sure whether the object which we learned will change." She also agrees that mathematics is about thinking and grasping real world phenomena. She argued, "by enhancing what really mathematics is, you will be more creative inusing various types of representation presented in this world." Regarding strategy of solving problem, she believes that she will not always be successful to find the solution of the problem even though the concepts and mathematical procedures that she used are correct. In this case, she asserted, "I often encounter this matter primarily when solving a problem which demands higher-order thinking. I am not sure whether this is caused by my lack knowledge or, indeed, there is no perfect solution to that problem."

Ana's belief about teaching mathematics also aligned with problem-solving. Except for her idea of choosing a suitable reward for motivating students to learn mathematical problem solving which is likely instrumental, Ana showed consistent view in putting learner as a focus of learning in instructional practice as indicated by the other cases. Her idea such as, "providing guidance on the form of representation the students develop is more fruitful than letting them think of strategy based on the form of representation, like figure shown in this option, will make students more freely express their idea","the way to motivate students learn central angle, even though they do not find any application in real life is by convincing them that the thinking skill used to find the relationship between central angle and circular angle is much more important in their future life", and "I will provide students opportunity to discuss ideas emerging from their thinking and determine which one is the best to clarify their misunderstanding" support her problem-solving view.

With regards to belief in how students should learn mathematics, Ana's view seemed to align with problem-solving view too. When asked to complete statement about learning strategy on being good in solving a mathematical problem, she said, "student should attempt to solve a problem by their own strategies based on their knowledge and experience, so they will get used of thinking first rather than practicing many times on similar problem. If they only attempt to apply what teachers teach, higher order thinking will not appear on their idea". Besides, Ana also suggests students to learn all strategies provided since she argued, 'if students do not learn all the strategies, they will not likely find which strategy best for solving problem." Surprisingly, Ana is the only one participant who agrees to permit students to use calculator. She said, "using calculator does not influence students' understanding of basic procedures or concepts, though and even it is recommended for facilitating students to solve problem which emphasizes on the process, not results, like in statistical problem." Despite all such indicate that her view seemed to support problem-solving view, her final conclusion regarding learning mathematics is rather platonist. She said, "students need to solve problem by their own way using their experience, and of course, need aims from teacher when experiencing difficulties, and especially about mathematical formula, they need to understand how such formula is constructed."

Ana's knowledge about problem solving seemed better than Toro and Yati, particularly on problem-solving content knowledge. Unlike their idea, Ana claimed that not all mathematics questions are problem for students. She said, "it depends on students' ability and whether the questions embody something challenging and interesting." However, again, Ana, as well as Toro and Yati, regarded that question 'find the area bounded by curves $y=x^2+1$, $x=-4$, $x=5$, and x-axis' is a problem for her students. She also expressed her good understanding on the types of problem. When interviewed about the idea to exemplify open-ended problems, she said, "I will reverse the question, like if the area of a rectangle is 225 cm^2, what are its probable measure? Draw as many as possible irregular plane with the area of 500 cm^2." Ana's pedagogical knowledge seemed also sufficient to hold problem-solving instruction. She told, "teaching problem-solving, you need to begin with asking students understand the problem such as selecting relevant and irrelevant information, listing given and not given but needed, encouraging them to think creatively on any ideas in attempting to find precise mathematical model, giving feedback on their strategies and finally asking them to consider whether the solution they found make sense or not."

4. Conclusion

Our findings indicate although our participants espoused their view of the relationship between mathematics and real life as well as the success of problem-solving strategy in a problem solving or platonist view, no participant who defined mathematics in a problem-solving view. Yati and Ana only view mathematics as a static discipline with regards to platonist view, while toro defined it as a discipline studying numbers and formula. However, generally their view on teaching and learning mathematics is seemingly more constructivist which corresponds to at least platonist view particularly on Yati and Ana. this finding may indicate that belief aboutmathematics and aboutteaching and learning are two distinct belief sets, as suggested by Thompson [12]. Such result also is in line with the study of Lui&Bonner [13] suggesting that despite teachers may hold constructivist belief for mathematics teaching and learning, they may holdtraditional beliefs about what mathematics is.

Regarding knowledge, our findings point out that generally teachers hold insufficient knowledge primarily on the nature of problem like deciding when a problem is suitable for building concept or applying concept This is aligned with findings in the study of Siswono et al [14] reporting that teachers found many difficulties when dealing with the nature of problem and identifying problem-solving strategies. In search of a possible interaction between teachers' beliefs and knowledge about mathematical problem solving, we found that, seemingly, Toro shows less conceptual knowledge about problem solving, while Ana shows sufficiency. This finding is consistentwith expectations that a problem-solvingbelief, would be associated with skillsin conceptual knowledge as Lui & Bonner [13] explained that teachers with aconstructivist beliefs had strong skills in conceptualknowledge.

To conclude, we highlight that all the teachers did not show a high consistency in responding views of their mathematics-related beliefs. Beside, they showed weaknesses primarily on problem-solving content knowledge. It is also noticed that teachers' beliefs have a strong relationship with teachers' knowledge about problem-solving. In particular, the instrumental teacher's beliefs were consistent with his insufficient knowledge about problem solving, while both platonist and problem-solving teacher's beliefs were consistent with their sufficient knowledge of either content or pedagogical problem solving. A possible further explanation of such relationships may be investigated from their observed teaching practice to examine whether teachers' belief and knowledge consistent with their practice and how each belief and knowledge influence their practice.

5. References

[1] Anderson J White P and Sullivan P 2005 *Mathematics education research journal* **17** 9-38
[2] Guskey T R 2002 *Teachers and Teaching: Theory and practice* **8** 381-91.
[3] Fennema E, Carpenter T P and Peterson P 1989 Teachers' decision making and cognitively guided instruction: a new paradigm for curriculum development *School mathematics: The Challenge to change* ed N. F. Ellerton& M. A.Clements (Geelong: Deakin University Press) p 174-187
[4] Raymond A 1997 *Journal for Research in MathematicsEducation* **28** 550-76.
[5] Bray W S 2011 *Journal for Research in Mathematics Education* **42** 2-38
[6] Purnomo Y W Suryadi D and Darwis S 2016 *Journal of Elementary Education* **8** 629
[7] Franke M L, Kazemi E and Battey D 2007 Mathematics teaching and classroom practice *Second handbook of research on mathematics teaching and learning* ed F. K Lester Jr (Charlotte, NC:Information Age) p 225–256
[8] Chapman O 2015 *LUMAT* **3** 19-36
[9] Ernest P 1991 *The philosophy of mathematics education* (London: The Falmer Press)
[10] Beswick K 2005 *Mathematics Education Research Journal* **17** 39–68
[11] Miles and Huberman 1994 *Qualitative Data Analysis: An Exposed Sourcebook 2nd* (London: SAGE Publication Ltd)
[12] Thompson A 1992 Teachers' beliefs and conceptions: a synthesis of the research *Handbook of research on mathematics teaching andlearning* ed D A Grouws (New York: Macmillan) pp 127-146
[13] Lui A M and Bonner S M *2016 Teaching and Teacher Education* **56** 1-13
[14] Siswono T Y E, Kohar A W, Kurniasari I and Astuti Y P 2016 *Journal of Physics: Conference Series* **693** 12015-33

Acknowledgment
We would like to thank Ministry of higher education and research on fundamental research grant in the second year 2016, the Rector of Universitas Negeri Surabaya (UNESA) and the Dean of the Faculty of Mathematics and Natural Science, UNESA.

The Use of Classroom Assessment to Explore Problem Solving Skills Based on Pre-Service Teachers' Cognitive Style Dimension in Basic Physics Course

Rahmawati[1,2*], Nuryani Y. Rustaman[2], Ida Hamidah[2], Dadi Rusdiana[2]

[1]Universitas Muhammadiyah Makassar, Jl. Sultan Alauddin No. 259, Makassar 90222, Indonesia
[2]Universitas Pendidikan Indonesia, Jl. Dr. Setiabudi No. 229, Bandung 40154, Indonesia

Email: rahmawati.syamsuddin@yahoo.co.id

Abstract. The aim of this study was to explore the use of assessment strategy which can measure problem solving skills of pre-service teachers based on their cognitive style in basic physics course. The sample consisted of 95 persons (male = 15, female = 75). This study used an exploratory research with observation techniques by interview, questionnaire, and test. The results indicated that the lecturer only used *paper-pencil test* assessment strategy to measure pre-service teachers' achievement and also used conventional learning strategy. It means that the lecturer did not measure pre-services' thinking process in learning, like problem solving skills. One of the factors which can influence student problem solving skills is cognitive style as an internal factor. Field Dependent (FD) and Field Independent (FI) are two cognitive styles which were measured with using Group Embedded Figure Test (GEFT) test. The result showed that 82% of pre-service teachers were FD cognitive style and only 18% of pre-service teachers had FI cognitive style. Furthermore, these findings became the fundamental design to develop a problem solving assessment model to measure pre-service teachers' problem solving skills and process in basic physics course.

1. Introduction

Assessment is systematic and varied process to get more information data and interpret it as a feedback in learning [1]. Furthermore, assessment strategy can help teacher to monitor student learning progress and help to improve learning process so that it is important to use relevant assessment [2]. In learning science, assessment should measure knowledge and concepts, science process skills, and higher-order thinking skills (like problem solving, critical thinking, logical thinking, and creative thinking skills) and use portfolio assessment and performance assessment for science process skills and scientific performance in learning science [3]. There is a problem in the field that many teachers or lecturers hope that their students can think critically and solve problems, but they just do not know how to design and implement assessment strategy of problem solving.

Problem solving is one of the activities using thinking process. Every student can be different between one to another in problem solving process. It could be caused by internal and external factors. Indeed, internal factor is related cognitive psychology, exactly cognitive style dimension [4]. The concept of cognitive style is closely connected to the idea of psychological differentiation [5]. It means that differences exist among individuals in relation with their cognitive structure and psychological function, the other name is called 'psychological individuality' [6]. The fundamental differences between individual FI and FD cognitive style can be observed to individual characteristics related to process information. Basically, FI persons perceive and process information analytically, while FD persons do it in global, holistic, and passive way [5]. Furthermore, FD persons have difficulty in separating an item from its context while FI persons can easily break up an organized field and separate relevant information from its context or discern 'signal' (what matter) from 'noise' (the incidental and peripheral) in confusing background. [6]. Finding of previous studies indicated that there are relationship between cognitive style and problem-solving skills of pre-service teachers to some learning materials, in particular the topic of mechanics basic concepts [5]. Result of that study showed pre-service teachers' problem-solving skills were statistically related to their FI/FD cognitive style. Based on finding of some previous studies, it means that cognitive style of persons have relationship with their problem-solving process, so that it is important to facilitate learners' thinking process in learning, like problem-solving skills. The effort that can be done to develop learners' problem-solving skills is by using learning strategy or assessment strategy because assessment is closely related to learning process. Using assessment strategy is formal effort to take more information related to knowledge, skills, and attitude of students as information for teachers to make a decision to improve learning process and student achievement [7].

This paper tried to explore the use of classroom assessment in basic physics course. In addition, GEFT test was administered to indicate FI/FD tendency of students. The findings of this study will be basic to develop assessment problem solving model related to student cognitive style.

2. Methodology

The type of this study was an exploratory research with observation techniques by interview, and questioners to pre-service teachers and lecturer. Subjects was 95 freshmen (male = 15, and female = 75) who have taken basic physics course in the first year 2015/2016. The students ranged in age from 18 to 21 years.

The instruments used in this research were questionnaire and GEFT test. Questionnaire was numerical rating scale that consisted of two aspects. Each aspect consisted of some statements about the important of problem-solving skills for students, and the assessment strategy used by lecturer in basic physics course. Observational techniques used in instruments were numerical rating scale for the first aspect. It consisted of strongly agree (SA), agree (A), cannot select (CS), disagree (DA), and strongly disagree (SD). In addition, the second aspect used technique of checklists with yes, cannot select, and no. After obtained data, those data was processed into percentage and then analyse data into qualitative and quantitative description. The instrument questionnaires were administered to subjects and the lecturer to obtain information about the use of classroom assessment and learning strategy in basic physics course.

Furthermore, test technique was used to measure cognitive style of pre-service teachers with using "Group Embedded Figure Test (GEFT)" instrument. GEFT is a psychometric measure whose reliability and validity has been supported by a growing body of studies. They are Salih *et al* [5], Herman *et al* [6], Reza *et al* [8], and Mohammad *et al* [9]. It consists of three sections; each section has simple figure embedded in large complex figures, which had to be traced. This instrument was administered to students. The first section which has a time limit of 2 minutes includes 7 items. This section is only as a warm-up to make them ready but the items in this section were not included in the total score. The real task began at the second section and into the third one. The second and the third section included a total of 18 items, 9 items for each section, which had to be traced in 12 minutes set within time limit of 6 minutes for each. These sections were given score 0 for the false answers and score 1 for the true answers so that the maximal score is 18 and the minimum is 0. After the second section had been administered, students stopped for more instruction on the third section and then went on. Those whose score above 12 out of 18 were labelled FI persons and those with score of 11 and less than 11 were branded as FD cognitive stylists.

3. Results and Discussions

3.1. Description about cognitive style of pre-service teachers

Figure 1 showed that the score of the participants related with cognitive style of pre-service teachers. GEFT score was ranged from 0 to 16. Dominant scores which ware obtained by students were score 4, 8, and 9. The analysis model of Howard and Laura [9] describes that persons who get scores between 0 and 11 are labeled FD cognitive stylist and they who get score more 11 until maximum score are called FI persons. From this analysis model, it can be explained that there were 78 students (82%) who get score in the range 0 – 11 as FD person. While they who get score 12 to 18 were 17 persons (18%) as FI.

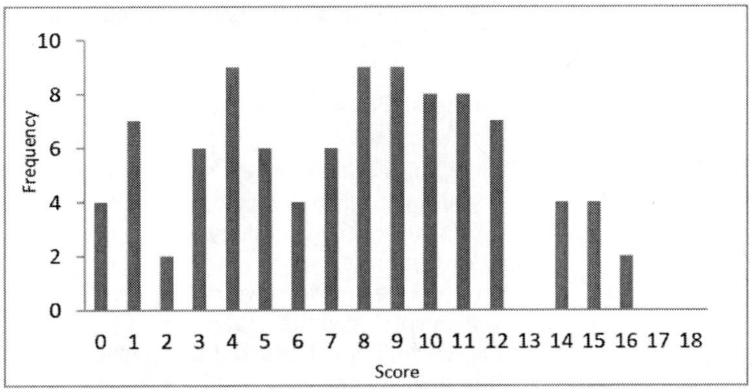

Figure 1. Distribution of student scores to GEFT test

Analysis result about obtaining test score of GEFT showed that the number of students who had FD cognitive style was more than the number of students with FI cognitive style. Based on this result, we can conclude that every student has cognitive style to get and process more information in the class. Therefore, it is important to consider the use of appropriate assessment and learning strategy with our student characteristic.

3.2. Description of questionnaire data analysis

Questionnaire instrument which were administered to pre-service teachers and lecturer aims to explore more information about assessment used by lecturer in basic physics course. These questionnaires consisted of two aspects. They were about the important of problem solving skills and assessment strategy used by lecturer. There are more statements used in the instruments which are showed by table 1 below.

Table 1. Statements used in the instrument to explore more information about the use of assessment and learning strategy

Aspects	Statements
The important of developing problem solving skills	1. Developing higher order thinking skills, like problem-solving skills, is an essential aspect for pre-service teachers 2. Developing problem-solving skills will add the load task for students 3. Developing problem-solving skills in basic physics course will spend more time 4. Problem-solving skills is needed by pre-service teachers 5. Every student who has characteristic can influence problem-solving

Table 1. Statements used in the instrument to explore more information about the use of assessment and learning strategy

Aspects	Statements
	skills of persons 6. It needs to habits for improving problem-solving skills 7. I am usually and able to solve problems 8. My problem solving skills depend on complexity of contexts 9. My problem solving skills depend on the situation of my feeling 10. My problem solving skills depend on the class situation 11. Problem solving skills should be implemented in the all courses, specific in basic physics course 12. Problem solving skills of pre-service teachers should be assessed for learning in the classroom
Assessment strategy used to develop thinking process of pre-service teachers	1. Do the kinds of tasks given by lecturer improve your thinking skills? 2. Do your tasks given by lecturer obtain feedback? 3. Does the assessment strategy by lecturer able to develop your thinking process? 4. Does the lecturer assess your task orientated to develop your problem solving skills? 5. Are your individual task and group task considered to obtain score in the final score? 6. Do you know about the assessment strategy used by lecturer? 7. Can your tasks help you to develop your thinking ability? 8. Does your lecturer assess your tasks oriented to develop thinking process? 9. Are you satisfied to your final score in basic physics course? 10. Do you hope to get comprehensive and continue assessment of your tasks?

The result of analysis from table 1 was presented into the figure 2 and 3 below. Figure 2 indicated about the student expectation to develop their problem solving skills by using appropriate assessment strategy of lecturer. It is important for them who will face more problems in their life. Therefore, pre-service teachers thought that problem solving skills was necessary to be tried by learning and assessment strategy.

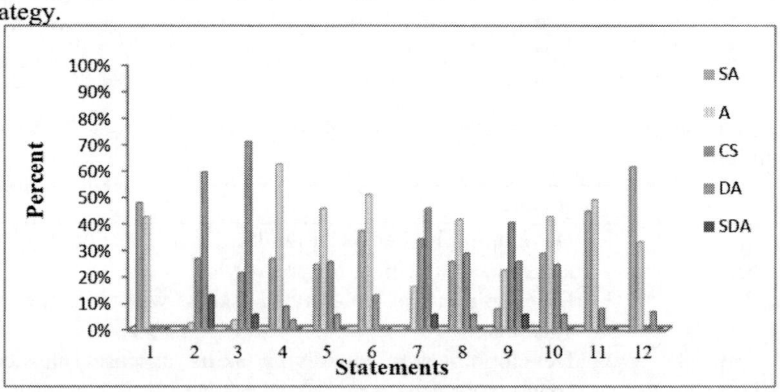

Figure 2. Percentage of student response to aspects of the important of developing problem solving skills in basic physics course

The explanation of above statement can be supported by figure 3. Figure 3 showed that there are some statements of students which indicated problem solving process has not been facilitated by lecturer in basic physics course. Furthermore, the authors claimed that lecturer in basic physics course still used summative assessment. There is no specific assessment model to measure problem solving process based on student cognitive style. Until now, the lecture only used assessment strategy which focused to student achievement using paper and pencil tests, like middle and final exam. In addition, the use of assessment strategy has not been able to support development of student problem solving skills even though problem solving skills can be trained gradually [11].

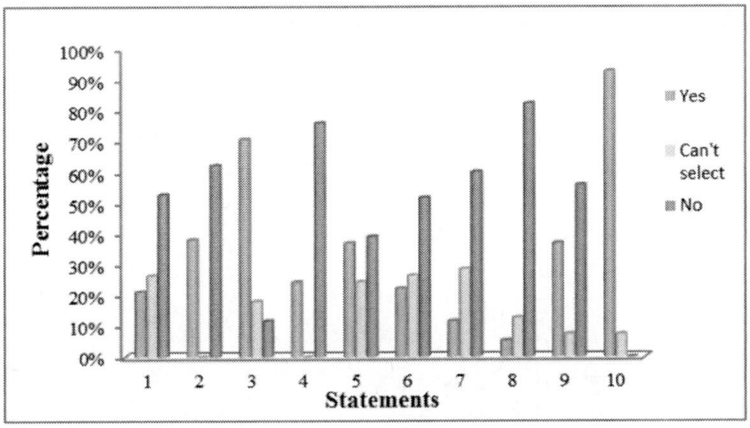

Figure 3. Percentage of student response to aspects of the use of assessment strategy by lecturer to develop thinking process in basic physics course

The effect of applying assessment strategy and learning model in basic physics course was that students felt difficult to solve problems when they faced to complex problems in basic physics course so that this situation made them have gotten low achievement (in grade A only 6% students, B = 26% students, C = 39% students, D = 12% student, and E = 12% students). From this statement, this is basic to design model of problem solving assessment with considering student cognitive style aspect so that this model can measure problem-solving process. In addition, it is effort to improve learning process and assessment strategy in other course.

4. Conclusions

Based on the result study, it can be concluded that using assessment strategy by lecturer has not been appropriate to develop problem-solving skills of students and did not consider psychology cognitive style of students in implementing assessment strategy. Even though, every student has the characteristic of individual cognitive style. The number of students with FD cognitive style more than FI cognitive style persons so that it is important to consider cognitive style variable in basic physics course. The effect of lecturer behavior was students have not been able to indicate their thinking process skills, specifically in problem solving process. In addition, this problem becomes the fundamental of designing model of problem solving assessment.

5. References

[1] National Research Council 1996 *National Science Education Standars* National (Washington, DC : Academy Press)
[2] Robert J M 2006 *Classroom Assessment and Grading That Work* (Alexandria, Virginia USA : Association for Supervision and Curriculum Development)

IOP Conf. Series: Journal of Physics: Conf. Series **812** (2017) 012047 doi:10.1088/1742-6596/812/1/012047

[3] David H J 2011 *Learning to Solve Problems A Handbook for Designing Problem-Solving Learning Environments* (New York : Routledge)
[4] Olivia N S 1997 *Teachers' and Students' Cognitive Style in Early Childhood Education* (London: Bergin and Garvey)
[5] Salih A and Erdat C 2007 *Research in Science & Technological Education* 167-78
[6] Herman A W, Chester A M, David R G and Philip W C 1977 *Review of Educational Research* **47** 1-64
[7] William J P 2011 *Classroom Assessment What Teachers Need to Know Sixt Edition* (Unites State of America: Pearson)
[8] Reza K and Mahmoud Q 2013 *Indian Jounal of Fundamental and Applied Life Sciences* **3** 600-19
[9] Mohammad K and Rosoul M H 2011 *Journal of Languange Teaching and Research* **2** 640-48
[10] Howard R D G and Laura J W 1994 *ERIC Journal* ED372711
[11] Peter G, Gunnar J and Margaretta E 2014 *European Journal of Engineering Education* 1-20

IOP Conf. Series: Journal of Physics: Conf. Series **812** (2017) 012048 doi:10.1088/1742-6596/812/1/012048

Student Obstacles in Ratio and Proportion Learning

W Andini[1] and Al Jupri[2]

[1]Department of Primary Education,Postgraduate School,UniversitasPendidikan IndonesiaJl. Dr.Setiabudhi No. 229, Bandung 40154, Indonesia
[2]Department of Mathematics Education, UniversitasPendidikan Indonesia Jl. Dr.Setiabudhi No. 229, Bandung 40154, Indonesia
Email: ullanandini@gmail.com

Abstract. Ratio and proportion are mathematics topics that are learned in elementary school. In the learning process, teachers often found students who have difficulties in learning these topics. Analysing the difficulties on the topics is important for teachers, as an attempt to doa reflection of the learning process and as a reference in devising a learning plan, including for anticipation of the possibility of these barriers that might appear in the next learning process. This research used as the qualitative method, and involved 49 students of grade 5 and one teacher. From the analysis of the data, i.e., student test results and observation of the learning, we found some obstacles that faced by students, such as students confused in understanding about ratio and proportion topic, students found it difficult in completing the problem which have different contexts, the learning material of ratio and proportion topic that presented in textbooks are still partial (not depth), and the lack of ability of the teacher in creating problems.

1. Introduction

Mathematics is a subject that must be learned by elementary student. It is because mathematics has important role in all of human life. This is in line with the opinion of Freudenthal [1] that mathematics as a human activity.

One of the branches of mathematics that must be taught in elementary school is ratio and proportion. Understanding and ability about these materials are needed in daily life, such as when we are shopping. When student have two choices, there are a food with weight 80gr and price Rp 6.400,00 and a food with weight 125gr and price Rp 9.000,00, buyer should use ratio and proportion so that he can buy the food with the large quantities but cheaper prices.

Besides used for activity connecting with mathematics, ratio and proportion also are used in a variety of other subject areas. For instances geography subject in senior high school. In that subject, ratio and proportion are used for the concept of population density. The concepts of ratio and proportion are important for some other subjects, such as in cartography (map-making); in the science to define various natural phenomena; in economics and statistic to calculate profit-and-loss and probability; and in technological studies for calculations in engineering, mechanics, robotics, computer science, among other [2].

By seeing the importance of ratio and proportion concepts, it is clear that these concepts are the materials that are required to be taught in elementary schools. That is because the ability of students about the concept of ratio and proportion in elementary school became the basis for students studying the concept at a higher level and applying it in their life.However, in ratio and proportion learning, students often have obstacles. One of study conducted to student grade 5 found that student had difficulty to solve ratio and proportion problems [3]. Other research in one of in Junior High School indicated most students got obstacles in solve simple problem of ratio and proportion, did not understand about the concepts, and had difficulty to solve problem with different context [4].

The obstacles that are faced by student did not occur if student have understood about ratio and proportion concept from elementary school. For getting the understanding about the concept, teacher must give student meaningful learning, so that student can construct their thought to understand the concept [5]. Teacher also must pay attention to student response in solving problems and the difficulties that students may face when work on the problems. Hence, making student responses prediction and good teaching plan is very important to help student avoid errors in learning process.

According to Piaget opinion [6] errors are not only the effect of ignorance, of uncertainty, of chance, as espoused by empiricist or behaviourist learning theories, but the effect of a previous piece of knowledge which was interesting and successful, but which now is revealed as false or simply un-adapted. Errors and student obstacles should be known its factors before learning process occur [7], so that student can get good achievement in ratio and proportion learning.

Many studies about ratio and proportion learning have done in math, but most studies were conducted for seeing the effect from one of learning models. Whereas, the study concerned student obstacles in ratio and proportion learning in math especially in elementary school is rarely. Therefore, it is important to know student obstacles in ratio and proportion learning and what factors are causing it. This study aimed to compare the predictions of student response and learning process with reality on the ground, to know strategies that student used to complete ratio and proportion problem, to find student obstacles in ratio and proportion learning, and to analyse what factors are causing it.

2. Experimental Method

This research was conducted at one of elementary school in Bandung West Java Province, Indonesia. There are 49 students as participants. They are students of grade 5, 27 males and 22 females.

This qualitative study was carried out in the following four procedures. Firstly, researchers made predictions about the learning process and student responses. Secondly, we collected data from textbooks, observation of learning process, and individual written test. Thirdly, we analysed data by comparing our predictions with data on collected from the learning and teaching process. Fourthly, we made conclusions.

3. Results and Discussion

Based on data analysis, we classify the results into three sections. The sections are seen from textbooks, learning process, and student test result.

Firstly, from analysing two available student textbooks, each book contains material of ratio and proportion that is less deep and still partial. Each book only gives a glimpse explanation about the material, it does not include all types of problems that often exist on a national exam. Incomplete material also makes students confused when solving other problems with a different context. It can actually be overcome by learning given to stimulate students to find a variety of strategy not only one strategy to solve the problems. Here, the teacher skills to design learning are important [8].

Secondly, comparison of prediction learning process performed by the teacher and the reality on the ground was almost the same. When describing the material, the teacher used concrete objects such as sticky-note. The advantages of using sticky-note are easy to use and the teacher can directly paste on the board so that students can see it. On the other hand, although the use of sticky note was quite effective, students tended to be passive because it was not actively involved directly in manipulating objects, because students only saw the teacher's explanation. The use of concrete objects was intended to help students recognize the concept of ratio and proportion, but the teacher also must use an effective method in teaching.

Thirdly, based on the analysis of student test data, there were some student responses that are similar to predictive strategies that have been made. There were two problems given by teacher, researchers made predictions of the strategy to be used by students in solving the matter given by the teacher. It is aimed for identifying student obstacles in solving problem that given by the teacher. As an illustration about the predictive strategies, student obstacles will be describedin the following section.

> "Problem 1:
> Andi's marbles: Beni's marbles is 3:5. If the number of marbles Andi = 27. How many Beni's marbles?"

For the Problem 1, we created 12 strategies that might be used by students in solving problems. Five strategies were made leading to the acquisition of the correct answer, while seven others that lead to the acquisition of the wrong answer. From the 12 predictions made, only four strategies were used students. Other students tackled the problems by using different strategies than expected, gave a direct answer without using the strategy, and did not give an answer. From 49 students, 37 students answered correctly by using strategies, two students answered correctly without the use strategy or just give a straight answer, and 10 other students answered incorrectly. As an illustration, the students' answers are presented below:

The first response is to use strategy A. This strategy was used by one student. Student worked with this strategy using integer arithmetic operations. Students counted the number of marbles in question by first dividing the total number of marbles at 27 with a number of comparison the number of marbles known, in this case the comparison numbers Andi that is 3. The result is multiplied by the ratio of the number of people that the number of Beni's marbles then. After that, it will be found that the number of Beni's marbles there are 45 items. Figure 1 is the result of the work of students who use this strategy.

Figure 1. The example of student answers of question problem 1 using strategy A

A second response to solve problem 1 is students worked with strategies G. There were three students working with this strategy. In resolving the problems, student gave wrong answer despite almost used ratio and proportion concept. Students did not understand numbers comparable or connection between comparison with the number of set. Figure 2 is the result of the work of students who use this strategy.

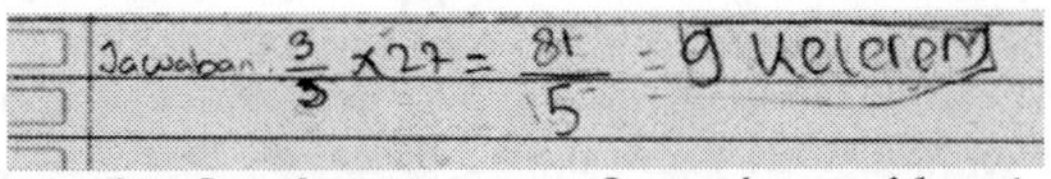

Figure 2. The example of student answers of question problem 1 using strategy G

The third response is the students did the problems with the strategy H. Student working with this strategy as many as 29 people. Students resolved the problem by using the concept of an appropriate. Figure 3 is the result of the work of students who use this strategy.

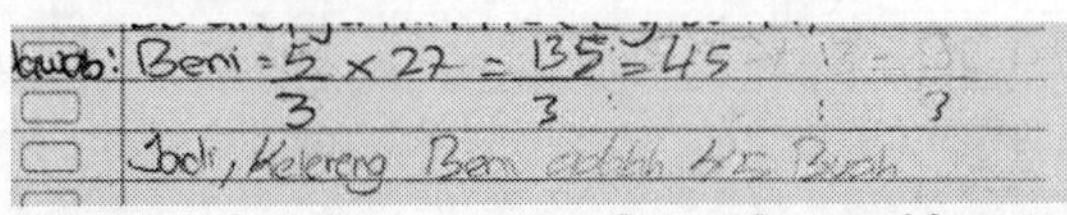

Figure 3. The example of student answersof question problem 1 using strategy H

The fourth response is the students did the problems with the strategy number J. There were three students working with this strategy. Students resolved the problem by using the concept of ratio and proportion but by comparing Beni's marbles with whole value. Figure 4 is the result of the work of students who use this strategy.

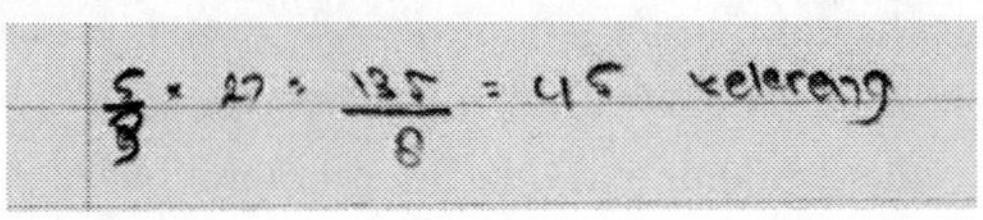

Figure 4. The example of student answersof question problem 1 using strategy J

Fifth, work on the problems the number 1, students were doing with the different strategies of the predictions that have been made. Student completed the problem with this strategy as many as seven people. Students work with this strategy as exemplified teachers. Figure 5 is the result of the work of students who use this strategy.

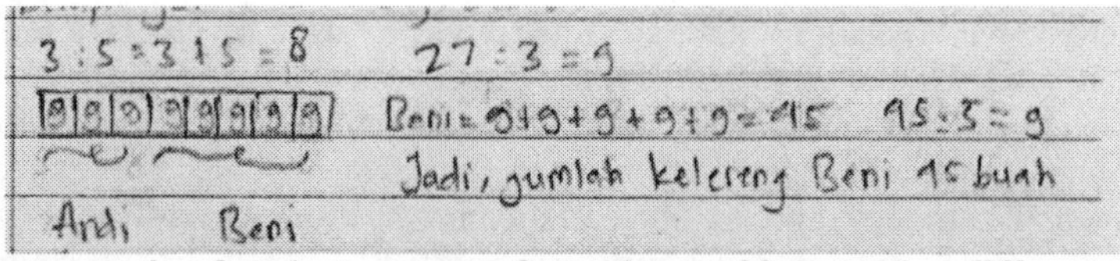

Figure 5. The example of student answersof question problem 1 using different strategy with prediction of response

Sixth, the students worked on Problem 1 with a different strategy from the predictions that have been made, but the response and the strategy was wrong. Student working with this strategy as many as one person. Students did not understand the question being asked. Figure 6 is the result of the work of students who use this strategy.

Figure 6. The answerof question problem 1 using different strategy with prediction of responsebut it is wrong strategy

Seventh, the student's response was to give an immediate response without the use of strategies to do it. Students who gave such a response there was one person.

Eighth, the students did not answer this question. There were three students who did not give this answer. Students who did not give this answer confused and did not understand the question.

"Problem 2:
Umi and Santi money ratio is7:4. Differences in the amount of money Umi and Santi isRp 2,250. How much money each of them?"

For problem 2, researchers created 15 strategies that may be used by students in solving problems. Three strategies were made leading to the acquisition of the correct answer, while ten others that lead to the acquisition of the wrong answer. From the 15 strategies that were made, only three strategies were used by students. Other students gave a straight answer without using its completion strategy. From 49 students, 26 students who answered correctly by using strategies, and 23 others answered with the wrong strategy or just give a straight answer wrong. The following description of the analysis.

The first response is to use strategy A. This strategy was used by 26 students. Workmanship about using this strategy was done by using the concept of comparison. All students who answered correctly in this matter used this strategy. This strategy was used all the students who answered correctly because most of the students' got difficulties in solving this matter. Then, the teacher explains the steps the process with this strategy. Figure 7 is the result of the work of students who use this strategy.

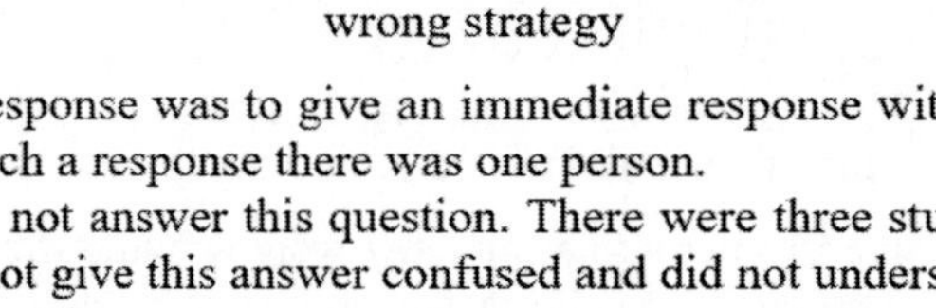

Figure 7. The example of student answers of question problem 2 using strategy A

The second response is the students tackle the problem by using strategy A, however the calculation is wrong and make wrong answers. This response was used by 18 students. Some students are less scrupulous in dividing or multiplying numbers. Figure 8 is the result of the work of students who use this strategy.

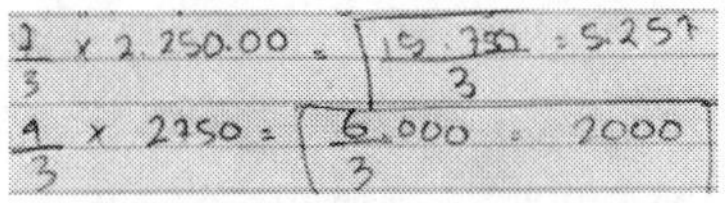

Figure 8. The example of student answersof question problem 2 using strategy A but error

Third, Students solved the problem by using strategy J. This response was raised by 2 students. Students work on the problems by comparing the total amount when the known is the difference. This makes students' wrong answers given. Figure 9 is the result of the work of students who use this strategy.

Figure 9. The example of student answersof question problem 2 using strategy J

The fourth response which raised is the student working on the problem by using strategy N. This response was used by 1 student. Student did the problems by directly multiplying comparison numbers Umi and Santi with the difference of the amount of their money. From this it is known that the students did not understand the concept of comparison the problems so that students cannot associate the concept of comparison that has been gained into the matter workmanship. Figure 10 is the result of the work of students who use this strategy.

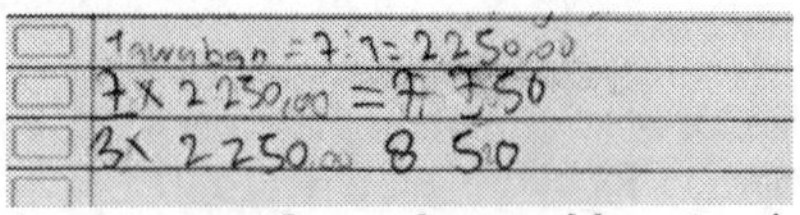

Figure 10. The answersof question problem 2 using strategy N

Fifth, students did not use any strategy in answering questions, students only gave the answer and that answer was the wrong answer. This response was raised by 2 students. Students gave a direct answer because students were confused because they did not understand the concept and the ways taught by teacher. Therefore, the students gave answers in a way to figure out the answer. Students gave answers that the results have a margin equal to the difference is unknown, that was Rp 2250.00.

From the observation and analysis of student test results that have been done, it can be seen that the students had some obstacles in resolving a matter of ratio and proportion. The difficulties include:

Students get difficulties in understanding the problem [9]. Studentscan not understand keyword in the problem and can not comprehending the sentence. It cause students confused in determining what strategy that they would use in solving problems.

Students have difficulty solving problems due to the weak capability of its prerequisites. The ability of such ratio and proportion prerequisite such as operations of integer (addition, subtraction, multiplication, division), fractions (including: simplifying fractions, comparing fractions, and fractional operations), and others.

Students have difficulty in applying the concept of comparison to the given problem. This can happen due to several factors, one of them is due to the teaching of teachers. Learning occurs without involving students to actively construct mind will make students' understanding to survive in the short term and quickly forgotten. In addition, from the observation, the teacher is showing two different ways, but the teacher did not explain to the students truly understand both ways, so that when students work on the problems in spite of the same shape, students' difficulties in completing.

Students have difficulty in the work on the problems that have a different context to the questions that have been learned. This is because students will work on the problems in accordance with what is exemplified by the teacher, so that when there are other contexts, students will be confusion in the work and can not imagine the context [10].

4. Conclussion

Based on the results, the ability of grade 5 students to solve ratio and proportion problem is still low. Most of students experienced obstacles. Students used strategies that given by the teacher, so that they get difficulties in completing the problems which have different contexts. The learning material of ratio and proportion that presented in textbooks is still partial (not depth). In addition, the lack of ability of the teacher in creating the problems also make student confuse in understanding the main purpose of the problems.

5. References

[1] Heuvel-PanhuizenM V D2003 *Educational Studies in Mathematics*(Netherlands: Kluwer Academic Publishers)**54** 9–35
[2] ChaimD B, KeretY, and IlanyB S 2012*Ratio and Proportion: Research and Teaching in Mathematics Teachers' Education (Pre- and In-Service Mathematics Teachers of Elementary and Middle School Classes)*(Rotterdam: Sense Publishers)
[3] RuizE F, and LupianezJ L 2009 *Electronic Journal of Research in Educational Psychology*7397-424
[4] Valindra D 2015 *Desain Didaktis Perbandingan Senilai dan Berbalik Nilai Pada Siswa SMP Ditinjau dari Learning Obstacle dan Learning Trajectory* (Bandung: Thesis Universitas Pendidikan Indonesia)
[5] Vallori A B 2014 *Journal of Education and Human Development.* **3** 199-209
[6] BrousseauG 2002*Theory of Didactical Situations in Mathematics New* (New York:Kluwer Academic Publishers)
[7] Sepeng P and Sigola S 2013 *Journal of Social Sciences***4**325-333
[8] Prasertcharoensuk T, Somprach K, and Ngang T K 2015 *Proc. Social and Behavioral Sciences* (Elsevier Ltd) **186** 566 – 572
[9] Phonapichat P, Wongwanich S, and Sujiva S 2014 *Proc. Social and Behavioral Sciences* (Elsevier Ltd)**116** 3169 – 3174
[10] Gooding S 2009 *Proc. of the British Society for Research into Learning Mathematics*,ed M. Joubertpp 31–36

Acknowledgments

We thank students and teacher for their help and participation.

The Implementation of Levels of Inquiry With Writing-To-Learn Assignment To Improve Vocational School Student's Science Literacy

R R Amarulloh*, S Utari and S Feranie

Sekolah Pascasarjana, Universitas Pendidikan Indonesia, Jl. Dr. Setiabudi No. 229, Bandung 40154, Indonesia

Email: rezaruhbani@gmail.com

Abstract. The aim of this study was to investigate the effectiveness of writing-to-learn assignment in a levels of inquiry learning to improve vocational school student's science literacy competence and knowledge on the subject of temperature and heat . This study used quasi experiment research methods. The data were obtained using 16 item of science literacy instrument with essay format. The result shows that there was a significant difference on the improvement of science literacy ability between the experimental class and control class. A significant difference occurred in the evaluating and designing experiments competency, interpretating data and science evidence competency, and procedural knowledge . Therefore it can be concluded that the implementation of levels of inquiry with writing-to-learn assignment can improve vocational student's science literacy competence and knowledge.

1. Introduction

The development of science and technology has been helping human to improve their quality of life, however this development is also carry negatives impact such as energy dissipation, destroying nature activities, etc. Hobson [1] stated "If an industrialized nation is democratic, its citizens must ultimately make the crucial choices about the uses of science and technology". This is imply that its neccesary for citizens to have a literacy of science and technology. Therefore, it will be important for us to consider the science literacy as one of the concerns and objectives in conducting science education in the basic education level or in higher education.

Based on the results of PISA (Programme for International Student assessment) in 2000, 2003, 2006 and 2009, the science literacy skill of Indonesian students are respectively at positions 38, 38, 50, and 60, significanly below the international average [2]. Based on the results of preliminary studies in one of the secondary schools in Ciamis, using the science literacy test instruments that are designed based on science literacy assessment PISA 2015, students science literacy skill is still quite low in both the competence and knowledge. This is indicate that the learning process doesnt facilitated students to achieve science literacy as the learning process tend to present science merely as a body of knowledge that students need to memorize. Therefore, its important to emphasize learning process that present science not merely as a body of knowledge, but also as a way to knowing about how the nature work. To achieve that goal, the learnig process has to give a sight to students how scientist do the science work, one way is to using inquiry learning in the science classroom, as Fang and Wei [3] stated that the literature in science education has been emphasized on inquiry activities in the development of science literacy.

Wenning [4]has proposed Levels of Inquiry as a sequences of lessons arranged hierarchically starting from the level of discovery learning, interactive demonstration, inquiry lesson, inquiry laboratory, realworld application, and hypothetical inquiry. Rohmi [5]have applied levels of inquiry on the theme of environmental pollution to improve science literacy. The results showed that there is an increased in students competence and knowledge of the science literacy after learning. However, this increase science literacy domain is not distributed evenly in every aspect, as smallest increase was in the epistemic knowledge and design and evaluate a scientific investigation competency.

Yore [6]stated that doing more hands on activity is not enough to improve meaningful learning, students need opportunities to strengthen his scientific experience and to contrast his understanding with the interpretation of the scientific establishment. This statement is similar to Glynn and Muth's [7]that state what is necessary to achieve science literacy is the emphasis on "minds on" in the study of science. One strategy that can be applied is to involve writing activities into science lessons. Peterson & Rochwerger [8]states that writing is a process that helps students to think more deeply about the ideas and information they encounter through reading, listening, watching and their exp9erience about the world around them. Baker et al. [9]states that the importance of writing skills of students to be attentive to all teachers, especially when using inquiry-based approach in science classes.

Keys et al. [10]stated that the nature of writing task for laboratory activity is depends on the nature of the activity of such labs. From the epistemological standpoint, traditional lab reports is appropriate for traditional laboratory activities. As for the Inquiry laboratory activities are still need to consider how is the appropriate writing form. One of suitable form is Science Writing Heuristic (SWH) which it is a tool to promote thinking, negotiate meaning, and writing about laboratory activities[11]. Science Writing Heuristic is a writing framework that designed to provide guidance for students to conduct laboratory investigations. SWH can be used as an alternative format of the laboratory reports but also can be used as a teaching model with the activity lab as a process of knowledge building and include arguments into inquiry based teaching [12].

Based on the preliminary study and the description above, researchers interested in studying the use of science writing heuristic in levels of inquiry strategy to improve students science literacy. In this study, science literacy is referring to the framework of assessment of science literacy PISA 2015 [13] which includes the competence and scientific knowledge. The problems in this research are:

1. How does the effect of use of levels of Inquiry with writing-to-Learn assignment to the improvement of students science literacy knowledge?
2. How does the effect of use of levels of Inquiry with writing-to-Learn assignment to the improvement of students science literacy competence?

2. Experimental Method

Considering the research needs and time constraints, the sample selection was not done randomly, but based group that has been formed previously, methods in such research is quasi experimental research methods [14]. The sample was 59 students of class XI at a vocational school in Kabupaten Bandung, West Java, distributed to the control class and experimental class. Control class was used the levels of inquiry learning on the subject of temperature and heat, while the experimental class was used the levels of inquiry learning with the writing-to-learn assignment. Both of class used the levels of inquiry proposed by Wenning [4]from discovery learning, interactive demonstration, inquiry lesson to inquiry lab. In the experimental group, after inquiry lab, students are asked to write down a personal meaning of laboratory activities that have been done following the students template proposed by keys [10]as presented at figure 1.

1. Beginning Ideas --What are my questions?
2. Tests-- What did I do?
3. Observations-- What did I see?
4. Claims -- What can I claim?
5. Evidence-- How do I know? Why am I making these claims?
6. Reading -- How do my ideas compare with other ideas?
7. Reflection – How have my ideas changed?

Figure 1. Student's Template

Then the students were asked to conduct discussions with his peers about the writings that have been made, then a representative student of the group presented the results of his writings in the classroom to get feedback from the teacher. After the lesson, the students were given the assignment to revise his writings at home by reference to the textbooks or online sources. The learning sequeces for both of class are presented in table 1.

Table 1: Learning sequences on control group and experimental group

LOI cycle	Day	Group	
		Control	Experimental
1	1	*Discovery Learning* *Interactive Demonstration*	*Discovery Learning* *Interactive Demonstration*
	2	*Inquiry Lesson* *Inquiry Lab* -	*Inquiry Lesson* *Inquiry Lab* *Writing-to-learn*
2	3	*Discovery Learning* *Interactive Demonstration*	*Discovery Learning* *Interactive Demonstration*
	4	*Inquiry Lesson* *Inquiry Lab* -	*Inquiry Lesson* *Inquiry Lab* *Writing-to-learn*

The test instrument has been matched to the PISA science literacy assessment, consists of 16 essay questions.Normalized gain value is used to determine the improvement of science literacy and then interpreted by using criteria proposed by Hake [15]. Based on normality test results, the n gain data are not normally distributed. thus, the hypothesis is tested by using "MannWhitney U Test". To determine the effect of the treatment to learning outcomes, we calculated effect size with cohen's formula.

3. Result and Discussion

3.1. The Improvement of Students' Science Literacy

Table 2 shows that there is an improvement of students' science literacy in the medium category both in experiment and control class. Both categorized as average improvement.

Table 2. Gain normalized of student's Science Literacy

Class	Pre-test	Post Test	N-Gain
Experiment	14,29	55,36	(0,48) Average
Control	17,64	43,25	(0,31) Average

To determine whether the differences in improvement of the control class and experimental class significant or not, then we used the statistical test analysis to test the hypothesis using MannWhitney test. Results of the hypothesis testing presented in the table 3 show that the differences catagorized as signifficant;

Table 3. Results of MannWhitney science literacy

Mean Rank of the control class (N=31)	Mean Rank of the experiment class (N=28)	Sig.	Decision
21,85	39,02	0,000	Significant

Based on the results of statistical hypothesis test nonparametric MannWhitney Asymp values obtained. Sig. (2tailed) or a pvalue of 0.000 where the value is smaller than the specified significance value (p. 0,000 <α. 0,05) with a 95% confidence level. Thus the null hypothesis was rejected, and the alternative hypothesis is accepted that that increase science literacy of students in the experimental group was significantly greater than the increase science literacy of students in the control group.

3.2. The improvement in the respective competence and knowledge of science literacy
Table 4 showed that there are average improvement on every aspect of science literacy knowledge both in experiment class and control class except on procedural knowledge of control group which its improvement categorized as low.

Table 4. Improved knowledge of each component of students' science literacy

No	Knowledge Type		Average Score		
			Initial test	Final test	N gain
1	Content	Experiment	27,97	56,32	0,39 (average)
		Control	24,19	50,53	0,35(average)
2	Procedural	Experiment	10,93	51,5	0,46(average)
		Control	18,14	39,11	0,26 (low)
3	Epistemic	Experiment	11,42	54,82	0,49(average)
		Control	12,9	45,48	0,37(average)

To determine whether the differences in improvement of the control class and experimental class on every aspect of science literacy knowledge significant or not, then we used the statistical test analysis to test the hypothesis using MannWhitney test. Results of the hypothesis testing presented in the following table;

Table 5. MannWhitney Test Results at each domain knowledge

Knowledge	Mean Rank of the control class (N=31)	Mean Rank of the experiment class (N=28)	Sig.	Decision
Content	27.31	33.09	0.186	Not significant
Procedural	20.74	40.25	0.000	Significant
Epistemic	25.92	34.52	0.053	Not significant

Table 5 show that the only significanly different improvement of the science literacy knowledge beetween control class and experiment class is on procedural knowledge. Beside the knowledge component, an analysis toward the components of science literacy competence was also conducted. Description of improvement of each competency is described in Table 6.

Table 6. The improvement of each component of competency of students' science literacy

No	Science Literacy Competency		Average Score (%)		
			Initial test	Final test	N-gain
1.	Explaining phenomena scientifically	Experiment	27,97	56,32	0,39
		Control	24,19	50,53	0,35
2.	Evaluating and designing experiments	Experiment	11,16	46,76	0,40
		Control	16,93	37,09	0,24
3.	Interpreting data and scientific evidence	Experiment	11,07	62,41	0,58
		Control	14,83	48,7	0,40

To determine whether the differences in improvement of the control class and experimental class on every aspect of science literacy competence significant or not, then we used the statistical test analysis to test the hypothesis using MannWhitney test. Results of the hypothesis testing presented in the table 7;

Table 7. MannWhitney Test Results at each domain of competency

Competency	Mean Rank of the control class (N=31)	Mean Rank of the experiment class (N=28)	Sig.	Decision
Explaining phenomena scientifically	27.31	33.09	0.186	Not significant
Evaluating and designing experiments	21.97	38.89	0.000	Significant
Interpreting data and scientific evidence	23.66	37.02	0.003	Significant

Table 7 show that there are significantly different improvement of the science literacy competency beetween control class and experiment class on evaluating and designing experiments competency and Interpreting data and scientific evidence competency. Results of the effect size calculation, shows that the treatment have a large effect to students science literacy. The calculation result presented in table 8;

Table 8. Effect size calculation result

knowledge	**d value**	**Category**
Procedural	10.76	Large effect
competence	**d value**	**Category**
Evaluating and designing experiments	4.23	Large effect
Interpreting data and scientific evidence	1.68	Large effect

3.3. Disscusion

Results of this study shows that the implementation of levels of inquiry strategies improve the science literacy of students. These findings are in line with previous studies that the learning lab by using inquiry strategy can improve science literacy of students [16]; [3]. With using levels of inquiry students are given the opportunity to be actively involved in the activities of scientific investigations so that they can develop the skills of their intellectual processes [17].

In the domain of competency, increase in scientific competence of students in the experimental group was significantly greater than the control group was in the evaluate and design a scientific investigation competency to and interpret the data and scientific facts competency, the effect of the treatment to both of these competencies are catagorized as large effect. Evaluating and designing scientific investigations competency are covering describe and assess the scientific investigation and propose ways to ask the question scientifically [13]On the writing-to-learn activities students are asked to write back what he did during the experiment, students were asked to create a proposition that describes what he did to answer the research question. Then students conduct a disscussion with his group, and then communicate the results of group discussion in front of the class. Then the results of the discussion group discussed jointly guided by the teacher to reflect on lab activities that have been done, such as the selection of tools and materials, experimental procedures, and the importance of controlling the experiment. Students are facilitated to interpret each of the steps and procedures of his experiment, in line with the statement of keys et al [10]that the activity in students template in writing-to-learn students reviewing procedures performed (both the procedure prepared by the teacher and the student) and observation for linking it directly with the research questions. Students are also encouraged to make an analysis of the experiments that have been done by comparing the experimental results that have been obtained to a written or online source in a hypothetical inference proposition, it can help students to improve their ability to design and evaluate experiments. Interpret the data and scientific facts competency covering analyze and evaluate data, questioned and argued in various representations and draw appropriate scientific conclusions [13]. On the writing-to-learn activities students are asked to make a claim from the experiment that they did, in addition students are also required to provide proof that lead students to make the claim, and then students compare the his claims to his peer, and also contrasting with other valid sources. In the final phase, students are asked to create a narrative about how his ideas change. Students are facilitated to make and reexamine the ideas and scientific claims are made, and the change of claims that have been made. This is similar to the claim of the Balgopal & Wallace [18]that the activity of writing to learn allows students to reexamine its ideas and modify the supporting facts for them to make a claim. This of course help students to be able to improve their competence in interpreting the data and scientific evidence.

The experimental group have significantly greater increase on procedural knowledge than the control group was. The effect of the treatment to this knowledge is catagorized as large effect. Procedural knowledge can be defined as knowledge of standard procedures used scientists to obtain reliable and valid data [13] On the writing-to-learn activities students are asked to to write a personal meaning and groups of procedures and the experimental results indicted. By doing that the students rewrite the ideas, designs, procedures, research results and conclusions that can be drawn from the investigation activities. This form a knowledge for students about the appropriate procedures and less appropriate procedures in order to answer the research question, as revealed by the Hand [19] that the new knowledge will be formed for the task of writing requires existing knowledge to be delivered in a way that has not been never been done before, and this led to a conceptual framework setting different topics.

4. Conclusion

Implementation of the levels of Inquiry strategy with the writing-to-Learn assignment can significantly improve the science literacy competence of students evaluating and designing scientific investigations, the competence to interpret the data and scientific facts, and procedural knowledge compared with levels of Inquiry learning only strategy. This research is empirical evidence that supports earlier studies, that the application of the levels of inquiry with the assignment of writing-to-learn can improve science literacy of students. A challenge for teachers is to integrate the activities of inquiry with the task of writing, that is necessary to train the students to be able to bring expected propositions. Teachers are expected to give an example of writing that shows these criteria on this training.

5. References

[1] Hobson A 2008 *The Surprising Effectiveness of College Science literacy Courses*. The Physics Teacher **46**
[2] Balitbang Kementrian Pendidikan dan Kebudayaan. 2011*Survei Internasional PISA*. [online].
[3] Fang Zand Wei Y 2010 Improving Middle School Students' Science LiteracyThrough Reading Infusion. *The Journal of Educational Research*
[4] Wenning C J 2005 *Journal Physics Teacher Education* Online **2**(3)
[5] Rohmi P 2015 *Penerapan Levels Of Inquiry untuk Meningkatkan Domain Kompetensi dan Pengetahuan Literasi Sains Siswa SMP pada Tema Pencemaran Lingkungan*. Tesis Sekolah Pascasarjana UPI. Tidak diterbitkan.
[6] Yore L D 2000 *Journal of Deaf Studies and Deaf Education* **5** (1) Winter
[7] Glynn S M and Muth K D 1994 *Journal Of Research In Science Teaching* **31**(9) 1057-073
[8] Peterson S S and Rochwerger 200 6*Cross-Curricular Literacy: Writing for Learning in a Science Program*. Voices from the Middle **14** (2)
[9] Baker, W et al. 2008 *A Journal of Educational Strategies, Issues and Ideas*
[10] Keys et al. 1999 *Journal Of Research In Science Teaching*
[11] Hand B, Wallace C W and Yang E 2004 *International Journal of Science Education* **26**(2) 131-49
[12] Xu H and Talanquer V 2012 *Journal of Chemical Education*
[13] OECD 2013 *PISA 2015 Draft Science Framework*
[14] Creswell J W 2009 *Research Design; Qualitative, Quantitative and Mixed Method Approach. Third Edition* (California: SAGE Publication)
[15] Hake R R 1999 *Analyzing Charge Gain Score*. Avalaible at http://www.physics .indiana.edu/Analyzing Change Gain.pdf
[16] Gormally et al. 2009 *International Journal for the Scholarship of Teaching and Learning*
[17] Wenning C J 2007 *Physics Teacher Education Online* **4**(2) 3-14
[18] Balgopal M and Wallace A 2013 Writing-to-Learn, Writing-to-Communicate & Science Literacy (*The American Biology Teacher* Vol. 75) (California : University of California Press)
[19] Hand et al. 2006 *International Journal of Science and Mathematics Education* 125Y143

Analysis of Students' Incorrect Answers on the Topic ofRoman Numerals in The Fourth-Grade of Primary School

D Selvianiresa[1], and Al Jupri[2]

[1]Department of Primary Education, Postgraduate School, UniversitasPendidikan Indonesia, Jl. Dr.Setiabudhi No. 229, Bandung 40154, Indonesia
[2]Department of Mathematics Education, UniversitasPendidikan Indonesia, Jl. Dr.Setiabudhi No. 229, Bandung 40154, Indonesia
Email: [1]dessiselvianiresa92@student.upi.edu

Abstract. Mathematics has an important role in solving problems related to mathematical symbols. Mathematics topics are related to the daily life of students, as well as mathematics in the primary school level. In this paper, we report the results of research that aims to understand the difficulties of primary school students towards mathematics, especially on the topic of Roman numerals. The subject of this research included 38 students of fourth grade students from a primary school in the city of Bandung. The results showed that the teacher is more likely do the learning to deliver topic directly to students, rather than having to relatethe learning with a real context. In addition, after the teacher gave some exercises to students, we observed that there are some students' answers that are unpredictable and made errors. Unexpected answers are caused by student's unconscientious, misunderstanding, or inability to memorize in seven Roman numeral in general. The results of this analysis can be used to develop teacher competence in the teaching and learning process in the future. In this way, we expect the teacher can prepare various student responses to the given problems in the learning process.

1. Introduction

Mathematics is formed from the daily human experiences. Mathematics has an important role in solving problems related to mathematical symbols.Age of children to be the first formal education is good start in the development of numeracy person skill. The development of numeracy skills of mathematics, language, working memory is developed rapidly in the period children and formal education atprimary school [1]. Mathematics skills of students at the primary school level would be an important start to be able to optimally develop their own skills. Teachers have an important role to develop students' skills to be able to solve the problems which is related to mathematics.

There is a primary school mathematics topic that essentially can be applied to students in daily life. However, most teachers prefer teaching to deliver the topic directly to students, rather than having to relate learning with real context, for example, the topic of Roman numerals. Teachers often have difficulties in looking for context to teach this topic, because the use of Roman numerals rare, andhard to find in daily life. Whereas, students will quickly understand mathematical concepts, when learning is relevant to students in daily life[2, 3, 4]. For example, even if it is still rare to use, teachers can use a concrete to the introduction in Roman numerals, as of books or novels, clock, street names.

In addition, mathematics has rules as the meaning of the concept of the material. As well as the Roman numerals, there are rules that students need to understand when working on the problems of Roman numerals. For example, firstly students must first memorize seven Roman numerals that will be used as the benchmark in each process. Students are also given the rules of how to writethe Roman numerals two or more numbers by addition and subtraction. This will affect their way of writing. When students do not understand it, it will affect the inability to write a Roman numeral. The students are still having difficulties in using the principles correctly [4]. Most students have difficulties in applying the rules of writing a Roman numeral. The core of the problems that occur at students is not memorizing rules of writing a Roman numeral [5]. Teachers can explain the Roman numeral directly or reception learning to achieve memorization learning.

In addition, teachers must be able to plan models, approaches, methods, techniques, and strategies that suitable to facilitate and enable students to understand the Roman numerals. Then, teachers must be able to predict student incorrect answers that will appear on the students in the Roman numerals topic, so the teacher has prepared and provided appropriate action to help students misconceptions. The teacher can predict the ability of their students, and is able to give a decision in determining the approach, and whatplan is suitable for students [3, 6]. Therefore, in the current study, we analyze students' incorrect answers in the Roman numerals topic, and the research questions of this study are the following: What are students' mistakes in learning the Roman numerals? What are the results of comparison between predictions of student answers and students' actual strategies? What are the factors of student mistakes?

2. Experimental Method

To address the research question, we conducted a qualitative study through classroom observation and an individual written test on Romal numerals. Data were collected through exploration ability of primary students on Roman numerals material. The subjects of the research were 38 fourth-grade students from one of primary school in the city of Bandung.

This study used a qualitative method with five procedures. First, we designed problems that would be used in the learning process. Second, wemade predictionsaboutthe learningprocess which would be carried out by the teacher and predictions about strategies which would be used by students to answer questions from the teacher. Third, we collected data from students' responses and the learning process. Fourth, we compared between predictions and the observation results. Finally, we analyzed data and gave recommendations about the results of analysis.

3. Result and Discussion

3.1. Analysis Strategy

Based on observations in the classroom, the learning model used by the teacher is a classical model. The teacher thought that the classical model is easier to use order to avoid misconceptions students' learning of mathematics. Submission of materials by teachers is a startbetter than the students must find and build their own understanding. This is reasonable, because skills such as the ability to compare numbers, the ability to recite the order, or the ability to recognize the number of a symbol is the development of numeracy students early [7,8]. For that, it would be better if the material is presented by the teacher.

Furthermore, the approach which is used by teachers when learning is similar to the prediction that the expository approach. The dominance of teachers not only for delivering the material, but there are times for to ask asking questions, and giving exercises to students. Expository learning, students are not only silently listening to the teacher, but the students were given training activities and asked questions [9].

For the methods which is used by teachers during lessons based on the observation that the prediction is similar to using lectures, question and answer, and exercise. The lecture method is used when the teacher presenting the material of writing two numbers along with the rules Romannumerals. This is not an issue when the teacher is using the lecture method in the delivery of the material, becausemethod is good if the teacher's lecture is really prepared welland understandably material by students. Furthermore, after the teacher delivering the material, the debriefing conducted between teachers and students. It gives an opportunity for students to be active and brave in expressing their opinions. The purpose of the activities of active students is for students to be proficient in math concepts.

Lastly, the students give an exercise to find student skill. The exercise can help students in providing an understanding and students will really understand what he learned. The counting activities is early to students is very influential on the development of skills. The counting activities which is do with their training provision. The exercise is very beneficial to the students aware of the numbers, the relationship with the student materials, and other skills related to the material being studied [10, 11].

The technique which is used by teacher was the expository technique, which is also in line with the predictions which is made. In addition, the strategy which is used by teachers in learning. There is different between prediction and students' actual strategies. Based on the prediction, researcher assumed that in explaining the Roman numerals will be easier to understand by student if the teacher which brings concrete objects, such as books or novels thatthe Roman numerals as the initial recognition. But, the reality showed that the teacher does not use any strategy in the classroom. Teacher is use directlymaterial in the Roman numerals, without any concrete objects or media.

3.2. Analysis Students' Response

Analysis the responses of the students which is earlier made is several predictions of students answers that will appear, which is then compared with the reality.

3.2.1. Analysis Problem 1

Based on observation, twenty eight students provided correct answers. Students have been corrected to writeRoman numerals 100 = C, 10 = X, and 3 = III. In addition, students have been to understand the rules subtraction that the smaller value is first written in Roman numerals. As 110, students have been to write XC, and 3 = III. So that, the answer is 93 = [100-10] + 3 = XCIII. But, more students is difficult and incorrect answers about problem 1. As an illustration, the students' answers were presented in Figure 1 below:

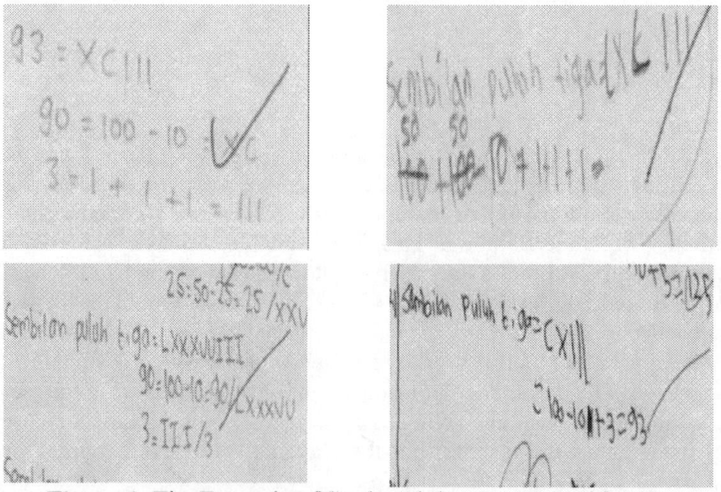

Figure 1. The Example of Students' Answer on Problem 1

Figure 1 shows that more students are incorrect answers in the rules Roman numerals. Six students similar with the prediction 93 = [100-10] + 3 = CXIII. Students have been correctto writing Roman numerals that the smaller value is first written in Roman numerals. But, the students have errors in exchange number is XCIII be CXIII.

From 38 students, one student similar with prediction is 50 + 10 = 93 + 10 + 10 + 5 + 5 + 3 = LXXXVVIII. Then, one student similar with prediction is 93 = [50-10] + 50 + 3 = LXLIII. The student have been correct to adding, but error to understand the rules five number. The teacher did not to explain about the rules it.

Based on analyzed, two students did not answer with the prediction researcher. One student is answer 93 = IXXXX. In addition, one student did not answer in problem 1.

3.2.2. Analysis Problem 2

Based on observation, twentysix students provided correct answer. Students have been corrected to write a Roman numerals 100=C, 10 = X, and 9=IX. In addition, students should understand the rules subtraction that the smaller value is first written in Roman numerals. As 99, students have been to write XC, and 9 = 10-1. So that, the answer is 99 = [100-10] + 9 = XCIX. But, more students is difficult and incorrect answers about problem 1. As an illustration, the students' answers were presented in Figure 2 below:

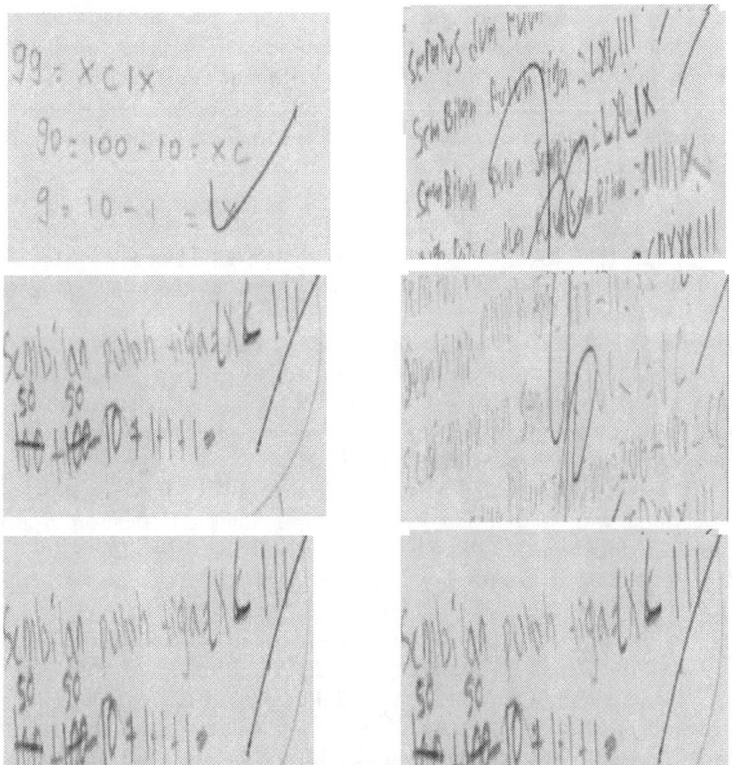

Figure 2. The Example of Students' Answer on Problem 2

Figure 2 shows that more students are incorrect answers in the rules Roman numerals. Two students similar with the prediction 99 = [100-10] +9 = CXIX. Students have been correctto writing Roman numerals, but the student did not understand about the rules of decrease. So that, the students have errors in exchange number is XCIX be CXIX.

From 38 students, four students are similar with prediction, such as99 = [50-10] + 50 + 9 = LXLIX. On observation, the teacher give to symbol that the answer is incorrect. However, the students areincorrectas the rules inRoman numerals.The rules in Roman numerals is X can be used to decrease

L and C. So,the student can be decreased by 50 or 100. But, the teacher explained about the rules V, L, D only be 1x written. So that,the teacher must explicate about the rules in Roman numerals.

In addition, one student response issimilar toprediction, i.e., 99 = 9 + 9 = IXIX. The student did not understand about 99 = 90 + 9, so the student answers9 = IX with 9 = IX. Other responses, two students is answer 99 = [100-1] = IC. Based on analysis, the students did not understand the rules who is explained by the teacher about decrease V and X. So that, C or 100 can not tosubtractionby 1. Problem 2, there is one student who did not answer.

Other responses, one student produced 99 = LXXXVVIX. Student is correct to writeRoman numerals LXXXVVIX = 50 + 10 + 10 + 10 + 5 + 5 + 9. But, the answer is a longer than the teacher answer 99 = [100-10] + 9 = XCIX.

Based on the analysis, one studentdid not answer according the prediction of researchers. One student answered90 = 90 + 9 = [9 + 10] +9 = IXXIX.

4. Conclusion

Based on the analysis between the predictions and observations, the teacheruseda classical model and the expository learning method. However, the teacher does not bring concrete objects that can helpstudents to understand and the benefits in learning Roman numerals. In addition, we found that some predictions are similar to student answers. Therefore, for the learning processes, we suggest the teacherto not only prepare the planning models/approaches/methods/techniques/learning strategies, but also to predict the responses of students.

5. References

[1] Cirino P T 2011 *Journal of Experimental Child Psychology* **108** pp713-33
[2] Hudojo H 2003 *Common Textbook:Pengembangan Kurikulum dan Pembelajaran Matematika* (Malang: UM Press) p 83
[3] Haylock and Fiona 2007 *Key Concepts in Teaching Primary Mathematics* (SAGE Publications) p 20
[4] Kusmaryono I and Hardi S 2016 The Effect of Constructivist Learning Using Scientific Approach on Mathematical Power and Conceptual Understanding of Students Grade IV*Journal of Physics*J. Phys.: Conf. Ser. 693 012019
[5] Wardono, S B Waluya, Scolastika M and Candra D 2016 Mathematics Literacy on Problem Based Learning with Indonesian Realistic Mathematics Education Approach Assisted E-Learning Edmodo *Journal of Physics*J. Phys.: Conf. Ser. 693 012014
[6] Blongkod S 2014 *Analisis Kesalahan Mengubah Bilangan As like Bilangan Romawi pada Siswa Kelas IV* (Gorontalo: Skripsi Universitas Negeri Gorontalo)
[7] Sumiaty E dan Endang D 2015 *Proc. Int. Seminar on Mathematics, Science, and Computer Science Education(Bandung)* (Bandung: FPMIPA UPI) p. 78-83.
[8] Desoete, A, CeulemansA, De Weerdt F and Pieters S2012 *British Journal of Educational Psychology* **82** p64-81
[9] Tim MKPBM Jurusan Pendidikan Matematika 2001 *Common Text book Strategi Pembelajaran Matematika Kontemporer* (Bandung: JICA UniversitasPendidikan Indonesia) p 171
[10] Barth H, La Mont K, Lipton J, Dehaene S, Kanwisher N and Spelke E2006 *International Journal of Cognitive Science* **98** pp199-222
[11] Andersson U and Lyxell B 2007 *Journal of Experimental Child Psychology* **96** p 197-228

Acknowledgments
We thank students and the teacher for their help and their active participation in this research.

IOP Conf. Series: Journal of Physics: Conf. Series **812** (2017) 012051 doi:10.1088/1742-6596/812/1/012051

Analysis of Students' Incorrect Answers at Scale Materials in The Fifth-Grade of Primary School

E N Amalina[1] and Al Jupri[2]

[1]Department of Primary Education, Postgraduate School, Universitas Pendidikan Indonesia, Jl. Dr. Setiabudhi No 229. Bandung 40132, Indonesia.
[2]Department of Mathematics Education, Faculty of Mathematics and Science Education, Universitas Pendidikan Indonesia, Jl. Dr. Setiabudhi No 229. Bandung 40132, Indonesia.
Email: erikanuramalina@student.upi.edu

Abstract. This research aims to analyze a comparison between predictions made by authors and the reality that occurs in the real classroom situation, and to analyze students' responses toward questions given by teachers on the topic of scale materials. This research used the descriptive-qualitative method, and involved 50 fifth-grade students from a primary school in the city of Bandung. The results of this research showed that there are some influences between learning methods and students' responses shown by the way students answer questions. In reality, there are many students' responses produced beyond the predictions of authors. Based on the results, we conclude: 1) incorrect answers made by students are not only influenced by the learning methods used by teachers, but also influenced by other complex factors, such as the lack of student understanding and reasoning; (2) teachers need to understand that every student has different perspectives in answering questions, so teachers should be more careful in assessing their students' answers.

1. Introduction

Mathematics is a tool to develop a way of thinking, is a deductive science that promote student thinking in a formal and abstract manner, but must be given to children from an elementary school that their way of thinking is still at the stage of concrete operations [1]. Therefore, a teacher will surely prepare the best strategy and methods of teaching mathematics to fit the students' thinking stage [4]. Strategies and methods which is chosen by teacher will greatly affect the way students in mathematical thinking.

According to Piaget [2], 'studying the cognitive structure of a person occurs because the process of assimilation and accommodation'. The assimilation is the process of acquiring new information that directly integrates with the information we already have, whereas the accommodation is the process of re-structuring the information that we already have with new information. So, learning is not only the process of delivering and collecting knowledge, but also the process of accommodating the old knowledge to new knowledge.

According to cognitive development, elementary students are at the stage of concrete operations, in which children will learn by manipulating concrete objects. Elementary students will have difficulty if they are asked to think abstractly [6], including when it is connected in the material about seeking scale. The teacher can provide concrete examples that commonly using the scale, for example on a map or drawing.

In addition to referring the theory of Piaget, this study uses the theory of Ausubel. The core learning theory of Ausubel [2] is meaningful learning. Meaningful learning is a process that links between new knowledge on relevant concepts contained in a person's cognitive structure. According to this theory, learning can be classified into two dimensions, namely a reception and discovery, where it gives students an emphasis on memorizing the material. By the reception dimension, we mean that the teacher directly communicates the material to students in its final form, then students are asked to memorize the material. By the invention dimension, we mean that the teacher initially provides learning at which the material must be found by the student, then the students memorize the material.

In responding problems, every student has definitely different ways, and mistakes that students may do would vary according to their respective capabilities [5]. Therefore, to analyze students' mistake in answering problems, the research questions of this study included: What are the results of the comparison between the predictions of researchers regarding the strategies which are used by the teacher in the learning process and the reality which occur in the learning process? What are the comparison between researchers predictions on students' responses in answering problems and actual reality that occur? How can we interpret students' mistakes from the perspective of the theories described earlier?

2. Experimental Method

To address the research question, we conducted a qualitative study through administering an individual written test on the topic of Scales [3]. Data were collected through exploration ability of primary students on scales materials. The subjects of the research were 50 fifth-grade students from one of primary school in the city of Bandung.

First of all, before learning process was happened, we made some predictions about learning strategies which would be used by teacher and predictions about strategies which would be used by students to answer questions from the teacher. In the analysis, we compared our predictions and the reality occurred in the learning and teaching process.

3. Result and Discussion

3.1 Analysis of Strategy

After making observations, we tried to compare the prediction of teaching method and the reality that occurs in the classroom. The comparison between predictions and the reality is presented in the Table 1.

Table 1. Comparison between Prediction Learning Model and Facts

Learning Model	Prediction	Fact
Model	Classical	Classical
Approach	Expository	Expository
Method	Lecture, Questions & Answers, Practice	Lecture, Questions & Answers, Practice
Technique	Expository	Expository
Strategy	Pictures and Map	x

Based on the observation, the learning model which is used by teacher is the classical model. This is similar to the prediction. The teacher used this model because this model is easier to use considering the number of students in a class, with 50 students. Because of that many students, the teacher is required to submit items as good and as comfortable as possible. Classical learning will make it easier for the teacher to organize the subject matter, because generally classical learning lesson material will be uniformly absorbed by students both the sequence and scope.

The learning approach used by the teacher was expository approach. This is consistent with the prediction. The reason of the teacher for using this approach is that because there are so many students but limited time to learn, so it is expected that using this approach can overcome these problems.This teaching approach is quite common in Indonesia [3].

The learning method which is used is the method of lecturer, discussion and exercises. This is consistent with the prediction. The reason of the teacher for using this method is that because the lecturer method is the application and narrative orally by the teacher to the class, using teaching aids to clarify the description given to the students. The lecturer method is often encountered in the processes of learning in schools starting from a low level to the college level, so that this method has been regarded as the practical method for teacher to make teaching and learning interactions. The use of this method can help the teacher to maximize her time and learning objectives.

Learning techniques which is used by teacher is the technique of expository. This is consistent with the predictions that have been made previously. The teacher using this technique because the number of students in the classroom quite a lot but limited time to learn, so that the use of this expository technique is suitable. Teacher did not use learning strategies. This is not consistent with the predictions that have been made.

3.2 Analysis Students Response to Questions by Teacher

After doing predictions about students' responses to questions which is provided by the teacher, the researcher analyzed the students 'response which is accordance with the results of observations that have been made. Beside of that, the researcher explained about the possible reasons why the students answer the question, how many students answered precisely and less precise, as well as the comparison between predictions and reality. Problems which is given by the teacher are 5 questions but the researcher analyzed at question number 1. The elaboration of predictions and reality is presented in the Table 2.

Table 2. Analysis of Students Response

Problems	Prediction(s)	Students' actual strategies	n/N
1. The distance from A to B on map is 5 cm. The actual distance of this is 60 km. What is the scale on the map?	a. Scale = 60 km = 6.000.000 cm = 5 : 6.000.000 = 1 : 1.200.000 b. Scale = 60 km = 600.000 cm = 5 : 600.000 = 1 : 120.000 c. Scale = 60 km = 60.000 cm = 5 : 60.000 = 1 : 12.000 d. Scale = 60 km = 6.000 cm = 5 : 6.000 = 1 : 1.200 e. Scale = 60 km = 600 cm = 5 : 600 = 1 : 120 f. Scale = 60 km = 60 cm = 5 : 60 = 1 : 12 g. Scale = 60 km = 6.000.000 cm = 5 : 600.000 = 1 : 120.000	- Response a exist - Response i exist - Response j exist - Response k exist - No student answers on the predictions. - Response b, c, d, e, f, g, h don't exist	39/50 2/50 3/50 2/50 4/50

	h. Scale = 60 km = 600.000 cm = 5 : 60.000 = 1 : 12.000 i. Scale = 60 km = 6.000.000 cm = 5 : 600.000 = 1 : 1.200.000 j. Scale = 60 km = 60.000.000 cm = 5 : 60.000.000 = 1 : 1.200.000 k. Scale = 60 km = 6.000.000 cm = 5 : 600.000 = 1 : 120.000

Note: n = number of students used kind of strategies; N = number of all students

3.2.1 Analysis Question Number 1

Based on observations of the student responses on problem 1, provided by the teacher, we found that almost all students answered correctly. The right answer is in accordance with the prediction of a response, which in fact as many as 39 students right in doing the calculation method or the answers. Students were able to change the unit of distance from kilometers to centimeters. Then, after changing the distance unit, students were able to compare the distance on the map with the actual distance. After that, the students were also able to simplify these comparisons in a simpler form. As in the ratio of 5: 6.000.000, students were able to simplify the comparison form. So the answer is 1: 1.200.000.

Continues on other responses, it was found that there are two students in accordance with the predictions i which is created by the researcher:

Scale = 60 km = 6.000.000 cm
= 5 : 600.000
= 1 : 1.200.000

Actually, the students have been able to change the unit of distance from kilometers to centimeters but when students compare the distance on the map with the actual distance, the student less scrupulous in rewriting the actual distance that should 6.000.000 to 600.000. In addition, students may feel confused by the large number of 0. However when simplifying this comparison is correct, students' answers are 1: 1.200.000. Some teachers may not consider this problem as a big problem, but if this problem is allowed to continue, the students will continue to have a misconception of matter on an ongoing basis. Therefore, early treatment is needed to help students reducing this type of mistakes. Furthermore, response prediction j answers appear by three students, as follows:

Scale = 60 km = 60.000.000 cm
= 5 : 60.000.000
= 1 : 1.200.000

If it is analyzed, it can be seen that students do not really understand how to change the unit of distance from kilometers to centimeters which has been described by teachers, by multiplying to 100.000. But in fact students take to 1.000.000. After that students write a comparison between distance on the map and the actual distance is 5: 60.000.000. Then, when simplify the comparison students answer is correct, it is 1: 1.200.000.

Furthermore, response prediction k answers appear by two students, is as follows:

Scale = 60 km = 6.000.000 cm
= 5 : 600.000
= 1 : 120.000

If it is analyzed, it can be seen that the students have learned how to change the unit of distance from kilometers to centimeters, but when students compare the distance on the map with the actual

IOP Conf. Series: Journal of Physics: Conf. Series **812** (2017) 012051 doi:10.1088/1742-6596/812/1/012051

distance the student less scrupulous in rewriting the actual distance that should 6.000.000 to 600.000. So that it causes some students write the wrong answer.

From the analysis of students' answers on problem 1, there are four students who answered beyond predictions. The examples of students' responses are as follows:

- *Scale = 60 km = 60.000.000 cm*
  = 5 : 60.000.000
  = 1 : 12.000.000
- *Scale = 60 km = 6.000.000 cm*
  = 5 : 6.000.000
  = 1 : 6.000.000
- *Scale = 60 km = 60.000.000 cm*
  = 8 : 60.000.000
  = 1 : 48.000.000
- *Scale = 60 km = 6.000.000 cm*
  = 5 : 6.000.000
  = 1 : 12

In the first answer, students were not yet understand how to change the unit of distance kilometers to centimeters so that the final answer becomes wrong. In the second answer, the students already know how to change the unit of distance from kilometers to centimeters. Students also have to do a comparison between the distance on the map with the actual distance, but the students are less scrupulous to simplify it so that the students' answer was wrong.

Later in the third answer, the students do not understand how to change the unit of distance kilometers to centimeters. In addition, students were careless in writing the distance on the map. Students write down the distance on a map that is 8 cm, while the matter is 5 cm. So the final answer becomes wrong.

In the last answer, finally, the students already know how to change the unit of distance from kilometers to centimeters. Students also have to do a comparison between the distance on the map with the actual distance, but the students are careless simplify it so that the students' answer was wrong.

Based on the predictions, made by the results of the students 'answers, there are seven predictions: b, c, d, e, f, g, h on question number 1 does not appear in the students' answers. Thus, of the 10 answers predictions that have been made, there are three predictions according to the prediction of the results of the students' answers.

In general, factors of students incorrect answers include because students do not understand how to change the unit of distance and students also are felt difficult in simplifying the comparison. Research has demonstrated that teacher instruction has a major impact on student learning [7]. In this study, the teacher gave problems to students, next the teacher told them the way to answer the problems. After that, the teacher gave students instruction to answer some similar problems. So that some students just duplicate the way from the teacher explanations without understanding why the answer like that. Indirectly it may influence students to do incorrect responses.

4. Conclusion

Based on the analysis of the predictions and observations that have been made, conclusions that can be drawn are as follows. Each method certainly has its advantages and disadvantages, it can be taken into consideration by the teacher in determining which method is most appropriate for use in learning. In addition to preparing methods, teachers are also expected to use the media as well as props to support the learning process so that students are more interested and more understand the material which is presented so that students are involved actively in the learning process.

Based on the results, the following are some recommendations: (1) The error responses of the students are not only influenced by the teaching methods used by the teacher, but also influenced by other complex factors; so we recommend for further research to assess students' mistakes by considering other factors; (2) Teachers need to understand that every student has a different way in answering a problem, so that teachers can be more prudent in assessing the students' answers.

5. References

[1] Hudojo H 2003 *Common Textbook (Edisi Revisi) Pengembangan Kurikulum dan Pembelajaran Matematika* (Malang: UM Press) p 40
[2] Dahar R W 2011 *Teori-Teori Belajar dan Pembelajaran* (Jakarta: Erlangga) p 79
[3] Jupri A, Drijvers P, & Van den Heuvel-Panhuizen, M 2014 *Mathematics Education Research Journal* **26** pp 683-710
[4] Majid A 2013 *Strategi Pembelajaran* (Bandung: PT. Remaja Rosdakarya) p 20
[5] Erman S *et al* 2003 *Strategi Pembelajaran Matematika Kontemporer* (Bandung: Jurusan Pendidikan Matematika FPMIPA UPI) p 74
[6] Tim MKPBM Jurusan Pendidikan Matematika 2001 *Common Textbook Strategi Pembelajaran Matematika Kontemporer* (Bandung: JICA Universitas Pendidikan Indonesia) p 55
[7] Jitendra Asha K 2012 *Journal of Hammill Institute on Disabilities* **36** pp 4 –8

Acknowledgements
We thank students and teacher who participated actively in this research.

Implementation of Performance Assessment in STEM (Science, Technology, Engineering, Mathematics) Education to Detect Science Process Skill

A Septiani* and N Y Rustaman

Biology Department, Universitas Pendidikan Indonesia, Jl. Dr.Setiabudi No. 229, Bandung 40154, Indonesia
Corresponding author : nuryanirustaman@upi.edu

Abstract. A descriptive study about the implementation of performance assessment in STEM based instruction was carried out to investigate the tenth grade of Vocational school students' science process skills during the teaching learning processes. A number of tenth grade agriculture students was involved as research subjects selected through cluster random sampling technique (n=35). Performance assessment was planned on skills during the teaching learning process through observation and on product resulted from their engineering practice design. The procedure conducted in this study included thinking phase (identifying problem and sharing idea), designing phase, construction phase, and evaluation phase. Data was collected through the use of science process skills (SPS) test, observation sheet on student activity, as well as tasks and rubrics for performance assessment during the instruction. Research findings show that the implementation of performance assessment in STEM education in planting media could detect students science process skills better from the observation individually compared through SPS test. It was also found that the result of performance assessment was diverse when it was correlated to each indicator of SPS (strong and positive; weak and positive).

1. Introduction

Education is very important aspect in human life. One of main efforts to improve the quality of education in Indonesia is curriculum development. Currently the development is applied by changing KTSP curriculum into the curriculum of 2013. Permendikbud No. 65 of 2013 states that the assessment of the curriculum 2013 uses *authentic assessment* that assess students' readiness, learning process, and their achievement entirely.

The demands of curriculum 2013 assessment, which not only assess aspect of knowledge, but also aspect of the psychomotor and attitude, as well as a through assessment during the learning process, it is necessary to apply that emphasizes the assessment process. One sort of assessment for learning that may be applied in the curriculum 2013 is the *performance assessment*.

Experts and some education researchers [11; 9; 13] stated that the assessment of the performance assessment is authentic assessment to assess the student's ability to apply the concepts to real situations, assess the ability of scientific work in learning science, can train students in having a positive mental attitude, and skillfully solve any problems. This assessment not only be used to assess learning outcomes or products, but also can be used as part of learning [5].

The use of performance assessment can not be separated from the learning process. One suitable learning approach using performance assessment is the study of STEM (Science Technology

Engineering Mathematics). With the performance assessment in STEM education, we can assess every phase in STEM approach. There are indicators to assess those phase in performance assessment that implemented in this research. Futhermore, this performance assessment completed with indicators that can assess science process skills. With indicators that held in performance assessment, allows the teacher to conduct an assessment during the learning process.

In STEM education, after the students finish making the project, it will be tested whether it is as expected or not. If not, then it will be redesigns. This process is implemented because STEM based instruction emphasizes mostly on engineering stage, but still intersects with the scientific processes [10].

In this study, the assessment of performance is implemented in the preparation of planting media. This subject is a contextual concept in daily life. There is a problem or issue that can be made from this material to be solved through STEM based instruction, such as the reduction of planting area in urban zone. The study was conducted in a vocational high school which enable implementing STEM based instruction. This is because the vocational education has a purpose as stated in Government Law No. 20 of 2003, which is to produce learners who are experts in their chosen fields to compete professionally, both in terms of knowledge and technology [6].

2. Method

The method used in this research is descriptive method. This study aimed to uncover the science process skills early on during the process of implementation of performance assessment on the subjects of preparation of the planting medium.

This research was conducted at only one class, without the control class. In determining the working group. Implementation assessment starts from the beginning until the end of the lesson. Assessment includes the discussion process, the implementation of the trial, during the observation (21 days to see effectivity of planting media that have been made previously), reporting and presentation. The results obtained in each activity are analyzed, to be concluded.

Participants in this research were 35 students from vocational high school 3 Baleendah, Bandung regency, with competence and expertise of Agribusiness and Tissue Culture Technique. Subject of this research were 10th grade students. Sample was taken using random cluster sampling technique.

Data obtained is qualitatively and quantitatively analyzed. In addition, the performed data analysis using SPSS Statistics 23 to show regression and Pearson correlation test on the variables of data, consisted of task 1, task 2, task 3 results and all the SPS indicators.

3. Instrumentation

The instruments used in this study include observation sheets, task rubric of performance assessment, and test on science process skills. Observation sheets used during the learning process, especially on the phase of constructing. Observation sheet was used directly during students performance. While the task and the rubric is used to analyze the science process skills indicators contained in the report, whereas the science process skills test was administered to students at the end of the lesson. The test result was then compared to the results of observations made previously.

The performance assessment given consists of three tasks. The first task was the elaboration of constructing and designing phases. Indicators of science process skills assessed on the first task is asking investigable questions, testing hypothesize, and designing experiments. The second task is the implementation of the stage for covering the manufacture of products in accordance with the design that was planned earlier. Indicators of science process skills (SPS) that were assessed at this stage were observing, classifying, using the tools and materials as well as conducting experiments. The third task was the implementation of the test phase. SPS indicators assessed at the third task included communicating, interpreting, predicting, and applying concept.

While SPS indicators that can be determined through tests including interpretation, predicting, communicating, asking questions, applying the concept, planning experiments, classification, and hypothesize. The tests were given consisted of 20 questions that represent indicators of SPS has been described previously, with five possible answers. Each correct answer was given a score of 1 while another false one was given a score of 0.

4. Results

The data resulted from this study was obtained in two ways: by direct observation (i.e. for indicators of observing, classifying, using apartus and equipment, and doing experiments); and through reports made by each student (i.e. for indicators of questioning, hypothesizing, interpretation, predicting, communicating and applying the concept). To ensure, the result data was also supported by SPS test result. The whole profile of science process skills of students as in Figure 1.

According to the data obtained, it can be seen that the achievement of students in answering SPS questions is still low, as seen in Figure 1, while in practical study, the student ability of SPS is good. On observation results, the lowest SPS indicator of the ability of students is predicting (53%) while the highest is asking questions (89%).

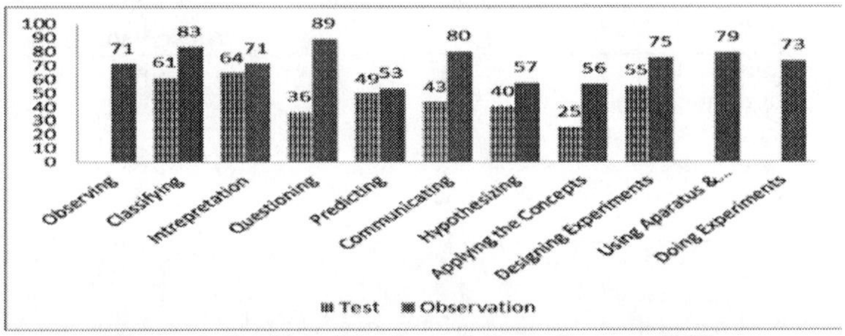

Figure 1. Profile of Students Science Process Skills

In contrast with the results obtained by students in answering questions, the ability of students who are lowest on SPS indicators of applying concept (25%), while the highest for the interpretation SPS indicators (64%). In the matter given to the student, not given the matter for observing indicators, using aparatus and equipment, as well as doing experiments, because these three indicators can only be expressed directly.

Contributions assessment of performance in revealing the ability of SPS students can be seen in Figure 2. In the first task, through group discussions, performance assessment can uncover some indicators SPS owned by the students, such as questioning, hypothesizing, and designing experiments. Measurement of indicators seen asking questions of formulation of the problem created by the students, while designing experiments judging by the way the students identify the variables, determine the tools and materials, and determine the procedure. The first task assessment is based on students' worksheets.

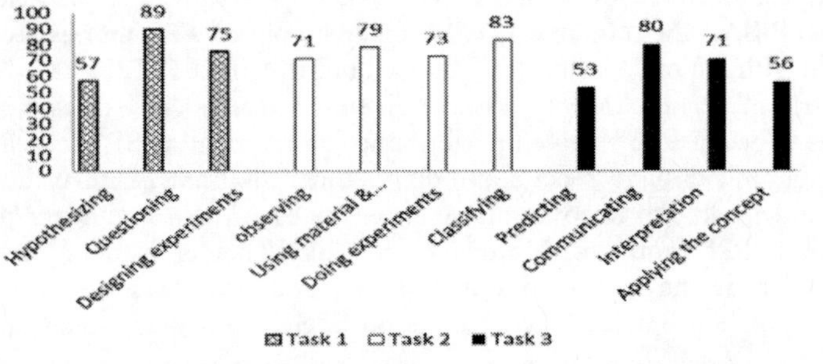

Figure 2. Contributions Assessment of Performance in SPS Revealing

In the second task, which revealed SPS includes indicators observing, classifying, using aparatus and equipment, as well as doing experiments. Measurement indicators observed by looking at the way students make a description of the characteristics of planting media used, while classifying measured by how students distinguish planting media group organic and inorganic. The second task assessment based on the observations made by researchers.

In the third task, which revealed SPS include communicating, interpretation, applying the concept, and predicting. Measurement communicating by looking at how the students in presenting the observed data. Indicator interpretation is measured by looking at how students make conclusions, applying the concept measured by the analysis of data and predicting measured by observing at how students provide design ideas improvement based on experiments that have been done before.

The correlation that exists between performance assessment with science process skills are shown in Table 1. The data shows that the assessment of the performance have a strong relationship and significant impact on the results of science process skills in observation, but has a weak relationship to the test results.

Table 1. Correlation Relationships Performance Assessment on Science Process Skills

		Performance Assessmen	SPS (observation)	SPS (Text)
Performance Assesment	Pearson Correlation	**1**	,860**	,283
	Sig-(2-tailed)		,000	,100
	N	35	35	35

5. Discussion

The Based on data obtained stating that the performance was good at observation results are shown in Figure 1 supposedly because during the learning process there is feedback from teacher to students, as stated by Earl [4] that assessment for learning can help students understand the subject. Besides the learning experience directly in accordance with the nature of science that not only trains students in knowledge but also skills and attitudes [3].

The achievement of science process skills of students is still low, especially in questioning due to lack of familiarity SPS students in answering questions like that. Students in Indonesia are generally more accustomed to answer the questions that are more knowledge given [7]. This is supported by the TIMSS results show that Indonesian students excel when working on theoretical problems and memorization. But collapsed when faced with questions that reveal aspects of high level. This fact shows that the students have not been able to use the concepts he learned and use logic thinking (reasoning) in resolving the matter. Additionally these results show that learners Indonesia is still lacking in conceptual understanding.

As the students are not used to the process of scientific method in practical activities, it will result in evidence the low achievement of students SPS test. Low achievement is also evident from the results of TIMSS and PISA, where Indonesia was ranked 40 out of 42 countries took part in TIMSS in 2011 and was ranked 64th out of 65 countries took part in PISA in 2012 [2].

The implementation of performance assessment given to students can express science process skills of the students. This is because in a given performance assessment as SPS indicators can be viewed directly or can be analyzed through reports and daily student journals. Unlike the SPS tests given at the end of the lesson, less able to uncover the science process skills of students because students are not used to work on the SPS problems. Moreover, most likely students guess the answers to the SPS tests for a given problem in the form of multiple choice, and also weaknesses in the given problem, there is still a matter of concept load, whereas good SPS is a matter which does not include the concept.

The amount of the acquisition value lower on tests of SPS with indicators predicting, hypothesizing, communicating, designing experiments (particularly in the identification of variables), as well as asking questions supposedly due in achieving the indicators necessary thinking skills, it is also reinforced by still many students who were at the level of concrete thought. In theory, the students age of tenth grade (15-16 years) is already at the level of thinking formally, but the level of intellectual development not only determined just by age, but is also influenced by the maturity, the experience of physical, logical-mathematical experience, social transmission and process the balance or the regulatory process itself [8; 1].

Based on Figure 2, the indicator is still low of SPS achievements are in task 1 and task 3. In the first task, the indicator is still low are hypothesized. In hypothesize, students must first understand how to specify the variables involved in the experiment, which is then connected between the two variables that had been predetermined. The hypothesis is also an approximate cause of the matter [9]. It takes skill to think in formulating hypotheses. It is also associated with cross-cutting concept of cause and effect arising from planting media used to plant growth and physical and chemical properties of growing media.

As for the third task, which is still low are predicting and apply concepts. Both of these indicators are very skilled thinking. This is because the indicator predicted, students tried to propose the type of planting media or the other based on the pattern that has happened before. In applying the concept, students should be able to link the results obtained with the preparation of the concept of growing media, plant growth, and hydroponics.

The thing that is contributing to the low achievement of students in terms of SPS and SPS indicators that require intellectual or thinking skills, the study habits of students who put more emphasis on manipulative skills (hands on) without taking part in developing the intellectual or thinking skills of students (minds on). This finding is consistent with the study conducted by Sumartini [12], that the vocational students by learning with a problem based learning has good reasoning power than vocational students who undergo conventional learning (hands-on only).

Based on the data obtained, it is known also that performance assessment conducted in this study can work well in every phase of STEM. This can be seen through the SPS indicator of students that can be detected in every phase. In addition, it can be seen also through the correlation between the performance assessment with SPS observation that shows a positive value.

Strong correlation relationship with the observation of science process skills for activities supposely undertaken students aiming to develop science process skills. This is because in learning activities, students are trained to perform lab activities are in line with the scientific method. This activity helps students understand the material preparation of the growing medium, because in the practical work, students become more active and students also have direct experience, so that the material presented will be long remembered by the students. This is consistent with the theory of the Edgar Dale's Cone of experience.

6. Conclusion

Based on the data obtained, it can be concluded that the assessment of the performance characteristics used are divided into two, namely votes directly and indirectly. Comparison of the results of tests and observation is that the observation of the student better able to describe the actual ability. Now, every task given to the students in this study were able to uncover the representative of indicators science process skills, and based on the correlation analysis, assessment of performance has a strong correlation to the results of science process skills in observation, and a weak relationship to the test results.

7. References

[1] Dahar R W 1996 *Teori-Teori Belajar* (Bandung: Erlangga)
[2] Dikdasmen 2014 *Gawat Darurat Pendidikan di Indonesia* (Indonesia: Kemendikbud RI)
[3] Direktorat PSMP 2013 *Hakikat IPA* (Indonesia: Kemendikbud RI)
[4] Earl L M 2006 *Rethinking classroom assessment with purpose in mind* (Canada: Curriculum Services)
[5] Haka N B 2013 *Penerapan Asesmen Kinerja untuk Meningkatkan Kemampuan Habits of Mind dan Penguasaan Konsep Biologi Siswa Kelas XI* (Bandung)

[6] Kemendikbud 2003 *Undang-Undang No. 20 Tahun 200 (Indonesia: Kemendikbud)*
[7] Nikmah N S and Masduki 2016 *Penggunaan Peta Konsep Dalam Penilaian Pemahaman Konsep Lingkaran* Proc on Nasional Penelitian Matematika dan Pembelajaran I (Indonesia: Universitas Muhammadiyah Surakarta_Press)
[8] Rezba R J, Sprague C and Fiel R 2006 *Learning and Assessing Science Process Skills* (Iowa : Hunt Publishing Company)
[9] Rustaman N Y 2003 *Strategi Belajar Mengajar Biologi* (Indonesia: Universitas Pendidikan Indonesia)
[10] Suwarma I R 2015 A Research on STEM Education Theory and Practices Method in Japan and Indonesia Using Multiple Intelligences Approach (Japan: Shizuoka University Repository)
[11] Stiggin R J 1994 *Student-Centered Classroom Assessment* (New York: Mac Millan College Publishing Company)
[12] Sumartini T S 2015 *J.Pendidikan Matematika* Vol **5 (1)**
[13] Wulan A R 2008 *Penilaian Kinerja dan Portofolio Pada Pembelajaran Biologi* (Indonesia: Universitas Pendidikan Indonesia

Acknowledgments
The author expresses her thanks to Mrs. Nuryani Y. Rustaman, for her great contribution and support to this study, and to reviewers for comments and input.

Development Scaffolding for Construction of Evaluation Instrument Training Program on The Cognitive Domain For Senior High School Physics Teachers and The Same Level

W Arif[1*], A Suhandi[2], I Kaniawati[3], and A Setiawan[4]

[1]Program Studi S3 Pendidikan IPA, Universitas Pendidikan Indonesia, Indonesia
[2,3]Departemen Pendidikan Fisika, Universitas Pendidikan Indonesia, Indonesia
[4]Departemen Pendidikan Teknik Mesin, Universitas Pendidikan Indonesia, Indonesia
Jl. Setiabudhi No. 229, Isola, Sukasari, Isola, Sukasari, Kota Bandung, Jawa Barat, 40154, Indonesia

Email: wa_upi@yahoo.com

Abstract. The development of scaffolding for evaluation instrument construction training program on the cognitive domain for senior high school physics teacher and the same level that is specified in the test instrument has been done. This development was motivated by the low ability of the majority of physics teachers in constructing the physics learning achievement test. This situation not in accordance with the demands of Permendiknas RI no. 16 tahun 2007 concerning the standard of academic qualifications and competence of teachers, stating that teachers should have a good ability to develop instruments for assessment and evaluation of process and learning outcomes. Based on the preliminary study results, it can be seen that the main cause of the inability of teachers in developing physics achievement test is because they do not good understand of the indicators for each aspect of cognitive domains. Scaffolding development is done by using the research and development methods formulated by Thiagarajan which includes define, design and develope steps. Develop step includes build the scaffolding, validation of scaffolding by experts and the limited pilot implementations on the training activities. From the build scaffolding step, resulted the scaffolding for the construction of test instruments training program which include the process steps; description of indicators, operationalization of indicators, construction the itemsframework (items scenarios), construction the items stem, construction the items and checking the items. The results of the validation by three validator indicates that the built scaffolding are suitable for use in the construction of physics achievement test training program, especially for novice. The limited pilot implementation of the built scaffolding conducted in training activities attended by 10 senior high school physics teachers in Garut district. The results of the limited pilot implementation shows that the built scaffolding have a medium effectiveness in improving the ability of senior high school physics teachers in constructing the physic achievement test instrument that is characterized by more than 70% of trainees achieve scores of test instruments construction of about 80 or more.

1. Introduction

In a ministerial regulation of national education of the Republic of Indonesia number 16 of 2007 on standard academic qualifications and competence of teachers stated that one of the pedagogical competence that must be held by teachers is competence in conducting the assessment and evaluation of learning process and product, which described in the competence of senior high school teacher as follows: 1) Understanding the principles of assessment and evaluation of the learning process and learning outcomes in accordance with the characteristics of the subjects of teaching, 2) determine aspects of the learning process and the learning outcomes that are important to be assessed and evaluated in accordance with the characteristics of the subjects of teaching, 3) Determine the assessment and evaluation procedure of the learning process and learning outcomes, 4) Develop assessment and evaluation instruments of learning processes and learning outcomes, 5) Administer the assessment of the learning process and the learning outcomes on an ongoing basis using a variety of instruments, 6) Analyze the results of the assessment of the learning process and the learning outcomes for a variety of purposes, and 7) Perform evaluation learning process and learning outcomes. From these rules it appears that one of the competencies required of teachers is developing assessment and evaluation instruments of learning processes and learning outcomes [1].

One domain physics learning outcomes should be evaluated at any time after the implementation of learning is the cognitive domain. Cognitive domain is a domain that is important is to be achieved learners after participating in learning. There are six categories of the dimensions of the cognitive processes that should be owned by the students as the acquisition of the learning process, namely remember, understand, apply, analyze, evaluate), and create [2,3,4]. In connection with the evaluation of this cognitive learning outcomes, teachers are required to have a high ability in constructing an item of cognitive test instruments. Unfortunately, the level of ability of teachers in constructing the instrument items of cognitive test is still not in line with expectations. Results of research conducted by researcher showed that the ability of the majority of teachers in constructing the instrument items of cognitive test is relatively low. The results of further research showed that some of the things that cause the low ability of teachers in construct an evaluation instrument learning outcomes include: (1) the majority of teachers stated only know some of the aspects of cognitive domain of learning outcomes according to Bloom's revised by Anderson, (2) most of teachers expressed only know several of all indicators of every aspect Bloom's cognitive domain revised by Anderson, and (3) the majority of teachers expressed do not understand the indicators of every aspect Bloom's cognitive domain revised by Anderson [5]. Ignorance and lack of understanding that have an impact on mistakes in the operationalization process in every aspect indicators. Another thing which is also the cause of the failure of teachers in constructing item test instrument is a result they do not take steps that systematically, they do not start with the construction of a frame or scenario of item test and they did not check the suitability of constructed item test with indicators of cognitive aspects.

This situation can not be allowed to continue to drag on, because it will have an effect on the poor quality of the learning process and results of Physics held. Need to quickly find the right solution so teachers can have pedagogic competence especially those associated with the ability to construct cognitive domain instrument test items, as expected. One effort that can be done to improve the ability of teachers to construct cognitive instruments test item is through in-service training activities. Actually they have completed the training activities on evaluation of learning held by government agencies that have the duties and functions as an institution guarantor of the quality of teachers, yet they feel that the training they follow not specifically oriented towards development instrument test item for measuring the competence of learning outcomes in cognitive domain, but the evaluation of learning in general. On the basis of the factors that cause the failure of teachers in constructing cognitive domain instrument test item as proposed in above, it is necessary to develop a scaffolding that contains stage process of preparing an instrument test item that can provide guidance and directives systematically to teachers in constructing a valid instrument test item. At least on the scaffolding there should be activity of cognitive indicators review, construction frame test item and checking items are constructed. Scaffolds are the supporting structures provided by those people or tools to promote learning [6]. Stone argues that scaffolds allow students to achieve a higher level of understanding within their Zone of Proximal Development. Sinaga et al was designed the scaffolding to improve the pre-service physics teachers' ability in writing teaching materials [7].

This article describes the results of the development of scaffolding for construction cognitive instrument test item, the results of expert judgement and results of limited test about using the scaffolding which was developed on in-service training activities.

2. Experiment

The method used in the development of scaffolding for construction cognitive domain instrument test item is a method of research and development (R & D) consist of a three stage process that are define, design, and develop. Define stage is done to obtain an overview of urgent needs related to improvement of professional competence of Physics teachers. At this stage, the activities are policy studies, field studies and literature studies. The design stage is done to design a training program for construction cognitive instrument test item based on the results from need assessment. One of the things that are designed at this stage is the scaffolding for construction cognitive instruments test item that will be used in the in-service training about construction of cognitive domain instruments test item for senior high school physics teachers. Develop stage conducted to produce scaffolding for in-service training activities based on the draft which has been made at the design stage. This includes creating a stage develop scaffolding, scaffolding validation, and limited testing about use of scaffolding in construction activities on in-service training construction cognitive domain instruments test item. Validation scaffolding conducted by three experts in the field of education evaluation, while the use of scaffolding in a limited test on in-service training activities conducted on 10 teachers of Physics in one district in West Java as research subjects. Limited test conducted by using experimental research design as shown in Figure 1.

Pretest	Treatment	Posttest
O	X	O

Figure 1. Chart design of experiments to limited test use of the scaffolding in- service training

In the chart above, O is the ability test to construct cognitive instrument test item. To measure the ability of teachers to construct cognitive instrument test item at before and after treatment used construction skills test cognitive ability test instrument that covers 19 indicators of aspects of cognitive domain. To determine the level of teacher's ability to construct a cognitive instrument test item was used levelling guide as shown in Table 1.

Table 1. Levelling guide for teacher's ability to construct cognitive instrument test item

Levels of ability	Range of average scores (AS)
High	AS≥ 80
Moderate	60 ≤AS< 80
Low	AS< 60

While to determine the effectiveness of the use of scaffolding in promoting teachers ability on constructing cognitive instrument test item used criteria as in Table 2. [8]

Table 2. Criteria for the effectiveness of the use of scaffolding in enhancing the teachers ability in constructing cognitive instrument test item

Criteria of effectiveness	The number of teachers (N) who achieve a score ≥ 80 on a scale of 100
High	N≥75%
Moderate	50%≤ N<75%
Low	N<50%

3. Result and Discussion

Figure 2 shows the schematic of the scaffolding for construction an instruments test item resulted from the development stage. Scaffolding was built includes five phases of construction activity of instruments test item of cognitive domain.

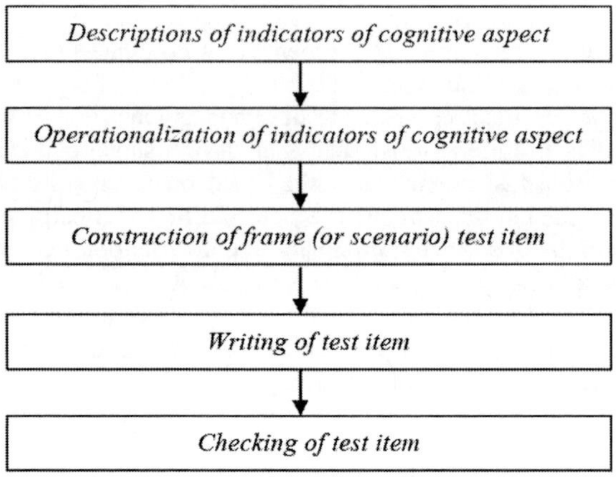

Figure 2. Scaffolding for construction of instrument test item of cognitive domain.

In practice, the use of scaffolding in the construction process of instruments test items of cognitive domain on the in-service training activities assisted by using a matrix as shown in Table 3.

Table 3. Matrix for the use of scaffolding in the construction process of the cognitive domain instrument test item

The test form	Cognitive aspects and indicators of aspects	Description of indicators	Evaluated Physics Content	Operationalization of indicators	Construction of frame (or scenario) items test	Writing test item	Checking of constructed test item
....	....	....	....	....	....	....	....

To fill the matrix using scaffolding during the construction process of the cognitive domain instrument test item, used the following provisions:

1. Column form of test filled in the forms/types of problems that will be created, for example, an objective test questions multiple choice or essay form about a subjective test.
2. Column cognitive aspect/indicator of aspects is filled with the cognitive aspects and indicators of cognitive aspects to be measured, for example aspects of understanding (C2) and the indicator explain (exemplifying).
3. Column description of the indicator filled in as indicators of cognitive aspects that will be measured in accordance described in the revised Bloom's taxonomy Anderson.
4. Column content filled with physics content and concepts were evaluated, for example, a static fluid material concept of hydrostatic pressure.
5. Column operationalization of aspect indicator is filled with the elaboration of indicators in a more specific aspects which include the concept being evaluated.

6. Making the construction of frame is filled with the exposure of each part consisting of questions about the stimulus, stem question and answer option (for about the multiple choice test form) according to the indicators and content /concepts were evaluated.
7. Column writing of items stocked with complete editorial matter within the framework of questions that have been created, along with an answer key.
8. Column checking items stocked by affixing yes and no on any part of matter is checked, the stimulus created accuracy, conformity with indicators stem matter aspect, homogeneity option answers (to questions on multiple choice form)

The results of the expert judgements on scaffolding developed indicate that: 1) Scaffolding for construction of cognitive domain instruments test item suitable for use in activities in- service training for physics teachers who do not have expertise in constructing the cognitive instruments test items, 2) activities stage in scaffolding construction of the cognitive instrument item ability test developed, assessed will be able to guide teachers in stages in constructing the test items cognitive ability towards problem valid, 3) scaffolding for construction of cognitive instrument test item suitable for learning adults, and 4) Scaffolding for construction cognitive instruments test item suitable for independent practice in which they work.

Table 4. Changes in the quantity of teachers at every level of ability to construct instrument test item before and after implementation of the scaffolding in the activities of in-service training.

Pretest	Posttest		
	Low Level	Moderate Level	High Level
Low Level (8)	-	3	5
ModerateLevel (2)	-	-	2
High Level (0)	-	-	-

The results of limited test about using the scaffolding suggest that the use of scaffolding which was developed on in-service training activities of construction cognitive instrument test item indicates that the change of the level of ability to construct cognitive instruments test item before and after implementation of the scaffolding on in-service training activities, as shown in Table 4. At the pre-test most of teachers are at the low level of the ability to construct instrument test item, a small fraction of teachers are at moderate levels, and none of the teachers are at a high level. At the post-test, teachers who initially had low level ability change to the moderate level of being 3 persons and change to the high level of being 5 persons. While the teachers who initially had a moderate level changed to a high level of being 2 persons.

In Table 4 it appears that at the post-test there are seven teachers (70%) who are at high-level capabilities (its means achieving test scores about constructing the instrument test items equal to 80 or more). Based on the criteria for the effectiveness of the use of scaffolding as described above, it means that the use of scaffolding on in-service training activities have the moderate effectiveness in enhancing the physics teachers ability to construct cognitive domain instrument test. These results indicate that the built scaffolding quite powerful in the enhancing ability to construct cognitive instrument test items among high school physics teacher. Systematic stages, slowly and have been able to trace give the guidance to novice to be able to construct instruments test item according to the measured aspects indicators, so that it meets the criteria valid test item.

4. Conclusion

Scaffolding for the construction cognitive domain instrument test item has been developed consisting of stages of the process: description of the cognitive domain indicators, operationalization of the cognitive indicator, construction of frame (or scenario) test items, writing test items, and checking the constructed test items. The result of the limited test suggest that the use of scaffolding which was developed in the activities of in-service training about construction of cognitive instrument test item

IOP Conf. Series: Journal of Physics: Conf. Series **812** (2017) 012053 doi:10.1088/1742-6596/812/1/012053

which followed by 10 subjects Physics teacher, showed that the use of scaffolding which was developed have the moderate effectiveness in enhancing ability to construct cognitive instrument tests item among teachers Physics became the subject of this experiment. The use of built scaffolding can change the level of ability to construct cognitive instruments test items towards higher ability, previously the majority of teachers are at a low level turn out to be at the moderate and high levels.

5.References

[1] Depdiknas 2007 *Permendiknas Republik Indoinesia nomor 16 Tahun 2007 tentang Standar Kualifikasi Akademik dan Kompetensi Guru* (Jakarta: Depdiknas)

[2] Anderson L W, Krathwohl D R, Airasian P W, Cruikshank K A, Mayer R E, Pintrich P R, Raths J and Wittrock M C 2001 *A Taxonomy for Learning, Teaching, and Assessing: A Revision of Bloom's Taxonomy of Educational Objectives* (United States: Addison Wesley Longman, Inc.)

[3] Ari A 2012 *Educational Sciences: Theory and Practice* **11**(2) 767-72

[4] Tutkun O F, Guzel D, Koroglu M, Ilhan A 2012 *The Online Journal of Counselling and Education* **1**(Issue 3) 23-30

[5] Arif W, Suhandi A, Kaniawati I, Setiawan A 2016. *Prosiding SNIPS 2016* (Bandung: Fisika ITB) p 165

[6] Stone C A 1993 *What is missing in the metaphor of scaffolding?* In Forman, E. A. & Minick, N. & Stone, C. A. (Eds.), *Contexts for learning: Socio-cultural dynamics in children's development* (New York : Oxford University Press) p 169

[7] Sinaga P, Suhandi A and Liliasari 2015 *International Journal of Instruction* **8**(1) 19-34

[8] Suhandi A, Wibowo F C 2012 *Jurnal pendidikan Fisika Indonesia* **8**(1) 1-7

Science Teachers' Understanding of Scientific Inquiry In Teacher Professional Development

Y H Adisendjaja, N Y Rustaman, S Redjeki and D Satori

Biology Department, Universitas Pendidikan Indonesia, Jl. Dr.Setiabudi No. 229, Bandung 40154, Indonesia
Corresponding author: ayusufhilmi@yahoo.com

Abstract. Inquiry is a main goal of science education reform around the world. This study investigated science teachers' understanding of scientific inquiry in teacher professional development program. The content of the program was focused on the nature of science and scientific inquiry. The program was conducted once in a week, every Saturday for 4 weeks, so it took about 30 hours. Twenty five science teachers from 3 districts with 5-25 years' experience were followed this program. Views About Scientific Inquiry modified was administered to all participants before and after TPD. VASI consists of 8 questions: 1) Scientific investigations all begin with a question and do not necessarily test a hypothesis, 2) There is no single set or sequence of steps followed in all investigations, 3) Inquiry procedures are guided by the question asked, 4) All scientists performing the same procedures may not get the same results, 5) Inquiry procedures can influence results, 6) Research conclusions must be consistent with the data collected, 7) Scientific data are not the same as scientific evidence, and 8) Explanations are developed from a combination of collected data and what is already known. Then, all responses are categorized as *informed*, *partially informed*, and *naive*. Results indicated that most of science teachers were not have good understanding of scientific inquiry. 30 hours teacher professional programs led to small measurable enhancements in teachers' understanding of scientific inquiry. Based on these findings, preservice and in-service program should focus on science education reform include scientific inquiry.

1. Introduction

Inquiry-based learning is a main focus in science education include science curriculum in Indonesia. Science education program inquiry-based is a main focus of National Science Education Standards and Project 2061's Benchmarks for Science Literacy (Colburn, 2000). Even though inquiry is considered as a science teaching process, many teachers do not apply in their classroom. Lederman (2009) stated that inquiry is a complex concept and have facets andnuances so teachers confuses about inquiry meaning. Reasons for teachers do not apply inquiry in their classroom are their believe that inquiry just for students with upper ability, teachers need more time to prepare inquiry-based teaching, inquiry is difficult to manage, and there is a tend to teach science fact, and the objective of science is to prepare for continuous learning (Colburn, 2000). The research of science teachers at Bandung area Indonesia showed the similar results. Teachers' views that the main objective of science learning was concept understanding so the students passed in national exam, inquiry teaching need more time, and teachers believe that their students could not yet to follow inquiry-based learning (Adisendjaja, 2010).
Basic abilities that students need to do inquiry are propose a natural phenomenon question, plan and do investigation, use equipment's to collect data, use data to develop explanation, communicate explanation investigation and explanation (NSES, NRC, 1996). These components indicate that

scientific investigation begin with a question then plan investigation to answer question followed by data collecting and analyses to make an evidence-based conclusion.

Science education reform around the world emphasize on science development that consistent to scientific practices and constructivist perspective (Hodson, 1998; Driver et al., 1996; NRC, 1996; AAAS, 1993). The National Science Education Standards [NSES] (NRC, 1996) stated that students must develop understanding what science is and what is not science, science do and not science do, and how science contribute to culture. Without values understanding and assumption of knowledge and its process, learners develop a small image science that consists of isolated facts and avoid contexts that science relevant and can be applied (Lederman, 1998; Schwab, 1962). Reform endeavor should continue to valid science curriculum from philosophical aspects (Hodson, 1988) and develop learner epistemology' of science (Schwartz et al. 2004).

Scientific inquiry refers to characteristics of scientific knowledge development includes consensus in development, receiving, and using of scientific knowledge (Schwartz, et al. 2004). Inquiry needs assumption identification, use logical and analytical thinking and judge alternative explanation (NRC,1996). Scientific inquiry consists of many methods in scientific investigation (myth of scientific method), scientific investigation understanding in relation to present knowledge, argumentation role in new knowledge development and receiving, data anomaly understanding, the difference between data and evidence, group review role and communication in negotiated scientific knowledge in social order (Buxner, 2014).Students should understand scientific inquiry aspects include (Lederman, et al. 2014): 1) scientific investigation do not need test a hypothesis, begins with question, 2) many methods in scientific investigation, 3) scientific procedure is guided by a question, 4) scientists perform the same procedures will not reach the same result, 5) inquiry procedure affects results, 6) conclusion should be consistent to collected data, 7) scientific data different from scientific evidence, 8) explanation is developed of collected data and what has been known.

2. Problem

Based on background information above, the problem is "how teachers understanding's about scientific inquiry before and after teacher professional development?"

3. Method

The impact of Teacher Professional Development(TPD)on teachers understanding' of scientific inquiry is the objective of this research. So the assessment was applied before and after TPD as pretest and posttest. Science teachers' views were explored by modified Views About Scientific Inquiry Questionnaire (VASI) (Lederman, 2009; Lederman *et al.* (2014). The rule of pretest and posttest referred to Lederman, *et al.* (2002), then, all answer were checked then grouped to informed, partially informed and naive category refers to Abd-El-Khalick & Akerson (2009). Some science teachers were interviewed to avoid misinterpretation during checking the answer so can lead to misleading conclusion. After finish check all responses, all answers were classified to informed if the answer parallel to constructivist view' about science; partially informed if part of an answer parallel to constructivist view' but still has unparalleled to science contemporer; naive category if answer was unparalleled to science contemporer view'. Then, each category was percentage.

4. Participant

Twenty five science teachers consist of 23 women and 2 men participated in TPD come from three districts at Bandung area West Java Indonesia. Their experience teaching was between 5-30 years. Twenty teachers graduated from science teachers' bachelors, 2 teachers have master degree in science education, 2 teachers in environmental science, and one teacher from social science. TPD was conducted at Biology Education Department, Indonesia University of Education, in every Saturday for 4 weeks from 08.00 am-17.00 pm exclude homework 2.5 hours each week.

5. Results and Discussion

The results before and after TPD are summarized at table 1. Before TPD, science teachers' knowledge about scientific inquiry was not satisfaction. They did not understand what is scientific investigation, kinds of investigation or not, is a method viewed as a scientific or not, the relationship among question, procedure, and conclusion, the differences between scientific data and scientific evidence.
After TPD, nearly all the percentage of scientific inquiry aspects increased for informed category, except for scientific evidence, method is viewed as a scientific procedure, and data collected to support conclusion did not improve. The increment was around 5% to 60%, and no one aspect of scientific inquiry decreased. The biggest increasing was the relationship between same question and same procedure with conclusion was 60% followed experiment as a method and scientific data were 40%. Other aspects increasing were 5% to 30.0%. The increasing result for informed category after TPD did not mean the decreasing of naive category. The results showed that decreasing nearly in all aspects of scientific inquiry except scientific evidence and information were needed to support their finding. Both aspects in naive category increased 5% and 10% after TPD. It is showed in Figure 1. Even though naive category decreased, it did not mean change to be informed category, part of naive category change to partially informed category.
Some aspects decreased for partially informed were experiment as a method, conclusion for the same question and the same procedure, scientific data and evidence. Two aspects of scientific inquiry increased for partially informed category were scientific investigation (5%) and scientific investigation can followed more than one method (30%), a method was viewed as scientific (75%), scientific question (15%), conclusion for the same question but different procedure (25%), and conclusion consistent with data (15%). Decreasing of naive category followed by increasing of partially informed indicated that respondents have not understood yet about scientific inquiry.

5.1. Scientific investigation

Scientific investigation consists of four questions, no one of respondents give answer in informed category for question 1a, 1c, and 1d. Only 25% respondents answered for 1b question in informed category (Table 1). These results indicate that science teachers have not yet understand about scientific investigation. Respondents answered that an investigation was scientific if followed "scientific method" with orderly steps. All science teachers stated that investigation (in discourse) was not scientific because it did not follow scientific methods and did not do experiment. This statement is a main misconception. A scientific investigation begin with question and answer question through data analysis and then give explanation based on evidence that collected through observation (Bell, Maeng, Peters, 2013). In relation to scientific investigation, only 25% respondents answered correctly to kinds of method. Lederman (2004, 2009)defines experiment if there is a manipulative variable and planned to find out cause and effect relationship. Experiment includes manipulation or observe natural world to test hypothesis (Lawson, 1995).

The first question related to scientific investigation consists of four sub questions, the first three questions can be answered by 70% respondents appropriately. However, they could not give accurate reason. Ninety percent of science teachers did not give answer to the question that was viewed as scientific investigation, so nearly all (90%) respondents belong to naive category. These results showed that science teacher not understand that an investigation is scientific. It means there was a tendency that respondents did not understand characteristics of investigations.To develop scientific understanding of learners, they should have experience and guidance to do explicit reflective model context to topic (Abd-El-Khalick & Lederman, 2000). Ifteachers have understanding and can teach science inquiry-based, they can help students learning inquiry-based science as curriculum suggested. Teachers should help students to learn inquiry through problem-based investigation to learn concepts specific. To train teachers in inquiry, they need actively engaged in laboratory activity to answer a question (McBride et al., 2004).

Table1. Science teachers' understanding on scientific inquiry before and after TPD

No	Aspectof scientific Inquiry	Category (%)					
		Informed		Partially Informed		Naive	
		Pre-	Post-	Pre-	Post-	Pre-	Post-
1.	Is that investigation scientific? (Yes: 90.0%, No: 10.0%)	0.0	10.0	75.0	80.0	25.0	10.0
	Is that investigation an experiment? (Yes: 30,0%; No: 70,0%).	25.0	65.0	40.0	20.0	35.0	15.0
	Is scientific investigation can follow more than one method? (Yes: 85.0%; No: 15.0%).	0.0	15.0	35.0	65.0	65.0	20.0
	Whether that method can be viewed scientific? (No response: 90.0%; Yes: 10%)	0.0	20.0	5.0	80.0	95.0	0.0
2.	Is scientific investigation always begins with scientific question? (Yes: 70.0%; Not: 20.0%; Yes/No: 10.0%).	10.0	25.0	25.0	40.0	65.0	35.0
3.	If question and procedure are the same, are scientists reaching the same conclusion? (Same conclusion: 10.0%; different: 90.0%)	15.0	75.0	45.0	25.0	40.0	0.0
	If questions are the same but procedures different, whether scientists reach the same conclusions? (Same conclusion: 55.0%; Different: 30.0%; Same/Different: 15.0%)	0.0	5.0	15.0	40.0	85.0	55.0
4.	Are scientific data same with scientific evidence? (Same: 5.0%; Different: 95.0%)						
	Scientific data	40.0	80.0	20.0	0.0	40.0	20.0
	Scientific evidence	35.0	40.0	20.0	20.0	45.0	40.0
5.	Questions guide procedure. A better than B (75.0%). B better than A (10.0%). Not A or B (15.0%).	35.0	40.0	20.0	20.0	45.0	40.0
6.	Conclusions must consistent with data collected. (Conclusion A: 95.0%; B: 0.0%; C: 5.0%).	90.0	95.0	0.0	0.0	10.0	5.0
		40.0	45.0	10.0	15.0	50.0	40.0
	reason refers data						
7.	Best construction and reasons	50.0	80.0	40.0	15.0	10.0	5.0
	Information needed	0.0	5.0	55.0	40.0	45.0	55.0

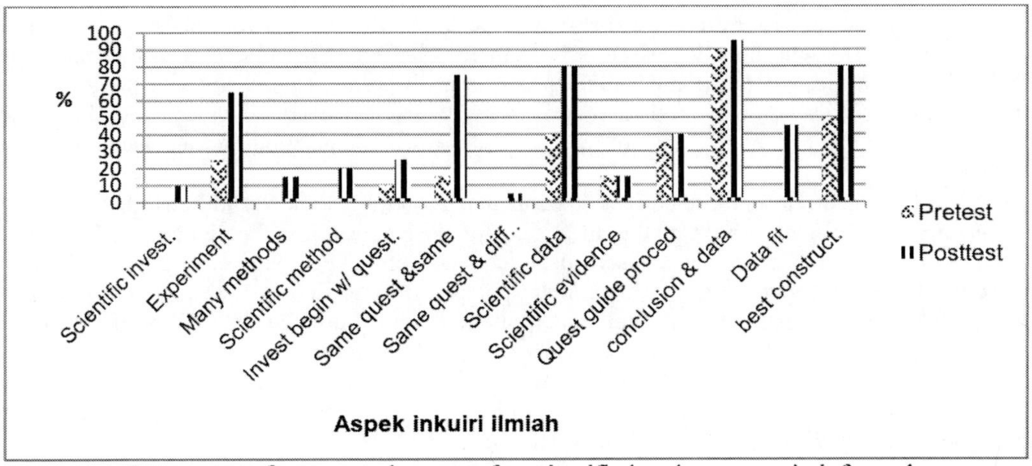

Figure 1. Percentage of pretest and posttest for scientific inquiry aspects in informed category

All aspects of scientific investigation in informed category increased around 10-40%. The biggest increasing was experiment as a method. It is showed by their answered in posttest that investigation in discourse was not an experiment. However, this understanding did not occur to all respondents, only 65% respondents answered in informed category. Respondents still lack of understanding for scientific of investigation and scientific definition. Respondent understanding' about method was limited to only one method, experiment so when facing to other method respondent can not answer. This view as is caused by scientific method understanding, where experiment is one step of scientific method.

5.2. Scientific investigation begins with question

Seventy percents respondents answered that scientific investigation begin with question and 20.0% did not need scientific question and 10.0% answered may begin or not begin with scientific question. From 70% respondents who answered that scientific knowledge begin with scientific question, only 10% included to informed category. Before TPD, only 10% teachers belong to informed category and most of them did not understand the role of question in an investigation.

Respondents' answer on whether scientific investigation begins with question was not satisfied. There was 30% decreasing in naive category from 65% to 35%, however it was not change to informed but it was changed to partially informed category. These results showed that respondents have not understood yet about the role of question in an investigation. Question roles were to identify reasons to do an investigation, to search information and to synthesize of finding and evaluate conclusion (Lewis, 2014). Many of respondents answered that hypothesis must be answered in an investigation, not a question. This view was possibly come from scientific method misconception.

After TPD, 40% respondents answered that investigation begin with scientific question to guide, direct, and focus investigation to reach conclusion. An investigation depends on question proposed (NRC, 2000). Students should have experience to design and to do different investigation include laboratory experiment and field observation (NRC, 2012). By doing these activities, teachers and students will understands that investigation have more than a method and direct experiences.

5.3. The same procedures may not results same conclusion

To the question whether scientists will get a same conclusion if they have same questions and same procedure? Ninety percents respondents answered will not get the same conclusion and only 15% of them belong to informed category. However, if the procedures are different, 55.0% respondents answered that scientists will get the same conclusion, 30% answered will get different conclusion and the rest (15%) may get the same or different conclusion and from those answered no one belong to informed category.

The next question is the relation between procedure and conclusion consists of two questions (Table 1). The first questions increased 60% from 15% before TPD to be 75% in informed category. However, for the second question where the procedures different, respondents have not yet determined the conclusion. It was showed by their answer that only 5% increased. Naive category decreased 30% from 85% to 55%, however it was not increasing informed category but it was changing to partially informed category.

After TPD their reason more correct that conclusion possibly different although their question and procedures were same because of the differences of their education, experiences, ability, social and culture, scientists' point of view, and scientists' interest. BeforeTPD many of respondents have known that conclusion not only depend on data, many factors affect conclusion. Data can be interpreted in many ways so scientists have different interpretation from the same data (Osborne, et al., 2003). Scientists have same data and follow same procedure can reach different conclusion Lederman, et al., 2014). The most reasons that respondent from informed category, stated were the differences of experience, education, and culture. Scientist can be affected by culture, believe, interest, background, previous theory (Ozgelen, Yilmaz-Tuzun & Hanuscin, 2013).

5.4. Procedure affects results

When respondents were asked how the conclusion if question was the same but procedures different,? Their answered are same conclusion were 55%; different conclusion were 30% andsame/different: 15%. These results showed that respondents have not understand yet that procedure will affect conclusion. Procedure was chosen affect results differently. Variable optimization, methods, variable measurement and method of data analysis affect results (Lederman, *et al.*, 2014). After TPD, the increasing for this aspect only 5%, most of answers belong to partially informed. This result indicated that respondents did not understand what factors will affect an investigation conclusion. To improve respondents' understanding on this problem, teachers and student need more exercise in different methods of investigation include laboratory experiment and field observation.

5.5. Procedure inquiry guided by question

To evaluate respondents' understanding about relation between procedure and research question, respondents asked to choose one of two design A or B that agree with research question and they asked to explain their reasons.

Seventy five percent of science teachers exactly chose design Athat presenting in table 1 question no 5, however their reason only 35% belong to informed category. The rest 20% include to partially informed category and 45.0% for naive category. The increasing after TPD was only 5%, from 35% to 40% for informed category. These results indicated that science teacher did not understand how to construct an experimental design and procedures.They could not make a design that appropriate with research question.Even though scientists present different procedure to answer the same question, procedures must answer research question (Lederman, et al., 2014).

5.6. Scientific data and scientific evidence

Ninety five respondents answered that scientific data and scientific evidence are differ, however their answered only 40% and 15.0% belong to informed category for data and evidence respectively. Forty percentsfor data and 70.0% for evidence belong to naive category. After TPD, improvement happened on scientific data from 40% to 80% for informed category. To scientific evidence there was no improvement. These results showed that respondents have not yet understand about scientific evidence. Respondents have not yet understood the differences between scientific data and scientific evidence. It is important to remember that data are theory-laden, it is not separate from previous theory or information (Crowther, Lederman & Lederman, 2005).Therefore the differences of scientific data and evidence should be discussed explicitly and need reflection after practical work finish.

5.7. Conclusion must consistent with data collected

Ninety five percent respondents exactly draw conclusion and 90.0% of this belong to informed category. However, if their conclusion should indicate or referred data only 40% belong to informed category and 50% to naive category. These results indicated that science teachers have not yet understand how to read data or relate data and their conclusion. This finding same as previous result that only half high school students able to identify, describe and explain based on data (Sadler, Chambers & Zeidler, 2004). After TPD, the increasing only 5% from 40% to 45% for informed category. They think that all data must support their conclusion. If there is an anomaly data or a data did not agree with their thinking even though most data are appropriate, they thing data were not support their conclusion. They had a notion that all data must support their conclusion, no data deviate.

5.8. Explanation developed of combination data collected and previous information

For the last question about relationship between investigation results and information were needed to support their finding, all respondent exactly chose conclusion and the best structure of possible *Dinosaurus* fossil. However, if their choice and conclusion relate to their reason only 50.0% of respondents belong to informed category. If their choice relate to what information was needed to support their conclusion, no one gave properly answered. Fifty five of respondents belong to partially informed and 45% belong to naive category. After TPD, increasing only 5% in informed category.

6. Conclusions

Before TPD, respondents' views on scientific inquiry can be summarized as follows, 1) ninety percent respondents did not give answer to the question whether a method is viewed as scientific, 2) ninety percent respondents give answer that even do the same question and the same procedures, scientist will not have the same question, 3) ninety five respondents agree that conclusion should be consistent with data collected.

After TPD, nearly all the percentage of scientific inquiry aspects increased for informed category, except for scientific evidence, method is viewed as scientific, and data collected to support conclusion did not improve. The increment was around 5% to 60%, and no one aspect of scientific inquiry decreased. The biggest increasing was the relationship between same question and same procedure with conclusion was 60% followed experiment as a method of inquiry and scientific data were 40%. Other aspects increasing were 5% to 30%. The increasing for informed category after TPD did not mean the decreasing of naive category.

7. REFERENCES

[1] Abdul E K F and Akerson V 2009 *Int.J.of Science Education* **31(16)** 2161-84.
[2] Abdul E K F and Lederman N G 2000 *Int.J of Science Education* **22 (7)** 665-701.
[3] Adisendjaja Y H 2010 Field Study: Studi Pendahuluan tentang Pembelajaran IPA di SMP. Laporan Penelitian. Tidak diterbitkan.
[4] American A A S 1993 *Benchmarks for science literacy, Project 2061* (New York:Oxford University Press)
[5] Bell R, Maeng J L and Peters E E 2013 *The J.of Mathematics and Science* **13** 5-25.
[6] Buxner S R 2014*J. of Astronomy and Earth Sciences Education* **1 (1)** 53-68.
[7] Colburn A 2000 *An Inquiry Primer.Science Scope* **23** (6) 42-4.
[8] Crowther D T, Lederman N G and Lederman J S 2005 *Science and Children NSTA* **43** 50-4
[9] Driver R, Leach J, Millar R and Scott P 1996 *Young People's Image of Science (*Buckingham: Open University Press)
[10] Hodson D 1988 *Science Education* **72** 19–40
[11] Lawson A E 1995 *Science Teaching and the Development of Thinking* (California: Wadsworth Publishing Company)

[12] Lederman N G 2004 *Syntax of nature of science within inquiry and science instruction* N Lederman (London: Kluwer) pp 301-17
[13] Lederman J S, Lederman N G, Bartos S A, Bartels S L, Meyer A A and Schwartz R S 2014 *J. of Research in Science Teaching* **51**(1) 65-83
[14] Lederman N G, Abd E K F, Bell R L and Schwartz R S 2002 *J.of Research in Science Teaching* **39** (6) 497-521
[15] Lederman J S 2009 *Teaching scientific inquiry: Exploration, directed, guided and open-ended levels* (New York: Brown Publisher) pp 8-20
[16] Lewis K G 2014 *Developing Questioning, Center for Teaching Effectiveness* (Austin:The University of Texas) pp 1-7
[17] McBride J, Bhatti M I, Hannan M A, Feinberg M 2004 *J. of Physics Education* **39** (5) 1-6.
[18] National R C 1996 *National Science Education Standards* (Washington DC: National Academy Press)
[19] National R C 2012 *A framework for K-12 science education: Practices, crosscutting concepts, and core ideas* (Washington DC: National Academy Press)
[20] Osborne J, Collins S, Ratcliffe M, Millar R and Duschl R 2003 *J.of Research in Science Teaching* ,**40** (7) 692-720.
[21] Ozgelen S, Yilmaz T O and Hanuscin, D L 2012 *Resource Science Education* **43** pp 1551-70.
[22] Sadler T D, Chambers F W and Zeidler D 2004 *Int.J. of Science Education* **26** (4) 387-409.
[23] Schwartz R. S, Lederman N G and Crawford B A 2004 *Science Teacher Education* **88** (4) 610-645.

Types of Reasoning in Framing Based Plant Anatomy and It Relation to Spatial Thinking

Ermayanti[1,2], N Y Rustaman[3] and A Rahmat[3]

[1] Student of Science Education Study Program, Graduate School, Universitas Pendidikan Indonesia, Jl. Dr. Setiabudhi No.229, Bandung 40154, Indonesia
[2] Biology Education Program, Universitas Sriwijaya, Jl. Palembang-Prabumulih Km. 32 Ogan Ilir, Palembang 30128, Indonesia
[3] Department of Biology Education, Universitas Pendidikan Indonesia, Jl. Dr. Setiabudhi No.229 Bandung 40154, Indonesia
Email: ema_antik@yahoo.co.id

Abstract. A study about the spatial framing in plant anatomy course was conducted to investigate the role of framing in improving student reasoning in understanding the structure and function of plant and its relation to spatial thinking. A number of biology students (n=35) at State University in South Sumatra, Indonesia was involved as participants in this study. Data was collected using reasoning test based on Marzano framework and spatial thinking instrument test that had been developed and based on expert judgment. Data obtained was processed by calculating the average and percentage (%) on each indicator. Research findings show that there was improvement in reasoning and spatial thinking of students after having experience through framing based learning with an average N-gain 51.9 and 46.6 (moderate category). Learning of framing based plant anatomy course improved the students reasoning at each indicator. Students reasoning before and after learning based framing instruction are: (1) comparing (41.1 and 87.4); (2) classifying (14.7 and 77.1); (3) inducing (28.6 and 64.6); (4) deducing (15.7 and 55.7); (5) analyzing error (21.9 and 40.0); (6) constructing support (33.3 and 54.3); (7) abstracting (31.4 and 34.3); and (8) analyzing perspective (38.3 and 47.1). Analysis of the relationship between reasoning related concepts of plant anatomy and spatial thinking showed r=0.454 (p=0.00*>0.01) (significant correlation). Further it was found that students' spatial thinking including generating a representation (i); maintaining a representations in working memory (ii); scanning the representation (iii); and transforming of representation (iv) are factors that improved student reasoning.

1. Introduction

Plant anatomy is one of the compulsory course for students of Biology Education. Demands of plant anatomy curriculum required students to understand the structure and function of cells, tissues or organs of plants which are three-dimensional (3D) structures. Plant anatomy generally includes a variety of topics: the anatomical structure of the cell, the ground tissue (parenchyma, collenchyma and sclerenchyma), dermal, vascular and secretory tissue, and the organs of plants such as roots, stems, leaves, flowers, fruits and seeds [1]. In addition, students should also be guided to understand the structure of plant anatomy in two-dimensional (2D) and three-dimensional (3D). For this purpose, it is indispensable spatial thinking ability (spatial thinking) for each student [2]. Spatial thinking is associated with problem solving through the coordinated use of space, problem representation and reasoning process3. Spatial thinking allows one to externalize relationship by creating a spatial representation in a variety of media, forms, 2D graphs, tree diagrams or relationships, 3D scale models and other forms of structure3. Spatial thinking can be developed based on: (1) the concept of space; (2) the representation; and (3) the process of reasoning [3],[4].

The concept of space is generally relate to the conceptual framework and data analysis can be integrated, relevant and whole structure. Thinking about the space in various disciplines can be defined in various forms. In this study, the concept of space associated with making a proper incision in location of the tissues to be observed, using a right microscope scale, positioning the object properly with the focus of observation, identifying 3D-plant tissue, drawing or visualizing 2D and 3D tissues (representation). It is important to recognize the position of a tissue among other tissues to recognize the color, shape, size or other attributes, as well as to construct 2D to 3D structure or vice versa. Representation is the ability to describe or stimulate some ideas, concepts or objects [3] and a tool to visualization that potential and understanding were exploited to improve the quality of science education [5]. The spatial representation are important in all branches of biology, where 3D representation are crucial for understanding the phenomena [6]. Thinking spatially uses representation to help remember, understand, reason and communicate about the properties of and relations between objects presented. The spatial thinking process begins by distinguishing and coding of spatial features [3]. Reasoning process includes manipulating process, interpreting and explaining structured information. The powerful features of spatial thinking are to change, manipulate, operate on representation. Some of spatial transformation are: perspectives changing, orientation, the shape change, resize, and reconfiguration [3],[5],[7].

The previous research showed that visuospatial based learning of plant anatomy can improved students reasoning ability [8]. There is a correlation between spatial visualization, logical thinking and concepts of objects with different scales [9]. Moreover, involving students in building models of the 3D structure of the cell will improve the students' understanding on the structure and function of cells [10]. The studies showed that spatial visualization was required more in receiving spatial information. Thus the space visualization involves various spatial ability and the ability for mental rotation, manipulating and rotation as well as the construction of the stimulus 2D into 3D. Meanwhile the representation in this study is the ability of students to make representations, managing representations in working memory, scanning the representation and the ability to transform representation. This is done by using the concepts of plant anatomy. However, some research has not revealed specifically how to improve spatial thinking of students in plant anatomy concepts. How reframe of spatial thinking student that has roles in improving the student reasoning related to the plant anatomy concept.

Research on framing in learning has been done by several researchers. However, the framing in education and the learning process is still very rarely conducted [11]. Framing is an ongoing process that is dynamic, where people continue to frame how to understand "what is happening", in a small adjustment of the scheme [12]. Frame is individual's sense of “what is it that's going on here?” [13]. Framing, was more often used in journalistic discourse, that is related to how the media interpretation and frame cases or events reported. But in education, it was described that the framing as cognitive strategies to sequence and synthesize an information. Framing strategy is a visual arrangement that enables a substantial amount of information to be put in a form of grid, framework, spatial or matrix [14],[15]. Framing involves making connections of main ideas and the relationship between them, it might aid students' organization and comprehension of structural knowledge, and remembering as well [15]. This showed that the framing creates connections of main ideas and relation between them, which may help to organizing and understanding the structure of knowledge and the students remember ability. Framing refers to the process of meta-communicative formation of social reality in the classroom. A learning context has been framed when someone uses meta-communicative signals that help to construct what someone do in it, so that create a "frame" in activities can be interpreted [11],[16].

Some research on framing shows that dynamics of framing on process of practice biology learning and social reality context on biology learning process influence the scientific arguing ability of students [12],[17], resolving the cognitive pressure18, transferring and creating the ability to explain

[16] for the prospective teacher [19]. From these studies, it can be seen that framing can be performed on the learning process in the classroom or in the biology laboratory activities. However, the studies of how framing the situation in the classroom and laboratory activities when students use reasoning and solve problems related to spatial concept of plant anatomy has never been done previously. Therefore problem statement in this research is how to improve the reasoning ability of students to the concept of framing based plant anatomy? This paper focuses on investigating the role of framing to improve spatial thinking in learning plant anatomy. This paper describes how the learning of plant anatomy in the class and the laboratory occur are framed to promote spatial thinking. It was predicted that spatial thinking would be promoted by plant anatomy learning based framing. Spatial thinking will improve student reasoning related to the concept of plant anatomy. Focus of this paper discussion is on how to investigate an overview the role of framing in improving student reasoning in understanding the structure and function of plant anatomy and it relation with spatial thinking.

2. Methods

2.1. Participants
This descriptive research was conducted at the Biology Education Study Program at a State University in South Sumatra, Indonesia. The activity of this study involved 35 students (33 females and 3 males). Participants of this research were the third semester students who take plant anatomy course. Instruction in plant anatomy course was generally consistent with a framing. In most lesson, student receiving information from the lecturers, identification of important concept, verbal representation, scanning 2D and 3D structure, constructing 2D to 3D picture or vice versa, constructing 3D model of plant tissues and refining concept.

2.2. Instrument test and procedure
Two instruments used in this research are spatial thinking instrument and reasoning of plant anatomy concepts instrument. These instruments were developed thoroughly, which were then validated by experts and through field testing. Spatial thinking instrument was specifically designed for the study in the form of multiple choice test items. To give a comprehensive measure spatial thinking, the instrument employed four indicators, such as; (1) generate a representation; (2) manage and maintain a representation in working memory; (3) scan a maintained representation in working memory; (4) transform a representation with rotation or view the object from different perspective [3],[20]. Meanwhile, the reasoning of plant anatomy concepts instrument was developed based on Marzano's framework, that are: comparing ability, classify, inducing, deducing, analyzing an error, constructing a support, abstracting, and analyzing a perspective [21].

Participants were tested with spatial thinking and reasoning of plant anatomy concepts instrument at the beginning and at the end of the framing based plant anatomy learning. Framing based plant anatomy learning used to improve spatial thinking of Biology students on plant anatomy concept. Dynamics of framing was adopted and modified from framing term [3],[13],[16],[18]. Dynamics of framing was specifically designed: concepts questions, spatial related concepts question, direction sentences [22] and also examples of 2D and 3D plant anatomy picture that had been constructed well (worked examples). Dynamics of framing directed the students to think spatially on plant anatomy concepts. The questions used consisted of non-spatial questions about anatomy concept and spatial relating concepts questions. Directing sentences used to guide spatial thinking process of students gradually. Worked examples used to reframe spatial thinking process of students while constructing 2D to 3D or vice versa, transforming representation and imagine the tissues structure from different perspective.

2.3. Data analysis
Test was administered at the beginning and the end of ground tissues concept learning. The Data was analyzed both quantitatively and qualitatively. Quantitative data obtained with calculated the average

or percentage on every indicator of spatial thinking and reasoning related concepts. Criteria of spatial thinking and the reasoning ability in plant anatomy was classified by reference and modified from [23] that is: ≤ 34 (very low); 35-50 (low); 51-65 (medium); 66-80 (high); ≥ 81 (very high). The improvement in reasoning and spatial thinking of student after having experience framing based learning was done with an average N-Gain [24]. Correlation between the scores on spatial thinking and reasoning ability was calculated with pearson correlation coefficients (SPSS 22). Qualitative data was analyzed from observation on framing based plant anatomy learning process.

3. Result and discussions

Research findings show that there was improvement in reasoning and spatial thinking of students after having experience framing based learning with an average N-Gain 51.9 and 46.6 (moderate category). Learning of framing based plant anatomy course improved the students reasoning at each indicator. Students reasoning before and after learning with framing based instruction are in Table 1.

Table 1. Percentage of the reasoning ability of basic tissues.

Indicator of Concepts Reasoning	Pretest	Criteria	Posttest	Criteria
Comparing	41.1	Low	87.4	Very High
Classifying	14.7	Very Low	77.4	High
Inducing	28.6	Very Low	64.6	High
Deducing	15.7	Very Low	55.7	High
Analyzing an error	21.9	Very Low	40.0	Low
Constructing Support	33.3	Very Low	54.3	Medium
Abstraction	31.4	Very Low	34.28	Low
Analyzing a Perspective	38.3	Low	47.14	Low

Note: Very high (75-100); High (67-74); Medium (51-60); Low (35-50); Very Low (≤34)

The data on Table 1 shows that the comparison of students reasoning ability before and after framing based learning instruction indicate most of each indicator moves from very low to high (or even very high), except on the last three indicators (medium for constructing support, low for abstraction, and still low for analyzing a perspective).

These research findings are similar and supported by other previous research findings, such as that framing on process of practical biology learning and social reality context on biology learning process influence the scientific arguing ability of student [12],[17], resolving the cognitive pressure and improve working memory capacity18, transferring and create the ability to explain [16], for the prospective teacher [19].

The higher score of the reasoning ability was comparing indicator (87.4), and the lowest score was abstraction indicator (34.4). The low score at posttest were analyzing an error, abstraction and analyzing a perspective. This study showed that the framing was not fully play role in improving reasoning, especially for that three things. According the observation results in the classroom, showed that student was difficulty in constructing the structure of 2D to 3D or otherwise and transforming representation. These results are related to the managing and transformation representation of spatial thinking indicator on posttest (Table 2). This study investigated the need to promote reasoning ability and spatial thinking. Research study on framing indicated that dynamics of framing on process of learning can resolving the cognitive pressure [18] and improve spatial working memory. In order to help undergraduate students to develop spatial thinking and reasoning ability, it will be designed and repaired dynamic framing. Repaired framing designed will be given at construction and transformation representation steps.

Table 2. Percentage of spatial thinking on basic tissues concept.

Spatial Thinking Indicator	Pretest	Posttest	Criteria
Result in a representation	32.3	85.7	Very High
Managing a representation	20.0	47.4	Low
Scaning a representation	21.1	76.6	Very High
Transforming a representation	13.3	32.9	Low

Note: Very high (75-100); High (67-74); Medium (51-60); Low (35-50); Very Low (≤34)

Based on Table 2, the highest spatial thinking was create a representation ability (85.7) and the lowest was the managing (47.4) and transforming representation ability (32.9). It is mean that on function of the framing test had not trained the construction and transforming representation ability fully. Result of the research was showed that students can't constructing a representation to the full. In plant anatomy, students must managing a representation ability to constructing two dimensional picture into three dimensional picture or otherwise. This is accordance to [3], Managing and maintaining a representation in working memory is students ability in manage the available representation and use it to solve the problem.

In plant anatomy, transforming a representation was created by plant anatomy material needs, rotating, viewing a tissue from different perspectives and from different positions. Students can know the shape, characteristic, and position of a tissue among other tissues from scanning and transforming a representation. Scanning a kept representation on working memory or fast thinking in representation, means to focus the attention on several parts of available representation [3]. In this research, scanning a representation was done by knowing parts of position and characteristic of plant tissues according to available representation. This study showed that most of student was difficulty to create transforming representations. This result have correlation with the low student reasoning at abstraction and analyzing a perspective (Table 1). These results are also supported by some results on previous studies on transformation representation that showed transformation representation related to spatial ability function that is a form of individual mental activity to create spatial image and manipulate a representation (changed with folding, rotating and compressing), to solve any practical problems and theoretical problems [25],[26],[27],[28],[29].

Analysis of the relationship between reasoning related concepts of plant anatomy and spatial thinking showed $r = 0.454$ ($p = 0.00^{*} > 0.01$) (significant correlation). Student with low spatial thinking have low reasoning related concepts, and student with high spatial thinking have high reasoning related concepts. Improving in spatial thinking will improve the reasoning related concepts. This study is appropriate with the previous study that student with low spatial abilities make more error [25], whereas student with high spatial abilities perform better in concept mastery [30],[31].

4. Conclusion

Based on this study show that there was improvement in reasoning and spatial thinking of students after having experience through framing based learning with an average N-Gain 51.9 and 46.6 (moderate category). Learning of framing based plant anatomy course improved the students reasoning at each indicator. Students reasoning before and after learning based framing instruction are (i) comparing 41.1 and 87.4; (ii) classifying 14.7 and 77.1; (iii) inducing 28.6 and 64.6; (iv) deducing 15.7 and 55.7; (v) analyzing error 21.9 and 40.0; (vi) constructing support 33.3 and 54.3; (vii) abstracting 31.4 and 34.3 and (viii) analyzing perspective 38.3 and 47.1. Analysis of the relationship between reasoning related concepts of plant anatomy and spatial thinking showed $r = 0.454$ ($p = 0.00^{*} > 0.01$) (significant correlation). Student spatial thinking including generating a representation (i); maintaining a representations in working memory (ii); scanning the representation (iii); and transforming of representation (iv) are factors that improved student reasoning.

In the study as indicated, most of students have low in managing and transformation representation that influence low reasoning in analyzing an error, abstraction and analyzing in perspective. These activities are necessary to facilitate the framing in constructing and transforming representation. Regarding the result, needs to repaired strategy of framing to improve spatial thinking process, specifically in constructing 2D into 3D or vice versa, and transforming representation with creating view of a tissue from different perspective, rotation, and scanning representation. Analysis of the relationship between reasoning related concepts of plant anatomy and spatial thinking showed significant correlation. Student with low spatial thinking have low reasoning related concepts, and student with high spatial thinking have high reasoning related concepts. This study as indicated, spatial thinking including generating a representation (i); maintaining a representations in working memory (ii); scanning the representation (iii); and transforming of representation (iv) are factors that improved student reasoning.

5. References

[1] Tasmania and Ermayanti 2013 *Syllabus of Plant Anatomy* (Palembang: FKIP Sriwijaya University) pp 1-30
[2] Ermayanti, Rustaman N Y and Rahmat A 2015 *Proc. National Conference Biology Education-2015 (Palembang),* ed Susanti *et al* (Palembang: FKIP Universitas Sriwijaya) pp 248-55
[3] National Research Council 2006 *Learning to think Spatially* (Washington DC: The National Academies Press) pp 10-25
[4] Diezmann C M and Lowrie T 2012 *International Journal of Science and Mathematics Education* **10** 1469-90
[5] Gilbert J K 2005 *Visualization in Science Education* (Netherlands: Springer) pp 1-5
[6] Bolotin M Mand Nashon S M 2012 *Protoplasm* **249** S25-S30
[7] Mathewson J H 1999 *Sci. Education* **83** 33-54
[8] Suprapto P K 2012 *Pengembangan Program Perkuliahan Anatomi Tumbuhan Berbasis Visuospasial melalui Representasi Mikroskopis Sistem Jaringan Tumbuhan untuk Meningkatkan Penalaran dan Penguasaan Konsep Calon Guru Biologi* (Bandung: Disertation, Universitas Pendidikan Indonesia)
[9] Jones M G, Gardner G, Taylor A R, Wiebe E and Forrester J 2010 *Research Science Education* **41** 357-68
[10] Lazarowitz R and Naim R 2013 *Science Education Technology* **22**
[11] Eagle R A 2006 *Journal of the Learning Sciences* **15** 451-98
[12] Berland L K and Hammer D 2012 *Research in Science Teaching* **49** 68-94
[13] Goffman E 1974 *Frame Analysis: An Essay on The Organization of Experience* (Boston: Northeastern University Press)
[14] Van P, Chao J, Reigeluth C M 1986 *Review of Education Reaserch* **56** 437-71
[15] Awofala A O A, Ariqbabu A A and Awofala A A 2013 *Acta Ditactia Napocensia* **6** 1-22
[16] Eagle R A, Nguyen P D and Mendelson A 2011 *Instr. Science* **39** 603-28
[17] Boerwinkel D J, Swierstra T and Waarlo A J 2014 *Science & Education* **23** 485-507
[18] Autin F and Croizet J C 2012 *Experimental Psychology* **141** 610-8
[19] Subramaniam K and Harrell P E 2013 *Sci. Teacher Education* **24** 1177-99
[20] Koslyn S M *Journal of Experimental Psychology: Human Perception and Performance* **4** 47-60
[21] Marzano R J, Pickering D and Mctighe J 1993 *Assessing Students Outcomes: Performance Assessment using The Dimensions of Learning Model* (USA: Association for Supervision and Curriculum Development) pp 67-83
[22] Injeong I 2007 *Aspect of Spatial Thinking in Geography Textbook Questions* (Texas: Ph.D. thesis, Texas A&M University)
[23] Bao L *et al* 2009 *Education Forum* **232** 586-7
[24] Meltzer D E 2002 *American Journal of Physic* **70** 639-54
[25] Hegarty M and Waller D A 2005 *Individual Differences in Spatial Abilities* (New York; Cambridge University Press) pp 121-69
[26] Hegarty M and Waller D A 2004 *Intelligence* **32** 175-91

[27] Koshevnikov M, Motes M and Hergarty M 2007 *Cognitive Science* **31** 549-79
[28] Hegarty M and Kozhevnikov M 1999 *Eduation Psychology* **91** 684-9
[29] Miyake A, Friedman N P, Rettinger D A, Shah P and Hergarty M 2001 *Exp. Psychol. Gen.* **130** 9621-40
[30] Yang E M, Andre T and Greenbowe T J 2003 *Int. J. Sci. Education* **25** 329-50
[31] Hoffler T N 2010 *Education Psychology Rev.* **22** 245-69

Acknowledgements
The author would like to thanks to expert judgment (1. Dr. Hj. Siti Sriyati, M.Si; 2. Dr. Hj. Sariwulan Diana, M.Si.; 3. Dr. Amprasto, M.Si and Dr. Rahmi Susanti, M.Si), that have been helped in this study. We also to thanks to head of Biology Education Program and the laboratory technician of Biology Education Program, FKIP Sriwijaya University, who helped us in this study.

Construction Of Critical Thinking Skills Test Instrument Related The Concept On Sound Wave

F Mabruroh[1*], and A Suhandi[2]

[1]Sekolah Pascasarjana Program Magister Universitas Pendidikan Indonesia, Jl. Dr. Setiabudi No. 229, Bandung 40154, Indonesia.
[2]Departemen Pendidikan Fisika, Universitas Pendidikan Indonesia, Jl. Dr. Setiabudi No. 229, Bandung 40154, Indonesia.

Email: faiza@student.upi.edu

Abstract. This study aimed to construct test instrument of critical thinking skills of high school students related the concept on sound wave. This research using a mixed methods with sequential exploratory design, consists of: 1) a preliminary study; 2) design and review of test instruments. The form of test instruments in essay questions, consist of 18 questions that was divided into 5 indicators and 8 sub-indicators of the critical thinking skills expressed by Ennis, with questions that are qualitative and contextual. Phases of preliminary study include: a) policy studies; b) survey to the school; c) and literature studies. Phases of the design and review of test instruments consist of two steps, namely a draft design of test instruments include: a) analysis of the depth of teaching materials; b) the selection of indicators and sub-indicators of critical thinking skills; c) analysis of indicators and sub-indicators of critical thinking skills; d) implementation of indicators and sub-indicators of critical thinking skills; and e) making the descriptions about the test instrument. In the next phase of the review test instruments, consist of: a) writing about the test instrument; b) validity test by experts; and c) revision of test instruments based on the validator.

1. Introduction

The main purpose of science education is to prepare students to understand the concept and improve the skills of thinking.One form of quality thinking is critical thinking. According to Ennis (1991) critical thinking is reflective thinking that is based on reasoning that is focused on deciding what to believe or do. Someone who thinks critically reasoning is a skilled person and have a tendency to trust and act in accordance with his reasoning. People who have the critical thinking skills to be able to evaluate, discriminate, and determine whether information, ideas of others or his own mind it was right or wrong. She will also be able to find alternative solutions to the problems encountered.

Critical thinking skills was developed in learning process with the assumption that most children can achieve critical thinking and thinking skills in children is always evolving. Similarly, thinking skills can be taught and learned. Based on a literature review of the results of previous studies, some strategies/methods/learning model can be used to construct critical thinking skills of students, among others learning based problem-solving[18],cooperative learning model of STAD *(Student Teams Achievement Division)*[21], learning to use interactive multimedia/virtual simulation [20], the learning cycle *(learning cycle)*[23] and so on. So the selection of strategies/methods/ learning model will greatly determine the success to construct critical thinking skills of students.

Efforts to assess the success of students in developing critical thinking skills must be supported by a measuring instrument that can measure such capabilities.Measurement is an important factor in education because through the measurement, teacher would be known exactly where students at one time in an activity.

Based on previous research the availability of measurement tools that can be used as guidelines in determining the level of critical thinking skills of students, especially in the subjects of physics is rarely, whereas critical thinking skills test instrument need to be developed in all the subject of physics. Previous research has developed a measuring tool critical thinking skills of students in the teaching material temperature and heat [19]. Thus, the present study teaching materials that will be developed is related to sound waves.

The learning process can construct critical thinking skills students will rely heavily on the use of strategies/methods/learning model applied by the teacher. Selection strategies/methods/model will greatly determine the final result of the increase in students' critical thinking skills. Thus, in this study applied learning models that they were problem based learning model, 5E learning cycle model, and the inquiry learning model.

Based on the above, it is necessary to develop the measurement tools of critical thinking skills related to teaching material sound waves to determine the qualification of students' critical thinking. The problems of the present study is, how to construct a critical thinking test instruments on teaching material sound waves that will be tested on high school students grade XII. The purpose of the present study is compiling test instruments critical thinking in teaching material sound waves that will be tested on high school students grade XII.

2. Experimental Method

This study used qualitative descriptive method to examine the characteristics of critical thinking skills instruments. This study tries to construct the critical thinking skills test instrument related the concept on sound waves. This study used mixed method with *sequential exploratory design.* This strategy is implemented with the aim to used the data and quantitative results to help interpreted qualitative findings.Based on the design, procedure to construct test instrument in this study was expressed by this chart below.

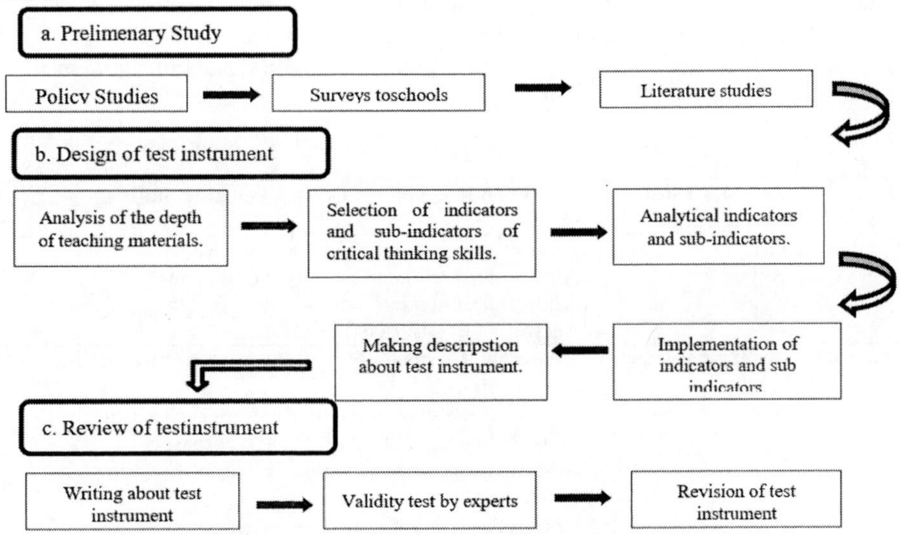

Figure 1. Procedure to constuct test instrument

Critical thinking skills test instrument was written in the form of open-ended with converging pattern because it forms a comprehensive assessment and construction of indicators can capture critical thinking skills [8], [15]. This essay test would be asked to students to express their response and put forward ideas or insights that exist in their minds so as to give a describe of the level of students' critical thinking skills. The level of difficulty that will be tested was varies according to the indicators and sub-indicators of critical thinking skills.

This instrument of critical thinking skills would be tested on four public high schools in Bandung city. Three schools was used learning methods and one school was used traditional method of learning. The students who followed the test was XII grade in each school.

3. Result and Discussion

This research was conducted to obtain critical thinking skills test instruments related the concept on sound waves. Construction of this test instruments using *sequential exploratory mixed method* design consisting of three phases, preliminary studies, designing and review of test instruments. Preliminary study in this research consisted of policy studies, field surveys and literature studies. Preliminary studies conducted by the authors to determine the considered problem. Critical thinking skills including high-level thinking skills required for the flyer to the students, making it one of the learning competencies of physics in the national education curriculum[2].

Based on the results of the study[17],[13] showed that the achievement of students' critical thinking skills, included the low category.This is due to the applied physics teaching method has not stimulate and cultivate students' critical thinking skills. Learning physics tends to be teacher centered, students are not active in building their own knowledge, and does not stimulate the reasoning power and thinking of students.

This is consistent with the results of a field survey conducted by the authors at some schools at Bandung city, through interviews with teachers of physics. The result was during and after the process of learning physics, the critical thinking skills of students have never been assessed by the teacher. Most teachers assess students' mastery of concepts, using test questions which refer to the tests in the National Exam. Most teachers do not understand about critical thinking skills assessment form and procedure. In addition, teachers have not had a test instrument that assess students' critical thinking skills especially physics content. Part of the teacher was not able to develop critical thinking skills instruments due to the limited knowledge on critical thinking skills assessment.

The author conducted the selection of indicators and sub-indicators of critical thinking skills expressed by Ennis (1985) adapted to teaching materials sound waves. The result of the selection of indicators and sub-indicators of critical thinking skills along with details of the number of questions that are designed were presented in Table 1.

Table 1. Indicators and Sub Indicators Critical Thinking Skills in Research

No.	Indicators of Critical Thinking Skills	Sub Indicators of Critical Thinking Skills	Number Problem
1	Elementary clarification	Analyze the argument: identify conclusions	1
		Analyze the argument: identify the reasons	1
		Asking clarifying questions	1
		Answering questions of clarification	1
2	Basis for the decision or basic support	Assessing the credibility of the source criteria: the existing procedure	1
		Assessing the credibility of the source based on the following criteria: the ability to give a reason	1
		Assessing reports observation based on the following criteria: Note of observation	1
		Assessing reports observation based on the following criteria: Competent or the suitability of the technology.	1

3	Inference	Hypothesis explanation: claims of general causal	1
		Induction: Activities of investigation especially aspects of experimental design	1
		Induction: Provide a reasonable assumption	1
		Induction: Generalization of the chart	2
		Make a statement of values, based on: the facts	1
		Make a statement of values, based on: their alternatives.	1
		Make a statement of values: the consequences	1
4	Advanced clarification	Assessing the definition: Definition of the report	1
5	Strategies and tactics	Interacting with others: A coherent strategy	1

Indicators and sub-indicators that have been analysed in order to know meaning of relationship between indicators and sub-indicators. This phase is carried out so that the indicators and sub-indicators of critical thinking skills that are general to be more specific and concrete that can be measured. Descriptions of test questions are based on the analysis and implementation of indicators and sub-indicators of critical thinking skills. Writing test questions refer to the description of the questions that have been created.As an example of the results of test instrument design presented in this below,

Indicator : Elementary clarification
Sub Indicator: Analyze the argument by: identifying the reason that are not explicity stated
Question :
Tora and Heru walk to school that was located in the center of city. When they arrived in front of the school, the ambulance was passed by while sirens, and from the opposite direction through the car of fire engine and sirens also rang. When the two cars passed each other in front of Tora and Heru, they heard a very loud buzzing sound and then it disappeared, and reappeared a few moments are repeated several times until the two cars apart.
Tora asked Heru, *"why is buzz happen?"*. Heru responded the question by saying that they had heard the buzz caused by increased amplitude of the sound waves emanating blend of the two cars.
From the reason ofHeru expressioned, identify the reasons why the sound of sirens can produce a very loud buzz!

The review of test instrument is the validity of the test. Validity of test instrument was performed with requested consideration by experts.Test instruments was validated by five experts consisting of two physics education experts and three physics content experts. The results of the test instrument validation of critical thinking skills presented in Table 2.

IOP Conf. Series: Journal of Physics: Conf. Series **812** (2017) 012056 doi:10.1088/1742-6596/812/1/012056

Table 2. Results of validation by experts

Problem Test Item	Aspects ratings					Suggestions /Improvements
	Suitability indicators / sub-indicators		Use item Problem			
	Corresponding (%)	Not available (%)	Not be Used (%)	Revision (%)	Can be Used (%)	
1	100	0	0	20	80	Add description by picture.
2	100	0	0	20	80	Repair the unit was used
3	100	0	0	0	100	-
4	80	20	0	40	60	Answer clarified
5	100	0	0	0	100	-
6	100	0	0	20	80	Answers clarified
7	100	0	0	40	60	Answer shortened
8	100	0	0	0	100	-
9	100	0	0	40	60	Add pictures
10	100	0	0	0	100	-
11	100	0	0	20	80	Clarified the question
12	100	0	0	20	80	Clarified the question
13	100	0	0	0	100	-
14	100	0	0	20	80	Add explanation of answer
15	80	20	0	40	60	Improvement questions by supporting data
16	80	20	0	0	100	-
17	80	20	0	20	80	Clarified the question
18	100	0	0	0	100	-

Conclusion:

- Validator 1: questions can be used after repair and analyzes each comment in the instrument
- Validator 2: the instrument can be used after the revision.
- Validator 3: the test instrument can be used after revision
- Validator 4: the test instrument can be used after revision the questions and answers should be equipped with a scoring rubric.
- Validator 5: need to be careful about the description conformity with questions and an answer key.

All validators provide that there was no problems that could not be used. Furthermore, the test instrument which consist of 18 items can be prepared to test and test of implementation to obtain the test instruments critical thinking skillsafter it was revised. The recommendation from experts is used to fix the test instrument that has been designed, so that the test instrument is really fit to be used in testing and implementation phase.

4. Conclusion

In this study, it has been prepared 18 items of critical thinking skills test instruments of high school students related concept on sound waves with five (5) indicators and eight (8) sub-indicators of critical thinking expressed by Ennis. Based on the analysis by validator that there was no problem that could not be used, but some questions need to be improved. The test instrument will be used in testing and implementation of test instruments, after it was revised.

5. References

[1] Alfianti, dkk. 2013 *Pengaruh Pembelajaran Kooperatif Model Reciprocal Teaching Dengan Teknik Example Non Example Terhadap Berpikir Kritis Dan Hasil Belajar Siswa (Siswa Kelas XI Man 2 JemberPancaran***2**(3) 187-200

[2] Anonim 2013 *Peraturan Pemerintah Nomor 64 Tahun 2013, tentang Standar Isi Pendidikan Dasar dan Menengah* (Jakarta: Depdiknas)

[3] Arthur L C 1988 *Developing Mind* (Association for Supervision and Curriculum Development: USA)

[4] Azismalayeri *et.al*. 2012 *Journal of Education And Practice*. **3** (10)

[5] Bowell T and Kemp G 2001*Critical Thinking* (New York: Routledge)

[6] Case R 2005 *Education Canada* (Spring 2005) **45**(2) 45-9.

[7] Creswell, John W 2016 *Research Design-Pendekatan Kualitatif, Kuantitatif, dan Mixed* (Yogyakarta: Pustaka Pelajar)

[8] Ennis R H 1993 *Critical Thinking Assessment-Theory into Practice* **32** (3)179-86.

[9] Ennis R H1996 *Critical Thinking* (New Jersey : Prentice-Hall Inc.)

[10] Facione P A 2001 *Test Manual: The Test Of Everyday Reasoning – A Measure of Thinking Skills* (MillbraeCA: Insight Assessment/The California Academic Press)

[11] Fisher A 2008 *Berpikir Kritis-Sebuah Pengantar* (Jakarta : Erlangga.)

[12] Feldman and Daniel A 2010 *Berpikir Kritis* (Jakarta : Indeks)

[13] Greenwald R & Ian J Quitadamo 2014 *The Journal of Undergraduate Neuroscience Education (JUNE), Spring 2014* **12**(2):A100-A06.

[14] Janes SP dkk. 2014 *Kemampuan Berpikir Kritis Siswa SMA Dalam Menyelesaikan Soal Uji Kompetensi Tertulis (UKT) Pada Olimpiade Sains Biologi SMP-SMA Se-Riau 2014.* Jurnal Pendidikan. Tersedia : http://www.unri.ac.id.

[15] Krathwohl D R 2002 *A Revision of Bloom's Taxonomy: An Overview.* Theory Into Practice **41**(4)

[16] Ku K Y 2009 *Assessing students' critical thinking performance: Urging for measurements using multi-response format. Thinking Skills and Creativity* **4**(2009), 70–6.

[17] Manao, H. (2013). *Pengaruh Pendekatan Pembelajaran Contextual Teaching And Learning (CTL) Terhadap Kemampuan Pemecahan Masalah Dan Kemampuan Berpikir Kritis Matematik Siswa Sekolah Menengah Pertama*. Tesis Magister UNIMED. Diakses Maret 2015. Tersedia : http://digilib.unimed.ac.id

[18] Martawijaya MA 2015 *Journal of EST* **1**(2)

[19] Masek & Yamin 2012 *International Scholarly Research Network*. Volume 2012, Article ID 759241, 6 pages.

[20] Ritdamaya, Desti. 2016 *Konstruksi Instrumen Tes Berpikir Kritis Terkait Materi Ajar Suhu dan Kalor*. Tesis Magister pada SPS UPI : Tidak Diterbitkan.

[21] Simon, N 2014 *Proceedings of Society for Information Technology & Teacher Education International Conference 2014* 453-59

[22] Soirwan, dkk. 2013 *Jurnal Pembelajaran Fisika*. **1**(3) Tersedia :www.unila.ac.id

[23] Thompson C 2011 *International Journal of Humanities and Social Science***1**(9) [Special Issue – July 2011].

[24] Udayani, dkk. 2014 *e-Journal MIMBAR PGSD Universitas Pendidikan Ganesha Jurusan PGSD* **2**(1)

The Integration of Javanese Indigenous Knowledge in Biology Learning Resources Development

D Anazifa* and R F Hadi

Student of Biology Education Program, Graduate School of Yogyakarta State University, Indonesia
*Corresponding author: rizqa2011@gmail.com

Abstract. The student's difficulties in learning and understanding Biology concepts are caused by the adoption of scientific phenomenon that not suitable with the environment they live in. Students who comes from the Javanese background sometimes find the Biology concepts hard to understand. Science content that comes from the West sometimes is not suitable with the student's background, because the cultural and geographical background that underlining the science development are different. It can potentially cause the clash in constructing knowledge of students. The proportion of western knowledge and indigenous knowledge has to be balanced, in order to give the scientific rationale of the natural phenomenon that faced by students in everyday life. The ethnoscience experienced by student is still in the form of concrete experience as a result of the interaction with the nature. As one of the largest tribe in Indonesia, Javanese has many unique cultures that can be adopted in science classroom especially in Biology class. The role on ethnoscience in the context of developing Biology learning resources is to connect the science concept with the real world situation. By considering indigenous knowledge as one of learning resources, teachers can start to adjust the Javanese indigenous knowledge into the curriculum. This paper is literature review which will present the background, rationale, and procedure in integrating Javanese indigenous knowledge into Biology classroom as learning resources. The integration of Javanese indigenous knowledge in Biology learning resources development is necessary in order to connect the Biology concept into real situation.

1. Introduction

The development of science and technology has affected to Biology material taught in school. Scientific phenomenon taught in Biology is adapted from the West that sometimes are different with the condition where the students come from. It is caused by the concepts are taken from the West that has different background with the students. The difference between the natural phenonemon presented in Biology teaching and learning with the natural phenomenon that usually faced by students can possibly caused the clash in building knowledge.

Learning Biology should be able to educate student to learn about their soroundings. It is called as contextual learning ehich students learn based on the reality they usually found in everyday life. Students need to know about the natural phenomenon in their environment and also learn about the ethnical prespective or traditonal knowledge that accompanies the phenomenon. But a lot of traditional knowledge is also being lost because of the communication gap, since neither children nor adults spend as much time in their communities. Because it is generally transmitted orally, it is susceptible to change, particularly when people move to new regions, or when people's lifestyles tend to be different from that of their ancestors.

Indonesia is an incredibly diverse country. It consists of many tribes and ethnics. One of the biggest tribes in Indonesia is Javanese tribes. It is centred in Central Java, Yogyakarta, and East Java province of Indonesia, but due to migrations it can also be found in other parts of the Indonesia and even in the other parts of the world. Javanese tribes has a variety of cultures. The majority of Javanese livelihood is farmers. The Javanese farmers has its own method in managing their farmland by using its indigenous knowledge called as *pranotomongso.* This paper will explain about the background, rationale, and procedure in integrating Javanese indigenous knowledge into Biology classroom as learning resources. The integration of Javanese indigenous knowledge in Biology learning resources development is necessary in order to connect the Biology concept into real situation.

2. Method

This study was a qualitative study. Data were collected using library study by analysing books, journals, and other sources from the internet. This study aims to analyse the background, rationale, and procedure in integrating Javanese indigenous knowledge into Biology classroom as learning resources.

3. Discussion

3.1. Indigenous Knowledge

Every culture has its own science and refers to the science in a given culture as its indigenous science [1]. In the last 25 years, biologist, ecologist, botanist, geologist, climatologist, astronomers, agriculturist, pharmacologist, and related working science have labours to develop approaches that are improving our ability to understand and mitigate the impact of human activity upon the environment [1]. Scientist has formalized a branch of biological and ecological science that has known as traditional ecological knowledge (TEK). Indigenous science is the interpretation of the nature phenomenon based on particular culture perspective, whereas traditional ecological knowledge is both the science of long-resident oral peoples and a biological sciences label for the growing literature which records and explores that knowledge [1].

Ogawa defines indigenous science is a culture-dependent collective rational perceiving of reality where collective means held in sufficiently similar form by many persons to allow effective communication, but independent of any particular mind or set of minds [1].Indigenous science, sometimes referred to as ethnoscience, has been described as the study of systems of knowledge developed by a given culture to classify the objects, activities, and events of its given universe [2]. Indigenous science interprets how the local world works through a particular cultural perspective. It has also been referred to as the unique, traditional, local knowledge existing within and developed around the specific conditions of women and men indigenous to a particular geographic area [3].

There are differentiation between formal knowledge and indigenous knowledge. Fromal knowledfe is developed within predominanty western- based education system. It is called as a formal because it is supported by written documents, rules and regulation, and technological infrastructure. On the other hand, the difinition of indogenous knowledge is the knowledge practiced and maintained by people with long histories of close interaction with natural phenomenon. Indigenous knowledge is fundamental to local decision-making regarding daily activities like hunting and gathering, fishing, agriculture, animal husbandry, water conservation, health, etc. Moreover, unlike formal scientific knowledge, indigenous knowledge is generally transferred as oral wisdom from one generation to the other, and is seldom, if ever, documented. Combining all forms of knowledge other than the formal ones as 'indigenous knowledge' would lead to its generalization and oversimplification, and may negate the outstanding contribution local knowledge can make to sustainable development [3].

In the recent years, indigenous knowledge has been applied in various filed including intercropping techniques, pest control, crop diversity, and seed varieties in agriculture; plant varieties, and fish breeding techniques in biology; traditional medicine in human healthcare; soil conservation, irrigation, and water conservation in natural resource management; and oral traditions and local languages in education. The realization of indigenous knowledge's contribution to these sectors has led to an increasing interest in it by academicians, and also policymakers.

3.2 Biology Learning Resources Development

Indigenous knowledge has been incorporated in modern applied sciences such as medicine, architecture, engineering, pharmacology, agronomy, animal husbandry, fish and wildlife management, nautical design, plant breeding, and military and political science. One example that can be used as one of Biology learning resources is pranotomongso system. *Pranotomongso* is a system that regulates the work of Javanese farmers based on the change of the season [4]. *Pranotomongso* consists of two words, *pranoto* which means determination and *mongso* which means season. *Pranotomongso* has been known by Javanesse farmer since longtime ago. They had known the planting time and the rest time for farmland. *Pranotomongso* can be used as a conservation mode in a traditonal way.

Pranotomongso is devided into 12 mangsa in a year There are kasa atau kaji (I), karo (II), katelu (III), kapat (IV), kalima (V), kanem (VI), kapitu (VII), kawolu (VIII), kasanga (IX), kasapuluh (X), dhesta (XI) dan saddha (XII). Pranotomongso is also divided into four big seasons. There are katigo, labuh, rendheng, and mareng. Every each big season is devided into three months. Katigo is devided into kaso (kasa), karo, dan katigo (katiga), while labuh is devided into kapat, kalimo (kalima), kanem. rendheng is devided into kapitu, kawolu (kawalu), kesongo (kasanga), whereas mareng is divided into kesepuluh (kadasa), apit lemah (hapitlemah), apit kayu (hapitkayu). Every months has own characteristic that can be noticed by the natural changing of the environment as a basic to start farming [5].

Pranotomongso is one of indigenous knoeledge relating to the agriculture. The application of pranotomongso shows that Javanese people are connected to the the environment since long time ago. Javanese people see the nature as asubject that they are bent down to teh nature. They also think that the change of the season is determined the actvities they should do in term of agriculture [6].

Pranotomongso had been used as a season determination since Hindu era in Java Island. In that time, people commonly used Saka calendar as a calendar system that made by Aji Saka, the king of Medang Kamulang, so that people called pranotomongso as mongso. As the Islamic era started to spread in Indonesia, Saka was changed into Hijra system that automatically affecting the usage of mangsa system. Javanese people started to leave the mangsa system [7]. Pranotomongso was standardized in June 22th 1856 by Sri Susuhunan Paku Buwana VII in Surakarta that divided into 12 mangsa or 12 months with 23 until 43 days in a month [8]. Pranotomongso was also written in Serat Centini written in 1820- 1822 by Paku Buwono V. It was published in The History of Java and Hien. In that time, pranotomongso was only used by Javanese people around Merapi Mountain and Merbabu Mountain. The standardized of pranotomongso aimed to strengthen the calendar system used to guide the farmer in managing their farmland based on the natural phenomenon that has been divided into mangsa.

Pranotomongso has cosmographic aspect and bioclimatology aspect. Both of them are underpinning the life od Indonesian in social aspect, economics aspect, and cultural aspects. Pranotomongso uses natural phenomenon as farming time direction. There are many natural phenomenon used in pranotomongso, such as plants conditions, bedidig phenomenon, the distinct temperature change in the beginning of dry season, animal behaviours, and astrological signs. The utilization of these natural phenomenon can be used as determinant to start farming.

There are many similarities between prantomongso and Georgian calendar system. The similarities are on the length of sun radiation, average temperature, average humidity, and evapotranspiration throughout the years that are relatively low. The similarities can also be seen through the cosmography method used in both calendar system. Indonesian had started to use this knowledge from the ninth century until Mataram sultanate as a direction in agriculture, economic, administration, and military.

Besides described as cosmography, pranotomongso is also explained as bioclimatology. Mangsa kasa, sotya murca ing embanan, depicts the fallen leaves. The natural phenomenon that can be seen during this season is that the leaves start to fall, the land begins to dry, and the water begins to shrink.

Mangsa Karo, bantala rengka, can be seen as the palawija plants and other plants such as randu and manggo trees start to grow, and the land begins to crack. Mangsa Katigo, suta manut ing bapa, means that lung-lungan plants start to spread. Mangsa Kapat, wasspo kumembang jroning kalbu, means that the sources begin to dry. This is the harvesting time for palawija plant and also the growing time to other plants such as uwi, gadung, kunci.Mangsa Kalimo, pancuran emas sumawur ing jagad, means that the weather begins to cold, tamarin trees start to grow, and animal such as snake and flies start to appear. It is also harvest time for mango. Mangsa Kanem, rasa mulyo kasucen, can be described as fruits season when rambutan, durian, and manggis begin to ripe. Mangsa Kapitu, wisa kentar ing maruto, means that it is a rainy season when farmers begins to plant paddy. Mangsa Kawolu, ajrah jroning kayun, means that it is a happy season when the weather is quite warm and the paddy field turns into green. Mangsa Kasongo, wedare wacana mulya, means that during this season there is a sound of cicada. Mangsa Kesepuluh, gedong minep jroning kalbu, means that it is a harvest time for paddy and nesting and hatching time for birds. Mangsa Dhesto, sotya sinarawedi, means that in this month, farmers start to harvest their paddy, pala pendhem, and tubers. Mangsa Soddha, tirta sah saking sasana, means that the weather begins to cold.

Pranotmongso shows the spirituality that shows respect and appreciation to the earth in which human live. Earh is an subject not an object. Pranotmongso shows us that human can not be sparated to the earth, so that human needs to know the place where they live in including the natural phenomenons that show the natural hange as a sign for the chhange of season. Pranotomongso also shows that Javanesse farmer dares to live from the earth spirituality by catching the natural phenomeneons. It is based on the mutal relationship between human and nature.

Pranotomongso is one of indigenous knowledge from Javanesse tribes that can be used as a reflection to others in order to learn the environmental change. Their understanding of natural phenomenon and natural behaviour is formulated into their argicultural system. Pranotomongso also helps farmes manage thir economic life by looking at their crops in every season. Pranotomongso is also called as ethnoscience that develop in Javanesse tribe. Ethnosisnce studies about the structure used in managing the environment that is consided as a concrete experience and what is important in the certain ethnic. In Javanesse tribe, ethnosicene is still in a form of concrete experience as a result of interaction between human and environment. The development and the utilization of ethoscince can be described as follows, it begins from the interaction between human and envrionment, then it becomes as an etnic knowledge that can be analysed through the ethnoscince prespective. Moderenization and rasionalizaionethnic knowledge through ethnoscince prespective is done afterwards. The next step is the development of ethnic knowledge and followed by the empowerment of ethnic knowledge. [7].

The application pranotomongso in agriculture is necessary that pranotomongso brings a local knowledge that is able to overcome agriculture problems such as the effective use of agricultural land and the high demand for food. Pranotomongso must be reinstated in the right place, so that local knowledge is still be used in the right proportion, such as the integration of pranotomongso in science learning and the guidance for local farmers in managing thier farmlands using pranotomongso.

Indigeous science, pranotomongso, can be used as learning resources in Biology teaching and learning, particullary in developing material for environment theme and ecosystem theme that includes conservation and environmental management. By using pranotomongso as a learning resources, it can teach students about conservation. In pranotomongso, there are environmental management concepts that can educate students to preserve the environmental functions. It can be usefull for farmes to manage and mantain thier plantation based on its characteristic towards the nature. In addition, by implementing pranotomongso, farmers also conducts the intercropping method in order to inhibit the soil damage. It also can improve sustainable production in plants and improve the ecological condition of the land.

Learning using indigenous knowledge is one of the examples of contextual learning. Students will learn through the biological phenomenons that usually happen in their environment they live in. In addition, learning using indigenous knowledge, it gives the opportunities to educators to modify the curriculum based. The indigenous knowlege can be adapted into the curriculum. The curriculum can be developed based on student's backgrounds. It also opens up the opportunities to make innovations in education based on local wisdom. Students can learn based to their own tradition [9].

Indigenous knowledge orlocal wisdom can be embedded and integrated in curricular programs using integrated model the existing curriculum, and or subjects that exist, even learning process. Curricular courses or subjects should be loaded with the values of local wisdom. This model requires the readiness and capability of school, the principal and teachers. School principals and teachers are required to be creative, full of initiative, and also rich in ideas. Teachers and principals must be clever and deft in order to describe the curriculum, to manage learning, and develop assessment. The advantages of this model are cheap and does not add the workload of principals, teachers or learner [10].

4. Conclussion

Pranotomongso, indiginous knowlede form Javanese tribes, can be devoleped as one of the Biology teaching and learning materials. The utilization of pranotomongso aims to preserve local wisdom and educate the students about conservation concepts of natural resources in the agricultural sector. *Pranotomongso* can also be used as a reference for the various natural phenomena that are expected to emerge in response to weather conditions or climate change. By using pranotomongso as lerning resources in Biology teaching and learning, it can edaucate people in uderstanding the biological phenomenons well, because people should be able to adapt with the environmental changing.

5. Reference

[1] Snively G and Corsiglia J 2000 *Discovering Indigenous Science: Implication for Science Education* (USA :John Wiley & Sons, Inc)
[2] Hardesty D L 1977 *Anthropology Hardesty* (New York: John Willey and Sons)
[3] Soni K 2007 *CINE project: Indigenous Knowledge and Importance* (Ahmedabad: Indian Institute of Management)
[4] Daldjoeni 1983 *Penanggalan Pertanian Jawa Pranatamangsa* (Yogyakarta: Proyek Javanologi)
[5] Hilmanto R 2010 *Etnoekologi* (Bandar Lampung: Penerbit Universitas Lampung)
[6] Dumadi J 2011 *Mikul Dhuwur Mendhem Jero: Menyelami Falsafah dan Kosmologi Jawa* (Yogyakarta: Pura Pustaka)
[7] Sarwanto and Budihartini R and Fitriana D 2010 *Identifikasi Sains Asli (Indigenous Science) Sistem Pranata Mangsa Melalui Kajian Etnosains*. Makalah Seminar Nasional Biologi FKIP UNS 2010
[8] Purwadi and Niken E 2007 *Upacara Pengantin Jawa* (Yogyakarta: Panji Pustaka)
[9] Sarwanto 2014 *J. Pendidikan Fisika Indonesia* ***10*** *15-21*
[10] Wagiran 2011 *J. Penelitian Dan Pengembangan* ***III*** *3* 85-100

Acknowledgments

We would like to acknowledge Biology Education Program, Graduate School of Yogyakarta State University for the support

IOP Conf. Series: Journal of Physics: Conf. Series **812** (2017) 012058 doi:10.1088/1742-6596/812/1/012058

Profile of Students' Critical Thinking Skill Measured by Science Virtual Test on Living Things and Environmental Sustainability Theme

N I Maulida[1], H Firman[2] and L Rusyati[1]

[1] International Program on Science Education Department, Universitas Pendidikan Indonesia, Jl. Dr.Setiabudi No. 229, Bandung 40154, Indonesia
[2] Chemistry Education Department, Universitas Pendidikan Indonesia, Jl. Dr.Setiabudi No. 229, Bandung 40154, Indonesia
Email: nur.ida@student.upi.edu

Abstract. The aims of this study are: (1) to investigate the level of students' critical thinking skill on living things and environmental sustainability theme for each Inch' critical thinking elements and overall, (2) to investigate the level of students' critical thinking skill on living things characteristic, biodiversity, energy resources, ecosystem, environmental pollution, and global warming topics. The research was conducted due to the important of critical thinking measurement to get the current skill description as the basic consideration for further critical thinking skill improvement in lower secondary science. The research method used was descriptive. 331 seventh grade students taken from five lower secondary schools in Cirebon were tested to get the critical thinking skill data by using Science Virtual Test as the instrument. Generally, the mean scores on eight Inch' critical thinking elements and overall score from descriptive statistic reveals a moderate attainments level. Students' critical thinking skill on biodiversity, energy resources, ecosystem, environmental pollution, and global warming topics are in moderate level. While students' critical thinking skill on living things characteristic is identified as high level. Students' experience in thinking critically during science learning process and the characteristic of the topic are emerged as the reason behind the students' critical thinking skill level on certain science topic.

1. Introduction

In this era, students are challenged with heaps of very much increasing information in everyday life, and coincidently human experiences and knowledge which are enormously flourishing. Considering diverse resources of information, selecting suitable materials amongst abundant sources is a difficult task for the student. Those challenges require high levels of intellectual and rational skills such as reflection, argument, understanding and evaluation, all of which are essential components of critical thinking [1]. Being focus on critical thinking, Paul also believed that acquisition of critical thinking skills is considered vital for students to face a multitude of challenges of adult life and function effectively in today's increasingly complex world [2]. Therefore, it is necessary for critical thinking to be formed in individuals.

In corresponds of the importance of critical thinking skill, researchers and educators have been working to increase demand of critical thinker by designing instructional program that focus on the acquisition and transfer of critical thinking skills [3]. Many studies have been done in order to find out the way and testing the effectiveness of their findings to be implemented on the lower up to higher education program.

From one side, we can realize that the development of critical thinking skills is often listed as the most important reason for formal education because the ability to think critically is essential for success in the contemporary world where the rate at which new knowledge is created is rapidly accelerating. But however, it also important to realize that development of critical thinking is not only focus on how it is developed and implemented during the learning process. It must also be supported by a testing process. As according to Jacob and Chase, critical thinking must be supported by evaluations or tests that reflect critical thinking, because the evaluation or testing is an integral part of learning [4]. In another supported side, it also known that the developments of learning and teaching process cannot be separated and highly depend on teacher and students' measured critical thinking skills [5]. In this sense, great responsibilities fall to the researcher in order to measure or evaluate students' critical thinking as the object for those developments.

Based on the importance of measuring critical thinking skill asserted on the theories above, this present study provides "Profiling" as the main focus. In academic perspectives, applying profiles is the process of identifying and representing a specific subject or to identify a subject as a member of a specific group or category and taking some form of decision based on this identification and representation [6]. Hence, profiling secondary students' critical thinking skill is expected to be a best way in figuring out the current condition of student's critical thinking skill in science. With current description and specific respondents, the stakeholders or educators in Indonesia can take an actual and on target actions for having a solution or generating critical thinking developments.

In specific, the critical thinking used in this study is adopted from Inch et al which stated that critical thinking embodies interconnected functions of generates purpose, raises question at issue, makes assumption, embodies a point of view, uses information, utilizes concepts, makes interpretation and inference, and generates implication and consequences [7]. At the end, by using Science Virtual Test as the critical thinking measurement, this study aims to find out the level of students' critical thinking skill on living things and environmental sustainability theme for each elements and overall; and to know the level of students' critical thinking skill on living things characteristic, biodiversity, energy resources, ecosystem, environmental pollution, and global warming topics.

2. Experimental method

This present study use descriptive research method, or non-experimental study. This method allows the author to collect and describe the data systematically, factual, and accurate about the characteristic of specific population from literature and in the field [8]. The respondents were 331 students enrolled in seven grade of lower secondary, taken from 5 public schools in Cirebon west Java. The sampling technique used was cluster sampling that allows the researcher to generate a more efficient probability sample which spread over a great distance [9]. Thus, the sample was taken either from city or regency area with the same percentage of respondents.

The instrument was Science Virtual Test by Firman and Rusyati, which has high reliability (Cronbanch's Alpha 0.651) and moderate difficulty level [10]. It contains 28 multiple choice test items that comprised eight elements of critical thinking skill: generates purpose, raises question at issue, makes assumption, embodies point of view, uses information, utilizes concept, makes interpretation and inferences, and generates implication and consequences [7]. In addition, those items was also made based on living things and environmental sustainability theme on VII grade science, encompassed living things characteristic, biodiversity, energy resources, ecosystem, environmental pollution, and global warming topic.

In order to determine the level of critical thinking skill, descriptive statistic data (Mean, Standard deviation, and score range) was analyzed and being categorized. The mean and the standard deviation of the data was used to categorize the students' attainments into low, moderate and high level [11]. Hence, the critical thinking attainments on this present study is categorized on these following score range; low (Score < 43.74), moderate ($43.74 \leq$ score < 74.59), and high (score≥ 74.59).

3. Results and discussion

3.1. Profile of students' critical thinking skill on eight critical thinking elements

The first objective of this study is to investigate the student's critical thinking skill level on overall and each element. The students' critical thinking attainments are summarized on Table 1.

Table 1. Students' attainments in eight critical thinking elements and overall critical thinking score.

Critical Thinking Elements	Mean	SD	Minimum Score	Maximum Score
1. Purpose	60.763	18.8897	0.0	100.0
2. Question at Issue	55.438	24.7846	0.0	100.0
3. Assumption	67.372	33.6025	0.0	100.0
4. Point of View	54.381	29.1884	0.0	100.0
5. Information	54.783	32.3040	0.0	100.0
6. Concepts	59.517	30.0264	0.0	100.0
7. Interpretation and Inference	61.631	38.0478	0.0	100.0
8. Implication and Consequences	59.315	28.8325	0.0	100.0
Overall Critical Thinking Score	59.169	15.4254	14.1	95.3

Table 1 shows that the students' critical thinking skill in generating purpose, raising question at issue, making assumption, embodying point of view, using information, utilizing concepts, making interpretation and inference, and generating implication and consequences are identified at moderate level. In another side, the overall critical thinking skill of the students on living things and environmental sustainability theme is also at moderate level (59.169 on 100 scale). Compared to the mean score on other critical thinking elements, making assumption appears as the highest score attained by the students. Meanwhile, the lowest score attained by the students is lied on embodying point of view.

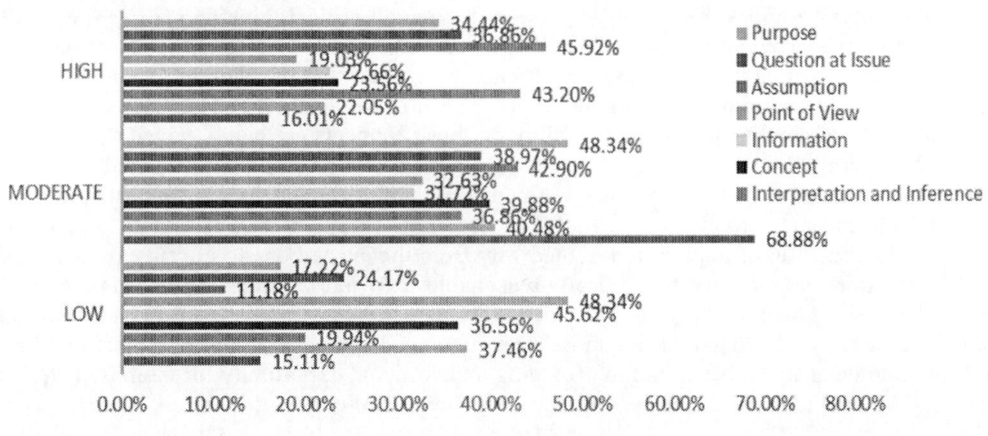

Figure 1. The Comparison of Students Frequency on Three Critical Thinking Attainments Categories for Overall and Each Elements.

In associated with the frequency distribution along defined range of score, the author also represents the percentage of the students' attainments frequency within low, moderate, and high attainments category as presented on the Figure 1. It shows a clear difference in the total student's attainments frequency percentage on three different attainment categories. In the overall critical thinking skill, the highest percentage is lied on moderate category level. It implies that most of the respondents (68.88% from 331) attained 43.74≤ score < 74.59 range in overall critical thinking skill. In line with it, most of the respondents also achieve moderate attainments category in generating purpose, raising question at issues, utilizing the concepts, and generating implication and consequences.

In the other hand, a high frequency on low attainments category is identified on embodying point of view. In another words, most of the students achieved score < 43.74 on the element. In contrast, a high frequency on high attainment category is lied on using information and making interpretation and inference ability. It means that most of the student achieved score ≥74.59 on those elements. It can be said that most of the respondents achieved score ≥74.59 on those elements.

As middle school students, the respondents are expected at least for being "practicing thinkers" [12]. As practicing thinkers students are expected for having enough skill in thinking to critique their own plan for systematic practice, and constructing a realistic critique of their powers of thought. Results of this study presents that students' critical thinking skill on living things and environmental sustainability theme is on moderate category, for the overall score fell above the 43.74 and below 74.59 as the cut off points. A moderate critical thinking category is characterized by mixed level analysis (beginning skill) for becoming knowledgeable of what it would take to systematically monitor the role in student's thinking about concepts, assumptions, inferences, implications, points of view, information, and etc.

Critical thinking is defined by eight elements that make it up. Whenever people think, we think for a certain purpose within a point of view based on assumptions leading to implication and consequences. People use concepts, ideas and theories to interpret data, facts, and experiences in order for answering questions, solving problems, and resolving issues [13]. The theory lead us to realize that a moderate level of students' critical thinking skill in this study is not a coincident thing. Ideally, a moderate overall critical thinking skill in science is also supported by at least at moderate level ability of the constructing eight thinking elements; generates purpose, raises question, uses information, utilizes concepts, makes inferences, makes assumption, generates implication, and embodies a point of view.

In line with the theory above, further exploration on every critical thinking elements attainment on science in this study revealed that the secondary students have moderate attainments in all critical thinking elements. It implies the suitability among the overall critical thinking skill as the result and the attainments on each element of thought ability as the foundation. This condition described 68.88% of the respondents of this study. At the end, it can be realized that the thinking elements are being structured in students mind during thinking process. Each of these structures has implications for the others [13]. Therefore, while the student's ability on those elements are being trained or fostered, it is expected that overall critical thinking skill of the students also can be improved, especially in science.

We noticed that such practice resulted in significant improvement in measures that are closely related to the original task (domain-specific CT measures) [14]. Therefore, science education program should include a specific of implicit redesigned way from the existed one to improve critical thinking skill. A suggestion that might be theoretically reasonable is by inuring the students' to be trained and evaluated for their critical thinking ability through digital technology media, as have been used on this present study; Science Virtual Test. It can be strongly considered since bringing critical thinking and digital technologies may be beneficial in providing an additional opportunity for interested students to achieve higher level of knowing and/or to practice the process of critical thinking skill [15].

3.2. Profile of students' critical thinking skill on science topics

The second objectives of this study is to investigate the level of students' critical thinking skill on the topics embodied on living things and environmental sustainability theme. Figure 2 represents visualization of the student' critical thinking attainments on each topic.

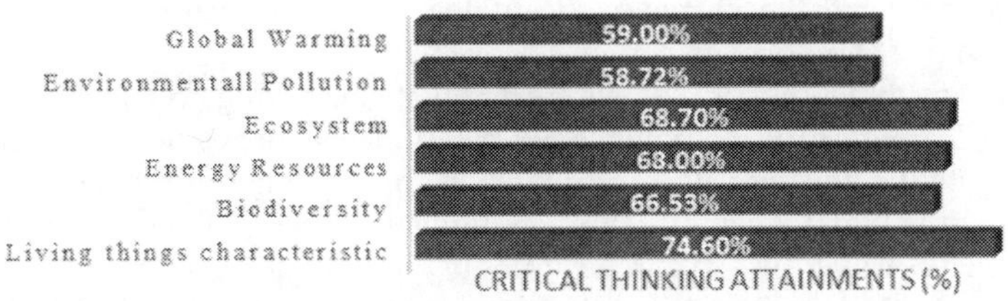

Figure 2. Students' Critical Thinking Attainments on Each Science Topics

Based on Figure 2 above, student's critical thinking skill on biodiversity, energy resources, ecosystem, environmental pollution, and global warming topic are at moderate level, or mixed level analysis (beginning skills). The lowest attainment is identified on environmental pollution topic. In contrast, student achieves highest critical thinking on living things characteristic topic, which also identified at high level, or commendable analysis level (skilled).

In this study, the result revealed that the student's critical thinking skill on biodiversity, energy resources, ecosystem, environmental pollution, and global warming topics are in moderate level. This result presents that science learning process on those elements are fair to support the student in thinking critically. Those topics also provide great opportunity for the teacher to train the students in thinking critically, due to the related popular issues. In another side, students who have high critical thinking skills on living things characteristics This conditions might be happened due to the characteristic of the topic which is common and observable in daily life. Those characteristic might be the reason for the students for having diver experiences. From the standpoint of the development of critical thinking skills some types of students appear to benefit more from engagement in diversity experiences than do other [16].

As the description of student's critical thinking skill on the topics, it is expected that the result above can be useful for the teacher to take it as consideration in developing learning activity. If we take the value of both student' attainments in each critical thinking element and on the science topics, it is important for teacher to train the students in strongly building their element of thought, as the process and foundation of critical thinking in science. It began more important since there is a strongly correlation among content in all subject and thinking. As Paul and Elder said that all contents is nothing more nor less than a mode of thinking (about something), a way of figuring something out, and a way of understanding something through thought [17]. Any subject or "content area" can therefore be understood as a mode of figuring out correct or reasonable answers to a certain body of questions.

In embedding critical thinking within science subject matters, a closer look at the CT instructional approaches adopted by studies included in a systematic reviews also confirm that the majority (over 75%) of the intervention studies targeted either the immersion or infusion approach [14]. This may indicate that embedding CT instruction within specific subject matter domains rather than teaching in separate courses is being considered as a more promising route to help students become critical thinkers.

4. Conclusion

In general, secondary student's critical thinking skill on living things and environmental sustainability theme is categorized on moderate level. it is characterized by mixed level analysis (beginning skill) for becoming knowledgeable of what it would take to systematically monitor the role in student's thinking about concepts, assumptions, inferences, implications, points of view, information, and etc.

Students' critical thinking skill on Biodiversity, Energy Resources, Ecosystem, Environmental Pollution, and Global Warming topics are in moderate level. While students' critical thinking skill on living things characteristic is identified as high level. Students' experience in thinking critically during science learning process and the characteristic of the topic are emerged as the reason behind the students' critical thinking skill level on certain science topic.

Based on the finding and discussion, it is expected that embodying critical thinking elements ability during the science learning process might help the students to get used in thinking critically. A suggestion that might be theoretically reasonable is by inuring the students' to be trained and evaluated for their critical thinking ability through digital technology media, as have been used on this present study; Science Virtual Test. It can be strongly considered since bringing critical thinking and digital technologies may be beneficial in providing an additional opportunity for interested students to achieve higher level of knowing and/or to practice the process of critical thinking skill.

5. References

[1] Nezami N R, Asgari M and Dinarvand H 2013 *Technical Journal of Engineering and Applied Sciences* **3** 2508-14
[2] Paul R W and Binker A J A 1990 *Critical Thinking: What Every Person Needs to Survive in A Rapidly Changing World* (California: Center for Critical Thinking and Moral Critique, Sonoma State University, Rohnert Park)
[3] Halpern D F 1998 *The American psychologist* **53** 449-55
[4] Jacob L C and Chase C L 1992 *Development and Using Test Effectively* (San Francisco: Jossey-Bass Publishers)
[5] Gedik H 2013 *Procedia Social and Behavioural Sciences* **93** pp 1020-024
[6] Ferraris V, Bosco F, Cafiero G, D'Angelo E and Suloyeva Y 2013 *Defining Profiling* (Social Science Research Network)
[7] Inch E S 2006 *Critical Thinking & Communication the Use of Reason in Argument* (United States of America: Pearson Education)
[8] McMillan J H and Schumacher A 2001 *Research in Education* (New York: Addison, Wesley Longman)
[9] Teddlie C and Yu F 2007 *Journal of Mixed Methods Research* **1** 77-100
[10] Firman H and Rusyati L 2015 *Development of Science Virtual Test as Student Centered Software to Measure Student' Critical Thinking Skill* (Unpublished Research Report)
[11] Arikunto S 2010 *Dasar- Dasar Evaluasi Pendidikan* (Jakarta: PT Bumi Aksara)
[12] Paul R and Elder L 2010 *Critical Thinking Development: A Stage Theory* (Tomales: The Critical Thinking Community)
[13] Paul R and Elder L 2007 *A Critical Thinker's Guide to Educational Fads* (Tomales: The Critical Thinking Community)
[14] Tiruneh D T, Verburgh A and Elen J 2014 *Higher Education Studies* **4**
[15] Carmichael E, Craigie D, Driscoll K, Farrell H, James B and Scoufis M 1998 *Critical Analysis – What is It? 2nd Edition* (University of Western Sydney Nepean: Kingswood, Sydney)
[16] Loes C, Pascarella E and Umbach P 2012 *Journal Higher Education* **83** 1-25
[17] Paul R and Elder L 1999 *A Foundation for the Logic of Teaching: Content is Thinking, Thinking is Content* (Online: The Critical Thinking Community)

Acknowledgment

The authors would like to express our sincere gratitude and appreciation for all the parties who have helped in conducting this present study, especially for all the teachers and the students who have been involved during the data collection process.

The Use of Correctional "NO!" Approach to Reduce Destructive Behavior on Autism Student of CANDA Educational Institution in Surakarta

N Anggraini

Magister of Special Education, Universitas Sebelas Maret, Jl. Ir. Sutami No. 36A, Surakarta 57125, Indonesia
Email: nurian.anggraini@gmail.com

Abstract. This research aims to reduce the destructive behavior such as throwing the learning materials on autism student by using correctional "NO!" approach in CANDA educational institution Surakarta. This research uses Single Subject Research (SSR) method with A-B design, it is baseline and intervention. Subject of this research is one autism student of CANDA educational institution named G.A.P. Data were collected through recording in direct observation in the form of recording events at the time of implementation baseline and intervention. Data were analyzed by simple descriptive statistical analysis and is displayed in graphical form. Based on the result of data analysis, it could be concluded that destructive behavior such as throwing the learning material on autism student was significantly reduced after given an intervention. Based on the research results, using correctional "NO!" approach can be used by teacher or therapist to reduce the destructive behavior on autism student.

1. Introduction

Autism is one of several classifications of children with special needs. Autism is considered to be the result of a neurological disorder that affect the function of the brain, it is not caused by psychological factors. At this point, there is no medical test diagnose for this disorder. Any diagnosis is based on observable characteristics, that is behavior that a person is exhibiting. Autism is a developmental disability significantly affecting verbal and nonverbal communication and social interaction. It is generally evident before the ages of three. This disorder affects a child's ability to communicate, understand language, play, and relate to other [1]. In education, the most important thing to note is the level of compliance student to a therapist or teacher. Focus of educational programs for autism students is on improving communication and social, academic, behavioral, and daily living skills. Behavior and communication problems that interfere with learning require the assistance of a knowledge professional trained in autism-specific teaching strategies. There are some characteristics that are indicative of the autism.

Generally, an individual with autism will have struggles in several categories, not just one. The characteristic on the domain of behavior are: Obsessions with objects, ideas or desires; Ritualistic or compulsive behavior patterns (sniffing, licking, watching objects fall, flapping arms, spinning, rocking, humming, tapping, sucking, rubbing clothes); Fascination with rotation; Play is often repetitive; Many and varied collections; Unusual attachment to objects; Quotes movies or video games; Difficulty transferring skills from one area to another; Perfectionism in certain areas; Frustration is expressed in unusual ways; Feels the need to fix or rearrange things; Transitioning from one activity to another is difficult; Difficulty attending to some tasks; Gross motor skills are developmentally behind peers (riding a bike, skating, running); Fine motor skills are developmentally behind peers (hand writing, tying shoes, scissors); Inability to perceive potentially dangerous situations; Extreme fear (phobia) for no apparent reason; Verbal outbursts; Unexpected movements (running out into the street); Difficulty sensing time (Knowing how long ten minutes is or three days or a week); Difficulty waiting for their turn (such as in a line); and Causes injury to self (biting, banging head). Some of this behavior, if not handled properly, will certainly disrupt the continuity of the learning process.

During the learning process for autism student, behavior is an important thing that must be controlled. This is because most autism student have a slightly different behavior than any other kids and if they can not behave properly, then the learning success will not be achieved. Some behaviors that are commonly seen in autism student is like aggressive behavior, destructive / disruptive behavior, improper behavior, self injury behavior, and so on. Basically, the behavior is an act caused by the organism itself. Behavior is the range of actions and mannerisms made by individuals, organisms, systems, or artificial entities in conjunction with themselves or their environment, which includes the other systems or organisms around as well as the (inanimate) physical environment. It is the response of the system or organism to various stimuli or inputs, whether internal or external, conscious or subconscious, overt or covert, and voluntary or involuntary. Aisyah (2015:1) also explains behavior is appreciation intact and person's reaction as a result of both internal and external stimuli that processed through cognitive, affective, and psychomotor.

In other to teaching autism student the classroom environment should be structured so that the program is consistent and predictable, it is also important to develop programs with parents so that learning activities, experiences, and approaches can be carried over into the home and community environment. Lots of strategies or approaches that can be used to change the behavior of autism students, one of which is through correctional "NO!" approach that can be using to reduce or to decrease the destructive behavior on them.

Correctional "NO!" approach is one of several approaches to handling an autism student behavior. The meaning of correctional "NO!" is the pronunciation of the word "NO!" as hard by the tone and volume, firmly, stomping, snapping (light / a little to heavy / strong). Pronunciation "NO!" on correctional "NO!" approach pronounced only once, but if necessary it can be repeated as needed if it is still necessary, but not with pauses (not consecutive, not concatenated). Pronunciation "NO!" May be prolonged for emphasis, and may also need to be accompanied by stomping. Principle of correctional "NO!" approach is do not let student make mistakes until its completed, therefore the therapist or teacher should be alert to deal with the behavior displayed by student. Correctional approach "NO!" Can be applied when the child is doing some bad behavior, such as:

- Aggressive behavior such as hitting, kicking, pinching, pulling, and so on.
- Destructive/ disruptive behavior such as throwing learning material, disrupts teaching material, and so on.
- Improper behavior such as touching others breast, and so on.
- Self-injury behavior such as biting his finger, hit his chest, banged his head against the wall, pinch his self, and so on.
- Endanger situation such as a child wanted to reach a pot of boiling water on the stove, would like striking to the outlet, and so on.

The use correctional "NO!" approach if necessary accompanied by the identification of problems to find the right intervention or behavior modification for specific behaviors.

2. Research methodology

This research was conducted in CANDA educational institution. The research method used is a Single Subject Research (SSR). Single Subject Research (SSR) is popular in the fields of special education and counseling. This research design is useful when the researcher is attempting to change the behavior of an individual or a small group of individuals and wishes to document that change [2]. The research design is the reversal design in form A-B, A for baseline and B for intervention. In this research, there are two variables that become the focus of research, the first is independent variable about correctional "NO!" approach and the dependent variable about destructive behavior such as throwing the learning material. Subject of this research is one autism student of CANDA educational institution named G.A.P. Data were collected through recording the frequency of behavior that occurs in direct observation at the time of implementation baseline (A) and intervention (A). The data that has been collected is summarized in event table for further analysis. Data were analyzed by simple descriptive statistical analysis in visual analysis, that is by describing the influence or change behavior is measured before and after treatment or intervention. Data were analyzed by comparing observation result in the phase of baseline (A) and intervention (B) and it is displayed in the form of a line graph.

3. Results and Discussion

The results of this research can be seen from the number of frequency counts that autism student perform destructive behavior such as throwing learning materials at baseline (A) and intervention (B). Teacher count or record the frequency of occurrence about destructive behavior to measurements during treatment at baseline (A), the destructive behavior observed directly without giving a treatment to student. The results that obtained during the baseline (A) can be seen in Table 1.

Table 1. Data of target behavior at baseline (A).

Date	Time (Start – stop)	Tally of target behavior	Total of tally
09 February 2016	13.35 – 14.55	~~IIII~~ III	8
10 February 2016	13.35 – 14.55	~~IIII~~ ~~IIII~~ II	12
11 February 2016	13.35 – 14.55	~~IIII~~ II	7
12 February 2016	13.35 – 14.55	~~IIII~~ IIII	9
Average			**9**

Subject : G.A.P
Observer : N.A
Time : Start at 13.35 – 14.55 (80 minutes)
Date : 09, 10, 11, 12 February 2016
Target of behavior : Throwing learning materials

At the time of intervention, teachers is giving a treatment when student is doing a destructive behavior such as throwing learning materials by using correctional "No"! approach. The results that obtained during the intervention (B) can be seen in the Table 2.

Table 2. Data of target behavior at intervention (B).

Date	Time (Start – stop)	Tally of target behavior	Total of tally
16 February 2016	13.35 – 14.55	~~IIII~~ II	6
17 February 2016	13.35 – 14.55	~~IIII~~ I	6
18 February 2016	13.35 – 14.55	III	3
19 February 2016	13.35 – 14.55	II	2
23 February 2016	13.35 – 14.55	IIII	4
24 February 2016	13.35 – 14.55	II	2
Average			**3.8**

Subject : G.A.P
Observer : N.A
Time : Start at 13.35 – 14.55 (80 minutes)
Date : 16, 17, 18, 19, 23, 24 February 2016
Target of behavior : Throwing learning materials

This can be seen from the difference of values obtained during the baseline (A) and intervention (B), it seemed that the frequency of occurrence the behavior that measured decline significantly. The average obtained at baseline is 9 and at intervention is 3.8. Based on the description of table that has been described above, researcher concluded that the use of correctional "NO!" approach have an effect to reducing the destructive behavior such as throwing learning materials on autism student.

Thing that often appears in the daily dealing with autism student is a behavioral problem. The disorders of the nervous system that they experienced cause unstable behavior or bad behavior. The main thing in training an autism student is to train the compliance and continued with change bad behavior and maintaining good behavior. The use of correctional "NO!" approach can train autism student's compliance level on therapist or teacher, and at the same time can also change they bad behavior bad habit. Use of the tone is firm and precise expression when applied approach is very helpful in the successful using this approach.

In this study, researchers wanted to change the bad behavior that often displayed by the subjects, it is destructive behaviors such as throwing learning materials. The behavior such as throwing learning materials is disturbing the learning process. Student can't receive the stimulation of learning if he kept throwing learning materials. Therefore, this behavior should get treatment as soon as possible so that the learning process can go on properly, one of the ways to overcome this destructive behavior is to use a correctional "NO!" approach.

The result of using correctional "NO!" approach to reduce destructive behavior such as throwing learning material on autism student is very successful. The analysis showed that the use of this approach to reduce destructive behaviors such as throwing a learning materials on autism student proven to give a good results. Changes can be seen in a decrease in the frequency of destructive behavior that looks at baseline (A) and intervention (B). The data generated can be seen in the Figure 1.

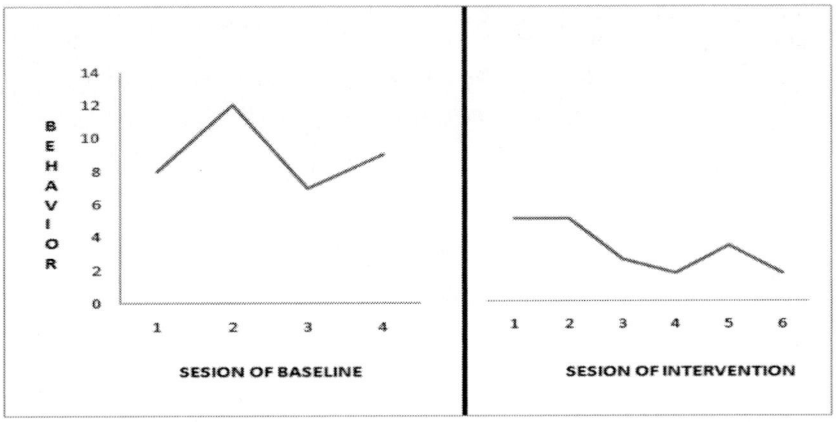

Figure 1. Graph of destructive behavior shown in the baseline and intervention phases.

Figure 1 shows the number of destructive behavior such as throwing the learning material at baseline and intervention phases. At baseline phase, obtained the amount of destructive behavior is quite high, but after being treated at intervention phase by using correctional "NO!" approach obtained reduction in the number of destructive behavior of autism student. Based on the results that obtained during the intervention (B) by using the correctional "NO!" approach it is obtained amount of emergence behavior much less when compared to the baseline (A). It could be a reference rate of successful use of this approach. In this research, the less behavior that appears it will be better anyway, because the thing is being measured is the destructive behaviors that are behaviors that should eliminated.

Giving an intervention by practicing correctional "NO!" approach to autism student when he was doing the bad behavior as destructive behavior managed to decrease the frequency of the number of events behaviors that may appear. It refers to the average value that obtained at baseline was 9 and at intervention was 3.8. A decrease in the number of destructive behavior has become the evidence of the successful use of this approach.

4. Conclusion

Based on the explanation of the research results to reduce the destructive behavior such as throwing learning material on autism student by using correctional "NO!" approach, the researcher obtained data that destructive behaviors such as throwing learning material on autism student decreased after getting a treatment or after given an intervention. This can be seen from the average of destructive behavior at baseline (A) and intervention (B), is from average value 9 to 3.8. Reduced the number of destructive behaviors that can be indicated as the positive effects of the use of correctional "NO!" approach, that destructive behavior such as throwing learning material can be reduces by using correctional "NO!" approach.

5. References

[1] Getskow V and Konczal B1996 *Kids with Special Needs* (California: The Learning Works) p 141
[2] Fraenkel J R and Wallen N E 2006 *How to Design and Evaluate Research in Education* (Boston: McGraw-Hill Companies) p 318
[3] Marliyah (privat communication)
[4] Sutadi R and Anwar A 2015 *Autism is Curable Inshaallah* (Bandung: Litaba) p 31
[5] Peeters T 2009 *Panduan Autisme Terlengkap* (Jakarta: Dian Rakyat)

Investigation of learning environment for arithmetic word problems by problem posing as sentence integration in Indonesian language

NHasanah[1, a)], Y Hayashi [1] and T Hirashima[1]

[1]Department of InformationEngineering, Hiroshima University, Japan
E-mail: nur@lel.hiroshima-u.ac.jp

Abstract. Arithmetic word problems remain one of the most difficult area of teaching mathematics. Learning by problem posing has been suggested as an effective way to improve students' understanding. However, the practice in usual classroom is difficult due to extra time needed for assessment and giving feedback to students' posed problems. To address this issue, we have developed a tablet PC software named Monsakun for learning by posing arithmetic word problems based on Triplet Structure Model. It uses the mechanism of sentence-integration, an efficient implementation of problem-posing that enables agent-assessment of posed problems. The learning environment has been used in actual Japanese elementary school classrooms and the effectiveness has been confirmed in previous researches. In this study, ten Indonesian elementary school students living in Japan participated in a learning session of problem posing using Monsakun in Indonesian language. We analyzed their learning activities and show that students were able to interact with the structure of simple word problem using this learning environment. The results of data analysis and questionnaire suggested that the use of Monsakun provides a way of creating an interactive and fun environment for learning by problem posing for Indonesian elementary school students.

1. Introduction

Arithmetic word problems remain one of the most difficult area of teaching mathematics. Indonesian Mathematics Teaching Monitoring and Evaluation result (P4TK ME) showed more than 50% teachers reported that most students have difficulty in solving word problems, due to not having a clear picture regarding the link between contextual daily state with the corresponding mathematical sentence, and not actively utilizing their mental state in problem solving [1].

Learning by problem posing, suggested as an important way to promote learner's understanding in arithmetic [2], involves generating new problems in addition to solving pre-formulated problems [3]. In Indonesia, practice of problem posing was introduced in National Conferences and Journal of Education [4] to improve the professionalism of mathematics teachers in conducting classroom. The application of new Competence-Based Curriculum (CBC/KBK) in Indonesia in 2004 then started new teaching method called Indonesian Realistic Mathematics Education (IRME/PRMI), in which problem posing method is encouraged as a means to improve students' problem development and problem solving ability [5]. Since then, various researches in problem posing practice has been conducted in Indonesia, reporting increased motivation and active performance in classroom and learning completeness percentage. However, reports of students in accustomed to problem posing task facing difficulty to carry out the learning and teacher needed extra time to evaluate the posed problems were observed [9].

In this study, we present a learning environment for posing arithmetic word problem of simple addition and subtraction called Monsakun, which require students to create problems instead of solving it, and to distinguish between necessary and extraneous information in the process of creating problem. Figure 1 shows the interface of Monsakun. In each assignment learners are provided with a set of sentence cards and requirements of the target word problem, then they pose problem by selecting and arranging appropriate cards. In this activity, learners do not make their own statements of problem, but they are required to distinguish the role of each sentence card and integrate them into problem, a process called "problem-posing as sentence-integration" [7]. Monsakun is developed based on Triplet Structure Model [8] and four schemas/story types or arithmetic word problem [9]. An arithmetic word problem is defined as consisting of two "independent quantity sentences" which describe numbers of objects and one "relative quantity sentence" which describe relation between the independent quantity sentences. Practical uses of Monsakun in several Japanese elementary schools have been reported [7][10] and in-depth analysis of students' log data were also conducted [11][12]. An experimental use of Monsakun for Indonesian and English university students showed their acceptance of our arithmetic model [13].

Our problem posing learning environment provides a way for students to touch and work directly with the structure of arithmetic word problem. Moreover, the dummy/distractor sentence cards give students opportunity to distinguish extraneous and necessary information in a word problem [13], a practice rarely seen in classroom situation. It addresses the issue of assessment and feedback time faced by the teachers by automatic agent-assessment method. The problem of math and language concept comprehension faced by some students is addressed by providing a closed problem space using simple sentence cards, which students can easily select and arrange to create a problem. The closed problem space also enables the recording and analysis of students' log data, which can give insight to their comprehension of arithmetic concepts and processes.

This research investigates the introduction of Triplet Structure Model for arithmetic word problem and Monsakun experimental use by Indonesian elementary school children living in Japan. With respect to the drawback of problem posing practice in Indonesian classroom as explained above and the usefulness of Monsakun practice in Japanese classroom situation, we believe that this software has a potential to provide a meaningful activity for Indonesian teachers as well as students in carrying out problem posing practice from early school grade.

To establish the position of this research in related field, we have investigated 70 literature/researches regarding problem posing in Indonesia in the past 17 years (1999-2015), including conference papers, journal papers, theses, and books. Among them, only 20% researches were conducted in elementary school, showing that this method of learning is not often investigated in low grade school compared to higher school. Moreover, we only found 3 researches of problem posing using interactive software, all of them used in middle/high school. Thus, our contribution is to pioneer the study for the implementation of interactive learning environment for problem posing in elementary school grade.

2. Experimental Method

Our target subjects were ten Indonesian children (age 9-11) who lived in Japan and went to Japanese elementary school. The age range were chosen because they have learned basic mathematics/arithmetic concept of addition and subtraction in Indonesian, and they have sufficient language ability.

The experiment was carried out in one day and divided into four sections. Firstly, students were given an extraneous problem test which consist of 20 simple addition or subtraction word problems. Second, the researcher, under supervision of an expert, conducted a teaching activity to introduce the Triplet Structure Model. Afterwards, students were given a task to pose problems using provided simple sentences. The third section was Monsakun use, followed by questionnaire.

The instruments used in this study are: (1) Extraneous problem test to measure students' arithmetic ability, (2) analysis of Monsakun log data to find out students' problem posing performance, and (3) questionnaire result to investigate students' acceptance of our learning model.

3. Results and Discussion

3.1. Extraneous Problem Test

The first section of this study was extraneous problem test. Students were randomly given one of two types of test consisting of 20 simple addition/subtraction word problems containing extra/superfluous information to be answered in 30 minutes. The material of the test was consulted beforehand to an Indonesian elementary school mathematics teacher as an expert in this field, and both type of the test have the same level of difficulty. Students were asked to read the sentences carefully, crossed out unnecessary sentences, write the corresponding calculation formula and the answer for the problem. An example of the test is provided in figure 3.

Although the instruction was to cross out the extra information, not all students did it well, so a score is given if the final answer is right. The mean score for the extraneous problem test was 70.5% of 20 problems (M = 14.1, SD = 4.84). As this type of word problem is generally not included in the Indonesian textbooks, it was the first time that the students encountered this exercise, and the mean score shows they had enough ability to distinguish extraneous information in simple word problems. As a comparison, extraneous problem test performed by Japanese elementary school students in a study of Monsakun practical use [14] resulted in mean score of 82% for high performance students (n=40) and 14% for low performance students (n=38).

3.2. Teaching of Problem Posing based on Triplet Structure Model

The second section was conducting teaching activity to introduce the concept of simple sentence problems based on Triplet Structure Model, four schema (story type) of arithmetic word problem, and exercise of problem posing in addition and subtraction. The teaching was necessary since according to the expert, Indonesian students in lower grade of elementary school are usually taught to solve word problems by paying attention to specific keywords such as "sum", "total of" for addition or "less", "decreased by" for subtraction, while the Triplet Structure Model put emphasis on the type of sentence and relation between them instead of focusing on keywords.

To begin the teaching activity, the translation from a conventional word problem to simple sentences are introduced. For example, figure 2 shows combination story problem translated into three sentences. Then, two types of sentences in Triplet Structure Model and types of story problem was explained, and then students were asked to construct simple sentences and identify their types. The target of this activity was to give students' understanding of the construction of a simple word problem in our model. Similar to the teaching in Japanese elementary school using Monsakun [10], the explanation was conducted using paper cards and whiteboard. Students were actively involved in this section by answering questions and offering opinions.

After the four story types have been explained, students were given a task to pose problems using provided simple sentences. This exercise aimed to confirm their understanding towards our model. An example of the exercise is shown in figure 4. Fifteen sentences consist of 6 independent quantity sentences and 9 relative quantity sentences was provided. Five sentences have unknown quantity. From the sentences, it is possible to pose four types of story with multiple combination of correct problem. Students were asked to pose one of each, and random students were appointed to explain their answer in front of the class. Students were also actively involved in this exercise.

3.3. Monsakun Experimental Use and Log Data Analysis

In the third section of this study, the Indonesian students used Monsakun to practice problem posing of addition and subtraction. 90 minutes of total time (excluding break time) were allocated for the students to work through all assignments from Level 1 to Level 5. Scene of students' experience with

Monsakun are shown in figure 5. We found that students posed 2,222 problems in total, which averages in 2.46 problems per minute. Without the use of a fixed problem space system like Monsakun, it is thought that students won't be able to pose such many problems in a short time in classroom situation. Furthermore, teachers' problem of needing extra time to grade students' problem exercise in conventional class is solved by automatic assessment system in this learning environment.

After the experiment, the log data was analyzed and presented in figure 6. Level 2 and 4 were omitted from the analysis because they only contain 3 assignments from one story type, while Level 1, 3 and 5 contain 12 assignments from all story types. The average of steps and mistakes at Level 5 is higher than ones at Level 1 and 3. This is the same trends as of Japanese students, which shows that posing problems with reverse thinking problems and calculation formula (Level 5) is difficult for both Japanese and Indonesian elementary school students [16].

The average steps and mistakes show increasing rate from Level 1 to Level 5, a trend also observed in previous practical use of Monsakun in Japanese elementary school students [10]. However, the average number is considered high compared to either the Japanese elementary school students [15]. We assume that the difference came from Indonesian students' short exposure to this new learning method (one experimental session consisting 60 minutes teaching and 90 minutes of system use), while the Japanese students had longer exposure (9 classroom sessions: 35 minutes teaching & 10 minutes system use, totaling 315 minutes teaching and 90 minutes of system use). Nevertheless, the Indonesian students expressed high interest during the Monsakun use.

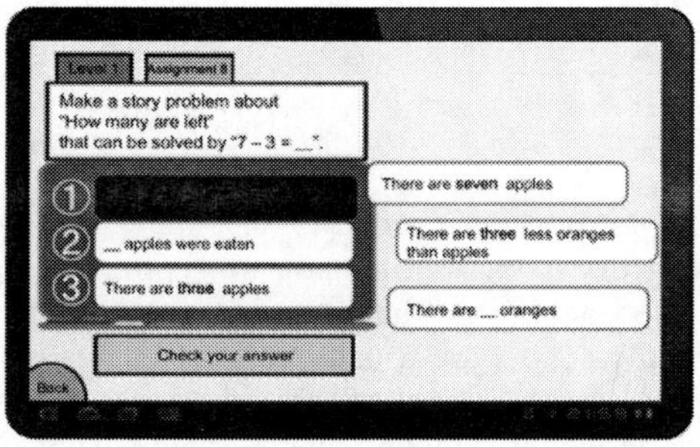

Figure 1. Monsakun interface.

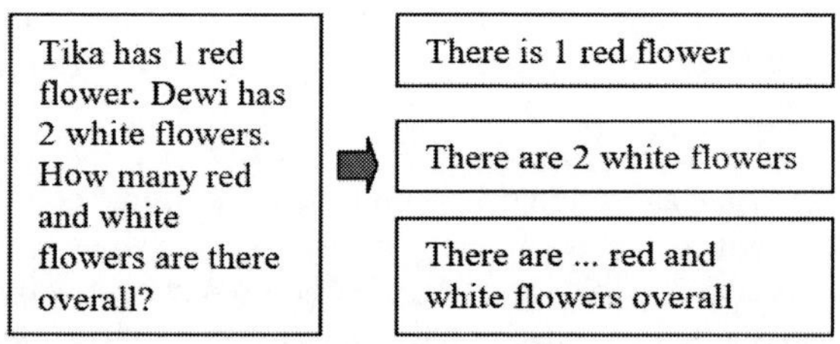

Figure 2. Example of translation from conventional word problem to three simple sentences.

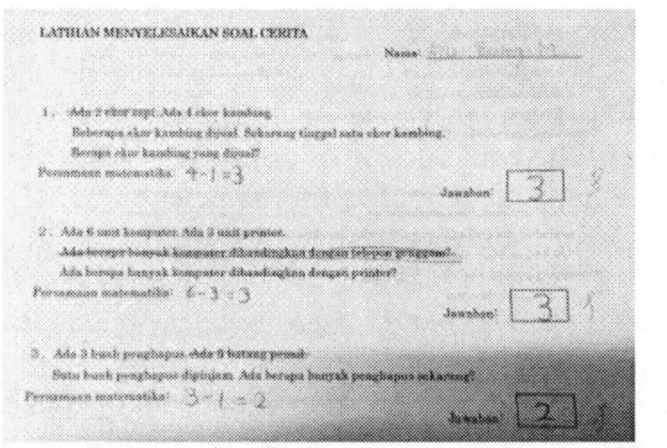

Figure 3. Example of extraneous problem test (in Indonesian language).

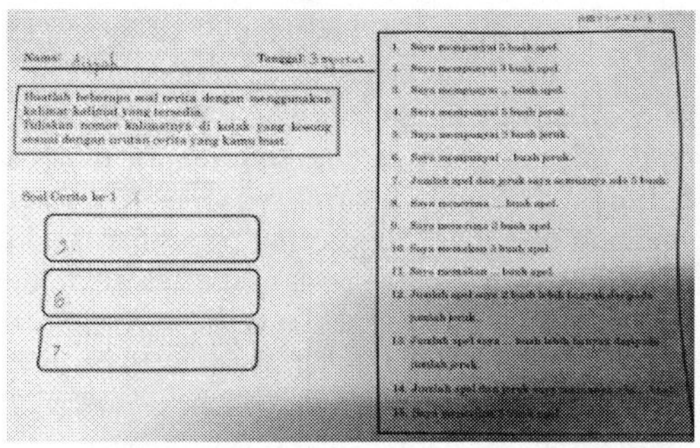

Figure 4. Example of student's answer in a problem posing task (in Indonesian language).

Figure 5. Scene of Indonesian students used Monsakun in Indonesian language.

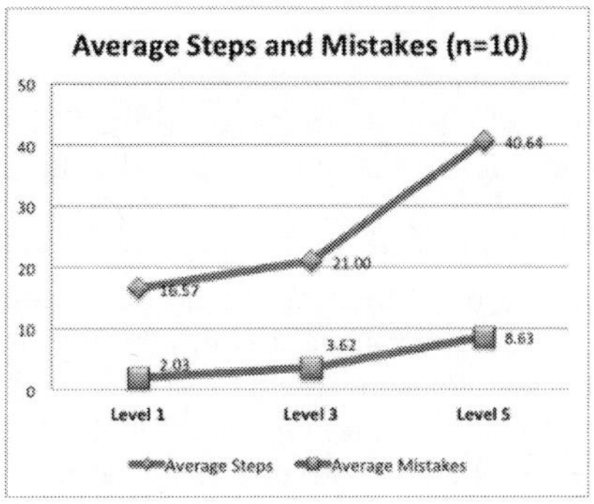

Figure 6. Average steps and mistakes for assignments in Level 1, Level 3, and Level 5.

The conventional teaching method in mathematics conducted one-way by the teacher rarely provides chance for students to become active learners. In contrast to this, we observed students initiating discussion among themselves or with the teacher during this section regarding the assignments and how to pose the correct problem in different story type. We believe that the use of Monsakun encourage this interaction, which in turn promote a deeper understanding about the structure of word problem.

3.4. Questionnaire

Table 1 shows the questionnaire result of Monsakun experience use for Indonesian children (n=10). We observed that most students found posing arithmetic word problems are enjoyable (Q1), however, posing problems were not easy (Q2). Students thought that posing problems are useful for arithmetics (Q3, Q4). Monsakun was fairly easy to use for them (Q5), but they needed more effort to understand the assignments and feedbacks (Q6, Q7). They expressed interests to attend classes that use problem posing activities (Q8). Based on the questionnaire results, we concluded that Monsakun is acceptable and useful as problem posing learning environment for Indonesian elementary school students.

Table 1. Questionnaire about Monsakun use.

	Questions	Strongly Agree	Agree	Weakly Agree	Weakly Disagree
1	Do you enjoy posing arithmetic word problems?	**70%**	30%	0%	0%
2	Are arithmetic word problems easy to pose?	20%	**30%**	**30%**	20%
3	Do you think that posing problems is a good learning method for arithmetic?	**50%**	20%	20%	10%
4	Do you think that posing problems made it easier to solve problems?	**40%**	**50%**	10%	0%
5	Do you think that it easy to use MONSAKUN?	**40%**	**50%**	10%	0%
6	Were the assignments and sentences easy to understand?	10%	**40%**	**50%**	0%
7	Were the feedbacks easy to understand?	**30%**	20%	**30%**	20%
8	Would you like to attend arithmetic classes where problem posing is used?	20%	**30%**	**40%**	10%

4. Conclusion

In this study, we investigated the introduction of Triplet Structure Model for arithmetic word problem and Monsakun experimental use by Indonesian elementary school students living in Japan. In the extraneous problem test, we checked that Indonesian students had good enough ability in solving word problems compared with Japanese students.

The questionnaire result shows that even though posing problem is not too easy, they enjoyed posing problems with Monsakun. Students were able to pose a lot of word problems. This number of problems could not be achieved in usual classroom situation. The averages of steps and mistakes at Level 5 are higher than ones at Level 1 and 3. This trend are also observed in the previous practical use of Monsakun by Japanese elementary school students.

The result of data analysis and questionnaire shows that problem posing by sentence integration based on Triplet Structure Model is accepted by Indonesian students. Furthermore, the use of Monsakun could provide a way of creating an interactive and fun environment for learning by problem posing for Indonesian elementary school students.

The limitation in this study are the small number of subjects and the short investigation/observation time. Ideally, the proposal of a new learning system is preferably done in multiple sessions spread in several weeks, including statistically measurable pre-test and post-test. However, we hope that this research will encourage others to continue, verify, expand, or apply it for further improvement.

5. References

[1] Raharjo M 2008 *Pembelajaran Soal Cerita Berkait Penjumlahan dan Pengurangan di SD* (Yogyakarta: PPPPTK Matematika)

[2] Polya G 1957 *How to Solve it: A New Aspect of Mathematical Method* (Princeton Univ. Press)

[3] Silver EA and Cai J 1996 An analysis of arithmetic problem posing by middle school students *J. Rsrch in Math. Ed.*275 521-539

[4] Suryanto 1998 *Problem Posing dalam Pembelajaran Matematika* (Malang: Program Pascasarjana IKIP)

[5] Suharta I 2003 *J. Pendidikan & Pengajaran IKIP Negeri Singaraja***XXXVI**pp 137-155

[6] Siswono TYE 2004 *Mendorong berpikir kreatif siswa melalui pengajuan masalah (problem posing)* Konf. Nas. Mtk Univ. Udayana Denpasar **XI** pp 23-27

[7] Hirashima T and Kurayama M 2011 *Proc. Artificial Inteligence in Education* vol 6738 (Auckland: Springer) pp 123-130

[8] Hirashima T, Yamamoto S and Hayashi Y 2014 *Triplet structure model of arithmetical word problems for learning by problem-posing* Proc. HCII2014(LNCS 8522) pp 42-50

[9] Riley MS, Greeno JG and Heller JI 1983 Development of children's problem-solving ability in arithmetic *The Development of Mathematical Thinking*,ed Ginsburg H (Academic Press) pp 153-196

[10] Yamamoto S, Kanbe T, Yoshida Y, Maeda K and Hirashima T 2013 *Learning by problem-posing with online connected media tablets* Proc. HIMI/HCII 2013 **III** pp 165-174

[11] Hasanah N, Hayashi Y and Hirashima T 2014 *Revealing students' thinking process in problem-posing exercises: analysis of first sentence selection* Proc. ICCE2014 135-137

[12] Supianto AA, Yusuke H and Tsukasa H 2016 *Rsrch. & Practice in Tech. Enhanced Learning***11.1** 14

[13] Hasanah N, Hayashi Y and Hirashima T 2014 *Utilization analysis of Monsakun in multiple languages as validation of triplet structure model of arithmetical word problems* Proc. SIG-ALST **72** pp 5-9.

[14] Muth D 1992 *Contemp. Ed. Psych.***17** 278-285

[15] Yokoyama T, Hirashima T, Okamoto M and Takeuchi A 2005 *Development of an environment for learning by problem-posing as integration of sentences* Proc. JSAI **19**

[16] Hasanah N, Hayashi Y and Hirashima T 2015 Investigation of students' performance in monsakun problem posing activity based on the triplet structure model of arithmetical word problems*Proc. Int. Conf. on Computer in Education 2015*,ed H Ogata et al (China: Asia-Pacific Society for Computers in Education) pp 27-36

Profile of Prospective Physics Teachers on Assessment Literacy

R Efendi[1*], N Y Rustaman[2] and I Kaniawati[1]

[1]Department of Physics Education, Universitas Pendidikan Indonesia Jl. Dr. Setiabudhi 229 Bandung, Indonesia, 40154
[2]Department of Science Education, Graduate School, Universitas Pendidikan Indonesia Jl. Dr. Setiabudhi 229 Bandung, Indonesia, 40154

Email: ridwanefendi@upi.edu

Abstract. A study about assessment literacy of prospective Physics teachers was conducted with the involvement of 45 prospective physics teachers. Data collected by using test consisted of seven competencies. The profile of prospective physics teachers on assessment literacy determined in descriptive statistics, in the form of respondent average values. Research finding shows that prospective physics teachers were weak at all competency areas. The average values of the Choosing assessment methods appropriate for instructional decisions is the highest average values and the average values of the communicating assessment results to students, parents, other lay audiences, and other educators is the lowest average values. In depth study to detect the reason underlined the results was still in progress so far, as another aspect was planned to be administered on the next semester.

1. Introduction

The term assessment literacy is not commonly know among educator [1]. A simple definition of assessment literacy refer to an understanding of the principles of sound assessment [2]. Previous inadequacies in teacher training in educational assessment highlight the need for teachers to be assessment literate [3].

Simply put literate assessment means: 1) have knowledge of the content and nature are studied; 2) has knowledge of learners and learning and a desire to help learners develop, improve and do their best; 3) have the skills to choose and make the task of assessment; 4) have knowledge of the criteria and standard assessment tasks are appropriate; 5) have the skills and expertise in analyzing and evaluating the use of assessment information; 6) has expertise in giving feedback that is appropriate to the targeted [4].

Two reasons why the concept of literacy assessment into the limelight lately. First, the standards-based improvements have been made expectations of students to learn more clear and increasing need to measure to determine whether students have reached the learning expectations. Second, literacy assessments into a support for using other forms of assessment are different [5]. Although both the attention on expectations of more formal learning and using alternative assessment has increased the burden on teachers and principles to understand how student learning adequately evaluated and what is supposed to provide the resulting information. This study aimed to identify profile of prospective Physics teachers on assessment literacy.

2. Research Method

This study used a descriptive survey research which involved atotal of fourty five prospective physics teachers. The respondents are prospective physics teachers who have passed the course of evaluation of learning. Assessment literacy of prospective physics teacher was collected by using Assessment Literacy Inventory (ALI). The test was adopted and adaptation from Metler[6]. The test consisted of seven competencies, *id est.* (1) choosing assessment methods appropriate for instructional decisions; (2) developing assessment methods appropriate for instructional decisions; (3) administering, scoring, and interpreting the results of both externally produced and teacher-produced assessment methods; (4) using assessment results when making decisions about individual students, planning instruction, developing curriculum, and improving schools; (5) developing valid student grading procedures; (6) communicating assessment results to students, parents, other lay audiences, and other educators; and (7) recognizing unethical, illegal, and other inappropriate methods and uses of assessment information. The profile of prospective physics teachers on assessment literacy determined in descriptive statistics, in the form of respondent average values.

3. Result and Discussion

The profile of prospective physics teachers on assessment literacy obtained from their response to ALI. Based on responses to seven competencies of ALI, data showing the profile of prospective physics teachers can be seen in Table 1.

Table 1. The Profile of Prospective Physics Teachers on Assessment Literacy

No	Competencies	Items	N	Minimum	Maximum	Average	Sd
1	Choosing assessment methods appropriate for instructional decisions	1,8,15,22,29	45	0	5	2,48	1,07
2	Developing assessment methods appropriate for instructional decisions	2,9,16,23,30	45	0	5	1,91	1,03
3	Administering, scoring, and interpreting the results of both externally produced and teacher-produced assessment methods	3,10,17,24,31	45	0	5	2,37	0,97
4	Using assessment results when making decisions about individual students, planning instruction, developing curriculum, and improving schools	4,11,18,25,32	45	0	5	1,76	0,99
5	Developing valid student grading procedures	5,12,19,26,33	45	0	5	1,33	1,08
6	Communicating assessment results to students, parents, other lay audiences, and other educators	6,13,20,27,34	45	0	5	1,26	0,95
7	Recognizing unethical, illegal, and other	7,14,21,28,35	45	0	5	1,78	0,99

Table 1. The Profile of Prospective Physics Teachers on Assessment Literacy

No	Competencies	Items	N	Minimum	Maximum	Average	Sd
	inappropriate methods and uses of assessment information						
		AVERAGE				1,84	1,01

Table 1 shows that in general the average achievement of literacy assessment of prospective physics teachers is low (1.84/5.00) is still far from the maximum standard. These results indicate that "Literacy assessment of prospective physics teachers is still weak". Prospective physics teachers show competence most excellent on competence 1 (M = 2.48 / 5:00) and the lowest in the competence 6 (M = 1.26 / 5:00). This result could be explained by the fact that these prospective physics teachers were surveyed after they have taken their assessment course where they were study varied assessment methods and were provided with the experience to choose assessment methods for assess learning target certainly as requirements in their course. The competence 6 should have been developed by the prospective physics teachers since they had lessons in their course on reporting assessment data. The participans are still prospective teachers, so they have had no hands-on experience in applying the lessons they had for this area of competency.

Results of this study are generally in line with the previous finding [7]. It was Also found that prospective teachers are weak on competency 6, communicating assessment results to students, parents, other lay audiences, and other educators. Thus a need to continue to develop the competence of student teachers of physics in the seventh literacy competency assessment.

4. Summary

Literacy assessment of prosepective physics teacher is still weak in every literacy competency assessment. Of the overall competence, competence chooses the method of assessment is the most excellent competencies possessed by prospective physics teachers. While communicating competency assessment results to students, parents, educators and lay audience are the weakest competence controlled by prospective physics teacher. The achievement of literacy assessment of prospective physics teachers results of this study should be used as input to the education of prospective teachers of physics in developing the literacy assessment in educational programs at the university.

5. References

[1] Eddy 2009 In *OnCUE Journal* **3**(1) 3-25.
[2] Popham, W J 2004 In *Educational Leadership* **62** (1): 82–83.
[3] Popham 2011In *The Teacher Educator* **46**(4) 265-73.
[4] Sadler D R 1998 *InAssessment in Education*, 5, 77-84
[5] Webb N 2002 Assessment Literacy in a Standards-Based Urban Education Setting.
[6] Mertler C A and Campbell C 2005 *The annual meeting of the American Educational Research Association,* Montreal, Quebec, Canada.
[7] Balagtas, Dacanay, Dizon, & Duque 2010 *Basis for Capability Building Program, in The Assesment Handbook* **4**(1) 1-19.

Acknowledgments
This study is supported by scholarship for DIKTI. We thank for all participans who had been contributed in this studied.

The Workshop Program on Authentic Assessment for Science Teachers

N Y Rustaman[1], D Rusdiana[2], R Efendi[2] and W Liliawati[2]

[1] Department of Biology Education, Universitas Pendidikan Indonesia, Jl. Dr. Setiabudhi No. 229, Bandung 40154, Indonesia
[2] Department of of Physics Education, Universitas Pendidikan Indonesia, Jl. Dr. Setiabudhi No. 229, Bandung 40154, Indonesia
Email: nuryanirustaman@upi.edu

Abstract. A study on implementing authentic assessment program through workshop was conducted to investigate the improvement of the competence of science teachers in designing performance assessment in real life situation at school level context. A number of junior high school science teachers and students as participants were involved in this study. Data was collected through questionnaire, observation sheets, and pre-and post-test during 4 day workshop. This workshop had facilitated them direct experience with seventh grade junior high school students during try out. Science teachers worked in group of four and communicated each other by think-pair share in cooperative learning approach. Research findings show that generally the science teachers' involvement and their competence in authentic assessment improved. Their knowledge about the nature of assessment in relation to the nature of science and its instruction was improved, but still have problem in integrating their design performance assessment to be implemented in their lesson plan. The 7th grade students enjoyed participating in the science activities, and performed well the scientific processes planned by group of science teachers. The response of science teachers towards the workshop was positive. They could design the task and rubrics for science activities, and revised them after the implementation towards the students. By participating in this workshop they have direct experience in designing and trying out their ability within their professional community in real situation towards their real students in junior high school.

1. Introduction

The Workshop program on authentic assessment was carried out as follow up of previous research results [1]. The science teachers had seriously supervised the prospective teachers in science teaching during teaching practice at Junior high schools [2]. Nevertheless they did not have enough ability in applying assessment during science instruction as they had less understanding about the nature of assessment compared to the prospective science teachers been supervised as a demand of the 2013 Curriculum in Indonesia [2]. It had been found too that most of science teachers had poor understanding about the nature of science and its component in order to be implemented in their science learning [3; 4]. Another research on prospective science teacher resulted in very limited ability in preparing rubrics [5]. Actually since the end of twentieth century there has been a tendency that standard for classroom assessment need to be improved [6]. Therefore it is very urgent to focus the study into the essential aspect of science epistemology [7], the characteristics of learning science [8] to improve the ability of science teachers regarding the theory and the practice of authentic assessment in science teaching as demanded by any curriculum [9].

In general training for science teachers on instruction in relation to assessment are very rare in Indonesia and many countries as well as in teacher education [10]. Mostly during science instruction, and in science teacher training/workshop, they are being explained how to conduct, but not being modelled or even never being direct experienced by the participants [11]. The workshop program on authentic assessment for science teachers gave theoretical aspect by observational learning model [12]. In this developmental study, the workshop program focused on interactive small group activities. They practice directly in small group of four using cooperative learning approach (think-pair share type). Later after practicing within their own groups how to implement their rubrics planned, they have the opportunity to try out their rubrics towards the junior high school students in order to be completed or revised [13].

By direct experience in planning and implementing their plans in groups, the junior high school science teachers can communicate each other, share ideas, and construct their capacity to enhance performance assessment before, during, and after science instructions [13]. In short they have authentic experience on the essence of the purpose of assessment, namely: assessment for learning, assessment of learning and assessment as learning [6]. The originality of the workshop program on authentic assessment is that: "having direct experience to generate the standard competence into indicators and/or learning objectives in science which includes the nature of science [7] and the nature of assessment in science instruction to empower them with performance assessment in real context with their students among their community [1][8][9][14].

2. Research methodology

2.1. Participants

A number of junior high school science teachers (n=30) as participants, divided into seven groups of 4-5, and 21 seventh grade students from one state junior high school at LPMP Bali Province in Denpasar. The science teacher participants were from nine "districts" in Bali. There were more science teachers with science background in biology major than in physic major, and it is relevant with the number of basic competences to be implemented in seventh grade.

2.2. Instrument and procedure

Data collected through pre-test and post-test results which were taken before and after the workshop. The instrument for the two tests was similar, consists of 20 test items with five statements (strongly agree, agree, neutral, disagree, and strongly disagree). The observation sheet was prepared to assess the quality of the tasks and rubrics planned by the groups of science teachers during the implementation of science activities. Documents from each group were collected after being presented and revised based on other groups' comments or questions as input.

Two sessions presentation from the facilitators was delivered. First session was about nature of science, its instruction and assessment [13], the second session was about formulating indicator and learning objectives to be integrated into lesson plan. Discussion using think-pair-share type of cooperative learning approach was then implemented during the workshop. It was used in sharing experience to formulate the indicators and the learning objectives derived from the basic competence selected from 2013 curriculum for the 7th grade school science. And it was used too in planning tasks in worksheet activity and its rubrics within each group of science teachers.

After having been limited tried out within the group, each science teachers group prepared the equipment and local materials by themselves for the next day try out. During the implementation of their own task and rubrics, the members of each group chose and took role in the try out, such as one member act as the teacher who explains what to do, other members as an observer using their rubrics, as photographer using hand-phone cameras. The number of seventh grade students involved in the workshop trial out was distributed into seven groups of three and faced each group activity for about one up to two sessions of 40 minutes.

The research team members observed the try out session using observation rubrics prepared before and took photographs of each group of students as documentation during the try out. The interactive direct experience of students and the science teachers in groups was then followed up by interactive communication between the science teachers and the students, as well as between the students and the

IOP Conf. Series: Journal of Physics: Conf. Series **812** (2017) 012062 doi:10.1088/1742-6596/812/1/012062

research team, and between the team and the group of science teachers through separated reflection activities which resulted in input for revisions.

2.3. Data analysis
The task in science activities prepared and revised by each group of science teachers were checked and interpreted. The documents obtained from seven groups of science teachers were then analysed each effectiveness using the rubrics. Each group of science teachers' finding using their own rubric was checked and interpreted thoroughly. Later it was scored based on the rubric prepared by the team. The pre-test and post test results were scored individually.

3. Results and discussion
Research findings show that there was improvement science teachers understanding about the authentic assessment in its relationship with science instructions. The plans of each group of science teachers were recorded in Table 1.

Table 1. Plan of science teachers worksheet, rubrics and observation & product from each group.

Group	Topics of YHS Science Activity	Lesson Plan	Alternative Assessment		Observation about	
			Task	Rubrics	SP Skills	Product
1	Classification: Identifying objects	No LS	Not clear	Complete	Obs, com	complete
2	Classification of Living things: Classifying Plants	No LS, complete Worksheet+Obs sheet		Complete	Obs, cla,	in-complete
3**	Chemistry & Physical Changes: separation of mixed materials	No LS	using lab equipment	Simple	Obs.	complete
4	Organization of Life: identifying Cell	Complete	Clear	Complete	Obs, com	incomplete
5	Temperature and its change: Using suitable thermometer	No LS, Worksheet & Observation sheet		Need ex-planation	Obs, select , measure	complete
6**	Interaction of biotic & abiotic factors:	Complete	Not clear	Need ex-planation	Obs, com, int.	incomplete
7	Measuring using specific equipment	No LS, complete Worksheet+Obs sheet		Complex	Obs, com., measure,	complete

Notes: YHS= Junior High School; Obs= observing; **) using sight only during observation
com= communicating; cla= classifying; int=interpretation, conclusion

Based on the information in Table 1, it can be stated that most of the group cannot complete the lesson plans. They just can plan worksheet for science activities and its rubrics. Science process skills mostly are observation, but it still not completes observation, because two of the group still using sight only. Group 2 and Group 7 have worked hard to prepare complete worksheet, observation sheet and its rubrics. Unfortunately they did not have enough time to prepare their lesson plans. They spent most of the time to complete and revise the rubrics.

Table 2. Results of task & try-out rubric for each group.

Group	Clear Descriptor of Science Activity for Students	Observation Result using Rubric*	Achievement on		Scoring guide in Rubrics
			Concept	SP Skills	
1	Clear, limited number of materials to be observed	100% relevant; between 2-3	Moderate	Good	General and complete
2	Clear, supported by materials observed by the students	80% relevant, mostly 3, but time allocation	Good	Good	Specific and complete
3	Separation of mixture	80% relevant, between 2-3	good,	Too simple	Not complete
4	Identifying Cell	100% relevant; between 2-3	Moderate	Obs., draw.	Complete
5	Using suitable thermometer	80% relevant; mostly 3, but target	Good	Limited	Not complete
6	Not clear, especially taking notes about characteristics of living things in nature	40% relevant; only 2-3 for time allocation	Moderate, not focus	Limited	Not specific, no example to fill in
7	Clear, too many objects to measure	100% relevant: mostly 3, but time allocation	Good	Good	very complicated, need interpretation

*designed by the research team

The quality of worksheet and observation sheet developed and achieved by each group of science teachers can be seen in Table 2. In Table 2, the results of using holistic and analytical rubrics are stated as "observation result using rubric". Meanwhile achievement on science concept and scientific skills are also stated qualitatively, as well as the characteristic of scoring guide in rubrics prepared by each group of science teachers. The scoring guide in rubrics was categorized as complete/incomplete; too general, not specific or specific; very complicated.

Data in Table 2 shows that based on "Observation Result using Rubric*" more than 80% the science teacher participants within their group were able to prepare tasks and rubrics properly (except group 6), and the range of score applied to their rubrics were mostly between 2-3 (from 3 scale). It means that the science teacher participants had understanding and ability to plan activity based performance assessment in the form of tasks and rubrics. They began to realize the relationship between assessment and instruction, which means they have understood the purpose of assessment for learning [6].

Working in group among science teachers can remind them that in order to develop performance assessment in real world (authentic assessment) they can develop their professional development in their learning community such as in MGMP [14]. By doing observation upon the science activities conducted by the student, they learn collaboratively and actively [14]. They can realize that there was a tendency that assessment in science education has moved from what is easy to assess towards what is highly value and transferable knowledge /skills [15].

They have direct experience and exploration in preparing task and rubrics for science activity performance assessment. By giving opportunity to their student to have learning experience, and at the same time they learn how to use the rubric. This experience helps them to understand the meaning of assessment for learning. The evidence collected during the authentic assessment workshop supported inferences concerning the attainment of learning objectives [8]. Assessment in this sense act as formative assessment which is used to inform teaching and to influence learning [16].

From the expression of the student during science activity and Forum group discussion (FGD), it was found that they were exciting doing science activities, as it derives them to the concept to master [8]. Further it was found too, that some of the seventh grade students are very creative in using science equipment and very critical in making inference of what they observe through the microscope. So active science need active assessment procedure [8]. it is relevant with the statement that "... authentic assessment creates assessment procedures that are closely modelled on real world tasks and require the same conceptual, physical, and social abilities as the real world task [8]. Further it was stated too that the process of active assessment development consists of five stages: empirical description, definition of aims, tool development, rubric development, and assessment piloting.

The findings that the rubrics were qualitatively assessed, and were not all complete due to the vague (not clear) of the task can be understood, as the assessor should comprehend and practice again-and again in applying alternative assessment [17]. For the first experience, using holistic rubrics is good enough, later they can develop and practice the analytic ones. Assessing scientific skills, especially in measuring objects using specific equipment will need precision and accuracy [9][18][19].

4. Conclusion

Based on the results and research finding as stated in the discussion, and refers to the research objective, it can be concluded as follows. The workshop program on authentic assessment for science teachers is effective to enhance science teacher knowledge and skills in delivering science learning with scientific approach. as well as to implement their task and rubric on science activity plans during the workshop through direct experience. They can improve their experience how to implement scientific approach in learning science, by directing the small group of students to conduct science activities and assessing their performance (skills and product such as worksheet fill-in report) and come to conclusion expected. Furthermore the workshop program is effective in enhancing science teacher ability in planning task and rubrics through direct experience and reflective activity. They work within the group that challenge their mind to comprehend new skills embedded with science knowledge which has epistemology meaning through social interaction and construction.

5. References

[1] Rustaman N Y, Rusdiana D, Efendi R and Liliawati W 2015 *Proc. Int. Seminar on Mathematics, Science, and Computer Science Education (MSCEIS) II (Bandung)* (Bandung: FPMIPA UPI)

[2] Rustaman N Y, Rusdiana D and Efendi R 2015 Studi tentang sikap dan kompetensi asesmen otentik guru sains dan pengembangannya *Laporan Penelitian Unggulan Pergurutan Tinggi Tahun ke 1*

[3] Adisendjaja Y H 2010 Field study: Studi pendahuluan tentang pembelajaran IPA di SMP *Laporan Penelitian*

[4] Adisendjaja Y H, Rustaman N, Satori D and Sri Redjeki 2013 Pengaruh pembelajaran berbasis hakikat sains terhadap penguasaan konsep keterampilan proses sains, kemampuan berpikir kritis dan persepsi siswa terhadap hakikat sains *Laporan Penelitian Mandiri*

[5] Efendi R and Rustaman N Y 2015 *Proc. International Seminar (MSCEIS) II (Bandung)* (College Park: American Institute of Physics)

[6] Black P and William D 1998 *Inside Black Box: Raising Standards The Classroom Assessment* vol 80 (Phi Delta) pp 139-48

[7] McComas *et al* (ed) 2000 *The Nature of Science in Science Education: Rationales and Strategies Dordrecht* (The Netherlands: Kluwer) pp 83-120

[8] Hanauer D I, Hatfull G F and Jacob-Sera D 2009 *Active Assessment: Assessing Scientific Inquiry* (New York: Springer-Science-Business Media)

[9] Doran R, Chan F, Tamir P and Lenbart C 2002 *Science Educator's Guide to Laboratory Assessment* (Arlington: NSTA press)

[10] Alkharusi K and Al-Musawai 2011 *Asia-Pasific Journal of Teacher Education* **39** 113-23

[11] Adisendjaja Y H 2016 *Pandangan Mahasiswa Calon Guru Biologi dan Guru IPA Peserta Pengembangan Profesional Guru tentang Hakikat Sains dan Inkuiri Imiah* (Bandung: Disertation UPI)
[12] Bandura A 1977 *Social Learning Theory* (New Jersey: Prentice-Hall)
[13] Rustaman N Y, Rusdiana D, Efeendi R and Liliawati W 2016 Panduan workshop asesmen otentik bagi guru sains *LPMP Provinsi Bali* (Denpasar: Draft)
[14] Roberts S M and Pruit E Z 2009 *Schools as Professional Learning Communities: Collaborative Activities and Strategies for Professional Development* 2nd Edition, ed D Alpert and C Harris (Thousand Oaks: Corwin Press)
[15] National Research Council (NRC) 2011 *Classroom Assessment and the National Science Education Standards 1 in Atkin j m, Black p and Coffey* (Eds.) (Washington DC: The National Academy Press)
[16] National Research Council (NRC) 1996 *Assessment in Science Education in National Science Education Standard* (Washington DC: National Academy Press)
[17] Zainul A 2001 *Alternative Assessment* (Jakarta: Pusat Antar Universitas, Dirjen DIKTI, Depdiknas)
[18] Koh H K 2011 *Teaching Education* **22** 255-76
[19] Mertler C A 2009 *SAGE Publications* **12** 101–13

Acknowledgments
This research is funded by KEMRISTEK DIKTI in Hibah Unggulan Perguruan Tinggi (HUPT) of the fiscal year 2016 (Nomor: 048/SP2H/LT/DPRM/II/2016). We also thank to Dr Ramlawati, M Si from Universitas Negeri Makasar, Dr A A Rai Istri Sudiatmika, M.Pd from Universitas Pendidikan Ganesa for their active participation as facilitators, Dr. I Made Alit Mariana, M Pd from LPMP Provinsi Bali, science teachers and students who have provided valuable comments.

IOP Conf. Series: Journal of Physics: Conf. Series **812** (2017) 012063 doi:10.1088/1742-6596/812/1/012063

Experiment of Enzyme Kinetics Using Guided Inquiry Model for Enhancing Generic Science Skills

N Amida[1], F.M.T. Supriyanti[2], and Liliasari[2]

[1]SekolahPascaSarjana, UniversitasPendidikan Indonesia, Dr.Setiabudi Street No 229, Bandung 40154, Indonesia
[2]Departemen Pendidikan Kimia, UniversitasPendidikan Indonesia, Dr.SetiabudiStreetNo 229, Bandung 40154, Indonesia

Abstract. This study aims to enhance generic science skills of students using guided inquiry model through experiments of enzyme kinetics. This study used quasi-experimental methods, with pretest-posttestnonequivalent control group design. Subjects of this study were chemistry students enrolled in biochemistry lab course, consisted of 18 students in experimental class and 19 students in control class. Instrument in this study were essay test that involves 5 indicators of generic science skills (i.e. direct observation, causality, symbolic language, mathematical modeling, and concepts formation) and also student worksheets. The results showed that the experiments of kinetics enzyme using guided inquiry model have been enhance generic science skills in high category with a value of <g> average of 0.77. Four indicators classified in the high category are direct observation, causality, symbolic language, and mathematical modeling with the value of <g>0,73; 0,70; 0,96; dan 0,85. Meanwhile, indicator of concepts formation in the medium category with a value of <g> 0.62

1. Introduction

Inquiry is the process of discovering knowledge based on search and discovery through a systematic process of thinking. Inquiry can beoccured in the process ofstudents learning and experiment. Inquiry conducted in the experiment can improve skill of thinking, concepts understanding, skill of cognitive, argue ability, learning active[1,2,3]and reduce misconceptions[4]. One of experiment performed by students in biochemistry lab. The purpose of this course are understanding of the basic concepts of biochemistry and develop laboratory skills. Based on the results of interviews conducted with a professor biochemistry at one of the universities at Bandung, current experiments with semi-open inquiry model. It means using local materials in the optimization of enzyme activity to conducted the best reagents and conditions. However, it was difficult for the students, because of the lack of experience in open inquiry and not all students were able to understand the purpose of the experiment. Students tend to remember the work steps contained of previous student reports. Therefore, there were still many students who need to be guided, them it was used inkuri guided models. Guided Inquiry Model develop analytical skills, cognitive process and social processes better than the open inquiry[5]. Stages of guided inquiry lab are observation, manipulation, generalizations, and applications[6]. Stage of guided inquiry will create an environment that motivate students to active learning by providing opportunities for them to build meaning and find the concept, therefore establishing a framework of an order high thinking skills, experiments in laboratory[7,8,9,10]. Framework for higher-order thinking skills, also known as scientific thinking or generic science skills.

Generic science skills are abilities of intellectual combination result of complex interactions between science knowledge and skills[11]. Generic science skills can be used to study various concepts and solve of problems in science[11]. Therefore, generic science skills are commonly used in a variety of scientific work, and can be used as a basic in doing experiments. To identify generic science skills, according to the indicators formulated by Brotosiswoyo[13]. Indicators generic science skills included, 1) direct and indirect observation, 2) sense of scale, 3) symbolic language, 4) framework logic obey the principle, 5) logical inference, 6) causality, 7) mathematical modeling and 8) concepts formation.
Enzyme kinetics experiment were used in this study. Enzyme kinetics associated with qualitative test of the rate reaction catalyzed by enzyme. The enzyme used in the experiment was polyphenol oxidase (PPO). PPO accepted from natural materials known as local materials. Eggplant was used as a local materials. In the qualitative test of enzyme kinetics can use guided inquiry model. This study aims to enhance generic science skills of students using guided inquiry model through experiments of enzyme kinetics. Indicators of generic science skillstested were direct observation, causality, symbolic language, mathematical modeling, and concepts formation.

2. Research Methodology

This study used quasi-experimental methods, with pretest-posttest nonequivalent control group design. Subjects of this study were chemistry students enrolled in biochemistry lab course, consisted of 18 students in the experimental class and 19 students in control class. Instrument in this study were essay test and student worksheets.The essay test was used to measure the generic science skills of students in enzyme kinetics experiments. The research procedure is describe in Figure 1.

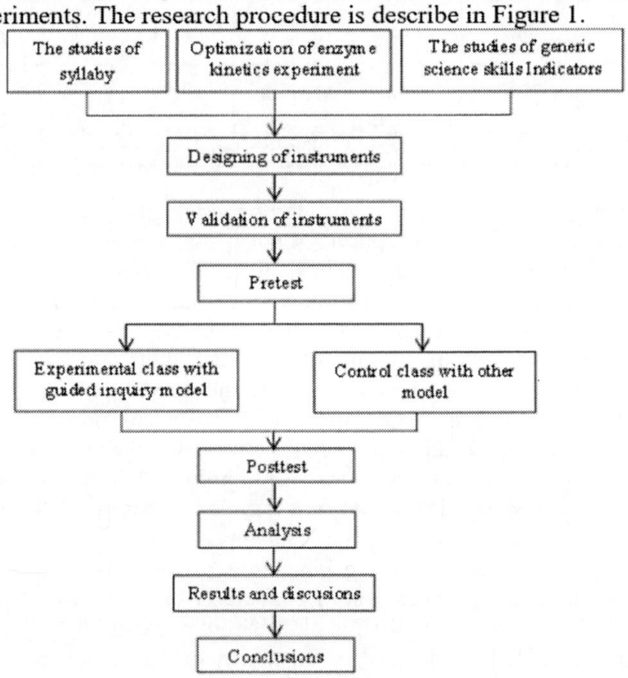

Figure 1. Research procedure

3. Resultsand Discussion

There were 8 aims of enzyme kinetics experiments: 1) isolating the enzyme PPO from local materials, 2) determine the type of substrate, 3) determine the optimum concentration of enzyme on concentration of substrate remains , 4) determine the optimum concentration of substrate on concentration of enzyme remains, 5) determine of optimum temperature on enzymatic reaction of PPO , 6) determine of optimum pH on enzymatic reaction of PPO, 7) determine the most influential inhibitor against enzymatic reaction rate of PPO, 8) determine the type of inhibition that occurs in PPO enzymatic reaction. In the experimental class, enzyme kinetics experiment used guided inquiry model. To enhanced of generic science skills of students, they helped by worksheets. Generic science skills of students were assess based of pretest and posttest. An increase of generic science skillsbased on the value of normalized gain<g> obtained. The value showed in each indicatorsin both classes can be seen in Table 1.

Table 1 Analysisof studentsgeneric science skills in enzyme kineticsexperiment

Indicators	Experimental class				Control class			
	Pretest Average	Posttest Average	<g> Average	Category	Pretest Average	Posttest Average	<g> Average	Category
Direct observation	50,00	86,11	0,73	High	45,61	74,56	0,53	Medium
Causality	64,49	89,13	0,70	High	60,87	77,12	0,42	Medium
symbolic language	71,19	98,15	0,96	High	71,82	92,98	0,75	High
mathematical modeling	27,50	88,89	0,85	High	20,84	68,37	0,60	Medium
concepts formation	59,83	84,83	0,62	Medium	58,50	74,70	0,39	Medium
Average	54,60	89,42	0,77	High	51,44	77,55	0,54	Medium

Based on analysis of research results, it can be stated generally that two classes of study have been increased in generic science skills. The generic science skills improvement that occurred in the experimental class was higher than in the control class, this is evidenced by the value of <g> respectively for 0,77 and 0,54. Increased generic science skills of students in experimental class because of the model used in enzyme kinetics experiments. In the experimental class, enzyme kinetics experiments using a model of guided inquiry and other model in the control class. Based on statistical test between the experimental class and controlclass, the increase of generic science skills has significantly different. This proves that the experiments of enzyme kinetics using guided inquiry model can improved generic science skills of student better than other model. The stages that exist in the guided inquiry model were able to train students in improving laboratory skills[9] as well as generic science skills. Stages of guided inquiry model was developed based on the phases of Wenning[6], included observation, manipulation, generalization, and application.

Stages of guided inquiry conducted in enzyme kinetics experiment begins with observation. At the observation, students observe the phenomenon of enzyme kinetics that associate with live everyday. At this stage, it will be assess direct observation skill.Direct observation skill would be increased of this stage. Direct observation skill rated based on the ability of students in finding related phenomena enzyme kinetics of PPO in everyday live, finding a relationship phenomenon enzyme kinetics with the concepts, and determining the tools and materials enzyme kinetics experiments. The enzyme kinetics experiments using guided inquiry model have value of <g> 0,73 on direct observation skill in high category. Increased direct observation of students in experimental class at high and medium category. Based on the test results statistically significant differences between the high and medium category ($p=0.002$) at level of significance 5%. This proves that experiments of enzyme kinetics were using guided inquiry lab suitable for increasing the direct observation of students in the high category.

Stages of guided inquiry after observation was manipulation. At this stage, student determining influential variables in the experiment, and then prove the influence of these variables[5]. The ability to express the relationship between two or more variables in the phenomenon of enzyme kinetics and estimating the cause was known from generic science skills indicator causality. There were a series of relationships between the various factors of the observed phenomena[11]. Causality skill rated based on the ability of students in finding related phenomena enzyme kinetics of PPO in everyday live, finding a relationship phenomenon enzyme kinetics with the concepts, determining the tools and materials enzyme kinetics experiments, analysis of data enzyme kinetics experiments. The stages of manipulation in determining variables involved in experiments of enzyme kinetics, generic science skill indicator causality would be better. This evidenced by the value of <g> 0,70 in the causal laws in enzyme kinetics experiments using guided inquiry model is classified in high category. Statistical tests carried out showed that there are significant differences between experimental class and control class class(p=0.000) at a level of significance 5%.

Causality trained by worksheetin found of association phenomenon enzyme kinetics with concepts, determine the tools and materials used in experiments of enzyme kinetics, conducting experiments of enzyme kinetics by connecting all the variables, for example, the type of suitable substrates,varying substrate that observed brown color changes were more significant. With a variety (type of substrate, temperature, pH, and the inhibitor) undertaken by students in the experiments of enzyme kinetics ,causality of students should be trained. Increased causality student experimental class at the high and medium category. Based on the test results statistically significant differences between the high and medium category with (p=0.000) at level of significance 5%. This proved that experiments of enzyme kinetics using guided inquiry model was suitable for increasing the causality of students in the high category.

There were no significant difference between experimental class and control class for the symbolic language and mathematical modeling occurs. Kinetics enzyme experiment was a procedural knowledge, regarding the activities of how to do something[14]. While the symbolic language and mathematical modeling is a conceptual knowledge. Symbolic language and mathematical modeling obtained by the students during the course of biochemistry. With the guided inquiry model, students will form a construction of thought and conduct a recall of concepts that have been obtained[11]. Symbolic language and mathematical modeling skills trained by worksheets, where was described the data in tables and graphic. The increased of symbolic language and mathematical modeling are high and low categories at experimental class. Based on the test results statistically significant differences between high and low categories (p=0.003) at significance level 5%. This proved that enzyme kinetics experiment using guided inquiry model lends itself was enhancing symbolic language and mathematical modeling of students in the high category.

From the generalization, students will be have a conclusion to find the concept of the relationship of each variable. Generic science skills that can be observed in this case that concepts formation . With the conclusion obtained by students through the stages of generalizing the generic science skills indicator concepts formation will be better. Concepts formation skill trained by worksheets, where was completing all of question. This evidenced by generic science skills indicator concepts formation in experiments of enzyme kinetics using guided inquiry model value of <g> 0.62 were classified in medium category and statistical tests carried out showed that there are significant differences between experimental class and control class (p=0.012) at level of significance 5%.

The experiments of kinetics enzyme using guided inquiry modelincreased concepts formation study of student in high, medium, and low category. Based on the test results statistically significant differences between the high and medium category (p=0.003), between medium and low categories (p=0.003) and between high and low categories (p=0.000) at atlevel of significance 5%. This proved that the enzyme kinetics experiments using guided inquiry model suitable for increasing concepts formation of students in the high category.Concepts formation skill assisted by the student worksheet.

The application of guided inquiry model on experiments of enzyme kinetics based on local materials improved generic science skills overall students in high category. With four indicators of generic

IOP Conf. Series: Journal of Physics: Conf. Series **812** (2017) 012063 doi:10.1088/1742-6596/812/1/012063

science skills classified as high category that direct observation, causality, symbolic language, and mathematical modeling. While one of indicator in generic science skills concepts formation in medium category . Application of guided inquiry model of enzyme kinetics experiments based on local materials suitable for improving science generic skills of students particularly high category on indicators of direct observation, the causality, symbolic language, and mathematical modelling.

4. conclusions

The results show that experiments of kinetics enzyme using guided inquiry model have been enhance generic science skills in high category with of average <g> value 0.79. Four indicators classified in the high category. There are direct observation, causality, symbolic language, and mathematical modeling with <g> value 0,73; 0,70; 0,96; dan 0,89. Meanwhile, indicator of concepts formation in the medium category with<g> value 0.60.

5. references

[1] Arias H, Lazo L, and Canas F 2014 *J. Chil. Chem. Soc* **59** (4) 2747-52
[2] Supasornand S & Lordkam A -*Social and Behavioral Sciences***116**
[3] Simonson S.R, and Shadle S.E 2013 *Journal of STEM Education* **14** (1) 56-62
[4] Sesen B.A.,and Tarhan L 2013*Res SciEdu* **43** pp 413-35
[5] Wenning C.J 2010 *J. Phys. Tchr. Educ* **5**(3) Winter 11-20
[6] Wenning C.J 2011 *Journal Physic Teacher Education* **6**(2) Summer 9-16
[7] Gaddis B.A, and Scoffstall A.M. 2007 *Journal of Chemical Education* **84** (5) 848-51
[8] Bailey C.P., Minderhout V., and Loertscher L 2012 *Biochemistry and Molecular Biology Education* **40** (1) 1-7
[9] Fakayode S.O 2014 *Anal Bioanal Chem* **406** 1267-71
[10] Simonson S.R., and Shadle S.E 2013 *Journal of STEM Education*, **14** (1) 56-62
[11] Tawil M and Liliasari 2014 *Keterampilan-keterampilan Sains dan Implementasi nya dalam Pembelajaran IPA* (Makasar: BadanPenerbit UNM)
[12] Bailey C.P., Minderhout V., and Loertscher L 2012 *Biochemistry and Molecular Biology Education* **40** (1) 1-7
[13] Brotosiswoyo B.S 2012 *Kiat Pembelajaran MIPA dan Kiat Pembelajaran Fisika di Perguruan Tinggi* (Jakarta: Departemen Pendidikan Nasional Kualifikasi Nasional Indonesia)
[14] Anderson L.W, (Ed.), Krathwohl D.R, (Ed.), Airasian P.W, Cruikshank K.A, Mayer R.E., Pintrich P.R, Raths J, and Wittrock M.C 2001 *A taxonomy for learning, teaching, and assessing: A revision of Bloom's Taxonomy of Educational Objectives* Complete edition (New York: Longman)

Developing Chemistry Teacher's Ability to Design Inquiry-based Lab through Scaffolding type of Teacher Training Program

A G Gumilar[1] , Y Sunarya[2,*], M Arifin[2]

[1]Chemistry Teacher in SMA Negeri 3 Sumedang, Jl. Cipadung 54, Sumedang, Indonesia
[2]Department of Chemistry Education, Universitas Pendidikan Indonesia, Jl. Dr. Setiabudhi 229, Bandung 40154, Indonesia
*Email: yayan_sunarya@upi.edu

Abstract. This study aims to develop chemistry teacher's ability in designing inquiry-based lab. Preliminary study showed that the chemistry teacher ability to design inquiry-based lab need to be developed. Teacher's training program with scaffolding strategy was selected to improve that ability. Twenty four chemistry teachers, who were involved in the Chemistry Teacher Council(CTC, also called MGMP Kimia) in Sumedang, and 60 students of class XI from two different schools were participated in this study. The instruments used were the test of designing inquiry-based lab, rubric to evaluate student worksheet (LKS) and rating scale questionnaire. The results showed that the ability to design inquiry-based lab increased with N-gain average was 0.8 or at the high category. The ability to assess the phenomenon of contextual and making steering questions had a highest mean N-gain is 0.9 at high category. Teachersgave positive responses about the training program generally, indicated by percentage of teacher's response in strong and very strong criteria. Although teacher training with scaffolding strategy was newly applied in CTC, it helped developing teacher's ability in designing inquiry-based lab and gave confidence to teachers to implement in their classroom.

1. Introduction

One of learning models that is recommended in the National Curriculum of 2013 is inquiry. Inquiry defined as multidirectional activities where the students make observations, ask questions, gather information from a variety of sources, planning and carrying out the investigation, using evidence to explain the question, use the tool to search, collect and interpret data; propose answers, questions, and predictions; and communicate the findings [1].

In lab-based inquiry learning, students are encouraged to use more high-level thinking skills, including the ability to think creatively [2]. Despite using more time than conventional lab, lab-based inquiry could encourage students to design their own steps of investigation based on the problems that arise after their observations of the phenomenon.

Teachers should be able to design appropriate learning model according to the times [3]. Therefore as one of the studentsdetermining factorsin achieving certain competence, teacher needs to be improved to mention his ability on a regular basis.

The results of interviews with some of chemistry teachers in Sumedang, it wasrevealed that laboratory workin chemistrylearning is usually done with another approach.Chemistry lab is often limited to verifying a concept that has been taught previously. The results from interview, indicating that teachers are still experiencing difficulties in designing inquiry-based lab, and requires the existence of activities that can improve their competence.

This is shown in research that is often difficult to equip teachers and train inquiry to students because of the low ability of the teacher's inquiry [4]. The ability of teachers in packaging materials and develop LKS inquiry is still low and needs to be improved through various means [5]. Weak ability of teachers in designing the lab as one of the causes of the scarcity of teachers doing practical work in chemistry learning [6].

One of the alleged effective training model in developing the pedagogical competence of teachers is the Scaffolding type teacher training [7]. This type of training consists of four steps, namely the teacher does it, the class does it, the group does it, and the individual does it.The aims of this paper is to define development of chemistry teacher's ability to design inquiry-based lab using scaffolding type of teacher training.

1.1. Teacher Ability to Design Inquiry-based Lab

There are seven aspects that could be expected to support the development capabilities of designing lab [8]: (1) determine the purpose of laboratory activities, (2) determine the type of trial, (3) determine the tools and laboratory materials, (4) determine the test series and describes the diagram, (5) to plan their own experimental procedures, (6) develop inquiry-based worksheets, and (7) designing evaluation laboratory activities.

1.2. Scaffolding Strategy

Scaffolding is a phrase that is used as a framework to illustrate how teachers can guide students through a learning task [9]. Scaffolding as a teaching strategy derived from Vygotsky's sociocultural theory and the concept of the zone of proximal development (ZPD). People who are more expert or more knowledgeable other (MKO) will set up a scaffold or guidance to facilitate the development of students, so that students 'ability awakened by students' prior knowledge and ability to internalize new information. Activities are prepared only located above the stage to do the students themselves. MKO set up scaffolding so that students can complete tasks throughout the ZPD (with guidance) that can't be resolved without guidance. Scaffolding is temporary, that is, when the ability of students has increased the scaffolding gradually discontinued, so that students can complete assignments independently.

Six major elements in the teaching process scaffolding, namely: (1) Establish a special purpose (sharing specific goal); (2) a comprehensive approach to the task (whole task approach); (3) Assistance is given on time (immediate availability of help); (4) Guidance directional (intention assisting); (5) Assistance at optimal level (the optimal level of help); (6) Setting up a model expert (conveying an expert model), the model expert can be prepared in an explicit example of task as experts to complete the task.

1.3. Teacher Training with Scaffolding Strategy

Scaffolding strategy has been used in teacher training program within few decade. One model of scaffolding in teacher training was Smith et. al (2013) model [10]. It has three objectives: diagnosis, responsiveness, and handover to independence.

Strategy scaffolding on teacher training in this paper implemented by the teacher does it, the class does it, the group does it, and the individual does it. The pattern of these measures is presented in Figure 1.

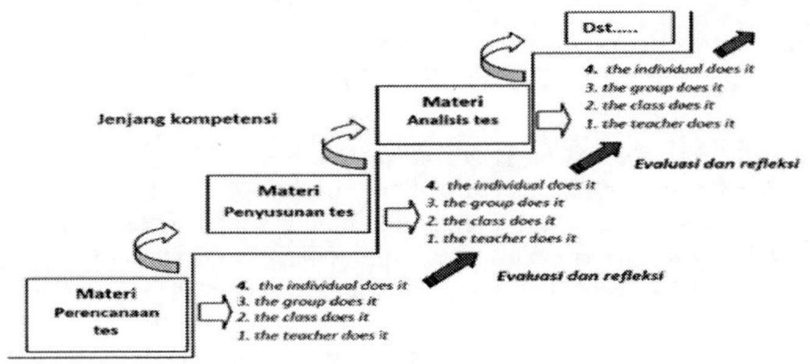

Figure 1. Pattern use in Teacher Training with Scaffolding Strategy [7]

- The first step, trainer give troubleshooting steps (the teacher does it), not with material explanation. In his capacity as a trainer scaffolding learning MKO (the more knowledgeable other), that is, those who have more competence than the participants. MKO will develop participants to achieve their potential competence. The coach gives a true example of solving the problem.
- The second step, namely the class does it. At this stage the responsibility handed over to the class coach, the coach only acts as a companion and container participants' opinions. The results are evaluated and reflected the working class to control the absorption (understanding) of the participants. Evaluation work is intended to determine whether the work meets the standards prescribed or not. If according to the evaluation results turned out to work does not meet the standards then do a reflection about solving problems at once to correct his mistakes.
- The third step is the group does it, the task is given to small groups. Problem solving is done with intensive interaction and communication among members of the group. The group had trouble can ask for help coach. The work groups were evaluated and are reflected as a second step. Until this third step means that each participant has three times gain experience in solving the problem.
- The fourth step, is the individual does it. Each participant was given the task / problem to do / be solved individually. In the final step, participants are required to demonstrate more independent completion of its task. The ability of the participants in completing the task at this stage illustrates that the competence of participants has increased one level higher than before. It also means that the participant has reached its potential competence.

2. Experimental Method

Subjects in this study were high school chemistry teachers in Sumedang who are members of Congress Subject Teacher (MGMPs) High School Chemistry. Customized training activities with programs planned by MGMPs. Teacher training activities conducted on 6-7 May 2015, housed in one of the high schools in KabupatenSumedang. The number of teachers involved in training activities as many as 24 people consisting of 20 women and 4 men.

The method applied in this study using mixed methods. Selection of mixed methods, because there is a need to examine more deeply not only in terms of quantitative, but also qualitative side. In this research will reveal the circumstances on the ground as it should be. The study design used is the *One Group Pre and Post Test.*

Some of the instruments used in this study include:

(1). Pre and posttest to measure the ability of teachers in designing inquiry-based lab.

(2). Questionnaire attitude scale to determine the response of teachers to the implementation of the scaffolding type training.

3. Result and Discussion

3.1. Implementation of Scaffolding Strategy in Teacher Traning

Implementation of teacher training type of scaffolding held over two days with 16 hours of training models (1 hour of training = 45 minutes). Earmarking time is based on the complexity of the material and understanding of the early participants. The material discussed in training tailored to the needs of participants who netted through a preliminary study. Table 1 illustrates the need for training materials.

Table 1. Analyse for Training Materials

Components in a preliminary study	Components in Design Lab Capabilities	Training Materials needed
• Understanding of Inquiry • Stages of inquiry-based lab • The role of teachers in the lab • The learning experience provided to students • Reflection on the ability to design lab • Evaluate the lab • Expectations of training	• Curriculum Analysing • Assess Contextual Phenomenon • Identify tools and materials • Develop procedures • Establish a steering question • Evaluate the lab	• Understanding of inquiry-based Lab • Characteristics of inquiry-based Practicum • Strategies to design inquiry-based lab • How to evaluate lab

In accordance with the characteristics of the type of scaffolding training, training events are on task. In order for the participants to understand the material, do not through the delivery of content, but through tasks that have been prepared. In the first step (the teacher does it) on any matter, the facilitator presents the expected goal after studying the tasks given. In addition, the facilitator provides techniques that can be performed by the participants in order to complete the task. Once is enough, then the next step is the class does it. Domination activeness of participants in the training started on this step. Discussions occur more live with their differences of opinion among the group at this stage. Discussions also occurred in the step group does it, in the respective groups occurred discussion in order to complete the task. The last step is the individual does it, here teachers working on each task based on their understanding of each stage.

3.2. Teachers Ability in Designing Inquiry-based Lab

The ability of teachers in designing lab examined included six abilities as measured from the change before and after the training type of scaffolding. The results of tests the ability of teachers in designing inquiry-based lab work can be seen in Table 2.

According to Table 2, it appears that the value of N-gain the ability of teachers in designing the lab generally increased with the medium category, except the ability to analyse the curriculum is still low category. The average value of the N-gain with the high category in two ability, i.e the ability to assess the phenomenon of contextual and prepare questions directional capability.

The results of observations made during the training appears that the teachers are mastering the ability to analyze the curriculum. Although there are differences in the curriculum used by SMA in Sumedang, but another workshop about curriculum undertaken by all members of the Chemistry

MGMPs very good. This is indicated by the data that teachers are able to analyze the core competence and basic competences contained in the curriculum of 2013.

Table 2. Increase of Teacher Ability to Design Inquiry-based Lab

No.	Indicator	Mean		Gain	N-gain	Category
		Pretest	**Posttest**			
1.	Curriculum Analysing	2,7	2,8	0,1	0,2	Low
2.	Assess Contextual Phenomenon	2,0	2,9	0,9	0,9	High
3.	Identify tools and materials	2,6	2,9	0,3	0,6	Moderate
4.	Develop procedures	2,1	2,4	0,3	0,7	Moderate
5.	Establish steering questions	2,1	2,9	0,8	0,9	High
6.	Evaluate the lab	2,4	2,6	0,2	0,3	Moderate

The situation is the same as the previous study [6] there are many teachers who do not carry out lab work, especially that presents a contextual phenomenon, for various reasons. One of the efforts that the teachers were able to examine the contextual phenomenon is often read books, expand the horizons and get to know the neighborhood. The lack of mastery in chemical materials also affect the ability of teachers to assess contextual phenomenon.

3.3. Teacher responses against the type Scaffolding Training

Teacher training with scaffolding strategy was first conducted in the Chemistry MGMPs in Sumedang, but feedback from teachers who attended the training was very positive. This is shown by data from the analysis of questionnaires that were distributed to the trainees. Data analysis of the questionnaire are presented in Table 3.

Table 3. Percentage Responses to the Teacher Training

No	Statement	Percentage of responses (%)	Criterion
1	Statement of 1	75	Strong
2	Statement of-2	84	Very strong
3	Statement of-3	85	Very strong
4	Statement of-4	80	Very strong
5	Statement of-5	86	Very strong
6	Statement of-6	75	Strong
7	Statement of-7	81	Very strong
8	Statement of-8	75	Strong
9	Statement of-9	88	Very strong
10	Statement of-10	75	Strong
11	Statement of-11	75	Strong
12	Statement of-12	75	Strong
13	Statement of-13	80	Strong
14	Statement of-14	75	Strong
15	Statement of-15	83	Very strong
16	Statement of-16	75	Strong
17	Statement of-17	75	Strong
18	Statement of-18	75	Strong
19	Statement of-19	75	Strong
20	Statement of-20	75	Strong
21	Statement of-21	75	Strong
22	Statement of-22	86	Very strong
23	Statement of-23	83	Very strong
24	Statement of-24	83	Very strong

Based on Table 3 it can be observed that the scaffolding type training is done to get a positive response from the participants. This is evident in the percentage criteria of teacher responses after training showed strong and very strong criteria. On the last topic, which is the follow-up training of teachers is very strong response. This means that the teachers are very confident to be able to implement the results of in-class training.

4. Conclusion

In this paper, we have reported the development of Chemistry teacher's ability to design inquiry-based lab through Scaffolding type of teacher training.Six teacher's ability to design inquiry-based lab has increased after training, with two abilitieswere attained highest gain. Moreover,the teacher training with scaffolding strategy was responded positively.

5. Acknowledgments

The author would like to thank the District Education Office Sumedang who have allowed the author conducted research. To the teachers who are members of the Chemistry MGMPs Sumedang District are willing to take the time to follow the training was also delivered thank you profusely.

6. References

[1] Hook V *et. al.* 2009 *Electronic Journal of Science Education.* **13**

[2] Suhandi A 2012 *Bahan Kuliah Prodi Pendidikan IPA SPs UPI* [Online] [http://file.upi.edu/Direktori/SPS/PRODI.PENDIDIKAN_IPA/196908171994031-ANDI_SUHANDI/Lab_verifikasi_Vs_lab_berbasis_inkuiri.pdf]

[3] Oon-Seng T 2009 *Problem-Based Learning and Creativity* (Singapore :Cengage Learning Asia Pte Ltd)

[4] Tresnagalih I 2011 *Kemampuan Inkuiri dan Sikap Ilmiah Guru Biologi SMA di Kota Bandung dalam Melaksanakan Kegiatan Inkuiri Ilmiah.* (Thesis) Graduate School, UPI, Bandung.

[5] Marissa R 2012 *Analisis Kemampuan Mengemas Materi dan Mengembangkan LKS Inkuiri pada Guru Biologi SMA di Kota Bandung* (Thesis). Graduate School, UPI, Bandung.

[6] Salirawati D 2011 *Materi Pelatihan Kepala Laboratorium Kimia Bagi Guru-Guru Kimia Kabupaten Kulon Progo* (Yogyakarta: Laboratorium Kimia FMIPA UNY)

[7] JalmoT , Rustaman N 2010 *Forum Kependidikan,* **30**

[8] Wiyanto 2006 *Jurnal Pendidikan dan Pengajaran IKIP Negeri Singaraja* **No. 2 TH. XXXIX** Singaraja : IKIP NegeriSingaraja.

[9] Nusu A 2010 *Scaffolding dalam Program Pengajaran Mikro Kimia.* (Dissertation). Graduate School, UPI, Bandung.

[10] Rahman B, Abdurrahman A, Kadaryanto B, Rusminto N E 2015*Australian Journal of Teacher Education,* **40**(11) http://dx.doi.org/10.14221/ajte.2015v40n11.4

IOP Conf. Series: Journal of Physics: Conf. Series **812** (2017) 012065 doi:10.1088/1742-6596/812/1/012065

Implementation of Creative Problem Solving Model to Improve The High School Student's Metacognitive

A. Effendi[12]
[1]School of Graduate Studies, UniversitasPendidikan Indonesia, Jalan Dr. Setiabudi No. 229, Bandung 40154, Jawa Barat Indonesia.
[2]Department of Mathematics Education.UniversitasGaluh,Jalan R.E. Martadinata No.150, Ciamis 46274, West Java Indonesia
Email: adang.effendi@yahoo.co.id

Abstract. This Research is quasi-experimented study with 3x2 factorial and nonequivalent control-group design. The population in this study were all 10th grade students at one of the Senior High School in Ciamis. Furthermore, two sample groups randomly selected (experimental class and control class) with a purposive sampling technique. Each sample group divided into high, medium, and low level based on students' mathematical prior knowledge. The experimental class used Creative Problem Solving models but the control class used conventional models. The instrument used in this study were the metacognitive ability test. The differences of metacognitive ability improvement based on students' mathematical prior knowledge and applied learning model was tested by two ways ANOVA at significance level 0.05, after prerequisites testing are met. Based on this research, it is known that (1) Students' metacognitive ability improvement that has been acquired the Creative Problem Solving model is significantly better than students who acquired conventional learning; (2) There are significantly differences in metacognitive abilities improvement among students who obtain the Creative Problem Solving model with students who received conventional learning in terms of students' mathematical prior knowledge level, high, medium, and low. Metacognitive abilities improvement of Experimented students who have a high and a medium mathematical prior knowledge level, are significantly better than the improvement of Control students' metacognitive abilities that have a high and a medium mathematical prior knowledge level. However, the increase in metacognitive abilities of students who have lower mathematical prior knowledge level in the experimental class and the control class did not differ significantly.

1. Introduction

Metacognitive skills in general is a person's consciousness to be knowledge about the process and results of thinking (cognition) and the ability to control and evaluate their own cognitive processes. Metacognitive skills have an important role in learning (Flavel in [1]. It is this ability which is key to the success of students. In solving the problem (Schoenfeld; Gourgey in [2]), which means that students who have the ability metacognitive low will lead to failure in solving problems , while students who have good metacognitive skills will enhance their problem-solving ability [3].

Another advantage of the metacognitive ability is its role in the success of student learning and is closely related to intelligence (Borkowski, et al. In [1]). These capabilities include common knowledge that can be used for various tasks which enables the use of strategies, the level of effectiveness of the strategy, and self-knowledge (Wildfire, 2013 [4]). Students who display the metacognitive skills, in solving mathematical tasks have better learning outcomes than students who do not show the metacognitive ability (Kramarski and Mizirachi in [5]). Students who have good metacognitive abilities, can find a cognitive style that suits his character (Brown; Rahman and Philips in [6])

When a student uses metacognitive skills, in the learning process going on an activity that involves a reflective process to what the students themselves. Students are trained to explain what was on his mind, will be able to explain it in detail, is also capable of assessing reasoning and problem-solving strategies, and be able to correct the misconception that understands (Web, et al. In [5]). Thus, the increase in metacognitive skills can train students to proficiently communicate, solve problems, control yourself, set the environment, as well as adept at selecting and implementing strategies to improve cognitive abilities.

The study habits led students are able to analyze the learning needs, formulating learning goals, and design a learning program. In addition, students are also able to select and implement strategies, monitor and evaluate whether the strategy has been implemented correctly, check the results, and reflect and get feedback [7] (Siswono 2004 [8]) describe the creative process, known as "Creative Problem solving (CPS) "in three main steps, namely to understand the problems, generate ideas and plan of action. Understanding of issues covering the steps to find purpose, find the data or facts and find problems as the target of questions.

2. Research Method

2.1 Design

This research is a quasi-experimental or quasi-experiment conducted using two research groups, the experimental group (using model Creative Problem Solving) and the control group (using conventional model, a model which is commonly carried out in the schools using the curriculum in 2013) .As consideration of the use of this study design is used because the sample group has been formed previously, meaning that the researchers did not classify the sample into groups randomly as it may interfere with the effectiveness of learning.

Therefore, the researchers used a purposive sampling techniques, the samples are used according to the purpose of research, is also a modest sample, in the sense that the researchers did not classify a random sample, but researchers only scrambles class groups only (Ruseffendi, 2005).

To view the Creative Problem Solving model implementation to metacognitive skills of students, each sample group was divided into three categories based on the capability initially, the initial capability of high, medium and low, so in this study is used a factorial design 3 x 2 factorial design this can be seen in the following table.

Table 13 x 2 Factorial Design

Early Mathematical Ability	Learning	
	Experiment (E)	Control (C)
High Initial Capability (T)	TE	TK
Initial capabilities Medium (S)	SE	SK
Low Initial Capability (R)	RE	RK

The variables used in this study consisted of independent variables and the dependent variable. The independent variables in this study the model of learning (Creative Problem Solving and conventional) and the beginning of students' mathematical abilities (a category of high, medium and low), while the dependent variable that metacognitive abilities of students.

In addition to a factorial design, this study also research design "non equivalent control-group design", namely the experimental group and the control group were selected without the procedure of random, but random class, then both groups were given pre-test and post-test, but only the experimental group is given treatment (Creswell, 2010 [10]). The study design is illustrated as follows:

Experiment group	0	X	0
Control group	0		0

Figure 1. Design of Research

With O : Pre-test / Post-test students' metacognitive skills

X : Creative Problem Solving learning model

2.2 Population and Sample Research

The location of the research is SMAN 1Ciamis, with consideration is one of the schools that have implemented the curriculum of 2013. However, only class X MIA who are already using the curriculum in 2013, so that the population used in this study were all students of class X in Lesson 2014 / 2015 and research samples taken two classes X MIA.

2.3 Research Instruments

The research instrument used in this study consisted of metacognitive skills tests. The research instrument used, developed through several stages: 1) the stage of manufacture of the instrument; 2) the validation phase a team of experts; and 3) the test phase instruments (tests metacognitive skills).

2.4 Metacognitive Ability Test

Metacognitive ability test is made in the form of a description. The written test consists of a pre-test and post-test. These instruments are given to all students in the experimental class and control class, with the question of pre-test and post-test made equivalent (relatively similar). The purpose of the pre-test is to see the homogeneity of the initial capabilities of students. While the purpose of the post-test was to determine student learning outcomes after the implementation of Creative Problem Solving learning model. Furthermore, from the pre-test and post-test visits gain or increase in metacognitive skills of students.

3. Results and Discussion

Results

Overview average difference in terms of each level early mathematical ability of students in both classes can be seen in the following figure.

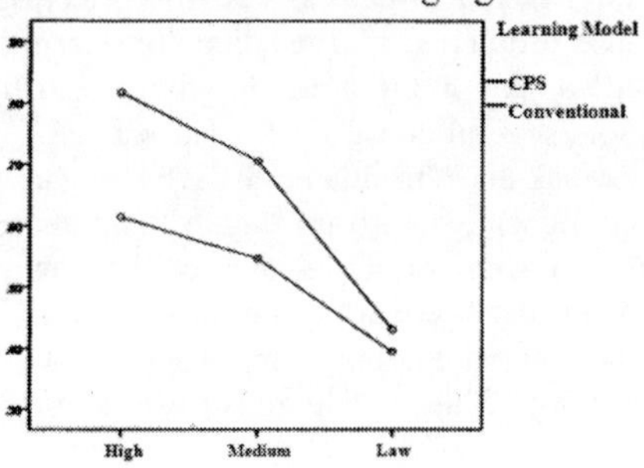

Figure 2. Graph N-Gain Metacognitive Ability in terms of early mathematical ability of students in the class CPS and Conventional

IOP Conf. Series: Journal of Physics: Conf. Series **812** (2017) 012065 doi:10.1088/1742-6596/812/1/012065

To see the effect of the interaction between the learning model and early mathematical ability of students to increase metacognitive skills, it can be seen that the value of the variable Model F * KAM (representing the interaction between the learning model and early mathematical ability of students) is 0849 with a value of significance is 0.434. The significance value greater than 0.05, so H0 is accepted. This means there is no effect of interaction between the learning model used by beginning students' mathematical abilities.

Discussion

Implementation of learning in this study held six meetings in the classroom. The time allotted at each meeting is 2 x 45 minutes. Observations in particular through student activity observation format are directly given to the experimental class, the class that implements the learning model Creative Problem Solving (CPS).

The initial stage of learning, the teacher gives motivation to the students through questions and answers and give some examples of applications of trigonometry in everyday life. Then the teacher describes the steps that will be undertaken student learning and divide students into small groups of 5-6 people with early mathematical ability composition of high, medium and low. She reminded the material prerequisites by providing some questions, in order to make the perception of student understanding. Furthermore, teachers share teaching materials to each student in the group. In the process of learning, the teacher only act as facilitators, teachers guide students to identify problems in teaching materials and describes the need to resolve problems that are given, such as source books, ruler, bow, and the paper's puzzle. After all the students understand what he should do, the teacher asks the students to work on all the problems that are given by way of discussion and learning together with members of his group.

By the time students discuss with each group, teachers around and observe the discussion and providing assistance with scaffolding techniques when needed. Thus, the students really think to understand the problem, identify the characteristics of the given problem, thinking about what he should do, to believe that they really can solve the problems given.

The last fifteen minutes, the teacher asks the class representatives to convey the results of the discussion in front of the class, with the aim of equalizing the perception and understanding of a concept and the resolution of a problem. Teachers also asked the students to criticize his work, delivered the rebuttal, or ask students deliver solutions with different strategies. Thus, students are trained to share ideas, express opinions, delivered the rebuttal, and train students to communicate in front of the class.

Furthermore, teachers with students conclude the matter and reflect on what was learned. Reflection teachers do is to evaluate the learning that has been carried out and correct mistakes in their next meeting, while the reflection of the student aims to make the perception of what they have learned so that students can apply the concept to a wider problem. At the end of the activity, the teacher informs the material for the next meeting.

During the learning process, student's activities is quite conducive and effective. The first meeting, several students had dared ask about understanding in solving the problems, it's just that the learning conditions are still stiff, especially at the time of presentation, students pay more attention without daring to ask. 2 x 45 minutes was still lacking, so that when the presentation, several questions last unassisted.

The second to the sixth meeting, some students began to actively ask and formulated the idea, began to increase student motivation, enthusiastic students in problem solving and his group began to increase. When solving the problems given, students help each other, teach, exchange opinions and cooperate with each other with group members as well as with other groups. Students also were active and enthusiastic in solving the problem, discuss, wondered answer, ask for clarification when discussing friend, give a different opinion, as well as answers to refute a friend by giving a reason. This is what adds to the learning environment more conducive and vibrant, although there are some students who are often silent, embarrassed provide ideas, just write without comment or give opinions.

Based on the observations of researchers, at first the problems of students in solving problems such as less accurate in reading and understanding the problems, students can not describe or illustrate problems in the form of diagrams or sketches, and can not take advantage of what is known as the capital to solve the problem. However, at the next meeting, with scaffolding techniques, these problems can be overcome with good.

Heterogeneous grouping students by ability level makes students more low initial mathematical helped in solving the problem. Only, they look more silent without comment or simply write the results of group discussion without actively contribute ideas, they look more relying on friends in the group who had prior knowledge of mathematical level high and medium.

The opinion of most students about learning with CPS models is quite good. They argue that by learning discussion, faster job completion, and if it does not understand the problem, they can share ideas with their friends. However, some students found it very tiring learning model, they want more explanations beforehand as learning materials are commonly carried out in school rather than have to find out for yourself.

4. Conclusion

Based on data analysis and discussion of the results of research that has been described, obtained the following conclusions. Increasing metacognitive skills students acquire Creative Problem Solving learning model is significantly better than the increase in metacognitive skills of students who received conventional learning.There is a difference significant increase in metacognitive skills among students who obtain Creative Problem Solving learning model with students who received conventional learning in terms of early mathematical ability of students are high-level, medium, and low. Increasing metacognitive skills of students who have a high level of mathematical ability early and being in class CPS is significantly better than the increase in metacognitive skills of students who have a high level of mathematical ability early and being in a conventional classroom. However, the increase in metacognitive skills of students who have a low level of mathematical ability early in the CPS classes did not differ significantly with an increase in metacognitive skills of students who have a low level of mathematical ability early in the conventional classroom. Thus, Creative Problem Solving Learning model is successfully applied to students who have a high level of mathematical ability early and moderate. However, less suitable for students who have a low level of mathematical ability early.

5. References

[1] Livingstone JA 2005 Met*acognition : Anoverview* [Online: http://www.gse.buffalo.edu/fas/sg/huell/cep564/metacog.html]
[2] NoolN R 2012 *Int.Conf. on Education and Management Innovation*(Singapore) vol 30 (Singapore: IACSIT press Singapore) pp 302-6
[3] Yoong W K 2002 Helping Your Students to Become Metacognitive in Mathematics*Mathematics Newsletter*12(5)
[4] Wildan M 2013*KecerdasanMetakognitifPadaKurikulum 2013* [Online" http://gurupembelajaran.blogspot.co.id/2013/09/kecerdasan-metakognitif-pada-kurikulum.html]
[5] Jbeili I 2012 I*nt. J. Res. Educ***32** 45–71
[6] Sholihah I M2012 *Jurnal Pendidikan Biologi* **4** 31-9
[7] Sumarmo U 2013*KemandirianBelajar: Apa, Mengapa, danBagaimanaDikembangkanpadaPesertaDidik*(Bandung: Kumpulan

MakalahBerpikirdanDisposisiMatematikSertaPembelajarannyaUniversitasPendidikan Indonesia) pp. 108–121.
[8] SiswonoT Y E2004 *Pengembangan Kriteria Tingkat Berpikir Kreatif Siswa dalam Matematika.* (Surabaya: Jurusan Matematika FMIPA UNESA)
[9] Ruseffendi E T 1993*StatistikaDasaruntukPenelitianPendidikan* (Bandung: IKIP Bandung Press)
[10] Creswell JW 2010*Research Design: Pendekatan Kualitatif, Kuantitatif, dan Mixed [Terjemahan]* (Yogyakarta: : Pustaka Pelajar)

Acknowledgement
The writer would like to express his great thanks to Ministry of Research and Higher Education Indonesia in giving funding for this research.

The Bacterial Endospore Stain on Schaeffer Fulton using Variation of Methylene Blue Solution

A Oktari*[1], Y Supriatin[1], M Kamal[1] and H Syafrullah[2]

[1] Department of Medical Analyst, School of Analyst Bakti Asih, Padasuka Street No. 233 Pasirlayung, Bandung 40192, Indonesia
[2] Medical School of Dharma Husada, Jakarta Street No. 75 Antapani, Bandung 40282, Indonesia
Corresponding author : nio80zahra@gmail.com

Abstract. Endospores staining is the type of staining to recognize the presence spore in bacterial vegetative cells. The bacterial endospores need a staining which can penetrate wall thickness of spore bacteria. A method of endospores staining is Schaeffer Fulton method that used Malachite Green. It is an alkaline substance staining that can staining the spore bacteria. In this research, it have found the alternative staining that can replace Malachite Green solution in spore bacterial stain. The alternative staining used is Methylene Blue solution (0,5 %, 0,7%, and 1% concentration) with pH variation (10, 11, and 12), and varyous heating time (3, 4, and 5 minutes). The all treatments staining have been effect on bacterial spores staining results. The warming time greatly affect the dye to penetrate the walls of bacterial spores, this can be seen in the results with various concentration at pH 10, indicates that the not long warm-up time 3 and 4 minutes, bacterial spores are not stained, while in the longer heating time is 5 minutes bacterial spores stained. This is caused because the longer heating time can make the pores of spore wall is open so that can facilitate the dye to get into the bacterial spores.

1. Introduction

The bacteria is an organism that was the most numerous and widespread compared the others. Bacteria are generally unicellular organisms, prokaryotic, and microscopic. Bacteria also have a chromosome, ribosomes, and some other species have a granular food, gas vacuoles, and magnestom. Some bacteria are able to form themselves into endospores which enable them to survive in extreme environments [1].

Bacterial spores are bacterial mechanism that is intentionally set in an attempt to secure themselves to the adverse effects of the external environment. Bacterial spores have the same functionality as Amoeba cysts, because the bacteria in the form of spores and cysts of Amoeba in the form of a second phase in which the microorganisms that changes shape to protect themselves against unfavorable external factors eksospora [2].

At the bacterial spore staining, there are two methods commonly used, the method of Schaeffer Fulton and Klein methods. The difference of these two methods is the dye used. In the method of Schaeffer Fulton dyes used are Malachite Green and Safranin, while in the Klein method using dye Carbol Fuchsin and Methylene Blue. Schaeffer Fulton method is a method often used by laboratory technicians, due to staining time faster than the method of Klein [2]. So in this study, researchers prefer to use the method of Schaeffer Fulton commonly used in coloring endospores.

In endospores staining method Schaeffer Fulton, that used an alkaline solution, namely Malachite Green and Safranin. Malachite Green used has a pH of 11.2 (it is alkaline). Dye Malachite Green and Safranin can work well in bacteria because of the alkaline (chromophoric component positively charged), while the bacterial cytoplasm is basophilic, it causes the bacteria can absorb dyes well [3].

Methylene Blue is one dye Thiazine frequently used, because they are economical and easy to obtain. It is a basic dye which is important in the process of dyeing leather, cloth and cotton cloth. The use of Methylene Blue may cause some effects, such as gastrointestinal irritation if ingested, cause cyanosis inhalation and skin irritation if touched by the skin. Methylene blue is an aromatic hydrocarbon compounds with molecular formula $C_{16}H_{18}N_3SCl$ has many uses in different fields, such as biology and chemistry. At room temperature appears as a solid, odorless, dark, and green colored powder [4].

Methylene Blue solution was also used as a dye in chromoendoscopy, smear staining bacteria, and is also used to examine RNA or DNA under a microscope. This solution is also an alkaline solution, which has a pH of 11.4, and is soluble in water and alcohol, have the same properties as malachite green so that researchers use it as an alternative to the dye staining method Shcaeffer Fulton endospores. From the results of the preliminary test, a solution of Methylene Blue with a concentration of 0.5%, 07%, and 1% can pierce or penetrate the bacterial endospores so endospores appear blue.

2. Experimental Method

Preparation of the solution pH variation begins 250 mL $Na_2B_4O_7$ 0.025 M NaOH is added with 27 ml 0.20 M NaOH, then add up to 500 mL with distilled water (for pH 10). The next solution for pH 11 begins 250 mL of 0.50 M $NaHCO_3$ was added with 57 ml of 0.20 M NaOH, then add up to 500 mL with distilled water. The last solution for pH 12 begins 250 mL 0.20 M KCl was added with 30 mL of 0.20 M NaOH, then add up to 500 mL with distilled water.

Preparation the solution of Methylene Blue 0.5% begins with dillute 0.1 grams of this dye with a pH buffer (10, 11, and 12) 20 mL. Then for Methylene Blue 0.7% begins with dillute 0.14 grams of this dye with a pH buffer (10, 11, and 12) 20 mL.The last step for Metylene Blue 1% begins with dillute 0.2 grams of this dye with a pH buffer (10, 11, and 12) 20 mL.The procedure of Schaeffer Fulton method using Malachite Green 5% begins prepare bacterial smear on the objek glass, fixation over a low heat and Malachite green 5% added over the heat fixed bacterial smear then heated over a steam bath for few minutes. This will soften the hard outer coverings of the spore and the primary stain gets stick to the spore. When taken from the steam bath followed by further cooling hardens the outer layer of the spore, during this stage both the spore and vegetative cells appear as green in color, then added water and H_2SO_4 1% as decolorizing agent for vegetative cells. The last added dye Safranin 0,5%, when counterstained with this dye, vegetative cells are easily stained with Safranin, and the cells appear in red or pink color. Chek using microscope 1000x.

The procedure of Schaeffer Fulton method using variation concentration of Methylene Blue (0,5%, 0,7%, and 1%) as the same as standard procedure but the difference is Malachite green replaced by variation concentration of Methylen Blue (0,5%, 0,7%, and 1%).The subject of bacteria in this experiment are *Bacillus substilis* and *Clostridium tetani* that can produce endospores with the different type.

3. Result and Discussion

3.1. Result

Staining of bacterial spores is staining using Malachite Green solution of 5% and 0.5% Safranin, which results in the coloring will appear green on the spores, as well as red in the vegetative cells. At Schaeffer Fulton method that is widely used in painting endospores, endospore first stained with Malachite Green by a heating process, this solution is a powerful dye to penetrate into endospores. After treatment Malachite Green, the cell culture is washed with water and then covered with paint safranin, this technique will produce endospores and green in color pink in vegetative cells [5].

In endospores staining methods Schaffer Fulton, that used an alkaline solution (Malachite Green and Safranin). Dye Malachite Green and Safranin can work well in bacteria because of the alkaline (chromophore component positively charged), while the bacterial cytoplasm is basophilic so there was the attraction between the components chromophore in a dye with bacterial cells, it causes bacteria can absorb dyes well [6].

In this study, researchers used a dye that is wet is Methylene Blue as an alternative to dye Malachite Green. Researchers varying the solution of Methylene Blue with a concentration of 0.5%, 0.7% and 1%, then varying the pH of a solution of Methylene Blue, and longer heating time at 3-5 minutes. Based on the results of research and observation and then analyzed with a statistical calculation Two-Way ANOVA test done on the variation of Methylene Blue concentration (0.5%, 0.7%, 1%), the pH variation 10, 11, 12, and the variation of time heating 3 minutes, 4 minutes and 5 minutes, showed a statistically significant difference, where F count > F table. Because of the significant differences in staining results are then followed a further test using LSD test. After further test (LSD), the data obtained from each treatment on a modified staining bacterial spores *Bacillus subtilis* and *Clostridium tetani* is there is a difference in each treatment.

In *Bacillus subtilis* bacterial spores staining with a variety of used give different results. At a concentration of 0.5% (Fig.1a), the same coloring treatment results with the standard is in the variation of pH 10 with variation of time (5 minutes), variations in pH 12 with variation of time (3 minutes, 4 minutes and 5 minutes). While the results of different treatments with standard staining is in the variation of pH 11 with variation of time (3 minutes, 4 minutes and 5 minutes). At a concentration of 0.7% (Fig.1b) results equal treatment with standard staining is in the variation of pH 10 with variation of time (4 minutes and 5 minutes), and variations in pH 11 with a variation of the time (3 minutes and 5 minutes). While the results of treatment of different staining with the standard is in the variation of pH 10 with a variation of the time (3 minutes), variations in pH 11 with variation of time (4 minutes), and at various pH 12 with a variation of the time (3 minutes, 4 minutes and 5 minutes).

At a concentration of 1% (Fig.2a) results the same treatment staining with the standard is in the variation of pH 10 with variations (4 minutes, 5 minutes), and variations in pH 11 with variation of time (4 minutes and 5 minutes). While the results of treatment of different staining with the standard is in the variation of pH 10 with a variation of the time (3 minutes), variations in pH 11 with a variation of the time (3 minutes), and variations in pH 12 with a variation of the time (3 minutes, 4 minutes and 5 minutes).

From observations obtained in the modification staining bacteria *Bacillus subtilis* spores, it can be concluded that all the coloring treatment effect on bacterial spores staining results. As we know that the warming time greatly affect the dye to penetrate the walls of bacterial spores [7], this can be seen in the results of staining with various concentration (0.5%, 0.7% and 1%) at pH 10, indicates that the not long warm-up time 3 minutes and 4 minutes, bacterial spores are not stained, while in the longer heating time is 5 minutes bacterial spores stained. This is caused because the heating time is longer make the pores of the spore wall is open so as to facilitate the dye to get into the bacterial spores.

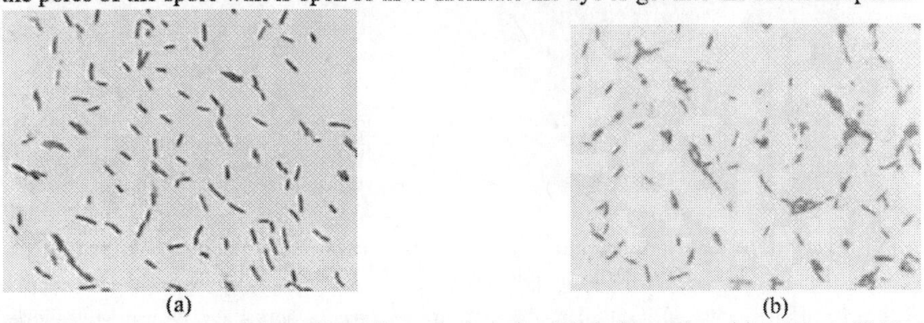

Figure 1. The endospore staining of *Bacillus subtilis* using Methylene Blue (a) 0,5% at pH 12 and 3 minutes; (b) 0,7% at pH 10 and 5 minutes

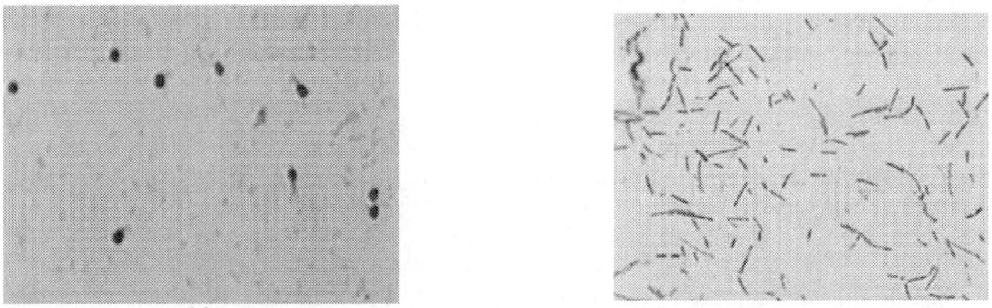

(a) (b)

Figure 2. The endospore staining using Methylene Blue (a) *Bacillus subtilis* at 1%, pH 11 and 5 minutes; (b) *Clostridium tetani* at 0,5%, pH 11 and 5 minutes

In *Clostridium tetani* bacterial spores staining with a variety of used give different results. At a concentration of 0.5% (Fig.2b) results equal treatment with standard staining is in the variation of pH 10 with variation of time (4 minutes and 5 minutes), variations in pH 11 with variation (time of 3 minutes, 4 minutes and 5 minutes), and the pH 12 with a variation of the time (3 minutes, 4 minutes and 5 minutes). While the results of different treatments with standard staining is in the variation of pH 10 with a variation of the time (3 minutes), variations in pH 12 with variation of time (3 minutes).

At a concentration of 0.7% (Fig.3a) results equal treatment with standard staining is in the variation of pH 10 with variation of time (4 minutes, and 5 minutes), variations in pH 11 with variation of time (3 minutes, 4 minutes and 5 minutes). While the results of different treatments with standard staining is in the variation of pH 10 with a variation of the time (3 minutes), and variations in pH 12 with variation of time (3 minutes, 4 minutes and 5 minutes). Meanwhile, at a concentration of 1% (Fig.3b) by varying the pH (10, 11, and 12) and the variation of heating time (3 minutes, 4 minutes and 5 minutes), the results obtained do not have the same standards.

From observations obtained in the modification staining the bacteria *Clostridium tetani* spores it can be concluded that, all the coloring treatment effect on bacterial spores staining results. As we know that the warming time greatly affect the dye to penetrate the walls of bacterial spores, this can be seen in the results of staining with various concentration (0.5%, 0.7% and 1%) at pH 10, indicates that the warm-up time is not long (3 minutes) of bacterial spores are not stained, while in the longer heating time (4 minutes and 5 minutes) stained bacterial spores. This is caused because the heating time is longer make the pores of the spore wall is open so as to facilitate the dye to get into the bacterial spores [8].

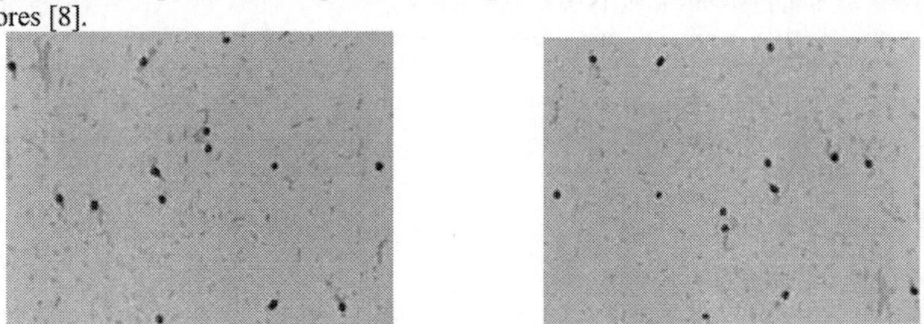

(a) (b)

Figure 3. The endospore staining of *Clostridium tetani* using Methylene Blue (a) 0,7% at pH 10 and 3 minutes; (b) 1% at pH 11 and 4 minutes

pH of a solution is very influential in the staining bacteria, where the solution Methylene blue is used in the coloring of bacterial spores must be alkaline, because the nature of the bacterial cytoplasm is basophilic [9], so in this study the researchers varying the pH of the solution Methylene Blue (pH

10, pH 11, and pH 12) to determine the optimum pH which can provide the same results with standard staining. From the observations obtained, variations Methylene Blue solution with a pH (10, 11, and 12) is influenced by the concentration of Methylene Blue solution (0.5%, 0.7% and 1%).

The concentration of the solution is also affected by variations in pH of the solution used. As an example can be seen at a concentration of 0.7%, at pH 10 and 11 look good bacteria spores, whereas at pH 12 of bacterial spores does not look good, because the use of high concentration and pH of the solution is more alkaline (strong bases) then dye methylene blue is more concentrated, so that when given a second dye with Safranin [10], vegetative cells appear red bluish and purplish red. So that the bacterial spores and vegetative cells does not look good.

4. Conclusion

Optimal concentration using Methylene Blue as a coloring alternative to Malachite Green in staining the bacterium *Bacillus subtilis* spores are at a concentration of 0.5% at pH 12 and the heating time for 3 minutes. While in staining the bacterium *Clostridium tetani* spores are at a concentration of 0.5% at pH 11 and the heating time 3 minutes.

5. References

[1] Zubaidah and Elok 2006 *Mikrobiologi* (Malang :Universitas Brawijaya Press)
[2] Waluyo and General L 2004 *Mikrobiologi* (Malang :UMM Press)
[3] Volk and Wheeler 2009 *Basic Microbiology 5th edition* (Jakarta: Erlangga)
[4] Sutedjo M 2011 *Soil Microbiology* (Jakarta: Rhineka Reserved)
[5] Sumarsih S 2004 *Test Activities lipolytic Some Bacteria Results Isolation of the Port of Tanjung Perak and Lipase Production of Selected Strain* (Surabaya: JIPTUNAIR)
[6] Ruth M, I 2009 *Photodegradation Methylene Blue with UV light and ZnO Catalyst* Thesis. (Bukit Jimbaran : University of Udayana)
[7] Albert and Bacillus 2008 *Cellular and Molecular Biology* (USA: Caister Academic)
[8] Dwidjoseputro 2010 *Fundamentals of Microbiology* 2 (Jakarta: Djambatan)
[9] Noirot P 2007 *The Replication of Chromosome Bacillus subtilis, Bacillus: Cellular and Molecular Biology* (USA: Caister Academic Press)
[10] Pelczar M J 2007 *Fundamentals of Microbiology* (Jakarta : UI Press)

Acknowledgments

The authors thanks to the Microbiology Laboratory of Health Polytechnic Bandung and the staff of the School Analysts Bakti Asih Bandung who have helped during the research process. Acknowledgements are also delivered to other parties who have a lot of feedback and suggestions on this study.

IOP Conf. Series: Journal of Physics: Conf. Series **812** (2017) 012067 doi:10.1088/1742-6596/812/1/012067

The Increase of Critical Thinking Skills through Mathematical Investigation Approach

N Sumarna[1], Wahyudin[2], T Herman[2]

[1]Faculty of Teacher Education and Training, Universitas Halu Oleo, Indonesia
[2]Department of Mathematics Education, Faculty of Mathematics and Science Education, Universitas Pendidikan Indonesia, Jl. Dr. Setiabudhi No 229 Bandung 40132, Indonesia
Email: nanafkipunhalu@gmail.com

Abstract. Some research findings on critical thinking skills of prospective elementary teachers, showed a response that is not optimal. On the other hand, critical thinking skills will lead a student in the process of analysis, evaluation and synthesis in solving a mathematical problem. This study attempts to perform an alternative solution with a focus on mathematics learning conditions that is held in the lecture room through mathematical investigation approach. This research method was Quasi-Experimental design with pre-test post-test design. Data analysis using a mixed method with Embedded design. Subjects were regular students enrolled in 2014 at the study program of education of primary school teachers. The number of research subjects were 111 students consisting of 56 students in the experimental group and 55 students in the control group. The results of the study showed that (1) there is a significant difference in the improvement of critical thinking ability of students who receive learning through mathematical investigation approach when compared with students studying through expository approach, and (2) there is no interaction effect between prior knowledge of mathematics and learning factors (mathematical investigation and expository) to increase of critical thinking skills of students.

1. Introduction

Critical thinking has been an important issue for many years. The word critical means to understand the people and things around us and analyses our own thinking processes. Unfortunately, criticizing ability is used just as limiting the thinking process of a person [1]. Construction of learning more teacher-centered response to the condition where students only receive the information without considering their ability to think. It has undercut the ability of students in an effort to do the analysis and synthesis. The main principle of learning is to provide a great opportunity so that students can develop the ability to think and manage this capability so that there is a meaningful learning. One type of thinking skills and become the focus of this research is the ability to think critically. In this context, [2] pointed out that "Critical thinking is the intellectually disciplined process of actively and skillfully conceptualizing, applying, analyzing, synthesizing, and/or evaluating information gathered from, or generated by, observation, experience, reflection, reasoning, or communication, as a guide to belief and action. This statement emphasizes that the ability to think critically is an intellectual process that guides a person in action. [3]views critical thinking as a mental process that involves a high quality and high level of thinking for problem-solving and decision-making. Terms such as higher level thinking and reflective thinking have often been used interchangeably with the term critical thinking throughout the literature ([4], [5], [6], [7]). However, while there are many definitions of the term critical thinking, [8] suggested that critical thinking resulted from the interaction of a set of dispositive toward critical thinking: seeking a clear statement of the questions, seeking reasons, trying to be well-informed, and trying to remain relevant the main point.

Expert opinion given above shows how important the role of critical thinking skills in a person, in particular for the development of the capacity of a student, as a subject as well as in a learning process. While different conceptualizations of critical thinking have shed considerable light on the nature and meaning of critical thinking, at the same time, it is often acknowledged that critical thinking skills require time to develop. In fact, [9] also emphasizes that the students should not only employ critical thinking skills in class, but also be able to activate them in real-life situations and to recognize situations when these skills should be used. But the condition of students' critical thinking skills, looks very varied

Some studies indicate it. For example, [10] found that the critical thinking skills of teacher candidates of elementary mathematics were medium level but not high enough. Learning only a content-based curriculum, children cannot become better thinkers able to give reasons for their conclusions, to think flexibly and creatively, to solve problems and make good decisions [11]. On the other hand, [12] conducted a study through the selection of learning content and found that students' critical thinking can be successfully encouraged and developed in elementary mathematics education by the selection of content (tasks), the solution of which emphasizes skills of critical thinking in all students, and that we should not make any limitations regarding the classification of students by their overall achievement in school.

On the other hand, mathematical investigation is an approach to learning that is focused on the exploration of a topic in depth and make connections between different representations. These efforts provide great opportunities, based on the ability of each student, to be able to solve the problems of open-ended, finding patterns, self-discovery, reducing the role of the teacher, using its own methods, divergent thinking, and so as an attempt to solve the problem. The Cockcroft report (1982) cited by [13] stated that the idea of investigation is fundamental both to the study of mathematics itself and Also to an understanding of the ways in the which mathematicscan be used to extend knowledge and to solve problems in very manyfields.

The investigative approach is broken down into steps in Maths Charter Working Mathematically (n.d.) and Kanasa (n.d.) (cited in [14]) has illustrated different methods of investigating and problem solving in a visual way which will add meaning for many students. The ability to pose questions, collect data, investigate and discover answers, solve problems, describe, share and elaborate on findings, evaluate, reflect on and judge the accuracy of answers, draw conclusions, revise and test models of investigating and reject or defend ideas are skills that should be stressed in the modern mathematics classroom. These processes need to take place in small groups and in the classroom as a whole. Furthermore, [15] explains that the stages of mathematical investigation consists of five phases: entry, goal setting (or problem posing), attack (or problem solving), review and extension. At the entry phase, the students should understand the task. There are generally two approaches to goal setting: students may pose a specific problem or they may decide to search for any pattern. The latter approach usually leads to specializing, formulating and testing of conjectures, and generalizing, in the attack phase. This is the process of investigation.

Context of this study, researchers applied an a mathematical investigation approach in order to develop students' critical thinking skills. The emphasis on the aspect of mathematical investigation will establish a process for students to think mathematically, and supported the process of interaction between students and lecturers as well as a source of learning as a process of communication and cooperation, are expected to have impact on the ability of mathematical understanding and thinking skills of students. The questions were: (1) Is the mathematical investigation approach can improve the ability of Critical thinking skills of prospective elementary teachers?, (2) Is there an interaction effect between the learning factor (mathematical investigation and expository) and prior knowledge of mathematics in improving the Critical thinking skills of prospective elementary teachers?

2. Experimental Method

Research design using Quasi-Experimental type design with pre-posttest design [16]. Based on this type, the researchers apply a pre-and posttest design, which is divided into two study groups: control group and the treatment group. Both groups were subjected to pretests as supporting material to perform the treatment, then do post-test to know the difference between the two groups. The experimental group using investigative approach to learning mathematics (mathematical investigation), while the control group using expository approach. These different approaches are arranged in a construction of instructional design each time the lecture.

Data were collected through a process of pretest and posttest in both study groups (experimental and control). Pre-test and post-test question contains a set of items in which there are indicators of critical thinking. Indicators of critical thinking abilities that are used in this study, are: (1) identify, select, classify into categories and describe attributes (2) organize the parts to form something new or unique whole, (3) integrate ideas, information and theory to design solutions, (4) summing up and explain the meaning to give the meaning of each section

Number of items in pre-test consist of 4 items and post-test consists of 5 items, whole items which of the measured variables of the study. Measurement of study variables using a score of 0-4 in each category have been determined under the rubric and scoring techniques.

Both these instruments do the testing process to test the validity and reliability of the instrument. Validity test using Pearson correlation test models. While the reliability of the instruments this study refers to internal consistency reliability with the test model used was Cronbach Alpha coefficients. Descriptive analysis of data using descriptive statistical measures such as mean, standard deviation, and curves. ANOVA test was used to test the hypothesis.

3. Result and Discussion

3.1 Result

The increase of critical thinking skills of research subjects is obtained through the ability to analyze, generalization, synthesis, justify, and resolve non-routine which is operationally constructed as an indicator of research and used as the measurement criteria of critical thinking skills. Based on the observation of these indicators, the results obtained in the following Table 1.

Table 1. Means, Minimum, and Maximum Observed Scores, and Standard Deviation in the critical thinking abilities

Group		*N*	*Minimum*	*Maksimum*	*Mean*	*Standard Deviation*
Control	Pre-test score (Maximum score = 16)	55	2	8	3.42	3.01
	Post-test score (Maximum score = 20)	55	4	18	8.42	1.64
	Gain value	55	0.0	0.89	0.28	1.93
Experiment	Pre-test score (Maximum score = 16)	56	2	7	6.78	3.74
	Post-test score (Maximum score = 20)	56	16	20	12.48	3.98
	Gain value	56	0.0	1.00	0.42	0.25

The distribution of the data (Table 1) provide a response that observed in the experimental group showed results higher scores on aspects of post-test and the value of the gain. This is an indication that the application of mathematical investigative approach produces positive impact on critical thinking skills of research subjects.

Table 2. Results of the Analysis of Variance (*one way*)
Tests of Between-Subjects Effects

Dependent Variable: gain value

Source	Type III Sum of Squares	df	Mean Square	F	Sig.	Partial Eta Squared
Corrected Model	,513[a]	1	,513	9,943	,002	,084
Intercept	13,978	1	13,978	271,029	,000	,713
learn_factors	,513	1	,513	9,943	,002	,084
Error	5,622	109	,052			
Total	20,162	111				
Corrected Total	6,135	110				

a. R Squared = ,084 (Adjusted R Squared = ,075)

Results from one way ANOVA analysis in the Table 2, shows that the mathematical investigation approach gives a positive response in providing enhanced critical thinking of research subjects, based on the F value of 9.943 with sig. 0.002<0.05 as the recommended value. Partial eta squared value generated by the model were high at 0.084 < 0.14, so it is classified in the category of medium (Cohen and Miles & Shevlin, (2001) cited in [17]). Value partial eta squared explains that the mathematical investigation approach able to deliver effect as 0.084 in contributing to a unique variant of the critical thinking. While the value of Adjusted R Squared generated model of the main effect of 0.075, which means 7,5% of the variance of the total variance in critical thinking described by mathematical investigation. Using partial eta-square value of this, researchers calculated effect size obtained by the size of Cohen's d and generate value 0.302. the value of the effect size of this is in the category small and the equivalent of percentile standing at 62.0 and percent of nonoverlap at 21.3% [18].

Table 3. Results of the Analysis of Variance (two way)
Tests of Between-Subjects Effects

Dependent Variable: gain value

Source	Type III Sum of Squares	df	Mean Square	F	Sig.	Partial Eta Squared
Corrected Model	,749[a]	5	,150	2,922	,016	,122
Intercept	8,255	1	8,255	160,954	,000	,605
learn_factors	,335	1	,335	6,534	,012	,059
prior_knowledge	,105	2	,053	1,025	,362	,019
learn_factors * prior_knowledge	,095	2	,048	,929	,398	,017
Error	5,385	105	,051			
Total	20,162	111				
Corrected Total	6,135	110				

a. R Squared = ,122 (Adjusted R Squared = ,080)

Then, the result of two-way ANOVA in the Table 3 illustrates that there is no interaction effect between the factors of learning as applied in this study and the prior knowledge of mathematics to increase the critical thinking of the study subjects were confirmed by F value of ***learn_factors * prior_knowledge*** amounting to 0.929with sig. 0.398>0.05 as the recommended value. Value partial eta squared of 0.017 in the category of small and use a value of partial eta squared, the calculated effect size with the size of Cohen's d and obtained a value of 0.131. Effect size is in the category of small and the equivalent of percentile standing at 54.0 and percent of nonoverlap at 7.7%. Based on these measures, we conclude that the effect of the interaction between learning factors and prior knowledge of mathematics to the improvement of critical thinking is quite small and is considered insignificant. This condition is confirmed by the interaction profile plot "learn_factors * prior_knowledge" below.

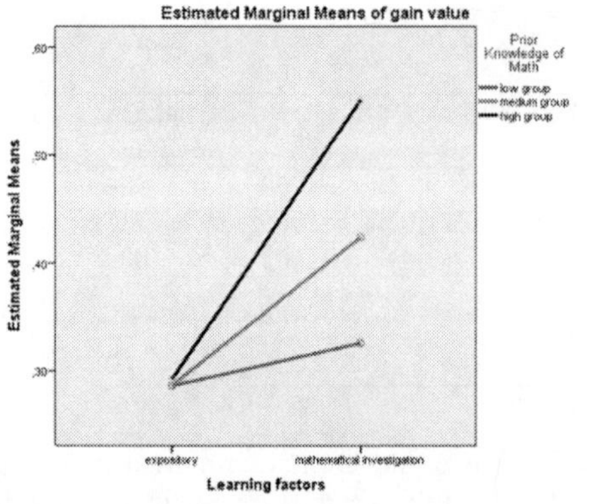

Figure 1a. Profile plot interaction with learning factors of as a horizontal axis

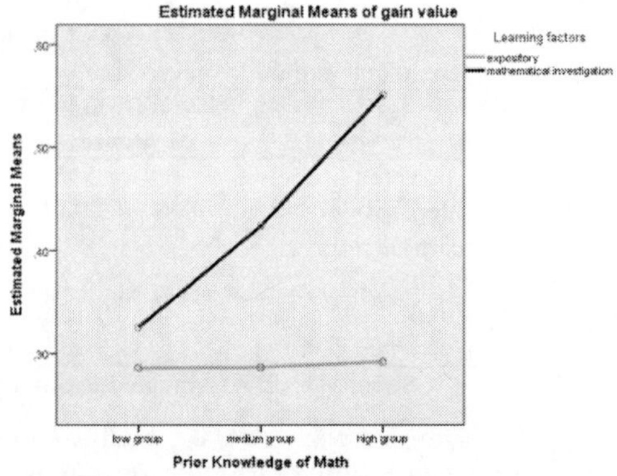

Figure 1b. profile plot interaction with prior knowledge of mathematics as a horizontal axis

In Figure. 1a makes clear that the movement of the line of the domain mathematical investigation to the domain ekspository a pattern of positive slope, meaning that there is an increase in the value of the gain of the reasoning abilities of higher mathematics at the mathematical investigation for any prior knowledge of mathematics, when compared to expository. In Figure.1b illustrates that the movement of the two lines that show gain value of the increase in critical thinking generated by factors of learning in each category prior knowledge of mathematics, seen not happen crossovers (intersection of the two lines). Expository lines symbolize movement has always been at the bottom of the movement of the line that symbolizes the mathematical investigation.

Based on the above description, the conclusion that there is no interaction effect that occurs in both the independent variables studied (learning factors and prior knowledge of mathematics) to the dependent variable (the critical thinking).

3.2 Discussion

The results of a one-way ANOVA test showed that there was the increase critical thinking skills significantly of students having learning through a mathematical investigation than learning by learning through expository approach.

The results of this, not in spite of attributes inherent in mathematical investigation. One is as a process of mathematical thinking. In the process of mathematical thinking, identified four basic processes. [19] stated that we identified four fundamental processes, in two pairs, and Showed how thinking mathematically very often proceeds by alternating between them: (a) specializing – trying special cases, looking at examples. (b) generalizing - looking for patterns and relationships, (c) conjecturing – predicting relationships and results, and (d) convincing – finding and communicating reasons why something is true. Viewed from the side of the mathematical investigation, based on the conclusions of various literature, concluded that the mathematical investigation is a process that includes 4 (four) stages of the process of thinking, namely: specialization, filing of alleged (conjecturing), justification and generalization [20].

Thus, critical thinking skills generated in this study are not detached from the learning process with the mathematical investigation approach in increasing critical thinking skills, because in the process of critical thinking will involve the mathematical thinking that is based on four fundamental processes, namely specializing, generalizing, conjecturing, and convincing, which developed from a low level mathematical thinking skills to a high level of mathematical thinking skills. The position of critical thinking is at a high level thinking skills that require the above process. Critical thinking is ... ... "*... the intellectually disciplined process of actively and skillfully conceptualizing, applying, synthesizing or evaluating information gathered from, or generated by observation, experience, reflection, reasoning, or communication as a guide to belief or action* [2].

As a comparison, researchers present a problem which is used to measure students' critical thinking skills following

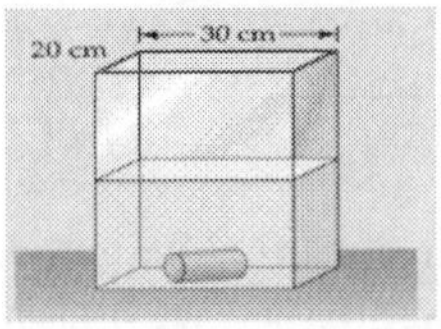

A cylinder with a 2-centimeter radius and a height of 10 cm is submerged in a tank of water that is 20 cm wide and 30 cm long (see Figure). How much does the water level rise? (Source: [21], p. 287)

And the answer sheets of students from the experimental and control group, which both have the same initial knowledge of mathematics, which both are in the medium category.

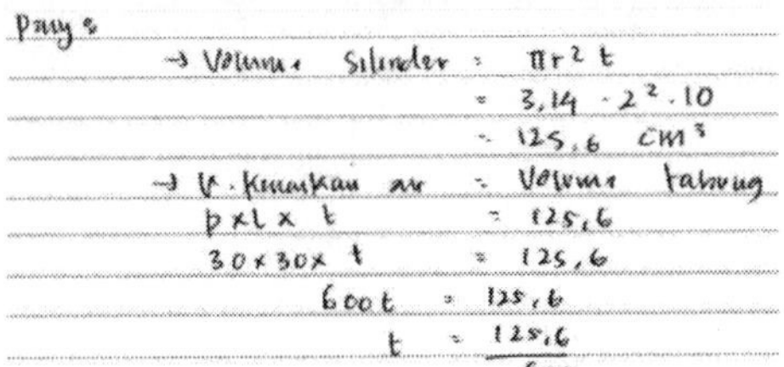

Fig.2a The answer sheets of M070 (exp. group)

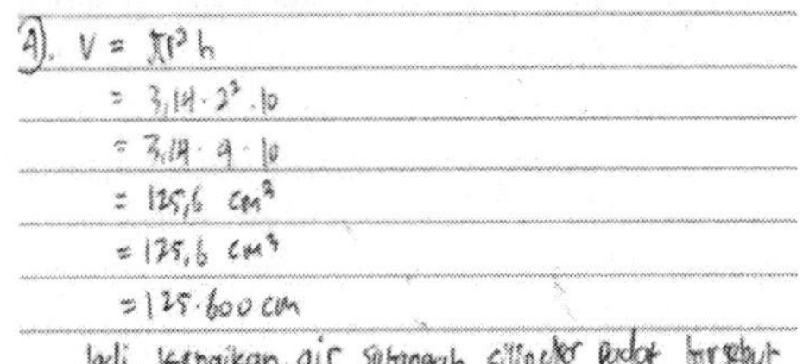

Figure.2b The answer sheets of N034 (cont. Group)

Response M070 (Figure 2a) looks give the correct response, despite the procedural errors, the errors included figures for the value of l (width) by 30, should have been 20 as an attitude that is less critical. It seems that M070 is able to provide an indicator of critical thinking skills such as being able to identify and describe the attributes; identify the relationship between the parts of the whole piece; arranging the parts to form something new or unique whole; and is able to integrate the ideas, information and theory to design a solution. These results lead to the correct answer.

Answer N034 (Fig 2b) shows incompetence bring up the indicators for the measurement of critical thinking skills, such as identifying the relationship between the parts involved; identify the relationship between the parts of the whole piece; organize the section to form a the flow solutions and integrate all the information into the design of the solution. The answer N034 portray their difficulties in the process of mathematical thinking. This condition is seen as N034 difficult to identify the relationship between the parts that exist, it is actually N034 difficulties in the Specializing and generalizing, especially on the search patterns and relationships to develop a solution. The end result, it is easy to guess there will be an error response

The answer sheet above, is an example put forward by researchers in demonstrating differences in the response of the two-group research. In the experimental group, showed an increase in critical thinking skills more than in the control group, where the influence of the investigation mathematically according to the effect size obtained by the value of partial eta squared of 0.084 and d Cohen's of 0.302 indicates the level of impact strength both at medium level. In line with this conclusion, [22] have reported a synthesis of the results of quantitative (meta-analysis) of the studies that evaluated the impact of the use of thinking skills program and approach to cognition, achievement and attitudes. Higgins and colleagues examined the results of 29 studies from around the world, mostly from the United States and Britain, which are used in elementary schools and secondary in literacy, math, and science. The results of the analysis of Higgins et al. supports the conclusion that the intervention skills that are effective in improving students' thinking skills, achievements in their fields, and motivation. Furthermore, [23] found that a group of students who receive instruction critical thinking skills (critical thinking skills instruction) explicitly designed in a study, concluded the increase of the value of the gain is significant in the analysis of their arguments than the group do not receive instructions explicitly critical thinking skills.

Other findings in the study is no interaction effect between the factors of learning (mathematical investigation and expository) and prior knowledge of mathematics (in categories: low, medium, high)

to increase the critical thinking. Profile plots (fig.1a & fig.1b) showed no intersection of the lines and the existence of the movement increase the ability of critical thinking of research subjects, which is moving toward the higher, from the domain of expository to the domain of mathematical investigation for all types of prior knowledge of mathematics. As it noted previously, one of the factors considered in the process of improving a person's ability critical thinking is the position of the subject in the context of prior knowledge of mathematics he had.

The context of the absence of these interactions can be identified from the prior knowledge of mathematics of the research subjects. Description of the answer sheet presented earlier and simple statistics, illustrates the significant difference in the increase critical thinking skills between the experimental and control, seen at the beginning of the study subjects were capable of medium and high. But in subjects with lower initial ability, showed no significant differences. In this condition, looks very realistic to see the urgency of giving mathematical investigation. Learning of mathematics like this will load the investigation activity, investigation task, investigation work or investigation process as well as covering also aspects of problem solving, filing problems, reasoning as well as a process of mathematical thinking.

4. Conclusion

The conclusion of this study stated that: (1) Investigation of mathematics as an approach to learning is able to give a positive response to the increasing of critical thinking, and (2) There is no interaction effect of the factors of learning and prior knowledge of mathematics to the increased ability of critical thinking. One of the implications of the implementation of the mathematical investigation is to provide opportunities and demands for teachers to train themselves in mastering the material to be taught, thus indirectly mathematical investigation serves as a "math lab" for teachers. It is as a consequence of other attributes inherent in the investigation of mathematics, that is, as an activity that requires a thought process, ranging from understanding the problem, the process of mathematical thinking (specialization, conjecturing, justify, generalization) and then ends with a process of review and extension. The condition occurs because of the mathematical investigations are confronted with problems of non-routine, where these conditions provide the demand for teachers to practice doing some investigative tasks that contain activity of investigation before it is taught.

5. References

[1] Chaffee J 1988 *Thinking critically (2nd ed)* (Boston: Houghton MifflinMA)
[2] Scriven M, & Paul R 1987 *The 8th Annual International Conference on Critical Thinking and Education Reform* [on line: www.criticalthinking.org/pages/defining-critical-thinking/766]
[3] Al-Mubaid H 2014 *International Journal of Advanced Corporate Learning* **7** 34-37
[4] Crenshaw P, Hale E, and Harper S 2011 *Journal of College Teaching & Learning* **8** 13-22
[5] Geertsen H R 2003 *Teaching Sociology* **31** 1-19
[6] Ness R 2015 *American Journal of Public Health* **105** 114-118
[7] Wallace B, Berry A, and Cave D 2009 *Teaching problem solving and thinking skills through science* (Oxon: Routledge Abington)
[8] Ennis R H 1986 A taxonomy of critical thinking skills *Critical Thinking*,ed J B Baron and R G Strenberg (Philadelphia: Franklin Institute press)
[9] Swartz R 1992 Critical thinking, the curriculum, and the problem of transfer In D Perkins J Bishop and J Lochhead (Eds.)*Thinking: The Second International Conference* (Erlbaum, Hillsdale, NJ) pp 261-284
[10] Aktaş G S and Ünlü M 2014 *Procedia - Social and Behavioral Sciences* **93** 831 – 835
[11] Massa S 2014 *Procedia - Social and Behavioral Sciences* **141** 387 – 392
[12] Maričića s and Špijunovićb K 2015 *Procedia - Social and Behavioral Sciences* **180** 653 – 659
[13] Jaworski B 1996 *Investigating Mathematics Teaching: A Constructivist Enquiry (No. 5)* (London: Taylor & Francis Ltd.)

[14] Quinnell L 2010 *The Australian Mathematics Teacher* **66** 35-40
[15] Yeo J B W and Yeap BH2009 *Mathematical Investigation: Task, Process and Activity*. [online: http://math.nie.edu.sg/bwjyeo/Publication/]
[16] Creswell John W 2008 *Educational Research: Planning, Conducting, and Evaluating Quantitative and Qualitative Research* (New Jersey: Pearson, Merrill Prentice Hall)
[17] Watson Peter 2016 *Rules of thumb on magnitudes of effect sizes MRC Cognition and Brain Sciences Unit* [on line: http://imaging.mrc-cbu.cam.ac.uk/statswiki/FAQ/effectSize]
[18] Carson C 2011 *The effective use of effect size indices in institutional research* [online: http://rlrw.bnu.edu.cn/uploads/ 2011420132315861.pdf]
[19] Stacey K 2007 What is mathematical thinking an d why is it important? *Progress Report of the APEC Project: Collaborative Studies on Innovations for Teaching and Learning Mathematics in Different Cultures (II) (Lesson Study Focusing on Mathematical Thinking)* (Center for Research on International Cooperation in Educational Development (CRICED) University of Tsukuba) pp 39-48
[20] Yeo J B W and Yeap B H 2009 Solving mathematical problems by investigation *Mathematical problem solving: Association of Mathematics Educators Yearbook*,ed B Kaur, B H Yeap and M Kapur (World Scientific, Singapore) pp 118-136
[21] Aufmann R N, Lockwood J S, Nation R D and Clegg D K 2008 *Mathematical Thinking and Quantitative Reasoning* (Boston: Houghton Mifflin Company)
[22] Higgins S, Hall E, Baumfield V and Moseley D 2005 *A Meta-Analysis of the Impact of the Implementation of Thinking Skills Approaches on Pupils* (Research in Education Library, London: EPPI-Centre, Social Science Research)
[23] Bensley D *et al* 2010 *Teaching and assessing critical thinking skills for argument analysis in psychology* (Education Resources Information Center ERIC No. EJ883208)

A DFT and TDDFT Study of PCM Effect on N3 Dye Absorption in Ethanol Solution

C D D Sundari[1,2,*], M A Martoprawiro[1], and A L Ivansyah[1]

[1]Inorganic and Physical Chemistry Division, Department of Chemistry, Faculty of Mathematic and Natural Science, Institut Teknologi Bandung, Jl.Ganesha No. 10, Bandung, West Java, 40132, Indonesia
[2]Program Studi Pendidikan Kimia, Universitas Islam Negeri Sunan Gunung Djati (UIN SGD), Jl. A.H. Nasution No.105 Bandung, West Java, 40614, Indonesia.
*E-mail: citra@uinsgd.ac.id

Abstract. Dye Sensitized Solar Cell (DSSC) is one kind of low-cost third generation solar cell which employed organic or complex dye molecule as photon harvester. $[Ru(H_2dcbpy)_2(SCN)_2]$, known as N3 dye, one of the best sensitizer for DSSC, was the focus of this research. All calculations were carried out with Gaussian03, using B3LYP hybrid functional DFT and TDDFT for ground state geometry optimization and excited states calculations respectively. Basis setused in all calculations was 3-21G* for all elements. Polarized Continuum Model was used for modelling the complex in ethanol solution. Combination of 3-21G* basis set with PCM in N3 dye calculation achieve the better approach regarding to experimental absorption spectra. The last two maximum absorption peaks dominated by charge transfer excitation are red shifted in ethanol solution with respect to the gas phase. Electronic polarization between complex and solvent caused the structure to relax, permitting a greater charge separation.

1. Introduction

Since founded in 1990s, dye-sensitized solar cell (DSSC) has attracted many researchers' attention and the number of publication related to this research topic increasing quite rapidly in recent years. DSSC attracts many interests because this new kind of molecular solar cell achieved lower cost-efficiency ratio than the first and second generation solar cell, moreover, the development of DSSC still have wide area to be investigated in order to achieve higher efficiency while maintaining the low cost production [1]. Many research subtopics arises from DSSC investigation, such as finding new sensitizer, new semiconductor, new electrolyte system, electrolyte additives, controlling the semiconductor pore size, etc., both experimentally and computationally [2-7].

DSSC is a solar cell that employed a dye (usually an organic or metal complex molecule) as a photon catcher, called sensitizer. DSSC provides an efficient source of renewable energy, environmentally friendly, and quite economical [8]. In the DSSC cells, the properties of components such as the sensitizer absorption spectra (UV-Vis spectra) [9] and redox properties [10] play an important role in the photovoltaic performance of DSSC cells. Solar energy absorbed by the sensitizer is used to inject electrons into the semiconductor photoanode then transferred to the conductor part. Hence, sensitizer material has a critical influence to the overall cell efficiency.

One of the best sensitizerscomes from ruthenium polypyridyl complexes and its derivatives, such as N3 dye and black dye [11]. N3 dye is a famous name for dithiocyanatobis(2,2-bipyridine-4,4-

dicarboxylicacid)ruthenium (II) [Ru(H_2dcbpy)$_2$($SCN)_2$], having an overall cell efficiency up to 11,18% [11] and can be enhanced by modifying the electrolyte systems [12]. However, computational study to investigate the electronic structure, photonic properties, and excitation mechanism of this material is still not fully understood.

2. Computational Details

All calculations were carried out with Gaussian03 [13] program package using Becke, three-parameter, Lee-Yang-Parr (B3LYP) hybrid functional [14] density functional theory (DFT). Ground state geometry optimization of neutral form of N3 dye were carried out using 3-21G* basis set for all elements with no symmetry applied. Time-dependent density functional theory (TD-DFT) excited state calculations were determined based on geometry optimization results in both the gas phase and ethanol solution. The approach to embed N3 molecule in a dielectric medium in order to include the solvent polarization effect was used in calculation for N3 dye in ethanol solution. The cavity was defined as the union of a series of interlocking atomic spheres using Tomasi's Polarized Continuum Model (PCM) for the ground state geometry optimization and TD-DFT calculations for 70 singlet to singlet transition states. Mixed (GENECP) basis set with PCM ethanol solvent was used to predict the absorption spectra of N3 dye. Visualization was carried out using Chemcraft 1.6 [15].

3. Results and Discussion

3.1. Molecular Structure

The optimized geometry of N3 dye is shown in figure1. Ruthenium (II) ion is coordinated to four N atoms from two 2,2-bipyridine 4,4-dicarboxylic acid ligands and to two N atoms from two SCN^- ligands forming an octahedral-like molecular shape in the center. Comparison of the calculated structure parameter with experimental data was presented in table 1. The calculated structure parameters were slightly different with the experimental result because of the different condition of measurement with the model in computational study. Experimental structure parameter is usually obtained from a single crystal (solid phase) while the calculated values were for gas and aqueous phase in ethanol solvent. However, the experimental value still can be used to make a rough comparison. From the table above, it can be seen that the calculated structure parameters were in quite good agreement with the experimental data. Values from each method were slightly different with each other. Interatomic distance values after the solvation effect applied were longer than the values in gas phase in order to overcome intermolecular interactions between N3 molecule (solute) and ethanol solvent.

3.2. Excitation Energies of N3Gas Phase and N3 in Ethanol Solution

70 singlet to singlet transition states were calculated for both N3 optimized structure (gas phase and solution phase) and excitation energies for wavelength larger than 380 nm using both methods were presented in table 2. A detailed analysis of several N3 dye molecular orbital composition are presented in table 3.

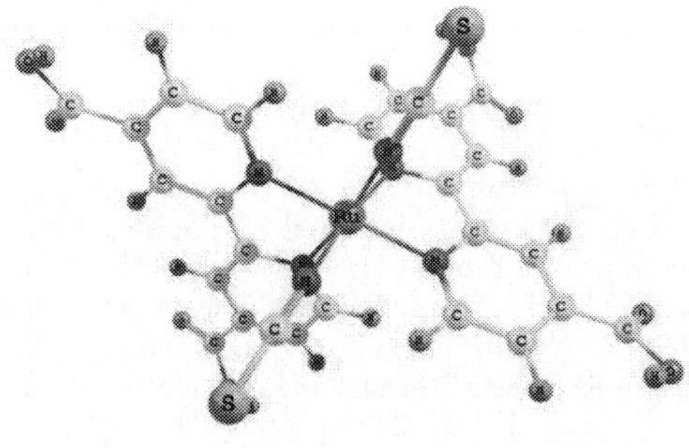

Figure 1 N3 dye optimized geometry obtained using 3-21G* basis set in ethanol solution.

Table 1 Structure parameter comparison between experimental [16] and computational results.

Structure parameter	Expt [16]	N3 Gas phase	N3 in Ethanol Solution
R (Ru-N1)	2.030	2.073	2.072
R (Ru-N2)	2.036	2.074	2.083
R (Ru-N3)	2.058	2.075	2.082
R (Ru-N4)	2.048	2.072	2.072
R (Ru-N5)	2.046	2.049	2.078
R (Ru-N6)	2.048	2.048	2.078
< N1 – Ru – N2	79.8	78.8	78.8
< N2 – Ru – N3	90.6	92.4	90.9
< N3 –Ru – N4	79.1	78.8	78.8
< N2 – Ru – N4	97.8	99.7	98.5

Table 1 Excitation energies for λ>380 nm, oscillator strength (*f*)> 0.01, and %composition>10%

State	N3 Gas Phase			N3 in Ethanol Solution		
	E (eV)	*f*	Composition	*E (eV)*	*f*	Composition
1	0.9178	0.0154	87% (HOMO to LUMO)	1.8154	0.0326	95% (HOMO to LUMO)
2	1.1969	0.0195	90.3% (HOMO-2 to LUMO+1)	1.8439	0.0143	85% (HOMO to LUMO+1)
3	1.3133	0.0367	43.8% (HOMO-3 to LUMO), 24.3% (HOMO-1 to LUMO+1), 17.3% (HOMO to LUMO+1)	2.1554	0.1091	55.3% (HOMO-2 to LUMO+1), 40.4% (HOMO-1 to LUMO+1)
4	1.3955	0.0367	91.3% (HOMO-3 to LUMO+1)	2.2480	0.0206	52.2% (HOMO-2 to LUMO+1), 27.1% (HOMO-1 to LUMO), 11% (HOMO to LUMO+3)
5	1.6770	0.0333	95.6% (HOMO to LUMO+2)	2.5992	0.0559	93.8% (HOMO to LUMO+2)
6	1.7565	0.0421	89% (HOMO to LUMO+3)	2.6718	0.0469	94.8% (HOMO-3 to LUMO+1)
7	1.8098	0.0147	92.3% (HOMO-1 to LUMO+2)	2.6946	0.0794	73.5% (HOMO to LUMO+3), 12.8% (HOMO-3 to LUMO)
8	2.0223	0.0238	64.2% (HOMO-3 to LUMO+3), 21.5% (HOMO-1 to LUMO+4)	2.8546	0.0477	77.1% (HOMO-2 to LUMO+2), 15.2% (HOMO to LUMO+4)
9	2.0237	0.0258	51.2% (HOMO to LUMO+5), 21.1% (HOMO-3 to LUMO+3),	3.0831	0.0368	91.5% (HOMO-4 to LUMO)
			19% (HOMO-3 to LUMO+2)	3.1299	0.0488	90.1% (HOMO-5 to LUMO+1)

10	2.7082	0.1843	70.7% (HOMO-6 to LUMO), 22.2% (HOMO-5 to LUMO+1)	3.1576	0.0103	62.3% (HOMO-5 to LUMO), 18.0% (HOMO-6 to UMO+1), 11.8% (HOMO-4 to LUMO+1)

Table 2 Percent composition of the highest occupied and lowest unoccupied orbitals of N3 dye in terms of Ru, bipyridine-dicarboxylic ligand, and SCN^- ligand for N3.

	N3 Gas Phase								N3 in Ethanol Solution							
MO	MO No.	% Ru			% Bipyridine		% SCN		MO No.	% Ru			% Bipyridine		% SCN	
		4d	5s 5p	other	ring 1	ring 2	1	2		4d	5s 5p	other	ring 1	ring 2	1	2
LUMO+3	181	0.6	0	5.5	46.6	46.9	0.2	0.2	181	0,5	0	4,4	46,2	48,5	0,2	0,2
LUMO+2	180	0.1	0	0.8	49.6	49.3	0.1	0.1	180	0,1	0	0,8	50,7	48,3	0	0
LUMO+1	179	1.5	0.1	13.6	41.5	39.9	1.7	1.8	179	1	0,1	10,5	42,1	44,8	0,7	0,7
LUMO	178	0.6	0.1	6	45	47	0.7	0.6	178	0,5	0,1	5,6	47,9	45,2	0,4	0,4
HOMO	177	2.5	0	22.2	5	4.5	32.2	33.6	177	5,7	0	49,1	3,9	3,9	18,7	18,7
HOMO-1	176	1.4	0	13	3	2.9	34.9	44.8	176	4,1	0	36,1	5,3	5,4	23,8	25,3
HOMO-2	175	1.2	0.1	11.2	3.1	3.2	45.6	35.6	175	4,3	0,1	38,3	3,7	3,7	25,6	24,4
HOMO-3	174	0	0.1	1.1	0.8	1	49.5	47.6	174	0	0	0,3	0,4	0,4	49,5	49,4
HOMO-4	173	7.6	0	65.6	5.2	5.1	8.3	8.2	173	5,5	0,1	48,1	5,9	5,9	18,7	15,9
HOMO-5	172	6.9	0.1	60.1	10.3	10.3	5.9	6.3	172	4,4	0,1	39	7,7	7,7	18,9	22,1
HOMO-6	171	7.2	0.1	64.1	8.4	8.4	5.9	5.7	171	4,8	0	42	5,9	5,8	20,7	20,7

It can be inferred from table 2 and table 3 that the gas phase N3 dye excitation energies with wavelength larger than 380 nm were mainly charge transfer. It can be seen from the large metal (Ru) contribution in mostly HOMO orbital that known to be the source of donor electron to LUMO orbital which is dominated by ligand (bipyridine ring) orbital contribution. From HOMO to HOMO-6, Ru atomic orbital contribution was larger than 10% except for HOMO-3. Donor electron in HOMO-4 to HOMO-6, where Ru contribution was larger than 67%, will be transferred to LUMO, LUMO+1, and LUMO+3 where the bipyridine ligands orbital contribution was larger than 81%. HOMO to HOMO-3 orbital, originated from more than 65% contribution of SCN^- ligands atomic orbital, will transfer its electron to LUMO, LUMO+1, LUMO+2, LUMO+3, LUMO+4, and LUMO+5 orbitals (dominated by bipyridine ligands atomic orbital) via Ru contribution ranging from to 2.1% to 35.9% from the HOMO-LUMO excitation pairs.

Calculation of N3 dye in ethanol solution used the same calculation parameter as gas phase, except the inclusion of PCM effect. 70 singlet to singlet transition states were calculated using the optimized geometry with PCM effect. Excitation energies for wavelength larger than 380 nm were also presented in table 2, and the detailed analysis of several solution phase N3 dye molecular orbital composition are also presented in table 3.

The excitation energies results from the addition of solvent effect are blue-shifted. This effect was expected to happen because when a molecule passes from gas phase into solution, the structure will relax to permit greater charge separation due to electronic polarization. The composition of molecular orbital in gas phase and in ethanol solution, if compared, will results in several difference of atomic orbital contribution. The solvent and solute interaction results in "shifting" composition of Ru and SCN^- ligand in HOMO to HOMO-2. In the case of SCN^-, the orbital composition in these orbitals is decreased.

Inspite of its role as donor electron, the decrease of SCN^- orbital contribution in HOMOs might be the reason of blue-shifting phenomena in excitation energies due to decreasing the donor electron strength, so the excitation became more difficult than before. This atomic orbital contribution difference was indirectly resulted from the interaction between molecule and solvent. When molecule interacts with solvent, polar-polar and polar-nonpolar interactions were occurred. This intermolecular interaction could make the interatomic bond in N3 molecule change, some become longer, some become shorter than its initial state. It caused the atomic orbital contribution to change.

HOMO and LUMO orbital energy levels of gas phase N3 were -0,191840 Hartree and -0,131480 Hartree, while HOMO and LUMO orbital energy levels of N3 in ethanol solvent were -0,201030 Hartree and -0,104770 Hartree. HOMO orbital stabilization and LUMO orbital destabilization occured when N3 dye moves to solvent environment.

3.3. Absorption Spectra

N3 dye absorption spectra obtained from gas phase calculation (figure2) using 3-21G* basis set was overestimating the absorbed wavelength compared to the experimental data. Hence, calculation results of the molecule in gas phase were not representative enough to model the system. Since the experiment data was obtained from measuring the N3 dye in ethanol solution, the next approach was to carried out calculation in ethanol solution. This approach was done with the inclusion of polarized continuum model (PCM) in the calculation. The result was presented in figure2.

In ethanol solution, calculated N3 dye absorption peaks shifted to shorter wavelength due to solute-solvent dielectric field. The calculated absorption spectra of N3 dye in ethanol solution have four absorption peaks for both basis sets used, as well as with the experimental absorption spectra. Peak with the largest wavelength located at 533-543 nm wavelengthand 576 nm for the results of experimental measurementsand calculations results using 3-21G* basis sets in solution, respectively. Wavelength absorption peak with the second largest in the experimental absorption spectrum lies at a wavelength between 403 to 409 nm. This peak was close to two absorption peaks in each calculation, it is almost in the middle of both peaks as if it is the result of the superposition of the two peaks. This peaks were at 391 nm and 455 nm on the results using the 3-21G* basis set in solution. Meanwhile, the third largest experimental absorption peak lies at a wavelength of 328 nm. This peak adjacent to the fourth peak from the calculation results in which lies at 285 nm for 3-21G* basis set in ethanol solution.

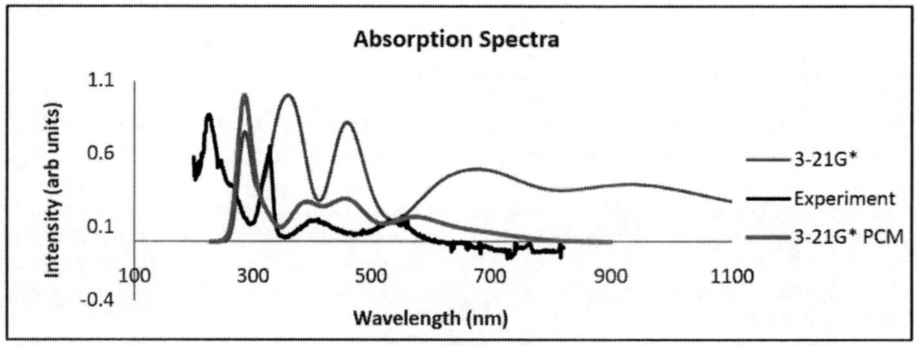

Figure 2 Calculated N3 dye absorption spectra using 3-21G* in gas phase (blue line) and 3-21G* in ethanol solution (brown line).

4. Conclusion

Computational study of N3 dye has been carried out by using B3LYP exchange correlation 3-21G* basis set and comparing its properties in gas phase and in ethanol solution (using PCM model). The calculation results and the comparison have been discussed above. 3-21G* basis set combined with

PCM solvation effect yields better approximation to molecular structure, excitation energies, and absorption spectra. Electronic excitation in N3 dye both in gas phase and in ethanol solution were mainly charge transfer. Solvation effect results in greater charge separation due to structure relaxationin order to overcome intermolecular interaction between N3 dye and solvent molecules.

5. Acknowledgements
The authors thank Prof. Hideaki Kasai for helpful discussion during the course of this research and Department of Applied Physics, Osaka University for research facilities. We also acknowledged financial support from Japan Student Services Organization (JASSO) for research funding.

References

[1] Grätzel M 2009 *Accounts of Chemical Research* **42** 1788-98
[2] Han W S, Han J K, Kim H Y, Choi M J, Kang Y S, Pac C and Kang S O 2011 *Inorg. Chem.***50** 3271-80
[3] McCall K L, Jennings J R, Wang H, Morandeira A, Peter L M, Durrant J R, Yellowless L J and Robertson N 2011 *Eur. J. Inorg. Chem.* **2011** 589-96
[4] Robertson N 2006 *Angew. Chem. Int. Ed.* **45** 2338-45
[5] Pastore M and De Angelis F 2012 *Phys. Chem. Chem. Phys.* **14** 920-8
[6] Hwang D, Jo S M, Kim D Y, Armel V, MacFarlane D R and Jang S Y 2011 *ACS Appl. Mater. Interfaces* **3** 1521-7
[7] Shi Y, Zhan C, Wang L, Ma B, Gao R, Zhu Y and Qiu Y 2009 *Phys Chem Chem Phys.* **11** 4230-5
[8] Wang M, Moon S and Zhou D 2010 *Advanced Functional Materials* **20** 1821-6
[9] Hagfeldt A and Gratzel M 2000 *Acc. Chem. Res.* **33**, 269-77
[10] Reynal A and Palomares E 2011 *Eur. J. Inorg. Chem.* **2011** 4509-26
[11] Nazeeruddin M K, De Angelis F, Fantacci S, Selloni A, Viscardi G, Liska P, Ito S, Takeru B, and Grätzel M 2005 *J. Am. Chem. Soc.* **127** 16835-47
[12] Yella A 2011 *Science* **334** 629
[13] Gaussian 03, Revision B.03, M. J. Frisch, G. W. Trucks, H. B. Schlegel, G. E. Scuseria, M. A. Robb, J. R. Cheeseman, J. A. Montgomery, Jr., T. Vreven, K. N. Kudin, J. C. Burant, J. M. Millam, S. S. Iyengar, J. Tomasi, V. Barone, B. Mennucci, M. Cossi, G. Scalmani, N. Rega, G. A. Petersson, H. Nakatsuji, M. Hada, M. Ehara, K. Toyota, R. Fukuda, J. Hasegawa, M. Ishida, T. Nakajima, Y. Honda, O. Kitao, H. Nakai, M. Klene, X. Li, J. E. Knox, H. P. Hratchian, J. B. Cross, V. Bakken, C. Adamo, J. Jaramillo, R. Gomperts, R. E. Stratmann, O. Yazyev, A. J. Austin, R. Cammi, C. Pomelli, J. W. Ochterski, P. Y. Ayala, K. Morokuma, G. A. Voth, P. Salvador, J. J. Dannenberg, V. G. Zakrzewski, S. Dapprich, A. D. Daniels, M. C. Strain, O. Farkas, D. K. Malick, A. D. Rabuck, K. Raghavachari, J. B. Foresman, J. V. Ortiz, Q. Cui, A. G. Baboul, S. Clifford, J. Cioslowski, B. B. Stefanov, G. Liu, A. Liashenko, P. Piskorz, I. Komaromi, R. L. Martin, D. J. Fox, T. Keith, M. A. Al-Laham, C. Y. Peng, A. Nanayakkara, M. Challacombe, P. M. W. Gill, B. Johnson, W. Chen, M. W. Wong, C. Gonzalez, and J. A. Pople, Gaussian, Inc., Wallingford CT, 2003.
[14] (a) Lee C, Yang W and Parr RG 1988 *Phys. Rev. B.* **37** 785
(b) Becke A D 1988 *Phys. Rev. A* **38** 3098
[15] Chemcraft, Version 1.7 (Build 382) 2013; Bluesnap, Inc.: Waltham, Massachusetts
[16] Fantacci S, De Angelis F, Sgamellotti A 2003 *J.Am.Chem. Soc.* **125** 4381-7

IOP Conf. Series: Journal of Physics: Conf. Series **812** (2017) 012069 doi:10.1088/1742-6596/812/1/012069

Improve Student Understanding Ability Through Gamification in Instructional Media Based Explicit Instruction

N Firdausi, H W Prabawa and H Sutarno

Computer Science Education Department, Universitas Pendidikan Indonesia, Jl. Dr. Setiabudi No. 229, Bandung 40154, Indonesia
Email: nuruly.firdausi@student.upi.edu

Abstract. In an effort to maximize a student's academic growth, one of the tools available to educators is the explicit instruction. Explicit instruction is marked by a series of support or scaffold, where the students will be guided through the learning process with a clear statement of purpose and a reason for learning new skills, a clear explanation and demonstration of learning targets, supported and practiced with independent feedback until mastery has been achieved. The technology development trend of todays, requires an adjustment in the development of learning object that supports the achievement of explicit instruction targets. This is where the gamification position is. In the role as a pedagogical strategy, the use of gamification preformance study class is still relatively new. Gamification not only use the game elements and game design techniques in non-game contexts, but also to empower and engage learners with the ability of motivation on learning approach and maintains a relaxed atmosphere. With using Reseach and Development methods, this paper presents the integration of technology (which in this case using the concept of gamification) in explicit instruction settings and the impact on the improvement of students' understanding.

1. Introduction

Vocational High School (SMK), according to article 15 of Undang-Undang No.20 of 2003, aimed at preparing students to be able to work in a certain field, which give priority to the development of students' abilities to certain types of work. But, the observations in the field, as stated in Document of Vocational Curriculum 2006, part of Rational curriculum, indicating, mostly graduates of vocational less able to adjust to the changes/developments in science and technology that are not easily trained back up the shortfall. The findings seem to indicate that learning in SMK has not appeared touching on self-development in the adaptability of learners. The study also noted that most graduates obtain vocational school can not be absorbed in the workforce because their capacity is not in accordance with the demands of the working world. The condition was also mapped in the document released by the World Bank in 2011 [1]. The document put the quality of facilities as the main problems of learning in vocational (by 29%), followed by the quality of teaching (by 23%), then a row is a problem with regard to general skills (13%), curriculum and cycle length (9%), relevance (8%), teacher skills and specific skills (4%).

Those facts led to allegations that the achievement of competency standards of vocational students are not fully met by the learning device available today. This paper raised the issue with regard to efforts to improve the ability of vocational students' understanding through the use of gamification in setting explicit learning. Understanding ability is a basic capital for student to be able to better interact with the material being taught concepts before ultimately concept implemented or developed in accordance with the needs of the problems faced.

Gamification is using elements of game mechanics to provide practical solutions by way of building interest (engagement) certain groups [2]. Gamification not only use the game elements and game design techniques in non-game contexts [3], but also to empower and engage learners with the ability of motivation on learning approach and maintains a relaxed atmosphere. In general, gamification contains some elements of which [4] : a) *games based* (these elements are intended to create a system in which learners or players involved in the abstract challenge, defined by rules, interactivity and feedback to produce measurable results ideally bring out the emotional reactions); b) *game mechanics* (level, earnings badge, point system, score and time challenge is the element used in the gamification); c) *game aesthetics* (how aesthetic experience felt by a person strongly influence or willingness to accept gamification); d) *game thinking* (the idea of thinking about everyday experiences such as jogging turn it into an activity that has elements of competition, cooperation, exploration and storytelling); e) *engagement* (the purpose of the gamification is to get someone's attention and engage the user in a process that's been made); f) *presence of people* (this can be learners, consumers, or the player. Individuals who will be involved in the process of creating and who will be motivated to take action)

2. Method

This research was conducted under research and development method, that includes several phases, which are analysing, designing, developing, implementing and evaluating. Learning media implemented to 37 students and 5 teachers that coming from one of the SMK in Bandung. The material presented in learning media is web apllication publication (especially with regard to the concept of domain and hosting, the principle of elections domain and hosting, and using cPanel application) that obtained in the even semester of the year 2015/2016. Technique of data collection implemented in this study was a test and questionnaire. Test instrument was designed as multiple choice questions with rationality, organized according to conceptual mastery indicators. While, questionnaires were used adopt the LORI instrument which cover interaction usability, accessibility, reusability and standards compliance.

3. Result and discussion

3.1. Explicit instuction syntax and gamification

Explicit Instruction model was first introduced by Rosenshine and Steven in 1986. The theory behind this model are the behavioral theory, research on teacher effectiveness, and social learning theory. In general, syntax of explicit instruction [5] are presented in Table 1.

Table 1. Explicit instruction syntax.

Phase 1 (*A Clear Goal*)	Setting the Stage for learning Clear explanation of what to do
Phase 2 (*Some Show & Tell*)	Modeling the process (showing)
Phase 3 *(Plenty of Practice)*	Guided Practice Independent Practice (when teacher is confident students will be successful) Assessment/Closure (informal or formal)

Based on three phases syntax (as presented in Table 1), instructional media focused on the first and second phases. In the third phase, independent practice activities managed manually because they relate to applications outside of instructional media. General overview of the media flow is presented in Figure 1. Multimedia developed use gamification elements as shown by Table 2.

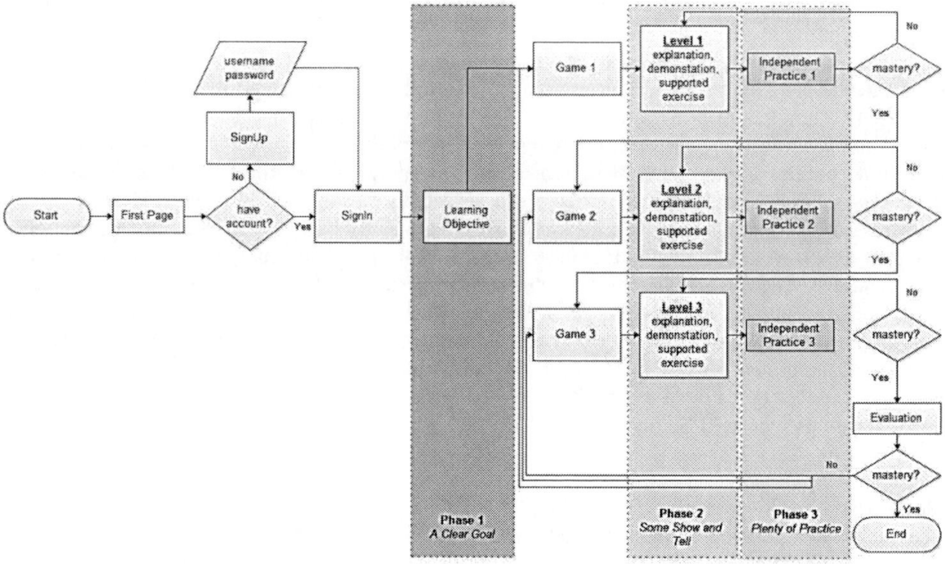

Figure 1. Explicit instruction syntax media with gamification flow chart.

Table 2. Gamification concept design.

Gamification Element	Information			
Game Mechanics	Level	Level 1 Task : Complete the exercises independently form five multiple choice questions (easy problem)	Level 2 Task : Complete the exercises independently form five multiple choice questions. (medium problem)	Level 3 Task : Complete the exercises independently form five multiple choice questions (difficult problem)
Game Aesthetics	Aesthetic element in media is the image color. The title of the game is a sky adventure, then the background of the game in the form of a stage at the top of skyscrapers. The blue color is more dominant because customized with the color of the sky			
Game Thinking	Moving to the right, to the left and jump where required to take the red key where the player must face the enemy advance. To deal with the enemy, players must put a bomb that exploded near the enemy and the enemy is lost, then the player must find a strategy to take the red key			
Engangement	Reward	Level 1 Reward : Getting 20 points for the correct answer. If values> 60, can continue to level 2.	Level 2 Reward : Getting 20 points for the correct answer. If values> 60, can proceed to level 2.	Level 3 Reward : Getting 20 points for the correct answer. If values> 60, can proceed to final evaluation.
Presence of people	Only one player who can play			

3.2. The impact of gamification in improving student's understanding ability

The assessment of media feasibility (before use in the classroom learning) is done through expert judgement that conducted by 2 media expert and 3 teachers, then media tested to 37 student. The summary of assessment shown by Table 2.

Table 3. The result of explicit instruction media assessing by expert.

Aspect							Aspect Rate
Interaction Usability	Accesibillity	Reusability	Standards Compliance	Software Engineering	Learning Aspect	Visual Communication	
0.836	0.815	0.903	0.850	0.817	0.873	0.887	0.854

From Table 3, obtained information that media aspect rate is 0.854. It's means, media has become reasonably well established in provide literature of web application publication. To further strengthen the research, conducted interviews to some students especially to further explore the potential of the media at the same weaknesses that have been developed. In general, given the response of students showed a positive influence, means consciously recognize that students obtain student outcomes (both within each level or in the final evaluation) is a real condition of their mastery abilities in web application publication material. So, it can be said that, gamification is an alternative in making the learning process more interesting, fun and effective.

However, the following questions should be considered whether the concept of gamification is used successfully to build engagement for students or not. Do students really interested and engaged in the learning process with the concept of the game was given? How to challenge and objective can improve students' proficiency and skills? How instructor/teacher to monitor the progress of the development of students as an evaluation?

4. Conclusion

In the learning process, gamification is an alternative in making the learning process more interesting, fun and effective. Despite using game mechanics, the use of gamification does not mean to make a game. The most important is the right concept, clear goals and capabilities in building engangement for students to learn.

Based on all the discussions that have been described, it can be concluded that the media gamification using explicit instruction give a positive impact on students' understanding. In addition, student responses regarding media gamification using explicit instruction also showed positive results.

5. References

[1] OECD/Asian Development Bank 2015 *Reviews of national policies for education - education in indonesia : rising to the challenge* (Paris: OECD Publishing)

[2] Vianna Y, Vianna M and Medina B 2014 *Recreating Companies Through Games Recreating Companies Through Games* 1ª edition vol 3 (Rio de Janeiro: MJV Press, Gamification, Inc)

[3] Werbach K and Hunter D 2012 *For the Win: How Game Thinking can Revolutionize Your Business* (Philadelphia: Wharton Digital Press)

[4] Kapp K M 2012 *The Gamification of Learning and Instruction: Game-Based Methods and Strategies for Training and Education* 1st edition (San Francisco: Pfeiffer & Company)

[5] Goeke K 2009 *Explicit Instruction: Strategies for Meaningful Direct Teaching* (Chicago: Merrill)

[6] Deterding S, Dixon D, Khaled R and Nacke L 2011 *Proceedings of the 15th International Academic MindTrek Conference: Envisioning Future Media Environments(Tampere)* (Tampere: MindTrek)

Acknowledgments
Grateful to the Computer Science Education Department of Universitas Pendidikan Indonesia who have given support to the writing of this manuscript.

IOP Conf. Series: Journal of Physics: Conf. Series **812** (2017) 012070 doi:10.1088/1742-6596/812/1/012070

Computational modeling of inclusion complex of r/s-omeprazole with β-cyclodextrin using oniom2 method

A. L. Ivansyah[a]*, M.A. Martoprawiro[a], Buchari[b]

[a]Inorganic and Physical Chemistry Division, Department of Chemistry, Faculty of Mathematic and Natural Science, Institut Teknologi Bandung, Jalan Ganesha No. 10, Bandung, West Java, 40132, Indonesia.

[b]Analytical Chemistry Division, Department of Chemistry, Faculty of Mathematic and Natural Science, Institut Teknologi Bandung, Jalan Ganesha No. 10, Bandung, West Java, 40132, Indonesia.

*)Corresponding author: attharquantum@gmail.com

Abstract. Host-guest inclusion complexes between R/S-omeprazole and β-cyclodextrin has been modeled using ONIOM2 (B3LYP/6-31g(d):PM3) method. Based on the value of binding energy obtained from the computational modeling, it was found that inclusion complex of R-omeprazole with β-cyclodextrin is more stable than the inclusion complex of S-omeprazole with β-cyclodextrin, moreover R/S-omeprazole can form stable inclusion complex with β-siklodekstrin by ratio of host : guest equal to 2 : 1. Based on the value of thermodynamic parameters, the formation of inclusion complex between R/S-omeprazole with β-cyclodextrin is an enthalpy driven process.

1. Introduction

Cyclodextrin is a cyclic-organic compound derived from starch transformation through enzymatic process. Cyclodextrin is widely used in many application due to its ability in forming inclusion complex with various large sized molecules [1–3]. Even though the main application of cyclodextrin is in the field of pharmacy, cosmetics, food, and agrochemical industry, cyclodextrin can also used in analytical chemistry application, especially in chiral compounds separation [4–6]. For example, cyclodextrin can be used as CSP in HPLC and CS in CE for S-omeprazole enantiomer separation from R-omeprazole and its racemic mixture [7].

By reducing gastric acid production, omeprazole is widely used to cure symtoms/syndromes caused by excess gastric acid, i.e. GERD (*gastroesophageal reflux disease)*, peptic ulcer, dan Zollinger–Ellison syndrome [8]. This compound with IUPAC name 5-metoksi-2-[(4-metoksi-3,5-dimetilpiridin-2-il)metilsulfinil]-1H-benzimidazol, has molecular structure composed of a substituted pyridine ring attached to benzimidazole through sulphoxide group [9]. The chiral center of this compound located in the sulphoxide group as shown in Figure 1.

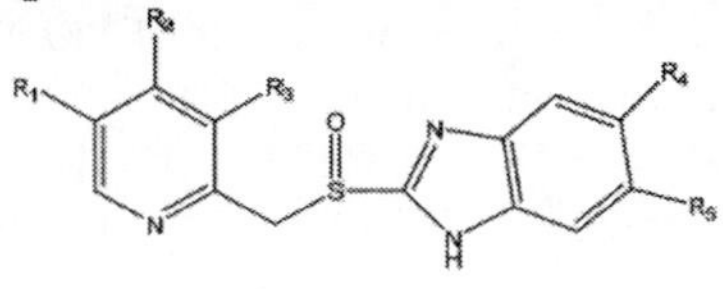

Figure 1. Omeprazole structure (R1 and R3 are methyl group, R2 and R4 are metoxy group, and R5 is H atom)

In nature, omeprazole exist as two enantiomers: S-enatiomer and R-enatiomer. However, S-omeprazole has higher effectivity in curing the syndrome compared with R-omeprazole and its rasemic mixture [9]. Meanwhile, omeprazole is normally obtained as rasemic mixture. Therefore, a certain method is needed to obtain a more effective drug (S-omeprazole), either by asymetric synthesis or by enantiomer separation [10]. S-omeprazole can be separated from R-omeprazole by using HPLC (*High Performance Liquid Chromatography*) or CE (*Capillary Electrophoresis*). HPLC employed CSP (*Chiral Stationary Phase*), while CE utilize CS (*Chiral Selector*) to obtain S-omeprazole from its rasemic mixture.

Some researcher have been doing experimental research about host-guest inclusion complex and reported the results, especially the inclusion complex between R/S-omeprazole with β-cyclodextrin [6,7,9-13]. However, due to the limitations of experimental method, molecular modelling can be used as alternative solution to predict and to explain experimental findings related to chiral recognition by β-cyclodextrin. Nowadays, various theories have been used by researchers to model host-guest inclusion complex, especially inclusion complex between a certain guest molecule and cyclodextrin host molecule. Among those theories, molecular mechanics [14,15], molecular dynamics [16-19], semiempirical methods (such as AM1, PM3) [20-22], hybrid ONIOM method [23-26], have been widely used by researchers to make a model of the inclusion complex between guest molecule and cyclodextrin (and its derivative) host molecule. The calculation method that is most frequently used i.e. hybrid ONIOM method developed by Morokuma et al. [23,24], because hybrid ONIOM method have a good accuracy as well as lower computational cost compared with ab initio and DFT methods. Most of researchers modeled host-guest inclusion complex with ratio host : guest = 1 : 1. However, Rajendiran et al [27] reported that omeprazole and β-cyclodextrin are able to form inclusion complex with ratio of host : guest = 2 : 1 experimentally. Rajendiran et al has not futher investigate about this inclusion complex formation using molecular modeling.

In this research, the interaction in the inclusion complex of R/S-omeprazol with β-siklodekstrin was studied using ONIOM (B3LYP/6-31g(d):PM3) method. The purposes of this research were to found the reason why and how chiral recognition of R/S-omeprazole by β-cyclodextrin occured, and to investigate the main driving force that lead to the inclusion complex formation, also to explain and obtain better understanding about the results reported by Rajendiran et al. In order to study chiral recognition mechanism of R/S-omeprazole by β-cyclodextrin, interaction energy between R/S-omeprazole with β-cyclodextrin was first determined using PM3 method, then continued with ONIOM (B3LYP/6-31g(d):PM3) method to observe interaction energy difference between R-omeprazole with β-cyclodextrin and S-omeprazol with β- cyclodextrin.

2. Computational details

In this molecular modelling, β-cyclodextrin act as host molecule and omeprazole enantiomers act as guest molecule. The structures of β-cyclodextrin used in this research were CD-HH, CD-HT, CD-TT, and CD. CD-HH is a symbol for β-cyclodextrin dimer in which 2'-hydroxyl and 3'- hydroxyl functional group in both β-cyclodextrin facing one another forming 180° angle towards the origin of Cartesian coordinate. CD-TT is a symbol for β-cyclodextrin dimer in which 6'-hydroxyl functional group in both β-cyclodextrin facing one another forming 180° angle towards the origin of Cartesian coordinate. CD-HT is a symbol for β-cyclodextrin dimer in which 6'-hydroxyl functional group in one β-cyclodextrin facing with 2'-hydroxyl and 3'-hydroxyl functional group in another β-cyclodextrin forming 180° angle towards the origin of Cartesian coordinate.

Each guest and host molecule was optimized using PM3 method with quantum mechanics Gaussian 09 Rev. D.01 [28] software until all eigen values from Hessian matrix were positive. The optimized structure of guest and host molecule then used for docking calculations using Autodock ver 4.2 [29]. The resulting inclusion complex structures from docking calculation were calculated with

Gaussian 09 Rev. D.01 using semiempiric PM3 method in vacuum (gas phase) condition. Several model sample structures from docking step were calculated using PM3 method, that is sample structure with minimum energy and sample structure with the most frequencies. The resulting optimized structures were further optimized with ONIOM method where β-cyclodextrin molecules which has structure CD, CD-HH, CD-HT, and CD-TT treated as low layer (calculated using PM3 method) and R/S-omeprazole molecule treated as high layer (calculated using DFT method with B3LYP exchange correlation functional and basis set of 6-31g(d)). Analysis of ONIOM (B3LYP/6-31g(d) : PM3) calculation results was performed by involving ΔE (binding energy), ΔH, ΔG, ΔS analysis, DEF (Deformation Energy) guest and host molecule, HOMO energy, LUMO energy, energy gap between HOMO and LUMO, μ (chemical potential), ω (electrophilicity), dipole moment, χ (electronegativity), η (hardness), Mulliken charges, and molecular orbital analysis to obtain further information about the character of the inclusion complexes between R/S-omeprazole with β-cyclodextrin.

3. Results and discussions

ONIOM2 method has been used by several researchers to model host-guest inclusion complexes. This method has relative low cost with good level of accuracy [23-26]. Shi et. al. [26] modeled the inclusion complex between ethyl-3-hydroxybutyrate with permethylated β-cyclodextrin using ratio of host : guest = 1 : 1. The modeling was carried out using quantum semiempiric PM3 method which is further calculated using ONIOM2 method. In ONIOM2 method performed by Shi et. al., guest molecule of ethyl-3-butyrate treated as high layer using DFT method with B3LYP as exchange correlation functional, while host permethylated β-cyclodextrin treated as low layer using quantum semiempiric PM3 method.

The result of ONIOM2 calculation indicate that one unit of β-cyclodextrin in CD-TT-R-OMZ and CD-TT-S-OMZ inclusion complex were distorted. Before the formation of inclusion complex occured, both β-cyclodextrin unit in CD-TT conformation were facing one another forming 180° angle towards the origin of Cartesian coordinate. However, after the formation of inclusion complex took place, both β-cyclodextrin unit were not in the same position and angle anymore. When guest R/S-omeprazole molecule forming inclusion complex with all β-cyclodextrin conformations, the geometry of R/S-omeprazole will undergo several changes. Changes on geometry of R/S-omeprazole can be seen from the shift in structure parameters as shown by Table 1.

The conformation of R/S-omeprazole undergo significant changes during the formation of inclusion complexes, where the molecule of R/S-omeprazole form a specific conformation to form the most stable inclusion complex with host molecule of CD, CD-HH, CD-HT, and CD-TT. These conformation changes in R/S-omeprazole structure indicate that R/S-omeprazole guest molecule tend to be flexible when entering the cavity of β-cyclodextrin in the inclusion complex formation process. According to research reported by Rekharsky et. al. [30], the flexibility of guest and host molecule is one of the important factor leading to the inclusion complex formation because host and guest molecule are able to adjust its conformation in such a way to ensure the inclusion complex formation and to increase entropy of host-guest inclusion complex formation. Based on the shift in interatomic distances and bond angles for atoms listed in Table 1, it can be inferred qualitatively that R-omeprazole was more flexible than S-omeprazole when forming the inclusion complex with β-cyclodextrin.

Table 1. Geometry structure parameter of R/S-Omeprazole before and after form inclusion complex

Distance between atoms	FREE OMZ		CD-R-OMZ	CD-HH-R-OMZ	CD-HT-R-OMZ	CD-TT-R-OMZ	CD-S-OMZ	CD-HH-S-OMZ	CD-HT-S-OMZ	CD-TT-S-OMZ
	R-OMZ	S-OMZ								
C21-C25	14.798	14.798	14.854	14.488	14.866	15.005	14.779	14.808	14.925	14.849
C4-C10	4.111	4.111	4.104	4.085	4.109	4.122	4.069	4.106	4.107	4.09
C21-C10	8.247	8.247	8.175	8.202	8.222	8.21	8.058	8.156	8.209	8.252
C3-C25	9.066	9.066	9.115	9.14	9.185	9.304	9.134	9.214	9.243	9.074
C3-C10	2.725	2.725	2.75	2.734	2.747	2.797	2.767	2.765	2.756	2.69
Angle (°)										
C3-S1-C10	95.542	95.542	97.012	96.21	97.332	99.128	97.580	97.916	97.772	94.593
N11-C10-S1	127.65	127.65	126.495	126.612	125.832	127.478	127.233	127.806	126.069	127.382
C4-C3-S1	110.27	110.27	107.605	110.348	108.463	106.594	105.432	106.573	107.617	110.224
Dihedral (°)										
N11-C10-S1-C3	64.05	64.05	64.575	56.649	58.222	61.981	65.702	50.141	62.642	71.032
C10-S1-C3-C4	177.89	177.89	177.248	160.815	179.543	168.435	163.314	173.736	175.421	171.242
N12-C10-S1-C3	119.33	119.33	120.818	130.612	130.237	127.438	118.916	139.787	126.662	111.308
S1-C3-C4-C5	90.42	90.42	90.593	90.462	86.732	99.505	98.632	96.033	95.044	142.543

Table2. ΔE, ΔH, ΔG, ΔS value of inclusion complex of R/S-omeprazole with β-cyclodextrin

	ΔE (kcal/mol)	ΔH (kcal/mol)	ΔG (kcal/mol)	ΔS (cal/mol.K)
CD-R-OMZ	-9.844	-10.125	8.165	-61.345
CD-HH-R-OMZ	-14.768	-15.830	7.895	-79.573
CD-HT-R-OMZ	-17.706	-19.463	7.870	-91.678
CD-TT-R-OMZ	-14.221	-15.343	8.013	-78.335
CD-S-OMZ	-8.849	-9.353	10.010	-64.943
CD-HH-S-OMZ	-11.310	-12.093	10.314	-75.155
CD-HT-S-OMZ	-15.741	-17.027	8.422	-85.353
CD-TT-S-OMZ	-13.296	-14.604	9.304	-80.189

Based on the value of ΔE, ΔH, ΔG, and ΔS obtained from computational modeling using ONIOM2 method, which is shown by Table 2, several points of conclusions can be inferred, i.e. inclusion complex formed from R-omeprazole with β-cyclodextrin is more stable than of S-omeprazole; Inclusion complex formed from R/S-omeprazole with β-cyclodextrin dimer (CD-HH, CD-HT, and CD-TT) is more stable than inclusion complex formed from R/S-omeprazole with monomeric β-cyclodextrin (CD); formation of inclusion complex between R-omeprazole with β-cyclodextrin is more exothermic than of S-omeprazole; formation of inclusion complex between R/S-omeprazole with β-cyclodextrin dimer (CD-HH, CD-HT, and CD-TT) is more exothermic than with monomer of β-cyclodextrin (CD); even though the values of ΔG formation were positive, the formation of inclusion complex between R-omeprazole and β-cyclodextrin is more spontaneous than of S-omeprazole at 1 atm and 298.15 K, also the formation of inclusion complex between R/S-omeprazole with β-cyclodextrin dimer (CD-HH, CD-HT, and CD-TT) is more spontaneous than with monomer of β-cyclodextrin (CD) at 1 atm and 298.15 K; and formation of inclusion complex between R-omeprazole with β-cyclodextrin is more favored than S-omeprazole with β-cyclodextrin because the more positive ΔS value. Based on ΔH and ΔS value obtained from computational modeling using quantum semiempiric PM3 and ONIOM2 method, it was found that the spontaneous formation of inclusion complex R/S-omeprazole is enthalpy driven process because both ΔH and ΔS values are negative. These negative values of ΔH and ΔS obtained from calculation indicate that the spontanity of complex formation is temperature dependent and the main cause of negative ΔG value will be the negative value of ΔH.

4. Conclusions

The results of computational modeling from this research confirms the research reported by Hancu et al. [31] and Figueiras et al. [11] that the S-enantiomer of omeprazole would be eluted first before the R-omeprazole due to the inclusion complexes of R-omeprazole is more strongly bound to the CSP or CS β-cyclodextrin compared with the S-omeprazole. This is due to the inclusion complex formed between R-omeprazole with β-cyclodextrin is more stable, more exothermic, more spontaneous, and more preferably than the inclusion complex of S-omeprazole with β-cyclodextrin based on the value of ΔE, ΔH, ΔG, ΔS obtained from the computational modeling. Computational modeling results also confirm the research reported by Rajendiran et al. [27] that the R/S-omeprazole can form stable inclusion complexes with a ratio of host:guest of 2:1. The results of the analysis of thermodynamic parameters such as ΔH and ΔS show that the inclusion complex of R/S omeprazole with β-cyclodextrin has a negative value of ΔH and ΔS indicating that the formation of inclusion complex is an enthalpy driven process.

References

[1] Sinha V R, Anitha R, Ghosh S, Nanda A and Kumria R 2005 *J. Pharm. Sci.* **94** 676

[2] Anselmi C, Centini M, Ricci M, Buonocore A, Granata P, Tsuno T and Facino R M 2006 *J. Pharm. Biomed. Anal.* **40**, 875
[3] Yap K L, Liu X, Thenmozhiyal J C and Ho P C 2005 *Eur. J. Pharm. Sci.* **25** 49
[4] Cass Q B, Lima V V, Oliveira R V, Cassiano N M, Degani A L G and Pedrazolli Jr. J 2003 *J. Chromatogr. B* **798** 275
[5] Orlando R M and Bonato P S 2003 *J. Chromatogr. B* **795** 227
[6] Nevado J J B, Penalvo G C and Dorado R M R 2005 *Anal. Chim. Acta* **533** 127
[7] Bonato P S and Paias F O 2004 *J. Braz. Chem. Soc.* **15** 318-323
[8] Cirilli R, Ferretti R, Gallinella B, De Santis E, Zanitti L and La Torre F 2008 *J. Chromatogr. A* **1177** 105
[9] Olsson J, Stegander F, Marlin N, Wan H and Blomberg L G 2006 *J. Chromatogr. A* **1129** 291
[10] Enmark M, Samuelsson J, Forssen P and Fornstedt T 2012 *J. Chromatogr. A* **1240** 123
[11] Figueiras A, Sarraguca J M G, Pais A A C C, Veiga J F and Carvalho R A 2008 *J. Incl. Phenom. Macrocycl. Chem.* **62** 345
[12] Redondo J, Capdevila A, Latorre I and Bertran J 2012 *J. Incl. Phenom. Macrocycl. Chem.* **73** 225
[13] Lu Y, Guo T, Qi J, Zhang J and Wu W 2012 *AAPS PharmSciTech* **13** 1222
[14] Tőkés B 2008 *J. Biochem. Biophys. Methods* **70** 1276
[15] Tablet C and Hillebrand M 2008 *Spectrochim. Acta. A. Mol. Biomol. Spectrosc.* **70** 740
[16] Zhou Z M, Li X, Chen X P, Fang M and Dong X 2010 *Talanta* **82** 775
[17] Raffaini G 2009 *J. Phys. Chem. B* **113** 9110
[18] Gotsev M G and Ivanov P M 2009 *J. Phys. Chem. B* **113** 5752
[19] Lipkowitz K B and Stoehr C M 1996 *Chirality* **8** 341
[20] Żuchowski G, Zborowski K and Bojarski J 2000 *Chem. Anal.* **45:6** 767
[21] Zborowski K and Żuchowski G 2002 *Chirality* **14** 632
[22] Issaraseriruk N, Shitangkoon A and Aree T 2010 *J. Mol. Graph. Model.* **28** 506
[23] Maseras F and Morokuma K 1995 *J. Comput. Chem.* **16** 1170
[24] Dapprich S, Komáromi I, Byun K S, Morokuma K and Frisch M J 1999 *J. Mol. Struct. THEOCHEM* **461–462** 1
[25] Attoui Yahia O and Khatmi D E 2009 *J. Mol. Struct. THEOCHEM* **912** 38
[26] Shi J H, Hu Y and Ding Z J 2011 *Comput. Theor. Chem.* **973** 62
[27] Rajendiran N and Venkatesh G 2015 *Spectrochim. Acta - Part Mol. Biomol. Spectrosc.* **137** 832
[28] Gaussian 09, Revision D.01, Frisch, M. J.; Trucks, G. W.; Schlegel, H. B.; Scuseria, G. E.; Robb, M. A.; Cheeseman, J. R.; Scalmani, G.; Barone, V.; Mennucci, B.; Petersson, G. A.; Nakatsuji, H.; Caricato, M.; Li, X.; Hratchian, H. P.; Izmaylov, A. F.; Bloino, J.; Zheng, G.; Sonnenberg, J. L.; Hada, M.; Ehara, M.; Toyota, K.; Fukuda, R.; Hasegawa, J.; Ishida, M.; Nakajima, T.; Honda, Y.; Kitao, O.; Nakai, H.; Vreven, T.; Montgomery, J. A., Jr.; Peralta, J. E.; Ogliaro, F.; Bearpark, M.; Heyd, J. J.; Brothers, E.; Kudin, K. N.; Staroverov, V. N.; Kobayashi, R.; Normand, J.; Raghavachari, K.; Rendell, A.; Burant, J. C.; Iyengar, S. S.; Tomasi, J.; Cossi, M.; Rega, N.; Millam, J. M.; Klene, M.; Knox, J. E.; Cross, J. B.; Bakken, V.; Adamo, C.; Jaramillo, J.; Gomperts, R.; Stratmann, R. E.; Yazyev, O.; Austin, A. J.; Cammi, R.; Pomelli, C.; Ochterski, J. W.; Martin, R. L.; Morokuma, K.; Zakrzewski, V. G.; Voth, G. A.; Salvador, P.; Dannenberg, J. J.; Dapprich, S.; Daniels, A. D.; Farkas, Ö.; Foresman, J. B.; Ortiz, J. V.; Cioslowski, J.; Fox, D. J. Gaussian, Inc., Wallingford CT, 2009.
[29] Morris G M, Huey R, Lindstrom W, Sanner M F, Belew R K, Goodsell D S and Olson A J 2009 *J. Computational Chemistry* **16** 2785
[30] Rekharsky M V 1995 *J. Am. Chem. Soc.* **117** 8830
[31] Hancu G, Papp L A and Rusu A 2014 *Chromatographia* **78** 279

The Use of Concept Map as a Consolidation Phase Based STAD to Enhance Students' Comprehension about Environmental Pollution

O. F. Nugroho[a], **D. T. Chandra**[b], **and Y. Sanjaya**[c]

Universitas Pendidikan Indonesia
[a,b,c)] Department of Science Education, School of Postgraduate Studies, Universitas Pendidikan Indonesia
[*)]oktianfajarnugroho@gmail.com

Abstract. The purpose of this study was to improve students' concept comprehension using concept map as a consolidation phase based STAD. This study was conducted by randomized control group pretest-posttest. Data was collected by using an instrument test to evaluate the effect of concept map as a consolidation phase based STAD on students'understanding about environmental pollution. Data was analyzed using normalized gain (n-gain) and independent t-test. The n-gain analysis shows the increased of students's understanding about environmental pollution at experimental group arehigher than at the control group. The result of this study showed that students' comprehension at the experimental class (0,53) higher compared to the control group (0,23). Whilst the t-test analysis shows that there is a significant effect of mapping concept as a consolidation phase based STAD towards students' concept comprehension. It can be concluded that the implementation of mapping concept based STAD may improve the students's understanding on science concept.

1. Introduction

There are several things that hang on experiences and adding visualization in the teaching learning practice for the student in the class that can improve the students' soft skill. Visualize the concept of the science materials can increase students interest to science. This thing has already implemented in some schools, such as concept map as assessment. Visualize the concept of science is also benefit during the teaching activities. Meanwhile there is another way using concept map, which is concept map as a consolidation phase.

According to Stegemann and Sutton-Brady(McNamara, *et al.*, 2010) provide a useful summary of the benefits of posters as identified by the literature. Several studies have reported that students feel comfortable with the concept of concept map. The map concept at an intervention session is an enjoyable and rewarding experience for students.Constructing a concept map is an effective of developing both research skills and creative abilities". According to Huddle(McNamara, et al., 2010), it was written that "Concept map allows students to become active learners and encourages deeper learning".

According to Bloom's Taxonomy proposed (Anderson, 2001) reveal the mastery is the ability of these terms, like being able to capture a material that is presented in a form that can be understood and be able to provide interpretation and classify. Mastery and understand of the concept of a person's level of learning outcomes that can define or describe a piece of information with their own words, students are required not merely to remember the material by using their own words. Mastery and understand the concepts is essential in teaching and learning activities because without concept, the learning process will be hampered.

Environmental pollution is a topic to be raised in the research of concept map as a consolidation phase bases STAD. There are various kinds of environmental pollution on earth, such as water pollution, air pollution, land pollution, etc. Those kinds of pollution materials explained to all students in the class that hopefully can be explored by students during the activity. This activity will increase of students' skills in finding the problem of environmental pollution, mention the characteristics of each environmental pollution, etc.

The concept map is a learning tool that has been used in learning and have a positive impact on the quality of student learning (Gerstner &Bogner, 2009; Lalor, 2014; Lian, 2009) and has been positively received in higher education (Lalor, 2014). A number of educational journals found that the various aspects of the application of a higher concept maps required the development of appropriate reviews to help students.

Based on Dahar (1996) said that the aim of concept map refers to investigate students behavior, knowing students way of learning, investigate students' missconception, and evaluation. Refers to Yamin (2006) concept map has some characteristics which are concept map is a representation of students mind, ilustration from two dimension that show the relationship between some concept to another concept, every concept map has different score each other, and the form of concept map is hierarchy.

Map concept seems to offer the most valuable contribution to student learning where the task concept map reflects actions taken to practice the discipline (Lalor, 2014). So, for example, in teaching physiology, students are encouraged to build concept maps actively integrating components subject and identify the causal relationship to reflect the desired learning outcomes (Lian, 2009).While the map concept is often a tendency to map the results of students in providing accurate and simple way to record student progress, it should be stressed that this approach has the potential of losing important information (Lalor, 2014)

Map concept is a completive approach that help to develop visual presentation of complex and allow reflection on the students learn new material (Gerstner &Bogner, 2009).Some knowledge structure is more open than the others to the assimilation of new knowledge (Lalor, 2014) and will be a determining factor in the success of the intervention of concept map, especially for certain students as their reference to assimilate or rejection of new content (Hay 2007 in the journal of Conradty&Bogner, 2012). An additional reason to focus on the concept map concept map is a tool that is commonly found in educational theory (as summarized in the journal Amadieu, et al., 2009), so as to study the conceptual gives access to a greater understanding to the students and find out why certain interventions can improve learning for more effective results.

STAD is a coorporate teaching method which was developed by Slavin. STAD is strategy in which small groups of learners with different levels of ability work together to accomplish a shared learning. It is considered as one of the simplest cooperative learning. The students was placed in small group or teams. Students will get more chance to share with other students. The reason why STAD was implemented because STAD can provide student to share their knowledge.

2. Experimental Method

The design that used in this research is quantitative research. In this research the test was conducted to evaluate students' prior knowledge or pretest (T1), Concept map as a consolidation phase based STAD (X), and post-test (T2). The type of method in this research isRandomized Control Group Pretest-Posttes Design. This method is chosen because using two classes which is one is control group and another one is experimental group (Arikunto, 2010).

Table 1.Randomized control group pretest-posttest design

Group	Pretest	Implementation	Posttest
Control group	T1	C	T2
Experimental group	T1	X	T2

From the table above show that in the experiment group did the intervention such as making concept map as a consolidation phase based STAD. Population in this reseach are all students of SMPN 1 CurugTangerang. Sample of this research are two classes of 7th grade from the population. Random cluster sampling was used as the sampling technique (Frankel & Wallen, 2007). Two clasess consist of experiment and control group. Experiment group was used concept map as a consolidation phase based STAD.

3. Result and Discussion

The result of implementation about the effectiveness of science learning using concept map as consolidation phase base STAD towards students' concept comprehension about environmental pollution. Five periods have already implemented in this research.

Mastery of the concept of data obtained throught a written test with a shaped of multiple choice question with 4 answer choices, with amounts to about 20 questions. Pretest is used to obtained students' concept comprhension about environmental pollution. After intervention using concept map (only in experiment group), there will be a posttest to measure students' improvement about environmenral pollution. The calculation results are summarized in the Table 1.

Table 2. Result of Research Experiment

	Experiment Group		Control group	
	Pretest	**Posttest**	**Pretest**	**Posttest**
N	39	39	36	39
Mean	67	84,63	48,36	64,36
The highest score	90	100	70	80
The lowest score	40	75	35	45
Gain	18,88		11,79	
N-Gain	0,53 (Medium)		0,21 (Low)	
Normality test	Sig. ≥0,05 = Normal			
Sig.	0,153		0,079	
Conclusion	Normal			
Homogenity test	Sig. ≥0,05 = Homogen			
Sig.	0,102			
Conclusion	Homogen			
Independent-Sample T-test	Sig. (2-tailed) ≥0,05 = H_0 Accepted			
Sig.	0,00			
Conclusion	H_0 Rejected, H_1 Accepted			

Based on the table shown that mean of students' concept comprehension in pretest (eperiment group) is lower than posttest which is 67 < 84,63 meanwhile in control group the pretest 48,36 > 64,36means that mean of experiment group is higher than control group. This table also show the results of high score and low score of students' concept comprehension. In pretest, the highest score of students is 90 and the lowest score of students is 40. Meanwhile in posttest, the highest score of students is 100 and the lowest score is 45.

The result of normality test of concept comprehension in experimental group is 0,152 and in control group is 0,079. Using significant interpretation (α) 0,05 so it can be concluded that the value of significant from normality test both of pretest and posttest is lower than significant. This result shows that the distribution of pretest and posttest score is normal.

The result of homogenity test of concept comprehension in pretest and posttest is 0,102. Using significant interpretation (α) 0,05 so it can be concluded that the value of significant from homogenity test both of pretest and posttest is higher than significant. This result shows that the distribution of pretest and posttest score is homogen. Based on the result of normality test and homogenity test is shown that the data of pretest and posttest of concept comprehension is not normal and homogen, and after that is hypothesis testing. By using a software of IBM SPSS 20 for windows is obtained that the value of significant is 0,00. This data is lower than α=0,05, it can be concluded **H_0 is not accepted, H_1 is accepted**, means that there is a significant effect of concept comprehension as a consolidation phase based STAD towards students' concept comprehension.

Concept map implementation as a treatment of the research was successfully conducted, the pretest as well as the posttest has been conducted as well. According to the result of compare mean analysis that there is improvement of students result between pretest and posttest. Based on the result the average amount of students' pretest in experimental group is 67 and the result of the average amount of students' posttest is 84,63 and the average amount of students' pretest in control group is 48,36 and the result of the average amount of students' posttest is 64,36 . Thus, it indicates that the implementation of conept map as a consolidation phase based STAD has positive impact in improving students' concept comprehension.

The result of n-gain average in experimental group is 0,53 and in control group is 0,21. Based on the result of gain average indicates that the quality of improvement of students' concept comprehension after implementation is categorized medium in experimental group and categorized low in control group. According to Slameto (2003) there are 2 factors that can influence the learning achievement they are; internal and external factor. The internal factor is determined by physical, physical factors and psychological factors (intelligence, interests, talents, and motivations. Concept map provide a physical activity to the students, the students actively involved in raising the prior knowledge and integrated it with another knowledge. Besides, the students not only having a physical activity they also had finding the best solutions to overcome the problem.

4. Conclusion

The implementation of concept map as a consolidation phase can improve student concept comprehension about environmental pollution. This study was calculated by SPSS 20 to measure student improvement. The results of experiment group is higher than control group. Cooperative learning based on STAD gave the best impact for student concept comprehension because students learned by sharing in the small group. Concept map helps teacher to analize student understanding and missconception about environmental pollution. Thus, concept map as a consolidation phase is the best way to do before doing the evaluation.

5. References

[1]Aderhold J, Davydov V Yu, Fedler F, Klausing H, Mistele D, Rotter T, Semchinova O, Stemmer J and Graul J 2001 *J. Cryst. Growth* **222** 701
[2]Arifin, Z 2009 *Evaluasi Pembelajaran* (Bandung: PT RemajaRosdakarya)
[3]Arikunto S 2012 *Dasar-dasar Evaluasi Pendidikan* (Jakarta: BumiAksara)
[4]Arikunto S 2010 *Prosedur Penelitian* (Jakarta: PT Rineka Cipta)
[5]Amadieu, F, *et all.* 2009 *Effect of Prior Knowledge and Concept Map Structure on Disorientation, Cognitive Load, and Learning, Learning and Instruction* 19, 376-386 DOI: 10.1016/j.learninstruc.2009.02.005
[6]Anderson O. W., Krathwohl D.R 2001 *A Taxonomy for Learning, Teaching, and Assessing A Revision of Bloom's Taxonomy of Educational Objectives (*New York: Longman)
[7]Conradty C & Bogner F. X. 2012 *Knowledge Presented in Concept Maps:Correlation with Conventional Cognitive Knowledge Test, Educational Studies* 38(3), 341-354 DOI: org/10.1080/03055698.2011.643100
[8]Gerstner S & Bogner F. X 2009 *International Journal of Science Education* 32(7), 849-870, doi: 10.1080/09500690902803604
[9]Lalor S B 2014 *International Journal of Science Education* 36(5), 839-864, doi: 10.1080/09500693.2013.829255
[10]Lian M W S 2009 *Reserach in Science Education* 39(3), 337-352 DOI: org/10.1080/03055698.2009.643100
[11]Martinez G, *et al.* 2012 *Journal Science Educatuion Technology* 22, 204-214 DOI: 10.1007/s10956-012-9386-8

[12]Munandar & Utami S C 1992 *Mengembangkan Bahan dan Kreatifitas Anak Sekolah (petunjuk bagi para guru dan orang tua)* (Jakarta: PT Gramedia)
[13] Munandar & Utami S C 2002 *Strategi Mewujudkan Potensi Kreatif dan Bakat* (Jakarta: PT Gramedia)

[14]Stewart M 2012 *Assessment and Evaluation in Higher Education* 37(3), 349-368 DOI: org/ 10.1080/02602938.2010.534764
[15]Sugiyono 2008 *Metode Penelitian Kuantitatif, Kualitatif dan R&D* (Bandung: Alfabeta)

Acknowledgments
The author would like to convey a gratitude to Universitas Pendidikan Indonesia for support this research. This paper was implemented in one of junior high school in Tangerang, Indonesia.

IOP Conf. Series: Journal of Physics: Conf. Series **812** (2017) 012072 doi:10.1088/1742-6596/812/1/012072

Performance Assessment of IT Governance with Balanced Score Card and COBIT 4.1 of Universitas Pendidikan Indonesia

N Y Wijayanti, W Setiawan and R A Sukamto

Departement of Computer Science Education, Universitas Pendidikan Indonesia, Jl. Dr. Setiabudi No. 229, Bandung 40154, Indonesia
Email: nurayuniar@gmail.com

Abstract. Information technology's application has become an important daily support for all sectors. Educational institutions, including Universitas Pendidikan Indonesia (UPI), enable information technology as the main asset to increase its qualities and global's competitive power. By the importances of using information technology for almost every scope, measurement is needed to identify how optimal the IT governance is. Based on these facts, the purposes of this reaseacrh are identify the IT governance's performance assessment indicators, discover the scores based on the indicators, and analyse IT governance's performance in UPI. This research is using the combination of Balanced Score Card (BSC) and COBIT 4.1 as the framework to establish assessment indicators in questionnaire's form.By combining both methods, the final scores of IT governance's performance will represent UPI's business goals and objectives in all sectors. This research used 26 COBIT's processes as assessment indicator of IT performance from the maping 15 IT and business goals of COBIT, and 17 UPI's strategic plans. The final score are 3.80 for financial perspective, 3.63 for customer perspective, 3.62 for internal business process perspective, and 3.72 for learning and growth perspective. With these scores, then the final result is each perspectives of Balanced Score Card's current maturity levels are at level 4, which is IT process criticality is regularly defined with full support and agreement from the relevant business process owners.

1. Introduction

Information technology or IT's application has become an important daily support for all sectors. IT ease and maximize target's achievements, both directly and indirectly. Including education institution, like university, enabling IT as the main asset to increase their qualities, both academic and non-academic such as financial, planning and developing, studenthood, and ICT (Information, Communication, and Technology). To have a global competitiveness, universities must have a significant excellence and not limited to academic sectors only.

Universitas Pendidikan Indonesia (UPI) has a vision to be a leading and outstanding university, also has strategics and goals to achieve its quality and Sglobal competitiveness that relevant to national education goals. IT's importance to UPI showed by the establishment of a directorate of IT's applications and supports, Direktorat TIK (ICT Directorate) on July 6th 2007. Also, based on UPI's 2011-2015 Strategic Plan, one of UPI's strategies to achieve goals and objectives is to leverage ICT on learning and management systems.

By knowing the IT's importance, it's necessary to have a measurement or assessment of IT governance's performance in UPI. Beside that, knowledge about the harmonization if strategics, business processes, an IT also affect the easeness of achieving business goals and objetives.

Balanced Score Card (BSC) is one of performance assessment and management system's approach [1]. BSC enables four perspectives at once in assessment, which are financial, customer, internal business process, and learning and growth. Control Objectives for Information and Related Technology (COBIT) is a comprehensive tool to measure IT governance's quality. COBIT unites and bridges the management's requirements of gap between business risks, control requirements, and IT's technical problems, and provide the references of best business practices that including all IT items and its connection with company's business process and explain it by logic activities' structures that can be managed and controlled effectively [2]. ISACA has been published five versions of COBIT, and this research use COBIT 4.1 that has four domains which are Plan and Organise (PO), Acquire and Implementation (AI), Deliver and Support (DS), and Monitor and Evaluate (ME).

Lot of companies have measured their IT governance, like PT Panin Sekuritas, Tbk and PT Pos Indonesia. PT Panin used IT BSC and gap analysis methods, but separatedly [3]. Otherwise, PT Pos Indonesia combines BSC and COBIT 4.1 methods, but specializing it to focus on performance indicators only [4]. Education institution, including Universitas Singaperbangsa Karawang also have measured information system's performance with COBIT 4.1, but only with the PO domain [5]. All three companies of above only measured the IT division, so that other sectors' IT performance haven't been measured yet.

Based on those facts, this research will use the combination of BSC and COBIT 4.1 methods with all BSC's perspectives and COBIT's domain by assessment indicators' maping from vision, missions, and strategic plans of UPI. Then, those indicators will measure all of working areas, so that the scores of performance will represent business goals and objectives of UPI on all sectors.

2. Literatures

2.1. IT Governance

Weill dan Ross stated that IT governance is specialized by decision making side and accountable framework's implementation to increase IT use based on the desired condition [6]. Here are the elements and domains of IT governance.

- Elements, to implement a proper IT governance needs the framework of these three main elements are structure, process and communication as rational mechanism
- Domains, a long with the breadth of research, there are few forms of IT governance's objectives/domains, which are IT value and alignment, risk management, accountability and performance measurement

2.2. BSC (Balanced Score Card).

BSC is is one of performance assessment and management system's approach [1]. The main element of BSC is scorecard, which is a record that illustrates key measurements with numbers that ease the company's executives to evaluare what happened on their work areas. And the balanced element reflects the concern about short term financial and budgetary issues only. A long with the developing, realized that both of those issues lead to more important issues such as customer development, customer development, changing markets, standards of service, and organisational learning, which never be payed attentionally or even be abandoned at all. Based on those thoughts, Kaplan and Norton formulated a model that consists of four quadrans/perspectives to represent key components, time span, and perspectives drom organisation's strategies [7]. Those four perspectives are financial, customer, internal business process, and learning and growth.

2.3. COBUT (Control Objectives for Information and Related Technology) Version 4.1

COBIT 4.1 is a Process Reference Model (PRM) of self-assessment programme. COBIT 4.1 provide the definition of processes with a life cycle form that describe the relationship between them, including the goals and achievements of each process. COBIT 4.1 has four domains and 34 processes. To conduct the IT governance performance's assessment, one of the indicators is maturity level that regulated by maturity models. Maturity models will be maped to questions form. Then, the scores of each maturity models will be represented by Sembiring (2013) with six levels of maturity level, which described in Table 1. BSC and COBIT 4.1 can be maped according to the connections between the BSC's perspectives, COBIT's business and IT goals, and processes.

Table 1. Maturity models' representations.

Level	Meanings	Details
0	Non-existent	Management processes are not applied at all.
1	Initial	Processes are ad hocand disorganised.
2	Repeatable but intuitive	Processes follow a regular pattern.
3	Defined process	Processes are documented and communicated.
4	Managed and Measureable	Processes are monitored and measured.
5	Optimized	Good practices are followed and automated

3. Methodology

This research consists four phase, which are preparation, maping, data collection, and analysis phase. The details of research design is showed in Figure 1.

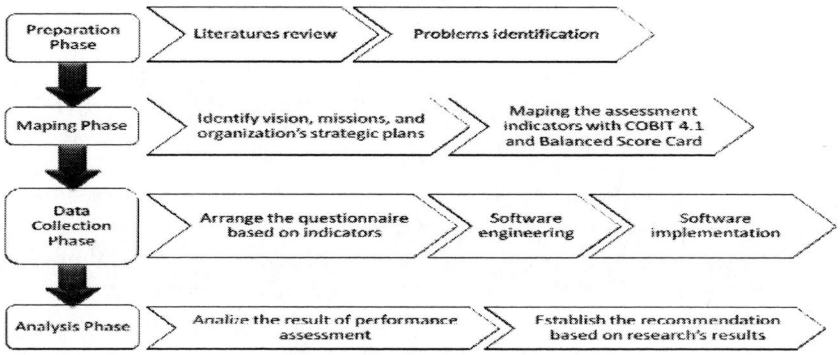

Figure 1. Research design.

Used method consists a few accurate data and information collection process, which are:

- Exploration and literatures review
- Observation
- Questionaire

BSC and COBIT is used on this research as a framework to maping the assessment indicators with a few steps showed in Figure [4].

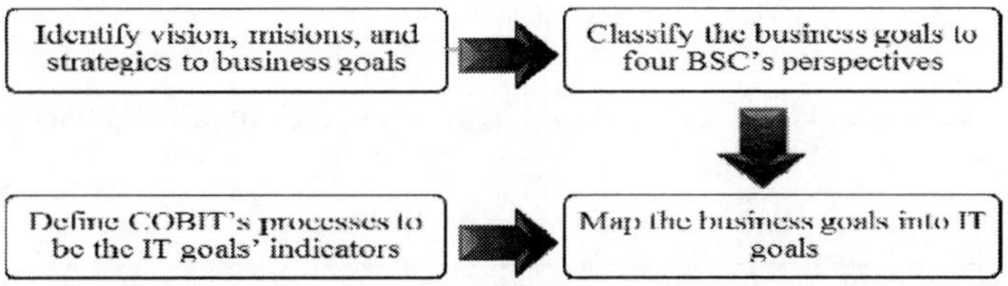

Figure 2. Steps of maping the assessment indicators.

4. Results

4.1. Maping The Assessment Indicators with BSC and COBIT 4.1 Methods

4.1.1. Maping UPI's business goals to COBIT's business goals. By selecting on priorities, there are 18 UPI's selected business goals based on vision, mission, and strategic plans. Then, map those business goals to COBIT's business goals. The result of maping are:

- Improve corporate governance and transparency.
- Improve customer orientation and service.
- Offer competitive products and services.
- Achieve cost optimisation of service delivery.
- Improve and maintain business process functionality.
- Provide compliance with external laws, regulations and contracts.
- Manage product and business innovation.
- Acquire and maintain skilled and motivated personnel.

4.1.2. *COBIT's IT goals and processes*. From the maping of UPI's business goals to COBIT's business goals, there are 8 business goals to be the assessment indicators. The grouping is showed on Table 2.

Table 2. Assessment indicator's maping

BSC's Perspectives	COBIT's Business Goals	COBIT's IT Goals	COBIT's Processes
Financial	1	ITG 2	PO10, ME1, ME4
		ITG 18	PO9
Customer	2	ITG 3	PO8, DS8, DS13
		ITG 23	DS3, DS4
	3	ITG 5	PO7, AI3
	4	ITG 24	PO5, DS6
Internal Business Process	5	ITG 6	AI1
		ITG 7	PO3
		ITG 11	AI4, AI7
	6	ITG 19	DS5
		ITG 21	DS12, ME2
		ITG 27	DS11, ME3
Learning and Growth	7	ITG 25	PO8
		ITG 28	ME3
	8	ITG 9	PO7, AI5
Amount of Processes of COBIT 4.1			26

4.2. IT Governance's performance

4.2.1. *Performance's score.* Details of each COBIT's processes in financial perspective showed in Table 3.

Table 3. BSC's perspective scores.

Perspective	COBIT's Prosesses	Average	Sum
Financial	PO10	3.78	3.80
	ME1	4.11	
	ME4	3.63	
	PO9	3.67	
Customer	PO8	3.89	3.63
	DS8	3.74	
	DS13	3.59	
	DS3	3.26	
	DS4	4.04	
	PO7	3.93	
	AI3	3.30	
	PO5	3.48	
	DS6	3.48	
Internal Business Process	AI1	3.81	3.62
	PO3	3.78	
	AI4	3.56	
	AI7	3.85	
	DS5	3.67	
	DS12	3.19	
	ME2	3.56	
	DS11	3.93	
	ME2	3.26	
Learning & Growth	PO8	3.74	3.72
	ME3	3.56	
	PO7	4.00	
	AI5	3.59	

The final score is each perspectives of BSC's current maturity levels are at level 4, which is IT process criticality is regularly defined with full support and agreement from the relevant business process owners.

4.2.2. Recommendations (The IT Governance Institute, 2007)

- The lowest score process on financial perspectives is ME4 (provide IT governance) which is 3.63, so that UPI needs to focus on preparing board reports on IT strategy, performance and risks, and responding to governance requirements in line with board directions. The achievement goals that can be tried are (1) stablishing an IT governance framework integrated into corporate governance and (2) obtaining independent assurance over the IT governance status
- The lowest score process on financial perspectives is DS3 (manage performance and capacity) which is 3.26, so that UPI needs to focus on meeting response time requirements of SLAs, minimising downtime, andmaking continuous IT performance and capacity improvements through monitoring and measurement. The achievement goals that can be tried are planning and providing

system capacity and availability, monitoring and reporting system performance, and modelling and forecasting system performance

- The lowest score process on financial perspectives is DS12 (manage the physical environment) which is 3.19, so that UPI needs to focus on providing and maintaining a suitable physical environment to protect IT assets from access, damage or theft. The achievement goals that can be tried are implementing physical security measures and electing and managing facilities
- The lowest score process on financial perspectives is ME3 (ensure compliance with external requirements) which is 3.56, so that UPI needs to focus on identifying all applicable laws, regulations and contracts and the corresponding level of IT compliance and optimising IT processes to reduce the risk of non-compliance. The achievement goals that can be tried are (1) identifying legal, regulatory and contractual requirements related to IT, (2) assessing the impact of compliance requirements, and (3) monitoring and reporting on compliance with these requirements.

5. Conclusions

- The assessment indicators to measure IT governance's performance on this research with BSC andf COBIT 4.1 methods are maped based on UPI's business goals. The results of maping are four BSC's perspectives, 8 COBIT's business goals, 15 COBIT's IT goals, and 26 COBIT's processes.
- The calculation of each BSC's perspectives is obtained from the average scores from the 26 respondents' answers in UPI, which are 3.80 for financial perspective, 3.63 for customer perspective, 3.62 for internal business process perspective and 3.72 for learning and growth perspective.
- The final score is each perspectives of BSC's current maturity levels are at level 4, which is IT process criticality is regularly defined with full support and agreement from the relevant business process owners. These are improvement focus on each BSC's perspective in UPI based on the research's results are (1) ME4 (provide IT governance) process for financial perspective, (2) DS3 (manage performance and capacity) process for customer perspective, (3) DS12 (manage the physical environment) process for internal business process perspective and (4) ME3 (ensure compliance with external requirements) process for learning and growth perspective.

6. Recommendations

- The continued research must be done with a bigger populations and samples to represent wider areas and higher number of accuration.
- Additional methods are needed to support the research's analysis, such as Key Performance Indicator (KPI), as is and to be scores, relatives scores, gap analysis, etc.

7. References

[1] Keyes J 2005 Aligning IT with corporate strategy implementing *The IT Balanced Score Card* (Boca Raton: Auerbach Publications)
[2] Surbakti 2012 *Jurnal Teknologi Informasi* **7** 1-14
[3] Benny, Saragih H and Reza B 2013 *Jurnal Teknik dan Ilmu Komputer* **3** 59-76
[4] Daniela A 2015 IT *Performance Management untuk Mengukur Kinerja Perusahaan dengan Metode Balanced Score Card* (Bandung: Skripsi UPI)

[5] Hendriadi A A and Jajuli M 2012 *Pengukuran Kinerja Sistem Informasi Akademik dengan Menggunakan Kerangka Kerja Cobit 4.1 pada Domain Plan and Organise di Universitas Singaperbangsa Karawang* vol 10 (Majalah Ilmiah Solusi Unsika)
[6] Rahmayuni I and PP I Y 2014 *Jurnal Momentum* **16** 88-94
[7] Mackay A 2012 Shareholder and stakeholder approaches to strategic performance measurement using the balanced scorecard *A Practitioner's Guide to the Balanced Scorecard* (Leeds: Leeds University).
[8] The IT Governance Institute 2007 *COBIT 4.1: Framework, Control Objectives, Management Guidelines, and Maturity Models* (Rolling Meadows: The IT Governance Institute)

Statistical Modelling for Dropped Out School Children (DOSC) in East Nusa Tenggara Province Indonesia

R D Guntur[1], and M Lobo[1]

[1]Department of Mathematics, Faculty of Science and Engineering, Universitas Nusa Cendana, Jl. Adisucipto Penfui, Kupang 85001, Indonesia.
Email: robertusdoleguntur@gmail.com

Abstract. A research has been carried out to investigate the characteristics of reasons for DOSC and to determine the statistical model explaining factors which influence on the DOSC in the age group 7 – 18 years in East Nusa Tenggara (ENT) Province. Primary data of out of school children had been collected throughout interviews using prepared questionnaires in three selected districts. Data was then analysed using descriptive and logistic regression method. The analysis shows that from the 341 samples, there were 194DOSC. The majority of them were males, lived in the countryside, had farmer parents, had family size of 5, and had mothers with only primary education level. The main reasons of children to drop out from the primary and junior education levels were the inabilities of paying the school fees and the willingness to work in the farms to help their parents. For senior education level, it was because of the unaffordable school tuitions and no desire of children in having good education. Both partial and simultaneous parameter tests in the logistic regression model show that children who lived in countryside, from poor families, males were the three factors that significantly affected the number of DOSC in the group age with odds ratio values 2.48; 2.37; 1.97 respectively.

1. Introduction

The issue of dropped out school children (DOSC) has been a public concern recently. Many studies have been conducted to investigate the complexity of the DOSC problem both in national and international level. Data shows that many children have no access to education. Globally, there were 61 million children in the primary education age group did not go to school in 2010 [1]. As a developing country Indonesia also faces similar problem. Around two and a half million children in the age group 7 -15 years were out of school [2].

East Nusa Tenggara Province (ENT) is one among the seven provinces in the eastern part of Indonesia that contribute to the large number of the out of school children in Indonesia. Most of out of school children in the area are males and live in the country side. A study conducted previously [3] shows that there were 795 out of school children in 10350 children in the junior high school age group in ENT.

A great number of investigations on DOSC have been conducted in many countries to explain factors contributing to dropping out of school. Rose [4] found that factors such as financial issues, teenage pregnancy, and family issues may have contributed to the number of students left school before graduation. Meanwhile Arun [5] indicated that physical disorders were the main reason for children in Kerala India for dropping out of school. Poverty has resulted in children have to work to help their families, as a result children might left school. Research conducted in The Philippines byRickey [6]found that children particularly those who lived in rural areas and who worked either part or full time have bad impacts on their schooling status. They were more likely to drop out of school. Then another research showed that gender issue is one of the crucial problems in basic education for developing countries. The attendance rate between male and female might be significantly different. The number of females who are out of school in urban and rural areas was two times higher than that of males [7]

Logistic regression is employed to describe the relationship between binary dependent variable and independent variables. It assumed that dependent variable has a binomial distribution [8]. In this case the dependent variable is the state of the children's drop out with category 'yes' and 'no'. The current study is to examine the characteristic of DOSC and to find a statistical model that explains factors that contribute to the high number of DOSC in the age group 7 -18 years old in the province of East Nusa Tenggara.

2. Experimental Method

This research employed primary data which had been collected throughout interviews using prepared questionnaires in three selected districts; East Sumba, Sikka, and Kupang City during 6 months from March to August 2016. The participants of the study were children who were out of schools and it was selected using purposive sampling.

The dependent variable of the study was the state of the children's drop out with category 'yes' and 'no'. The independent variables are individual characteristics as gender (male and female), and household characteristics as location of household (urban and rural), number of family members, education level of mother (no education, primary school, package A, junior secondary school, package B, senior secondary school, package C, academy, university), the occupation of household head (Civil Servant, Private employee, Labour, Farmer (Owner), Businessman/Entrepreneur, Army/Police, Pensioner, Fisherman, Craftsman , Driver, Motorcycle driver, and others) ,Income (the first category, the second category, the third category, the fourth category), and the reasons for drop out of school.

Initially, descriptive analysis was undertaken to observe the characteristics of DOSC based on individuals' and households' characteristics. A binary logistic regression model is then constructed by employing a stepwise method. To evaluate predictors in the regression model, parameter estimation significant test and model suitability test are carried out [8][9]. As the final step, model interpretation considering the scales of independent variables is conducted.

3. Result and Discussion

3.1 Characteristics of DOSC in East Nusa Tenggara Province.

Figure 1 shows the distribution of DOSC based on gender and location of household. From 341 research samples, 194 children were found to be dropped out from school. The number was dominated by male students (59%) whereas their counterpart female children were about 41%. With reference to location of households, the majority of children who were dropped out from school living at rural area at 65 %, whilst there were about 35% of them living in urban region.

Figure 1. The Proportion of dropped out School Children Based on Gender and Location of Households

This results is in line with a study conducted by Parikh A and Sadoulet E [14] using data survey in Brazil with 28,819 children involvement shows that children who live in rural areas spent most of their time to help their farmer parents. These children were at risk of being drop out of school.

With regard to the occupation of household head, it can be clearly seen in Table 1 that the majority of DOSC is from farmer family, which is 46.9%. Then, it is followed by laborer family and businessman family with 14.9 % and 9.8 % respectively. Meanwhile the proportion of DOSC from fisherman and craftsman is the same at 5.2 %. In general, most of DOSC in East Sumba and Sikka District is from farmer family at 19.1% and 23.2 consecutively. However, in Kupang City district, the highest percentage of DOSC is from businessman family at 5.7 %, and the proportion of DOSC from farmer and labor families in the region is the same at 4.6 %.

Looking at the table into more detail, it is evident that the majority of DOSC is from family which has income in category two at 37.90% followed by family in the third category at 28.5%. Meanwhile, the proportion of DOSC who has family in the first and fourth category is 14.70% and 19.10% respectively.

Table 1. The Percentage of DOSC based on Occupation of Parents, Income of Household, and Districts.

	Districts			Total
	East Sumba	Sikka	Kupang City	
Occupation of Parents				
Farmer	19.1	23.2	4.6	46.9
Laborer	4.6	5.7	4.6	14.9
Businessman	2.6	1.5	5.7	9.8
Fisherman	2.1	2.1	1	5.2
Craftsman	2.1	2.1	1	5.2
Civil Servant	3.1	0.5	0.5	4.1
Driver	1	0.5	1.5	3
Motorcycle Driver	0	2.1	0.5	2.6
Household wife	1	0	1	2
Pensioner	1	0	1	2
Others	1.5	1.3	1.5	4.3
Total	38.1	39	22.9	100
Income				
The First Category	10.00	1.20	3.50	14.70
The Second Category	14.10	13.20	10.60	37.90
The Third Category	8.20	10.90	9.40	28.50
The Fourth Category	6.50	6.70	5.90	19.10
Total	38.70	32.00	29.30	100.00

With reference to the highest level of education obtained by mothers as shown in Table 2 most of DOSC (39.2%) have mothers with primary education level followed by mothers with no education in 37.6% and mothers with junior high school education level in 10.8%. DOSC with mothers who have senior high school education was about 7.7%. It is interesting to note that with refer to districts the majority of DOSC in East Sumba and Kupang districts has mothers who mostly graduated from elementary school level at 13.9% and 12.4% respectively, in Sikka district is from family with mothers who has no diploma at 19.6%.

The result also shows that DOSC were mostly from families with 5 members with the percentage of 23.20%. This particularly happened in Sikka district with the proportion of 10.38%. . However, the situation is quite different in other districts. In East Sumba district for example, the DOSC was mostly from families with 7 members at about 9.76% and City of Kupang was families with 6 members at 6.70%.

The level of education of mothers plays an important role in supporting children education. The result of descriptive method of the research indicates that most of drop out children in ENT are from family with mothers who just finish elementary education level. This is in line with what stated by

Mike [13] that education level of parents is highly important in reducing the number of out of school children.

Table 2. The Percentage of DOSC based on the highest Level of Mother's Education, Family Size, and Districts

	Districts			Total
	East Sumba	Sikka	Kupang City	
The Highest Diploma of Mother				
No Diploma	13.4	19.6	4.6	37.6
Elementary School	13.9	12.9	12.4	39.2
Package A	0	0.5	0	0.5
Junior High School	4.6	3.1	3.1	10.8
Package B	0	0	0.5	0.5
Senior High School	3.6	1.5	2.6	7.7
Package C				0
Diploma	0.5	0.5	0.5	1.5
Bacherlor Degree	2.1	0	0	2.1
Family Size				
2	1.03	2.06	2.06	5.15
3	2.06	4.12	1.55	7.73
4	1.55	5.67	2.06	9.28
5	8.76	10.31	4.12	23.20
6	3.61	5.67	6.70	15.98
7	8.76	6.19	2.06	17.01
8	2.06	2.58	1.55	6.19
9	3.61	1.55	1.03	6.19
10	3.09		2.58	5.67
11	1.03			1.03
12	2.06			2.06
14	0.52			0.52

3.2 Reasons for dropped out from schools

Data in Table 3 shows that the main reason for dropping out from schools is quite different amongst three categories of students and between genders. It is clearly seen that in senior high school group there were 2 reasons that caused children to be dropped out from schools; can't afford to pay school fees and unwillingness to go to school which is recorded at 38.90% and 20.60% respectively. The reason to help parents in farm was also found as a significant factor that caused DOSC with percentage 19%.

At this level, the surroundings in children's residences and getting married were the reasons that contributed significantly to the number of DOSC at about 7%. The economy factor was the main reason for both male and female children to be dropped out from school with the proportion of 21.4% and 17.50% respectively. The reason for work to help the economy of the families for male children was about one and a half higher than that of their counterparts' female children with the proportion of 11.9% and 7.10% respectively.

In Junior High School group, the proportion of DOSC because of the inability to afford school fees was the highest at level of 28.3% followed by the reason to work for supporting families finance at 24.50%. In this group, the main reason for male children to be dropped out was school fees at 26.40% whereas the reason for female children was working for helping the families' economy. The health reason contributed to the number of DOSC for both male and female children at 3.8%. The same proportion was contributed by the number of female DOSC because of having families.

In primary education group, economic factor that led to the inability in paying school fees was the main reason contributed to the number of DOSC with the percentage 46.70%. Working to help families contributed about 26.70% to the number of DOSC. Surprisingly, the percentage for reasons for DOSC due to health issues and location of school which is far away from house is the same which is 6.70%.

Table 3. Proportion of reasons for dropped out school children based on the age group of the children

Reasons	Senior High School group			Junior High School group			Primary School group		
	Males	Females	Total	Males	Females	Total	Males	Females	Total
Can't afford school fees	21.40	17.50	38.90	26.40	1.90	28.30	20.00	26.70	46.70
Work to help families	11.90	7.10	19.00	15.10	9.40	24.50	26.70	0.00	26.70
School far from residence	0.80	0.00	0.80	0.00	7.50	7.50	6.70	6.70	13.30
Incapable to learn in school	0.80	0.00	0.80	1.90	0.00	1.90			
The effect of surroundings in residence	3.20	4.00	7.10	3.80	0.00	3.80			
Parents' health				0.00	1.90	1.90			
Experience health issues	2.40	2.40	4.80	3.80	3.80	7.50	6.70	6.70	13.30
Unwilling to go to school	14.30	6.30	20.60	18.90	1.90	20.80			
Married	0.00	7.90	7.90	0.00	3.80	3.80			
Total	54.8 (69)	45.2 (57)	100 **(126)**	69.8 (37)	30.2 (16)	100 **(53)**	60 (9)	40 (6)	100 **(15)**

3.3 Statistical Inference

The research was aiming at analyzing factors that related to DOSC based upon individual and household characteristics in age group 7-18 years. A logistic regression method is employed to estimate the probability of a child being drop out of school by controlling some particular variables. The application of stepwise method has been carried out to determine variables that can be input into each model to produce the best binary logistic regression model under criteria in Peng, et al [10]. The results of the suitable model for the data are displayed in Table 4.

Table 4. Best selected model for DOSC

Variables	β	Odds Ratio	Sig.
Gender_Male	0.68	1.974	0.003
Location_Rural	0.908	2.479	0
Income			
The First Category	0.055	1.057	0.889
The Second Category	0.863	2.369	0.008
The Third Category	0.438	1.549	0.19
Constant	-1.01	0.364	0.001

Table 4 shows that variables that affect the DOSC in ENT are gender, location of households, and the income level of parents with alpha 1%. It is clear from the table that Variable income is the most significant factor affecting the DOSC occurrence in ENT. Results show that its odds ratio is 2.37 which imply that if we compare children from family which has income level on the fourth level, children who are from the second category level are 2.37 times higher to be drop out of school. This is understandable since children in this group have financial issue which in turn will hinder the children in school. As a consequence, children might drop out from school.

The result of the research is in agreement with a study conducted by Sabates, R., et all [11]using multivariate logistic method towards 9047 children age 6 – 15 years in six districts of Bangladesh. The results show that children who were out of school and who have never been at school were from low income and low education level families. Moore, et all. [12] in his study also found that the welfare level of families was related very closely to children schooling status. Furthermore, his census in 2007 in United States showed that children from very low level income tended to be dropped out from school, had no academic achievement and had health, mental behavior problems. Moreover, Arun, et all [5] found that financial problems has become the biggest issue in most cases of drop out children.

4. Conclusion

The results of the research show that the characteristics of drop out school children in ENT are males, lived in the country side, had farmer parents, had family size of 5, and had mothers with only primary education level male children. The main reason for children to drop out in all education levels is inabilities of paying the school fees. Factors such as location of households, income of family, and gender might have contributed to the number of drop out school children in East Nusa Tenggara Province Indonesia.

5. References

[1] UNESCO 2012 *UIS Fact Sheet: Reaching out of school children is crucial for development* (*UNESCO Institute for Statistics* No. 18)
[2] UNICEF 2012 Global *Initiative on Out of School Children Indonesia Country Study*
[3] Guntur R and Lobo M 2015 *Asia-Pacific Collaborative Education Journal* **11** p 23-36
[4] Rose S 2013 *International Refereed Research Journal* **4** pp 81-86
[5] Arun N R, Kishore K and Shaji S 2012 *Indian Journal of Psychological Medicine* **34** 318-323
[6] Rickey L 2009 *The Determinants of Child Labour and Schooling in the Philippines* (California: Thesis Stanford University).
[7] Dursin R 2001 *Indonesia: Education Still a Male Domain* (*Asia Times Online*) [online: http://www.atimes.com/se-asia/CG18Ae02.html]
[8] Powers D and Xie Y 2000 *Statistical method for categorical data analysis* (San Diego: Academic Press, A Harcourt Science and Technology Company)
[9] Agresti A 2007 *An Introduction to Categorical Data Analysis* (Florida: Wiley series in probability and Mathematical statistics, John Wiley and Sons)
[10] Peng C J L K and Ingersoll G 2002 *The Journal of Educational ResearchIndiana University-Bloomington* **96** 3-14
[11] Sabates R, Hossain A and Lewin K M 2010 School drop out in bangladesh: new insights from longitudinal evidence *Research Monograph No. 49*
[12] Moore K A, Redd Z, Burkhauser M, Mbwana K and Collins A 2009 Children in poverty: trends, consequences, and policy options *Child Trends Research Brief, 4301 Connecticut Avenue, NW, Suite 350, Washington*
[13] Mike I, Nakajjo A and Isoke D 2008 Socioeconomic determinants of primary school dropout: the logistic model analysis *Economic Policy Research Centre No. 54*
[14] Parikh A and Sadoulet E 2005 The effect of parents' occupation on child labour and school attendance in brazil *Department of Agricultural & Resource Economics, University of California, Berkeley. Paper 1000. CUDARE Working Papers*

Acknowledgments
This research was supported by Research Grant "Competition Grant Program 2016", Ministry of Research, Technology and Higher Education, Indonesia.

Development of Guided Inquiry-Based Student Lab Worksheet on the Making of Pineapple Flavoring

G. Dwiyanti[1,a)], A. Suryatna, and I. Taibah

[1]Department of Chemical Education, FPMIPA, University of Indonesia
[a)] gebi.dwiyanti@upi.edu

Abstract. The aim of this research was to develop guided inquiry based student lab worksheet on making pineapple flavour and knowing the quality of worksheet that is being developed. Research methods that is being conducted is research and development that is limited by a preliminary studies (literature studies, field surveys, and preparation of the initial product) and development of the model (within limited testing). The results from analyze the books sources and fields survey showed that the characteristic of esterification lab worksheet that currently available still in the direct instruction form (cookbook). The optimization result of making pineapple flavour experiment that was conducted are the ethanol volume 3 mL, butyric acid volume 2 mL, sulfuric acid 5 drops, saturated $NaHCO_3$ solution volume 9 mL, and temperature of heating was 80°C. The characteristic of guided inquiry based student lab worksheet that was developed contained phenomenon and instructions that suitable with inquiry stages to guide the students in doing the experiment of making pineapple flavour. The evaluation of designated teachers and lecturers of the developed student worksheet were very good (96,08%). Lab-experiment feasibility achieved by using guided inquiry based student lab worksheets that is being developed based on the inquiry stages that conducted by student were found very good (97,50%) and accomplishment based on students' answer of the tasks in the worksheet were found very good (83,84%). Students' responses of the experiments using the developed worksheet are found very good (81,84%).

1. Preliminary

The chemistry has the nature as a product and as a process. The aim of the 2013n education curriculum is to prepare people of Indonesia to have the ability to live as individuals and citizens who has belief, is productive, creative, innovative, and affective also be able to contribute to the society, nation, state, and world civilization [1]. One of the methods that correspond to the nature of the chemistry and can be used to achieve the objectives of the curriculum in 2013 is the experiment method. In the learning with experiment method students not only get the concept (of the product), but students are also invited to experience the process in got the concept, so students later be able to apply the concepts they earn to contribute to society. Activities of students in learning with experiment methods can improve cognitive ability, metacognitive, psychomotor and attitude [2].

The 2013 education curriculum applies the scientific approach and learning by the pattern called 5 M which are observing, asking, collecting data, associating and communicating. One model that can be used is the inquiry model because the pattern of inquiry learning is similar to the 5 M pattern, which are questioning problems, formulating hypotheses, collecting data, analyzing the data, and making conclusions.

Learning by inquiry learning model is a student-centered learning because learning is compiled based on the principle of search and discovery through the process of thinking. [3] divides the inquiry learning model into three categories; *guided inquiry, inquiry-based design,* and *open-ended inquiry.* According to [4] learning by inquiry model can be implemented with the stages of observation, manipulation, generalization, verification, and applications. Type of inquiry appropriate to apply to the high school level learning is guided inquiry, because the guided inquiry learning provided more guidance for students, making it easier for students in solving a given problem.

One of the factors holders of the implementation of learning with experiment method is the availability of lab-worksheet.. According to [5] lab-worksheet is divided into expository lab-worksheet , guided inquiry lab-worksheet, and problem-based lab-worksheet differentiated based on the results, approaches and procedures. The results of analyze of the book sources showed that characteristic of esterfication lab worksheet that currently available still in the direct instruction form (cookbook). Cookbook lab worksheet does not make students actively participate in learning because students simply follow the procedures indicated on the lab worksheet. Therefore we need a lab worksheets to make students play an active role, namely guided inquiry based lab-work-sheet.

A number of studies about the advantages of the use of guided inquiry based lab worksheet has been done by [6] on improving students' thinking skills, [7] on students' memory enhancement, [8] on improving students' understanding significantly, and [9] concerning the improvement of the level of student understanding.

So that learning process can be more meaningfully implemented and can equip students to be able to contribute in the society in the future, then the material presented in the learning process should be based on everyday life. Esters have many uses in everyday life, such as a solvent for cosmetics and adhesives, as an ingredient in perfumes, and as food flavorings [10]. Based on the analysis of the authors, in West Java alone, especially in the area of Bandung many food industries that require the supply of food flavoring, such as mango banana, or pineapple flavoring. Therefore, it is necessary for students to do lab works on making food flavoring. Based on preliminary studies that has been done, esterification lab works in schools is still rare, and worksheets that are in school or in the student handbook are still in the form of *a cookbook* that is not yet appropriate for the learning to be applied.

Based on the problem analysis written above, it is necessary for a research concerning the development of students lab worksheet based on guided inquiry to be conducted. The aim of the research is to develop lab worksheet based on guided inquiry and to determine the quality of the lab worksheet. Authors conducted this research to develop the kind of lab worksheet mentioned above on making pineapple flavoring.

2. Methodology

The method used in this study is a research and development according to [11]. The stages of research are as follows: 1. A preliminary study and; 2. Development of a model (limited trial). Sources of data in this study are 13 high school chemistry books circulating in the city of Bandung, 7 high school chemistry teachers in Bandung and Cimahi, three lecturers from the Department of Chemistry Education FPMIPA-UPI, and 40 high school students. The instruments used in the form of sheet analysis worksheets, interview guides, design optimization, observation sheets enforceability stage, inkuiti, students' answers assessment rubrics, worksheets assessment sheet by teachers and lecturers, as well as the student questionnaire responses.

3. Results and Discussion

3.1. Results of Preliminary Study

3.1.1. Characteristics of Lab Worksheet Used in High Schools. Based on the research literature study on the characteristics of esterification lab-worksheet circulating in the field showed that of 13 books were examined only 6 books that contain lab esterification procedures and all procedures are still in the form of direct instruction or cookbook.

3.1.2. Lab Works in High School and kinds of Lab Worksheet Used. Based on the results of a field survey of the implementation and usage of esterification lab-worksheet at 10 high school in Bandung and Cimahi, there are only 7 school that doing experiment of esterification and lab- worksheets that are used are still in the form of direct instruction or cookbook.

3.1.3. Lab Worksheet Characteristics Developed in Conducted Research. To make a guided inquiry-based lab worksheets required an optimized reference lab procedure. Lab procedures that are used as reference material in the manufacture of guided inquiry-based student lab worksheet are created by [12]. Optimization is done to optimize the technical ethanol volume, volume of butyric acid, the amount of concentrated sulfuric acid droplets, heating time, the volume of a saturated solution of $NaHCO_3$, and the heating temperature.After the optimization of several variables that have been made, then the optimum results obtained from the experiment of making pineapple flavoring as contained in Table 1.

TABLE 1. Results of optimization on the making of pineapple flavoring

Ethanol Volume (ml)	Butyric acid volume (mL)	The amount of sulfuric acid droplets (drops)	Heating Time (min)	Volume $NaHCO_3$ Saturated (mL)	Heating temperature (° C)
3	2	5	5	9	80

Characteristics developed lab-worksheet containing: 1. Title of the identity of the experiments to be performed; 2. The phenomenon that contains a picture of a situation / problem that can lead students to a question about the problems that must be resolved through lab work to be performed; 3. Direction in formulating the problem; 4. Referrals in making hypotheses; 5. Tutorial experiment containing questions for the design of experiment,direction to determining the material , tools , tool sets and experimental procedure.; 6. Referral to experiment; 7. Referral to data analysis contains questions that lead students to load the relationship between the data obtained from the students experiment with the concepts that have been owned by the students, so that with the analysis of this data will facilitate students in making conclusions; 8. Tutorial prove the hypothesis; and 9. Referral to make conclusions.

3.2. Research Findings on Model Development.

3.2.1. Assessment of Teachers and Lecturers on Developed Guided Inquiry-Based Lab Worksheet on Making Pineapple Flavoring. Assessment of teachers and lecturers on the quality of developed guided inquir based lab worksheet. Assessment is done by seven high school Chemistry teacher in Bandung and Cimahi and 3 lecturers in the Department of Chemistry Education FPMIPA UPI. Results of the assessment of teachers and lecturers against developed guided inquiry based student lab worksheet can be seen in Figure 1.

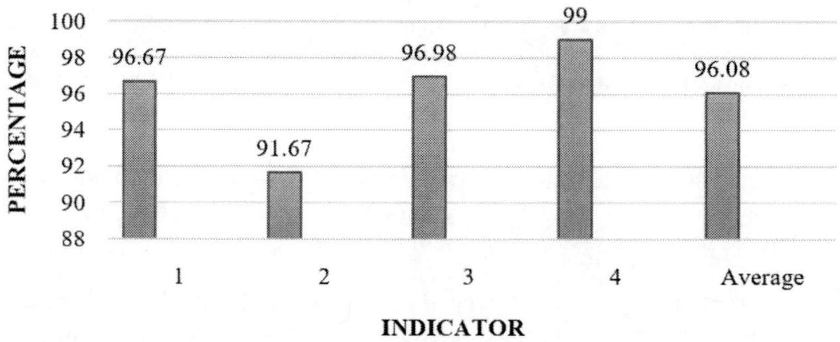

FIGURE 1. Diagram assessment of teachers and lecturers on developed guided inquiry-based student lab worksheet. 1. Conformity with the core competencies and basic competencies; 2. Compliance with the esterification material; 3. The effectiveness of the sentence in developed lab worksheet; 4. The layout and appearance of developed lab worksheets

Based on the diagram in Figure 1 can be seen the results of assessment of teachers and lecturers on developed lab worksheet with an average of 96.08%, which according to [13] belongs to the category of very powerful, meaning the developed lab worksheet has very good compliance to core competencies and basic competencies also esterification teaching material, sentences used are very effective, and the layout and appearance are very good.

3.2.2. Lab Experiment Feasibility. Lab-experiment feasibility using the worksheets were developed aiming to see the extent of adherence to lab work done by the students with the help of direction in developed lab worksheets .. Lab-experiment feasibility using developed lab worksheets viewed from two aspects of the assessment, the feasibility of the stages of inquiry and the students' answers to tasks that exist in the developed lab worksheet.

a. *Feasibility of the Stages of Inquiry.* To determine the experiment enforceability of use worksheets that were developed based on the stages of the inquiry conducted by the students is done by observing the practical activities undertaken by students. Therefore, placed one observer in each group of students. This is done so that any observer can observe student activities carefully and focus.Figure 2 shows the results of feasibility of the stages of inquiry.

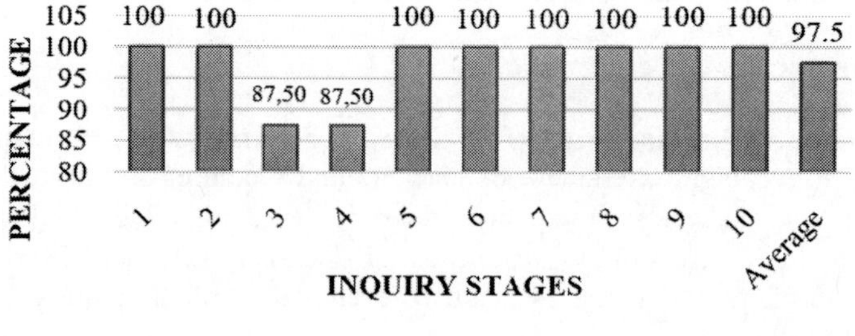

FIGURE 2. Diagram of the assessment results of feasibility of the stages of inquiry 1. Formulate the problem; 2. Make a hypothesis; 3. Choose the material; 4. Select the tool; 5. Make a sketch of tools set; 6. Conducting the experiment; 7. Write the observations; 8. Analysis of observation data; 9. Proving the hypothesis; 10. Make conclusions.

Based on the diagram in Figure 2 can be seen on the observation of the feasibility of the stages of inquiry conducted by the students gained an average of 97.5% according to [13] belongs to the category of very strong, meaning feasibility of the stages of inquiry is very good and lab- worksheets developed capable of directing students conduct inquiry stage.

b. *Students' answers to the task given in lab worksheet.* Results of the assessment of the students' answers to the tasks that exist in the developed lab worksheet shown in Figure 3.

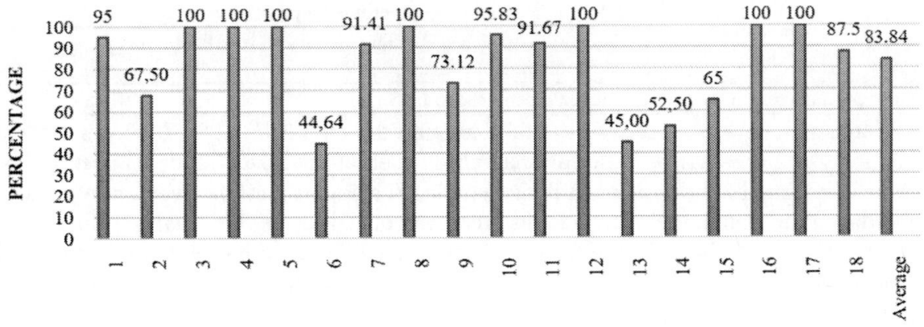

FIGURE 3. Diagram assessment students' answers to tasks given in the developed lab-worksheet 1. Formulating the problem; 2. Making a hypothesis; 3. Answering the question of the direction of the experiment 1; 4. Answering questions referrals experiment 2; 5. Answering questions referrals experiment 3; 6. Answering a question the direction of the experiment 4; 7. Selecting experimental tools; 8. Drawing tools set developed lab; 9. Designing an experiment; 10. Filling the observation table 1; 11. Filling the observation table 2; 12. Answering the question of data analysis 1; 13. Answering the question of data analysis 2; 14. Answering question 3 data analysis; 15. Answering question 4 data analysis; 1 6. Answering the question of data analysis 5; 17. Proving the hypothesis; 18. Making conclusions

Based on the diagram in Figure 3, it can be seen the average score on the assessment of the students' answers to the tasks contained in the developed lab worksheets is 83.84%. According to [13], the result belongs to the category of very powerful.That is feasibility of the experiment use developed lab worksheet based on assessment of the students' answers to the tasks contained in the developed lab-worksheet has been very good and the developed lab-worksheet capable of guiding students to discover concepts esterification.

3.2.3. Students responses to the use of the guided inquiry-based lab worksheet. The results of the students' response to use of the developed lab worksheets can be seen in Table 2.

TABLE 2. Percentage of Student Responses

Aspect measured	No.	Statement	Percentage (%)
The response of students to the esterification material	1.	Esterification teaching material can be understood easily	83.12
	2.	Esterification teaching material interesting to learn	88.75
	3.	Esterification material related to everyday life	78.75
The response of students to the implementation of experiment using developed lab worksheet	4.	The ease in formulating the problem	77,50
	5.	Ease in making hypotheses	81.25
	6.	Ease in determining the tools and materials	83.12

7.	Ease of sketching tools	78.75
8.	The ease of making an experimental procedure	78.75
9.	Ease of doing experiment	81.25
10.	The ease in filling the observation table	85.62
11.	Ease in analyzing the data	73.12
12.	Ease in making inferences	76.25
13.	Satisfaction in formulating the problem	88.75
14.	Satisfaction in making hypotheses	79.38
15.	Satisfaction in making the experimental procedure	85.00
16.	Satisfaction in using based guided inquiry lab-worksheet to the understanding of concept.	90.00
The average percentage of aspects of the students' response to LKS (%)		81.84

In Table 2 looks percentage score for ease of analyzing the data is 73.12%. It shows students are not too easy in analyzing the data because for analyzing observational data students need to connect this data with the concepts that have been students have until students can discover a new concept. This is in line with the values obtained in the students answered questions on data analysis (Figure 3).

Of all aspects of Table 2 shows the average response of students to the implementation of developed lab-worksheet get a percentage score of 81.84%, which according to [13] classified as category is very strong or very good.

4. Conclusions

Based on the results of research and discussion explained above, conclusions reached are as follows:

1. Characteristics of lab worksheet of esterification circulating in the school are still in the form of direct instruction *(cookbook).*
2. Optimization results of experiment procedure of making pineapple flavoring are the volume of ethanol is about 3 mL, the volume of butyric acid by 2 mL, the amount of sulfuric acid droplets as much as 5 drops, the heating time for 5 minutes, the volume of solution of saturated $NaHCO_3$ as much as 9 mL, and the heating temperature of 80 ° C.
3. Characteristics of guided inquiry based student lab worksheets on making pineapple flavoring are as follows:
   a. Contains phenomenon that guide students to formulate the problem regarding the making pineapple flavoring.
   b. Contains direction that correspond to the stages of inquiry to guide students doing making pineapple flavoring.
4. Assessment of teachers and lecturers against guided inquiry based student lab-worksheet on making pineapple flavoring is in accordance with core competence and basic competence, is in accordance with the esterification, the use of the phrase in the lab-worksheet has been very effective and has the layout and appearance of lab-worksheet has been excellent.
5. Feasibility of the stage of inquiry using guided inquiry based student lab-worksheet on making pineapple flavoring that conducted by student were found very good and accomplishment based on students answer of the task in the lab-worksheet were found very good.
6. Students' response to the implementation of the guided inquiry based student lab-worksheet on making pineapple flavoring were found very good.

References

[1] Kemendikbud 2013 *Permendikbud No. 69 Year 2013 About Standard Process SD/MI, SMP/MTs, SMA/MA, and SMK/MAK.* Jakarta: Ministry of Education and Culture.
[2] Hofstein, A., 2004 The Laboratory in Chemistry Education: Thirty Years of Experience with Developments, Implementation, and Research. *Chem. Educ. Res. Pract.,* 5, 247-254
[3] Mohrig, *et al.* 2009 Synthesis and Hydrogenation of Disubstituted Chalcones. A Guided-Inquiry Organic Chemistry Project. *Journal of Chemistry Education.* 86(2). 234

IOP Conf. Series: Journal of Physics: Conf. Series **812** (2017) 012074 doi:10.1088/1742-6596/812/1/012074

[4] Wenning, C.J. 2011 The Levels of Inquiry Model of Science Teaching. *J.Phys.Tchr.Educ.* 6(2), 9-16
[5] Johnstone, A. H. and Al-Shuaili, A 2001 *J of U. Chem. Ed.*, 5, 42-51.
[6] Wulandari, AD, *et al.* 2013 *Journal of chemical education research and practice,* vol. 1, No. 1, p. 18-26
[7] Domin, D.S. 2007 *Chemistry Education Research and Practice,* 8 (2), p. 140-152.
[8] Sessen, B. A. dan Tarhan, L. 2013 *Research Science and Education* .43, p. 413-435.
[9] Wang, H.P., *et.al* 2014 *Journal of Modern Education Review, USA,* 4 (2), p. 112-118.
[10] Toon, TY and Kwong, CL 2004 *Chemistry Matters for GCE 'O' Level.* Singapore: Federal
[11] Sukmadinata N S 2012 *Methods of Education (*Bandung: Youth Rosdakarya)
[12] Rahayu, I. 2009 *Chemistry for Class XII High School / Madrasah Aliyah Program Natural Sciences (*Jakarta: Ministry of National Education)
[13] Riduwan 2007 *The scale of measurement of the study variables (*Bandung: Alfabeta)

Prospective Primary School Teachers' Understanding on States Of Matter and Their Changes

W Sopandi[1,2], A Latip[3] and A Sujana[4]

[1] Elementary Education Program, Graduate School, Universitas Pendidikan Indonesia, Jl. Dr. Setiabudi No. 229, Bandung 40154, Indonesia
[2] Chemistry Education Department, Faculty of Mathematics and Science Education, Universitas Pendidikan Indonesia, Jl. Dr. Setiabudi No. 229, Bandung 40154, Indonesia
[3] Science Education Program, Graduate School, Universitas Pendidikan Indonesia, Jl. Dr. Setiabudi No. 229, Bandung 40154, Indonesia
[4] Primary School Teacher Education Program, Universitas Pendidikan Indonesia, Jl. Dr. Setiabudi No. 229, Bandung 40154, Indonesia
Email: wsopandi@upi.edu

Abstract. This study investigated Prospective Primary School Teachers (PPST)' conceptual understandings on states of matter and their changes. Data were collected from 37 students Primary School Teacher Education Program. The students were given open-ended questions that explored their understandings on states of matter and their changes divided into macroscopic, submicroscopic and symbolic levels. Their understandings were divided into categories: understand, partial understand, misconception, and do not understand. The results showed that almost all students understood that matters are composed of particles. But most of them did not understand that between the particles exists "particle-free space". Almost all students knew most of phenomena (macroscopic level) related to states of matter and their changes. However, most of them could not explain the phenomena at submicroscopic and symbolic levels. At these levels, their understandings were still largely in the categories of partial understanding, misconception, or do not understand. Given these results it can be concluded that most of the students did not have fully understandings on states of matter and their changes. These results imply the need for improving the teaching and learning quality on states of matter and their changes for PPST in Indonesia.

1. Introduction

Learning science subjects is knowing the various natural phenomena that exist in the environment and can explain them scientifically [1],[2]. According to the constructivist view, it is important for teachers to know what students already know or think before they teach their students scientific conception[3]. Students often come to class with a variety of knowledge which is different from scientific conceptions [4]. These misconceptions inhibit students to gain scientific conceptions. Therefore, it is very important for teachers to know what already exists in the student's cognitive structure and eliminate them to make students acquire scientific conception easier [5].

In chemistry subject, according to Taber most of the misconceptions is not the result of interaction between students and the environment outside the school but due to the interaction of students in a formal learning environment, namely due to students' misconceptions from earlier science learning, limitations in science models, errors in model applications and language statements used that are less precise [6]. Misconceptions caused by previous science learning will further affect on how students interpret and construct new concepts[7]. To avoid misconceptions in chemistry it needs good science teaching before they study chemistry specifically in high school, especially learning about the states of matter and their changes that are fundamental to study chemistry at higher levels [8]. The concepts about states of matter and their changes are very important basic concepts in order to understanding other science concepts [1],[9],[10]. Furthermore, the concept of states of matter and the changes are also a key component in science curriculum from primary school to middle school [10],[11],[12].

As known, chemistry subject is study about matter (properties, structure, composition) and its change [13],[14]. Based on chemistry point of view, the topic states of matter and their changes include three levels that need to be studied, namely macroscopic, submicroscopic and symbolic level [15]. Given this, after learning the topic students are expected to know the various phenomena of changes in states of matter (macroscopic level), to explain the phenomena scientifically (submicroscopic level) and symbolizes the phenomena (symbolic level). Good conceptual understandings about matter, substances and particles will be good prerequisite to understand subject matter of chemistry [8]. Without a good understanding of these concepts then most of the next study of chemistry can be as learning a mystery story. If this happens then study of chemistry can only be done by rote memory [3].

In other countries such as in Turkey the above concepts are studied by students ranging from grade 4 to grade 8 (10-14 years) [9]. Ozmen found that primary school students given early introduction of the particle concept had a better particle conception than that of students who are not given the introduction of the particle concept [9]. Research results of Sanmarti and Merino indicated that students aged 9-11 years (equivalent to primary school students ages) introduced about the particle concept have a better ability in describing the arrangement of particles in the process of physical changes as compared to that of students not given the introduction [2].

Previous research results indicated that students of various levels have misconceptions and difficulties in understanding the topic states of matter and changes [9],[10],[16],[17]. Misconceptions that occur in the topic caused by various factors, including, according to Ben-Zvi et al, the concept is too abstract and according to Gabel et al, the poor visualization capabilities of students [10]. In addition, students' misconceptions is also due to the learning process in schools, "school-made misconceptions"[18]. The students' difficulties to learn about the states of matter and the changes did not apart from the teachers' difficulties to understand particulate nature of matter composing all matter. Research results found by Håland in Norway showed that at least 80% of prospective teachers have problems to understand the concepts associated with matter, substances and particles [3]. Tatar in Turkey found that PPST have a number of misconceptions [19].

Teachers can be a source of the misconceptions for students in understanding science contens due to several things, including a lack of knowledge about the concepts to be taught [20], educational background speziallitation and teaching experience of teachers [21]. Actually teachers should be more than just mastery in subject matter but they should also be able to diagnose misconceptions and designing learning strategies to help students confront misconceptions and turn them into scientifically conceptions [22]. Based on the review of the literature, Ozmen conclude that there are three main reasons why chemistry is considered difficult by students [9]. The first is chemistry topics are abstract. Secondly many everyday terms used have different meanings from everyday meanings. The third is students' cognitive development stages who have not reached the formal operational stage and their low capacity of visualization. In addition to those three things, according to Johnstone chemistry also includes three levels of representation, macroscopic, submicroscopic and symbolic [18]. According to Hinton and Nakhleh students fully understanding of a phenomenon can be demonstrated by their ability to present and interpret such phenomena in the three levels [23].

Although several researches on the states of matter and the changes have been reported, but researches on the understanding of the topic categorized into macroscopic, submicroscopic and symbolic level has not been done. Given all above, the purpose of this study is to investigate on "how PPST in Indonesia understand about the states of matter and their changes at macroscopic, submicroscopic and symbolic level".

2. Research methodology

This descriptive study aimed to describe the conception of primary school teachers about the states of matter and their changes. The research was conducted for six months, from September 2015 until February 2016. The study involved 37 student teachers in a Primary School Teacher Education Program in West Java taking science as their major. They have already completed Basic Concepts of Chemistry, Biology, and Physics courses. The research instrument used is a multiple-choice test combined with essay, questionnaire about students' confidence level of their answer. This test is used to measure students' understanding levels about states of matter and their changes classified into macroscopic, submicroscopic and symbolic level. Students' answers in each level were then classified following the simplified categorization from Abraham [24], as follows:

- Understand: Respondents answer questions correctly includes all components in accordance with the scientific conception and the respondents were convinced by the answer.
- Partly Understand: Respondents answer questions correctly, but does not include all the components in accordance with the scientific concept and respondents undecided or are not sure of the answer.
- Misconception: Respondents answer questions with answers that are not logical with the wrong information.
- Do not understand: Respondents answer questions with the wrong answer or no answer at all.

3. Results and discussions

3.1. PPST' understanding on the discontinuous nature of matter

Table 1 below shows that the majority of PPST already understand and believe that any substance, whether in the form of solid, liquid and gas are composed by particles.

Table 1. PPST' understanding on the existence of particles of matter.

Levels of Understanding	The Concept on the Existence of Particle of Matter		
	Solid	Liquid	Gas
Understand	86.48%	97.29%	97.29%
Partly Understand	13.52%	2.71%	2.71%
Misconception	0%	0%	0%
Do not Understand	0%	0%	0%

In Table 2 below shows that PPST who already understand that there is a particle-free space between the particles composed solid, liquid and gas is still low (around 30%). Most of them have misconceptions, such as between particles of solids is air or other substances that fill it; between the particles of liquids is air, empty spaces and other liquids; between particles of gases is air and empty space. These results are consistent with previous researches which found that students/college students have misconceptions in understanding the existence of particle-free space, mostly referred to it as an empty space [1],[25],[26].

Table 2. PPST' understanding on the existence of "particle-free space" between particles of matter.

Levels of Understanding	The Concept on the Existence of "Particle-Free Space" between Particles of Matter		
	Solid	Liquid	Gas
Understand	24.32%	27.03%	24.32%
Partly Understand	18.92%	18.92%	16.22%
Misconception	37.84%	43.24%	40.54%
Do not Understand	18.92%	10.81%	18.92%

From the above findings, it seems that PPST have difficulty to accept the concept that between the particles of matter (solid, liquid or gas) is particles-free space. In the science history was known that ealier times scientific community needed up to 2,000 years to accept the discontinuous nature of matter or the concept that all matter are composed by the particles separated by particle-free spaces [27]. Only after discovery of vacuum, the concept of the discontinuous nature of the matter can be accepted by the scientific community. Based on the literature, the lack of time to introduce the concept causes many misconceptions related to particle concepts [9],[28],[29].

3.2. PPST' understanding on states of matter and their changes at macroscopic level

In Table 3 below shows that some PPST still have misconceptions regarding expansion, sublimation, and deposition phenomena. Misconceptions related to expansion such as solids, liquids and gases do not change the volume when heat is given. Misconception related to sublimation such as shrinking of mothballs involve evaporation process. And misconception related to deposition such as the formation of icebergs at the poles as the process of freezing and sublimate.

Table 3. PPST' understanding on states of matter and their changes at macroscopic level.

Levels of Understanding	Concept Labels						
	Expansion	Epavoration	Condensation	Melting	Freezing	Sublimation	Deposition
Understand	32.43%	100%	100%	100%	100%	97.3%	29.73%
Partly Understand	8.11%	0%	0%	0%	0%	0%	21.62%
Misconception	35.13%	0%	0%	0%	0%	2.7%	40.54%
Do not Understand	24.33%	0%	0%	0%	0%	0%	8.108%

3.3. PPST' understanding on states of matter and their changes at submicroscopic level

As known, representation at submicroscopic level is based on the particle theory used to explain real phenomena in the form of the motion of a particle such as electrons, ions, atoms and molecules [30]. In line with that, this study investigated PPST' understanding in verbally explaining the process of expansion (where the states of matter does not change but it changes size) and changes of states of matter such as evaporation, condensation, melting, freezing, sublimation, and deposition.

Based on Table 4 below shows that PPST who are able to explain all concepts at the submicroscopic level verbally are below 50%. Most of the students can be categorized as partly understand, misconception or do not understand.

Table 4. PPST' Understanding on states of matter and their changes at submicroscopic level verbally.

Levels of Understanding	Concept Labels						
	Expansion	Epavoration	Condensation	Melting	Freezing	Sublimation	Deposition
Understand	45.95%	46.85%	32.43%	40.54%	43.24%	33.33%	39.64%
Partly Understand	18.92%	13.51%	18.02%	18.02%	18.02%	15.32%	18.92%
Misconception	8.11%	19.82%	24.32%	17.12%	21.62%	17.12%	18.02%
Do not Understand	27.03%	19.82%	25.23%	24.32%	17.12%	34.23%	25.23%

Some misconceptions that arise are as follows:

Table 5. PPST' misconceptionson states of matter and their changes.

Expansion/ Changes in states of matter	PPST' Misconceptions
Expansion	Particles of solids, liquids, and gases enlarged when experiencing expansion
Evaporation	Water particles disappear when water evaporate
	Water particles swell when water turns into water vapor
Condensation	Water particles swell when it turns into dew
	Water particles do not have movement when water vapor condenses to water
Melting	Water particles are smaller when ice melts
	Water particles disappear when ice melts
	Water particles do not have movement when ice melts
Freezing	Water particles do not move in frozen ice
Sublimation	Particle of mothballs shrink and disappear
Deposition	Water particles do not move when vapor turns into solid

Those above misconceptions of PPST in explaining the phenomenon at the submicroscopic level are in line with previous research findings [9],[10],[16],[26]. According to Nussabaum and Novick, misconceptions that occur in the submicroscopic level is due to various factors, including everyday experience[1] and learning processes which present events and assumptions but those are not in accordance with the phenomenon known in everyday life [31].

PPST' understanding to explain the phenomena visually by using drawing were also investigated. Table 6 below shows that PPST who are able to draw particle model for all the labels concept is below 50%. Many of them still have misconceptions.

Table 6. PPST' understanding on states of matter and their changes at submicroscopic level visually.

Levels of Understanding	Concept Labels						
	Expansion	Epavoration	Condensation	Melting	Freezing	Sublimation	Deposition
Understand	15.32%	24.32%	24.32%	32.43%	32.43%	29.73%	8.11%
Partly Understand	33.33%	10.81%	8.11%	13.51%	10.81%	18.92%	16.22%
Misconception	17.12%	24.32%	21.62%	16.22%	16.22%	10.81%	10.81%
Do not Understand	34.23%	40.54%	45.95%	37.84%	40.54%	40.54%	64.86%

Misconceptions that occur include drawings of particle models that are changing the size (shrink and swell) as shown below:

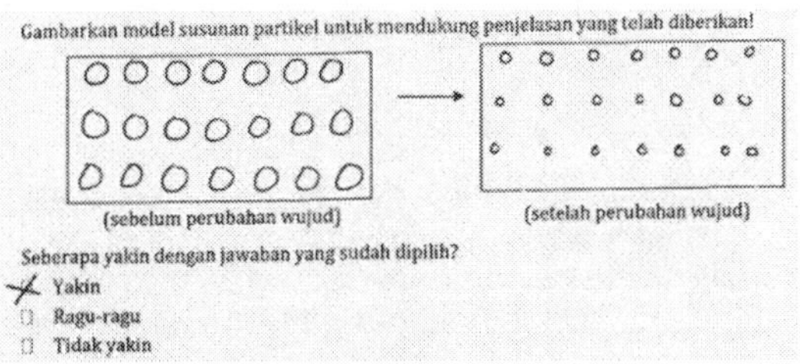

Figure 1. PPST' answer on condensation at submicroscopic level visually.

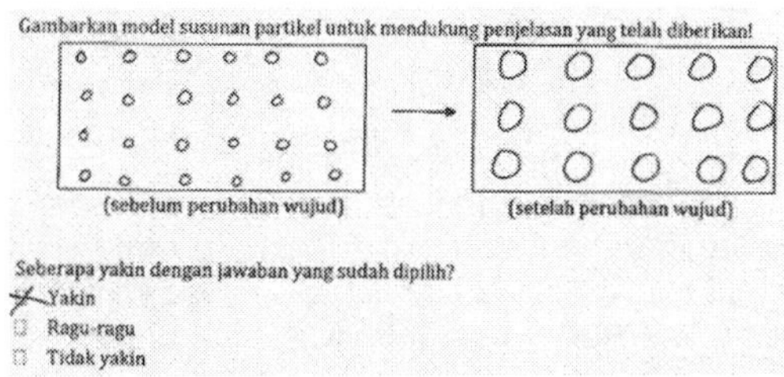

Figure 2. PPST' Answer on Evaporation at Submicroscopic Level Visually.

Figures 1 and 2 above belong to misconceptions because the drawings of particle size change (become smaller / become larger), particle size of solids, liquids, and gases should scientifically remain unchanging during the changing. These research results are consistent with previous studies which found that students/ college students assumed that particle size increases when there is a change form from solid to liquid and to gas [26],[32],[33]. Another misconception found in Figures 1 and 2 is that drawings of the distance between the particles before and after the changes are the same. The distance between the particles should be different before and after the change. In general, the distance between particles in solids is dense, in liquids is distant and in gases is very distant. Besides the changes of the distance between particles, drawings should also show the change in the arrangement of the particle when states of matter change. The difficulties of PPST in Indonesia to explain the phenomena at submicroscopic level is in line with several studies conducted in other countries which found that students of all levels, college students and even prospective student teachers have difficulty to explain the states of matter and their changes at the submicroscopic level [34],[35],[36],[37].

3.4. PPST' Understanding on States of Matter and Their Changes at Symbolic Level

Symbolic level is representation in the form of chemical symbols, formulas and reaction equation [30]. In this study, PPST' understanding level was measured by asking them to write symbols of states of matter such as solid (s), liquid (l) and gas (g) at different changes states of matter. PPST' understanding about changes states of matter at symbolic level shown in Table 7 below:

Table 7. PPST' understanding on changes states of matter at symbolic level.

Levels of Understanding	Concept Labels					
	Epavoration	Condensation	Melting	Freezing	Sublimation	Deposition
Understand	37.84%	35.14%	43.24%	45.95%	21.62%	5.41%
Partly Understand	27.03%	24.32%	43.24%	29.73%	16.22%	18.92%
Misconception	13.51%	10.81%	0.00%	0.00%	0.00%	0.00%
Do not Understand	21.62%	29.73%	13.51%	24.32%	62.16%	75.68%

Table 7 shows that the percentage of PPST who can write the symbolic level for all changes states of matter is still below 50%. On the concept evaporation and condensation there are still PPST who have misconceptions. Following is a misconception that occurs in writing the symbol of the evaporation.

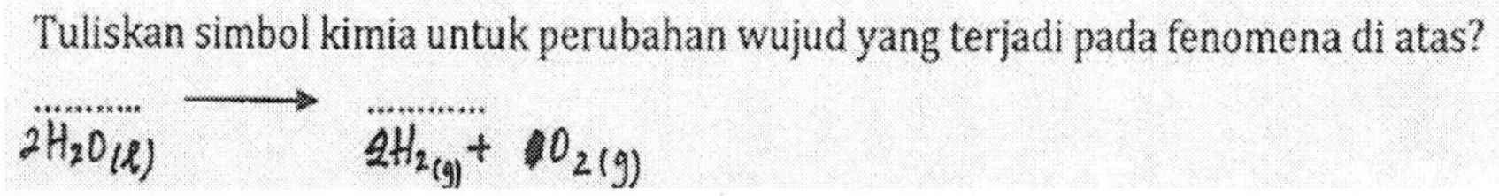

Figure 3. PPST' answer in writing symbols of evaporation.

This above symbolic reperesentation shows a misconception in which the evaporation process is regarded as the decomposition of H_2O into H_2 and O_2. Results were consistent with previous researches found that students of all levels, college students and prospective teachers assume that when the process of evaporation, the water molecules will break down into oxygen and hydrogen [38],[39]. As known, in the evaporation, water molecules are not changes, they exist before and after the evaporation.

4. Conclusion

The results of this study indicate that the Primary School Teacher Education Program has not provided its PPST with a full understanding related to the topic of states of matter and their changes. They did not have a full understanding about the discontinuous nature of matter. They already understood that matter is composed by particles, however, they did not understand the existence of the particle-free space that exists between the particles. Eventhough they already understood phenomena regarding to states of matter and their changes at macroscopic level, however, they still had difficulties to explain the phenomena at submicroscopic level both verbally and visually. And they also had difficulties to represent the phenomena at symbolic level. The results suggest the need for improving the quality of teaching and learning processes on states of matter and their changes at Primary School Teacher Education Program.

5. References

[1] Merritt J, Shwartz Y and Kracjik J 2007 *Middle School Students' Development of The Particle Model of Matter* (National Science Foundation)
[2] Merino C and Sanmarti N 2008 *Chemistry Education Research and Practice* **9** 196–207
[3] Håland B 2010 *NorDiNa* **6** 109-23
[4] Osborne R 1982 *Australian Science Teaching Journal* **28** pp 21-30
[5] Talanquer V 2006 *Journal of Chemical Education* **83** 811-6
[6] Taber K S 2001 *Chemistry Education: Research and Practice in Europe* **2** 123-58
[7] Boo H K and Watson J R 2001 *Science Education* **85** 568–85
[8] Johnson P 1998 *International Journal of Science Education* **20** 695-709
[9] Ozmen H 2011 *International Journal of Environmental & Science Education* **6** 99-121
[10] Özmen H 2013 A *Eurasian J. Physical & Chemical Education* **5** 81-110
[11] Snir J, Smith C L and Raz G 2003 *Science Education* **87** 794–830
[12] Treagust D F, Chandrasegaran A L, Crowley J, Yung B H W, Cheong I P and Othman J 2010 *International Journal of Science and Mathematics Education* **8** 141-64
[13] Chang R 2010 *Chemistry* Ed 10 (Boston: Mc Graw-Hill)
[14] Geller E 2003 *Dictionary of Chemistry* Ed 2 (New York: McGraw-Hill)
[15] Sopandi W 2006 *Pembelajaran Kimia yang Berorientasi pada Struktur: Perlukah?* (Bandung: FPMIPA UPI, Seminar Pendidikan Nasional Pendidikan IPA)
[16] Abdullah N, Ismail S and Surif J 2013 *Proc. Int. Conf. on Education and Educational Technologies (Malaysia)*

[17] Stojanovska M I, Soprajanov B and Petrusevski V 2012 *Creative Education* **3** 619-31
[18] Barke H D, Hazari A and Yitbarek S 2009 Chapter 2: Students' misconceptions and how to overcome them *Misconceptions in Chemistry: Addressing perceptions in chemical education* (Berlin: Springer-Verlag)
[19] Tatar E 2011 *Educational Research and Review* **6** 197-200
[20] Kolomuç A dan Tekin S 2011 *Eurasian Journal Physical Chemistry Education* **3** 84-101
[21] Sheehan M, Childs P E and Hayes 2015 *Pre-Service Irish Science Teachers' Misconceptions of Chemistry*
[22] Giuesepe M D and Fraser D 2012 *Curriculum Connection:All Grades and Subjects Myths and Misconceptions in Science Education* vol 3 (Cruble Online)
[23] Hinton M E and Nahkleh M B 1999 *Chemical Educator* **4** 1-29
[24] Çalik M and Ayas A 2005 *International Education Journal* **6** 30-41
[25] Griffiths A K and Preston K R 1992 *Journal of Research in Science Teaching* **29** 611-28
[26] Aydin A and Altuk Y G 2013 *International Education Studies* **6** 104-15
[27] Barke H D and Harsch G *Chemiedidaktik Heute: Lernprozesse in Theorie und Praxis* (Heidelberg: Springer-Verlag)
[28] Tytler R, Prain V and Peterson S 2007 *Research in Science Education* **37** 313–31
[29] Novick S and Nussbaum J 1978 *Science Education* **62** 273-81
[30] Treagust D F, Chittleborough G and Marmiala T 2003 *International Journal of Education* **25** 1353-68
[31] Harrison and Treagust D F 2002 The particulate nature of matter: Challenges in understanding the submicroscopic world *Chemical education: Towards Research-Based Practice* edited by J K Gilbert, Od Jong, R Justi, D F Treagust and J Hv Driel (Dordrecht: Kluwer Academic) pp 189–212
[32] Boz Y 2006 *Journal of Science Education and Technology* **15** 203-13
[33] Özmen H and Kenan O 2007 *Asia-Pacific Forum on Science Learning and Teaching* **8**
[34] Ayas and Özmen H 2002 *Bogazici University Journal of Education* **19** 45-60
[35] Yılmaz A and Alp E 2006 *Chemistry Education Research and Practice* **7** 22-31
[36] Flores-Camacho F, Gallegos-Cazares L, Garritz A and Garcia-Franco A 2007 *Science & Education* **16** 775-800
[37] Adbo K and Taber K S 2009 *International Journal of Science Education* **31** 757-86
[38] Valanides 2000 *Chemistry Education Research and Practice in Europe* **1** 249-62
[39] Othman J, Treagust D F and Chandrasegaran A L 2008 *International Journal of Science Education* **30** 1531-50

Acknowledgement
The authors wish to thank Elementary Education Program, Graduate School, Indonesia University of Education for the financial support for the development of the research project from which this paper originated.

IOP Conf. Series: Journal of Physics: Conf. Series **812** (2017) 012076 doi:10.1088/1742-6596/812/1/012076

Developing Online Course Portal to Improve Teachers' Competency in Creating Action Research (CAR) Proposal Using Learning Management System (LMS) Moodle

A A Muhtar

Science Teacher, Wira Bhakti Senior High School, Jl. Nani Wartabone, Bonebolango 96184, Indonesia

Email: agus.muhtar@yahoo.com

Abstract. Online course can offer flexible and easy way to improve teachers' competency in conducting education research, especially in classroom action research (CAR). Teachers can attend the course without physically present in the class. This research aims to (1) develop online course portal to improve teachers' competency in creating CAR proposal, and (2) produce proper online course portal validated and evaluated from four aspects: learning process, content, graphic user interface and programming. Online course in this research developed using Learning Management System (LMS) Moodle. The research model is using modified Borg & Gall Research and Development (R&D) started from preliminary studies, designing product, creating product, and evaluation. Product validated by three experts from three universities. Research subjects for field test are seven teachers as participants from different schools in several provinces in Indonesia. Based on expert validation and field test results, the product developed in this research categorized as "very good" in all aspects and it is suitable for teacher to improve their competency in creating CAR proposal. Online course portal produced in this research can be used as a proper model for online learning in creating CAR proposal.

1. Introduction

Referring to the Indonesia's Act of Teacher and Lecturer 2005 Number 14, one of the Teacher's Professional Competency is that the teacher must be able to conduct reflective actions [1]. Classroom action research (CAR) as one of the reflective actions can help teacher answer questions about the effectiveness of specific instructional strategies and identify changes that improve teaching and learning. Even though CAR is very important, it is rarely performed due to the teachers' lack of ability to conduct it. In the other word, teacher competency in CAR needs to be upgraded.

Teacher can improve their competency especially in conducting CAR by several ways, such as by participating in the teacher training or workshop [2]. But not all the teacher can attend training or workshop for several issues such as; this kind of training is rarely arranged, high cost and held in the other region far away from their home. Online course can be a solution for teacher to improve their competency without leaving their duty in the school. This research is conducted to solve these problems by developing free online course portal in Bahasa Indonesia. Teachers can access material in this online course anytime and anywhere using smart phone, tablet or laptop as long as they are connected to the internet.

An "online course" defines by Digital Learning Department, USA (2009) as a course, or grade-level course work, in which more than half of the course content is delivered electronically using the internet or other computer-based methods, conducted from a remote location through an online course learning management system or other online or electronic tools [3]. Online learning involve text that had merely been converted to electronic form and placed on a website [4]. To make a reliable online course, it must fulfill three qualities condition, (1) content and goal qualities, (2) instructional qualities, and (3) technical qualities [5].

Online course portal in this research developed using LMS Moodle and using free hosting in https://www.gnomio.com/. Learning Management System (LMS) is web applications that give educator to create a course web site that can be accessed by using a web browser [6]. LMS also offer a wide variety of tools that can make online course more effective. Moodle (Modular Object-Oriented Dynamic Learning Environment) is a free open-source LMS or e-Learning platform that serves educators and learners across the globe [7]. Moodle appears to present a clear advantage practically in all features. It can suport any teaching style or educational mode [8]. It enables users to create and join communities of practice, engage in reflective learning and collaborate with peers online [9].

It can be inferred from the previous explanation that teachers are still need to upgrade their competency in creating CAR proposal. Online course portal developed using Moodle is one of the alternative solutions to improve teacher competency. The research objectives are (1) develop free online course portal in the topic creating classroom action research proposal, and (2) know teachers' response about this online course portal from its learning process, content, graphic user interface and programming. This online course will give tremendous benefit for education development in Indonesia, especially to help teacher in remote area who want to improve their competency.

2. Model, design, analysis, and implementation

2.1. Model and design

The research model is using modified Borg & Gall Research and Development (R&D) [10]. This model consists of 4 development stage started from (1) preliminary studies, (2) designing product, (3) building product, and (4) evaluation. Output product developed of this research is online course portal to improve teacher competency in creating CAR proposal. All the design of this research is shown in figure 1.

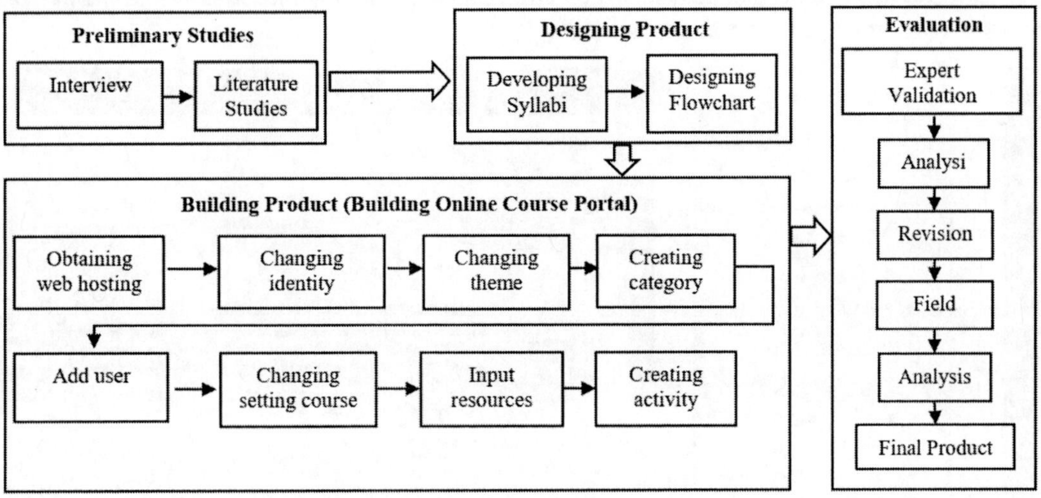

Figure 1. Research design consists of 4 development stage started from: (1) preliminary studies, (2) designing product, (3) building product, and (4) evaluation. (Adapted from Borg & Gall, 1983)

2.2. Analysis and implementation

Three experts validate the product quality in the four aspects; learning process, course content, graphic user interface, and programming using questionnaire that can be accessed in https://goo.gl/qfFWBN. Data from the expert validation then analyzed using descriptive qualitative method. Raw data in numbers/score from the questionnaire converted to interval data with scale five (5). Sukardjo (2008) formulated the conversion in the Table 1.

Tabel 1. Conversion of number/score to be interval data with scale five (5)

Category	Number/Score	
	Equation	Interval
Very Good	$X > x + 1,8\ SBi$	$5,00 < X \leq 4,206$
Good	$x + 0,60\ SBi < X \leq x + 1,8\ SBi$	$3,402 < X \leq 4,206$
Fair	$x - 0,60\ SBi < X \leq x + 0,80\ SBi$	$2,598 < X \leq 3,402$
Not Good	$x - 1,80\ SBi < X \leq x - 0,60\ SBi$	$1,794 < X \leq 2,598$
Really Not Good	$X < x - 1,80\ SBi$	$X \leq 1,794$

Where x is average ideal score = ½ (maximum ideal score + minimum ideal score), *Sbi* is standard deviation ideal score = 1/6 (maximum ideal score + minimum ideal score), and X is obtained score.

Experts validation and evaluation then used to improve the quality of product. After revised, the product is ready to be implemented in the field test. Seven (7) teachers as participants from different province in Indonesia were taken randomly as research sample in the field test. Participant questionnaire can be accessed in https://goo.gl/Ttzkac. Data obtained from the field test then analyzed again to complete the final product also using descriptive qualitative method. The final product can be implemented if in all four aspects categorized as "Good".

3. Result and discussion

3.1. Result

Preliminary studies show that teachers have difficulties to conduct CAR and online course can be an alternative solution to overcome this problem. The online course portal developed in this research can be accessed in https://oase.gnomio.com/. The online course portal interface is shown in Figure 2 and the architecture of the system is shown in figure 3.

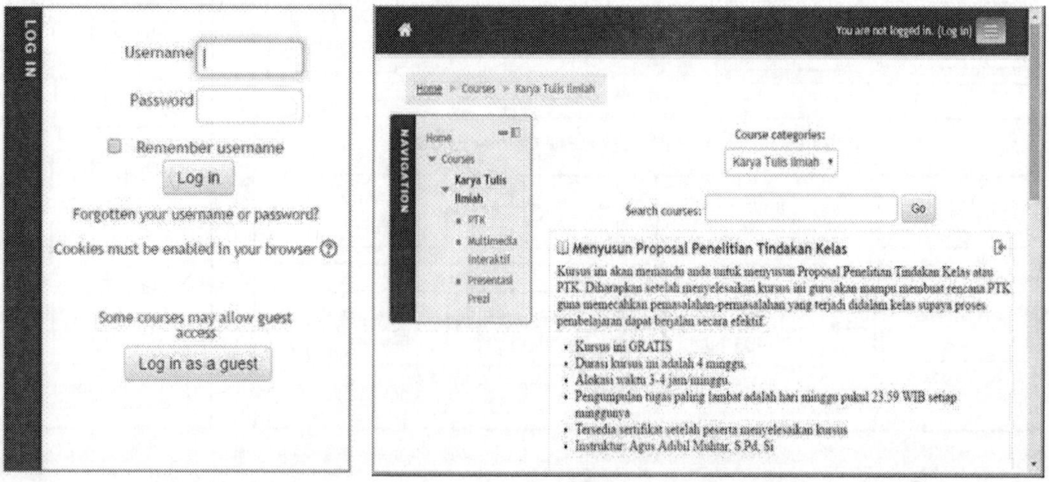

(a) (b)

Figure 2. The figure (a) and (b) shows online course portal interface.

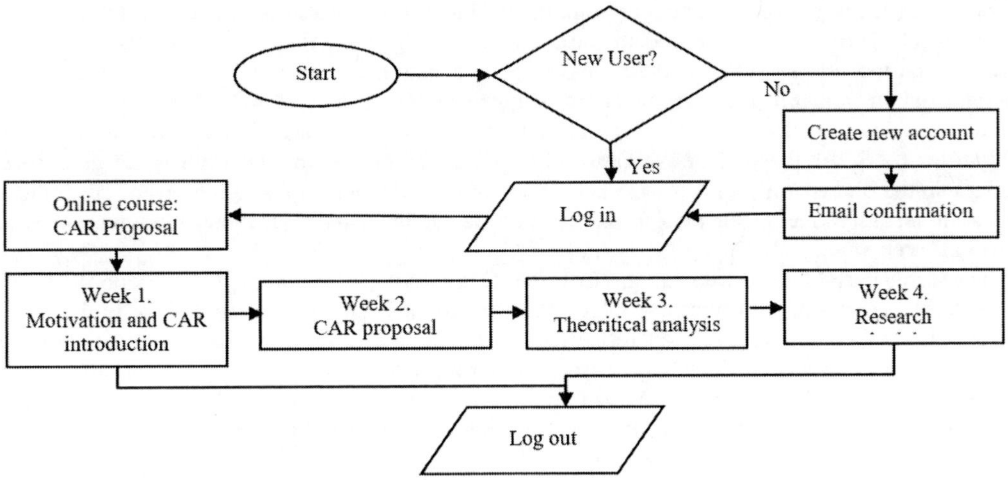

Figure 3. The figure shows architecture of online course portal system.

Online course portal system then validated and evaluated by expert before tested in the field. The validation and evaluation result can be seen in Table 2. After revised and upgraded based on the advice from expert, the product then tested to 7 teachers as participants from different province in Indonesia (Maluku, Gorontalo, East Java, Central Java, West Sumatra and Lampung) to see their teacher responses. Table 3 shows the result from field test.

Table 2. Expert validation result.

Validation Aspect	Average Score	Category
1. Learning process	4,60	Very Good
2. Course content	4,53	Very Good
3. Graphic user interface	4,47	Very Good
4. Programming	4,62	Very Good

Tabel 3. Teachers' response in using online course portal.

Validation Aspect	Average Score	Category
1. Learning process	4,68	Very Good
2. Course content	4,72	Very Good
3. Graphic user interface	4,56	Very Good
4. Programming	4,64	Very Good

3.2. Discussion

Based on expert validation and field test results, it can be inferred that online course portal developed in this research categorized as "very good" in all aspect and it can be implemented to improve teachers' competency in creating CAR proposal. Course content in this online course consist of several activity designed to be accomplished in four (4) weeks and teacher only have to spent 2-3 hours/week to finish the tasks. This flexible time management will give teacher opportunity to read material, discuss, and collaborate with other teacher. The topics on each week are: Week 1. Introduction and Motivation, Week 2. Title and Background, Week 3. Literature study, and Week 4. Research Methodology and Writing Research Proposal.

There are five activities on the first week; (1) introduction, (2) motivation, (3) reading about CAR introduction, (4) discussion forum about potential problem in the classroom for CAR, and (5) quiz. In the first week, participants are asked to introduce their self, write their motivation and expectation about this online course. They are also have to read material related to CAR introduction. This material contain valuable information about purpose and benefit of CAR, characteristic of CAR, explanation about the difference between CAR another educational research, and how to find potential problem for CAR. As the part of evaluation they are asked to respond each other and give positive feedback in the discussion forum. At the end of session, participants are asked to accomplish the quiz and submit an assignment. The quiz format is multiple choices with 5 questions about what they have learned in week 1.

The courses in the second week are divided into five activities; (1) reading CAR proposal structure, (2) CAR title discussion, (3) writing background for proposal, (4) the objectives, the limitation, and the benefits of the research, and (5) reflection. In the first activity, participants read material and explanation about structure of CAR proposal. After finished first activity, they are asked to make a forum discussion. Each participant has to submit at least one title for their CAR proposal and required to actively involve and collaborate with other participants to get feedback in the discussion about this title. Participants learn how to write proposal background and the objectives, limitation, and the benefits of the research in the third and fourth activity. They are also required to accomplish the quiz and submit their work in these activities. As a closure, participants discuss and share their thought about the problems and obstacles found in week 2 and how they solve these problems.

Participants learn more about CAR proposal in the third week. There are four activities in this week; (1) how to write theoretical base and literature, (2) how to write CAR framework, (3) discussion, and (4) quiz. To give optimum experience in activity one and two, participants are required to read material from external source portal and watch video from YouTube. They are also required to make some discussion about this week topic and submit assigntment in the pdf or doc format before the end of third week. Research methodology is delivered in the fourth week. This fourth week is the end of the course and all participants must submit their full CAR proposal. Participants also required submit the final survey about this course.

This product has been validated and evaluated by expert to make sure that the product is valid. Overall, evaluation from the expert categorized as "very good" and the online course ready to be implemented. The expert suggest that the graphic user interface navigation should be consistent and avoid using flashy color. For further development, they suggest to facilitate participants' different learning style and participants are required to finish all the activity in their current week before moving to the next week. Based on this evaluation, the product then updated and upgraded to meet expert expectation.

Seven teachers were selected randomly from different province in Indonesia as the field test participants. They show enthusiasm and gave positive response when participating on the online course. Participant from Gorontalo said *"I am so happy can join in this online course. I can learn and improve my competency without limited by time. I can learn something new without leaving my beloved family because I access this course from my home. This online course allows me to keep balance between my duty as mother and my career as an educator"*. Other participants from Maluku said *"This online course is really interesting and innovative. It can be accessed with easy, anytime and anywhere"*. In line with that, participants from central Java also commented *"This online course can be a solution for everyone who wants to gain their knowledge about CAR"*. On the other hand, participants also gave some advice; (1) add more course content, and trainer should give immediate feedback to the participants, (2) add more information about every activity. Based on these advice, the product then upgraded by added some content and notification for tutor to give immediate feedback. Overall, product evaluation from the participants categorized as "very good" and the online course ready to be used. However, this product has not been tested in the field for larger scale.

4. Conclusion

- Online course portal to improve teacher competency in creating CAR proposal developed using LMS Moodle and using modified Borg & Gall Research and Development (R&D) model, started from preliminary studies, designing product, creating product, and evaluation.
- Online course portal produced in this research is proper and can be used to improve teachers' competency in creating CAR proposal. Overall, product evaluation from expert and participants on the aspects learning process, content, graphic user interface and programming are "very good".

5. References

[1] Indonesian Republic 2005 *Teacher and Lecturer Law No. 14 of 2005*
[2] Indonesian Ministry of Education 2010 *Technical Guidance for Teacher Function and Credit Points* (Jakarta: Indonesian Ministry of Education)
[3] Digital Learning Department 2009 *Online Course Definition Available*
[4] Anderson T 2008 *The Theory and Practice of Online Learning* (Kanada: Athabasca University Press)
[5] Arsyad A 2009 *Learning Media* (Jakarta: Rajawali Pers)
[6] Singh J 2014 *How to Use Moodle 2.7*
[7] Dougiamas, Martin and Taylor P 2003 *World Conference on Educational Multimedia, Hypermedia and Telecommunications (Honolulu)*
[8] Aydin C C and Tirkes G 2010 *IEEE Educon Education engineering 2010 (Madrid)* pp 593-600.
[9] Pandey S R and Pandey S 2009 *Proc. of the International Conference on Academic Libraries (Delhi)* pp 249-54
[10] Borg W R and Gall M D 1983 *Educational Research an Introduction* 4th edition (New York: Longman)

Acknowledgments

The author thanks to Pujianto, M.Pd (UNY), Abdul Haris Odja, M.Pd (UNG), Septi Ambarwati, M.Pd, Si (UST) for many discussions and reviews. The author also thanks to participants from : 1) SMA Negeri 11 Ambon, 2) SMA PGRI 1 Maospati, 3) SMAN 1 Koto XI Tarusan, 4) SMAN 4 Bandar Lampung, 5) SMP PGRI 2 Somagede, 6) SMP Negeri 1 Kedung, and 7) MTsN Suwawa who have provided valuable support in this work.

Guided Inquiry Learning With Sea Water Battery Project

A Mashudi[1*]

[1]SMP Negeri 4 Juwana, Jl. Raya Tluwah-juwana, Pati, Jawa Tengah 59185 Indonesia
*Corresponding author's email : anwarmashudi@gmail.com

Abstract. Science learning process is expected to produce valuable product, innovative and real learning environment, and provide memorable learning experience. That orientation can be contained in Inquiry Based Learning. SMP N 4 Juwana is located close to the beach. That's why, Sea Water Battery Project is very suitable to be applied in learning activity as an effort to fulfill the renewable energy based on local wisdom. This study aims to increase interest, activity and achievement of students. Learning implementation stage, namely : Constructing Sea Water Battery project, observation, group presentations, and feedback. Sea Water Battery is renewable energy battery from materials easily found around the learner. The materials used are copper plate as the anode, zinc plate as the cathode and sea water as the electrolyte. Average score of students Interest on the first cycle 76, while on the second cycle 85. Average score of students Activity on the first cycle 76 and on the second cycle 86. Average score of students achievement on the first cycle 75, while on the second cycle 84. This learning process gave nurturant effect for students to keep innovating and construct engineering technology for the future.

1. Introduction

National Education System Law has mandated that the National Education aims to educate the nation and develop students' potency to become a noble man. Science as one of the national curriculum in national education that has an important role in achieving the goals of the National Education. Garbet (2011) states that Science relevant to find out about a systematic nature, so that the Science is not only knowledge mastery, such as : facts, concepts, or principles, but also a discovery process.

Indonesian education implements four approaches, namely 1) life skills -oriented education, 2) competency-based curriculum and learning, 3) production-based learning and 4) broad-based education. Life skills-oriented education, competency-based curriculum and learning process are expected to produce valuable products, demand real and innovative learning environment, and provide an integrative learning experience. The orientation can be contained in Project Based Learning.

Energy needed in Indonesia in particular and the world, increases generally because of population growth, economic growth and energy consumption patterns are constantly increased. Fossil energy, which has been the main source of energy, is very limited and continue to diminish. Alternative of energy is needed to overcome energy needs.

Maritime Ministry Data shows that Indonesia's land area is 1,910,931.32 km^2 and Indonesia's sea area is 3.544.743,9 km2 or 64.97% of Indonesia territory, so the Indonesian marine water resource is very abundant and can be used as an renewable alternative energy. SMP N 4 Juwana is located at 9 km from the Java Sea and 3 km from the River Juwana. Local wisdom on the abundance of marine water resource can be included in learning activity of electric current source chapter through Sea Water Battery Project Based Learning.

SMP N 4 Juwana students have problem to master electric current source chapter. It can be seen from their achievement, which down. Data from the last two years were as follows, (1) in the academic year 2013/2014, learning mastery is 75%. (2) in the academic year 2014/2015, learning mastery is 75%

Students' Interest and activity in learning were still low. It was marked from 36 students, only 7 who were active. Learning innovation is needed to improve the understanding of the concept, interest and activity of students on this chapter.

Observation and reflection from author and observer in the learning process of electric current source chapter, can be concluded that the factors that led to the lack of students understanding were as follows: 1) The learning method used was one direction, no feedback from students. 2) The learning process did not arouse interest and activity of students. 3) Students did not understand the basic competency. 4) Teachers did not optimize media around the students into educational media in the learning process.

The problem in this classroom action research are as follows, 1) How does the Sea Water Battery project design? 2) Can the Sea Water Battery Project Based Learning increase the students' interest? 3) Can Sea Water Battery Project Based Learning enhance the activity of students? 4) Can Sea Water Battery Project Based Learning improve mastery learning on the electric current source chapter?

The goals in this classroom action research are as follows, 1) Finding Out Sea Water Battery project design. 2) Finding Out whether the Sea Water Battery Project Based Learning can increase the interest of. 3) Finding Out whether Sea Water Battery Project Based Learning can enhance the activity of students. 4) Finding Out whether the Sea Water Battery Project Based Learning can improve learning mastery on on the electric current source chapter.

Bell (2010) states Project Based Learning is a learning model that involves a project in the learning process. Definition of Project Based Learning has been described above is in line with the exposure by Savery (2006) is as follows. 1) Project Based Learning is fueled and standards-based curriculum. 2) Project Based Learning Asks a question or poses a problem that each student can answer. 3) Project Based Learning Asks students to Investigate issues and topics addressing real-world problems while integrating subjects across the curriculum. 4) Project Based Learning is a model that fosters abstract, intellectual tasks to explore complex issues.

Project-Based Learning has five main steps, namely: (1) setting the theme of the project, (2) establishing the context of learning, (3) planning activity, (4) processing activity, and (5) implementing activity. Silver (2004).

Tobias (1994) says that the interest is more like a taste and sense of belonging to something or activity, without being told. Interest is basically the acceptance of a relationship between yourself with something outside of yourself. The stronger or close the relationship, the greater the interest.

According to Indonesian Dictionary (2005), Active means enterprising. Prince (2004) argues that the activity here is both physically and mentally. In learning activity, both activities must be interrelated. The link between the both will produce optimal learning activity

2. Model, Analysis, Design and Implementation

This Classroom Action Research was conducted in odd semester of 2015/2016 academic year, namely October to December 2015 at SMP N 4 Juwana at Tluwah street Juwana, Pati. The research subject is class IX A by the number of 34 students, boys were 18 students, while girls were 16 students. Class IX A was choosen as research subjects, because: (1) learning interest with good category is only 65%. (2) Students activity with good category is only 16 from 34. (3) learning mastery is only 57%.

There are two data sources in this Classroom Action Research, namely primary and secondary data sources. The primary data consists of 1) qualitative data (observation result of students' interest, observation result of students' activity , video documentation). 2) quantitative data is the achievement of students on the electric current source chapter. While the secondary data is the advice from the observer.

Data were collected by using an instrument as follows, 1) Interest of students data were collected

by questionnaire. 2) Learning activity was collected by observation sheet. 3) Students' achievement were collcted by written test in narrative form.\
Success indicator in this Classroom Action Research, namely: 1) Learning interest with a good category reaches 85% 2) Learning activity with a good category reaches 85%. 3) Learning mastery reaches 85%.

The materials needed in Sea Water Battery project such as: a) a copper plate as the anode, b) zinc plate as the cathode, c) sea water as an electrolyte, and d) a plastic tube. The materials used are very commonly found in the environment around the students. It proves that the materials around us can be used as a medium of education. Design Sea Water Battery as shown in Figure 1.

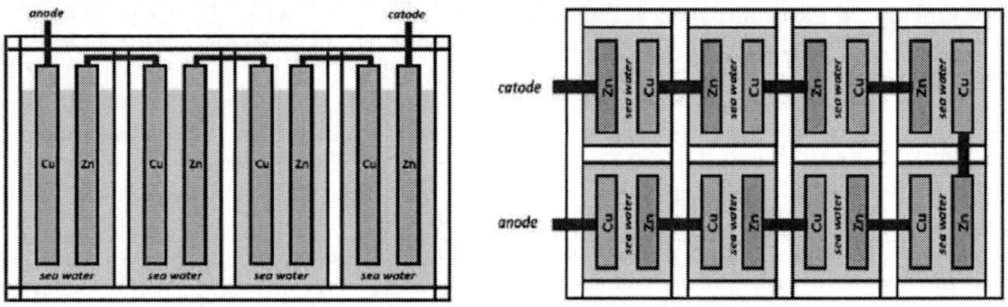

(a) (b)

Figure 1. Sea water battery project design (a) side view, (b) top view

2.1. Research Procedure

2.1.1. First Cycle Plan, namely : 1) arranging Syllabus Development and Learning Device of Basic Competency 3.3 Describing the element work principle and generated electric current well as its application in everyday life by learning indicators : a) Describing the composition and the electric element work. b) Developing the research instrument. c) Discussing learning device, research instrument used by the observer is Widyastuti Trirahayu, S.Si, M.Pd.

Implementation, namely : Teacher carried out learning based on the plan of Cycle I with the following main stages. 1) Preparimg the Sea Water Battery Project. 2) Group Presentation. 3) Feedback.4) Learning Evaluation.

Observation, peer teacher observed the learning process with the observation instrument of students' activity and teacher learning process observation. On this stage, researcher collected data in the learning process by using the instrument. (1) questionnaire to measure the interest of students. (2) Observation sheet to measure the activity of students. (3) Evaluation test to measure concept understanding of students on the electric current source chapter.

Reflection stage that is the result of observation was collected and analyzed. From the observation, teacher conducted reflection to determine deficiencies, obstacles and constraints during the learning process. The data obtained was used as a basis and reference for researcher to evaluate teacher and learners during the learning process. The result of the analysis was used to plan further action on the second cycle.

2.1.2. Second Cycle Plan, namely: 1) developing Syllabus Development and Learning device at basic competency 3.3 Describing the element work principle and the generated electric current as well as its application in everyday life. Learning indicator measures the voltage between the poles of the voltage source and clamp voltage (applied voltage).

Implementation, namely 1) Students were divided into several groups to construct materials that have been prepared to be a Sea Water Battery project. Teacher facilitates students in compiling Sea Water Battery. Students measure the voltage between the poles of the voltage source and clamp voltage (applied voltage) corresponding student activity sheet. 2) The results of each group were presented. 3) From the group presentations were given feedback from other groups. Teacher gave scantlings from the feedback of students. 4) Learning Evaluation.

Observation, peer teacher observed the learning process with the observation instrument of students' activity and teacher learning process observation. On this stage, researcher collected data in the learning process by using the instrument. (1) questionnaire to measure the interest of students. (2) Observation sheet to measure the activity of students. (3) Evaluation test to measure concept understanding of students in the indicator measuring the voltage between the poles of the voltage source and clamp voltage (applied voltage).

Reflection, the result of observation was collected and analyzed from the observation, the teacher hold a reflection to determine deficiencies, obstacles and constraints during the learning process. The data obtained was used as a basis and reference for researcher to evaluate subsequent learning.

3. Result and Discussion

Description of the initial condition of class IX A shows that (1) Learning interest with good category only 65%. (2) Students' activity with good category only 16 from 34 students. (3) Mastery learning is 57%.

3.1. Description First Cycle

Plan, researcher sets Syllabus Development and Learning Device of Basic Competencies 3.3 Describing the element work principle and the generated electric current as well as its application in everyday life by learning indicators : a) Describing the composition and the electric element work. b) Developing the research instrument. c) Discussing learning device, research instrument used by the observer.

Implementation : teacher carries out learning according the plan of first cycle with the following main stages.1) Preparing the Sea Water Battery Project. 2) Group Presentation. 3) Feedback. 4) Learning Evaluation first cycle. Figure execution of the first cycle is shown in Figure 2.

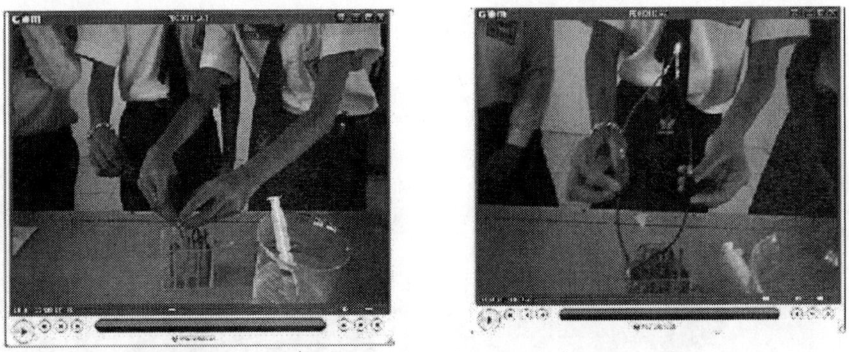

Figure 2. Video compilation of sea water battery project

Observation of first cycle was carried out during the activity. Observer followed the entire process of the actions performed in class IX A. The observation result of the implementation of learning on the first cycle can be seen as follows 1) The teacher had facilitated the students in the group to work together to design the Sea Water Battery Project. 2) Teacher had facilitated students in groups to communicate Sea Water Battery Project. 3) The teacher had guided the students to make inferences about the concept of an electric current source to train independence of thought. 4) Teacher had identified the students' problem to understand the material based on the debriefing and discussion in the classroom. 5) The teacher had guided the students to reflect on the learning difficulties to understand the material honestly and seriously.

3.2. Description Second Cycle

Plan: 1) developing Syllabus Development and Learning device at basic competency 3.3 Describing the element work principle and the generated electric current as well as its application in everyday life. Learning indicator measures the voltage between the poles of the voltage source and clamp voltage (applied voltage).

Implementation, namely : 1) Students were divided into several groups stringing materials that have been prepared to be a Sea Water Battery project. Teacher facilitated students in compiling Sea Water Battery. Students measured the voltage between the poles of the voltage source and clamp voltage (applied voltage) corresponding student activity sheet. 2) The results of each group were presented. 3) From the group presentations were given feedback from other groups. Teacher gave scantlings from the feedback of students. 4) Learning Evaluation in the second cycle. Video presentation of the group is shown in Figure 3.

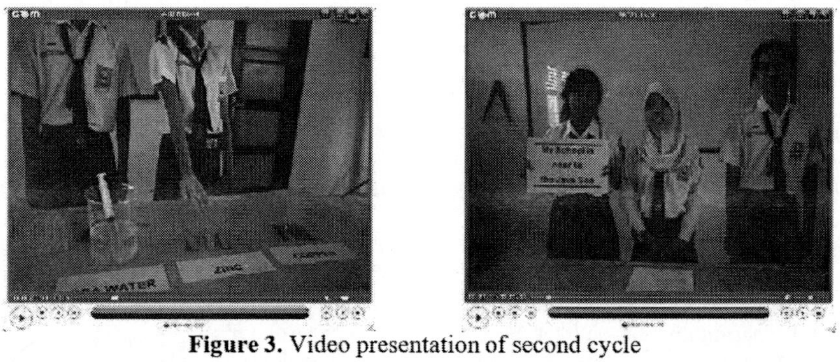

Figure 3. Video presentation of second cycle

3.3. Results

Learning Interest was measured using a questionnaire. The comparison results of the first cycle and second is shown in Figure 4.

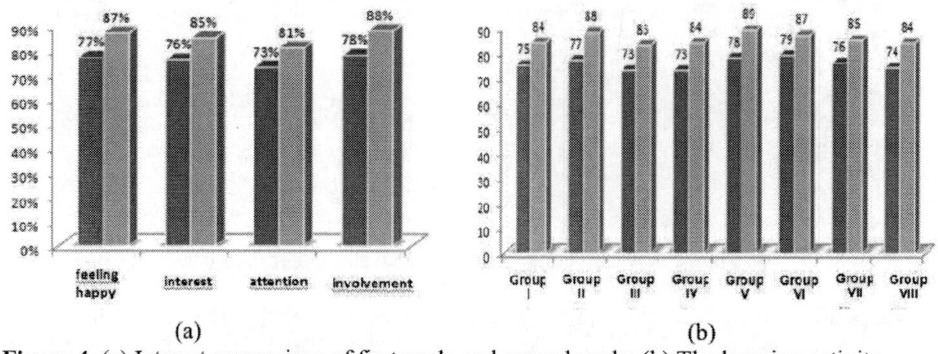

Figure 4. (a) Interest comparison of first cycle and second cycle; (b) The learning activity comparison of first cycle and second cycle

Students activity during Sea Water Battery Project Based Learning for each cycle can be seen in Figure 4b. The students activity is seen based on sheet format of students' process skills observation by observer. From the observation sheet is obtained students' activity through Sea Water Battery

Project Based Learning increased.

Knowing the learning achievement of the of electric current source concept, post test was performed using a briefing test. The comparison of learning achievement in the first cycle and the second cycle can be seen in Figure 5.

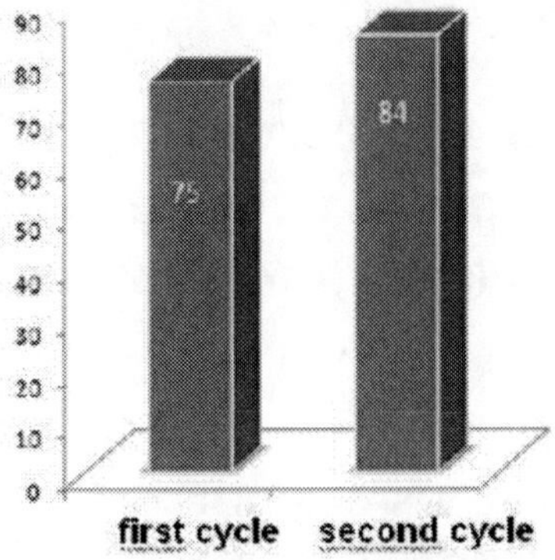

Figure 5. The average value of first and second cycle

4. Conclusion

Sea Water Battery is a renewable energy battery from materials easily found around the students The materials used are copper plate as the anode and zinc plate as the cathode as well as sea water as the electrolyte. Students Interest in Sea Water Battery project-based learning has increased, the first cycle has an average score of 76%, while the second cycle has an average score of 85%. Students activity on the first cycle has an average score of 76 and the second cycle has an average score of 86. The learning achievement of students has increased, the first cycle has an average value of 75, while the second cycle has an average value of 84. Suggestions for future research are the development of Sea Water Battery products more varied and Improve control of students in implementing the project to obtain a more optimal result.

5. References

[1] Bell S 2010 *A Journal of Educational Strategies, Issues and Ideas* **83** (2) 39-43
[2] Garbett D 2011 *Australian Journal of Teacher Education* **36** (6) 36-49
[3] Prince M 2004 *Journal of Engineering Education* **93** (3) 223-31
[4] Savery J R 2006 *Interdisciplinary Journal of Problem - Based Learning,* **1** (10 1-13
[5] Silver C E 2004 *Educational Psychology Review* **16** (3) 235-56.
[2] Tobias S 1994 *Review of Educational Research* **64** (1) 37-54

Bockron as a Medium of Learning in The Process of Inquiry based Learning to Improve Science Process Skills of Junior High School Students in Growth and Development Concept

D Mayasari, M.Pd

SMPN 3 Margahayu, Jl. Sadang No. 184, Margahayu, Bandung 40225, Indonesia
Email : diah_m@ymail.com

Abstract. Investigative research on Influence of bockron as a medium of learning in process of inquiry-based learning to the development of science process skills on the concept of growth and development. This research was done in an effort to follow up underdeveloped skills of observing, communicating andconclude on students. This research was conducted using classroom action research (PTK), which consisted of 3 cycles. Cycle 1 students observe differences in growth and development, cycle 2 students measure the growth rate,cycle 3 students observe factors that influence growth and development, In these three cycles is used as a planting medium bocron (bottles and dacron). It involves 8th grade junior high-school students of 14-15 years old as research subjects in six meetings. Indicators of process skill include observation, communication, interpretation and inference. Data is collected through students' work sheets, written tests and observation. Processing of the data to see N-Gain used Microsoft Excel 2007, and the results showed that an increase in science process skills with a value of medium N-Gain (0,63). Bokron learning medium easily and cheaply obtainable around the students, particularly those in urban areas is quite difficult to get land to be used as aplanting medium. In addition to observation of growth and development, bokron media can also be used to observe the motion in plants. The use bokronas a learning medium can train and develop science process skills, attitude and scientific method also gives students concrete experience of the processof growth and development in plants.

1. Introduction

Study and practice on science involves three main elements, namely attitude, process/method, and product. Process of invention and composition of science theories or concepts is carried through scientific method steps. A good science learning process ideally passes through process of how the concept is invented. However, when students use scientific method in learning process, they are required to understand components in scientific method. The components of scientific method are known as Science process skill. Science process skill (SPS) is very important to learn and master for everyone. If one has mastered process skill, he/she has mastered necessary skill for higher learning level, namely conducting research and solving problems [1].

Although there are some types of process skills, they are distinguished into two groups: basic process skill and integrated process skill. Included in basic process skill are skills of observing, classifying, measuring, interfering, interpreting, communicating, and predicting [1],[2],[3].The skills are needed by students to run the scientific method well. Process skill can also be a tool to train thinking ability. Thinking process is a part of intelligence. Intelligence can be developed by training thinking process. To train students' thinking process can be done by increasing student participation in learning. one learning can help students practice thinking is inquiry based learning.

Inquiry learning is designed to engage students directly into the scientific process. Schlenker research results, in Joyce and Weil [4], the inquiry showed that exercise can improve the understanding of science, productive in creative thinking, and students become skilled in obtaining and analyzing information. Cleaf [5] states that the inquiry is one of the strategies of student-centered teaching, which encourages students to investigate the problem and find information. The process is the same as the procedures used by scientists who investigate problems and find information. Wilson [6] states that the inquiry model is a model of the process of teaching based on learning theory and behavior. Inquiry is a way to teach students how to learn by using the skills, processes, attitudes and knowledge of rational thinking [7]. Inquiry approach can increase the science process skills [8],[9].

The learning process often does not go as expected because there are several factors that influence the behavior of the learning process. According Dollar and Miller [10] the effectiveness of the learning behavior is influenced by four things: motivation, students must want something; attention and know the targets, students should pay attention to something; businesses, students must do something; evaluation and consolidation of the results, students must obtain something. One way that learning to work effectively is to use instructional media. The media is important and necessary in the learning process. Because the media is anything that can deliver a message from the sender to the receiver of the message so that it can stimulate the mind, feelings, concerns and interests of students so that learning occurs. Learning media useful for: a.Overcoming the passive attitude of the students, b. Stimulating learning, c. Pose a direct interaction between students with the environment and reality, d. Allows students to learn on their own according to their ability and interest, e. Provide the same stimulus in students, f. Liken the experience of students, g. Raises the same perception in students. [1].

According to Setiawan *et al.* [11] at first the teachers have used a variety of media and simple homemade props. Unfortunately, the media and simple props was later abandoned because teachers prefer and depending on the use of modern learning media. The teacher becomes passive and less creative. For that is necessary to develop further the use of simple media from recycled materials and simple equipment. Not only will stimulate the creativity of teachers but also be able to function effectively is not inferior to modern media and can be superior if the user is right and appropriate.

Growth and development in seed plants is a process that is often observed to be used as a medium of learning. Soil is a natural medium which is used as a growth medium for seed plants. The use of soil as a growing medium on observations of growth and development of seed plants less effective. The roots of plants grown in soil cannot be observed, soil media also can contaminate the classroom or laboratory. The students who are in the urban areas will be quite difficult to find land. Because some of these shortcomings Then the media was replaced with cotton planting soil. Media planted cotton on observations of the growth and development also was less effective. When cotton exposed to water will deflate so that the growth of roots and stems difficult to observe. Therefore we need to develop ways and growing media that facilitate observation of the process of growth and development of seed plants.

2. Research methodology

This research was conducted by the method of action research or Classroom Action Research (CAR). According Wardhani et al [12] CAR is the research conducted by teachers in the class itself through self-reflection with the aim to improve its performance as a teacher, so that the result of increased student learning. The subjects of research involved are class 8th A SMPN 3 Margahayu Bandung District 2012-2013 school year, amounting to 22 boys and 20 girls. In order for this research directional, scope of the study is limited to the following points. Mastery of science process skills were measured: observation, interpretation, inference, communicate, [2],[13]. Topics were selected in this study refers to the Competency Standards and competency base curriculum KTSP 8th class, that is Competency Standards 1. Understand the different systems in the human life and basic competence 1.1 Analysing the importance of growth and development of living things.

The procedure is done by Classroom Action Research (CAR). CAR is an assessment process through a system of cycles of various learning activities. According to Joni [14], there are five stages in the implementation of the CAR, is: 1. To determine the focus of the research problem, 2. plan for an improvement actions, 3.To carry corrective action, observation, and interpretation, 4. Analysis and reflection, 5. Make planning a follow-up.

Implementation of research begins with the research held the pretest. Pretest activities filled test the ability of science process skills, research was conducted as many as four meetings. cycles 1 is students to observe the growth and development of seeds by using bocron and continued discussions in the classroom with a sub-topic material differences in growth and development. In cycle 2 students measure the length of roots, stems and leaves, followed by counting the growth rates. The measurement results then graphed. Cycle 3 students observe factors that influence growth and development. Observations were made from the cycle 1 until cycle 3. The variables measured were the development of science process skills, as well as increasing mastery of the concept of growth and development. All variables are measured and observed before and after the learning process. The measurement results in the form of student worksheets score data for the development of science process skills, written test score of concepts mastery of growth and development.

The research instrument consists of worksheets students to measure progress in the mastery of science process skills and a written test to measure the growth and development of the procurement concept. Written test mastery of the concept of growth and development carried out at the beginning and end of the sub-subject learning. This test is done to detect students' mastery of the concept of the material before and after the learning process.

3. Result and discussion

3.1. The development of bocron media

One of the growing media which interest to be developed to observe the process of growth and development of seed plants are Dacron. Dacron is a synthetic material that is popular in the needs of the textile as a material for filling dolls, bolsters and pillows. Dacron is a trade name of a textile fiber made by the company Du-Pont from England. Dacron is real name is polyethylene terephthalate which is a plastic polymer resin termoplast of polyester groups. When dacron was waterd, it would not be going down, flat that makes it easier for seed storage set appropriate experimental design. Dacron is also good at storing water so appropriate planting medium used. The media developed can be seen in the Fig. 1.

Bocron is used as a medium of learning to observe the growth and development of the seed plants, namely by entering the dacron rolls into a transparent plastic bottle, then the seeds that will be observed placed between dacron and bottles. Seeds are watered twice a day through the dacron. Growth and development of seeds is observed every day. Observations consist of measuring the length of the roots, stems and leaves, observing the timing of the roots, stems and leaves. Bocron also be used to observe the motion in plants with clear and easily.

Figure 1. Bocron as a learning media.

3.2. The impact of using bocron media on students' science process skills

On this research was obtained data of percentage score, science process skills and mastery of the concept of growth and development. Data of science process skill was obtained from the percentage score of student worksheet and mastery of the concept was obtained from the percentage score of written test about growth and development.

This research observes science process skills of observation, communication, interpretation and inference. N-gain observation, communication and inference has high category (0.77, 0.77 and 0.71 and interpretation has moderate category (0.61). N-gain from SPS indicator is illustrated in Figure 2.

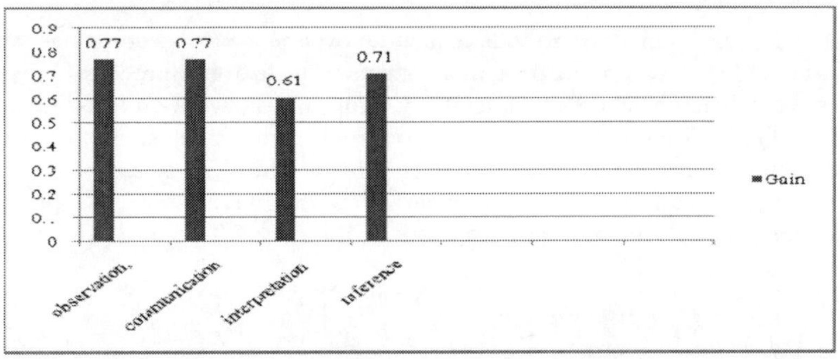

Figure 2. N-gain from observation, communication, interpretation and inference.

After brokron was used as a learning medium it can increase the percentage score of SPS, the mastery of the concept of growth and development. Increased percentage scores (N-Gain) are categorized as being (0.61) to high (0.77). In the learning process with Bocron media learning that based on this inquiry. students are invited to observe the growth and development of seed plants directly and easily. When students observe the growth and development of seed plants they conduct observation activities such as, began observing the appearance of roots, stems and leaves. Measuring changes in the length of roots, stems and leaves, observing the effect of the amount of water, light and storage location of the seeds in the process of growth and development of seed plants. Observation according Rezba [2] is the process of searching for information in a way perceived through one or more of the five senses. Therefore, in this study the students were given the opportunity to make observations using many senses to capture the stimulus in the form of a natural material object that is observing the growth and development of seed plants directly.

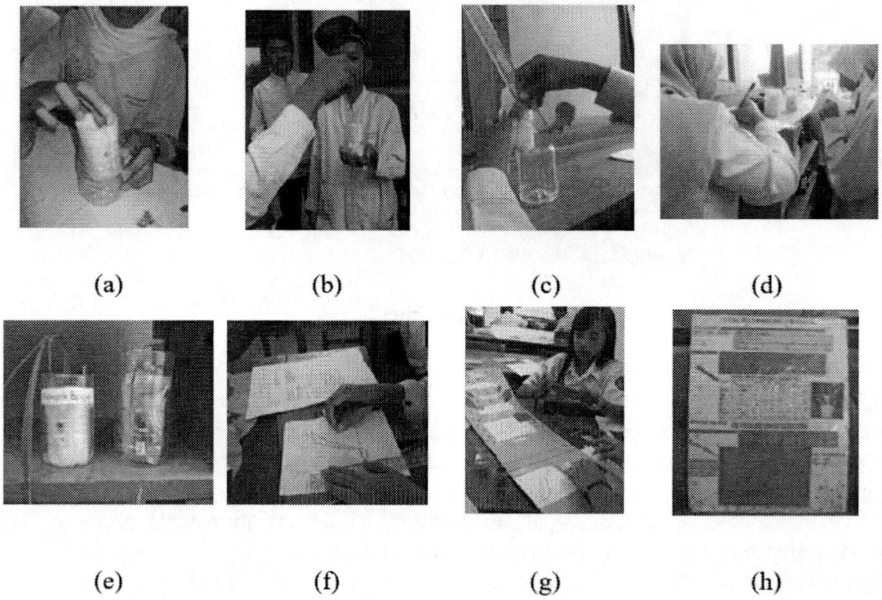

Figure 3. Photos (a) (b), (c), (d). Activities students are conducting observations of growth and development of the seed plants. Photos (e) of corn and beans are planted in bokron. Photos (f), (g) and (h) the student is making a report on the observation.

Observation activity also aims to provide concrete experience in accordance with the students' cognitive development (15), thus providing a more meaningful learning process [1]. Observing natural material object is very helping students understand terms the concept of growth and development. Catching stimuli natural material object then processed through a series of thought processes [15],[16],[17] the thought process is necessary science process skills. Science process skills greatly assist students in their concept mastery. This is evident from the increase in the procurement concept conveniently indicates generally increased the criteria N-Gain high (C1: 0.73; C4: 0.70; C5: 0.72; C6: 0,71).

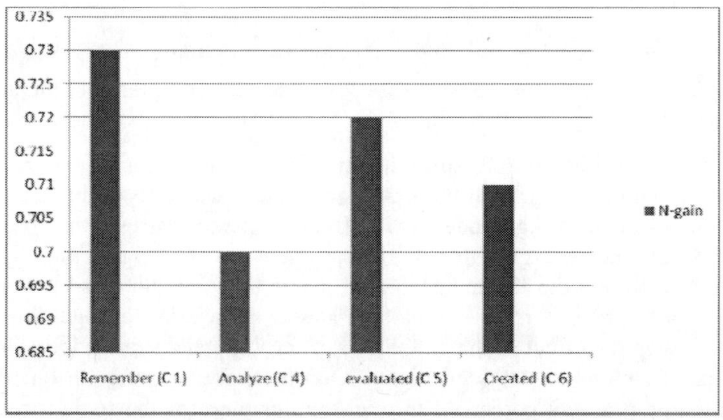

Figure 4. N-gain from remember, analyze, evaluated and created.

The next science process skills that students master important is communicating. Communicating is the ability of the delivery and reception symbols verbal and non-verbal [16], these capabilities will thrive when supported by the ability of interpretation. Interpretation is the ability to interpret observations by recording the observations and then link the observations and then find the pattern or regularity of a series of observations [13].

In this research to develop the skills deliver, receive and interpret the symbol of non-verbal students are trained to make reading and interpreting charts, tables and graphs from the observation of the growth and development of seed plants were planted on Bocron. Meanwhile, to develop capabilities deliver, receives and interprets verbal symbols, students are trained to present his exhibition boards in front of the classroom during class discussions. Such skills are indicators of the science process skills that is communication and interpretation. The results demonstrated through these activities there is a significant improvement in communication skills (50% pre, post 89%, N-Gain 0.77) and inference students (Pre 38%, Post 92%, N-Gain 0.85) .

Table 1. Percentage of Pre-Post Test Scores and N-Gain of Science Process Skills Indicators on growth and developments observations seed plants is by using bokron.

Indicator	Pre (%)	Post (%)
Observation	51	89
Communication	50	89
Interpretation	37	72
Inference	38	92

On observation of the growth and development activities by using Bocron students gain new information among other things, on how the process and the rate of growth and development of roots, stems and leaves, plant condition if the lack of water, how the light affects the position of the roots, stems and leaves. The next activity students are trained to make a conclusion (inference). Inference is a conclusion made using old information that is already known from past experience and new information is being observed through the senses through observation [2].

When students put their inference practice process and incorporate new information and nature patterns from the observation of the growth and development of seed plants with the knowledge that they have learned previously. Then student then interpret and explain the process of growth and development. Students detect patterns of interaction between the plant and its environment.

4. Conclusion

According to the research data results, it can be concluded that using Bocron as a learning media can improve students' science process skill and mastery of the concept of growth and development. Using the concrete media Bocron also facilitate students to easily learn the growth and development of seed plants.

5. References

[1] Departemen Pendidikan Nasional 2005 *Ilmu Pengetahuan Alam, Keterampilan Proses Ilmu Pengetahuan Alam* (Jakarta: Direktorat Jenderal Pendidikan Dasar, Direktorat Pendidikan Dasar Lanjutan Pertama)

[2] Rezba R J *et al* 2002 *Science Process Skill* (Iowa: Kendall/Hunt Publishing Company)

[3] Rustaman N Y, Dirdjosoemarto S, Yudianto S A, Achmad Y, Subekti R, Rochintaniawati D and Nurjhani M K 2003 *Strategi Belajar Mengajar Biologi* (Bandung: Common Textbook, JICA IMSTEP)

IOP Conf. Series: Journal of Physics: Conf. Series **812** (2017) 012078 doi:10.1088/1742-6596/812/1/012078

[4] Joyce B, Weil M and Calhoun E 2009 *Models of Teaching* (Yogyakarta : Pustaka Pelajar)
[5] Cleaf D W V 1991 *Action in Elementary Social Studies* (Singapore: Allya and Bacon)
[6] Trowbridge L W and R W Bybee 1990 *Becoming A Secondary School Teacher* (Merlbourne: Merril publishing company)
[7] Bruce W and Bruce J 1992 *Learning Social Studies Through Discrepant Event Inquiry* (Annapolis, MD: Alpha Publishing Company)
[8] Aksari Y D, Widoretno S and Santosa S 2012 *Peningkatan Keterampilan Proses Sains dan Hasil Belajar Siswa Melalui Penerapan Strategi Pembelajaran Guided Inquiry di SMP Negeri 26 Surakarta Kelas VIII-B Tahun Pelajaran 2011/ 2012* vol 1 (Surakarta: Universitas Sebelas Maret) pp 63-74
[9] Strom R K 2012 *Using Guided Inquiry to Improve Process Skills and Knowledge in Primary Science* (Montana State University)
[10] Syamsuddin and Abin Makmun 2000 *Psikologi Pendidikan* (Bandung: PT Remaja Rosdakarya)
[11] Setiawan et al 2009 *Komputer dan Media Pembelajaran* ed 13 (Jakarta: Universitas Terbuka)
[12] Wardhani et al 2007 *Penelitian Tindakan Kelas* ed 2 (Jakarta: Universitas Terbuka)
[13] Rustaman N Y 1990 *Kemampuan Klasifiaksi Logis Anak (Study tentang Kemapuan Abstraksi dan Inferensi Anak Usia Sekolah Dasar pada Kelompok Budarya Sunda)* (Bandung: Disertasi Pascasarjana Institut Keguruan Dan Ilmu Pendidikan Bandung)
[14] Joni and Raka 1998 *Penelitian Tindakan Kelas* (Jakarta: PCP PGSM Dikjen Dikti)
[15] Setiono K 2009 *Psikologi Perkembangan* ed 2 (Bandung: Widya Padjadjaran)
[16] Rakhmat J 1989 *Psikologi Komunikasi* ed revision (Bandung: Remadja Karya, CV)
[17] Solso R L, Maclin O H and Maclin M K 2008 *Psikologi Kognitif* ed 8 (Jakarta: Erlangga)

Variation on the similar-size disk tower of hanoi puzzle

S Zuchri

SMP Negeri 1 Batujajar.Jl. SMP No 12 Batujajar, Bandung Barat40561, Indonesia
Email: saefudinzuchri.71@gmail.com

ABSTRACT. The famous Tower of Hanoi puzzle was invented by Edouard Lucas in 1883. This puzzle proposed three pegs, and the number of disks with different size. The puzzle starts with the disks in a neat stack in ascending order of size on one peg, the smallest at the top. The objective of the puzzle is to move the entire stack to another peg, by following these simple rules: (1) only one disk can be moved at a time; (2) Each move consists of taking the upper disk from one of the stacks and placing it on top of another stack; (2) No disk is placed on the top of a smaller disk and the minimum number of move is the goal of this puzzle.Many variations have been proposed as exercises and challenges. Some have more than three pegs and some with colours. This paper poses a new variation. There are two or more disks with similar size. The goal is to move each stack of the disk from its initial location to its final location. As usual, disk must be moved one at a time and a disk can never sit above a disk of smaller. Let n be a number of disks and there are p similar size disks. The disks are labelled from 1 to $n - p + 1$ in increasing order of size so the disk with similar size has the same label. If m is the label of the similar disks, so $M_p(n; m)$ is the minimum number moves needed to move all the disks in original peg to destination peg. We have,

$$M_2(n; m) = 2^{n-1} + 2^{n-m-1} - 1$$
$$M_3(n; m) = 2^{n-2} + 2^{n-m-1} - 1$$
$$M_p(n; m) = 2^{n-p+2} + (p-1)2^{n-m-p+1} - 1$$

The number moves needed to move if there are p_1 similar size disks m_1 and p_2 similar size disks m_2 is

$$M_{p_1,p_2}(n; m_1, m_2) = 2^{n-p_1-p_2+2}[(p_1 2^{-m_1} + p_2 2^{-m_2}) - (2^{-m_1} + 2^{-m_2}) + 1] - 1$$

1. Introduction

The tower of Hanoi Puzzle was invented by French Matematician Édouard Lucas in 1883. He is a famous French number theorist and recreational mathematician. The tower of Hanoi Puzzle has been around for at least a hundred years and is a favorite subject of recreational mathematician.

There are three vertical pegs and some number n of disk, all with a hole in the center that allows them to be stacked on the pegs. The disks are all different size in increasing order. They are stacked in order on one of the pegs-original peg, with the biggest on the bottom and the smallest on the top. The objective of the puzzle is to move entire stack from original peg to destination peg, by following these simple rules: (1) only one disk can be moved at a time; (2)Each move concists of taking the upper disk from one of the stack and put it on top of another stack; (3) No disk is placed on the top of smaller disk. The third peg is used to temporarily hold disk during the process. The objective of the puzzle is to find the minimum move to transfor teh stack of disk from origin peg to destination peg.

We know that $2^n - 1$ is a unique minimum-move solution. It is easily described recursively: for $n > 0$, recursively move a tower of $n - 1$ disks from original peg to temporary peg; move disk nfrom original peg to destination peg; and then recursively move the tower of $n - 1$ disk from temporary peg to destination peg. If $M_0(n)$ is the number of moves made for n disks, we have

$$\begin{aligned} M_0(n) &= M_0(n-1) + 1 + M_0(n-1) \\ &= 2 \times M_0(n-1) + 1 \\ &= 2 \times [2 \times M_0(n-2) + 1] + 1 \end{aligned} \quad (1)$$

or

$$\begin{aligned} M_0(n) &= 4 \times M_0(n-2) + 2 + 1 \\ &= 8 \times M_0(n-3) + 4 + 2 + 1 \\ &= 2^{n-1} + .. + 8 + 4 + 2 + 1 \\ &= 2^n - 1 \end{aligned} \quad (2)$$

We can see that disk n need one move, disk $n-1$ need two move, disk $n-2$ need four move, and disk m need 2^{n-m} move.

Many variations have been proposed, with feature such as colored disk, multiple pegs, restricred moves. Lukas himself invented a version with five pegs and a stack of disks of four different color. Indeed Paul invented variation with four pegs. This paper will discuss two, three, and m similar size disks but still with three pegs.

2. Main Result

In this section we present the formula of minimum number move n disks from original peg to destination peg. It will be easier if we divided it by some cases. First, there are two similar size disks. There are at least three disk on the stack. The second is three similar size disks, there are more than three disks on the stack. From case 1 and case 2 we can see the pattern and make the formula for m similar size diks.

2.1Case 1:

Let n be a number of disks, two of the disks have similar size. The disks are labelled from 1 to $n-1$ in increasing order of size. The similar size disks have the same label. Tabel 1 is the number of move from the stack with two similar size that have label 2.

Its clear that if there are two similar size disks, the number of move that are needed for the disk is twice that it should be. Let n be a number of disks, m is the label of the the similar size disks for $1 < m < n$. The number of move needed for a disk with labelled m is 2^{n-m-1}. Because there are two disks with labelled m, so the number of move that needed for the disks with labelled m is twice than 2^{n-m-1}.

Table1. The Number Moves of Two Similar Size Disks at label 2

The Number of Move at Disk D_i	**The Number of The Disk**					
	3	4	5	6	...	n
D_1	2	4	8	16	...	2^{n-2}
D_2	2×1	2×2	2×4	2×8	...	$2 \times 2^{n-2}$
D_3	-	1	2	4	...	2^{n-4}
...		-	-	...	...	...
D_m		-	-	...	...	2^{n-m-1}
...		-	-	...	...	...
Total Steps	4	9	18	36	...	...

D_i is the number of move that have label i.

The total number of move for label m is,

$$M_2(n;m) = 1 + 2 + 4 + \cdots + 2(2^{n-m-1}) + \cdots + 2^{n-2}$$
$$= 1 + 2 + 4 + \cdots + 2^{n-2} + 2^{n-m-1}$$
$$= 2^{n-1} + 2^{n-m-1} - 1 \quad (3)$$

2.2Case 2:

Let n is the number of disks and there are three similar size disk, the disks on a stack have label from 1to $n-2$, in increasing order. The similar size of disk have the same label. If m is the label of the similar disks then the number of move of each disk can we see at Tabel 2.

Table2. The Number Moves of Three Similar Size Disks at label 2

The Number of Move at Disk D_i	**The Number of The Disk**				
	4	5	6	...	n
D_1	2	4	8	...	2^{n-3}
D_2	3×1	3×2	3×4	...	$3 \times 2^{n-4}$
D_3	-	1	2	...	2^{n-5}
...	-	-	...	...	...
D_m	-	-	...	...	2^{n-m-2}
...	-	-	...	...	...
Total Steps	5	11	23	...	...

D_i is the number of move that have label i.

Tabel 2 shows that if there are three similar size disks, the number of move for the similar disk is three times that it should be. The number of move that are needed for disk m is 2^{n-m-2}. So the number of move that needed for the disks with label m is three times than 2^{n-m-2}.

The total number of move is,

$$M_3(n;m) = 1 + 2 + 4 + \cdots + 3(2^{n-m-1}) + \cdots + 2^{n-3}$$
$$= 1 + 2 + 4 + \cdots + 2^{n-3} + 2.2^{n-m-2}$$
$$= 2^{n-2} + 2^{n-m-1} - 1 \quad (4)$$

Without Loss of Generality, we can find the total number of move m similar size disks. Let say n as a number of disks, and p as a number of similar size disks, $1 < p < n$. The disks are labelled from 1 to $n - p + 1$ in increasing order of size. If m is a label of similar size disks, the number of moves needed to move disk m is $2^{n-m-p+1}$. Because there are p disks label m, so the number of moves needed to move all the disk to destination peg are,

$$M_p(n;m) = 1 + 2 + 4 + \cdots + p(2^{n-m-p+1}) + \cdots + 2^{n-p+1}$$
$$= 2^{n-p+2} + (p-1)(2^{n-m-p+1}) \quad (5)$$

We can find if there are p_1, p_2 similar size disks where their disks have different label. Let n be a number of disks, the stack of disks have p_1 disks label m_1 and p_2 disks label m_2, for $p_1 + p_2 < n$.

The number of move that needed for a disk whom have label m_1 is $2^{n-p_1-p_2-m_1+2}$ and label m_2 is $2^{n-p_1-p_2-m_2+2}$. There are p_1 disks have label m_1 and p_2 disks have label m_2 , so the number of move that needed for the disks label m_1 and m_2 consecutively are $p_1 \times 2^{n-p_1-p_2-m_1+2}$ and $p_2 \times 2^{n-p_1-p_2-m_2+2}$

Table3. The Number Moves of Three Similar Size Disks at label 2 and 4

The Number of Move at Disk D_i	**The Number of The Disk**				
	7	8	9	...	n
D_1	8	16	32	...	$2^{n-p_1-p_2+1}$
D_2	2×4	2×8	2×16	...	$2 \times 2^{n-p_1-p_2}$
D_3	2	4	8	...	$2^{n-p_1-p_2-1}$
D_4	3×1	3×2	3×4	...	$3 \times 2^{n-p_1-p_2-2}$
...	-	-	...	...	...
D_m	-	-	...	...	2^{n-m-2}
...	-	-	...	...	...
D_{m_2}	...	...	...	...	$2^{n-p_1-p_2-m_2+2}$
...	-	-	...	...	...
Total Steps	21	42	84	...	...

D_i is the number of move that have label i.

The total number moves needed to move all disk are,

$$
\begin{aligned}
M_{p_1,p_2}(n; m_1, m_2) &= 1 + 2 + 4 + \cdots + p_1(2^{n-p_1-p_2-m_1+2}) + \cdots + p_2(2^{n-p_1-p_2-m_2+2}) + \cdots + 2^{n-p_1-p_2+1} \\
&= 2^{n-p_1-p_2+2} + (p_1 - 1)(2^{n-p_1-p_2-m_1+2}) + (p_2 - 1)(2^{n-p_1-p_2-m_2+2}) - 1 \\
&= 2^{n-p_1-p_2+2}[1 + (p_1 - 1)2^{-m_1} + (p_2 - 1)2^{-m_2}] - 1 \\
&= 2^{n-p_1-p_2+2}[(p_1 2^{-m_1} + p_2 2^{-m_2}) - (2^{-m_1} + 2^{-m_2}) + 1] - 1 \qquad (6)
\end{aligned}
$$

3. References

[1] HinzA M, *et al* 2013 *The Tower of Hanoi–Myths and Maths* (Switzerland: Birkhäuser Basel)
[2] Allouche J A, Astoorian D, Randall J and Shallit J 1994 *Mathematical Association of America* **101** 651-658
[3] Stockmeyer P K and Lunnon F 2008 *Proc. of International Conference on Fibonacci Numbers and Their Applications (Patras)* (Dodrecht: Kluwer Academic Publisher)
[4] Stockmeyer PK 1994 *Congressus Numerantium* **102** 3-12

IOP Conf. Series: Journal of Physics: Conf. Series **812** (2017) 012080 doi:10.1088/1742-6596/812/1/012080

Developing Save Your Food Kit *(Sayofu Kit)* to Support Inquiry, Improve Student Learning Outcomes at SMP Plus Hidayatul Mubtadiin and Public Awareness on Food Additives

J Astutik

Indonesian School of Kuala Lumpur, Lorong Tun Ismail No. 1, Kuala Lumpur 50480, Malaysia
E-mail: ummu.syauqee505@gmail.com

Abstract. Food additives are materials that can not be separated from the lives of students and the community. Based on the preliminary questionnaire, it indicates the lack of kit supporting material additives in some schools and communities. The research objectives of this development are (1) to develop Kit experiment *(SAYOFU KIT)* and supplementary books to improve student learning outcomes in the classroom and public awareness on food additives (2) to describe the feasibility and potential effectiveness of *SAYOFU KIT* developed (3) to analyze the practice of *SAYOFU* KIT and benefits for students and the community. This development study uses 4-D models Thiagarajan, et al (1974). through some stages, they are: defining, designing, developing and disseminating which involes the students and community. The developed SAYOFU KIT includes additives sample kit, borax test kit, curcumin test kit, formaldehyde test kit, modification heater to the identification of dyes and dye test paper. The study is conducted at SMP Plus Hidayatul Mubtadiin, and TKIT Al Uswah. The products are validated by experts and education practitioners. Qualitative data processing uses descriptive method, whereas quantitative data by using the *N-gain*.The average yield of expert validation of SAYOFU KIT with supplementary books 76.50% teacher's book and 76.30% student's book are eligible. The average yield of 96.81% validation of educational practitioners criteria, piloting a small group of 83.15%, and 82.89% field trials are very decent.The average yield on the student questionnaire responses SAYOFU kit and supplementary book is 87.6% with the criteria very well worth it. *N-Gain* 0:56 cognitive achievement with the criteria enough. The results of the public poll showed 95% feel the benefits SAYOFU kits for testing food. Based from description indicates that SAYOFU Kit developed feasible, practical, useful to support inquiry learning and improve student learning outcomes as well as public awareness of food additives.

1. Introduction

Material food additives can not be separated from everyday life is one of the materials science class VIII. Ideal learning science materials covers aspects of microscopic, macroscopic and symbolic [1]. If these aspects found, the science material will be easily learned and applied. Food additive teaching materials available today do not represent these three aspects much. Student activities on experiments in teaching materials are dominated by food composition analysis with many scientific terms so that these materials tend to be difficult to understand and apply because it reviews of symbolic aspect only.

Methods of learning and teaching materials of food additives should be investigated and developed more. This material is not separated from the daily life of current students, future and the impact on life in the community. According to Piaget in Olson & Hergenhahn, the optimal education requires a challenging experience for learners so that optimal process of assimilation and accommodation [2]. "Appropriate learning method, learning resources and learning experiences greatly affect the students 'competence' [3]. Method of inquiry is expected to accommodate and improve learning outcomes with real and meaningful experience. The availability of teaching materials are also expected to improve the reasoning ability of students in learning.

Teaching materials that support the process of inquiry are still limited. Activity is dominated by experimental analysis composition of food additives on food packaging. Curriculum resources VIII class 2013 contains experimental activities already [4]. However, experimental activities additives are only tested on dye textiles, while the issue of additives and other addictive are presented by information and analysis only. Grain material and concept maps are not much different from other books. Value divinity embedded in religious teaching students to associate with the material additives and addictive substances studied are still lacking.

Questionnaire preliminary study to parents and teachers from several areas indicate that the material teaching materials additives are needed. Respondents noted that variations have not done an experiment in learning the material additives yet. This material is dominated by the activities of the analysis of the composition of additives on food packaging. Exhaustiveness average grade of only 50-75% only. After studying this material, it showed no change in the attitude of the students in choosing the food consumed. The preliminary study questionnaire on 50 junior high school students of class IX Plus Hidayatul Mubtadiin showed that 65.22% of students did not like this material because a lot of memorizing the type and function of the term - a foreign term. Students who forgot the name of the food and non-food additives that need to watch reaches 40 to 71%, so they have not been able to apply the knowledge about additives in food consumption. The majority of parents also find it difficult to determine the safety and how to test food additives and feel uneasy with the widespread use of non-food additives on the market.

The existence of a kit to test food additives at this time is there but still need to be developed and disseminated in the learning material additives and in community life. The kits of non-food additives that are commercially available are relatively expensive and not so well known. Kit as a part of teaching materials will support the inquiry learning process. One characteristic of inquiry learning is to develop systematic thinking skills, logical, and critical [5]. They do demonstrate that the process of reasoning is needed in learning. So the inquiry teaching materials that support the process of reasoning is needed. Everyone should remember this sentence before eating or drinking *"save your food, make sure before you consume it"*. The abbreviated sentence SAYOFU message is supported by experimental tools, named SAYOFU KIT.

Suwarni states that the inquiry teaching materials effectively increase the cognitive, psychomotor, and affective student [6]. The results of the other development is the interactive teaching materials with guided inquiry approach in chemical materials of foodstuffs are also feasible to use [7]. Other studies based learning inquiry is already quite a lot and give positive results in improving the quality of learning of them stated [8]. Sulistina *et al*, Juntunen and Aksela also stated approach of inquiry can be applied to materials chemistry at all levels of school [9],[10]. Use of research in various fields of connecting the world outside the school can prepare for study of scientific inquiry better to improve the concept and process skills using inquiry method [11]. Design of project-based activities to the development of creativity and knowledge of the inquiry-based learning is very effective at secondary school [12].

Based on the above explanation, the development of teaching materials in the form of SAYOFU KIT and supplementary to additives material is made which can be used by eighth grade students and the community. The proceedings are designed in view of material and design of experiments varied. The objective research and development is to develop a proper and validated inquiry-based teaching material and determine the potential effectiveness in improving learning outcomes of student learning. Another goal is to increase public awareness of food additives and how to test using materials that are easy to obtain and inexpensive.

2. Model, analysis, design and implementation

2.1. Research and development model

The development model used in this study refers to the development of the research and development of 4D models of Thiagarajan *et al* which includes the step *define, design, develop* and *disseminate* [13]. In the *define* phase to do a needs analysis. These stages are complete made through a 5-step analysis: analysis front-end *(front-end analysis),* analysis of student *(learner analysis),*analysis of the task *(task analysis),* analyzes the concept *(concept analysis),* and goal setting *(Specifying Instructional Objectives*).

The front-end analysis aims in determining the problems encountered in learning based on the curriculum of 2013 and the facts on the ground so we need a material development material additives [14]. Analysis of students aims in determining the characteristics of the students to be adapted to the design of the development of teaching materials. Characteristics of students and the community analyzed will include background experience, the general attitude towards learning topics, as well as the media, formats and languages. Information about the characteristics of students and the public in accordance with the design development of teaching materials. The task analysis aims to identify the key skills that must be mastered and analyze it in order to be a skill that must be mastered. This analysis is done through a study of core competencies, basic competencies and learning indicators that are used to develop teaching materials additive material. Analysis of the concept aims to identify the main concepts in the material being taught additives which can not be separated from KI, KD and indicators of achievement of competencies. Formulation of learning objectives and task analysis is converting concepts into instructional objectives were developed from learning indicators.

Design stage is implemented by arranging *story board* that refers KD and indicators arranged on the *define* phase. This stage includes the preparation of materials, media selection, the selection of teaching materials floating format, the initial design. Penentukan early design phases covering material according to the material that has been determined, and the determination of the type KIT trial. Type kit is determined to make variations of the experiment could be done simply. The final outcome kit named SAYOFU KIT which consists of a kit of samples of additives, test of borax (paper, liquid and sticks), formalin (paper and liquid), for example paper test kit to test borax, formalin test kit, curcumin test kit, heater modifications and test paper dye.

Develop stage aims to produce a final draft of the revised based on input from the experts. The activities of this stage is the assessment of the experts and product trials. Prior to validation by the validator, the product must be validated by supervisor first, further validated by educational experts, subject matter experts, and practitioners. Validation expert *(expert appraisal)* aims to determine the feasibility or validity of the product before the final product is produced. The revised products are used to test the product in small groups and field trials. The end product will be used for field trials in the experimental class to determine the potential effectiveness of teaching materials to improve student learning outcomes as on. Products are also tested in people from various circles in Figure 1.

(a) (b) (c)

Figure 1. Testing SAYOFU KIT. (a) Testing products by students, (b) by Community in SMP Plus Hidayatul Mubtadiin and (c) TKIT Al Uswah.

2.2. Analysis

The type of data obtained in this research and development consists of quantitative and qualitative data from the questionnaires using *Likert* scale and column comments and suggestions. The qualitative data is obtained from the advice and input from education experts, subject matter experts, education professionals, students of class IX and class VIII Junior High School Plus Hidayatul Mubtadi'in Singosari Malang. The quantitative data is obtained from a questionnaire using *Likert* scale with a score of 1-5. Data obtained from expert validation is used to determine the feasibility and validity of teaching materials. The data from students, teachers and the community is used to determine the effectiveness, the attractiveness and teaching materials applicability.

Data collection instrument used in this study is in the form of questionnaire *checklist* and column suggestions and comments. The type of questionnaire used, namely the preliminary questionnaire (the needs of teachers and students), expert validation questionnaire, questionnaire testing small groups, and affective self-assessment questionnaire students. These preliminary questionnaire contains certain criteria of instructional materials needed by teachers and students. Questionnaire validation of experts comprising two types of assessment, namely the *Likert* scale and the truth of the concept of the material by subject matter experts using a questionnaire using yes and no options. Through this validation questionnaire can be used as teaching material improvement SAYOFU KIT developed.

Additive material cognitive matters are made according to the indicators achieved by competence, namely KD 3.7 and 4.7 grade VIII junior high school. Matters being used is validated by experts with the validity of 94.63% and a very decent criteria. The trials were conducted at about 44 students of class IX and continued with analysis using *IBM SPSS statistics 21*. The reliability obtained was 0.732 with the high category. The amount of cognitive questions are 30 questions that declared valid by modest and easy difficulty level.

Analytical techniques done is to calculate a score acquisition and converted in percentage form. The results of this quantitative data are used to get the conclusion of a pre-determined criteria. Data from preliminary study questionnaire is processed by the calculation of percentages for each item selection. As stated by Sudjana with the formula [15]: The determination of this percentage descriptive processed by counting the number of respondents divided by the total number of respondents and multiplied by 100 percent, Sudjana [15] as follows:

$$P = \frac{f}{N} x\ 100\% \qquad (1)$$

P = the percentage of poll respondents
f = number of respondents who answered item
N = the total number of respondents

Criteria for eligibility qualifications percentage calculations with predetermined criteria [16]. The questionnaire results with the percentage of 81% - 100% criteria are very decent (not revised), 61% - 80% criteria are eligible (not revised), 41% - 60% criteria are quite decent (not revised), 41% - 60% criteria are less worthy (revision), and 0% - 20% criteria are not worth (revised).

Analysis of scientific reasoning ability and cognitive learning outcomes form of multiple choice questions are conducted to determine improved results *pretest* to *posttest.* The counting uses the formula N -*gain*. Analysis of scientific reasoning ability and cognitive learning outcomes form of multiple choice questions are conducted to determine improved results *pretest* to *posttest.* Perhitunganya using the formula N -*gain* as follows:

$$N\ gain = \frac{<S_{post}> - <S_{pre}>}{S_{m\ ideal} - <S_{pre}>} \quad (2)$$

Furthermore, *the N gain* obtained is interpretated based on predetermined criteria. A high N-gain criteria if > 0.70, middle criteria if 0.30 to 0.7, while the low criteria if the N-gain <0.3 [17].

2.3. Design

Design of experiments performed in this study is at the stage of *develop* from the design of the 4-D, consisting of several stages: 1) validation expert *(expert appraisal)* by two lecturers, two science teachers JUNIOR HIGH SCHOOL, 2) pre-trials *(initial testing)* on at 8 junior high school students of class IX appointed, 3) field trials *(quantitative testing),* in class VIIIA SMP Plus Hidayatul Mubtadiin to test adherence and to get data from student questionnaire, continued to junior high school students grade VIIIB to test the potential effectiveness of teaching materials to improve student learning outcomes, 4) dissemination in the community involves teachers and parents of students kindergarten, elementary school, and junior high school. Research trial design in the classroom is *one group pretest-posttest design* as shown in Table 1 [13].

Table 1. Study design effectiveness test subjects.

Class	*Pre test*	treatment	*posttest*
VIII	O_1	x	O_2

Information:
O_1 = cognitive achievement test before treatment
O_2 = cognitive achievement test after treatment
X = learning inquiry-based teaching with SAYOFU KIT

2.4. Implementation

Research and development is done by 1) conducting *a pretest* about cognitive 2) implementing learning using teaching materials, 3) granting a matter *posttes* to measure whether there is an increase in cognitive achievement of students, 4) Socialising to the community through school kindergarten, elementary and junior high students. Quantitative data analysis of data is in the form of a percentage of student response to learning using teaching materials, the data *pretest* and *posttest* using cognitive tests. Furthermore, to determine student's cognitive learning outcome is calculated by the formula *N-gain.* Public response is obtained through qualitative data from the questionnaires. Implementation of SAYOFU KIT in research can be seen in Figure 2.

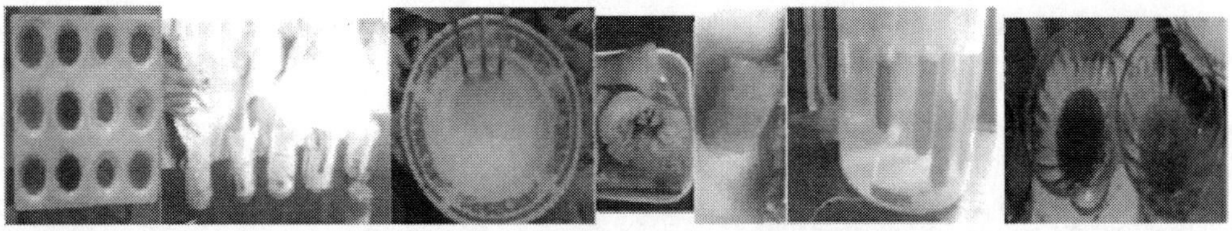
(a) (b) (c) (d) (e) (f)

Figure 2. Impelentation SAYOFU Kit for Experiments. (a) SAYOFU KIT used for identify natural and artificial colouring agents, (b) dye paper test, (c) Borax test on cracker, (d) curcumin test on yellow cassava fried, (e) modification heater used to identify colouring agents, (f) and formaldehyde test on meat.

3. Result and discussion

Prepared teaching materials are developed based on the Curriculum 2013 on the basis of competence 3.7 and 4.7 for teaching science to students of class VIII grade junior high school material additives and addictive substances in the form of books and SAYOFU KIT. The book consists of two kinds of books the teacher's and student's books. Teaching materials for teachers are not much different with the students teaching materials containing of four learnings, in each lesson there is an introduction, learning objectives, *lesson inquiry,* reflection, material comprehension, concept mapping, a summary of the material, tips, exercises. At the end of the book there is a summary of the material, concept mapping, a glossary and competency testing, concept maps, and a glossary. The difference between a teacher book and student book is at the teacher's guide, answer key, lesson plans and syllabi parts in the teacher's book.

Sayofu kit consists of 1) additives sampel kits consisting of three colors (red, yellow, and green) of the label to depend on the risk, 2) *heater* modification kits to test natural dyes and artificial through heating, 3) curcumin *test kit* , made from alkaline solution that can be used to test food color yellow, 4) borax test kit (paper, liquid, spray and stick) 5) formalin-borax *test kits* contain active *anthocyanins* from the skin of dragon fruit, hibiscus, or use ectract of sappan wood (*Caesalpinia sappan L*). 6) Dye paper test that works on the principle of paper chromatography. SAYOFU KIT and supplementary books developed can be seen in Figure 3.

(a) (b) (c) (d) (e) (f)

Figure 3. SAYOFU KIT and Supplementary Books. (a) SAYOFU KIT on box, (b) additives sampe kit, (c) Borax test kit, (d) curcumin test kit, (e) Formaldehyde test kit, and (f) Supplementary books for teacher and student.

The average yield of expert validation of the teaching materials criteria that include SAYOFU KIT with 76.50% teacher's book and 76.30% student's book are eligible. The average yield of 96.81% validation of educational practitioners criteria, piloting a small group of 83.15%, and 82.89% field trials are very decent. The average yield on the student questionnaire responses SAYOFU kit and supplementary book is 87.6% with the criteria very well worth it. From product testing, the highest scores obtained on the tips and reflection that is 97% and 93%. Results of the assessment *cognitive* learning outcomes *N-gain* of 0.56 with the medium category. Percentage outcomes 82,25 % of cognitive learning, 80% of learning outcomes affective, and psychomotor learning outcomes 85%. The results of the public poll showed 90% of respondents do not understand the additives and 95% feel the benefits SAYOFU kits for testing food, so they suggested to be produced, and developed further. The percentage of resesearch result can be seen in Figure 4.

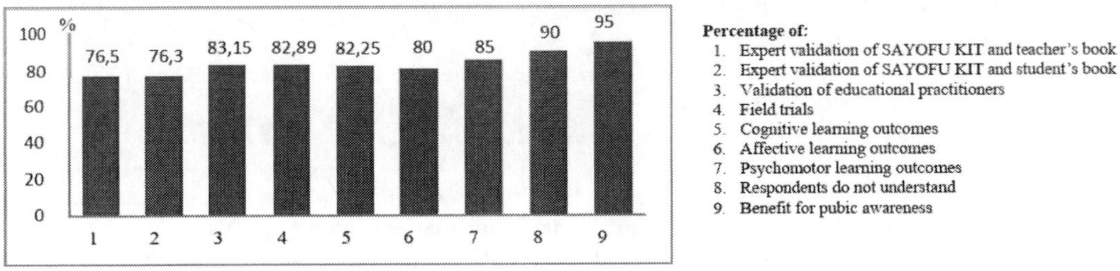

Figure 4. Result data of the research.

4. Conclusion

Based on the results of the validation and test of the product, it showed that the form of SAYOFU KIT teaching materials and supplementary developed are feasible to use for students and the community. Teaching material has the potential to improve student learning outcomes and increase public awareness as well as life skills to food additives. The results of research and development can be proceed in the *final* stage *of testing* is the use of *packaging* product by others and *disseminating* followed by a much larger scale. Teaching materials that have passed this phase will be determined its effectiveness, if proven effective, it can be used on a wider scale or distributed.

Further advice product development are (1) Developing kit other additives, such as test kits for MSG, cyclamate, and Rhodamine B because the current test kit price to test is still quite expensive; (2) developing supporting kits to another science materials or topics, because its existence is very helpful in the learning process; and (3) Using materials that are easy to obtain and use by the public.

5. References

[1] Chittleborough G and Treagust D F 2007 *Chemistry Education Research and Practice* **8** 274-92

[2] Hergenhahn B R and Olson M H 2008 *Theories of Learning* (Teori Belajar) 7th Edition, Translate by Tribowo B S (Jakarta: Media Grafika) pp 324-325

[3] Mulyasa 2014 *Pengembangan dan Implementasi Kurikulum 2013* (Bandung: PT Remaja Rosdakarya) pp 5-6

[4] Zubaidah S, Mahanal S, Yuliati L and Sigit D 2014 *Ilmu Pengetahuan Alam Kelas VIII* (Jakarta: Kemdikbud) pp 175-7

[5] Hosnan M 2014 *Pendekatan Saintifik dan Kontekstual dalam Pembelajaran Abad 21* (Bogor: Ghalia Indonesia)

[6] Suwarni E 2013 *Pengembangan Bahan Ajar Kimia Materi dan Perubahannya Berdasarkan SMK Kelas X Program Keahlian Teknik Batu dan Beton* (Malang: Thesis, Malang State University)

[7] Kartika D A 2013 *Pengembangan Bahan Ajar Interaktif dengan Pendekatan Inkuiri Terbimbing untuk SMK RSBI kelas X Program Keahlian SMKN 2 Bondowoso pada Mata Pelajaran Kimia Bahan Makanan* (Malang: Thesis, Malang State University)
[8] Nugroho C 2011 *Pengaruh Penerapan Strategi Pembelajaran Inkuiri dengan Teknik Praktikum pada Materi Pemelahan Sel terhadap Motivasi dan Hasil Belajar siswa Kelas XII IPA SMAN 8 Malang* (Malang: Thesis, Malang State University)
[9] Sulistina O, Dasna I Wvand Iskandar S M 2010 *Penggunaan Metode Pembelajaran Inkuiri Terbuka dan Terbimbing dalam Meningkatkan Hasil Belajar Kimia Siswa Kelas X SMA Laboratorium Malang* (Malang State University of Chemistry) pp 82-89
[10] Juntunen M and Aksela M 2013 *CEPS Journal* **3** 157-80
[11] Feldman A, Chapman A, Hernandez V, Ozalp P and Alshehri F 2012 *Journal of Science Education International* **23** 328-37
[12] Knezek G, Christensen R, Tyler-wood T and Periathiruvadi S 2013 *Science Education International* **24** 98-123
[13] Thiagarajan S, Semmel S D and Semmel M I 1974 *Instructional Development for Training Teachers of Exceptional Children* (Pennsylvania: Indiana University) pp 75-6
[14] Kemendikbud 2014 *Dokumen Kompetensi Dasar untuk Sekolah Menengah Pertama/Madrasah Tsanawiyah sebagai Salah Satu Perangkat Kelengkapan Dokumen Kurikulum 201* (Jakarta: Kemendikbud)
[15] Sudjana N 2011 *Penilaian Hasil Belajar Mengajar* (Bandung: PT Remaja Rosdakarya)
[16] Ridwan 2011 *Skala Pengukuran Variabel-Variabel Penelitian* (Bandung: Alfabeta)
[17] Hake R R 1998 *American Jornal of Physics* **66** 67-74

Acknowledgements
I would like to express my deepest appreciation to all those who provided me the possibility to complete this research. I would like to express my gratitude to Prof. Ir. Ari Purbayanto and Drs. H. Agustinus Suharto, M.Pd who allowed me to represent Indonesian School of Kuala Lumpur in Ki Hajar Dewantara Award. I would like to thank my advisors Dr. Munzil Arif, M.Si and Dr. Hadi Suwono, M.Pd for the useful comments, remarks, full effort in guiding me, and engagement through the learning process of this research. Furthermore I would like to thank Mr Muhamad Mustofa the headmaster of SMP Plus Hidayatul Mubtadiin and Mrs. Fitri the headmaster of TKIT Al Uswah Singosari for allowing research in their school also supporting on the way. Also, I like to thank the participants in my research, who have willingly shared their precious time during the process of research, their comment and advices. A special gratitude I give to Mrs. Astri Rahayu whose contribution in stimulating suggestions and encouragement, helped me to coordinate my research especially in writing this report. Furthermore, I would like to give my highest appreciation to the the staff of Indonesian school of Kuala Lumpur and SMP Plus Hidayatul Mubtadiin, who let me use all required equipment and the necessary materials to complete the research. A special thanks goes to all my friend, who help me to assemble the parts and gave suggestion about the SAYOFU KIT. I would like to thank my family and loved ones, Syaifudin Juri who have supported me throughout entire process, both by keeping me harmonious and helping me putting pieces together. I will be grateful forever for your love. Last but not least, many thanks go to anyone may I have forgotten. I do apologize and thank you as well.

IOP Conf. Series: Journal of Physics: Conf. Series **812** (2017) 012081 doi:10.1088/1742-6596/812/1/012081

Developing Critical Thinking of Middle School Students using Problem Based Learning 4 Core Areas (PBL4C) Model

R Haridza[1] and K E Irving[2]

[1]SMP Negeri 3 Pontianak, Jl. Kalimantan No. 123, Pontianak 78121, Indonesia
[2]College of Education and Human Ecology, The Ohio State University 333, Arps Hall 1945 North High Street Columbus, Ohio 43210, Amerika Serikat
Email: amallia.yuliana@gmail.com

Abstract. Traditional methods such as rote learning and memorization in teaching science create passive students in science classrooms. The impact of this continuous action for many decades is inactive learners who cannot develop higher order thinking skills. Based on the performance test, students' critical thinking skill in Public Middle School 3 Pontianak was in low level although their achievement score were higher than school standards. The purpose of this study is to develop critical thinking skills of middle school students using Problem Based Learning 4 Core Areas (PBL4C). The design of this research is classroom action research with two cycles. Data has been collected using observation checklist, rating scale, self and peer assessment. Research findings reveal that students experience development from 11.11% to 88.45% in identifying the problem correctly, 37.03% to 76.92% for sub skills distinguish knowledge and opinion, 18.51% to 65.38% for sub skills providing possible solution, 22.22% to 69.23% for sub skills making decision, and 11.11% to 69.23% for sub skills identifying the impact of the implementation of their solution. In conclusion, the findings indicate that development of students' critical thinking skills occurs when PBL4C model applied in science classroom. These findings suggest that teachers should act as facilitator in a classroom as well as should provide meaningful learning resources that can benefit students' critical thinking skills. On the other hand, students should practice constantly to offer a sharp, accurate and appropriate solution.

1. Introduction

Nowadays, the development of teaching approaches in physics education has been changed. Transforming from the rote learning, memorization and other traditional methods in teaching, currently students are expected to be more involved in the teaching and learning processes as well as graduate improved and ready to be successful in today's global economy. In order to achieve this idea, students must learn the essential skills, for instance problem solving and collaboration skills within their context of core course knowledge such as physics. This concern becomes the main focus in the framework for 21st century learning. This framework develops some skills such as learning and innovation skills, information, media, and technology skills, and life and career skills that must be mastered by students to be successful in work and life.

I have been teaching physics for nine years in a regular public school. The last four years of my teaching experience was in top school in West Kalimantan Province. Based on my experiences, I have found that during my teaching practice, most of my students are not able to express clear ideas and transfer their knowledge to solve real-world problems in physics class. For example, students understood about metal and non-metal classification, but they failed to identify the correct material from the available resources that they needed to design a safe electrical circuit. Another example, students mastered the methods of separating mixtures such as distillation, filtration, and chromatography, but they were unable to identify which was the best material to purify water. Students were unable to develop their critical thinking in solving problems and judged a decision using scientific reasons. Based on the performance test of those experiments, the critical thinking achievement of each aspect of students from Grade VII D Public Junior High School 3 Pontianak Year 2013 is provided into the table below.

Table 1. Critical thinking achievement of students from grade VII D Public Junior High School 3 Pontianak Year 2013.

No.	Aspect of critical thinking	Percentage of student's performance
1.	Identifying problem correctly	11.11 %
2.	Distinguish knowledge and opinion	37.03 %
3.	Providing possible solution	18.51 %
4.	Making decision	22.22 %
5.	Identify the impact of the implementation of their solution.	11.11 %

Based on the Indonesian school based curriculum (2006), the central education authority developed general competences and minimum content outlines while individual teachers developed their subject curricula, including formulating learning objectives, selecting content, teaching strategies as well as developing learning evaluations independently. This curriculum stressed the achievement of standardized competences that students had to achieve, and the development of life skills to prepare graduates to survive in life [1]. In order to develop an evaluation, teachers must determine the minimum standard score, which involves complexity of content knowledge, availability of learning resources, as well as students' intake. The calculation of minimal standard score is started from determining the minimal standard score of indicator, basic competency, and standard competency respectively. The final calculation will result in the minimum standard score of a course, in this case, science for grade VII. In Year 2012/2013, based on the calculation, the minimum standard score for students in science was 80 out of 100. For this reason, all students who achieved a score average of 80 out of 100 for each aspect of critical thinking were assumed to have adequately developed their critical thinking skills.

According to my observations and the result of the performance test, students' critical thinking currently was very low. This condition could be the impact of implementing traditional methods in teaching physics such as rote learning and memorization. Those methods were not providing enough space for students to develop their multi dimensional skill, mental processes as well as universal harmonious values. Students were not used to working cooperatively, appreciating their diversity, and effectively analyzing. In addition, they also were not fostered to evaluate evidence, arguments, and alternative points of view before judging a decision and solving various types of problems with both conventional and innovative methods. In other words, students were not able to transfer their knowledge to solve problems in real life scientifically. Therefore, teaching and learning activities at this point were not preparing students to adapt and survive as well as to succeed in work and life.

During my participation in the Customized Course for Indonesian Secondary Science and Mathematics Educators at RECSAM, Penang, Malaysia on September 2012, I encountered the new invention in teaching approach. I was very impressed with its features. The instructors integrated the content of the lesson and some events that usually happen in our daily life in promoting an instructional strategy or learning model. This model was developed and focused on four core areas of physics education. Those areas were multi dimensional skills, multi disciplinary content knowledge, thinking skills and universal harmonious values so that students were prepared to be able to solve their problems in daily life. Thus both knowledge and learning activities became a meaningful learning for students. Therefore, in order to help teachers to combine the content knowledge and real life problems in teaching physics, teachers could apply the PBL4C Model as one of the alternatives.

When I reviewed some literatures, there were ample evidences showing that the enactment of PBL4C model in physics classroom was able to develop students' critical thinking skills. For example, Larmer and Mergendoller documented that the use of PBL4C model in secondary school was more meaningful if students were asked to conduct real inquiry [2]. They further suggested that PBL4C model by inquiry had developed students' high order thinking skills such as critical thinking. The most fascinating aspect of PBL4C

Model is the context problem in lesson based on the real life problem. Numerous studies corroborated that PBL was an effective approach to cultivate critical thinking and problem solving skills [3],[4],[5]. This model is students' centered; developing students' accountability, responsibility and lifelong skills. It also promotes problem-solving skills, asks student to apply content realistically, and stimulates creative and critical thinking. Considering the benefits of this model, I would like to conduct classroom action research focusing on the implementation of PBL4C Model and investigate its impacts in the development of students' critical thinking skills. Therefore, this Classroom Action Research questions are as follow: How to develop students' critical thinking skills using Problem based Learning 4 Core Areas (PBL4C) Model? What are the problems that students have in developing their critical thinking skills?

2. Research question and purpose of this study

The purpose of the study is to investigate the development of critical thinking skills of middle school students using Problem Based Learning 4 Core Areas (PBL4C). This teaching method is based on the active learning and constructivism learning theory and inquiry based science education. The central focus of constructivism is the student as a self-governed creator of knowledge [6]. Constructivist teaching should actively engage students in relevant and interesting problems; provide opportunities for students to discuss problems with each other and the teacher; give students time to reflect and test their ideas; and encourage students to reflect on other students' ideas [7]. In addition, Savery and Duffy considered Problem Based Learning (PBL) as one of the best examples of constructivist-learning environment [8].

3. Methodology

Research design of this study is class action research with two cycles. The main purpose of this study is developing critical thinking of middle students using PBL4C model. This study is relevant to the central idea of action research which is to intervene in a deliberate way in the problematic situation in order to bring about changes, and even better, improvements in practice [9]. This study has been conducted in Public Junior High School 3 Pontianak (SMPN 3 Pontianak), West Borneo, Indonesia. SMPN 3 Pontianak is an academically strong middle school in West Borneo located in Pontianak which means this school only accepts students based on the high level of students' academic performance in elementary school. Based on results of the national examination Year 2013, five students from SMPN 3 Pontianak become top ten students who achieved the highest score in that province, thus this school is considered as the best middle school in West Borneo Province this year similar to the previous years. This study had been conducted specifically in a physics course for grade VIII Term I Year 2014/2015 and the topic was Force and Motion. The course meeting is twice a week, 120 and 80 minutes, from July to December every year. This class action research was conducted for 6 weeks with the topic force and motion. The study started on the third week of July 2014 and continued to the fourth week of September 2014. Students from grade VIII were the participants of this study. The number of students was 32 and consists of both female students and male students. Students from grade VIII usually have common age, economic, social, and intellectual ability background thus will reduce or minimize the impact of these backgrounds on their critical thinking skills.

3.1. Procedures

Research for this study consists of four main phases: (1) planning, (2) implementation, (3) observation, and (4) reflection.

3.1.1. Planning. In this section, the steps to be implemented in this action research are planned. The researcher reviews the standard competency and basic competency in Curriculum 2013. After reviewing those competencies, especially for the topics of force and motion, the researcher determines the study objectives, learning experience, indicator of achievement, and teaching methodology, in this case, PBL4C. The next step, the researcher will design the lesson plans comprised of learning objectives, two scenarios, learning resources, and instruments of assessment.

Integrating PBL4C in lesson plans in the science classroom is based on the main principle of PBL which is that "all lesson begins with a contextual problem" where it provides a new dimension, direction and motivation for learning [10]. Thus, the learning activities will start with a contextual problem as the first stage. The contextual problem is delivered to all groups in the classroom. This contextual problem provides the opportunities to examine and try out what they know and what they need to know, discover what did they learn, develop skills for achieving higher performance in teams, and improve the communication skills [10]. Students are given a worksheet and they have to fill in the blank columns to identify the problem and the fact list. In addition students need to list what they need to know to make inferences, and what they need to know to identify implicit assumptions and possible solutions based on the contextual problem.

The second stage is that students are given two scenarios. A scenario is a unique component of PBL that is especially suited to introduce challenge and thereby provide good opportunities to solve the problem. Appropriate management of change is by no means an inborn skill but one that needs knowledge, observation, communication, accommodation, personal practice and evaluations. In scenario 1, each group test their decision or model on how to solve the problem through experiment. They determine the strength and weaknesses of the decision or model of the problem and whether it works or not in solving the problem. In scenario 2, they present their plan or model, its strength and weaknesses on solving problem, and/or recommendation to solve the problem to their group as well as their classmates. This presentation covers the identified problem, questions, data collection, data analyses, possible solutions and recommendations based on the data analyses. In their plan, they not only learn the content knowledge, thinking processes, but also practice the skills and values in decision making to figure out the possible solution and used higher-order thinking skills [11].

Through these processes, students discuss, analyze, and argue, as well as provide possible solutions in order to obtain the mutual understanding about the contextual problem. Those processes develop students' ability to identify the problem correctly, analyze arguments, ask clarifying questions, judge the credibility of a source or observation report, make inferences, identify implicit assumptions, and decide strategies and tactics to solve the problem. Through these activities, students are able to improve their critical thinking on those sub skills.

3.1.2. Implementation. The lesson plan had been implemented in the classroom. The researcher promoted the problem about force in real life through a video. After all students understand about the main problem revealed through the video, the researcher promoted scenario 1 which was a real life problem. Students formed a group consisting of 4 members and discussed the possible solutions for the problem. Then, students designed their model using appropriate materials and then tested whether it works or not in solving the problem.

After students completed their tasks in scenario 1, the researcher continued to the scenario 2. In scenario 2, each group presented their solution in front of the class. They will showed their model to other students and explain if the model works properly in solving the problem or not, its strength as well as its limitation. Students also informed the group about suggestions what they will do in the future in designing the better model based on their experience in testing it. At this stage, the researcher provided self and peer assessment to obtain information about the developing students' critical thinking. Those instruments will be filled and collected after the class is completed.

3.1.3. Observing. While students designed, tested, and promoted their model, the researcher observed the specific elements of critical thinking that had been developed through an observation checklist.

The specific elements consist of four categories: (1) Elementary clarification, (2) Basic support of an argument, (3) Inferences, (4) Advanced clarification, (5) Strategies and tactics [12]. This observation started from the beginning of scenario 1 through the end of scenario 2.

3.1.4. Reflection. Based on the result of self-assessment, peer assessment, as well as observation, the researcher defined the number of students who develop their critical thinking for each sub skills. The development of critical thinking skills are determined by observing carefully the number of critical thinking sub-skills used in solving problem. These sub-skills are stated in the self-assessment form, peer assessment form, and rating scale. Based on those findings, the researcher conducted interviews with students in order to obtain the clear picture of the development of students' critical thinking in applying PBL4C model as well as reduced biases and minimize different interpretation between the researcher and students. The researcher also defined some common problems in applying PBL4C in the classroom and reduced it on the next second cycle.

3.2. Data collection

Related to the Research Questions, some instruments were developed to document any data needed in this research. The types of instrument are summarized in the following table:

Table 2. Research questions, data to be collected, and instrument that will be used to collect the data.

Research Question	Data to be collected	Instrument
How to develop students' critical thinking skills using Problem based Learning 4 Core Areas (PBL4C) Model?	Data that indicated students' attitude towards critical thinking skills: 1. Identifying problem of the experiment correctly; 2. Distinguish knowledge and opinion; 3. Providing possible solution; 4. Making decision; 5. Identifying the impact of the implementation of their solution.	Self assessment Peer assessment Rating scale Checklist
What are the problems that students have in developing their critical thinking skills?	Data that indicated students' difficulties in developing critical thinking skills. 1. Identifying problem of the experiment correctly; 2. Distinguish knowledge and opinion; 3. Providing possible solution; 4. Making decision; 5. Identifying the impact of the implementation of their solution.	Self assessment Peer assessment Rating scale Checklist

Data collected during this research study included self-assessment and peer assessment, observation checklist and rating scale. Self and peer assessment instruments had been filled by the students after they complete the scenario 2 at the end of meeting. The researcher observed the development of students' critical thinking through observation checklist and rating scale. The observation had been conducted at the beginning of scenario 1 through the end of scenario 2.

As a qualitative research study, data has been analysis qualitatively. Description based on the rich data yielded from triangulation using various instruments for such a data may provide a clear picture of the findings. The function of triangulation is to locate and reveal the understanding of the object under investigation from "different aspects of empirical reality" [13]. The common trends arising from the data will be segmented and labeled accordingly to build meaning and interpretations before reporting the findings and conclusions of this study.

4. Results and conclusion

SMPN 3 Pontianak is an academically strong junior high school in West Kalimantan located in Pontianak City which means this school only accepts students based on the level of students' academic performance in elementary school. Based on the results of the national examination Year 2013, five students from SMPN 3 Pontianak become top ten students who achieved the highest score in that province, thus this school is considered as the best junior high school in West Kalimantan Province this year similar to the previous years.

4.1. Description of the first cycle

Target observation of this research is students' activities in learning process in cognitive and affective aspects related to Newton's third law. Results from self-assessment and observation as well as learning objectives are appropriate with the research instruments in planning, main activities and closure. Results of the students' critical thinking skills achievement sub skill: Identifying problem of the experiment correctly. Data from peer assessment perform 57.69% students are able to identify the problem correctly. This finding is supported by the data from observation (checklist) that indicate 50% students are able to identify correctly. Therefore, there is an increasing critical thinking skills compare to the data from initial research. The achievement increases from 11.11% (initial research) to 50%. Data from observation also indicate the 73.07% students have sensitivity to the feelings, level knowledge of others in various level yet only 15.38% students look for explanation and reasons and uses and cites credible sources.

Results of the achievement of students' critical thinking skills sub skill: Distinguish knowledge and opinion. Data from peer assessment indicate that students perform better in providing facts from the problem compare to mentioning opinion about the problem. There are 38.46% students always able to provide facts from problem and 23.07% students always mention the problem when groups share their results. Data from observation (checklist) also indicate that 38.46% students look for explanation and reason when they distinguish knowledge and opinion. Therefore, there is a development of critical thinking skills compare to the data from initial research. There is a slightly increase of achievement from 37.03% (initial research) to 38.46%. Data from observation also indicate the all student use and cite credible sources yet there is only 30.76% students who are able to be open minded and keep to the main relevant point.

Results of the achievement of students' critical thinking skills sub skill: Provided possible solution. Data from peer assessment indicate that students perform better in what they need to solve the problem rather than mention possible solution to solve the problem. All students always able to mention what they need to solve the problem and 30.76% students always mention possible solution to solve the problem when groups share their results. Therefore, there is an increasing of critical thinking skills compare to the data from initial research. There is an enhancement of students' ability to provide possible solution from 18.51% (initial research) to 30.76%. Data from observation also indicate 73.07% students are sensitive to the feelings, level knowledge of others yet there is only 30.76% students who are able to look for alternative when they provide possible solution.

Results of the achievement of students' critical thinking skills sub skill: Making decision. Data from peer assessment indicate that students perform better in making a decision towards the problem rather than providing logical reason in making a decision. About 50% students always able to make a decision towards the problem yet only 15.38% students sometimes provide logical reason when they make the decision. This finding is supported by the data from observation; 53.84% students look for alternatives when they make a decision. Therefore, there is an increasing of critical thinking skills compare to the data from initial research. There is an enhancement of students' ability to make a decision from 22.22% (initial research) to 50%. Data from observation (checklist) also indicate that all students take a position of an issue yet only 42.30% who change a position of an issue with good reason(s).

Results of the achievement of students' critical thinking skills sub skill: Identify the impact of the implementation of their solution. Data from peer assessment indicate that students perform better in mentioning whether the solution work properly in solving problem and mentioning the weakness of their model compare to mentioning the strength of their model. There are 61.53% students always able to both mention whether the solution work properly in solving problem and mention the weakness and 19.23% students always mention the strength of their model when groups share their results. Therefore, there is a development of critical thinking skills compare to the data from initial research. There is a slightly increase of achievement from 11.11% (initial research) to 61.53%. Data from observation also indicate the all student take a position of an issue and sensitive to the feeling, level knowledge of others but only 57.69% students are open-minded.

Results from the first cycle reveal that there is a development of students' critical thinking skills for all aspects compare to the preliminary data. The largest development of students' critical thinking skills is in sub skill: Identifying the problem correctly. In these sub skills, students have significant development from 11.11% to 50%. Some other sub skills also have significant development for example making a decision towards a problem and identifying the impact of the implementation of their solution. However these developments are only occurred in basic level. In the sub skill: Making decision, about 50% students always able to make a decision towards the problem yet only 15.38% students sometimes provide logical reason when they make the decision. This data implies that students are able to make a decision of a problem but their proposed solution may not have scientific explanation or logical reason behind it.

Data from sub skill: Identifying the impact of the implementation of their solution also reveals a similar case. Students only master the basic level of critical thinking skills. 61.53% students always able to both mention whether the solution work properly in solving problem and mention the weakness yet only 19.23% students always mention the strength of their model when groups share their results. This finding implies that students are struggling to identify the strength of their model. They only focus in the final goal that is whether the solution works properly or not. They also give more attention to provide explanation and suggestion to overcome the weakness of their model rather than boost the strength of it. These findings disclose that students' critical thinking skills need to be improved to the advanced level in the next second cycle.

4.2. Description of the second cycle

Results from self-assessment and observation as well as learning objectives are appropriate with the research instruments in planning, main activities and closure. Results of the achievement of students' critical thinking skills sub skill: Identifying problem of the experiment correctly. Data from self-assessment perform 92.30% students are able to identify the problem correctly. This finding is supported by the data from observation (checklist) that indicate 88.45% students stated the main idea or thesis clearly, enthusiastically, and interestingly, for the audience. Therefore, there is an increasing critical thinking skills compare to the data from cycle 1. The achievement increases from 50% (initial research) to 88.45%.

Results of the achievement of students' critical thinking skills sub skill: Distinguish knowledge and opinion. Data from self-assessment indicate that 76.92 % students are excellent in recognizing the difference between knowledge and opinion. Data from observation (checklist) also indicate that 88.45% students give highly appropriate facts, excellent generalizations from facts, that support the thesis, sources of facts are credible, facts used well in making the argument. Therefore, there is a development of critical thinking skills compare to the data from cycle 1. There is an enhancement of achievement from 38.46% (cycle 1) to 76.92%.

Results of the achievement of students' critical thinking skills sub skill: Provided possible solution. Data from self-assessment indicate that 80.76% students are excellent in thinking of the possible solution toward the problem. Data from observation also indicate 65.38 % students actively look and encourage others to express points of view different from or opposing own. There is an enhancement of students' ability to provide possible solution from 30.76% (cycle 1) to 65.38%. However, data from checklist also reveal that 30.76% students mention all relevant and important alternative positions, present it fairly, and evaluate it properly. There are also 19.23% students attempting to rebut the alternative position but they are ineffective or incomplete. These findings imply that although students perform excellent in providing solution, they still need to develop the abilities to evaluate and rebut those possible solutions fairly.

Results of the achievement of students' critical thinking skills sub skill: Making decision. Data from observation indicate that 69.23% students give excellent supporting reasons, good diversity, directly applicable. Therefore, there is an increasing of critical thinking skills compare to the data from cycle 1. There is an enhancement of students' ability to make a decision from 50% (cycle 1) to 69.23%.

Results of the achievement of students' critical thinking skills sub skill: Identify the impact of the implementation of their solution. Data from self-assessment indicate that 69.23%s students are excellent in providing new information. Furthermore, 69.23 % students are very clear in organizing and enhancing the argument and keep the audience interest focused on the main issues. Therefore, there is a development of critical thinking skills compare to the data from cycle 1. There is an enhancement of achievement from 61.53% (cycle 1) to 69.23%.

Results from the second cycle reveal that there is a development of students' critical thinking skills for all aspects compare to the data from cycle 1. The largest development of students' critical thinking skills is in sub skill: Identifying the problem correctly. In these sub skills, students have significant development from 50% to 88.45%. The development for sub skills distinguish knowledge and opinion is from 38.46% to 76.92%. The development for sub skills providing possible solution is from 30.76% to 65.38%.

Some other sub skills also have slightly development for example making a decision towards a problem and identifying the impact of the implementation of their solution. The development for sub skills making decision is from 50% to 69.23%. The development for sub skills identifying the impact of the implementation of their solution is from 61.53% to 69.23%. Data from checklist also reveal that 30.76% students mention all relevant and important alternative positions, present it fairly, and evaluate it properly, 15.38% students mention some of the relevant alternative positions, portray it properly, and evaluate it properly, and 53.84% students mention alternative positions but they are either not portrayed fairly, not evaluate it properly, or not relevant to the thesis. There also 19.23% students attempt to rebut the alternative position but they are ineffective or incomplete, 80.76% students do not attempt to rebut the alternative position. This data implies that students are struggling to evaluate and rebut alternative solutions properly.

5. Conclusion

Based on the analysis and the things that have been raised in advance, it can be concluded as follows:

- PBL4C model is a learning approach that can stimulate students' critical thinking
- The role of teachers in science learning process using PBL4C model is as a facilitator and learning resources that can guide and direct students to find solutions with respect to the provided problems.
- The courage and the ability to think critically is the basis for students in using PBL4C model to be more successful.
- The results showed that the use of PBL4C model develop students' critical thinking skills significantly. Students experience development from 11.11% to 88.45% in identifying the problem correctly, from 37.03% to 76.92% for sub skills distinguish knowledge and opinion, from 18.51% to 65.38% for sub skills providing possible solution is, from 22.22% to 69.23% for sub skills making decision, and from 11.11% to 69.23% for sub skills identifying the impact of the implementation of their solution.

This research study can further proposed some suggestions as follows:

- Teachers should be a facilitator and learning resources that can help students to study the learning material.
- Teachers conduct an analysis of the various issues involved, so it can be solved immediately.

- Students should be more active in implementing PBL4C model to increase their critical thinking skills.
- Students should be able to analyze a sharp, accurate and appropriate for any problems that occur to immediately be able to find a solution.

6. References

[1] Parker L and Raihani R 2011 *Educational Management Administration and Leadership* **39,** 712-32
[2] Larmer J and Mergendoller J R 2010 *Interdisciplinary Journal of Problem Based Learning* **1** 34-7
[3] Brush T and Saye J 2000 Educational Technology Research and Development **48** 79-100
[4] Hmelo-Silver C E and Ferrari M 1997 *Journal for the Education of the Gifted* **20** 401-22
[5] Mergendoller J R, Maxwell N L and Bellisimo Y 2006 *Interdisciplinary Journal of Problem Based Learning* **1** 49 – 69
[6] Skolnick R 2009 *Case Study Teaching in High School Biology: Effects on Academic Achievement, Problem Solving Skills, Teamwork Skills and Science Attitudes* (California: Ph.D. dissertation, TUI University)
[7] Dethefs T 2002 *Relationship of Constructivist Learning Environment to Students' Attitudes and Achievement in High School Mathematics and Science* (Lincoln: Ph.D. dissertation, The University of Nebraska)
[8] Savery J R and Duffy T M 2001 *Educational Technology* **35** 31-8
[9] Burns A 2010 *Doing Action Research in English Language Teaching: A Guide for Practitioners* (New York: Routledge)
[10] Tat T B, Preechaporn W, Kin L C and Kheong F H 2011 *Problem-based learning the 4 core areas (PBL4C): Identifying and Redefining Values for Nurturing Creativity in Building a New Nation in Southeast Asia.* (UNESCO-APEID International Conference)
[11] Teoh B T, Preechaporn W and Leong C K 2010 *Problem Based Learning 4 Core Areas (PBL4C) in The Search of Excellence in Mathematics Education* (UNESCO-APEID Internattional Conference)
[12] Brookhart S M and Nitko A J 2008 *Assessment and Grading in Classroom* (Upper Saddle River: Pearson Merrill Prentice Hall)
[13] Denzin N K 1978 *The Researcher Act: A Theoretical Instruction to Sociological Methods* (New York: McGraw-Hill)

Acknowledgements
This research was supported by:
1. Public Middle School 3 Pontianak, West Borneo, Indonesia
2. Educational Authorities of Pontianak City, West Borneo, Indonesia
3. The Ohio State University, United States

We are thankful to our colleagues Dr. Adrian Rodgers, Executive Director of U.S./ Indonesia Teacher Consortium (USINTEC) and Dr. Edwina Carreon from The Ohio State University who provided expertise that greatly assisted the research, moderated this paper and in that line improved the manuscript significantly although they may not agree with all of the interpretations provided in this paper. We are also grateful to Dr. Theresia Enny Anggraini from Sanata Dharma University for assistance with translation acknowledgement and for her comments on earlier versions of the manuscript, although any errors are our own and should not tarnish the reputations of these esteemed professionals.

Inquiry-Based Integrated Science Education: Implementation of Local Content "Soil Washing" Project To Improve Junior High School Students' Environmental Literacy

Syifahayu

SMPN 2 Hulu Sungkai, Jl. Raya Tulung Buyut, Lampung Utara 34555, Indonesia
Email: syifahayu6@gmail.com

Abstract. The study was conducted based on teaching and learning problems led by conventional method that had been done in the process of learning science. It gave students lack opportunities to develop their competence and thinking skills. Consequently, the process of learning science was neglected. Students did not have opportunity to improve their critical attitude and creative thinking skills. To cope this problem, the study was conducted using Project-Based Learning model through inquiry–based science education about environment. The study also used an approach called *Sains Lingkungan and Teknologi masyarakat - "Saling Temas"* (Environmental science and Technology in Society) which promoted the local content in Lampung as a theme in integrated science teaching and learning. The study was a quasi-experimental with pretest-posttest control group design. Initially, the subjects were given a pre-test. The experimental group was given inquiry learning method while the control group was given conventional learning. After the learning process, the subjects of both groups were given post-test. Quantitative analysis was performed using the Mann-Whitney U-test and also a qualitative descriptive. Based on the result, environmental literacy skills of students who get inquiry learning strategy, with project-based learning model on the theme soil washing, showed significant differences. The experimental group is better than the control group. Data analysis showed the p-value or sig. (2-tailed) is $0.000 < \alpha = 0.05$ with the average N-gain of experimental group is 34.72 and control group is 16.40. Besides, the learning process becomes more meaningful.

1. Introduction

Learning science as part of process of education has significant role in developing individual function in this globalization era. One of the roles is to improve students 'ability to think creatively in learning process so that they can face the globalization era. The ability of such creative thinking leads students who can express and elaborate original concept or idea to solve any problem. This is in line with Munandar (1992) who stated that creativity or to think creatively as ability to see various possibilities of problem solving which is as the idea which still doesn't get attention so far.

It has been known that the process of learning science took place in junior high school generally provoked lack of attention to the students' thinking skills and its developmental process. This is due to the implementation of learning science which mostly occurred with conventional models namely a way of classical teaching. This learning process did not give students to get involved in the learning process. This model has many weaknesses such as the role of the teacher is still more dominant (teachers centered), students tend to be passive and just accept information by memorizing concepts without adequate understanding. It is because the learning process did not guide the students to think critically, creatively, practicing to find concepts or principles to develop their creativity. So that, the achievement of learning is limited only to the aspects of knowledge (cognitive) but it did not develop in the aspect of attitudes (affective) and skills (psychomotor).

Another fact, students today have tendency to step away from values of protecting the environment. It can be seen from their consumptive behavior yet less productive in protecting the environment. There are number of students who use both private and public facilities in bad way. Gradually, their responsibility to take care public facilities is decrease even they do not have it at all. They could not take the advantage of technology to solve their problem about the environment as well as the issues about it. Also, they do not employ the environmental science, such as throwing rubbish in its place, taking care of plants in the garden, and watering the flower, then, recycling the potentially useful materials. Thus, the application of utilizing environment is still considered low so that it affected their awareness toward the environment. Therefore, problem occurred in both school system and in society today. Moreover, it seem like the environmental awareness is no longer needed. Based on this condition, one of the efforts to cope this problem is by using analysis of environmental literacy.

2. Model, analysis, design and implementation

The study used a quasi-experimental research method with pretest-posttest control group design. Subjects in both classes were given a pre-test and then the experimental group was given inquiry learning and control group was given conventional learning. After the implementation of the study, subjects in the two classes were given post-test. The subjects were students of class VII SMPN 2 Hulu Sungkai, namely Class VIIB as control class that implemented conventional learning and class VII A as the experimental class implemented inquiry learning. The study used purposive sampling technique. The purpose of this technique is the study could be carried out effectively and efficiently, especially in terms of supervision, the condition of the subjects, research time, place, and procedural licensing. Based on that technique, sample of two classes obtained, namely class VII A as experimental group consisting 26 students and the class VII B as control group consisting 23 students.

By the assumption that experimental group and control group are homogenous, thus the design is likely:

Experimental group : O------X--------O.
Control Group : O O
O : pretest or post-test of problem solving ability
X : learning use discovery learning strategy
--- : Subjects are not grouped randomly

Whereas, qualitative methods played role in answering research questions derived from observations, interviews and observations that had been conducted and analyzed qualitatively by describing the findings that were obtained during research in the field.

Data process was calculated based on the normalized score-gain with the formula proposed by Cheng (Gunawan and Liliasari, (2012)) namely:

$$\text{N-gain} = \frac{S_{post} - S_{pre}}{S_{max} - S_{pre}} \times 100\% \quad (1)$$

Note:
S_{post} = Post test score
S_{pre} = Pretest score
S_{max} = maximum score gained

Table 1. N- Gain classification by Cheng (Gunawan and Lilisari 2012).

N- gain Classification	Description
N-gain > 0,70	High
0,3 <N-gain > 0,70	Middle
N-gain < 0,30	Low

2.1. Model project-based learning

Arends (Trianto, 2007: 1) explains that the learning model refers to the learning approach that will be used in learning process, including the learning goals, and the stages in the learning activities, learning environment and classroom management.Trianto (2007:2) states that learning model can be said as conceptual framework that defines systematic procedure in organizing learning experience in achieving the learning goals. The function of learning model is as the guidance for the designer of learning model and for the teacher as well in implementing learning model.The term of learning model has four special characteristics, namely: 1) logical theoretical rational that is prepared by the designer 2) The basic thoughts about what and how students learn, 3) behavior of teaching that is necessarily required for the model to be successful, 4) learning environment needed so that the learning objectives can be achieved (Sanjaya, 2006).

The syntax of learning model shows clearly what activities should be done both by teachers and students as well as particular assignment which is done by the students. Syntax of learning model has similar components such as attracting students' attention by motivating them to get involved in learning process both in the beginning and at the end of the learning process. Arranging the syntax of the learning model must concern several factors such as differences in the management of learning environments, differences in the role of the teachers and students as well, differences in physical spaces and differences in social class system. Accordingly, those differences should be understood by teachers in implementing the learning model so as to it can be applied properly.

Project-based learning is a learning process which provides more opportunities for the teachers to manage classroom learning by involving work project. Work Project includes complex tasks based on very challenging questions and problems and that require students to design, solve problems, make decisions, conduct investigations. Moreover, the task provides opportunities for students to work independently to in completing the task. Project-based learning has syntax: (1) starts with the essential question, (2) design a plan for the project, (3) creates a schedule, (4) monitor the students and the progress of the project, (5) assess the outcome (6) Evaluate the experiences.

Scientific inquiry refers to a variety of ways for scientists to study the universe and aims to provide an explanation based on the evidence obtained. Inquiry also refers to students' activity when they build a knowledge and understanding of scientific ideas, as scientists' attempt to understand the universe. (National Science Education Standards, the United States, p. 23)

According to Nuryani (2005: 17), inquiry in learning indicates that effective science learning depends on the availability, organizing materials, media and tools, and technologies. Accordingly,a learning that refers to the inquiry wants to be more scrutiny through continuous research. Bruce (Sahin, 2013), explains that *"inquiry-based learning is a teaching and learning approach that engages students to explore their own questions and interests, by (1) asking meaningful questions, (2) planning and executing investigation strategy, (3) gathering information from a variety sources, (4) discussing information in productive presentation, (5) reflecting on their own learning".*

Thus, it can be seen that the purpose of the inquiry learning is to develop the ability to think in a systematic, logical, and critical or develop intellectual abilities as part of the mental process. Hence, in inquiry learning students are not only required to master the subject matter, but also to know how they can use their potential. Students who are able to master the lessons only may not be able to develop their optimal thinking ability. Whereas, the students will be able to develop their thinking skills as they could master the subject.

Hollweg*et al* defines environmental literacy as "*The cognitive domain refers to the individual's knowledge of ecological concepts and processes that provide the foundations for comprehending human impact on natural systems: environmental issues and environmental action strategies; as well as the cognitive skills for analyzing environmental problems and for the use of environmental action strategies. The affective domain refers to the individual's environmental awareness and sensitivity; attitudes, values and worldview regarding the environment; locus-of-control (sense of ability to influence a situation through personal behavior, i.e. self-efficacy) and assumption of personal responsibility (sense of obligation toward the environment. i.e. personal commitment to environmentally corrective behaviors). Behavior is the ultimate expression of EL - the individual's EL should be reflected in his/her behavior concerning the environment. It can be inferred that developing EL is equivalent to developing responsible environmental behaviour, i.e. in the context of EL, knowledge, dispositions, and competencies enable and are expressed as behaviours (Daphne Goldman et.al., 2014:3).*

In brief, environmental literacy is the ability or skill in understanding the importance of protecting the environment both for current need and future generations as well. According to Amini environmental education are expected could guide students to behave and aware of the environment (Aryanti). It is an attempt to change behaviors and attitudes conducted by some sides in which in its implementation it can bring the classroom to the factual environment and in other way bring the environment to the class. This means that the learning would use the environment as a learning resource in order to achieve the learning objectives that have been stated. Good environmental management can guarantee the availability of natural resources which are important for public welfare and environment safety in the future.

2.2. Implementation local content soil washing project

In this study, the researcher used local content of soil washing project that came from *Ikatan Pelopor Pertanian Oranik Lampung - IPPOL* (Lampung Association of Organic Farming Pioneers) under responsible of the foundation Yabima Indonesia. IPPOL continuously conducts counseling, mentoring and coaching to Lampung organic farmers so that they become more productive and welfare through several superior programs. One of them is the training of making and using soil washing formula in soil washing project.

The soil washing technology is a technology that uses liquid (usually water and certain chemical additives) and mechanical processes to wash the soil (United States Environmental Protection Agency, 1996). Developed to improve conditions of the soil from the contaminants (hazardous substances) and pollutants using herbal ingredients which are easily found in Indonesia. Though, the technology itself is not so familiar and stills such a new thing in Indonesia, especially in our neighborhood area in North Lampung regency (an experimental procedure of soil washing is attached in appendix 5).

The materials needed to make solution of the soil washing formula are: (1) 10 kg of lime, (2) 5 kg of noni, (3) 10 kg of star fruits, (4) 40 liters of water, (5) 0.5 of citronella, (6) 0.5 kg of red betel, (7)2 kg of Limestone. All ingredients except water and limestone mashed or blended together. After that, add the limestone and stir well. Furthermore, put all the ingredients that have been mashed into a container which had been provided. Finally add water into the container of soil washing solution materials, Stir it all up well. For soil cultivation, 1 kg of coriander is necessarily added to soil washing formula. Then, close the container tightly. Open the container and stir it once in every 5 days in order to get perfect fermentation process. The fermentation process lasts in a month and after that the solution is ready to use.In consideration of the principle of usefulness for future generations, thus the researcher deliberately intended to introduce this soil washing technology in learning process so that students have a sensitivity and concern/awareness toward the environmental continuity.

The utilization of soil washing is to wash the soil from contaminants. It is similar to the use of medicinal plants, in which as many as 25-30 cc of dose mixed with 12 liters of water is sprayed to the ¼ ha soil. The soil is best sprayed 3 times a week before planting. If it is used as learning materials at school, so the portion produced by each group is 1/10 of the dose so as to students can easily apply it. The Media of the soil being tested also kept in some polybags.

Figure 1. Processing soil washing project.

3. Results and discussion

Based on analysis of N-gain score of students' environmental literacy skills, the data shows the increase classification (quality) of students' scorescompared with the ideal maximum score. The average of N-gain represents an increase of scientific literacy skills of students who were given inquiry learning using project-based learning model or students who were given conventional learning. The average N-gain of student's scientific literacy skills in experimental group and control group are presented in the following table:

Table 1. Average and N–gain classification of environmental literacy.

Group	Average N-gain	Classification
Control	16.40	Low
Experimental	34.72	Middle

Based on Table 1 above, it shows thatexperimental group with project-based learning model get higher average N-gain score. N-gain classification of control group considered as low grade, whereas the N-gain classification of experimental group considered as medium. This demonstrates the environmental literacy skill of experimental group is higher than the control group. It also indicates that the inquiry learning strategies using project-based learning model gives a good contribution to improve environmental literacy skills than conventional learning model does.

However, to ensure that the increase in students' environmental literacy skills who get the learning environment project-based learning in theme Soil Washing Project is better than students who get conventional learning needs advanced statistical tests. Statistical tests which is needed to prove the hypothesis "environmental literacy skills of experimental group is better than the control group" is namely the differential average N-gain scores Test. Although, N-gain score of the data must meet the requirements of normality and homogeneity before it is being tested.

The normality test of N-gain score is calculated by Kolmogorov-Smirnov test with SPSS 16. The results of the test are presented in the following table:

Table 2. Normality test of N-gain score.

Group	Kolmogorov-Smirnov			Conclusion
	Statistic	Df	Sig.	
Control	.133	26	.200*	Normal distribution data
Experimental	.197	23	.021	Non-Normal distribution data

Table 2 above shows that N-gain score of environmental literacy skills of students in experimental group has the Sig. <A = 0,05. it means Ho is rejected. This indicates that the N-score data of experimental group is non-normal distribution. Whereas, the N-gain of control group has the Sig. > A = 0,05,it means Ho is accepted. This indicates that the N-score data of control group is normal distribution. Based on the results, it can be concluded that N-gain score of experimental group considered as non-normal distribution and control group is normal distribution.

To prove the N-gain score of experimental group is better than the control group, it is necessarily conducted differential average N-gain scores Test using non-parametric test (Mann-Whitney U-test). The first hypothesis proposed in the research study, namely: there is different average on environmental literacy between the experimental group and control group.

To test the hypothesis proposed above, statistical hypothesis is formulated as follows:

- Ho : $\mu_1 = \mu_2$, There is no significant difference in improvement of environmental literacy between studentswho were given inquiry learning strategy with project-based learning model and students who were given conventional learning.
- Ha : $\mu_1 \neq \mu_2$, There is significant difference in improvement of environmental literacy between studentswho were given inquiry learning strategy with project-based learning model (experimental group) and students who were given conventional learning (control group).

Notes:

The improvement of math problem solving ability of the students who were used inquiry learning strategy.
μ_1= average N-gain of environmental literacy skill of students who get inquiry learning strategy (experimental group).
μ_2= average N-gain of environmental literacy skill of students who get conventional learning strategy (control group).

Here is the summary of the results of the differential average test of N-gain score at significance level $\alpha = 0.05$.

Table 3. The differential N-gain test average of environmental literacy.

Statistics	Value	Description	Conclusion
Mann-Whitney	75.500	Ho rejected	There are differences
Z	-4.480		
Asymp. Sig. (2-tailed)	.000		

Based on Mann-Whitney Test above, *p-value* or *sig.* (*2-tailed*) gained is 0.000 < α = 0.05, meaning that Ho is rejected. Thus, improvement of environmental literacy skills of students who get inquiry learning model using Project-Based Learning is better than students who get conventional learning. Therefore, the hypothesis which states there is significant difference in the average score between students' environmental literacy skill of experimental group and control group is proven.

Based on the research result, inquiry learning strategy on Project-Based learning model on theme Soil washing project gave good impact on the learning process in Grade VII SMPN 2 Hulu Sungkai. It can be seen from the improvement of students' ability or skill to find a concept and could complete their tasks, and they also could implement it in their daily life. In addition, students get their encouragement to express their opinion, ideas and questions. From the pre-test and post-test questions given by the teacher, there were various answers appeared from students. The result was generally the same. The students also got involved in checking their work so they could know the right and wrong answer.

Teachers also led the students to think creatively in order to be able to solve problems. At the beginning of the learning, there were only few students who dared to ask questions, to develop ideas or to comment on other students. So that teachers still have to motivate the students or to manage students to come forward and present the results of working group. When students are grouped to work together and discuss how to complete teacher instruction, their collaboration and cooperation still did not appear. There are several groups that are dominated by one or two students. Although, the learning activity had increased in the next meeting, collaboration between group members appeared. It can be seen from the involvement of students in completing and implementing the instructions provided by the teacher. The number of students who conveyed the idea alsoincreased known from variations of answers from each group in completing the worksheets provided by the teacher.

The result of analysis from pre-test and post-test also shows that there are differences on environmental literacy skill in each indicator:

Table 4. Average N-gain Indicator of environmental literacy.

No.	Indicator	Average N-gain	
		Control	Experimental
1	Awareness to the environment	16,60	32,48
2	Knowledge of the environment	16,17	43,96
3	The application of material science in the development of integrated environmental curriculum	20,26	29,54
4	The application of environmental knowledge	17,71	24,45
5	Initiative and works in environmental management	19,71	30,96

The table shows that the students' environmental literacy skills in the control group are low because the average value of N-gain is below 30%, but in the experimental group for indicator such as awareness on the environment, environmental knowledge, then the initiative and works in environmental management are categorized middle. In experimental group, indicator which was categorized as low is indicator of the application of material science in the development of integrated environmental curriculum and initiative and works in environmental management.

Based on results from the series of experiments which was conducted, students can find out that the use of chemical fertilizers continuously can reduce the fertility of the soil, because the fertilizer that is given only contain small amounts of elements needed by plants. Sometimes, the soil itself will be more excessed in amount of the particular elements while the other elements are declining in number. Then, for the improvement of soil composition, it needs to conduct either by using organic fertilizers or reducing the use of inorganic fertilizers.Students also know that the ideal soil conditions must be made up of 15 elements nitrogen, potassium, phosphorus, magnesium, manganese, iron and so on. With the criteria of soil pH ranges from 6 to 8 in which good for planting. Soil has unique properties in which it can remediate itself by the assistance of particular microorganisms. So that it can degrade the heavy metal elements which pollute the soil then turn into particles or molecular which is not dangerous for life.

Thus, students can understand the importance of preserving the environment, especially the preservation of their soil as part of their life in the future. The project is expected can affect the students' mindset in order to evoke a sense of belonging and care to the environment. Also, the project can arouse students' desire to develop their district later when they are adult. Addition, they should forget their dream to find a job in the city or become Indonesian labor force as it had been done by people before them.

This is one of the nature of meaningful learning and one of the objectives of this study. Thus, not only to improve the students' environmental literacy skills, but also to create moral and social values of students so that they can answer the issues and solve the problems that occurred wisely and appropriately in the future.

4. Conclusion

Based on the results of data analysis and discussion presented before, it is derived some conclusions as follows:

- Inquiry-based environmental education: the implementation of the project "soil washing" is proven could help increase the students' environmental literacy.
- The environmental literacy skill of students who use inquiry learning strategy with project-based learning model on the theme soil washing showed significant differences, the experimental group is better than the control group. The data analysis shows the *p-value* or *sig.* (2-tailed) is $0.000 < \alpha = 0.05$ with Average N-gain 34.72 for experimental group and 16.40 for control group.
- Students' learning Interest who used inquiry learning strategy with project-based learning model on the theme soil washing project has increased, it can be seen from the enthusiasm of the students when learning process took place.
- There is a correlation between students' environmental literacy skills and students' learning interest.
- Students can develop values that appear in the learning process such as moral and social values of the environment so that it could make meaningful learning.

5. References

[1] EPA 1996 *A Citizen's Guide to Soil Washing* (United States Environmental Protection Agency)
[2] Olson S, Loucks-Horsley S 2011 *Inkuiri dan Standar - Standar Pendidikan Sains Nasional,* Translated by Ismunandar, Agustiani E D, Astuti D R F (Bandung: Qitep in Science SEAMEO)
[3] Achmad and Rukaesih 2004 *Kimia Lingkungan* (Jakarta: Andi Yogyakarta)
[4] Capraro, Robert *et.al* 2013 *STEM Project Based Learning: An Integrated Science, Tevhnology, Engineering, and Mathematics (STEM) Approach,* 2nd Edition (Rotterdam: Sense Publisher Netherlands)
[5] Cresswell J W 2010 *Research Design Penelitian Kualitatif, Kuantitatif dan Mixed* (Yogyakarta: Pustaka Pelajar)
[6] Robert E 1995 *Critical Thinking* (New York: University of Illinois)
[7] Fitriyanti *et al* 2012 *Analisis Literasi Lingkungan Hidup Mahasiswa pada Mata Kuliah Ilmu Pengetahuan Lingkungan* (Jurnal Nasional Universitas Riau)

[8] Fraenkel J R and Wallen N E 1993 *How to Design and Evaluate Research in Education* (Singapore: McGraw Hill)
[9] Furqan 2013 *Statistika Terapan untuk Penelitian* (Bandung: Alfabeta)
[10] Daphne G 2014 *International Journal of Science and Education* **9** 369 – 83
[11] Kemendikbud 2013 *Lampiran B Peraturan Menteri Pendidikan dan Kebudayaan Nomor 68 tahun 2013* (Jakarta: Kemendikbud)
[12] Permendikbud 2013 *Standar Isi Kurikulum 2013 untuk SMP/ MTs* (Jakarta: Kemendikbud)
[13] Progresia 2013 *Buku Siswa Lebih Dekat dengan Zat* (Bandung: Qitep in Science SEAMEO)
[14] Puskur 2013 *Buku Siswa Elektronik IPA Terpadu Kurikulum 2013 Kelas VII Semester I* (Jakarta: Kemendikbud)
[15] Rustaman, Hendrawati and Toharuddin 2011 *Membangun Literasi Sains Peserta Didik* (Bandung: Humaniora)
[16] Rustaman N *et al* 2005 *Strategi Belajar Mengajar* (Bandung: Imstep JICA Jurusan Pendidikan Biologi UPI)
[17] Sugiyono 2015 *Metode Penelitian Pendidikan: Pendekatan Kuantitatif, Kualitatifdan R&D* (Bandung: Alfabeta)

IOP Conf. Series: Journal of Physics: Conf. Series **812** (2017) 012083 doi:10.1088/1742-6596/812/1/012083

Analyzing Concepts Mastery and Misconceptions About Evolution of Biology Major Students

L O L Putri*, T Rahman and D Priyandoko

Department of Biology Education, Postgraduate's School, Universitas Pendidikan Indonesia, Jl. Dr. Setiabudi No. 229, Bandung 40154, Indonesia
Corresponding author: lukitaolp@gmail.com

Abstract. The purpose of this study was to analyze the concepts' mastery and misconceptions of evolutions' subject. This is a descriptive research and involve 2 biology major students from 2 universities in Bandung. Sample of the study is a biology major student who has attended the evolutions and genetics' course. Using Diagnostic Question Cluster and interview as a instrument research. Result of the study indicated that concepts' mastery is on the low level (41,25% on first data and 48,25% on second data) and misconceptions is on the high level (61,25% on first data and 60,25% on second data. Some types of misconceptions only found in specific subconcepts. We concluded that low level of concepts' mastery lead to misconceptions, and it is also caused by low mastering of key concepts in evolution's theory. Content-based misconception has a highest percentage of all kind of misconceptions (27,25% in first data and 28,5% on second data), because the biology major student's don't understand the scientific concept correctly.

1. Introduction

Biology education's curriculum was compulsoring their students to possessed many biology topics, also encouraging their students, the biology major students, to become a high school biology teacher. As a biology teacher, the biology major students need a deep understanding about many biology topics. One of the topics which is need to be learned by the biology major students is evolution. Evolution is also one of the primary subjects in biology education's curriculum. In biology class organics evolution is the most discussed topics in biology's class. Organics evolution refers to the alteration of the living things [1, 2, 3, 4, 5]. There are three main aspects of organic evolution which is distinguished as (a) evolutionary devices (include DNA and gene), (b) evolutionary mechanism (include natural selection and reproductive success) and (c) the result of their combination in the form of macroevolution (include speciation and fossils) [1]. Evolution as a subjects or topics held a high percentage of low understanding [2] and misconception [3].

Students' conceptions divided as preconception and concept assimilation [4, 5, 6, 7]. Preconception refers to the idea or understanding which is held by student before they had learning process [4]. After they had the learning process, their understanding called concept assimilation, this concept assimilation, ideally, directed to the concept mastery. But students' apparently found hard to assimilating their conceptions after learning process, so misconceptions found as the lack-conceptions of the students. Misconceptions refers to the different conceptions [5], or the alternative conceptions [8, 9, 10]. Students who held misconceptions didn't realize it and they did nothing to change those misconceptions [4, 11, 7, 12]. Misconceptions caused by teachers' learning method [13, 14, 15, 16], different teachers' interpretation of the concepts [13, 17] and by the reference [18, 19, 13, 17].

Misconceptions of evolution's concept identified as common-sense misconception, content-based misconception, nature of science-based misconception, non-scientific misconception and vernacular misconception [10]. Table 1 shows the characteristics of misconceptions's kind which is found in this study.

Table 1. Characteristics of Misconceptions

Kind of misconceptions	Characteristics and descriptions
Common-sense misconception	The answer suggest reason by comparing 2 unequal evolution's phenomena
Content-based misconception	The answer is using uncorrect scientific term/ concepts about evolution but the students believe that the term/ concepts is the right concepts
Nature of science-based misconception	The answer is using sceptical-belief to the evolutions evidence
Non-scientific misconception	The answer is denying the scientific concepts using religious concept/ statement
Vernacular misconception	The answer is purposively exchange the terms because lack-understanding of evolutions' terms

Concepts' mastery and misconceptions can be measured by the Diagnostic Question Cluster (DQC) [20], type three-tier test [21, 22]. The DQC type three-tier test have 3 levels question, the first level was to identify the students understanding [20]. This first level question using a Bloom's taxonomy revision from the C3 to C4 [21]. On the second level we identify the certainty degree. This certainty degree used as a cosideration about the concept mastery and the misconceptions. On the third level, we identify the misconceptions reasoning [10]. The three tier test reputed better to identify misconception than the two tier test [22]. This study explains the analysis of the concept's mastery and the misconceptions about the evolution on biology major students.

2. Research Methodology

This study using a descriptive method. Descriptive method used to picture the situation [23], so we used this method to identify concepts' mastery and misconceptions of biology major students who already attended the evolutions course in college. We purposively choose two universities in Bandung as the sample of this study and involved 20 biology major students from each university. Data collection conducted two times by using the DQC's test, the second data collection conducted 4 weeks after the first test, so we can identify the consistency of the students' answer. We also use interview as the second and complement data.

3. Result and Discussion

The DQC's test type three-tier test result showed that each biology major students from two universities possessed different level of concepts' mastery and misconceptions. As a simplification to comparing two universities, we use code A and code B to describe comparation of those two universities. Biology major students from university code A just passed evolution's subject last semester, and biology major students with code B passed evolution's subject for two semesters ago. Different semester from this universities did not caused a remembrance factor [24] [25], because we tested the concept mastery and understanding not the recitations.

On Table 2, we can see the comparation of the concepts mastery's biology major students of two universities. Those data collected by DQC's test from the first tier. First data collection shows that biology major students from university with code A has a highest percentage (21,75%) of concept mastery than the biology major students from university with code B (19,5%). Second data collection shows that biology major students from university with code B has a highest percentage (29%) of concept mastery than the biology major students from university with code A (19,3%). Based on interview data, those biology major students passed the evolution's subject with high score from 85 to

100. From the DQC's test, this two biology students still held low concept mastery (41,25% to 48,25%) of the evolution's subjects from the first data collection. Although there's a enhancement percentage of biology major students code B from the first data collection to second data collection, it doesn't mean that the biology major students understanding the concept better on the second data [26]. The level of understanding and concepts' mastery must fill three requirement, (a) there's an alteration of the answer, (b) the subject can explain the alteration answer and (c) understand the key concepts [26]. From the interview data, 60% students remember the test's question, so the alteration of the answer also caused by the memory of the question [25]. Alteration of the answer is a factor that caused this enhancement percentage.

Table 2. Concepts' mastery of the biology major students

Code of the university	First data collection (%)	Second data collection (%)
A	21,75	19,3
B	19,5	29
TOTAL	41,25	48,25

The second tier from DQC's test is measuring the certanty response of the first tier (concept mastery). In this tier, we asked students about their answer's confidence. Certainty response help us to assessing students' answer and determine which students is mastering the concept, not mastering the concept and misconception [27, 28]. Table 3 shows the certainty response of the right answer. First data collection shows that only 50% students that certain with their right answer, also there is 24,5% which is guessing the right answer and 8% students not certain with their right answer. Although there is 50% students choose the right answer with certain, but it doesn't mean that those students did not possessed misconception [29].

Table 3. Certainty response of the right answer

Sample of the university	First data collection (%)			Second data collection (%)		
	Sure	Guess	Not sure	Sure	Guess	Not sure
A	45.5	6.5	1	33.5	9	2
B	4.5	18	7	37	10.5	4.5
TOTAL	50	24.5	8	70.5	19.5	6.5

Based on the Table 3, the certainty response shows indicate that sample A has decreased percentage of the sure options' answer (from 45,5% to 33,5%) and increased percentage of the gueess and not sure options' answer. Based on interview data, 60% of biology major students from the sample A did not remember the questions, they also thought that the second test was using another questions test. Those data indicate that the sample A had a low level of understanding and the concepts' mastery. The sample B has increased percentage of the sure options' answer and decreased percentage of the guess and not sure options' answer. Based on interview data, 80% of biology major students from the sample B alternated their answer to the most-choosen answer. It is also indicate the low level of understanding and the concepts' mastery.

The third tier of DQC's test used to determine the kind of misconception which is held by the biology major students. Table 2 shows that the concepts' mastery of the biology major students is on the low percentage, this low percentage indicated the low level of understanding about the evolutions' subject and lead to misconception [26, 30]. And there's a positive correlation between low level of understanding and misconception [31]. Based on the students' answer from third tier of DQC's test, this is the common answer of the student which is found by analyzing their answer. Common-sense misconception found when the subject believe that use and disuse theory immediately change the organism's genotipe. The content-based misconception found when the subject using a mutation's term to answer the use and disuse phenomena [32]. The nature of science-based misconception found when the subject believe that no evidence of the use and disuse theory in organism [2]. Non-scientific misconception and vernacular misconception found as unstable misconceptions, it found only on first or the second data collection only. The non-scientific misconception found when the subject held religious conception and reject the scientific misconception, in this study we also found a local wisdom conception which is held by the students. Vernacular misconception found when the subject using scientific term out of the right scientific terms. Table 4 below shows percentage of misconception which is held by biology major students. Content-based misconception possessed the highest percentage of misconception, it caused by low level of understanding the concept whenever the students learn the concept. Vernacular misconception possessed the lowest percentage of misconception, it is caused by the failure understanding of the key concept. Non-scientific misconception possessed high percentage on other study [26, 16, 31, 33, 10], but in this study we only found 3,5% misconception on first data collection and 3,5% on second data collection from DQC's test. There is dissimilarity data from the DQC's data and the interview data about the non-scientific misconception. Most students denying the evolutions concept and believe that evolution is not a feasible explanation [34]. We also found that there's students who believed that evolution never happen because God is created all thing perfectly. Those answer did not go out on the DQC's test, the students explain that they need to answer scientifically on the test because the test asked for the scientific answer.

Table 4. Misconceptions in biology major students

Kind of misconceptions	First data collection (%)	Second data collection (%)
Common-sense misconception	10,5	10,75
Content-based misconception	27,25	28,5
Nature of science-based misconception	15,5	14,75
Non-scientific misconception	4,5	2,75
Vernacular misconception	3,5	3,5
TOTAL	61,25	60,25

Based on previous explanations, we merged the concepts' mastery and misconceptions based on the evolutions' subject. Table 5 shows the concepts' mastery and misconceptions based on the evolutions subject. According to all subconcepts, the organic evolutions' theory has the highest percentage of concepts' mastery and the lowest percentage of misconceptions. It is indicated that higher level of understanding caused lower level of misconceptions. Based on organics evolutions' theory, it has the highest average percentage of concepts' mastery but also has the highest average percentage of misconceptions. Although there's a positive correlation between concepts' mastery and misconceptions, but rejection of evolution's theory also caused misconceptions [16, 35, 36, 9, 37, 32, 38]. Evolutionary devices subconcepts has the lower percentage of concepts' mastery and the higher percentage of misconceptions. Genetics variation subconcepts has the lowest percentage of concepts' mastery and the highest percentage of misconceptions. Based on previous subconcepts, this subconcepts also positively correlated from the low level of concepts' mastery to the high level of misconceptions.

Table 5. Concepts' mastery and misconceptions based on evolutions' subject

Subconcepts	Concepts' mastery		Misconceptions	
	First data collection (%)	Second data collection (%)	First data collection (%)	Second data collection (%)
Organics evolution's theory: Lamarck, Darwin, Neo-Darwinism	56,9	59,4	58,8	52,5
Evolutionary devices: DNA, Genes, Chromosomes	33,1	38,8	59,4	66,3
Genetics variation	26,3	45	65	71,3

4. Conclusions

The low level of concepts' mastery caused the misconceptions. Based on certainty response, the concepts' mastery is also on the lowest percentage. It shows that biology major students apparently not mastering the concepts and the key concepts. Misconceptions found in a high percentage. It indicates that biology major students held wrong conceptions about evolution. Organics evolution's theory subconcepts held the highest percentage of concepts' mastery and the lowest percentage of misconception. Genetics variation's subconcepts held the lowest percentage of concept mastery and the highest percentage of misconception.

5. References

[1] Alexander D 2009 *Creation or Evolution: Do We Have to Choose?* (Oxford: Monarch Book)
[2] Nadelson L S and Sountherland S 2012 *International Journal of Science Education* **34** 1637-66
[3] Miller K R 1999 *Finding Darwin's God: A Scientists' Search For Common Ground Between God and Evolution* (New York: Cliff Street Books)
[4] Dahar R W 1996 *Teori-teori Belajar* (Jakarta: Erlangga)
[5] Tekkaya C, Ozkan S and Asci Z 2001 *Egitim ve Bilim* **120** pp 29-36
[6] Fraser B J, Tobin K G and McRobbie C J 2014 *Springer International Handbooks of Education* **24**
[7] Kubiatko M and Prokop P 2009 *Orbis Scholae* **3**(2) 97-112
[8] Anderson D L, Fisher K M and Norman G J 2002 *J. of Research in Science Teaching* **39**(10) 957-78
[9] Sadler P M 1998 *J.l of Research in Science Teaching* **35**(3) 265-98
[10] Yasri P 2014 *Int. J.of Biology Education* **3**(3)
[11] Dahar R W 2011 *Teori-teori Belajar dan Aplikasinya* (Bandung: Erlangga)
[12] Hasan S D B and Kelley E L 1999 *Physics Education* **34**(5) 194-9
[13] Chaniarosi L F 2014 *J. MPBio PPs Unsyiah* **2**(2) 187-250
[14] Saehana S and Kasim S 2011 *Proc Seminar Nasional Penelitian, Pendidikan dan Penerapan MIPA*
[15] Athanasiou K and Mavrikaki E 2014 *Int. J.of Science Education* **36**(8) 1265-85
[16] Borgerding L A, Klein A, Gosh R and Eibel A 2015 *J. of Science Teacher Education* **26** 371-92
[17] Mustika A A, Hala Y and Arsal F 2014 *J.Sainsmat* **3**(2) 122-9
[18] Tshuma T and Sanders M 2014 *J..of Biological Education* **49**(4) 354-69
[19] Adisendjaja Y H and Romlah O 2007 *J. Pendidikan Biologi FMIPA UPI*
[20] Bayrak B K 2013 *Mevlana Int.J. of Education* **3**(2) 19-26
[21] Parker J M, Anderson C W, Heidenmann M, Merril J, Richmond G and Urban-Lurain M 2012 *CBE-Life Science Education* **11**
[22] Schaffer D L 2013 (Missouri: University of Missouri)
[23] Cresswell J W 2014 *Research Design* (Yogyakarta: Pustaka Belajar)
[24] Ausubel D P 1962 *Journal of General Psychology* **56**
[25] Karpicke J D and Roediger H L 2007 *J. of Memory and Language* **56**
[26] Battish B T, Hanegan N, Sudweeks R and Cates R 2009 *Int. J. of Science and Mathematics Education* **8** 845-68
[27] Lederman N G, Abd-El-Khalick F, Bell L R and Schwartz R S 2002 *J.of Research in Science Teaching* **39**(2)
[28] Wiles J 2010 *Science Education Review* **9** 18-27
[29] Kagan T and Sanders M 2013 *African J.l of Research in Mathematics Science and Technology Education* **17** 38-49
[30] Settlage J J 1994 *J. of Research in Science Teaching* **31**
[31] Bishop B A and Anderson C W 1990 *J. of Research in Science Teaching* **27**(5) pp 415-27
[32] Galli L M G and Meinardi E 2011 *Evolution: Education and Outreach* **4** pp 145-52

[33] Scott E 2004 *Evolution vs Creationism: An Introduction* (California:University of California Press)
[34] Pugh K J, Koskey K L K and Linnenbrick-Garcia L 2014 *J of Biological Education* **48**(1) 22-33
[35] Rice J W, Clough M P, Olson J K, Adams D C and Colbert J T 2015 *Evolution: Education and Outreach* **8(8)**
[36] Williams J D 2014 *J.of Biological Education* **49**(3) 322-33
[37] Smith M 2010 *Pedagogical Issues, Science and Education* **19** pp 539-71
[38] Nadelson L S 2008 *Evolution: Education and Outreach* **2**(3) pp 490-504

Analysis of Prospective Mathematics Teachers' Basic Teaching Skills (a Study of Mathematics Education Departement Students' Field Experience Program at STKIP Garut)

D. V. Rahayu

Department of Mathematics Education, Sekolah Tinggi Keguruan dan Ilmu Pendidikan, Jl. PahlawanSukagalih No. 32Sukagalih, Garut 44151, Indonesia
Email: Diar_math@yahoo.com

Abstract: This study was intended to figure out basic teaching skills ofMathematics Department Students of STKIP Garut at Field Experience Program in academic year 2014/2015. This study was qualitative research with analysis descriptive technique. Instrument used in this study was observation sheet to measure basic teaching mathematics skills. The result showed that ability of content mastery and explaining skill were in average category. Questioning skill, conducting variations skill and conducting assessment skill were in good category. Managing classroom skill and giving motivation skill were in poor category. Based on the result, it can be concluded that the students' basic teaching skills weren't optimal. It is recommended for the collegians to get lesson with appropriate strategy so that they can optimize their basic teaching skills.**Keywords**Basic Teaching Skill, Field Experience Program.

1. Introduction

Education has a very important role in developing a nation. The main factor that decides the quality of education is teacherso thatthe quality of teachers should be in the main focus to enhance the quality of education in our country. Professional teachers can be created through appropriate courses in every subject since they are prospective teachers. Act no 2 in 2012,about university/college, explicitly describe that courses inuniverisity/college,is expected to give change to the students to optimally develop cognitive, affective, and psycomotoric aspect. For the sake of this target, government has published Presidential Regulation no. 8 in 2012 about KerangkaKualifikasiNasional Indonesia (KKNI) which guides the arrangment of the graduated learning target in every educational grade in the country. The learning achivement for graduated bachelors is stated up to level 6.

Concerning that matter, The Indonesian Mathematical Society (IndoMS) has arranged Mathematic Department bachelor degree based on parameter of job field skill, cognitive job field skill, and managerial skill. Lesson achievement based on job field skills are: 1. to be able to plan, to implement, to evaluate innovatively through applying mathematicsdictatepedagogics concepts and mathematic knowledge and also to make use various learning sources, science and technology which are life skill oriented. 2. To be able to interpret and apply various available mathematics learning methods which are innovative and tested. 3. To be able to do fostering to students in Mathematics learning process. 4. To be able to arrange and perform research to produce alternative ways to solve problem in Mathematics learning and publish them [1][2].

Based on lesson achievement qualification of KKNI and IndoMS above, Students ofDepartment of Mathematic Education must have good basic teaching skill in teaching Mathematic. So that, students of university/ college shouldn't only have mathematic skill but also have classroom management skill. The course that makes it possible for students toapply that skills is Students' Field Experience Program (Program PengalamanLapangan/PPL). PPL is a course that must be taken by 7thsemester students in Department ofMathematic Education of STKIP Garut.

The importance of basic teaching skills is as stated in Act no. 14 in 2005 about teachers and lecturers, such thing like pedagogics competence and professionalism competence. Sanjaya [3] stated that teachers' basic teaching skillsare needed in order to make teachers be able to do their role in managing learning process so that learning process can run smoothly and efficiently. Furthermore, basic teaching skills are an absolute requirement for teacher to be able to implement various lesson strategies.

According to the explanation, the writer's interested in investigatingbasic teaching skills of Mathematics prospective teachers that joined PPL. In order to perform it, the writer did research entitled Analysis of Prospective Methematic Teachers' Basic Teaching Skills(a study of Matematics Education Departement Students' Field Experience Program at STKIP Garut).

2. Study Literature

Basic Teaching Skills is defined as measured and coherent activity performed by teachers in order to make students study [4]. Wragg [4] stated that basic teaching skill as "strategies that teachers use which facilitate pupils' learning and which are acknowledged by those competent to judge as being skills"

Basic teaching skill is involved with some fundamental skills that must be acquired and actualized by every teacher, lecturer, and instructor in doing their duty. As for teacher, in regulation of Minister for the Empowerment of State Apparaturs no.16 in 2009, it's stated that teachers should plan and perform quality learning process, evaluate learning result, and provide enrichment and remedial [5].Teachers are in charge in choosing and deciding materials, strategies, methods, media, and instruments for evaluation in performing learning process to achieve education result which has high quality and has been accepted by teachers' ethic code.

Regarding those issues above, every teacher should have got good basic teaching skills. Basic teaching skills consist of questioning skill, motivational skill, conducting variation skill, explaining skill, opening and closing learning skill, guiding small discussion group skill, classroom management skill, teaching individual skill [6].

Basic teaching skills in teaching mathematics involved with special mathematics' skills and they should be implemented by every teacher [7]. One of those characteristics is for example the existence of abstract concepts which require special explanation and appropriate samples so that explaining skills is one of basic teaching skills to be mastered by mathematics teachers [8]. In addition, questioning skill needs to be mastered by Mathematics teachers in order to be able to develop students' thinking level [9].

Shellard and Moyer [10] stated characteristic of effective mathematics teaching, for example teachers should make students "explain" and ask "why" in questions to students answers. Ball [11] told that in teaching mathematics teachers must be able to:

- Design systematic and accurate explanations so that students can use them
- Use systematic approaches and complete definitions
- Present ideas carefully, map things and graphic models, note symbols, operations or processes
- Interpret and make pedagogicassessments
- Respond students' inquiries and mathematics questions

- Make assessmentsof mathematics lesson material quality
- Question good mathematics questions and problems and then modify according to necessities
- Assess students' mathematic learning progress and give it to next step

According to definitions and some opinions above, the writer states that teaching mathematics need basic teaching skills which can accommodate mathematics characteristics. it is meant for teachers to pack the learning process to be good and interesting for students so that it can persuade them to study. Those skills are called basic teaching mathematic skills, such as:

- Ability to master mathematic contents
- Ability to explain concepts, definitions, formulas, and mathematics symbols
- Ability to question in order to explore mathematics ideas, to respond students' inquiries and questions
- Ability to perform variation of using various mathematic learning approaches or strategies which are effective depend on situations and conditions
- Ability to give motivation through mathematic connections
- Ability to manage classroom and facilitateboth individual and group discussions
- Ability to perform assessments

3. Research Method

This Research is a qualitative research with analysis-descriptive technique. The subjects of the research were 4 students of Mathematic Education Departmentat STKIP Garut which took PPL SMAN 16 Garut in 2015. Instruments used in this research is observation sheet of basic skills of teaching mathematics arranged by some writers with guided by Instrument Penilaian Kinerja Guru (IPKG)/ and some related journals. The data analysis technique used in this research were data reduction, data presentation, and conclusion drawing.As for deciding basic teaching skills mastery, it's based on the criteria stated by Furqan [12] in the table 1 below:

TABLE 1. To Percentage Category Assesment

Average Class Score	Category
$80\% < x \leq 100\%$	Very Good
$67\% < x \leq 80\%$	Good
$53\% < x \leq 67\%$	Average
$40\% < x \leq 53\%$	Poor
$x < 40\%$	Very Poor

4. Research Results and Discussion

Basic teaching skills which were observed and analysed in this research were 1. Ability to master mathematics contents, 2. Explaining skill 3. Questioning skill 4. Motivational skill 5. Skill of performing variation 6. Skill of managing classroom 7. Skill of conducting assessments. Analysis result for the seven basic teaching skill components can be seen in table 2 below:

Table 2. Assessments of every Component in Basic Teaching Mathematics Skills

Basic Teaching Skill Components	Skor Total	SkorMak	Nilai	Criteria
(1) Ability to master Mathematics contents	29	48	60,4%	Average
(2) Explaining skill	20	32	62,5%	Average
(3) Questioning skill	61	80	76,25%	Good
(4) Giving motivational skill	31	64	48,4%	Poor
(5) Conducting variations kill	35	48	72,9%	Good
(6) Managing class skill	74	144	51,4%	Poor
(7) Conducting assessment skill	24	32	75%	Good
Total	272	448		

Mostly, it can be concluded that Students of Mathematic Education Department who took PPL in SMAN 16 Garut had average basic teaching skill 60,71%. As for achievement and categories for each basic teaching skill components are:

4.1 *Ability to master mathematics content*
Assessments percentage of ability to master mathematics content had major role in affecting others achievement of skill components like explaining skills.

4.2 *Explaining Skill*
Explaining skill category described those studentswho were mathematics prospective teachers' skill was on average category with percentage assessments 62,5%. The skill has deep connection with the level of mathematics content mastery, the higher his mathematic content mastery, the better his explaining skill. It's showed by the average category of mathematic content mastery of trainee students.

4.3 *Questioning Skill*
Percentage assessmentsof collegians reached 78,25%. It's considered good. If it's generally inspected, the component of questioning skill was in good category. But if we check it through every indicator, one of five indicators was not fulfilled. It was to lead student to problem solving. It's caused by the lack of collegians' understanding about problem solving.

4.4 *Giving motivation skill*
Giving motivational skill of mathematics prospective teachers was in poor category with 48,4 % assessments. The poor score was affected by many factors. One of them was the lack of experience gained by the collegians in responding students' conditions. The case was obviously showed in the indicator of using material to make student think critically. In the practice, the collegians only focused on the delivering of material without directing students into thinking critically.

4.5 *Skill of conducting variations*
Skill of conducting variations of the mathematics prospective teachers was in good category with assessment percentage 72,9%. Based on analysis result of every indicator, they had conducted variations well especially in using various tools/media/learning sources effectively. In the practice, they had used learning media effectively. It is appropriate to learning objectives which were told in the beginning of learning process.

4.6 *Managing class skill/skill of classroom management*

The managing class skill of matematics teacher is in poor category percentage achievement 51,4%. Managing class skill in this research includes opening and closing lesson skill, guiding group discussion skill, and teaching small groups and individual skill with nine indicators. The lack skill of collegians in managing class was caused by the little experience they got and also for students; authority aspect of university students seems to be less than guide teachers.

4.7 Conducting assessment skill
Conducting assessment skill of collegians, mathematics prospective teachers, is in good category percentage achievement 75%. Assessment is a very important thing in measuring learning process result. Generally, the trainnee students had performed the assessment well and it could measure the achievement level of learning objectives.

5. Conclusion and Suggestion
Mostly, Basic teaching skills of Mathematics Education Department's students in PPL were in average categories with achievement percentage 60,71%. Based on the result, it can be concluded that the basic teaching skills of prospective teachers has not been optimal yet. They needs development. Responding to this case, the writer recommends that they need to get lesson with appropriate strategies so that they can optimize their basic teaching skills.

6. References
[1] Tim Penyusun KKNI Dikti2013*Kerangka Kualifikasi Nasional Indonesiadan Implikasinya pada Dunia Kerja dan Pendidikan Tinggi*(Jakarta: Direktorat Jenderal Pendidikan Tinggi)
[2] Tim Kurikulum Prodi S1IndoMS 2013 *RekomendasiCapaianPembelajaranserta StrukturKurikulum Minimal Prodi S1 Matematika, PendidikanMatematika, StatistikadanIlmuKomputer/TeknikInformatika*(Bandung: The Indonesian Mathematical Society-IndoMS)
[3] Sanjaya W 2009*StrategiPembelajaranBerorientasiStandar Proses Pendidikan*(Jakarta: Kencana)
[4] Kyriacou C 2007*Essential Teaching Skills (Third Edition)* (Cheltenham: Nelson Thomes Ltd)
[5] PeraturanMenteri Negara PendayagunaanAparatur Negara danReformasiBirokrasi No 16 Tahun 2009 TentangJabatanFungsional Guru danAngkaKreditmya.
[6] Usman M U 2002*Menjadi Guru Profesional*(Bandung: RemajaRosdakarya)
[7] Nasution M2014*JurnalKajianKeislamanStudiMultidisipliner***1**1-15.
[8] Frasetyana, *et al* 2015*JurnalElektronikPembelajaranMatematika***3***383-94.*
[9] MauigoaL and Tekene2006*Journal of Teachers' Work***3** 12-23
[10] Shellard E and Moyer P S 2002*What Principals Need to Know about Teaching Math*(Alexandria, VA: The National Association of Elementary School Principals)
[11] Ball DL 2003*What Mathematical Knowledge is Needed for Teaching Mathematics* (Michigan: Secretary's Summit on Mathematics)
[12] Mulyatun2014*Jurnal Pendidikan MIPA Phenomenon***4** 79-90

Ionic Liquids as a Basis Context for Developing High school Chemistry Teaching Materials

Hernani, A Mudzakir and O Sumarna

Chemistry Education Department, Universitas Pendidikan Indonesia, Jl. Dr. Setiabudhi No. 229, Bandung 40154, Indonesia
Email: Hernani.kimia@gmail.com

Abstract. This research aims to produce a map of connectedness highschool chemical content with the context of the modern chemical materials applications based on ionic liquids.The research method is content analysis of journal articles related to the ionic liquid materials and the textbooks of high school chemistry and textbooks of general chemistry at the university. The instrument used is the development format of basic text that connect and combine content and context.The results showed the connectedness between: (1) the context lubricants ionic liquid with the content of ionic bonding, covalent bonding, metal bonding, interaction between the particles of matter, the elements of main group, the elements of transition group, and the classification of macromolecules; (2) the context of fuel cell electrolite with the content of ionic bonding, covalent bonding, metal bonding, interaction between the particles of matter, Volta cell, and electrolysis cell; (3) the contect of nanocellulose with the content of ionic bonding, covalent bonding, metal bonding, interaction between the particles of matter, colloid, carbon compound, and the classification of macromolecules; and (4) the context of artificial muscle system with the content of ionic bond, covalent bond, metal bonding, interaction between the particles of matter, hydrocarbons, electrolytes and non-electrolytes, and the classification of macromolecules.Based on the result of this content analysis, the context of ionic liquid is predicted can be utilized for the enrichment of high school chemistry and has the potential to become teaching material's context of high school chemistry in the future.

1. Introduction

Scientific literacy accession is a sustainable and continuously process that growing throughout human life. Student science literacy assessment includes aspects of attitudes, values, basic skills, knowledge and understanding of science. Science literacy assessment for learning in school is only able to measure the seeds of literacy that exist within the students and not to measure the outcome of scientific literacy to be achieved by a person [1].

Chemical literacy as part of the unity of science literacy potentially to be developed at the high school level. According Cigdemoglu and Geban (2015, p.304) Chemical literacy is the students ability to use and connecting the information provided in the chemical problems and the ability of students to use their knowledge and chemistry skills to shows comprehensive information related to daily problems. Many life problems have a close relation within technology, so that the context of the related technology will affect the ability of students in literacy chemistry.

IOP Conf. Series: Journal of Physics: Conf. Series **812** (2017) 012085 doi:10.1088/1742-6596/812/1/012085

The definition of chemical literacy developed by obtaining an agreement among chemists, educators, and teachers in high school. As a result, the definition of chemical literacy was specified as it looked in Figure 1 (Cigdemoglu and Geban. 2015,p. 302).

1. Chemical content knowledge
   A chemically literate person understands the following:
   A. General chemical ideas
   - As an experimental discipline, chemists carry out scientific investigations, generalize findings, and propose theories to explain to the world.
   - Serves knowledge to other fields in order to explain phenomena

   B. Characteristics (key ideas) of chemistry
   - Explain the macroscopic level by means of the molecular structure of matter
   - Seek the dynamics of processes and reactions
   - Investigate the energy changes accompanied in a reaction
   - Understand/explain life by chemical processes and structures of living systems
   - Appreciate the contribution of scientific language to this discipline
2. Chemistry in context
   Chemically literate students are able to:
   - Understand the knowledge of chemistry in explaining everyday situations
   - Understanding of daily life chemistry, such as being user of new products/technology, a decision process, and involving in social argumentation on chemistry-related issues.
   - See the relatedness of innovations in chemistry and sociology.
3. Higher-order learning skills
   Students who are chemically literate ask questions, investigate relevant information when required. Additionally, he/she can evaluate pros/cons of debates.
4. Affective aspects
   Students who are literate have a fair and rationale perspective of chemistry and its applications. Furthermore, literate students show interest in issues of chemistry, specifically in a non-formal environment like mass media.

Figure 1. Definition of chemical literacy.

De Jong (2006) argues that the context is a situation or event that helps students to be able to acquire concepts, principles, laws and so on. Technological developments being one reference that can be used in selecting the context in order to broaden students knowledge through discussions of issues emerging technologies in the global world.

Modern technology ionic liquid material-based is a learning context that's globally and very interesting to be a topic in the chemistry learning with reasons: First, the attention of the international community of science and technology in the use of ionic liquids as a new generation of green solvent, electrolyte material and fluid engineering in early 21st century was greater, in line with the increasingly strong demands of the industry for the provision of new materials that are reliable, safe, and environmentally friendly for various purposes [2],[3]. In the electrochemistry, ionic liquids research directed as electrolyte materials in batteries, metal plating and sensor systems [4],[5],[6]. Ionic liquids are also directed as green solvents in chemical synthesis, catalysis and bio-catalytic [7],[8],[9]. In the engineering field, the ionic liquid is also used as a heat carrier fluid, lubricants, surfactants and the liquid crystals [10],[11],[12],[13],[14],[15]. Secondly, scientific explanation related to ionic liquid context contains a lot of facts, concepts, principles, laws, models, and theories that can be used to reinforcing high school chemistry content as a medium for developing thinking skills (process/competence) demanded by PISA (Programme for International Student Assessment).

Ionic liquids is a salt that in liquid phase at room temperature or below room temperature and melts overall shape composed of ions consisting of organic cations and inorganic anions [16]. Ionic liquid is a material that only consist of ionic species (cations and anions), doesn't contain neutral molecules, and relatively has low melting point between <100-150°C, although generally at room temperature [17]. In contrast to the molten salt, which usually has a higher melting point and viscosity, it is also highly corrosive. ionic liquids generally in liquid phase at room temperature, has a relatively lower viscosity and doesn't corrosive [18]. Ionic liquids have a very wide range of liquid phase; non-volatile; unburned (non-flammable); have a good thermal stability (in some cases have thermal stability up to 400°C); vapor pressure value can be ignored; able to dissolve many organic and inorganic compounds; and has a diverse miscibility properties by mixed with water or organic solvents [19].

Cation system in the ionic liquid (and liquid crystal ionic) generally an bulk organic cations, such as N-alkylammonium, P-alkylphosfonium, N-alkyl-pyridinium, S-alkylsulfonium, N-alkylpirolidinium, N,N-dialkylpirazolium and N,N-dialkylimidazolium. The biggest fraction of the research focused to the cation N,N-dialkilimidazolium 1, because of the enormous differences in physicochemical properties that can be provided [7].

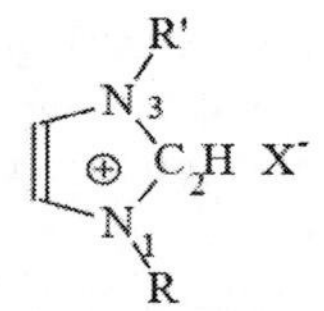

1

Figure 2. N,N-dialkylimidazolium 1 cation structure.

The properties of ionic liquids can be adjusted by changing the structure of the cation and anion. The properties of ionic liquids such as thermal properties, electrochemical stability, ionic conductivity and viscosity can be adjusted depending on the cation and anion are arranged to make the ionic liquid known as tailored-made solvents [20]. The melting point of salt ionic liquid 1-alkyl-3-methyl-imidazolium (with R = CH_3) for example, will decline with increasing length of the alkyl group R. After the minimum price achieved (usually in groups hexyl, heptyl, or octyl) melting point will begin to increase with increasing length of the R group [15].

Stability properties of ionic liquids at high temperatures, causing the ionic liquid can be used as a lubricant in the high temperature or low pressure process [11]. Lubricant can prevent or reduce wear and tear as a result of direct contact between two metal surfaces rub against each other so that the wear can be reduced. amount of power needed due to friction can be reduced and the heat generated by friction would be reduced (Lumas Multi Sarana, 2009).

Another utilization of the ionic liquids properties is Fuel Cell. Fuel cell is a power plant that converts chemical energy directly into electricity using hydrogen as fuel and oxygen as oxidant. Fuel Cell acts like a battery that is producing an electrical voltage. The output voltage is generated by the fuel cell is the direct voltage, ranging from 0,7V.

Excellence properties of ionic liquids can be used also to modify the cellulose in some applications nanocellulose. Nanocellulose is cellulose which has a diameter range between 1 to 100 nanometers and have a length about tens to hundreds nanometers, have a relatively high aspect ratio and has high properties of the crystalline structure [21].

Other applications that take advantage of the ionic liquids properties is Electron Active Polymer (EAP) as artificial muscle. In the manufacture of artificial muscles typically used for an actuator material which has properties such as artificial muscles. Actuator is a substance / material that can move when given a stimulus and movements resemble artificial muscles. Materials for the artificial muscle actuator typically uses a polymer known as EAP (Electro Activ Polymers) (Chen, et. al., 2015, p.106).

Based on the ionic liquids contexts review, the purpose of this research was to produce enrichment chemistry books for high school chemistry that can develop students literacy skills. These books will be constructed in order to enrich and improve the mastery of science and technology, skills, and shaping student's personality and other community, which serves as a passage for students and other parties related to the education field (Book Centre of Ministry Education, 2008).

The specific problem in this research is: How does the prospective of use of ionic liquids context in construction high school enrichment chemistry books? To answer these problem descriptive method used through analysis content steps from journal articles that related to the ionic liquids context, high school chemistry books , general chemistry textbooks and high school chemistry content standards curriculum in 2013.

2. Method

This research is an initial step of educational products development, such as determination (define) step, which is done by analyzing the related content of ionic liquids context in journals articles, analyze the chemicals content in high school chemistry books and general chemistry textbooks, as well as the content standard of high school chemistry subject based on curriculum 2013. the research instrument that used is the connectedness format of ionic liquids context and high school chemistry content.

The results of this analysis step will be used to define the design steps. The development step aim to formulating the basic text that the content and context already integrated and inseparable then evaluated the accuracy of its formulation. The process will be guidelines for constructing high school chemistry enrichment book that validated and legibility tested.

3. Result

The results in this research are presented in four types of applications related to ionic liquids context that derived from related journals articles, which includes Application at magnetic lubricating, electrolyte fuel cell, nanocellulose and artificial muscles. In all four application are discussing findings context from journals and it related content.

3.1. Magnetic media lubricant

3.1.1. Contexts analysis result. Findings from the journal articles, generates related terms that must be understood, namely: tribology, friction, wear, andlubrication. Tribology is the study of friction or interacting surfaces in relative motion. Frictionis the resistance to relative motion between two bodies in contact. The magnitude of friction force influenced by weight of the object and the surface roughness touching. Wearis removal (or displacement) of material from one body when subjected to contact and relative motion with another body. Lubricationis the act placing a lubricant between surface muatually shifted to reduce wear and friction [22].

Further explained that is necessary to lubricant used for magnetic media, the media that is normally used to record all data (audiovisual information). In magnetic recording, thin film media typically have their surfaces lubricated to reduce friction and wear resulting from contacts between the read/write magnetic head and media surface. In practice, to avoid adhesion related problems, lubrication has to be achieved with a molecularly thin lubricant film (Homola in Kondo) [23]. It is important that the lubricant remain on the media surface over the life of the file without being subject to desorption, spin off, or chemical degradation [23].

The characteristics which must be met for lubricant for magnetic thin film media with low friction and wear can be summarized as follows high viscosity, low vapor pressure (volatility), low surface tension (replenish surface with lubricant), some affinity to the surface (with polar end group), good thermal and oxidative stability to prevent degradation during storage and use, chemical inertness to metals, glasses and plastics and non-deposit forming to insure no damage to heads, guides, devices, etc., no toxicity and flammability (to prevent degradation in the presence of water), good boundary lubricant (forms easily sheared chemical film during asperity contact), and solubility in organic solvents [23].

The material chemical predicted meets the characteristics is the material based on ionic liquids, because the ionic liquid is a salt compound composed of ions (cation and anion), does not contain any neutral molecules specific, and has a melting poin is relatively low (<100°C). The cation system in ionic liquid is generally an organic cation with bulk properties. Moreover, the characteristics of ionic liquid can adjusted by changing the structure of the cation and anion [17].

Based on the novel research from Hayashi and Hamaguchi (2004) have found an ionic liquid that has magnetic properties and has the characteristics required as lubricant for magnetic media such as high thermal stability, non-volatile (low vapor pressure), high viscosity, and others [24]. This ionic liquid named [bmim] $FeCl_4$ synthesized from a mixture of compound 1-butyl-3-methylimidazolium dan $FeCl_3$. [bmim] $FeCl_4$ able to respond with strong magnetic field and the results of magnetic susceptibility testing indicate that [bmim] $FeCl_4$ has paramagnetic properties with magnetic susceptibility 40,6 x 10^{-6}emu/g [24]. The magnetic susceptibility is one of measurement method that is non-destructive and effectively used in knowing the existence of paramagnetic elements in the sample. The entire core, individual or sediment sample was placed in magnetic field that causes the sample magnetized and it is highly dependent on the amount of the magnetic substance contained in the sample. Susceptibility Meter type Bartington MS2 is an instrument for measuring magnetic susceptibility. This instrument detects the spin of unpaired electrons that are in the compound. Instruments MS2 or MS3 system can operate by generating a frequency and intensity of the magnetic field generated AC current carrying wire coiled around the sensor.

When a sample is placed near the sensor, the change in the magnetic field intensity generated can be detected by the system and converted into magnetic susceptibility readings for both values either positive is ferromagnetic and paramagnetic and negative is diamagnetic, with a resolution of 2 x 10^{-6} SI units. Measurements conducted are non-destructive or maintain the magnetic characteristics of the sample, and the relative does not affect the conductivity of the sample. Therefore, the magnetic susceptibility of the ionic liquid compound is a small positive and then classified into the metal that is paramagnetic [25]. With this findings are limited to the compound having paramagnetic properties, it is expected to be ferromagnetic compound can be synthesized in the future because of the strength of magnetic properties can make better quality lubricants in overcoming problems of friction and wear.

3.1.2. Content analysis result. The results of content analysis of high school chemistry and general chemistry textbooks mentions that the cation is a positively charged particle, which is formed from metal atoms lose valence electrons that are positively charged. By discussing the ionic liquid material, theses is widespread understanding of the cation is not only derived from the metal atom but also as positively charged organic molecules. Between the cation and anion occurred electrostatic attraction, forming an ionic bonding.

Other analysis of chemical content is the viscosity. Viscosity caused by the cohesive forces between like molecules. The cohesive force decreases as the temperature increases, because the force between molecules are declining. If a liquid has a strong intermolecular forces, the viscosity will be higher than the liquid has a weak intermolecular forces. Intermolecular forces are attractive force between molecules that include Van Der Waals forces, dipole-ion forces, and hydrogen bonding. The cohesion forces (the intermolecular attraction between like molecules) and adhesion forces (the intermolecular attraction between like molecules) causes the surface tension. Another phenomenon that arises when the phase change from liquid phase to gas (steam) phase or otherwise involves an equilibrium which is called vapor pressure. If a substance that has a low vapor pressure, the high boiling point. Boiling point is strongly influenced by intermolecular forces. The stronger the intermolecular forces then its boiling point will be higher and its means a low vapor pressure (non-volatile) [26].

The chemical content analysis of chemical elements subject matter give information that the aluminium alloys with magnesium or often called alnico can form a strong magnet. Metal that is highly resistant to corrosion are chromium (Cr), so that the chromium metal is often used as an alloy. Alloy formed from a mixture of cobalt (Co), chromium (Cr), and tungsten (W) the nature of which remains hard despite heated to very high temperatures. In addition, the alloy is formed from metal Co and Cr have ferromagnetic properties that can be applied to the magnetic coating on the magnetic recording medium. The phenomenon of ferromagnetic properties is only applicable if the materials in solid form. Not all of the elements that contain an unpaired electron in solid form having ferromagnetic properties. To form the ferromagnetic properties can be done by setting the exact distance between the paramagnetic ions, the ions so that each can held on to form a magnetic field. Metal that has no ferromagnetic properties (non-ferromagnetic), which lies between the ions are very close together, can be made is by being converted into a ferromagnetic alloy. This has been accomplished on chromium, where the addition of a cobalt cause Cr^{2+} ions in a geometric pattern of very strong interaction to form a magnetic field, so that the ferromagnetic alloy [27].

Another content chemical analysis of lubricant is polymer. Polymers are very large molecules (often called macromolecules) formed from the merger of small molecules called monomers, which is a recurring unit. Polymers can be chemically degraded to break down into simpler molecules, both natural and artificial [26]

3.2. Electrolyte fuel cell

3.2.1. Contexts analysis result. Findings from the journal articles provide information about the terms of the fuel cell. Fuel cells are electrochemical devices that convert chemical energy in fuels into electrical energy directly, promising power generation with high efficiency and low environmental impact [28]. Typically, a process of electricity generation from fuels involves several energy conversion steps, namely [29]:

- Combustion of fuel converts chemical energy of fuel into heat,
- This heat is then used to boil water and generate steam,
- Steam is used to run a turbine in a process that converts thermal energy into mechanical energy, and finally
- Mechanical energy is used to run a generator that generates electricity.

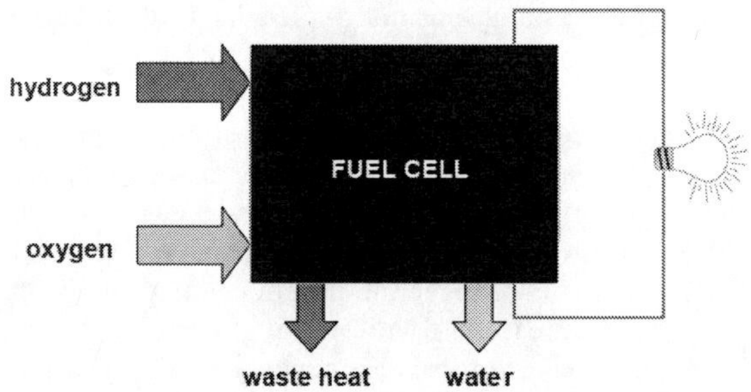

Figure 3. Fuel cell scheme.

A fuel cell in some aspects is similar to a battery, but it requires a constant supply of fuel and oxidant, and it electrode do not undergo chemical changes.

Fuel Cell could be work continuing as long as it fuel and oxidant are available. Typical reactants for fuel cells are hydrogen and oxygen; however, neither has to be in its pure form. Hydrogen may be present either in a mixture with other gases (such as CO_2, N_2, CO), or in hydrocarbons such as natural gas, CH_4, or even in liquid hydrocarbons such as methanol, CH_3OH. Ambient air contains enough oxygen to be used in fuel cells. Yet another difference between a fuel cell and a battery is that a fuel cell generates by-products: waste heat and water, and the system is required to manage those (a battery also generates some heat but at a much lower rate that usually does not require any special or additional equipment) [29].

Fuel cell contains two electrode, wich is anode and cathode that separated by electrolyte. Hydrogen supplied to the fuel cell through the anode, however Oxygen or ambient air supplied to the fuel cell through the cathode [28]. By the electrolyte hydrogen and oxygen couldn't be mixed. Electrolyte functioned as catalyzator that can decompose hydrogen to electron and proton. Protons flow through the electrolyte, while the electrons can not pass through the electrolyte, so that electrons would pile up at the anode, while in the cathode positively charged ions piled up. When the anode and the cathode is connected to an electrical conductor, there will be a flow of electrons from the anode to the cathode, so that there is an electricity. The chemical reactions that occurs in the fuel cell :

$$Anode : 2H_2 \rightarrow 4H^+ + 4e^-$$
$$Cathode : O_2 + 4H^+ + 4e^- \rightarrow 2H_2O$$

The use of hydrogen that contained in the fuel cell can be designed as a portable system that can be moved and has a light mass. Unlike the combustion system found on conventional transport equipment (use the piston in a car/motorcycle), the Fuel Cell has no moving parts or vibrating so will not causing pollution. Fuel Cell can be applied in various types of applications that use energy as in cell phones, personal computers, power station as in hospitals, office buildings, schools and also on vehicles such as motorcycles and cars.

3.2.2. Content analysis result. The results of high school chemistry content and general chemistry textbooks analysis explain that the electrochemical interpreted as a chemical changes that produce an electricity or chemical changes that produce an electricity [30]. Electrolytes are substances that can dissociate into ions, can be solids dissolved in a suitable medium or in molten form an ionic compound that can conduct electricity. Electrical energy is generated due to the movement of electrons in a solution or ionic liquids. Electrons move from the negatively charged electrode (anode) to the positively charged electrode (cathode) [26].

Another content analysis result explained that Volta cells or galvanic cells are electrochemical cells that can undergo oxidation-reduction reactions spontaneously to generate electricity [30]. In the Volta cell, two parts of the redox reaction are separated, it is done so in such there is a transfer of electrons in the system . In this way, useful electrical energy obtained. In some electrochemical cells spontaneous chemical reactions can occur with electrical energy drain, this process is called electrolysis [26]. Electrolysis cell consists of a container for reaction with the electrode material is immersed in an electrolyte and connected to a source of direct current. Inert electrodes are often used so that they do not react.

3.3. NanoCellulose

3.3.1. Contexts analysis result. The results of the study on several journal articles generate information that nanocellulose is cellulose molecule, being in nanometer size and has large surface area (150 m^2/g) (Bai *et al* in Tan *et al*) [21] . Nanocellulose divided into two kinds, namely cellulose nanofibrils (CNF) or better known as nanofibrillar cellulose (NFC) and cellulose nanocrystals (CNC) or better known as nanocrystalline cellulose (NCC) [31], [32]. Nanoselulosa can be obtained from plant cell walls (Klemm *et al* in Chirayil *et al*) [33] CNF can be obtained by mechanical method (Moon *et al* in Salas *et al*) [31], while CNC can be obtained by three main stages: pre-treatment of cellulose in which the most common method is acid hydrolysis to break down the amorphous part of cellulose, addition of anti-solvent to regenerate the cellulose and followed by treatment of sonication to disperse the CNC uniformly (Sheltami *et al* 2012; Man *et al* 2011; Lu *et al* 2013; in Tan *et al*) [21]. Moreover, nanocellulose can be obtained by dissolving cellulose in ionic liquids (Swatloski *et al* in Isik *et al*) [34]. Ionic liquids from the result of a merger between the cation 1-butyl-3-methylimidazolium and various types of anions have been investigated as a solvent of cellulose (Swatloski *et al* 2002; Gericke *et al* 2012 in Isik *et al*) [34].

The results of further studies on the context, it was found that nanocellulose very easy to form hydrogels (Hoffman, 2002; Jen *et al*1996; in Salas *et al*) [31]. Compared to CNC, CNF forms hydrogels more readily (Klouda and Mikos, 2008; Lee and Mooney, 2001; in Salas *et al*) [31]. Nanocellulose can also form a suspension in a non-polar dispersing medium by adding a surfactant to the system [35]. Nanocellulose can also form an emulsion systems, nano-cellulose is used as a stabilizer in the emulsion system [31], because nanocellulose containing hydrophilic group on its surface [35]. Hydrophilicity of nanocellulose can form an oil-in-water (O / W) emulsion (Oza and Frank in Salas *et al*) [31]. However, nanocellulose can also form a water-in-oil (W / O) emulsion by modifying the surface of nanocellulose (Lif *et al* 2010; Xhanari *et al* 2011; in Salas *et al*) [31]. In addition, nanocellulose can also form a stable foam. CNF and CNC can form a stable foam by modifying the surface of nanocellulose into a hydrophobic surface (Cervin *et al* in Salas *et al*) [31].

It was also found in several journal articles that nanocellulose applications in daily life is very broad. For example in the field of industrial electronic and optic devices (Bai *et al* 2009; Lu *et al* 2013; in Tan *et al*) [21], regenerative medicine (Lu *et al* 2013; Lu and Hseih 2010; in Tan *et al*) [21], a thickener in cosmetics, stabilizers in the food industry, enhancing the quality of the paper and furniture, as well as composite materials for construction (buildings) and vehicles (Mandal and Chakrabarty, 2011; Bai *et al* 2009; Haafiz *et al* 2013; Voronova *et al* 2012; in Tan *et al*) [21]. Nanoselulosa can also be used as coating materials, optical devices, stabilizers in the multiphase systems, and as a reinforcing material in the nanocomposite [31].

3.3.2. Content analysis result. The results of the study on several chemistry books for high school and chemistry books for universities produce information that the chemical content related to the context of nanocellulose is polymer content. Polymers are large molecules that formed from the simple repeating units. The term of polymer is derived from the Greek, *Poly* means "many", and *mer* meaning "section" [36]. Based on the many kinds of monomers making up the polymer, the polymer is divided into two. If a polymer was prepared only from a single monomer, then the polymer is classified as a homopolymer, whereas if a polymer prepared by more than one type of monomer then the polymer is classified as copolymer [36].

Another chemical content related to the context of nanocellulose is carbohydrate content. Carbohydrates are polyhydroxy aldehydes or ketones, or substances that yield such compounds on hydrolysis. Many, but not all, carbohydrates have the empirical formula $(CH_2O)n$, some also contain nitrogen, phosphorus, or sulfur. There are three major size classes of carbohydrates: monosaccharides, oligosaccharides, and polysaccharides (the word "saccharide" is derived from the Greek, meaning "sugar"). Monosaccharides, or simple sugars, consist of a single polyhydroxy aldehyde or ketone unit. Some of the monosaccharide units can form oligosaccharides, which are connected by bonds called glycosidic bond. In addition, a combined 20 or more monosaccharide units can form a polysaccharide. Some polysaccharides, such as cellulose, are linear chains; others, such as glycogen, are branched. Both glycogen and cellulose consist of recurring units of D-glucose, but they differ in the type of glycosidic linkage and consequently have strikingly different properties and biological roles [37].

Another chemical content related to the context of nanocellulose is about colloidal or colloidal dispersion. Colloidal dispersion is a collection of particles measuring between 1-1000 nm (the dispersed phase) is dispersed in the dispersing medium. The size of the colloidal particles is greater than the atomic and molecular size, so that the colloidal particles have physical and chemical properties that are different from atoms and molecules and also have different properties with macromolecules [38]. Colloidal dispersion is divided into eight types, which are summarized in the following table:

Table 1. Classification of colloidal dispersion [38].

Dispersed phase	Dispersion Medium	Name
Gas	Liquid	Foam
Gas	Solid	Solid Foam
Liquid	Gas	Liquid Aerosol
Liquid	Liquid	Emulsion
Liquid	Solid	Solid Emulsion
Solid	Gas	Solid Aerosol
Solid	Liquid	Sol, Suspension
Solid	Solid	Solid Suspension

Another chemical content related to the context of nanocellulose is the content of hydrogen bonds. The hydrogen bond is a special type of intermolecular forces between the dipoles hydrogen atom in a polar bond such as N-H, O-H, and F-H, with the atom electronegative O, N, and F [26]. Hydrogen bonds can be interpreted also as the bond between the H atoms in polar molecules and atoms with high electronegativity on polar molecules. Hydrogen bonding can occur in the polar molecule (called the intramolecular hydrogen bond), or it can also occur intermolecular polar (called intermolecular hydrogen bonding) [39].

3.4. Artificial muscle

3.4.1. Contexts analysis result. The results of the context research in several journal articles generate artificial muscle definition. Artificial muscle is a kind of actuator generally defined by analogy with the skeletal muscle. Like the natural muscle, any artificial muscle is characterized by its ability to contract in response to a chemical or physical stimulus [40]. Artificial muscles can be formed of a polymersmaterial that change shape or size in response to electrical stimulus are called electroactive polymers (EAP)

[41].The most attractive feature of EAPs is their ability to emulate biological musclesoffering resilience, toughness, large actuation strain, inherent vibration damping and soft with a quick response when stimulated electric field [42].

Further studies the in context from journal articles explaining thatpolymer artificial muscles have been divided into two major groups. In the first group, dimensional change (actuation) is in response to an electric field. These are commonly known as electronic or electric electroactive polymers (EAPs) [43]. Some of the technologies that fall under this category are :

Table 2. List of the leading electronic EAP materials [42].

Electronic EAP
• Dielektric elastomer EAP
• Elektroactif graft elastomers
• Electroactive-viscoelastikc elastomers
• Ferroelektric polymer
• Liquid Christal Elastomers (LCE)

The second group of polymer artificial muscles is a class of materials in which the presence and movement of ions is necessary to make actuation possible. This group is referred to as ionic EAPs. For the ions to be able to move, an electrolyte phase is necessary, which is often liquid; so these actuators are also known as wet EAPs [43]. Some of the technologies that fall under this category are :

Table 3. List of the leading Ionic EAP materials [42].

Ionic EAP
• Carbon Nanotube (CNT)
• Conductive Polymer (CP)
• Ionic Polymer Gels (IPG)
• Ionic Polymer Metalllic Composite (IPMC)

The electronic polymers be made to hold the induced displacement under activation of a DC voltage, allowing them to be considered for robotic applications. Also, these materials have a greater mechanical energy density and they can be operated in air with no major constraints. However, they require a high activation fields (>100-V/mm) close to the breakdown level. In contrast, ionic EAP materials require drive voltages as low as 1-2 Volts. However, there is a need to maintain their wetness, and except for conductive polymers it is difficult to sustain DC-induced displacements [44].

Also described in several articles that the technologies of artificial muscle polymer are being developed that can produce the same strain and high pressure using electrostatic forces, elektrostriksi, ion insertion, and changes in molecular conformation. Materials used include elastomers, conductive polymers, ionic polymers and carbon nanotubes. Other technologies using elastomer actuator dielectrik (Dielectric Elastromer Actuator), ferroelectric polymer and liquid crystal elastomers. In addition, alternative way another of obtaining actuation in the polymer is to use the movement of ions in the polymer [42].

3.4.2. Content Analysis Result. The result of the study on several chemistry books for high school and chemistry books for universities produce information that the chemical content related to the context of artificial muscle is polymer content. Polymers are large molecules formed from the simple repeating units. The term of polymer is derived from the Greek, Poly means "many" and mer means "section" [36].

Another chemical content related to the context of artificial muscle, that is:

- Stress, is defined as the change in force to the cross-sectional area which is subject to force
- Strain, is defined as the ratio of length change material to the original length due to a force with a direction parallel to the length change [45]
- Conjugation, is defined as a electron bonding cloud system formed by the interaction between the electron cloud on the double bond and a single bond : happen on molecule (generally organic molecules) containing a double bond and a single bond criss-cross. The interaction that occurs the form of clouds of electrons double bond delocalized (shifted) to the single bond.Example of simple compounds that have the conjugation bond is: H2C=CH-CH=CH2 [39].
- Elastomer,is defined as a polymer that has the ability to stretch and return to its original state quickly [36].
- Conductivity Electrical in Polymer, is a polymer that can conduct electricity if the polymer occurs electron delocalization in conjugation bond, have cis-tran isomer and has a heterocyclic ring [36]
- Electrostatic forces, is defined as the force generated by electrically charged objects [39].
- Covalent Bond, defined as the bond formed between atoms that share electrons [46].

4. Conclusion

Based on the context and content analysis, can be concluded that the relevant ionic liquids context such as (1) ionic liquid lubricants, (2) electrolyte fuel cell. (3) nano cellulose, and (4) the artificial muscle potentially to be used as material for enriching chemical bonds content. In addition the context (1), (3) and (4) the potentially to be used as material for enriching macromolecule/polymer content; the context (1) potentially to be used as material for enriching chemical elements content, the context (2) potentially to be used as material for enriching electrochemical content, the context (3) potentially to be used as material for enriching colloids content, and context (4) potentially to be used as material for enriching electrolytes and non electrolytes content.

5. References

[1] Shwartz Y, Ben-Zvi R and Hofstein A 2006 *Chemical Education Research and Practice* **7** 203-25
[2] Earle M J and Seddon K R 2000 *Pure and Applied Chemistry* **72** 1391-98
[3] Brennecke J F and Maginn E J 2001 *AIChE Journal* **47** 2384-9
[4] Blomgren G E 2003 *Journal of Power Sources* **119** 326-9
[5] Bhatt A I, May I, Volkovich V A, Hetherington M E, Lewin B, Thied R C and Ertok N 2002 *J. Chem. Society, Dalton Transaction* **24** 4532-4
[6] Buzzeo M C, Hardacre C and Compton R G 2004 *Analytical Chem.* **76** 4583-8
[7] Olivier-Bourbigou H and Magna L 2002 *J. Molecular Catalysis A: Chemical* **182-183** 419-37
[8] Vidis A, Ohlin A, Laurenczy A G, Küsters E, Sedelmeier G and Dyson P J 2005 *Advanced Synthesis and Catalysis* **347** 266-74
[9] Miao W and Chan T H 2006 *Accounts of Chem. Research* **39** 897-908
[10] Blake D M, Moens L, Hale M J, Price H, Kearney D and Herrmann U 2002 New Heat Transfer and Storage Fluids for Parabolic Trough Solar Thermal Electric Plants, *Proc. of the 11th SolarPACES International Symposium on concentrating Solar Power and Chemical Energy Technologies (Zurich)*
[11] Ye C, Liu W, Chen Y and Yu L 2001 *Chem. Communications* **21** 2244-5
[12] Merrigan T L, Bates E D, Dorman S C and Davis J E 2000 *Chem. Communications* **20** 2051-2
[13] Awad W A, Gilman J E, Nyden M, Harris R H, Sutto T E, Callahan J, Trulove P C, DeLong H C and Fox D M 2000 *Thermochimica Acta* **409** 3-11
[14] Holbrey J D and Seddon K R 1999 *J. Chem. Society, Dalton Transaction* **13** 2133-40
[15] Gordon C M, Holbrey J D, Kennedy A R and Seddon K R 1998 *J. Materials Chem.* **8** 2627-36

[16] Mudzakir A, Permanasari A and Mahiyudin 2007 *Proc. of the First International Seminar of Science Education, Science Education Program Graduate School (Bandung)* (Bandung: FPMIPA UPI)
[17] Hagiwara R and Ito Y 2000 *J. Fluorine Chem.* 105 221-7
[18] Toma G, Gotov B and Solcaniova E 2000 *Green Chem.* **2** 149-51
[19] Davis J H and Fox P A 2003 *Chem. Communications* **11** 1209-12
[20] Gordon C M 2003 Synthesis and purification of ionic liquid *Ionic Liquid in Synthesis* ed Wasserscheid P and Welton T (Frankfurt: Wiley Verlag)
[21] Tan X Y, Lai C W and Hamid S 2014 Facile Preparation of Highly Crystalline Nanocellulose By Using Ionic Liquid (Nanotechnology and Catalysis Research Centre) pp 1-5
[22] Shaffer S J 2013 *101 – Introduction to the Basics of Tribology* (Bruker Corporation)
[23] Kondo H 2011 Ionic liquid lubricant with ammonium salts for magnetic media *Applications of Ionic Liquids in Science and Technology* vol 20 ed Handy S pp 211–8
[24] Hayashi S and Hamaguchi H 2004 *Chemistry Letters* **33** 1590-1
[25] _____ 2013 *Magnetic suseptibility system* Bartington Instrument (Online: Bartington)
[26] Chang R 2010 *Chemistry* 10th edition (New York: McGraw-Hill)
[27] Brady J E *Asas dan Struktur Jilid I dan II* 1998 Translated by Sukmariah M, Kamianti A and Tilda S) (Jakarta: Binarupa Aksara) pp 361-71
[28] Williams M C 2004 *Fuel Cell Hanbook* (Morgatown: EG&G Technical Sevices inc.) pp 20-1
[29] Barbir F 2005 *PEM Fuel Cell: Theory and Practice* (London: Elsevier Academic Press) pp 1-6
[30] Whitten R E, Davis M L, Peck and Stanley G G 2004 *General Chemistry* seventh edition (USA: Thomson Brooks) pp 849-58
[31] Salas C, Nypelo T, Rodriguez-Abreu C, Carrillo C and Rojas O J 2014 *Current Opinion in Colloid and Interface Science* **19** 383-96
[32] Xu X 2015 *Applied Materials and Interface* **5** 2999-3009
[33] Chirayil C J, Mathew L and Thomas S 2014 *Reviews on Advenced Material Science* **37** 20-8
[34] Isik M, Sardon H and Mecerreyes D 2014 *International Journal of Molecular Sciences* **15** 11922-40
[35] Islam M T, Alam M M and Zoccola M 2013 *International Journal of Innovative Research in Science, Engineering and Technology* **2** 5444-51
[36] Stevens M P 2007 *Kimia Polimer* Translated by Iis Sopyan (Jakarta: PT. Pramita) pp 3-151
[37] Nelson D L and Cox M M 2005 *Lehninger Principle of Biochemistry* 4th edision (USA: University of Wisconsin Press) p 238
[38] Schramm L L, Emulsions, Foams and Suspensions 2005 *Fundamentals and Applications* (Weinheim: WILEY-VCH Verlag GmbH & Co. KgaA) pp 1-2
[39] HAM M 2009 Kamus Kimia (Bandung: PT Bumi Aksara) pp 145-185
[40] Tondu B 2015 *Actuators* **4** 336-52
[41] Cottinet P J, Guyomar D, Guiffard B, Lebrun L and Putson C 2010 Electrostrictive polymers as high-performance electroactive polymers for energy harvesting *Piezoelectric Ceramics*, ed Gomez E S chapter 10
[42] Bar-Cohen Y 2000 Electroactive polymers as artificiafchenl muscles: Potentials and challenges *Handbook on Biomimetics* section 11 chapter 8 "Motion"(Lubbock: NTS Inc.) ed Yoshihito Osada
[43] Mirfakhrai T, Madden J and Baughman R H 2007 *Materials Today* **10** 30-8
[44] Bar-Cohen Y 2001 *Electroactive Polymers As Artificial Muscles : Potentials And Challenges. Ballinghem* (WA : SPIE Press)
[45] Utami M B 2014 *Pengukuran Compressive Strength Benda Padat* (Universitas Airlangga)
[46] Toon T Y 2007 *Chemistry Matters* (Singapore: Marshall Cavendish Int (S) Pte Ltd) p 102

Acknowledgements
This research made possible by a grant from the Directorate General of Higher Education 2016 budgets, with the facilitation from LPPM UPI, FPMIPA management and the Department of Chemistry Education, and with help from students related to thesis preparation (1) Atep Rian Nurhadi, (2) Saefulloh , (3) Billy Oktora Abdillah Fauzi, and (4) Lutfi Lulul Ulum. For that, we extend many thanks for the cooperation that has been established, hopefully this research beneficial to the advancement of education.

IOP Conf. Series: Journal of Physics: Conf. Series **812** (2017) 012086 doi:10.1088/1742-6596/812/1/012086

Designing Adaptive-Content trough E-learning on Electromagnetic Concept

L Hakim[1]*, A Setiawan[2], and P Sinaga[3]

[1]Universitas PGRI Palembang, Department of Mathematics and Science Education, Palembang, Indonesia30263
[2]Universitas Pendidikan Indonesia, Department of Mechanical Engineering, Bandung, Indonesia40154
[3]Universitas Pendidikan Indonesia, Department of Physics Education, Bandung, Indonesia40154

Email: lukmanhakim1976@gmail.com

Abstract. Teacher competence development is a national education agenda. Although teachers haveadequate learning experience, based on *UKA* (Academic Competence Test) 2013 results, the content mastery of teachers is still low. In order to reach the maximum developmentof teacher, it is a must to consider the knowledge level of teachers andthe difficulty of content given. This study used a questionnaire given to 40 teachers but only 25 teachers who returned the questionnaire. According to the research, 82% of teachers stated that the electromagnetic is a difficult content. There are several factors why electro magnetic content is considered to be difficult by teachers such as it is abstract, uses a lot of mathematical equations, and correlation with other concepts and content material. From these results, adaptive e-learning design for teacher to learn electromagneticis created.

1. Introduction

Competence development of teachers is a national education agenda in improving the nationwide quality of education. Nationally, teachers have the professional competence standards that include the mastery of subject taught. Decent subject mastery will assist teachers in determining methods or strategies that can be conducted properly so that it can assist students in achieving the expected learning objectives.

In order to improve teacher professional competence, various activities have been carried out by the government such as teacher training. However, these activities have not been able to resolve the education problems occurred at national scale.According to Loucks-Horsley, science teacher training programs are less successful due to lack of attention to the background, experience, knowledge, beliefs and needs of teachers [1]. It can be seen from UKA (Academic Competence Test) of high school physics teacher in South Sumatra in 2013 that are grouped into three categories: high (> 75), moderate ($50 \leq x \leq 75$) and low (<50) (Source: P4TK Bandung).

Therefore, improving the effectiveness of teacher training in mastering physics content, the content given should be in accordance with the level of knowledge and experience of teachers. The level of

[1]L Hakim: to whom any correspondence should be addressed.

knowledge of physics can be seen from the difficulties of physics content given. The level of knowledge can be seen from the level of difficulty of the content provided. Difficult content is given only to the students with higher level of knowledge while is given lower level. Some difficult concepts according to the students is to identify the source of the electric field like magnet and currents [2], used to, graphs, math such as algebra and geometry, complex mathematics, too many laws, principles and formulas, integrate to examples daily life [3, 4, 5]. In addition, experiments, mathematical equations, graph and explanation of the concept at the same time and the concept domain that is associated with other concepts such as force, terrain, electricity and so on [6, 7] so that more concepts involved, the more difficult comprehended.

The level of knowledge is determined by teststo classify students based on certain criteria. It uses Blooms taxonomy obtained from the tests and classified into beginner, intermediate, advanced and expert [8, 9]. The concept of physics tends to be hierarchy[10] from the level of difficulty, if a teacher fails to answer questions at the beginner level, the teacher is grouped into beginner level because teacher does not master the basic concepts well. Therefore, the questions were designed based on novice teacher level of knowledge and thus the test scores can discriminate experts and beginners groups. Two adaptive learning system resources (TSAL) can help teachers construct the content of their subject by using students' personal information such as learning styles and learning behavior[11].

Therefore, the physics content design for teachers should be based on the use of mathematics, the correlation with other concepts and the depth or scope of content. Hence, teachers will learn the appropriate level of knowledge. Ability to customize the content learnt with the level of teacher content knowledge is referred to adaptive.

The purpose of this study is to determine teachers' difficulty in the electrical force and the electrical field as the basis in developing adaptive content based on the level of teacher knowledge. Less difficult level of concept or easy concepts is given to teachers at the novice group, while the concept with a higher level of difficulty is given to the expert groups of teachers. Grouping the teachers into expert and novices is based on test results.

2. Methods

The research was accompanied on July 2015 in Palembang. The questionnaires, which were utilized in, collect teachers' response toward the difficult content via Liekert Scales (4 scales). The questionnaire was given to teacher directly and then a few days later asked return or direct complete his/her response. A total of 40 copies of the questionnaires were given to teachers about teachers' difficulties in understanding the content of electromagnetic. The questionnaires use Liekert scale. Out of 40 copies of the questionnaire given, 25 teachers returned them. Design research to design content of adaptive e-learning based on difficult content for teachers such as below:

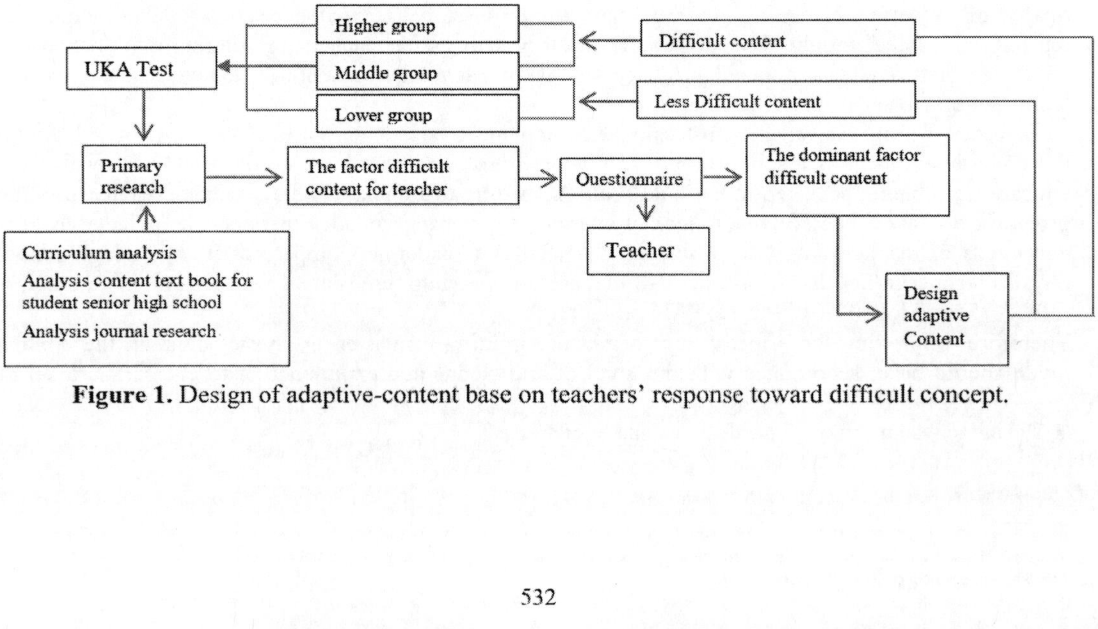

Figure 1. Design of adaptive-content base on teachers' response toward difficult concept.

3. Results and Discussions

The questionaires analysis gives result as follow:

Table 1. Teachers' response towards electro magnetic content.

The difficulties of electromagnetic concept	Response				% teachers' response
	SA[a]	*A*[b]	*TS*[c]	*STS*[d]	
The abstract concept	17	5	3	0	89
The use of higher level of math	10	8	4	3	75
Too many recitation and formula	12	8	5	0	82
Too many operations	15	8	2		88
Too theoretical	9	5	7	4	69
A lot of unfamiliar terms	16	5	4		87
A lot of sub-topic and terms used	14	8	3		86
Assosiated with other concepts	11	8	6		80
Average					82

[a]SA = Strongly Agree
[b]A = Agree
[c]DA = disagree
[d]SDA = strongly disagree

Based on Table 1, 82% of teachers or teachers in general responded that the electromagnetic concept is difficult [12]. Most of the teachers stated that the concept of electromagnetic is difficult because it is abstract. Abstract concept is a concept whose examples in everyday life are hardly found. To help students understand abstract concepts then a model concept is created so that it looks more real. Because the purpose of learning is to help students understand both abstract and concrete concepts, the content presented for the learners both expert and novice group is assisted with the model or illustrations to be more concrete. Then, the content presented does not distinguish the novice and experts group since it is an abstract concept.

Furthermore, the use of higher mathematics highlight used many mathematics calculations and formulas are employed in the electric force and electric field material. Learners have difficulty using mathematics in the electric force and electric field. A lot of sub topic and unfamiliar term cause the concept difficult. A lot of sub topic add load to learn so purpose more time to understand and unfamiliar term cause content strangecan it cause bore. The same thing is applied when dealing with the association of electromagnetic concepts with other concepts. Force two particles charge are hung by rope that effected gravitation more difficult than only force two particles on the floor because electrics force no effecting by gravitation. Therefore, based on these results, there are three major difficulties experienced by learners in learning electromagnet, namely the use of mathematics, concepts assosiation with other concepts and variables involved. Therefore, adaptive content are designed to accommodate these difficulties in planning the electromagnetic content covered

3.1. The use of mathematics

Integral: the use of integral in electrical magnet mainly atcontinuous charge electric field like line charge distribution, a ring of charge, balls and so forth. So, the concept of field caused by particular charge is given to the expert groups. Meanwhile, the electric field caused by point charges, the basic theory of electric field are given to the novice and expert groups. Thus, the participants can learn the basic theory of the electric field with a few mathapplications so that after completing the content. Consider the example of the application of integral in an electric field concept below.

Both groups receive the content of electric field caused by the point charge. The value of electric field can be measured by placing a very small test charge. It will cause the electrostatic force between the two charges:

$$\vec{F} = q_0 E \tag{1}$$

so

$$\vec{E} = \frac{\vec{F}}{q_0} = \frac{1}{q_0}\left(\frac{1}{4\pi\varepsilon_0}\frac{qq_0}{r^2}\right)\hat{r} = \frac{1}{4\pi\varepsilon_0}\frac{q}{r^2}\hat{r} \tag{2}$$

As for the continuous charge distribution such as line charge with 2l length and λ charge density, the value of field charge within r is as shown by figure 1 below:

$$E = E\cos\theta = \int \frac{1}{4\pi\varepsilon_0}\cos\theta \frac{dq}{l^2 + r^2} \tag{3}$$

Solving these equations requires the integral ability so more difficult than charge point and thus the electric field caused by continuous charge is only given to expert group.
In physics, vectors has two known quantities based on the direction of motion. The vector magnitude or vector have its own operating system differs from scalar magnitude operating systems. Vectors used in electric style content is as the following equation:

$$\vec{F} = \frac{qq_1\left(\vec{r} - \vec{r_1}\right)}{4\pi\varepsilon_0\left|\vec{r} - \vec{r_1}\right|^{3/2}} + \frac{qq_1\left(\vec{r} - \vec{r_2}\right)}{4\pi\varepsilon_0\left|\vec{r} - \vec{r_2}\right|^{3/2}} \tag{4}$$

3.2. The correlations with other concepts
Electromagnetic concept can be connected with other concepts such as the motion of particles in an electric field, the equilibrium particle influenced by the electric and gravitational forces. To ensure that participants have comprehended the concept of force and the electric field, in the novice group electromagnetic concept has not been linked with other concepts while at expert groups, the concept is linked with other concepts. The relation electrostatic force to gravitation force

$$\overrightarrow{F_g} = \overrightarrow{F_e} \tag{5}$$

The solution is given more difficult with present gravitation force than without gravitation

3.3. The scope or depth of material
Various applications of electromagnetic concept magnets in various technologies are given to the expert group. The reason is the expert group requires more insight to increase their knowledge. Concepts such as force and torque on the dipole are only given to expert group while the novice group only learns electric field on the dipole and thus the novice group learn adequate subject. In line with researchers who interested in doing research about electromagnetic concept such as [12, 13], they stated that electromagnetic was one of challenging concept and it could be aided by multimedia and computer simulations.They also stated that scope and depth of content on electromagnetic should be definitely focused to the content. Furthermore, Kaniawati et al. [14] utilized computer simulation as a part of e-learning in physics context to analyze the conceptions in physics learning because e-learning need adaptive-content that could facilitate students' understanding. Although the adaptive-content utilized by Kaniawati et al. [14] was in computer format then paper based, both the adaptive-contents could facilitate students in learning physics concepts.

4. Conclusions

There are three difficulties teachers had in comprehending electromagnetic concept namely excessive use of mathematical equations, correlation of electromagnetic concept with many other concepts, and scope of content that must be considered for adaptive content. The design content would be implemented in creating adaptive e-learning.

5. References

[1] Chval K, Abell S, Pareja E, Musikul K and Ritzka G 2008 *Eurasia Journal of Mathematics, Science & Technology Education* **4**(1) 31
[2] Huang Y, Yang B W, Adams R, Howell B, Zhang, J Zand Burbank K 2008 *Proceedings of The 2008 IAJC-IJME International Conference (New York City)* vol. 1 (New York: IAJC-IJME) p 209
[3] Tseng J C R, Chu H-C, Hwang G-J, Tsai C-C 2013 *Journal Computers & Education* **51** 776
[4] Roussel H M H 2012 *Advanced Electromagnetics* **1**(1) 65
[5] Erinosho, S Y 2013 *International Journal for Cross-Disciplinary Subjects in Education* **3**(3)1510
[4] Angell C, Guttersrud Ø, Henriksen E Kand Isnes A 2004 *Science Education* **88** 683 [7] Ornek F, Robinson W R and Haugan P 2008 *International Journal of Environmental and Science Education* **3**(1) 30
[5] Guisasola J, Almud'i J M and Zubimendi J L 2004 *Science Education* (doi: 10.1002/sce.10119)
[6] Bhaskaran and Swaminathan F 2014 *Research Journal of Applied Sciences, Engineering and Technology* **7**(16) 3298
[7] Nousiainen M and Koponen I 2010 *Nordina* **6**(2) 155
[8] Huang S-Land Shiu J-H 2012*Educational Technology & Society* **15**(3) 214
[9] Samsudin A, Suhandi A, Rusdiana, D, Kaniawati, I and CoştuB 2016 *Asia-Pacific Forum on Science Learning and Teaching***17**(1) 1
[10] Samsudin A, Suhandi A, Rusdiana, D and Kaniawati, I 2016 *Journal of Physics: Conference Series***739** 01
[11] Kaniawati I, Samsudin A, Hasopa Y, Sutrisno A D and Suhendi E 2016 *Journal of Physics: Conference Series***739** 01

Acknowledgment

This research was monetarily held by "Hibah Disertasi Doktor DIKTI" Research Grants in the monetary year 2016.

Genetic Pedagogical Content Knowledge (PCK) Ability Profile of Prospective Biology Teacher

W Purwianingsih*, E Muthmainnah and T Hidayat

Biology Department, Universitas Pendidikan Indonesia, Jl. Dr.Setiabudi No. 229, Bandung 40154, Indonesia
Corresponding author: widipurwianingsih@upi.edu

Abstract. Genetics is one of the topics or subject matter in biology that are considered difficult. Student difficulties of understanding genetics, can be caused by lack of understanding this concept and the way of teachers teach. Pedagogical Content Knowledge (PCK) is a way to understand the complex relationships between teaching and content taught through the use of specific teaching approaches. The aims of study was to analyze genetic PCK ability profile of prospective biology teacher.13 student of sixth semester Biology education department who learned Kapita Selekta Biologi SMA course, participated in this study. PCK development was measured by CoRes (Content Representation). Before students fill CoRes, students are tested mastery genetic concepts through a multiple-choice test with three tier-test. Data was obtained from the prior CoRes and its revisions, as well as the mastery concept in pre and post test. Results showed that pre-test of genetic mastery concepts average on 55.4% (low category) and beginning of the writing CoRes, student get 43.2% (Pra PCK). After students get lecture and simulating learning, the post-test increased to 63.8% (sufficient category) and PCK revision is also increase 58.1% (growing PCK). It can be concluded that mastery of subject matter could affects the ability of genetic PCK.

1. Introduction

Genetics is one of the topics or subject matter in biology that are considered difficult, both by students or teachers. This caused many sub topics in genetics contains concept that has character inaccessible, abstract and composed of a complicated structure. Student difficulties to learn genetics of which were related to the structure (gene, gamet, allel etc.) and process (cell division, crosses, etc.) [1]. For example, the concept of DNA, protein synthesis and heredity are concepts which have been considered difficult by students. Students have low understanding of these concepts and possibility have misconceptions. Examples of misconceptions is about character of gene. Students assume that one gene determines one trait, the gene is mutated always cause disease, and the gene is always expressed in the form of the phenotype [1]. Student difficulties of understanding the concepts of genetics, can be caused by lack of understanding of the concept and the way of teachers teach.

Briefly, PCK can be interpreted as an illustration of how a teacher to teach a subject to access what he knows about the subject and what he knew about the learner that he teaches, what he knew about the curriculum related to the subject and what he believes as way of good teaching in the context of [2]. Another understanding shows that PCK is a way to understand the complex relationships between teaching and content taught through the use of specific teaching approaches and it is developed through the process based practices in the classroom. Thus it can be said that PCK is also an integration between the knowledge of the subject matter (subject matter knowledge / SMK) with pedagogical knowledge (PK) held by teachers as a way of improving student learning [3]. Teacher PCK can be evaluated and measured using an instrument called CoRes (Content Representations) and PaP-eRs (Pedagogical and Professional experience repertoires)[3]. Analysis of the CoRes and PaP-eRs demonstrate the effectiveness of learning activities in accordance with the objectives to be achieved.

Canbazoglu [4] investigated the relationship between subject matter knowledge and PCK prospective science teachers. The results showed that the knowledge of the subject matter prospective teachers to the concept of the particulate nature of matter is very low. The low of subject matter knowledge of prospective teacher can impact on the process of transferring knowledge to students, teaching strategies, and assessment techniques. Thus, it can be said that subject matter knowledge of teacher or prospective teacher is closely connected with PCK.

Knowledge of teachers in learning activities in the classroom has a close relationship with how well students are learning [5]. The ability of understand content is the one of important goals in learning. To achieve this learning content must be formulated by teacher in proper PCK. Discussing Pedagogical Content Knowledge, understanding of concepts held by teachers is very important to know the extent of they understanding of the content material that will be taught. Teachers use a variety of its knowledge in conducting learning activities. Knowledge of a teacher is something very complex because it is influenced by the knowledge of life history, background experiences, emotions and goals [6].

CoRes can provide a picture of how a teacher conceptualize a particular subject matter. CoRes was developed by asking teachers to think about what they consider the "big ideas" related to learning activity on a topic on a certain level of education based on their experience in teaching [7]. The big idea that has become the horizontal axis is then examined with eight questions listed on the left side of the vertical axis cores, namely 1) What do you want students to learn from this idea ?; 2) Why is it important to know the students ?; 3) The other thing from this material that you know, but not yet known by the students; 4) Difficulties / limitations associated with this way of teaching material; 5) Knowledge of students' thinking affects you in teaching this material; 6) Other factors that affect the way you teach this material; 7) The procedure of teaching (and the specific reasons for its use); and 8) a specific way to ensure students' understanding or confusion regarding this matter. The study results between the vertical and horizontal axes can be specific information that affects the way content is taught explicitly [3]. Cores generally consists of a set of ideas about a particular topic, so that he can help the teacher because they can relate how, why, and what content to be taught, with the concept of what they consider important in learning.

Kapita Selekta Biologi SMA is one of course that develop skills and equip students apply the concept to pedagogy specificof Biology by considering the nature and characteristics of concept on appropriate pedagogy as implementation PCK. So it can be stated that in this course the student can be facilitated to integrate content knowledge and pedagogical knowledge as PCK. Genetics is one of the material covered in this course, so to analyze development of student PCK, research conducted through this course. The results of this study are expected to provide an overview of PCK student biology teachers especially in genetics.

2. Experimental Method

This study used descriptive method with triangulation techniques. Data consists of qualitative (CoRes) and quantitative (score mastering concept). Data taken simultaneously, compare the results, and then used to find the validity of each other. Participant in this study were 13 student at Department of Biology Education sixth semester UPI (Universitas Pendidikan Indonesia) who are conducting Kapita Selekta Biologi SMA course.

PCK development measured by CoRes by Loghran [2]. The big idea of CoRes has become the horizontal axis is then examined with eight questions listed on the left side of the vertical axis cores, namely 1) What do you want students to learn from this idea ?; 2) Why is it important to know the students ?; 3) The other thing from this material that you know, but not yet known by the students; 4) Difficulties / limitations associated with this way of teaching material; 5) Knowledge of students' thinking affects you in teaching this material; 6) Other factors that affect the way you teach this material; 7) The procedure of teaching (and the specific reasons for its use); and 8) a specific way to ensure students' understanding or confusion regarding this matter. CoRes has given before and after learning. In addition, to access the content knowledge of student biology teachers, they given test of genetics before and after learning.

Cores assessment done by coding refers to Widodo [8] as shown in Table 1. To test mastering concepts, we used 10 questions of Threetier-test pattern about genetics.

Table 1.Coding Category of PCK

No.	Total Score Coding	Category
1	0-9	Pre PCK (Level 1)
2	10-13	Growing PCK (Level 2)
3	14-18	Maturing PCK (Level 3)

Source: Adapted from [8]

To categorize or coding the answers of mastery concepts test, used criteria refer to Baird [9] (tabel 2.)

Table 2.Category of Answering Test

Category	Answer	Score	Code
Understood Concept	Right answer+ right reason + sure	3	UC
Less Understood Concept	Right answer+ right reason+ not sure	2	LUC
Not Understood Concept	False answer+ false reason+ not sure	0	NUC
Guess	False answer + right reason +not sure	1	False
	right answer + false reason+not sure	1	

Source: Baird [9]

Pretest and posttest results data from the 13 participants collected and calculated its N-gain. Similarly, the data CoRes before and after lectures calculated its N-gain. Gain normalized scores are categorized into: high, medium and low. Normal gain is calculated using the formula:

$$g=\frac{S_{pos}\text{-}S_{pre}}{S_{maks}\text{-}S_{pre}} \quad (1)$$

where:

S_{pre}= Score pretes;

S_{pos}=Score postes;

S_{maks}=Score ideal maksimum

To equalize the value of the test concept and value of PCK, both value is create per cent using the formula :

$$\frac{PCK\ score}{maximal\ PCK\ scores} \times 100\% \quad (2)$$

The value of test mastering genetic did the same way.

Table 3.Score categorization

Percentage (%)	Category
86-100	Very good
76-85	Good
60-75	Sufficient
55-59	Low
≤54	Very low

Source: Kind [11]

Table 3 show that score categorization have obtained the value of PCK and content value (%),this categorization refers by Kind [11].

3. Result and Discussion

Based on results of CoRes and mastery concepts, we get data shown in table 4 and clarified by Figure 1 and Figure 2.

Table 4. Result of mastering concept of genetics and PCK before and after the lecture

Aspect	N	Pre Test mean (%)	Category	Post test mean (%)	Category	N gain
Mastering Concept	13	55,4	Low	63,8	Sufficient	0,31
PCK	13	43,2	Pra	58,1	Growing	0,38

Table 4 show that level of mastery concepts before lecturing of prospective teacher are at low level (55.4 %) . Besides, the value of PCK prospective teachers before lecturing is still at level of pre PCK (43.2 %). Lecturing on Kapita Selekta Biologi SMA include strengthening concepts through lectures by professors, assignment of create the Concept Map, Pra RPP and simulation learning by the students followed by reflection. Such learning is expected to equip the PCK student mastery of concepts at once, and it is hoped that test results and student PCK can be improved. Based on data in Table 4, it can be stated that the mastery concepts and PCK of students increased respectively to 63.8 % (sufficient category) and 58.1 % (Growing category). From Table 4 we can also see N gain obtained on the mastery of concepts and PCK respectively by 0,31 and 0,38.

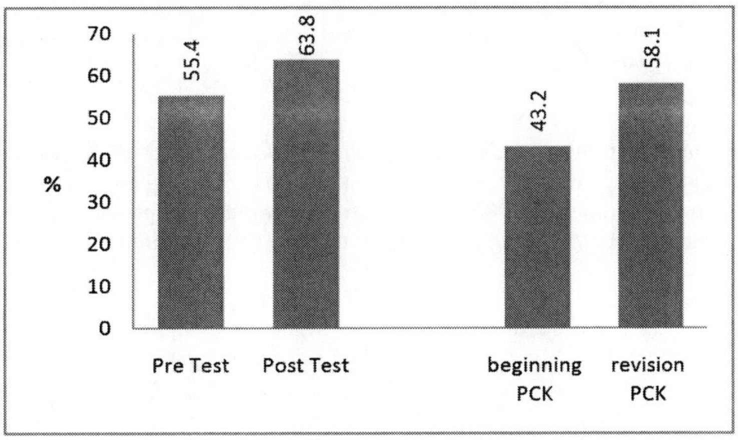

Figure 1. Score (%) of mastering concept & PCK, before & after lecture.

Figure 1 show that improvement of score mastering concept & PCK before and after lecturing. On Genetics material, it can be seen that an increase of ability PCK from category *Pre* to *Grow*, whereas on understanding concepts Genetic material is also an increase from *Less* to *Sufficient*. Based on the this data, it can be stated that ability of genetics PCK of biology prospective teacher is closely related to the mastery concept.

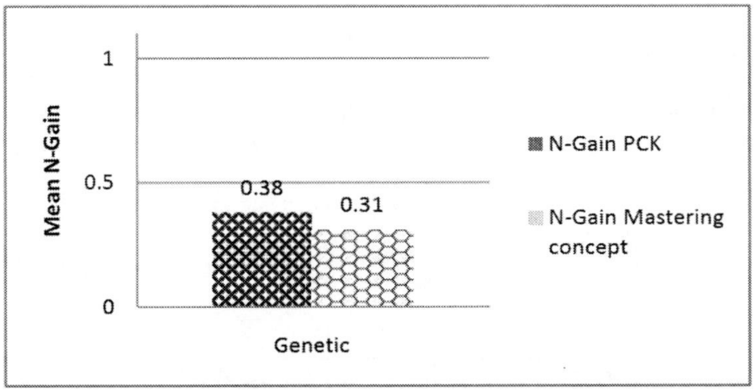

Figure 2. Comparison Values of N-gain mastering concepts and PCK

Figure 2 show that comparison value of N-gain mastering concept and PCK. Based on these results, it can be stated that when students are not mastering the concept correctly, the ability to plan learning in the form set forth in PCK cores will also be low. This is in accordance with the opinion of Baird [9] which states that the pre conception is defined as prior knowledge possessed by students before learning the subject. These preconceptions influenced by the culture of the social environment, and teaching in the classroom [9]. Students tend to use pre-conceptions they have to master new knowledge received later. From this research can also be seen that the key concepts of genetics that include structures and processes were not initially well controlled with students, but after learning mastery of this material is sufficiently increased.

4. Conclusion

General conclusion of this study. it can be stated that ability of genetics PCK of biology prospective teacher is closely related to the mastery concept. Based on this research can also be concluded that the factors that influence development of PCK include strengthening concept, assignment of directing the planning and implementation of learning and reflection of the learning simulation results.

5. Acknowledgment

The authors gratefully acknowledge the input of those academic, technical and support staff from the College of Biology Education, Universitas Pendidikan Indonesia, especially to Dr. Riandi, M.Si and Dr. Adi Rahmat, M.Si, as validator.

6. References

[1] Topcu M S and Pekmez E S 2009 *J. Of Turkish Science Education* **6**

[2] Rollnick M, Bennett J, Rhemtula M, Dharsey N and Ndlovu T 2008 *Int J. of Science Education* **30** (10) 1365-87

[3] Loughran J, Berryn A and Mulhall P 2012 *Understanding and Developing Science Teachers' Pedagogical Content Knowledge 2nd Edition* (Rotterdam: Sense Publishers)

[4] Canbazoğlu 2010 *Elementary Education Online* **9**(1) 275-91

[5] Marzano R J 2012 *Becoming a Reflective Teacher* (Virginia: Association for Supervision and curiculum development)

[6] Newsome J G 1999 *Pedagogical Content Knowledge: An Introduction and Orientation.* (Boston: Kluwer Academic Press) p 3-20

[7] Shulman L 1987 *Knowledge and teaching: Foundations of the new reform.* (United state: Harvard Educational Review*).* **57**(1) p 1–22

[8] Widodo A 2014 *J. Ilmiah Pendidikan Tahun 2014* **XXX 1** p 58-72

[9] Baird K A, Brodie M M, Bevins S C and Christol P G 2007 *Science teacher education* **27** 339-46
[10] Meltzer D E 2003 *American J. of Physics.* **70** 7
[11] Kind V 2009 A conflict in your head: An exploration of trainee science teachers' subject matter knowledge development and its impact on teacher self-confidence. *Int. J. of Science Education* **31(11) p** 1529–62

IOP Conf. Series: Journal of Physics: Conf. Series **812** (2017) 012088 doi:10.1088/1742-6596/812/1/012088

The Development of Monograph with 3-Dimentional Illustrations Titled "Augmented Chemistry: Hydrocarbon"as Learning Enrichment Materials

D Ernawati and J Ikhsan

Department of Mathematics Education, Universitas Negeri Yogyakarta, Jl. Colombo No. 1, Caturtunggal, Yogyakarta 55281, Indonesia
Email: dianerna15@gmail.com

Abstract. The development of 3D technology provides more advantages in education sectors. In chemistry, the 3D technology makes chemistry objects look more tangible. This research developed a monograph titled "Augmented Chemistry: Hydrocarbon" as learning enrichment materials. The development model consisted of 5 steps, which were the adaptation of the ADDIE model. The 3D objects of chemistry were built using the computer applications of ChemSketch, and Google SketchUp with ARPlugin. The 3D objects were displayed by relevant markers on the texts of the monograph from which the visualizations of the 3D objects appeared when they were captured by digital camera of laptop or smartphone, and were possibly viewed with free-rotation. Not only were 3D chemistry objects included in the monograph, but also graphics, videos, audios, and animations, which facilitated more fun learning for readers of the monograph. After the reviews by the experts of subject matter, of media, of instruction, and by peers, the monograph was revised, and then rated by chemistry teachers. The analysis of the data showed that the monograph titled "Augmented Chemistry: Hydrocarbon" was in the criteria of very good for the enrichment materials of Chemistry learning.

1. Introductions

The use of technology in education enables the learning process to be more active, motivating, attractive, meaningful, even brings new psychological influence to the student. Augmented Reality (AR) is an advanced technology that enables user to interact between virtual and real world in real time [1]. Augmented Reality can display the real world although the multimedia of text, graphic, video and audio that can improve learning experience of student [2]. The another advantage of the use of AR into learning was the ability of AR to display data and information in real time to enhance 3D perspective skill of users [3].

In chemistry, AR technology makes chemistry objects look more tangible, especially in learning and understanding molecular structures. Molecular structure can be visualized in 3-dimention, which adds real sensation and experience for students. Students enjoyed it and gained more knowledge of molecular structures. It substantially improves their spatial intuition and learns to better understand visual cues [4].

The purpose of this research is developing monograph titled "Augmented Chemistry: Hydrocarbons" as enrichment materials. "Augmented Chemistry" is the use of AR technology in chemistry. "Hidrokarbon" is chemistry concept. The book is meant to be intuitive and include 3D modeling, video, audio and the discussion of broader and deeper. With integrated AR technology into the book, that can solve disadvantage of 2-dimention representation on the book. So, it is expected that the book make hydrocarbon concepts is able to clear and more tangible.

2. Theoritical background

2.1. Chemistry learning

Learning is essentially a process of interaction between students with the environment, which changes the behavior of students going to a better direction. Learning interactions influced many factors both internal factors of the individual and the external factors of the environment. In learning, the most important of teacher's task is to condition the environmental in order to support behavior change for student [5].

Chemistry is an abstract and imaginative knowledge, but very useful for human life. Chemistry is the knowledge that involves skill and reasoning to searching for answers to the what, why, and how natural phenomena are interrelated [6].

Chemistry learning is giving learning experiences that involve the skills and reasoning to reveal how natural phenomena are interrelated. Chemistry learning can be effective if it takes place in interactive learning activities.

2.2. Learning media

Media is the mean for transmitting or delivering messages. Media is also to delivering content to student, to achive effective instruction [7]. Learning should manipulate the resource and processes, so the learning objective can be achived effectively. Therefore, teachers should use suitable instructional media to facilitate student as subject learning and teacher as the subject of educator [8].

Instructionall media are a variety of things that are used by teachers which can delivering the message, stimulating thoughts, feelings, and the willingness of students to encourage the learning process on the students.

2.3. Augmented reality (AR)

AR is real visualization methods that enable natural interaction to improve the user experience. The another advantage use AR was to improve 3D perceptive skills of the user by displaying the data in real-time information [3]. AR at simplifying the user's life by bringing the virtual information to immediate surroundings and the real-world environment, like live-video stream [9].

The advantages that are possessed by AR technology are the interaction seems so real, less expensive implementation, system delays shown fewer possibilities, user's imagination are more increased. In addition to the various advantages, AR technology still has some shortcomings, such as technology can only be used on a few specific Operating Systems, technology is currently developing more profitable Virtual Reality, users cannot distinguish between virtual objects and real objects with ease, the production facilities of the design as a whole are less supportive environments [10].

2.4. The role of 3-dimentional monograph

To increase knowledge and insight of students, teachers can teach students to read reference and enrichment books [11]. Enrichment books became one of the companion books to be a supplement for teaching materials that are still restricted to textbooks. enrichment books its enrich of certain aspects of the textbook, the material does not directly tied to the curriculum but still supports, as well as the target or group of readers cannot be classified explicitly [12]. The characteristics of enrichment books are the material does not have to be tied directly to the curriculum; target readers are not explicitly restricted although aspects of the look and content of eligibility envisaged readership; the presentation can be prose with pictures; generally use popular language; the material grow and develop cognitive, skills, attitudes, moral, and personality aspects.

These books contain broaden and deepen material, illustrations, sheet activities and exercises which support the current curriculum. Based on the contents of the book student can develop active thinking and motivate interest in independent study

3. Research methods

This development adopts ADDIE model. ADDIE model is a method of instructional design developed by Dick & Carry which consists of five stages.

3.1. Analysis

In this step analysis is the determination of the benefits and goals of monographs, a review of standard competence and basic competence of chemistry learning especially to hydrocarbon concept, analyzing basic competence became indicators and referencing.

3.2. Design

This step is about making monograph storyboard, preparation of resource materials, and manufacture marker by GIMP 2 and 3D molecular models by ChemSketch, and making research instruments.

3.3. Development

In this step, drafting monograph using Adobe InDesign; making object 3D by SketchUp; combine the Augmented Reality file, marker, and monographs; reviewed by the supervisor, expert of subject matter, of instruction, of media, and by peers to get a correction and suggestion. After that, product revised and print to assess by high school chemistry teacher.

3.4. Implementation

Rate the quality of products by 5 high school chemistry teachers using a quality assessment questionnaire of product and suggestion sheet. Assessment questionnaire consists of 6 aspects such as the feasibility of the subject matter, presentation, layout, language, software design and visual communication. These aspects are translated into 33 indicators.

3.5. Evaluation

The data from implementation stage is analyzed to determine the quality of products "Augmented Chemistry Hidrokarbon". The correction and suggestion data from chemistry teachers used to enhance final product "Augmented Chemistry Hidrokarbon".

4. Result and discussion

The result of this research was monograph entitled "Augmented Chemistry Hidrokarbon" provided 3D visualization that can be displayed through ARPlayer. Based on data, the average score of the book was 142.2 with percentage of ideality 86.67%. Teaching materials are feasible as learning sources if it has percentage of ideality at least 51% -75%. So the book can use as source of chemistry learning. The percentage of ideality for each aspect can be seen at Figure 1.

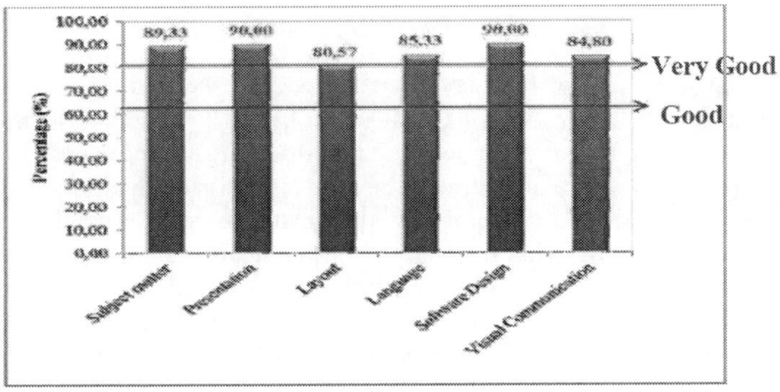

Figure 1. The percentage of ideality for each aspect.

Subject matter aspect consisted of suitability of the material with basic competence, the depth of the material, accuracy of the concepts and definitions, the breadth of material, actual, and the accuracy of the term. Reference [13] subject matter aspect which are generally assessed based on its relevance to the curriculum, the relevance of the material with the purpose of education, truth in terms of linguistics and humanities, and its compliance with the cognitive development of students.

Presentation aspects consisted of presenting examples of questions; exercise; conformity with the answer key practice questions; and cohernly presentation of learning indicators. Presentation aspects has very good quality and appropriate as learning materials for students. Presentation aspect which assessed in terms of the inclusion of learning objectives, learning stages, the attractiveness for students, easy to understand, ability to raise student activity, connectivity between the material and the availability of questions and exercises [13]. Materials of the monograph explained deductive, meaning that the material presented from easy to get a more difficult matter. This is expected to facilitate the learners in understanding the overall concept.

Layout aspect consisted of the suitability the size of the module; the letters color; layout front cover, back and spine; typeface, font variations, the field of print and margins; and the accuracy of the pictures, diagrams, and illustrations. The size of the book is A4 with 80 gsm of thickness. The size and the thickness selestion of the paper based on the effectiveness of using teaching materials in the classroom. The size of A4 is the standard size of ISO and 80 gsm is an ideal thickness of paper, so the paper does not easily tear if repeatedly inverted by students [14]. Layout design of cover, back and spine were harmonious. The colors used are orange. The aim is to attract student from the beginning through the cover books.Color is one of methods used as a guiding tool and attract the reader's attention [15]. Cover layout can be seen in Figure 2.

Figure 2. Cover layout.

Typing field were 1.5 cm from the bottom and 2.5 cm from the top, the right and the left edges of the paper. Line spacing was 1.5 cm. Font style use varied, but the majority used Times New Roman with 12 pt. Good font size for text (text book or handbook) is 12 pt [15]. Selection of the text colors, images, and illustrations in the book ware good so its increase the attractiveness of the reader when see the monograph. A book will be more attractive if they are equipped with colors and pictures. [16]

Language aspects of consisted of suitability the materials with knowledge level of students; communicative; and unambiguous. This book used communicative language according with the level knowledge of students, does not cause a double interpretation and used scientific language but not rigid, so it's flexible. That mean, not all parts of this book uses scientific language. It aims to facilitate students in understanding the material. The scientific language deemed not familiar described in the glossary. Language categorized good if it was arranged by Communicative sentence, using good and true Indonesian, does not cause a double interpretation, and does not offend the student[11].

Software design aspect consists of media effectiveness, reliability; compatibility; clarity of the instructions for use of media, creativity and innovation, usability, clarity workflow application usage; and opportunities to the development of science and technology. Augmented Reality technology has not been widely used in education, especially in the utilization combined with printed books. Quality teaching materials should be able to follow the development of science and technology [17], and in the use of media-based instructional technology, the software should be easy to use [18]. The use of AR Technology are very promising when it used in education.

Visual communication aspect consisted of communicative applications; suitability of audio; the attractiveness of the animation; function of navigation buttons; and clarity of illustration. This product ware combines text, graphics, animation, sound, and video. The 3D model visualization can be seen in Figure 3.

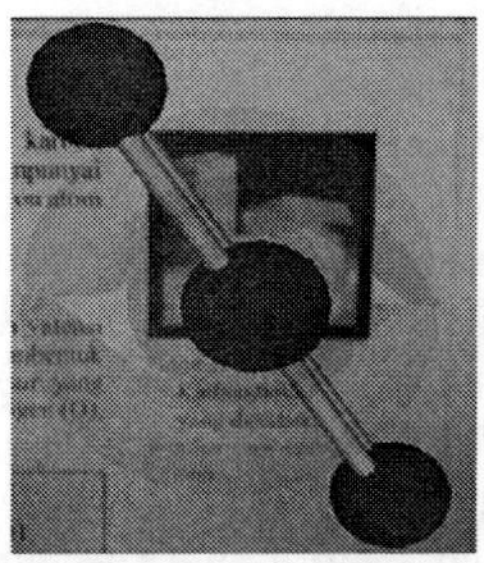

Figure 3. 3D model visualization

Varied media applications can present the information in a form that is fun, exciting, easy and clear to understand. Information will be easy to understand for as many senses, especially the eyes and ears, used to absorb the information [15]. The use of multimedia in this book is expected to increase the level of knowledge the students to understanding hydrocarbons concept.

5. Conclusion

This research produce of monographs "Augmented Chemistry Hydrocarbon" as a chemistry enrichment book. Monographs are include with 3D objects visualization that can be displayed using a computer, laptop, or smartphone at least version 4.0.3. Based on the assessment by senior high school chemistry teacher, it can be concluded that the monograph has a very good criteria to used as learning resources.

6. References

[1] Azuma R T 1997 *MIT Press Journal* **6** 355-85
[2] Neha T and Archana M 2015 *Journal of Engineering Education Transformation* 188-91
[3] Siltanen S 2012 *Theory and Applications of Marker-based Augmented Reality* (Espoo: VTT Technical Research Centre of Finland)
[4] Singhal S, Bagga S, Goyal P and Saxena V 2012 *International Journal of Computer Application* **49** 1-5
[5] Mulyasa E 2006 *Kurikulum Berbasis Kompetensi* (Bandung: PT Remaja Rosdakarya)
[6] Suyanti R D 2010 *Strategi Pembelajaran Kimia* (Yogyakarta: Graha Ilmu)
[7] Naz A A and Akbar R A 2010 *Journal of Elementary Education* **XVIII** 35-40
[8] Sadiman A S, Rahardjo R, Haryono A and Rahardjito 2011 *Media Pendidikan: Pengertian, Pengembangan dan Pemanfaatannya* (Jakarta: PT Rajawali Press)
[9] Furht B 2011 *Handbook of Augmented Reality* (New York: Springer)
[10] Patkar R S, Singh S P and Birje S V 2013 *International Journal of Advanced Research in Computer Science and Software Engineering* **3** 64-9
[11] Mendiknas 2008 *Peraturan Mentri Pendidikan Nasional RI No 2/ 2008 about Book* p 31
[12] Depdiknas 2007 *Pedoman Penilaian Buku Pengayaan Pengetahuan No 77B/A8.2/LL/2007* p 12
[13] Hanifah U 2014 *Jurnal At-Tajdid* **3** 99-121
[14] Purwanti I and Sunariningsih 2015 *Jurnal Pendidikan Bisnis dan Manajemen* **1** 11-23
[15] Arsyad A 2015 *Media Pembelajaran* Edisi Revisi (Jakarta: Rajawali Press)
[16] Sulistyani N H D, Jamzuri and Rahardjo D 2013 *Jurnal Pendidikan Fisika* **1** 164-72
[17] Situmorang M 2013 *Prosiding Semirata FMIPA Universitas Lampung (Lampung)* vol 1 237-41
[18] Saputro W and Purnama B E 2012 *Journal Speed-Sentra Penelitian Engineering dan Edukasi* **4** 60-7

Inductive-Deductive Approach to Improve Mathematical Problem Solving for Junior High School

Mariam Ar Rahmah

Email: mariamarrahmah@gmail.com

Abstract. This research is experimental quantitative about quasi experiment that emphasize on improve mathematical understanding and problem solving for Junior High School students by inductive-deductive implementation. The population of this research are all of students at IX degree students in Subang 2012/2013. Two of nine classes were chosen as sample for this research. The topic which used is probability including random occurrence, basics of chance, relative frequency, calculation of probability, determining the value of probability, the expected frequency and the combined odds of two events. The instrument that was used are test and non-test. Mathematical understanding and problem solving were used as a test methods. Meanwhile questionnaire, and observation sheet were used as non-test methods. The data was analyzed by Mann-Whitney and t-test. According to whole analyze in this research, can be concluded: 1) the student improvement of mathematical understanding by using inductive-deductive approach is in middle quality, 2) there is no significant difference at improvement mathematical understanding between experimental class and control class, 3) the improvement of student's ability in mathematical problem solving that use inductive-deductive approach has a low quality, 4) there is no significant difference at mathematical problem solving between experimental class and control class, 5) most of students has positive responses to mathematic learning by inductive-deductive approach, although the students have many problems when learning takes place.

1. Introduction

In 21st century is a globalization era, the era of resulting technology products with abundant quantity and quality that increasingly sophisticated and dissemination of information flow is increasingly and unstoppable. This has led to intense competition among individuals who have skills and ability to think critic, systematic, logic, creative, and able to communcate creative ideas that will be part of it.

To create human who meets the above characteristics, it can be achieved through education, such as through mathematics education. By studying mathematics as a whole, the students will be able to have ability of understanding, communicating, connecting, reasoning, problem solving, logical thinking, systematic thinking, critical thinking, and creative thinking. This is because mathematics is a means of thinking, so that mathematics can be regarded as a "vehicle" for developing ability to think logic and higher cognitive skills for children [1].

Nevertheless, the research that examined on this study was on mathematical problem solving aspects. This was because mathematical problem solving ability is an ability that considered important on learning, as proposed by Sabandar [2], that is mathematical problem solving is an ability that must be achieved, and improved mathematical thinking are the priority objectives of mathematics.

In addition to the opinion stated above, the importance of this ability can be seen from the regulation of the *Minister* of National *Education* of the Republic of Indonesia Number 20 of 2006 on the Standards Content [3] stated that the purpose of study mathematics is to make learners have the following capabilities:

- Understanding mathematics concepts, the relationship every concepts and apply concepts or algorithms as flexible, accurate, efficient, and precise in troubleshooting.
- Using the reasoning in the patterns and nature, perform mathematical manipulation in making generalizations, compile evidence, or explain mathematical ideas and statements.
- Problem solving, include the ability to understand the problem, contruct a mathematical model, solve the model and interpret the obtained solution.
- Communicating the ideas with symbols, tables, diagrams, or other media to clarify situation or problem.
- Having a good respect of the usefulness mathematics when solving the problem, also confidence in problem solving.

But the reality has found in the field that the level of achievement of mathematical problem solving ability is not satisfactory. It is based on research conducted by Nurhadiyati [4] to the junior high school students in the Bandung city. Generally, the results of mathematical problem solving ability of junior high school students is not satisfactory, around 30% - 50% from the ideal score. Similar opinion was also expressed by Ahmad [5] that is based on case studies in mathematics subject probability and statistics conducted on 41 students of second grade in SMP Negeri 2 Purwokerto, it was found that the students are still experiencing difficulties in resolving problems that related with the ability of understanding mathematical and mathematical problem solving.

The failure of students to achieve mathematical abilities above is not impossible causing the formation to negative course. According to Suherman [6] mathematics is a formation of affectives mathematics towards to the formation of cognitive area, although sometimes the opposite occurs. For example, a student who often feel able to solve mathematics problems, they enjoyed and desire to get more mathematics problems. Conversely, if they qite often can not feel disable, it will lead them to scare and shy. This is proven on the reality that most of students have negative affectives to mathematics. As stated by Muijis and Reynolds [1] mathematics is usually regarded as the most difficult subjects for children and adults. At school, many students seems to be uninterest on mathematics, and often there is doubt about the relevance of so much time spent for teaching mathematics.

These problems occur not only due to the students factors but also it could be occur because of insufficiency facilities and infrastructures when learning takes place, the educational environment not supporting, and also because teachers could not maximize their competency as teachers. From all factors, a slight incompetence teacher in teaching lead to an enormous impact on the students' lack on mathematical ability. Besides from Gage and Berliner's opinion [7], teachers should be able to play as a role, in charge of, and responsible for: (1) Planning, which should be prepare what is needed in teaching and learning process; (2) Implementing, which should be create a situation, leaders, stimulate, mobilize, and direct the KBM according to plan; and (3) Assessor, who should be collect, analyze, interpret, and make a consideration of teaching and learning process based on defined criteria.

In this condition, teacher not only prepare physical things (eg props) but also teachers have to prepare non-physical things, like mastery material that will be delivered through learning approach. The approach adopted is not based on self-interest, for example, practicality or the approach of "it" is the most controlled, but a teacher should use learning approaches that can stimulate interest and explore the knowledge of students, so it will impact on the emergence of students positive affectives towards mathematics. Such an approach is of course adapted to the material to be learned and the objectives to be achieved.

The ideal ways of learning that was expressed in Standard Process on the National Education Standards [8], is a process of learning in the educational unit organized in an interactive, inspiring, fun, challenging, motivating the students to actively participate and give enough room for innovation, creativity, and independence in accordance with their talents, interests, and physical and psychological development of learners.

Hamzah [9] suggested that in learning, students need to be active mentally, build a knowledge based on cognitive maturity they had. In other words, students are not expected to be like the little bottles ready to be filled with a variety of science in accordance with the will of the teacher. Meanwhile, Dahlan [10] also noted when the learning takes place, knowledge is not accepted passively. Knowledge gained through active activity in solving relationships, patterns, and make generalizations that are integrated in the new knowledge that obtained by the students, and learning is a social activity that occurs from the interaction of students with teachers and students with their peers. Such learning can be applied with inductive-deductive approach.

Inductive-deductive approach refers to activities undertaken by teachers so that teaching materials can be adapted by the students. Inductive-deductive approach according Mulyana [11] is the process of presenting a concept mathematical principle started by giving examples, followed by finding / construct the concept, constructing a conjecture, and ends with the givingexercise correspond to the stages of the concepts and principles that have been given. Through learning by using this approach students are trained to make generalizations.

To reachof making generalization stages, it needs capability to understand the relationship/linkages for given examples, problem solving plan, calculation process, and the process to re-examine the truth of the results obtained. These elements are an indicator of the ability of mathematical problemsolving. Based on the above, allegedly learning by using inductive-deductive approach could be improve students' mathematical problem solving ability.

2. Problem

Based on the introduction of the problems that have been described previously, the issues examined in this research are: (1) is the increase in mathematical problem solving ability of students to get math learning with inductive-deductive approaches are better than students who received conventional approaches? and (2) how is the students' affective using inductive-deductive approaches for mathematics learning?

3. Research Method

This research is a quasi-experimental research involving a group that has been formed to serve as the object of research with research design called non-equivalent control group design [12]. The population in this research were all students of third grade Junior High School in Subang Academic Year 2012/2013. Two of nine existing class was selected as research sample, that is the experimental group (the group that gets learning by using inductive-deductive approaches) and a control group (group with conventional learning). The formation of two classes aimed to determine the effect of learning mathematical problem solving ability. The instrument used is the ability tests, student questionnaire, and observation sheet.

4. Results And Discussion

Results and Discussion in this study are based on factors that were observed and found in the research include:

4.1 Description of Mathematical Problem Solving Ability Based on Learning (PSA)

Here are the results of descriptive statistics score students' mathematical problem solving ability.

Table 1 Descriptive statistic ofPSA score

Test	Experiment Group				Control Group			
	N	$\bar{X}$	S	$\%\bar{X}$	N	$\bar{X}$	S	$\%\bar{X}$
Pre-test	35	10,11	5,465	25,28	35	7,69	4,234	19,23
Post-test	35	14,66	4,065	36,65	35	11,17	5,366	27,93
Ideal Maximum Score: 40								

Based on Table 1 shows that the average score of the pre-test of mathematical problem solving ability to the experimental class is 10.11, while for the control group was 7.69 with early abilities difference in the second grade reached 6.05%. However, based on nonparametric statistical test Mann-U Whitne obtained Sig. (2-tailed) is 0.055> $\alpha = 0{,}05$ so that H0 is accepted. This means there is no difference in the average scores pre-test mathematical problem solving ability in the experimental class and control class.

While the average post-test score for an experimental class was 14.66 and the average for the control group was 11.17. Differences in the average scores in both classes is 3.52% with the highest average is in the experimental class. To prove that the mathematical problem solving ability score in experiment class is better than the control class there is a test of difference in the average score of N-gain using independent sample t-test, because the normal distribution of data and derived from a homogeneous variance. Here's a summary of the results obtained.

Table 2 Difference of Average Post test

t-test for Equality of Means		
T	**df**	**Sig. (2-tailed)**
3,152	68	0,003

Based on statistical tests above it can be concluded that there is the different mathematical problem solving ability between the students who got a math learning with inductive - deductive approach and conventional learning.

4.2 Enhance Mathematical Problem Solving Based Learning (PSA)

Table 3 below presents the average of mathematical problem solving abilities by learning N-gain.

Table 3 N-Gain Average and Clasification PSA

Class	**N-gain Average**	**Clasification**
Experiment	0,138	Low
Control	0,107	Low

From Table 3, although the classification of the enhancement in both classes is low, but the enhancement in mathematical problem solving ability of students in the experimental class is greater when compared with the increasing capability in the control class . To determine the significance then tested the nonparametric Mann - Whitney U because the results of the analysis show that the data are not normally distributed . Here's a summary of the results obtained.

Table 4 Difference of N-Gain Average Test

Statistic	Score	Explanation	Conclusion
Mann-Whitney U	521,000	Ho Accepted	Hipotesys rejected
Z	-1,081		
Asymp. Sig. (2-tailed)	0,280		
Asymp. Sig. (1-tailed)	0,140		

Thus, there are no significant differences between the average N-gain mathematical problem solving ability of students who got the inductive-deductive learning with students getting conventional learning. From the results, it can be concluded that there are no significant differences in mathematical problem solving enhancement abilities between students who study with inductive-deductive approach and conventional.

The discrepancy between the hypotheses are made with the results obtained made possible because of the inductive-deductive approach is less suitable for use in the classroom are the reasons:

First, students still unfamiliar implementing learning by using inductive-deductive approach. Workmanship in groups and provision of teaching materials about probability using tools to each student (in this case the coins, dice and cards bridge) was not enough to help the students to understand the material. Most of the students are still asking how to get a conclusion/ generalization of the examples above. Whereas Hudojo (Dahiana, 2010) states that, think math is a mental activity, which is in the process using generalizations. This indicates that the students thinking activity mathematically with inductive-deductive approach to the selected sample was not going according to what is expected. The students problems being unfamiliar with the way of learning using constructivist-based approach is confirmed by Dharmadasa (Muijis, D. and Reynolds, D., 2008), which suggests that a number of studies shows that many students thinks constructivist methods is quite difficult to be implemented. The same thing also expressed by Au and Carroll (Muijis, D. and Reynolds, D., 2008) stating that the teacher sees constructivist methods are burdensome and alarming effect on classroom discipline. They are not sure about the provision of appropriate materials, promoting experimentation, and started constructing the child's knowledge.

Second, there is lack of researcher that studies the class while the class is relatively large resulted not all students can be handled optimally. As a result, the students who have not been handled by researchers, conclusions/ formulas that have been made in the teaching materials is not based on an understanding of the premises above, it is obtained from the copying conclusion at the end of learning.

Third, the classic problem, is the limited time. Although researchers have designed a time as possible so that learning takes place in line with expectations, but the reality is different. Researchers do not have enough time to discuss the problems of mathematical problem solving for probabilities material, so there are still many students who are not accustomed to do problem solving mathematical problems contained in the post-test. This problem is recognized by Dharmadasa (Muijis, D. and Reynolds, D., 2008) based on the results of an informal interview to the implementation of programs designed to introducing constructivist methods shows that teachers consider a constructivist approach as a challenge and a concept hard to capture in a short time. These constraints may cause some differenciesof hipotesys and the result obtained.

5. Conclusions andRecommendation

Based on the results of research and discussion, it could be produced some conclusions as follows: (1) there is no significant difference between comprehension ability of mathematical problems solving in both of experimental group and control group; and (2) most of students have positive affective on learning mathematics by using an inductive-deductive approaches, even though in reality students have problems during and after the learning takes place.

The conclusion has been stated above, implies that: (1) learning mathematics by using inductive-deductive approaches could not improve mathematical problem solving skills significantly compared to conventional learning, and (2) learning mathematics by using inductive-deductive train students to socialize, communicate, and be unyielding before the goals are reached.

The recommendation after the course of this study are as follows: (1) Learning mathematics by using inductive-deductive can be applied in schools with high clusters and (2) teaching mathematics using inductive-deductive could be tested in a shorter period of time.

6. References

[1] MuijisD and Reynolds D2008*Effective Teaching (TeoridanAplikasi)*(Yogyakarta: PustakaPelajar)
[2] Kurniawan 2010 (Tidak ada di Referensi)
[3] RegulationNomor 22Tahun 2006tentangStandar Isi (Jakarta: Depdikbud)
[4] Anriani N*2011PembelajarandenganPendekatan Resource Based Learning untukMeningkatkanKemampuanPenalarandanPemecahanMasalahSiswa SMP Kelas VIII*(Bandung: Thesis PPS UPI)
[5] Ahmad2005*KemampuanPemahamandanPemecahanMasalahMatematikaSiswa SLTP dengan Model PembelajaranBerbasihMasalah* (Bandung: Thesis PPS UPI)
[6] Suherman., *et al*2003*StrategiPembelajaranMatematikaKontemporer*(Bandung: FPMIPA)
[7] Makmun AS2007P*sikologiKependidikan (PerangkatSistemPengajaranModul)* (Bandung: Rosda)
[8] National Education Standards (2009) (Tidak ada referensi)
[9] Hamzah2001*PembelajaranMatematikamenurutTeoriBelajarKonstruktivisme* [Online:http://depdiknas.go.id/jurnal/40/pembelajaran]
[10] Dahlan J A2004*MeningkatkanKemampuanPenalarandanPemahamanMatematikSiswaSekolahLanjutan Tingkat PertamamelaluiPendekatan Open Ended* (Bandung: Dissertation PPS UPI)
[11] Mulyana T2005Up*ayaMeningkatkanKemampuanBerpikirKreatifMatematikaSiswa SMA Jurusan IPA MelaluiPembelajarandenganPendekatanInduktif-Deduktif*(Bandung: Thesis PPS UPI)
[12] RuseffendiET2005 *Dasar-dasarPenelitianPendidikandanBidang Non EksaktaLainnya (*Bandung: Tarsito)
[13] Hafriani2004*MengembangkanKemampuanPemecahanMasalahMatematikamelalui Problem Centered Learning* (Bandung: Thesis PPS UPI) (Tidak ada di badan teks)

What do they know about Heat and Heat Conduction? A case study to excavate Pre-service Physics Teachers' Mental Model in Heat and Heat Conduction

I M Sari

Department of Physics Education, Mathematics and Science Education Faculty, Universitas Pendidikan Indonesia. Jl. Dr. Setiabudhi No. 229 Bandung, 40154 Indonesia

Email: ikams@upi.edu

Abstract. Teacher plays a crucial role in Education. Helping students construct scientifically mental model is one of obligation of Physics Education Department of Teacher Education Institute that produce physics teacher. Excavating students' mental model is necessary to be done in physics education. This research was first to identify 23 physics students' mental model of heat and heat conduction. A series of semi-structured interviews was conducted to excavate the students' understanding of heat and mental models on heat conduction. The students who involved in this study come from different level from sophomore to master degree in Physics Education Department. This study adopted a constant comparison method to obtain the patterns of the participants' responses through the students' writing, drawing and verbal utterances. The framework for assessing mental model and the instruments were adopted and adapted from Chiou and Anderson (2010). We also compared the students' understanding of heat and mental models on heat conduction. The result shows that Heat is treated as Intrinsic property, material substances, and caloric flow. None of students expressed heat as transfer of thermal energy. Moreover, there are two kinds of students' fundamental component of mental model in heat conduction were found: medium and molecules. Students understanding of heat and fundamental components of mental model in heat conduction are not resulted from running mental model.

1. Introduction

One of challenges in teaching physics is to help students develop a scientifically compatible understanding of heat phenomena based on their existing ideas and beliefs [1]. Understanding that exist in someone thinking will help in describing the phenomena and process the understanding to analyze the new phenomena in the form of internal representation [2]. Internal representation that is exist in the students' mind based on his/her understanding of a concept is called Mental Model. Mental model can be run in explain and predict the new phenomena. Moreover, the central goal in physics education is to help students construct a compatible scientifically mental model [3]

Excavating students' mental model in certain topics, provide a constructive feedback to stakeholders of physics education particularly to probe the reasons behind misconception. This feedback is used to improve the quality of physics learning [2]. The concept of heat and heat transfer is an essential concept and all physics students should master this concept as a basic concept. Moreover, as candidate of Physics' teachers, Students must have a scientifically understanding about what heat is and how the mechanism of heat conduction since it is listed in the national curricula. Heat and its transfer become essential topics both in Junior High School and in Senior High School.

Mental models refer to individuals' internal, mental representations of external, physical phenomena or systems [1] [4]. Meanwhile, Johnson-Laird [5] stated that mental models are structural analogues of the world; propositional representations are strings of symbols that correspond to natural language; and images are perceptual correlates of models from a particular point of view. Human beings understand the world by constructing mental models, i.e., working models (models that predict and explain) in their minds [6]. Someone who runs mental model, he/she will be able to predict and explain the new phenomena. Reinfried & Tempelman [7] listed mental models as constructions of knowledge and understanding that play a central role in the theory of learning of cognitive constructivism, which represents the epistemological position this study is based on.

Students' understanding of Heat conceptions has been extensively studied since the 1970s due to the difficulties to be studied as heat represent emergent processes [1]. Most of students treated as substance rather than as a process. As listed in page 829-830, Chiou [1] comprised four patterns the students' interpretative framework of heat as it can be found based on the past studies on misconceptions or conceptual change. The four patterns of heat conceptions are: (1). Heat is treated as an intrinsic property of a substance. Coldness is property of ice and hotness is property of wood- is an example of students response in pattern (1); (2). Heat is treated as material substance. For this pattern, hotness, which can be envisioned as a group of particles and can move from a hot object toward a cold object; the opposite of hotness particle is coldness, which can move from a cold object to a hot object; (3).Heat is treated as non material entity, caloric flow, which propagates from object at higher temperature to object at lower temperature; (4). Heat is treated as transfer of thermal energy (the total kinetic energy of all the particles in a substance) due to a temperature difference. The fourth framework represents a scientifically accepted view. The framework represents the developmental stages of peoples' conceptions of heat, developing from naïve view toward a more scientific one.

2. Research Method

2.1 Research purpose
The aim of this study was to identify pre-service physics teachers' mental model in heat conduction concept and investigated the relationship between the mental models' found and students' understanding about heat.

2.2 Research method
A series of semi-structured interviews were held to probe students' understanding of heat and students' mental model in heat conduction concept. The students who were involved in this study varies from sophomore to senior and some of them are Physics' Education master students. They were first asked to allow the researcher to record along the interview. Students were given a set of questions related to heat, heat conduction, mechanism in heat conduction, examples of the phenomena related to heat conduction which they found in the daily life. They then asked to predict the heat conduction phenomena and follow up questions were then probed to strengthening the participants' prediction and the reasons why the participants make the prediction. At the end of the interviews, the participants were given an opportunity to reflect and/or to change the answers.

There were three interview protocols: first interview protocol consisted of basic concept of heat, heat conduction, samples of phenomena heat conduction in daily life and the mechanism underlying heat conduction. The main purpose of questioning of the mechanism underlying heat conduction was to probe what happened in heat conduction based on the their existing idea. Moreover, the question whether there are students who tend to encompass the microscopic level of heat conduction, become the interestingly to be investigated.

The second interview protocol contained one interview-about-event question. The students' were asked to predict and explain the temperature condition of two kinds rod when it was connected to the iron circular disk. The iron circular disk was heated by the flame as shown in Figure 1.

2.3 Analysis Data
The data of mental model were analyzed by constant comparison method. After that, the participants' responses were analyzed and compared to Chiou's framework [1]. Three rounds of data analysis were conducted to construct each participant's mental model [3]. The first round focused to the participants drawing of phenomena; the second round focused to spatial arrangement and the last round was examining the reliability of patterns the students' responses.

Two lecturers who have expertise in thermal physics were asked to recode the participants' understanding of heat and mental models in heat conduction. The inter-coder reliability was calculated by Kappa coefficient which was 0.86 ($p<0.05$) and this was showing an excellent agreement between coders.

3. Result and Discussion

3.1. Students' understanding of heat
Based on the four patterns of students' interpretative frameworks of heat that stated by Chiou [1], The pre-service physics teachers' conceptions of heat can be seen in Fig. 2. As can be seen that there was only one student who treated heat as property intrinsic. He/she explained that heat is a form of energy. It can be converted to others form. Heat is a property of object, likewise the potential energy. He/she agreed that heat is owned by person, ballpoint, etc. He/she did not notice the temperature difference.

Furthermore, none of students answered that heat is an energy that exist as transfer of thermal energy due to the temperature difference. Most of students (more than 70%), treated heat as Caloric flow. Students' responses are homogeny that heat as Energy that is move due to the temperature difference. If they continued to be asked about the energy itself, they failed to explain in detail about the Energy. Its responses are closest to the definition written in high school text book of Physics.

In the other hand, there are two students who treated heat as combination of others framework. First student, combine the framework of intrinsic property and caloric flow. He/she explained that heat is energy that transferred due to the temperature difference when it is connected/touched each other. In addition, in the last part, he/she explained that heat is owned by each substance. The higher the temperature, the more heat is contained and vice versa. Second student, He/she expressed heat as combination of fourth framework and third framework. He/she stated that heat is transfer of average kinetic energy from the particle with high energy toward particle with low energy, whereas average kinetic energy is a definition for temperature. He/she pretended as having a scientifically understanding about heat due to the use of energy term. But, he/she could not determine the difference between temperature and heat.

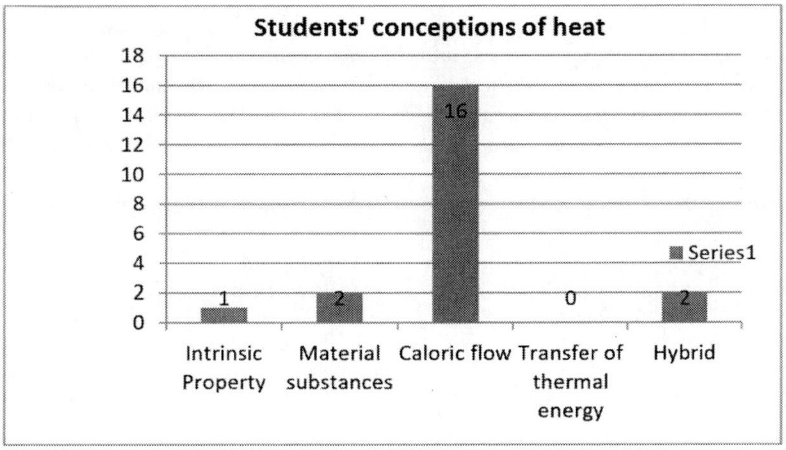

Figure 1. Students' understanding about heat

3.2. Students' mental model in heat conduction

Heat conduction is a dynamic process of how heat is transferred through the solid substance. Students' fundamental components of mental model in Heat Conduction were found as described in Figure 2.

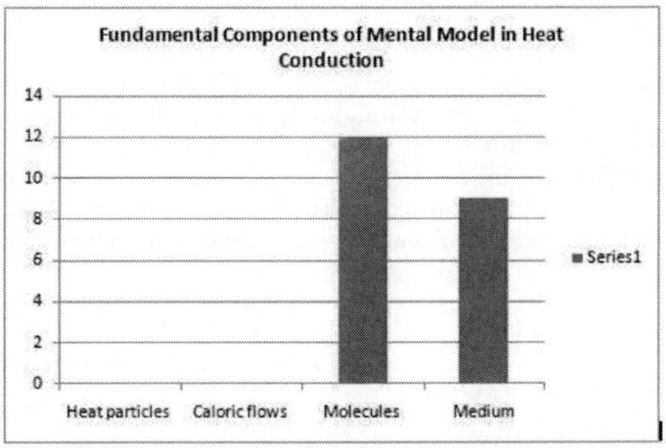

Figure 2. Fundamental components of Mental Model in Heat Conduction

There are two kinds of students' fundamental components of mental model in heat conduction based on the interviews. First component mental model is Molecules. Students, who chose molecules as a fundamental component mental model in heat conduction, explain the underlying mechanism of heat conduction as: "heat conduction proceeds as molecular collisions". For Example, Student A01 mentioned that heat conduction as impact of molecular collision:

I : Please describe the mechanism of heat conduction!

A01 : The mechanism ya? Ooo... when the metal is being heated... the particles in the metal... moving faster...

I : What kind of moving?

A01 : Hmmm... The particles arrangement in metal is dense, so when it is heated, the particles will be vibrated. As hotter, then they more vibrate... they collide each other and influence the next particles to vibrate, so the energy propagates

The second fundamental component of mental model in heat conduction is Medium. Most of students who chose this component, explain the heat conduction as process of heat transfer when the medium did not move with the heat. There is one student who additionally explained that in heat conduction involved the temperature difference, from the higher temperature toward lower temperature. He/she did not explain much more about the mechanism in microscopic or in molecular. Another student, A12 who is in this category, stated that heat conduction is process of energy transfer without medium (heat is directly propagates from one object toward another object without mediator).

I : Please describe the mechanism of heat conduction!
A12 : The process? ... Process in which energy transfer directly...
I : Directly? Could you explain?
A12 : Directly transfer is without mediator, for example .... Eee.... From the flame, flame directly touch the pan.... Flame is hot, it has an energy, directly touch the pan...
I : Emh... OK... so it is directly energy... is there any limitation for your answer?
A12 : No..
I : So, each energy that directly transfer, it is called heat conduction?
A12 : Yes mam.... For example, iron... in ironing... it is conduction

The results show that the fundamental component of mental model in heat conduction is majority of medium. All participants firstly answer that heat conduction is heat transfer in medium. Some students additionally explain with interaction molecular, but others only explain in medium. These findings are different with Chiou's finding [1], that he found that there are three kinds of fundamental component in heat conduction, i.e. Heat particles, Caloric flows, and Molecules.

4. The Relationship of fundamental component of mental model in heat conduction and students' understanding of heat

Ideally, students who have a scientifically understanding of heat, they should also have a scientifically accepted of mental model in heat conduction. Table 1 shows the relationship between students' conception of heat and Fundamental components mental model in heat conduction.

Table 1. *Relationship between students' conception and Fundamental components of mental model*

Students' conceptions of heat	Fundamental components of mental model in Heat Conduction	
	Medium	Molecules
Intrinsic Property	-	1
Material Substances	1	1
Caloric Flow	6	10
Transfer of Thermal Energy	-	-
Hybrid	2	-

They did not run their mental model to explain heat, they were just memorized the sentences of definition about heat. It is worth noticing that students in high level, they should have a better understanding of heat from the microscopic view, that heat is transfer of energy thermal due to the temperature difference. What kind of thermal energy, they should involve the total kinetic energy of all particles, such as atoms and molecules, in a substance.

5. Conclusion

Based on this finding, we can conclude that students' understanding of heat is still dominated by the definition written in the textbook. None of students chose Heat as transfer of thermal energy. The participants' fundamental components of mental model in heat conduction were medium and molecules. Students' understanding of heat and heat conduction is not a result from running mental model, otherwise their thinking is still fragmented. This finding led us to improve the quality of learning especially in delivering heat concept as a dynamic and essential concept.

6. References

[1] Chiou G L and Anderson O R 2010 *Science Education Journal* **94** 825-54

[2] Sari I M and Saepuzaman D 2015 *Proc. Int. Conf. on Innovation in Engineering and Vocational Education (ICIEVE) Atlantis Press doi: 10.2991/icieve-15.2016.48 (Bandung)*

[3] Chiou G L 2013 Reappraising the relationships between physics student's mental models and predictions: an example of heat convection *Physical Review Special Topics-Physics Education Research* **9**

[4] Vosniadou and Brewer 1994 Mental models of the day/night cycle *Cognitive Science* **18** (1) 123-84

[5] Johnson-Laird P N 1983 *Mental models: Toward a cognitive science of language, inference, and consciousness* (England: Cambridge University Press)

[6] Greca I M and Moreira M A 1997 The kinds of mental representations—models, propositions and images—used by college physics students regarding concept of field *International Journal of Science Education* **19** (6) pp 711-24

[7] Reinfried S & Tempelman S 2014 The impact of secondary school students' preconceptions on the evolution of their mental models of the greenhouse effect and global warming *International Journal of Science Education* **36** (2) pp 304-33

Acknowledgments
This study was supported by the Universitas Pendidikan Indonesia, under scheme: Hibah penelitian dan penguatan kompetensi , contracts No. 3858/UN40/LT/2015.

IOP Conf. Series: Journal of Physics: Conf. Series **812** (2017) 012091 doi:10.1088/1742-6596/812/1/012091

Isoquinoline Alkaloids from *Erythrinapoeppigiana* (Leguminosae) and Cytotoxic Activity Against Breast Cancer Cells Line MCF-7 In Silico

T Herlina*, R Mardianingrum, S Gaffar, and U Supratman

Department of Chemistry, Faculty Mathematics and Natural Sciences, UniversitasPadjadjaran ,Jatinangor 45363, Indonesia
*E-mail:tati.herlina@unpad.ac.id

Abstract. *Erythrinapoeppigiana*(Leguminosae) is a higher plant that has been used as a folk for the treatment of infection, fever, and inflammation. In the course of our continuing search for novel cytotoxic compounds from genus *Erythrina*, the methanol extract of *E. poeppigiana* showed a significant cytotoxic activity against breast cancer cells line MCF-7 in silico. The compounds in methanol extract of the *E. poeppigiana* was separated using a bioassay-guided fractionation. By using a cytotoxic activity to follow separation, the methylene chloride was separated by several column chromatography techniques on silica gel and ODS to yield three active compounds (**1-3**). The chemical structures of active compounds were determined on the basis of spectroscopic evidence and comparison with those identical compounds that previously reported and identified as a 10,11-dihydroxyerysodine (**1**) 6,7-dihydro-17-hydroxyerysotrine (**2**) 6,7-dihydro-11-methoxyerysotrine (**3**). Compounds (**1-3**) showed cytotoxic activity inhibits EGFR 2 against breast cancer cell line MCF-7 in silico molecular docking method with bond Gibbs free energy (ΔG) (kcal/mol) and inhibition constants (Ki) (nM) of value (-8.61121, $4.84x10^{-7}$); (-8.1145, $1.12x10^{-6}$); and (-7.3394, $4.14x10^{-6}$), respectively.

1. Introduction

Breast cancer is the most commonly diagnosed invasive non-skin malignancy and second leading cause (after lung cancer) of cancer related deaths in the women. Both epidemiological and laboratory studies show that environmental and behavioural factors are more important than genetic factors in determining overall cancer frequency among populations. Currently, systemic cytotoxic chemotherapy approaches, controlling and treating breast cancer, are being used which are not only less effective but are also non-selective and highly toxic to normal tissues [1].

Recently, the acceptance of traditional medicines as an alternative form of healthcare has increased among all socio-economic groups of the population in Indonesia. For these reasons, medicinal plants have become the focus of intense study, in term of validation of their traditional uses, through the determination of their actual pharmacological effects [2]. Recently, the attention has been focused on medicinal plants to provide new anti-cancer agents. *Erythrinapoeppigiana* (Leguminosae) is a famous medicinal plant widely distributed in tropical and subtropical region of the world. This plant is locally known as "dadapbelendung" in Indonesia and the leaves of *E.poeppigiana*are used as fork medicinal to treatment infection, inflammation, and fever. Previous studies have shown that the stem bark of *E. poeppigiana* contain isoquinoline, α-erythroidine, 8-oxo- α-erythroidine, and 8-oxo-β-erythroidine alkaloid [3,4]. As part of our continuing search for novel anti-cancer compounds from *E.*

*poeppigiana*plant, the isolation alkaloid and test the cytotoxic activity against breast cancer cell line MCF-7 that its mechanism of action inhibits the Epidermal Growth Factor Receptor 2 (EGFR 2) by molecular docking in silico method has been described during the present investigations. In short of our experimental procedure, we extract and isolate alkaloid, using common methods. Detailed method can be found in many references [5].

2. Experimental Method

2.1. Plant Materials

The leaves of *E.poeppigiana* were freshly collected in September, 2014, in Bandung District, West Java, Indonesia. The plant was identified by a staff at the Laboratory of Plant Taxonomy, Department of Biology, UniversitasPadjadjaran, Bandung, Indonesia, and a voucher specimen has been deposited at the herbarium.

2.2. Extraction and isolation compounds (1, 2, and 3).

The dried leaves (2.6 kg) of *E. poeppigiana* were extraction of methanol acidified with 1% hydrochloric acid to a pH of 2-3. Furthermore, fraction of acidic water was basified with ammonium hydroxide until pH 9-10 and solvent methylene chloride (1:1). Methylene chloride fraction (11.2 g) was separated using column chromatography silica gel G60 (70-320 mesh) with an eluent of chloroform: methanol (9.8: 0.2) in isocratic, obtained four fractions (F1-4) positive alkaloid. Fraction F1 is purified using column chromatography silica gel with chloroform: methanol (9:1) in isocratic system, obtained four fractions (F1A-D). F1B fraction was purified using silica gel column chromatography with chloroform: methanol (0.25%), obtained three fractions (F1B1-3). F1B2 fraction was separated using preparative thin layer chromatography with eluent ethyl acetate: acetone (9:1), obtained compound **1** (4.5 mg). F2 fraction was separated using silica gel column chromatography with chloroform: methanol (2%), obtained four fractions (F2A-D). F2A fraction separated using preparative thin layer chromatography with methylene chloride eluent: chloroform: methanol (1: 0.9: 0.1), obtained compound **2** (8.1 mg). F2B fraction separated using silica gel chromatography column with isocratic solvent methylene chloride: chloroform: methanol (1:0.9:0.1) obtained four fractions (F2B1-3). F2B2 fraction separated using preparative thin layer chromatography with solvent methylene chloride: methanol (9:1), obtained compound **3** (10.4 mg).

2.3. Preparation of Ligands

Ligands are drawn in 2D using software MarvinSketch 5.2, and then do the geometry optimization protonated ligand with a pH of 7.4, save in the form .mrv. Searching for the determination of confirmation structure and save as a file in the form of pdb and moles [6].

2.4. Validation of Docking Method

ArgusLab program validated to get a reliable method. Receptors that are used are obtained from the tuberculosis receptor protein banks data. Docking method is used to be good if Root Mean Square Deviation (MRSD) value less than or equal to 2 [7].

2.5. Docking Using ArgusDock

Docking compound with epidermal growth factor receptor 2 (EGFR 2) use the software ArgusLab . In the process of docking used each grid box receptors that are valid on previous validation process.

3. Result and Discussion

3.*1. Characteristics of Compounds (1-3)*

10,11-dihydroxyerysodine (1): yellow oily; Uvλmax (MeOH) nm 237 (ε 3846) ; 278 (ε 1009); IR (KBr) Vmax 3400, 2923, 1641, and 1382 cm^{-1}; HR-TOFMS $[M\text{-}H]^+$ m/z ; (298.1017); ^{1}H-NMR (CD_3OD): δ_H8.12 (1H, d, 5), 7.55 (1H, d, 5), 3.58 (1H, m), 2.14 (1H, m), 2.12 (1H, m), 5.32 (1H, bs), 4.54 (1H,m), 4.40 (1H, m), 6.80(1H, d, 5), 6.79(1H, d, 5), 7.24 (1H, s), 7.45(1H, s), 3.99 (3H, s),

3.73(3H, s), 4.02 (1H,s); ^{13}C- NMR (CD_3OD): δ_c121.9, 130.4, 73.3, 42.7, 64.8, 140.9, 119.4, 37.8, 120.5, 113.1, 127.1, 129.4, 106.2, 148.4, 148.3, 108.6, 56.2, 56.3 ppm.
6,7-dyhydro-17-hydroxyerysotrine (2)[8].
6,7-dihydro-11-methoxyerysotrine (3): white powder; ^{1}H-NMR (CD_3OD): δ_H 3.81 (3H, *s*), 3.50 (3H, *s*), 3.74 (3H,s), 3.49 (3H,s), δ_H 6.66 (1H, d, *J*=10, H-1), 6.56 ppm (1H, d, *J*=10, H-2), 3.81(3H, s, H-3), 1.59 (1H, brs, H-6), 2.89 (2H, m, H-7), 3.17 (1H, m, H-8), 3.83(1H, m),3.28 (2H, m, H-10), 3.55 (1H, m, H-11), 3.49 ppm (3H, s, 11-OCH_3), 3.74 (3H, s, 15-OCH_3), 3.50 ppm (3H, s, 16-OCH_3).

3.2. Docking of Compounds (1-3)

This program should be added hydrogen ArgusLabreceptors, because receptor in pdb is not hydrogen bonds. An 3W32 native ligand and a ligand copy docked with EGFR2 reseptor using ArgusDock method values 1.917605 A RMSD obtained using the grid method box X = 20.75 ; Y = 14:25 , and Z = 17:25 with a grid resolution of 0.4 A on condition of flexible ligands. RMSD smaller value than 2 indicates that a valid method of docking, so it can be used for docking of compounds **1**, **2**, and **3**. Results docking compounds **1**,**2**, and **3** of the EGFR 2 receptor compared with canertinib compounds which are anticancer drugs work on the receptor EGFR 2. Results docking form of the Gibbs free energy and inhibition constants (Table 1) .

Table 1.Gibbs free energy of binding and inhibition constants of compounds (1, 2, and 3)

Compounds	ΔG (kcal/mol)	Ki (nM)
10,11-dihydroxyerysodine (**1**)	-8,61121	4.84x10^{-7}
6,7-dihydro-17-hydroxyerysotrine (**2**)	-8.11457	1.12 x 10^{-6}
6,7-dihydro-11-methoxyerysotrine (**3**)	-7,3394	4,14.10^{-6}
Canertinib	-8.11505	1.12 x 10^{-6}

Results of docking (Table 1) showed that compounds 1, **2**, and **3**have a Gibbs free energy which is almost equal to the compound canertinib. Compounds **1**,**2**, and **3** can be predicted to have similar activity such as anticancer canertinib working on EGFR 2 receptor. The lower of value of a Gibbs free energy released during the interaction between compounds **1**, **2**, and **3** with a receptor showed that stability and strength ofinteraction of non- covalent bonding cause easily get into cells and interfere with DNA replication or the metabolic process to cell death.

ΔG value is proportional to a constant value inhibition. Value inhibition constants (Ki) illustrate the affinity of compounds (1, 2, and 3) and decomposition. The smaller of Ki value, the equilibrium of reaction tends towards formation of compound-receptor complexes (Purnomo, 2013). The complex compound-receptor binding affinity is said to have a good if it has a Ki value in the nanomolar scale. Docking results showed that compounds (**1**, **2**, and **3**) have an estimated value of Ki in the nanomolar scale.The smaller value of the inhibition constants, reaction equilibrium tends toward complex formation of compounds **1**, **2**, and **3** with a receptors. The results showed that a docking 3D visualization of the interaction between all the compounds **1**, **2**, and **3** with EGFR 2 reseptor (Figure 2). Compounds **1**, **2**, and **3** through the docking process showed the properties as inhibitors of EGFR 2 receptor. Epidermal growth factor (EGF) receptor (EGFR) has most an important role in the growth and development of breast cancer MCF-7 [9].

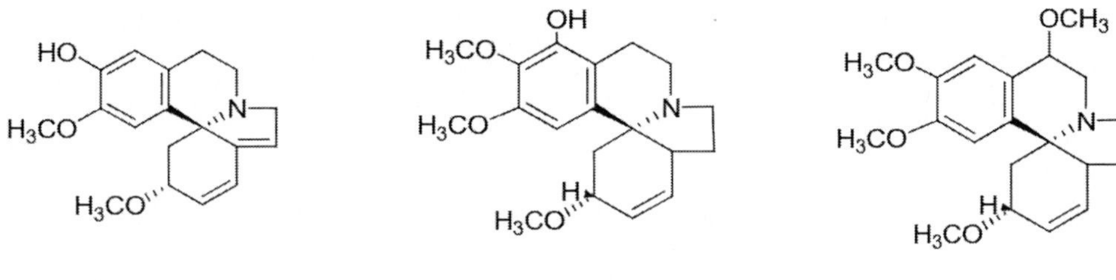

1 2 3

Figure 1. Structure Compounds 1, 2, and 3

Figure 2 showed that compound **1** (purple), compound **2** (yellow), compound **3** (light blue) and a canertinib (green) through a process of docking in the area would bind of target compound (black) which are known to have properties as an inhibitor of EGFR 2.

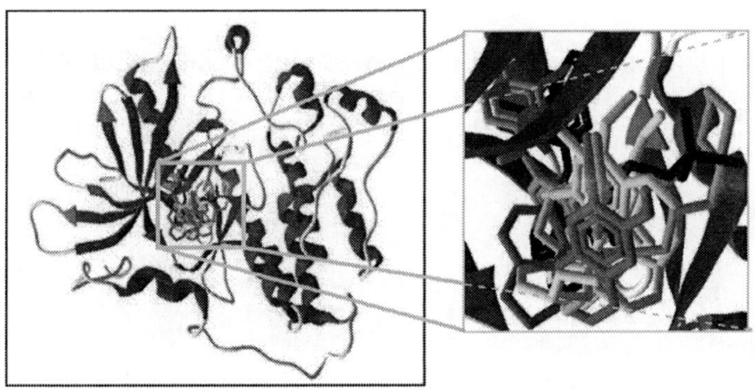

Figure 2. Visualization of an interaction between compounds 1 (purple) , 2 (yellow) , 3 (blue pale), and canertinib (green) with EGFR 2 reseptor.

4. Conclusion

Three alkaloids isoquinoline: 10,11-dihydroxyerysodine (1), 6,7-dihydro-17-hydroxyerysotrine (2), and 6,7-dihydro-11-methoxyerysotrine (3) isolated from leaves of*E.poeppigiana* . Molecular docking showed that compounds 1, 2, and 3 have a Gibbs free energy which is almost equal to canertinib. Compounds (1-3) showed cytotoxic activity inhibits EGFR 2 against breast cancer cell line MCF-7 in silico using molecular docking method with bond Gibbs free energy (ΔG) (kcal/mol) and inhibition constants (Ki) (nM) of value (-8.61121, 4.84×10^{-7}); (-8.1145, 1.12×10^{-6}); and (-7.3394, 4.14×10^{-6}), respectively. Compounds 1,2 , and 3 can be predicted to have similar activity such as anticancer canertinib working on EGFR 2 receptor. Compounds 1,2, and 3 are predicted to have a cytotoxic activity against cell MCF - 7 breast cancer using molecular docking method.

5. References

[1] Tyagi A K, Agarwal C, Chan D C F and Agarwa K 2004 Synergistic anti-cancer effect of silibinin with conventional cytotoxic agents doxorubicin, cisplatin and carboplatin against human breast carcinoma MCF-7 and MDA-MB468 cells *Oncology Reports* **11** pp. 493-9

[2] Lima M R F, Luna J S, Santos A F and Andrade M C C 2006 *Journal of Ethnopharmacology* ***105*** pp. 137-47

[3] Herlina T, Harneti D, Supratman U, Subarnas A, Supriatna, and Hayashi H 2003 *Mathematica et NaturaActa* **3** pp. 5-6

[4] Djiogue S M, Halabalaki X, Alexi D, Njamen Z T 2014 *Planta Med.* **80**, pp.861-9

[5] Nandiyanto A B D, Sofiani D, Permatasari N, Sucahya T N, Wiryani A S, Purnamasari A, Rusli A, and Prima E C 2016 *Indonesian Journal of Science and Technology* **1** pp 132-55.

[6] Purnomo H 2013 Computational chemistry test of anticancer compound in silico (Yogyakarta : Student Library)

[7] Kartasasmita R and Emran 2009 *Indo J. Chem* **9** pp 297-302

[8] Mardianingrum R, Herlina T and Supratman U 2015 *Chemica et NaturaActa* **3** pp. 90-3

[9] Garcia R, Franklin R A, McCubrey J A 2006 *Cell Cycle* **5** pp. 1840-6

Acknowledgments
Authors are grateful to Directorate General for Higher Education, Ministry of National Education of Republic Indonesia, for Research Grant of HibahKompetensi (2016) as financial support.

IOP Conf. Series: Journal of Physics: Conf. Series **812** (2017) 012092 doi:10.1088/1742-6596/812/1/012092

Students' mental model on heat convection concept and its relation with students conception on heat and temperature

R Amalia, I M Sari and P Sinaga

Department of Physics Education, Mathematics and Science Education Faculty, Universitas Pendidikan Indonesia. Jln. Dr. Setiabudhi No. 229 Bandung, 40154.
Corresponding author's e-mail: rifdahlia@gmail.com

Abstract. This research depended by previous studies that only to find out the misconceptions of students without figuring out the mechanism of the misconceptions. The mechanism of misconceptions can be studied more deeply with mental models. The purpose of this study was to find students 'mental models of heat convection and its relation with students conception on heat and temperature. The method used in this study is exploratory mixed method design that implemented in one of the high schools in Bandung. The results showed that 7 mental models of heat convection in Chiou's study (2013), only first model (diffusion-based convention), third model (evenly distributed convection) and fifth model (warmness topped convection II) were found and model hybrid convection as a new mental model. In addition, no specific relationship between mental models and categories of students' conceptions on heat and temperature.

1. Introduction

Senge [1] explained that to be a learning organization, schools should be able to motivate five core things of forminglearning organization. E.g : (1) Systems Thinking, (2) Personal Mastery, (3) Mental Model, (4) Building a Shared Vision, and (5) Team Learning. One of them is mental model that has not been frequently applied in teaching. As educators, teachers need to revealstudents' initial information to help students' understanding of more complex topics. The students' initial information can be seen from students' mental models, because mental models can be affected by previously acquired knowledge. This relates to Vosniadou& Brewer's opinion[2] regarding mental models, beliefs and mental models emerged out from previously acquired knowledge structures, and give effect to the new knowledge. According to Chiou [3] to help students building their scientific knowledge, educators need to know the students' mental models.

Mental models represent ideas in individuals mind are used to describe and explain the phenomenon. Mental models can provide important information about the origin of students' knowledge structures (specifically and theoretically framework) Vosniadou & Brewer [2]. According to Al-Diban& Ifenthaler [4] mental models are basic cognitive constructs that describe complex learning and problem-solving processes. Mental models are important in understanding the development process of the students in learning. Arslan& Durikan [5] said mental model is the structure of daily experience and the learning process, they can reform unwittingly when needed.

The concept of heat and heat convection is an essential concept and all high school students should master this concept as a basic concept [6]. The characteristic of heat convection is dynamic interaction. Convection is the process by which heat carried from one place to another through the movement of the fluid. When the heat convection occurs, the thermal energy move along with the heated fluid. Convection is the flow of heat transfer fluid accompanied by fluid movement [7]. Although the heating of water with a pan or hot air in the room, the process is the same. If the fluid is heated from below, the speed of the molecules increases; the density of the fluid to be reduced and pushed by the cooler fluid so that the density of the fluid cooler took place at the very bottom. Here, convection currents keep fluid to be heated [8].It is also stated by Griffith, & Brosing [9] when heating the air inside the house, the warm air has a lower density than cold air, so the air from radiators or baseboard heaters tend to move upwards (upper house /roof). The warm air makes the air currents distribute energy around the room. Warm air move upwards along the wall near a source of heat, and the cold air moves down along the wall opposite. This is called natural convection. When the fluid is heated, for example, the air above the fire, enlarged volume of the fluid, and its density decreases. If it is associated with the Law of Archimedes, cooler fluid and its density is greater, pushing the fluid warmer and push it to the top. When the warmer fluid moves up, the cooler fluid replaced. This colder fluid will be warmer and pushed upward. Therefore, a continuous flow of heat formed during receiving heat. When the heat is moved by the displacement of gas or liquid, can be said to be heat transferred by convection. Fluid flow can be called by convection currents [7]. Convection involves the physical movement of a substance (such as water or air) from thermal contact with a system for thermal contact with other systems. Substances move brings the internal energy. You can see the rising heat energy in the form of flames and hot air over a fire in a campfire [10].

Lee's study [11] just to find out misconception and how to eliminate these misconceptions on pre-service elementary school teachers about the concept of heat and heat transfer, this study according to Mansyur's [12], he said that study in Indonesia for cognitive science is more dominant in the discussion of misconceptions. Most studies examined how students have misconceptions, but they not explored in more depth mechanism of the misconceptions, the actual mechanism of the misconceptions can be studied more deeply with mental models. One of the basic competencies of high school students in class X on the temperature, heat and heat convection concept is to analyze the impact of heat and heat transfer in everyday activity. Therefore this studyto investigate mental models of high school students on convection heat and its relations with students conception on heat and temperature.

2. Research Method

2.1 Research purpose
The aim of this study was to identify students' mental model on heat convection concept and investigated the relationship between the mental models' found and students' conception on heat and temperature.

2.2 Research method
This research method is mixed methods (both quantitative and qualitative). The study design used in the study is exploratory design, "Exploratory mixed methods design emphasizes qualitative data than quantitative data and quantitative data to build qualitative data" [13]. In this study, the first is collecting quantitative data followed by collecting qualitative data, the results of quantitative data is intended to help the results of qualitative data.

2.3 Participant
In quantitative data, 28participants were students of class XI IPA at high schools in Bandung. Then in the qualitative data, 30% of 28 students were selected to be the subjects of mental model identification.

2.4 Instrument
Collecting data used tests and interviews. The instrument test used is a three-tier test about temperature and heat concept. The first tier contains 5 options, namely A, B, C, D, and E, this tier

using instruments Thermal Evaluation Concept [14] [15] as many as 30 questions that are adopted and adapted in the numbers 1 to 15, 17 to 27, and 29 to 32. As for the two other questions made by the researcher is at number 16 and 28, each question has been validated by experts and physical education specialists as much as three experts. To test the students' conceptions researchers added two others tier, which are: second tier contains the reason of students choiceon first tier, and third tier is confidence index (if the student is sure to choice and reason). The level of confidence in the three-tier test provides an opportunity to assess the nature and strength of their misconceptions [16].

Interview's instrument is adopted and adapted from Chiou [3]. Three rounds of interviews were conducted to investigate not only the understanding of the students but also their mental model of heat convection. Each student spends an average of 30 to 60 minutes per student to complete the three stages of the interview. The first phase aims to investigate students' understanding of heat convection as well as some basic scientific concepts about temperature, heat and heat convection. Student responses in this interview was expected to provide a foundation and a framework to guide the interpretation of their mental models. In a second phase, some generative questions designed to investigate the students' mental models about heat convection. Then, follow-up questions were given to investigate how students simulate the process of heat convection. In the third phase, several questions were posed to probe the participants' reflections on their previous responses.

2.5 Analysis Data

Quantitative data analysis conducted in accordance with the table category analysis answers three tier test [17]. There are four categories will we find ; scientific knowledge, error, lack of knowledge and misconceptions. It is from the analysis of the answers to each question. Researchers summing the results of the analysis of each question on the fourth categories of each student. The results of the analysis can be categorized based on the categories of students' conceptions in Table 1. The results of the analysis of quantitative data associated with mental models of qualitative data analysis later.

Table 1. The categories of students' conception

No	The categories of students' conception	Types of answers
1	Dominantscientific knowledge	Scientific knowledge51%-100% from all the answers
2	Dominantlack of knowledge	Lack of knowledge51%-100% from all the answers
3	Dominant misconception	Misconception 51%-100% from all the answers
4	Misconceptionand lack of knowledge are equal	Misconception and lack of knowledge 48%-52% from all the answers

Qualitative data analysis using the constant comparison method. Then analyze the results of interviews and compared with the results in Chiou [3]. 3 phases of interviews were analyzed to find out the students' mental models. The first phase is from the image of each student about the heat convection. Then, the focus is placed on the explanation of the temperature difference of the fluid as a result of convection currents. Then the last phase of testing is the reliability of each category of students. 2 deal physics lecturer from mental models that formed was tested by calculating Kappa coefficient. Kappa coefficient calculation is calculated using 2x2 contingency tables [18].

3. Result and Discussion

3.1. Student' Conception on heat and temperature

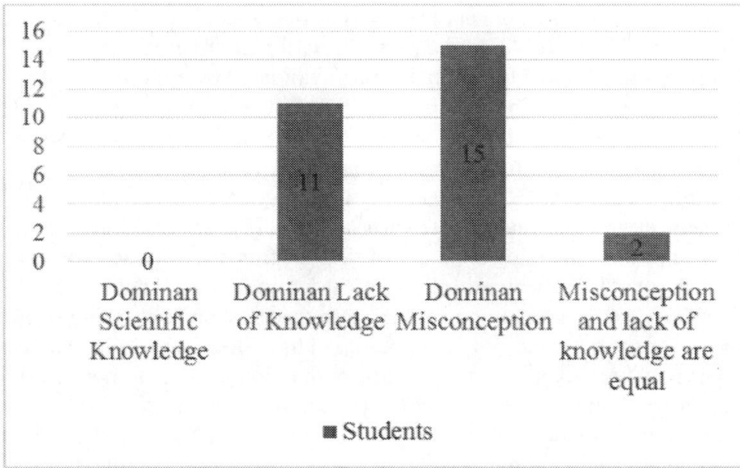

Figure 1. Students conception on heat and temperature

Based on table 1, there are4 categories based on students' answers;dominant scientific knowledge, dominant lack of knowledge, dominant misconception, and misconceptions and lack of knowledge are equal. 30% students of each category in the picture 1 are 4 students dominant lack of knowledge (IY, SSA, AZ, and MZA), 5 students dominant misconceptions (FMA, RMR, PAB, NRS, and BA) and 1 students misconceptions and lack of knowledge are qual (CNP). No students were included in dominant scientific knowledge,then only 10 students were investigated in mental models.

3.2. *Students' Mental Model*

From 7 mental model in Chiou's study [3] which scientific accepted is model 7. Howeverin this study, there was no student in model 7. 3 mental models by Chiou, G. L [3] and 1 mental models that the new findings. Categorizing the mental model of heat convection reliability was examined by calculating theKappa coefficient, which was 1, showing excellent agreement between the two experts.

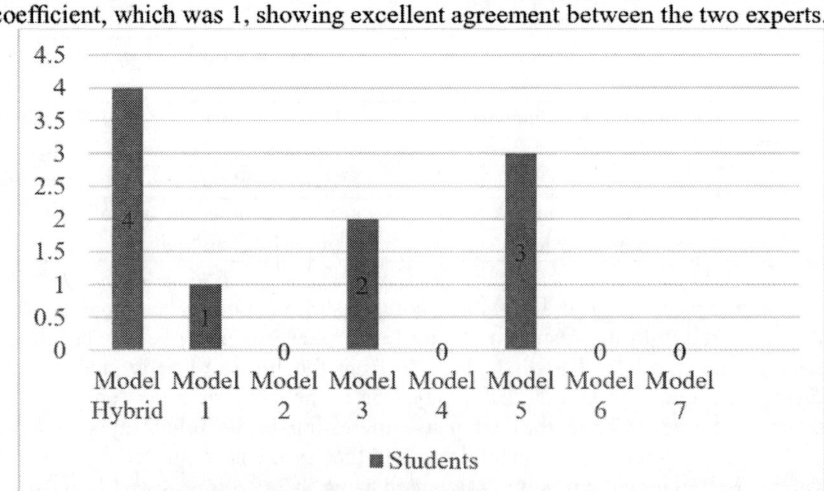

Figure 2. Students' mental model on heat convection

3.2.1. *Model hybrid convection*

In hybrid convection model, when heated, the molecules enlarged so its density decreases, then move up gradually and not back down. Liquid at the top have the highest temperature than other parts of the liquid. For example, when students BA in the interwiew about mechanism of heat transfer on tube I, he replied:

BA: When heated, volume increases, its density decreases, thus evaporate andmove up

There are four students in this model BA, MZA, RMR, and SSA. MZA has a similar thought with BA, MZA and BA thought thatevaporation occurs after reaching top of liquid. However, RMR thoughtafter reaches the top, boiling occurs.

3.2.2. Model 1: Difussion-based convection

Characteristics of the model 1 is diffusion mechanism that occurs when heating. Only one student is included in the model 1, he is FMA. When the interwiew about mechanism of heat transfer on tube I, he replied:

FMA: The first,heated at the bottom and move upward gradually

R: Why does it move up?

FMA: Because of heat at the bottom

In this model, students ignore the liquid as a heat carrier. The highest temperature in the liquid is bottom of the tube and gradually decreases to the top of the tube.

3.2.3. Model 3 : Evenly distributed convection

In model 3, students assume that the heat changes molecules near a heat source becomes larger so its density decreases and moves upward and then immediately turned down and so on, and the temperature in the liquidis the same. As shown in snippet interviews between researchers and AZ:

Q: Describe the process of heat transfer that occurs in the tube!

AZ: Because the heat from here (bottom) spread up, (drawing current convection) going around so continue.

Q: What about the temperature in different parts of the water when convection occurs? (e.g. bottom = A, the middle = B, the top = C)

AZ: A = B = C

However, CNP had different ideas, she thought the current convectionfrom bottom of the tube to the top of the tube, it happens continuously until boiling and then after boiling evaporation. CNP also thought that the temperature in each part of the liquid on the top tube has a high temperature and gradually decreases to the bottom of the tube.

3.2.4. Model 5 : Warmness-topped convection (II)

There are three students in model 5, they are IY, NRS, and PAB. They thought that the bottom of the liquid had the highest temperature and gradually decreases upward. When asked about the mechanism of heat transfer when heated at the bottom, IY answers that because it is heated, the volume of molecules in the bottom of liquid enlarged so its density decreases ($\rho = m / V$), then the molecules are moving up to the top and moving down to the bottom so that make the convection current. When the molecule moves up to the top, the top displace the molecules move. NRS have the same answer with IY, but PAB has a different answer to both. PAB thought that when the molecule at above of liquid, its density enlarged because of boil. From the top, spread straight down. Later, all parts of the liquid will be hot.

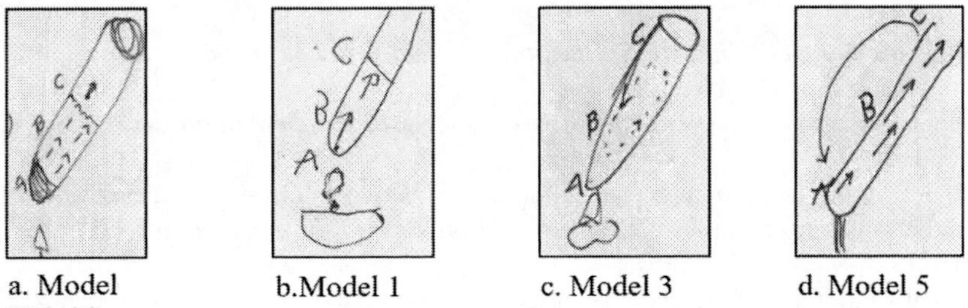

a. Model Hybrid b.Model 1 c. Model 3 d. Model 5

Figure 3. Drawings of students in each mental model

3.3. Relationship between students Mental Model and Student' Conception

Table 2. Relationship between students mental model and students' conception

Mental Model	Students' conception			Total
	DominanLack of Knowledge	**DominanMisconception**	**Misconception and Lack of Knowledgeare equal**	
Model X: ***Hybrid convection***	MZA, SSA	BA, RMR	-	4
Model 1: ***Difussion-based convection***	-	FMA	-	1
Model 3: ***Evenly distributed convection***	AZ	-	CNP	2
Model 5: ***Warmness-topped convection (II)***	IY	NRS, PAB	-	3
Total	4	5	1	10

From Table 2, students in dominant lack of knowledge category has a different mental model, there are model hybrid, model 3 and model 5. And students in dominant misconceptions has a different mental models too, there are model 1, model hybrid and also model 5. And the latter, studentin misconseprions and lack of knowledge are equal has model 3. This result proving that no specific relation between mental models and students conception.

4. Conclusion

From 7 mental models about heat convection inChiou [3], model 1 (diffusion-based convention), model 3 (evenly distributed convection) and model 5 (warmness topped convection II) that found in this study, and new mental model as a hybrid convection model. Among the four mental models no mental models in scientific accepted.

No specific relationship between mental models and categories of students' conceptions. Students are included in dominant lack of knowledge have mental models ranging from a hybrid convection model, model 3 (evenly distributed convection) and model 5 (warmness topped convection II). While students are included in dominant misconceptions have a varies model, they are in model 1, hybrid convection model and also model 5 (warmness topped convection II). And the last of students belonging to the category 4 has a model 3 (evenly distributed convection).

5. References

[1] Senge P M 1990 *The fifth discipline: the art and practice of the learning organization* (New York: Currency Doubleday)

[2] Vosniadou S and BrewerWF1994 Mental models of the day/ night cycle *Cognitive Science* **18** 123-183

[3] ChiouG L 2013Reappraising the relationships between physics students' mental models and predictions: an example of heat convection *Phys. Rev. Spec. Top: Phys. Edu. R***9** 0101191-01011915

[4] Al-DibanS and IfenthalerD2011Comparison of two analysis approaches for measuring

externalized mental models*Edu. Technol. & Soc***14** 16–30
[5] ArslanAS and DurikanU2016Pre-service teachers' mental models of basic astronomy concepts *Sci. Edu. I***27** 88-116
[6] Sari IM and Saepuzzaman D 2015 Physics students' mental model and prediction inheat convection concept*ICIE VE*224-227
[7] CutnellJD and Johnson K W 2007*Physics, Seventh Edition* (New Jersey: John Wiler & Sons)
[8] Hewitt P G 1992*Conceptual Physics, Seventh Edition (*New York: Harper Collins College Publisher)
[9] Griffith W T and BrosingJ W 2009 *The Physics of Everyday Phenomena: A Conceptual Introduction to Physics. Seventh Edition* (New York: The McGraw-Hill Companies)
[10] Bauer W and Westfall G D 2014 *University Physics, Second Edition* (New York: The McGraw-Hill Companies)
[11] Lee C K 2014A conceptual change model for teaching heat energy, heat transfer and insulation *Sci. Edu. I***25** 417-437
[12] Mansyur J(2010). *Kajian fenomenografi aspek-aspek model mental subyek lintas level akademik dalam problem solving konsep dasar mekanika.* (Tesis). Sekolah Pascasarjana, Universitas Pendidikan Indonesia, Bandung.
[13] CreswellJ W 2008 *Educational Research. Planning, Conducting, and Evaluating Quantitative and Qualitative Research, Third Edition* (USA: Pearson Prentice Hall)
[14] Yeo S and Zadnik M 2001Introductory thermal concept evaluation: assessing students' understanding*The. Phys. Teac.* **39**496-504
[15] Alwan A A2011Misconception of heat and temperature among physics students *Proc. Soc. and Behav. Sci.* **12**600-614
[16] Gurel D K, Eryilmaz A and McDermott L C 2015A review and comparison of diagnostic instruments to identify students' misconceptions in science *Eur. J. Math. Sci. Tech. Edu.* **11**989-1008
[17] Kaltakci D and Didis N 2007Identification of pre-service physics teachers' misconceptions on gravity concept: a study with 3-tier misconception test. *AIP Publishing* 499-500
[18] Murti B 2011*Validitas Dan Reliabilitas Pengukuran*Matrikulasi Program Studi Doktoral, Fakultas Kedokteran (Semarang: UNS)

IOP Conf. Series: Journal of Physics: Conf. Series **812** (2017) 012093 doi:10.1088/1742-6596/812/1/012093

Expert Strategies in Solving Algebraic Structure Sense Problems: The Case of Quadratic Equations

Al Jupri[1] and R Sispiyati[1]

[1]Department of Mathematics Education, Faculty of Mathematics and Science Education, Universitas Pendidikan Indonesia
Jl. Dr. Setiabudhi No. 229, Bandung 40154, Indonesia
Email: aljupri@upi.edu

Abstract. Structure sense, an intuitive ability towards symbolic expressions, including skills to interpret, to manipulate, and to perceive symbols in different roles, is considered as a key success in learning algebra. In this article, we report results of three phases of a case study on solving algebraic structure sense problems aiming at testing the appropriateness of algebraic structure sense tasks and at investigating expert strategies dealing with the tasks. First, we developed three tasks on quadratic equations based on the characteristics of structure sense for high school algebra. Next, we validated the tasks to seven experts. In the validation process, we requested these experts to solve each task using two different strategies. Finally, we analyzing expert solution strategies in the light of structure sense characteristics. We found that even if eventual expert strategies are in line with the characteristics of structure sense; some of their initial solution strategies used standard procedures which might pay less attention to algebraic structures. This finding suggests that experts have reconsidered their procedural work and have provided more efficient solution strategies. For further investigation, we consider to test the tasks to high school algebra students and to see whether they produce similar results as experts.

1. Introduction

Structure sense is considered as an important collection of abilities to deal with algebra ([1], [2], [3]), a core topic in secondary school mathematics that should be mastered by students for advanced study or professional work [4]. Performing this structure sense ability in, for instance, solving algebra problems shows a relational rather than only an instrumental understanding [5].

The lens of structure sense has been used in several studies to investigate student algebraic proficiency in algebra (e.g., [1], [2], [6]). In Indonesia, however, the use of this lens is still scarce where its use is limited only for interpreting student difficulties [7], but not for developing algebra tasks to assess student algebraic proficiency in, for instance, high school level.

Taking this into account, we carried out a small-scale case study on developing and solving algebraic structure sense problems. This case study aims to validate designed algebraic structure sense tasks for high school algebra and to investigate expert strategies dealing with the tasks. To address this aim, the idea of structure sense is used as a main theoretical framework.

The term structure sense, coined by Linchevski and Livneh [8], initially was used to describe students' difficulties with using knowledge of arithmetic structures in the context of learning initial algebra. This term, then was refined and developed by Hoch and Dreyfus [1], refers to a collection of abilities towards symbolic expressions, including skills to interpret, to manipulate and to perceive symbols.

For the purpose of this study, we distinguish between the use of standard procedures and structure sense strategies when dealing with algebra problems. Someone displays the use of standard procedures if s/he can, for instance, solve equations using certain algebraic procedures without considering the efficiency of the procedures (e.g., [1]). For example, to solve the equation $(3x-2)^2 - 5(3x-2) - 6 = 0$, rather than using a substitution $(3x-2)$ to obtain a simpler equation that can be solved using a more efficient factorization strategy, someone inefficiently transforms the equation into $9x^2 - 27x + 8 = 0$ and uses the quadratic formula to solve it.

Someone displays structure sense strategies when dealing with algebra problems if s/he can: (1) recognize a familiar structure in its simplest form; (2) deal with a compound term as a single entity; and (3) choose appropriate manipulations to make best use of a structure ([1], [2], [3]). For example, to solve the equation $(x-5)^4 - (x+5)^4 = 0$ by using structure sense strategies, someone can deal with $(x-5)^2$ and $(x+5)^2$ as single entities, can use these as substitutions to obtain a simpler equation, and can manipulate the equation into $((x-5)^2 + (x+5)^2)\,((x-5)^2 - (x+5)^2) = 0$.

2. Experimental Method

This case study – a part of a larger qualitative study on investigating symbol sense and structure sense abilities – aims to validate the appropriateness of algebraic structure sense tasks and to investigate expert strategies when dealing with the tasks. To do so, first, we developed three algebra tasks on quadratic equations based on the characteristics of structure sense for high school algebra (see Table 1). Next, the tasks were theoretically validated to seven experts, including four mathematics lecturers having mathematics education background, and three mathematics lecturers having no mathematics education background. In this validation process, we requested experts to solve each task using two strategies and to give comments and suggestions whether the tasks are appropriate for senior high school students or not. Finally, we analyzed expert solution strategies dealing with the problems in the light of structure sense characteristics ([1], [2], [3]).

Table 1. Structure sense tasks on solving quadratic equations

No	Structure sense characteristics	Tasks
1.	Recognize a familiar structure in its simplest form	$64 - x = 8 + \sqrt{x}$.
2.	Deal with a compound term as a single entity	$(x^2 - 2x)^2 - (x-2)^2 = 0$.
3.	Choose appropriate manipulations to make best use of a structure	$(x^2 - 4x)^2 - 2x^2 + 8x - 15 = 0$.

3. Results and Discussion

Table 2 summarizes expert initial strategies dealing with algebraic structure sense tasks on quadratic equations. For example, three experts used standard procedures and four experts used structure sense strategies for solving Task 1. We found that all seven experts, not influenced by having or having no mathematics education background, solved the three tasks correctly using either structure sense strategies or standard algebraic procedures.

Table 2. Expert initial strategies solving algebraic structure sense tasks

Tasks	Expert strategies	
	#Standard procedure	#Structure sense strategy
1	3	4
2	2	5
3	2	5

For the Task 1, all experts considered that this task is an appropriate and a creative task for assessing senior high school skills dealing with equations related to quadratic equations. Concerning solution strategies for solving the task, i.e., solving the equation $64 - x = 8 + \sqrt{x}$, four experts used structure sense strategies, such as recognizing the structure of $64 - x$ as a $8^2 - (\sqrt{x})^2$ and therefore solved it as follows: $8^2 - (\sqrt{x})^2 = 8 + \sqrt{x} \Leftrightarrow (8 - \sqrt{x})(8 + \sqrt{x}) = 8 + \sqrt{x}$ $\Leftrightarrow (8 + \sqrt{x})(7 - \sqrt{x}) = 0$, and so $x = 49$ (see Figure 1 left part); whereas three expert used standard procedures, such as by simplifying the equation into $56 - x = \sqrt{x}$ and then squaring both sides (see Figure 1 right part). In the second occasion, six experts used structure sense strategies, and one provided no solution. We consider that the structure sense strategies may include graphical solutions of the equation because perceiving a symbolic equation as a graphical representation shows a part of collection of abilities dealing with symbols ([1]). We found one expert used a graphical solution to the equation, i.e., the solution of the equation is an intersection point between graphs $f(x) = 64 - x$ and $g(x) = 8 + \sqrt{x}$. This suggests that most of the experts had reconsidered their thinking strategies when they were required to provide more than one solution strategy.

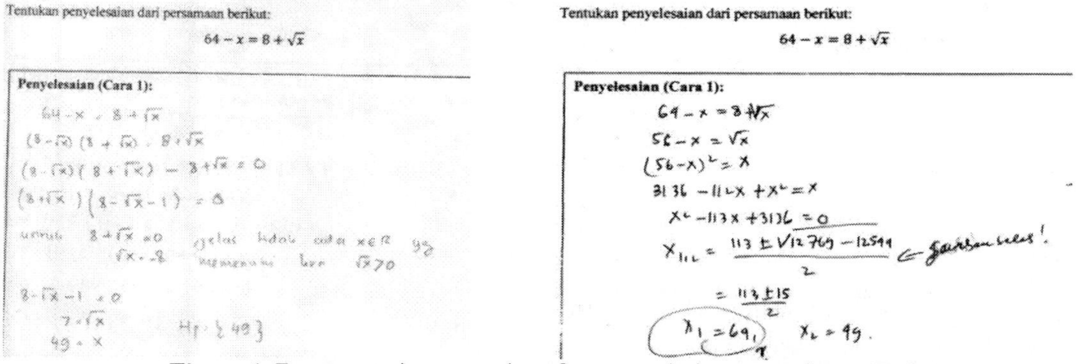

Figure 1. Representative examples of expert solution strategies on Task 1

Task 2, i.e., solving the equation $(x^2 - 2x)^2 - (x - 2)^2 = 0$, is considered by the seven experts as a good and appropriate problem for assessing student skills in dealing with quadratic equations having compound terms. By recognizing the structure of the equation as $a^2 - b^2 = 0$, where $a = (x^2 - 2x)$ and $b = (x - 2)$, someone can solve this equation in the following manner: $(x^2 - 2x)^2 - (x - 2)^2 = 0 \Leftrightarrow (x^2 - x - 2)(x^2 - 3x + 2) = 0 \Leftrightarrow x = -1,\ x = 1$, or $x = 2$. For this task, five experts used structure sense strategies and two experts used standard procedures in the first occasion of solution processes. All experts can eventually produce structure sense strategies in either the first or the second occasion. This finding suggests, again, the experts can change their thinking strategies in a versatile manner. Figure 2, left part, shows an example of structure sense strategies, and the right part shows an example of the use of standard procedures, i.e., using Horner's method to solve, in this case, a polynomial equation of the degree four.

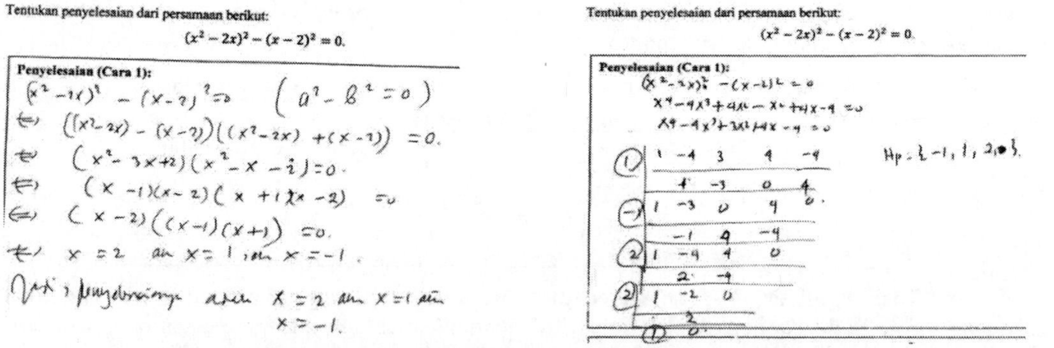

Figure 2. Representative examples of expert solution strategies on Task 2

All seven experts considered the equation in Task 3, i.e., $(x^2 - 4x)^2 - 2x^2 + 8x - 15 = 0$, as a good task for investigating high school students' skills dealing with equations related to quadratic one. We found that five experts used structure sense strategies and two experts used standard procedures in the first occasion of solution processes. The use of structure sense strategies is shown by using the substitution, for instance $y = (x^2 - 4x)$, to simplify the equation into $y^2 - 2y - 15 = 0$, to factorize it into $(y-5)(y+3) = 0$, and to obtain $y = 5$ or $y = -3$. The use of standard procedures is demonstrated by, for instance, expanding the equation, rewriting it into $x^4 - 8x^3 + 14x^2 + 8x - 15 = 0$, and solving it using the Horner's method. Figure 3, left and right parts respectively, shows an example of the use of structure sense strategies and of the standard procedures. Even if six experts eventually used structure sense strategies in the second occasion, still one expert used standard algebraic procedures for solving the equation in both occasions.

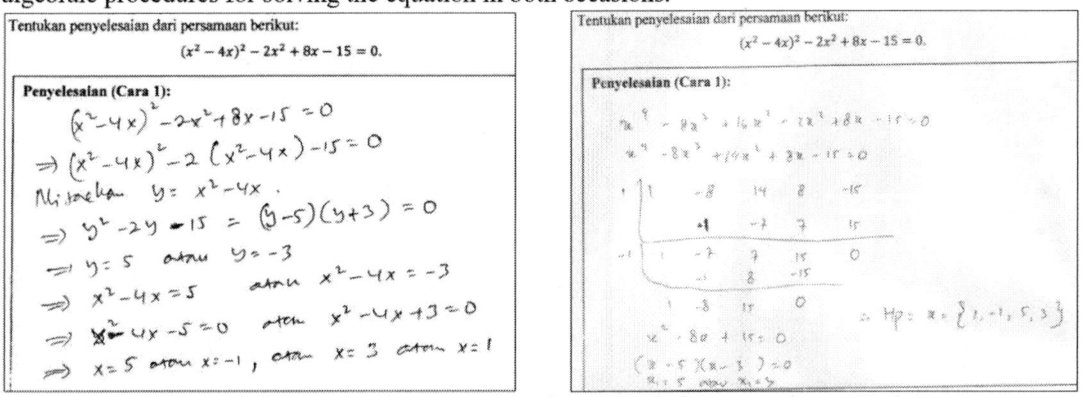

Figure 3. Representative examples of expert solution strategies on Task 3

4. Conclusion

From the results described in the previous section, we draw the following two conclusions. First, the designed algebraic structure sense tasks are considered by experts as appropriate for assessing high school student algebraic proficiency and understanding on the topic of quadratic equations. In other words, the designed tasks are appropriate to structure sense characteristics and are theoretically valid for assessing student skills and understanding on solving quadratic equations. Second, notwithstanding eventual expert strategies in solving equations are in line with the characteristics of structure sense, some of their initial solution strategies, as shown in their written work, used standard procedures. This suggests that initially some experts pay less attention to the structure of the equations, but then they pay more attention after being asked to provide more than one solution strategy. This finding indicates that experts have reconsidered their initial procedural work, and in a flexible manner can provide more efficient solution strategies. For further investigation, we wonder whether high school students will perform similarly as experts when dealing with the designed tasks.

5. References

[1] Hoch M and Dreyfus T 2006 *Proc. of the 30th Conference of the International Group for the Psychology of Mathematics Education,*ed Novotná, H Moraová, M Krátká, and N Stehlíková (Prague: PME) pp 305-312

[2] Hoch M and Dreyfus T 2010 *Proc. of the 6th Congress of the European Society for Research in Mathematics* (Lyon: Institut National de Recherche Pédagogique) pp. 529-538

[3] Novotná J and Hoch M 2008 *Mathematics Education Research Journal* **20** 93-104

[4] Katz V K 2007 *Algebra: Gateway to a Technological Future* (Washington DC: The Mathematical Association of America) pp 1– 45

[5] Skemp R 1976 *Mathematics Teaching* **77** 20-26

[6] Stiphout I M V 2011 *The Development of Algebraic Proficiency* (Eindhoven: Ph. D Thesis Eindhoven University of Technology)
[7] Jupri A, Drijvers A P, and Panhuizen M V D H 2014 *Mathematics Education* **9** 39-55
[8] Linchevski L and Livneh D 1999 *Educational Studies in Mathematics* **40** 173-196

Acknowledgments
This study was funded by Ministry of Research, Technology and Higher Education Indonesia through Fundamental research scheme, year 2016. We thank seven mathematics lecturers, as experts, for their active participation.

IOP Conf. Series: Journal of Physics: Conf. Series **812** (2017) 012094 doi:10.1088/1742-6596/812/1/012094

Synthesisofc-lifepo$_4$composite by solid state reaction method

I Rahayu[1,*], S Hidayat[2], A R Noviyanti[1], D Rakhmawaty[1], and E Ernawati[1]

[1]Department of Chemistry Faculty of Mathematics and Natural Sciences Universitas Padjadjaran Bandung-Sumedang, Km 21 Jatinangor, phone/fax (022)7794391

[2]Department of Physics Faculty of Mathematics and Natural Sciences Universitas Padjadjaran, Bandung-Sumedang, Km 21 Jatinangor

*Corresponding author: Email: iman.rahayu@unpad.ac.id

Abstract. In this research, the enhancement of $LiFePO_4$conductivity was conducted by doping method with carbon materials. Carbon-based materials were obtained from the mixture of sucrose, and the precursor of LiH_2PO_4 and α-Fe_2O_3 was synthesized by solid state reaction. Sintering temperature was varied at 700°C, 800°C, 900°C and 1,000°C. The resultshowed that C-$LiFePO_4$could be synthesized by using solid state reaction method. Based on the XRD and FTIR spectrums, C-$LiFePO_4$can be identified as the type of crystal, characterized by the appearance of sharp signal on (011), (211) and typical peak of $LiFePO_4$materials. The result of conductivity measurement from C-$LiFePO_4$at sintering temperature of 900°C and 1,000°C was $2x10^{-4}$ S/cm and $4x10^{-4}$S/cm, respectively. The conductivity value at sintering temperature of 700°C and 800°C was very small ($<10^{-6}$ S/cm), which cannot be measuredby theexisting equipment.

1. Introduction

$LiFePO_4$ has been known as lithium battery cathode material that has a high theoretical capacity (170 mAh/g), a stable potential voltage of 3.5 V vsLi|Li+, and environmental friendliness. However, $LiFePO_4$ have weaknesses, those are a slow kinetic diffusion of lithium ion and a low electronic intrinsic conductivity. Therefore, it needs a modification to improve its electrochemical performance [1,2].

One of the attempts to increase the low kinetic diffusion of lithium ion is coating the surface of $LiFePO_4$ with carbon, copper, and silver [3]. Enhancing electronic conductivity can be conducted by minimizing the particle size with modification in the sintering process. One of the methods for synthesizing $LiFePO_4$ is solid state reaction. The solid state reaction is more effective than sol-gel method. In addition, doping process with carbon can be conducted simultaneously in the solid state reaction. Therefore, with solid state reaction[3, 4], the results of synthesizing $LiFePO_4$ are expected to produce a material with a fairly high conductivity.

The purpose of this study was to synthesize C-$LiFePO_4$composite by solid state reaction method.The present method used carbon for enhancing and getting excellent $LiFePO_4$ material. Carbon was selected because carbon is easily produced. [5]

2. Experimental Method

The materials used are aquades, LiH_2PO_4 p.a. (Aldrich, 99%), α-Fe_2O_3 p.a. (Aldrich, 99%), sucrose ($C_{12}H_{22}O_{11}$), N-metilpirolidinon (NMP), polivinildene fluoride (PVDF), and nitrogen gas (N_2).

The initial step for synthesizing C-$LiFePO_4$ was preparing the precursor from the material LiH_2PO_4, α-Fe_2O_3 and sucrose. Those three precursors were made by the molar ratio Li: Fe: P: C= 1:1:1:1,2. After weighing those three precursors and mixed it into mortar containing aquades, the mixture was crushed by ball mill for 2 hours. After that, drying process was conductedat temperature 80°C, using oven. After the three precursors were dried, it was heated at temperature of 520°C for six hours in furnace without flowing N_2 gas, then the powder was crushed again in mortar for 2 hours. The powder which had subtle was heated with temperature variation of 700°C, 800°C, 900°C and 1,000°C in the furnace for 16 hours with flowing N_2 gas. The physical characteristics of cathode materials were analysed by using the x-ray diffractometer(XRD), infrared spectrophotometer (FTIR).

The production of Film C-$LiFePO_4$ was conducted by mixing C-$LiFePO_4$, and PVDF as an adhesive with weight composition of 92%:8% in total weight of 2 gram. The mixture was put in 5 mL of NMP solvent while stirred by magnetic stirrer for 16 hours until the homogenous $LiFePO_4$ liquid were produced. The production of $LiFePO_4$ film is conducted by printing thehomogenous liquid of $LiFePO_4$ on glass substrates with size of 2cm x 2cm.The film of $LiFePO_4$ was dried in the oven at temperature of 80°C for 4 hours. Then, each thickness and conductivity of the dried film was measured by four line probe method.

3. Result and Discussion

During the calcination process, areduction inthe mass ofthe precursorfrom28.8992gram to23.4556gram was observed. This is caused by the releasing of water on calcinations process which occurred at temperature of 100°C until 300°C, chemical reaction which occurred on calcinations process was written in equation (1) and (2)[2].

$$2LiH_2PO_4 \rightarrow Li_2H_2P_2O7 + H_2O \quad (1)$$

$$Fe_2O_3.2H_2O \rightarrow Fe_2O_3 + 2H_2O \quad (2)$$

During the sintering, there were threestages of microstructural processes which contained expansion on the initial condition, there was no reaction yet, and particle composition did not change. During theinitialstage of sintering, rearrangement occurred, there was a little moves or particle rotation to enhance the number of contacts between the particles andthe formation oflinkagesbetween the particles, C-$LiFePO_4$composite materialswere stillred, it indicated that there was no reaction of formation a perfect C-$LiFePO_4$composite.

In second stage, size of linkages between particles grew and its porosity decreased due to the particles which come closer each other. On this stage, grain growth started, formed the pore, the distance between the particles became smaller, and shrinkage occurred as shown in Figure 1(b),and (c), C-$LiFePO_4$composite begun to form and characterized by changing into blackish, it indicated that reaction stage was at intermediate stage.

On this stage, the pore became a round pore eventually, size of particle increased and the rate of shrinkage pores became smaller as shown inFigure 1(d),C-$LiFePO_4$compositematerialhad beenformed then compaction and hardening occurred. It showed that effect of the higher sintering temperature would change form structures material of C-$LiFePO_4$composite became more solid and the color turn into blakish. The reaction on the sintering process $Li_2H_2P_2O_7$+2FeO (Fe_3O_4+Fe)→2$LiFePO_4$+H_2O[2].

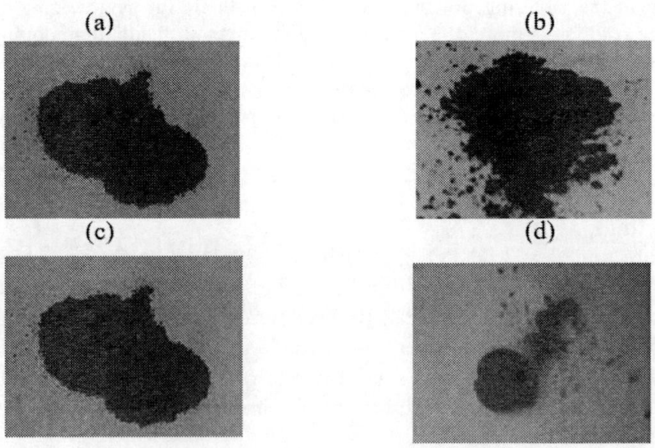

Figure 1. Powder as the result of *sintering* (a) 700°C (b) 800°C (c) 900°C (d) 1000°C

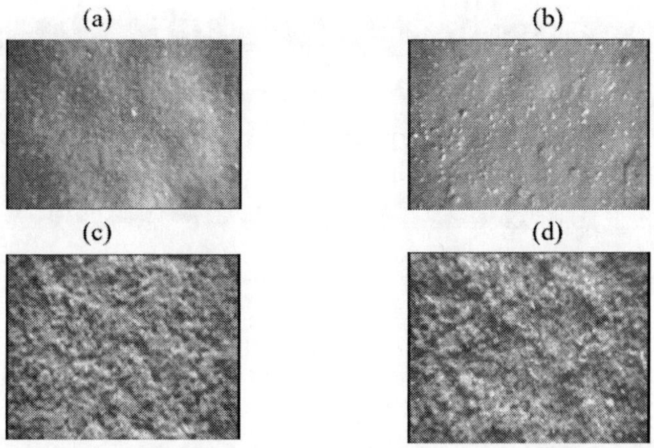

Figure 2. Profile of C-$LiFePO_4$ surface with 100x of magnification, formed in *sintering* film (a)700°C (b) 800°C (c) 900°C (d) 1,000°C

Figure 2 shows the surface of C-$LiFePO_4$cathode material on the addition of the sintering temperature (a) 700°C (b) 800°C (c) 900°C (d) 1000°C. The surface of $LiFePO_4$cathode on heating the sintering temperature of 700°C looked still subtle but the heating at sintering temperature of 800°C looked to start rough and coarsening occurred, then hardening occurred on heating at the sintering temperature of 900-1000°C. It occurred due to the heating of the sintering temperature, the higher sintering temperature, then the more formation of the perfect C-$LiFePO_4$ would be formed, otherwise at the temperature of 700-800°C, C-$LiFePO_4$had not formed completely. It showed that the higher sintering temperature, then the more perfect C-$LiFePO_4$would be formed.

Each thickness and conductivity of dried film would be measured by four line probe method. Table 1 showed that increasing of conductivity occurred along with the enhancing of the sintering temperature.It occurred because the enhancing of the sintering temperature then the crystal structure had been formed, which the bonding between the powder particles bind each other in these conditions. The conductivity of C-$LiFePO_4$composite materials at temperature of 700-800°C was not detected on the equipment, the ability of the equipment was only capable of reading up to 10^{-6} S/cm.

Table 1. The conductivity ofC-$LiFePO_4$

The sintering temperature /oC	Conductivity/S/cm
700	-
800	-
900	2 x10^{-4}
1000	4 x10^{-4}

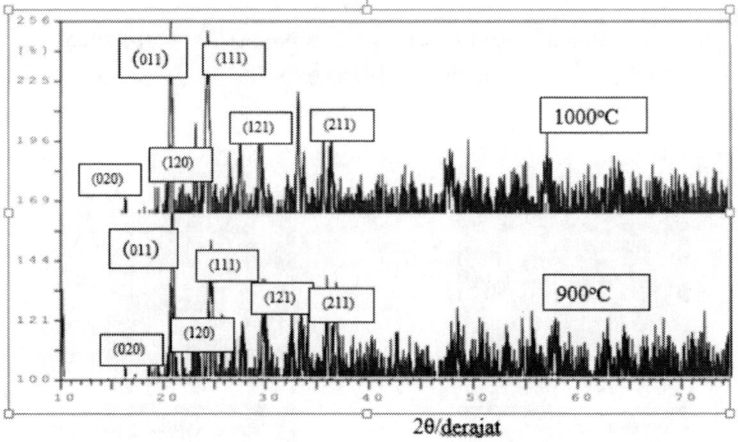

Figure 3. The result of XRDC-$LiFePO_4$

Figure 3 shows that x-ray diffractogram pattern from the variation of sintering temperature. All the products were identified as the type of olivine crystal $LiFePO_4$, marked by the appearance of sharp signal on (020), (011), (120), (101), (111) and (211) but the relativeintensitiesofthe peaknumberswere different amongsamples, due to the effecton the size ofthe crystal. There were alsosignals thatdid not match and it was possibility of the existence of impurities.

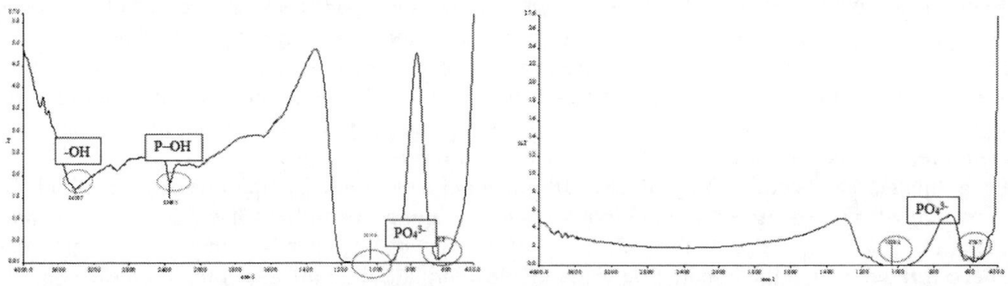

Figure 4. Spectrum infrared C-$LiFePO_4$sintering at(a)700 and (b) 800°C

To determine the functional groups contained in the C-$LiFePO_4$, the characterization by FTIR. Figure 4.(a) is an FTIR spectrum measurement results of C-$LiFePO_4$ sintering 700°C. On the spectrum, formed in wave numbers 3419,7 cm^{-1}, 2349,1 cm^{-1}, 1039,4 cm^{-1}, and 650,9 cm^{-1} that each identifies -OH stretch vibration, Un-deprotonated P-OH , and shows PO_4^{3-} functional groups. Figure 5(b) is an FTIR spectrum measurement results of C-$LiFePO_4$ sintering 800°C. On the spectrum, formed in wave numbers 2344,4 cm^{-1}, 1039,4 cm^{-1}, and 579,1 cm^{-1} identifies each functional group PO_4^{3-}. At 800°C sintering visible loss of OH and P-OH, the high temperature heating leads to loss of impurities such as OH. From the results of FTIR, two samples identified as PO_4^{3-} functional groups.

IOP Conf. Series: Journal of Physics: Conf. Series **812** (2017) 012094 doi:10.1088/1742-6596/812/1/012094

4. Conclusion

C-$LiFePO_4$composite materialcan be synthesized by using solid chemical reaction method.Based on theXRD and FTIR characterization, it can be identified as C-$LiFePO_4$which has the olivine crystal structure with the appearance of sharp signal on (011), (020), (011), (120), (101), (111) and (211). The composite materials of C-$LiFePO_4$on the sintering temperatures of 900°C and 1,000°C were 2 x10^{-4} S/cm and 4 x10^{-4} S/cm, respectively. The sintering temperature of 700°C and 800°C provided a very small conductivity (<10^{-6} S/cm). Therefore, it cannot be measuredby theexisting equipment.

5. References

[1] Myounggu P, Zhang X, Myoungdo Chung, Gregory B. Less, Ann Marie Sastry 2010 *J. of Power Sources* **195** 7904
[2] Gaojian, Jianjun Li, He X, Jiang C, Wan C 2011 *Int. J. Electrochem. Sci.* 2818-25
[3] Drozd V, Liu G Q, Liu, R S, Kuo, H T, Shen C H, Shy D S, Xing, X K 2009 *J. of Alloys and Compounds* **487** 58–63
[4] Zhang, S S, Allen J L, Xu K, Jow T R. 2005 *J. of Power Sources* **147** 234
[5] Nandiyanto A B D, Fadhlulloh M A, Rahman T, Mudzakir A 2016 *IOP Conf. Proc.* **128** (1),012042

Acknowledgement

Authors are grateful to Directorate General for Higher Education, Ministry of National Education of Republic Indonesia, for Research Grant of Penelitian Unggulan Perguruan Tinggi (2015) as financial support.

Finding The Most Important Actor in Online Crowd by Social Network Analysis

I Yuliana[1], P I Santosa[2], N A Setiawan[2] and Sukirman[1]

[1] Department of Informatics Engineering Education, Universitas Muhammadiyah Surakarta (UMS), Jl. Ahmad Yani No. 1, Pabelan, Surakarta 57162, Indonesia
[2] Departement of Electrical Engineering, Universitas Gadjah Mada, Jl. Grafika No.2, Sleman 55281, Yogyakarta, Indonesia
E-mail: irma.yuliana@ums.ac.id

Abstract. Billion of people create trillions of connections through social media every single day. The increasing use of social media has led to dramatic changes in the of way science, government, healthcare, entertainment and enterprise operate. Large-scale participation in Technology-Mediated Social Participation (TMSP) system has opened up incredible new opportunities to deploy online crowd. This descriptive-correlational research used social network analysis (SNA) on data gathered from Fanpage Facebook of Greenpeace Indonesia related to important critical issues, the bushfires in 2015. SNA identifies relations on each member by sociometrics parameter such as three centrality (degree, closeness and betweenesse) for measuring and finding the most important actor in the online community. This paper use Fruchterman Rein-gold algorithm to visualize the online community in a graph, while Clauset-Newman-Moore is a technique to identify groups in community. As the result found 3735 vertices related to actors, 6927 edges as relation , 14 main actors in size order and 22 groups in Greenpeace Indonesia online community. This research contributes to organize some information for Greenpeace Indonesia managing their potency in online community to identify human behaviour.

1. Introduction

The growing impact of social media (Facebook, Twitter, blogs, wikis, Youtube, Flickr, etc.) is changing the way people work, relate, think, and react [1][2][3]. It has great potential to solve national priorities problem such as disaster response, health/wellness, community safety, energy sustainability, citizen science, open government, etc. Ensuring universal access in usability, huge scalability, and reliability during peak usage are very challenging, so that require interdisciplinary research. There a lot of information in online crowd, that need to be analyzed specifically to get the trusted and useful ones.

The online crowd makes people confused about trusted informatin, ass seen on Fanpage Facebook of Greenpeace Indonesia which a non-profit global campaigning organization that acts to ensure the ability of Earth to nurture life in all its diversity. October 2015 was the peak time where Indonesia's rainforests were on fire, threatening human health, endangered species with devastating global consequences, around half a million people have been struck by acute respiratory infections, and at least 10 have died. People might give their opinion on the page, they were demanding government response because this was a humanitarian crisis should be national disaster. The online crowd content could be in positive or negative opinion. It will be very useful when Greenpeace can manage their environment issues by social media and heading it to be national policy. The research goal is to reveal some hidden information in online crowd by Social Network Analysis (SNA).

SNA focuses on relationships among social entities. It is used widely in the social and behavioral sciences, as well as in political science, economics, organizational science, animal behavior, and industrial engineering. The social network perspective, which used in this research, has been developed over the last sixty years by researchers in psychology, sociology, statistics, and anthropology [1][2][3]. The goal of this research is to map the online crowd of social media in Fanpage Facebook of Greenpeace Indonesia by detecting some actors that have strong influences to the group (Greenpeace Indonesia). Furthermore, community detection is used to identify more specifically actors that found by SNA.

2. Literature Review

Social network analysis is a method to map and measure relationships and communication among people, groups, organizations, computers or entity that processes the information [4][5]. The relationships were visualized by graph so it will be easier to be analyzed. Social networks can be generated to analyze soccer games patterns in a video [6]. It constructed two social network for both team based on the interactions between the players. The strong connections between the players of different positions belong to the winner team. In culinary, social network analysis methods are used even in recipes around the world to link the reviews the historical background of a region through the hallmark of traditional culinary offerings [7]. Subsequently another research applying SNA to identify terrorism with a particular community and visited website by the criminal [8][9] Other studies utilizing the connectivity through graph theory and SNA in disaster management [10][11], political campaign [12], trip planning travel [13], monitoring traffic congestion [14], economy and business[15][16][17].

Graph in SNA represents interaction between the individual and his role in the dissemination of information and innovation. One of the SNA method development is community analysis. Basically, online community in social media is formed from a large number of small groups or meta-community. Identification of a small group which is also a meta-community can be based on many things, such as profession, educational background, preferences and even groceries [18][19]. Identification of this group is very useful to determine the structure of the community, which in turn is able to identify individual profiles.

3. Method and material

The descriptive-correlational research material that used in this study was taken from Greenpeace Indonesia online community by SocialNet Importer in October 2015. The online crowd consist of many opinios that are not entirely true, it depends on the actor who spread and the content. SNA detects the relations of each member in a group even the content of the topic.

The stages were carried out after literatur study in this research includes several steps as shown on figure 1.

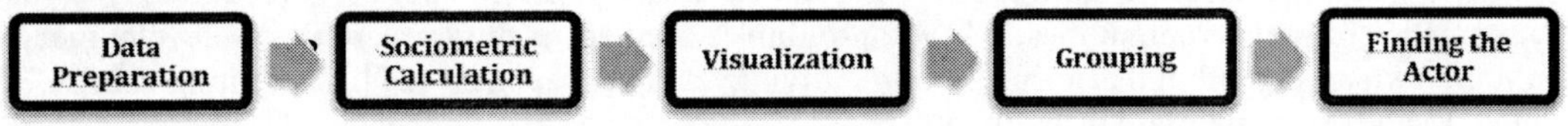

Figure 1. Research method.

Data preparation is very importan to cleanse the data from multiple and redundant relationships and node. NodeXL as an open source adds in Excel has brought many advantages in the process[20][21][22]. Node is representation of an actor in real world. Unique node indicates that there is only one actor in the online community. While SNA is a series of process of sociometrik calculation [23] and visualization by Fruchterman-Reingold [24]. The detected nodes will be grouped by Clauset-Newman-More (CNM) method based on the modularity [19]. The important actors that shown in visualization as a whole online community will be cross checked on each it group or meta-community..

3.1 Social network analysis - sociometry

Sociometry is a quantitative method to measure social relationships among the people, that developed by psychotherapist Jacob L. Moreno in his research of the relationship between social structures and psychological well-being [1]. In this section we introduce three types sociometric centrality that commonly used for SNA that are degree, closeness, and betweennes. The centralities assume that relation are directed and weighted social networks.

Degree Centrality, in case of weighted and directed network in a graph, an actor's degree is defined by the number of another actors that directly connected to him as in-degree, vice versa out-degree. Node size that formed by actor depends on the metric. The most influence actor can be genarated from the biggest value in-degree centrality [25]. Where d_i is the degree (number of adjacent edges) of node v_i.

$$C_d(v_i) = d_i^{in} \quad \textit{(prestige)} \quad (1)$$

Closeness centrality directly relates to the geodesic distance (or the cardinality of the shortest path) between two actors. It has a reflection of the whole connectivity in the network structure [23]. where d(i,j) defines the geodesic distance between two actors i and j, n is number of nodes/ actor.

$$C_c(i) = \frac{n-1}{\sum_{j-1}^{n} d(i,j)} \quad (2)$$

Betweenness centrality depicts the position of an actor in between, so he can control other two actors, which do not have direct connectivity between them. This metric is important in global centrality measure, which investigates the strength of connectivity between the actors in the network [23]. where, *pjk* is the number of geodesic paths between two actors j and k, and *pjk(i)* is the number of geodesic paths between *j* and *k* that contains actor *i*.

$$C_B(i) = \sum_{j<k} \frac{p_{jk}(i)}{p_{jk}} \quad (3)$$

3.2 Community detection

Clauset-Newman-Moore (CNM) is a community method detection address this problem with modularity with an estimated time of less O (nlog2n). Modularity is the property of a network and specifically aims to divide a network into a meta-community [26]. Basically a community formed from smaller or meta-communities. CNM algorithm is applied to this research as a development of Girvan-Newman algorithm that focuses on identifying edge betweenness [27]. Computing using CNM limited to node scale no more than 500,000.

4. Result and discussion

SNA mapped out the online community graph as a whole, depicted from figure 1 that found 3735 vertices and dan 6927 edges. Overall density is 0.000486287 from the maximum number is 1.The number indicates that the node connectivity is very rare. The average in-degree and out-degree totally are same that is 1,855, while every node has its size that depends on in-degree centrality. People prestige shown by the in-degree centrality. Prestige indicates that the actor have many follower and link like on his page so he has a strong influence to the other. The betweenness centrality average is 7530,996, the number is vary and it related with control and manipulation information in a group, so it quite important metric compare to the others. While maximum closeness centrality is 0.0001760 that indicates the distance among the actor is quite nearby in range 0 – 1. The three sociometrics calculate and put the 14 importan actors in order from the biggest size in-degree and betweenesse centrality, while the small one in closeness centrality.

Figure 2 shows that every vertice that connected by edge to another vertices.Vertice that have strong influences, indicated by the biggest node on the graph and appear more clearly and has extra size than the other. Greenpeace Indonesia laid right in the middle of the graph and shown as red dot. On Figure 3,the edges also shows the intense relationship among the node that showed by a thick line. There are 14 important actors, depend on the size of vertex, that is more than 50. Each of these actors are include in the groups (meta-community). The most important actor according to the calculation and visualization is still Greenpeace Indonesia, because of the fan page characteristic that need the administrator keep on exist on the online community to make it dinamic. On the other hand, groups detected has their own important actors and the combine from the first. Groups that found by modularity declare that every small group of vertices that has strong relationship will make their own community.

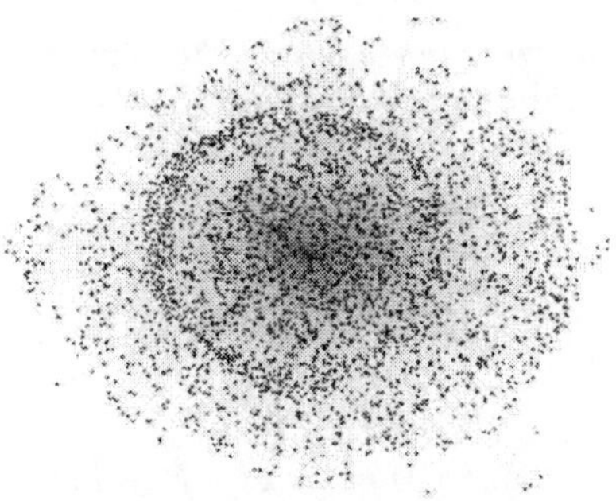

Figure 2. Online community graph.

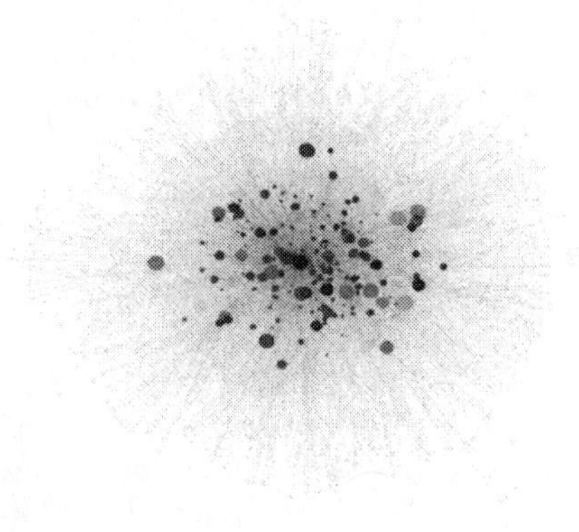

Figure 3. The important actors indicated by node size.

This research was conducted on finding some important actors by sociometric on Social Network Analysis. Greenpeace is the main actor of all centrality, it has the highest value on every metric calculation. The most important metric is betweenness centrality that gain 13039778,534 belongs to Greenpeace Indonesia.While 22 meta-communities are found as community detection result, by modularity clustering, with an important actor in each groups. Most of them are the same as the important actor of the SNA result.

5. Conclusion

Finding the most important actor in online crowd could be done by social network analysis. In this research the results obtained can be said to be ideal, because the main actor is the administrator of fanpage Greenpeace itself. The application of methods in SNA needs to be deepened, for example by eigenvalue analysis and page rank, as well as depictions of the graph and the detection method of community. Besides sampling too short need to be extended in order to obtain more accurate data.

6. References

[1] John S 2000 *Social Network Analysis A Handbook* 2nd Edition (London: Sage Publication Ltd)
[2] Nemiche M, Cavero V and Loez R P 2012 *Understanding Social Behavior Evolutions through Agent-Based Modeling*
[3] Tsvetovat M and Kouznetsov A 2011 *Social Network Analysis for Startups* (California: O'Reilly Media, Inc)
[4] Song J, Kim M, Kim H and Kim K 2010 *A Framework: Workflow-Based Social Network Discovery and Analysis*
[5] Aggarwal C C 2011 *Social Network Data Analytics* (New York: Springer)
[6] Park K J and Yilmaz A 2010 *Int. Conf. Pattern Recognition (Istanbul)* pp 3935–3938
[7] Kular D K, Menezes R and Ribeiro E 2011 *Proc. 2011 IEEE 1st Int. Netw. Sci. Work. (New York)* pp 38–45
[8] Tayebi M A and Glasser U 2012 *IEEE/ACM Int. Conf. Adv. Soc. Networks Anal. Min. (Istanbul)*
[9] Yang C C and Ng T D 2007 *IEEE Intell. Secur. Informatics (New Brunswick)* pp 55–8
[10] Zelenkauskaite, Bessis N, Sotiriadis S and Asimakopoulou E 2012 *Fourth Int. Conf. Intell. Netw. Collab. Syst. (Barcelona)* pp 503–08
[11] Cheong F and It B 2011 *Pacific ASIA Conference on Information Systems Proc.(Brisbane)*
[12] Ya-ting L 2011 *The Social Network Analysis of Political Blogs in People* pp 5441–4
[13] Ahmedi L, Rrmoku K and Sylejmani K 2012 *Tourist Tour Planning Supported by Social Network Analysis*
[14] Dimokas N 2010 *Social Network Analysis Concepts in the Design of Wireless Ad Hoc Network Protocols* pp 23–29
[15] Alamsyah A and Rahardjo B 2013 Financial Fraud Detection using Social Network Analysis (San Fransisco: Researchgate)
[16] Jackson M O 2010 *An Overview of Social Networks and Economic* pp 1–96
[17] Wei K and Wen W *2011 Research on Emergency Information Management Based on the Social Network Analysis üü A Case Analysis of Panic Buying of Salt* pp 1302–10
[18] Berger-wolf T and Kempe D A 2007 *Framework For Community Identification in Dynamic Social Networks* (San Jose: ACM)
[19] Clauset, Newman M E J and Moore *C Finding Community Structure in Very Large Networks* pp 1–6
[20] Smith M, Milic-Frayling N, Shneiderman B, Mendes R E, Leskovec J, Dunne C 2010 *NodeXL: a Free and Open Network Overview, Discovery and Exploration Add-in for Excel*
[21] Hansen M A S and Shneiderman D L B 2011 *Analyzing Social Media Network with NodeXL Insight From a Connected World* (Burlington: Elsevier)
[22] Hansen D L, Rotman D, Bonsignore E, Milic-Frayling N, Rodrigues E M, Smith M and . Shneiderman B 2012 *Do You Know the Way to SNA?: A Process Model for Analyzing and Visualizing Social Media* pp 304–13
[23] Analysis D D and Raya V 2013 *Social Network Analysis, Methods and Measurements Calculations* pp 2–5

[24] Kobourov S 2013 Force-directed drawing algorithms *Handbook Graph Draw* pp 383–408
[25] Rachman Z A and Maharani W 2013 *Int. Conf. Inf. Commun. Technol. (Guilin)* pp 72–6
[26] Malliaros F D and Vazirgiannis, M 2013 *Clustering and Community Detection in Directed Networks*
[27] Girvan M and Newman M E J 2002 *Proc. of the National Academy of Science of the United States of America (Washington)*

Acknowledgments
This study is based on research conducted on the bushfires disaster that almost happen annually in Indonesia. We are so grateful to Greenpeace Indonesia that has shared valuable information about the organization.

Improving Middle School Students' Quantitative Literacy through Inquiry Lab and Group Investigation

N S M Aisya[1]*, B Supriatno[2], Saefudin[2], S Anggraeni[2]

[1]Magister Student of Universitas Pendidikan Indonesia, Dr. Setiabudi Street No. 229, Bandung40154, Indonesia.
[2]Biology Education Department of Universitas Pendidikan Indonesia, Dr. Setiabudi Street No. 229, Bandung40154, Indonesia.
Corresponding author: nurasyifa.ma@gmail.com

Abstract. The purpose of this study was to analyze the application of metacognitive strategies learning based Vee Diagram through Inquiry Lab and Group Investigation toward students' quantitative literacy. This study compared two treatments on learning activity in middle school. The metacognitive strategies have applied to the content of environmental pollution at 7th grade. This study used a quantitative approach with quasi-experimental method. The research sample were the 7th grade students, involves 27 students in the experimental through Inquiry Lab and 27 students in the experimental through Group Investigation. The instruments that used in this research were pretest and posttest quantitative literacy skills, learning step observation sheets, and the questionnaire of teachers and students responses. As the result, N-gain average of pretest and posttest increased in both experimental groups. The average of posttest score was 61,11 for the Inquiry Lab and 54,01 to the Group Investigation. The average score of N-gain quantitative literacy skill of Inquiry Lab class was 0,492 and Group Investigation class was 0,426. Both classes of experiments showed an average N-gain in the medium category. The data has been analyzed statistically by using SPSS ver.23 and the results showed that although both the learning model can develop quantitative literacy, but there is not significantly different of improving students' quantitative literacy between Inquiry Lab and Group Investigation in environmental pollution material.

1. Introduction

Based on characteristic of biology learning, biology of the twenty-first century is an increasingly turn into quantitative science [1]. Not only mathematician that needs the quantitative data, almost all of the field uses mathematics on their works . Mathematics is an essential aspect which it's currently still separated by other subjects [5]. Mathematics itself serves as the basis of quantitative literacy. In using mathematics, students are required to have a sensitivity of numerical. Aside from that, the use of numerical data train to critical thinking activities and metacognitive skill. Quantitative literacy is a skill, knowledge, beliefs, habits of thought, dispositions, communications capabilities, and problem-solving skills needed to engage effectively in quantitative issues in the field of work and life. Quantitative literacy is a habit of minds or habits of thinking to create a meaning of numerical information [3].

Nowdays, much of the information presented in numerical data sets. Numbers have long been important in the management of life, but they have never been so ubiquitous as they are now. The new circumstances arrived suddenly with the coming of technology and their application to gathering, processing and disseminating quantitative information. But recently, there is a problem in our education implementation which after leaving high school, students will already experience changes in the curriculum. The curriculum has taken into quantitative aspects and the students were not ready to follow the system [4]. In this stage, student confronted into the gap between the curiculum require and the unqualified quantitative literacy skill of the students. Students were not familiar with mathematical applications that demanding quantitative reasoning. Therefore, the quantitative literacy is necessary implanted in the middle school to reduce the gap between curiculum require and the quantitative skill of the students.

The low achievement of the literacy skills of quantitative supposedly occurred because the learning process is not maximized to developing literacy quantitative of students. Based on the findings of [5]. learning outcomes are designed by high school teachers through the syllabus that made by National Education Standards Agency. On the syllabus contains the assessment of learning outcomes that are less material to facilitate students in developing literacy skills. This could be a problem in our educational process. So, we need to develop quantitative literacy of students at the school, at least in the subjects of biology.

As we know, biology learning required the students to learn by direct experience and contextual learning, so that students can reveal phenomena around them. To uncover the phenomenon, students need a strategy that bring students to think about how the phenomenon can be revealed, then they could develop into a form of new knowledge. Strategies like these are known as metacognitive strategies. Cognitive strategy is needed to help students figure out what should he know and what he did not know [6]. Learning with metacognitive strategies can be implemented through a variety of ways. We could apply this strategy through inquiry learning model in which students are directed to the learning process [7]. According [8], the long term inquiry-based learning is known could affect the quantitative literacy of students significantly.

Besides Inquiry Lab, there are also various learning models began to be used to maximize educational purposes. [9] found that learning with partner through discussions was good to do to learn something. The students can easily understand difficult concepts if they are able to discuss the issues with friends [10]. This can be done through group investigation [10]. Group Investigation (GI) could put the students to learning actively and train the quantitative literacy skill. So, through this research, we want to know how the learning models can give the impact of students quantitative literacy skill. Of the facts and issues that have been described previously, this research proposed.

2. Method

This study used a quantitative approach with quasi-experimental method that aims to see the effect of Inquiry Lab and Group Investigation learning models on material environmental pollution toward the students' quantitative literacy. The experimental research groups consist of an experimental class that applied Inquiry Lab learning model and an experimental class that applied Group Investigation learning model. Both of the learning model was modified by Vee Diagram metacognition aspects. Each research groups was given an early test, then treated them with different models, and given a final test. The results of the pretest and posttest of each group were compared.

The study was conducted at a secondary public school in Majalengka on the second semester in the academic year 2015/2016. This research involves 54 students from 7th grade, divided into two experimental classes (27 students in the Inquiry Lab class and 27 students in the Group Investigation class). Subjects of this study were taken by purposive sampling. The instruments that used in this study were pretest and posttest quantitative literacy skills in the form of multiple choice questions, observation sheets of learning process, and questionnaire responses of teachers and students.

The data were analyzed by using SPSS/PC version 23 statistical program, two different t-tests were performed: Paired Samples t-test was conducted to determine whether there was a significant difference between pretest and posttest results in each group, in-group analysis, as a result of the instructional methods used in the study. Independent Samples t-test was performed to identify whether quantitative literacy skill of all Inquiry Lab and Group Investigation were equal or at least similar before the study, and to find out whether students significant changes arise between groups as a result of the instructional methods used. Significance level was decided by taking p values into consideration p>0.05, meant there was not a meaningful difference, p<0.05 meant there was a meaningful difference.

3. Finding and Result

3.1. *Analysis of students' quantitative literacy skills*

The result of pretests, posttest and N-Gain of quantitative literacy skills between the Inquiry Lab class and the Group Investigation class can be seen on Table 1.

Table 1. Descriptive statistics of students' quantitative literacy in the study.

Variable	N	Data	Inquiry Lab			Group Investigation		
			Pretest	Posttest	N-gain	Pretest	Posttest	N-gain
Students' quantitative literacy	27	$\bar{x}$	22,67	61,11	0,49	21,60	54,01	0,42
		SD	7,80	24,13	0,27	8,42	17,51	0,19
		%	68, 02	61,11		51,84	58,92	

According to the results presented in Table 1, the average pretest was 22.67 for the Inquiry Lab class and 21.60 for the Group Investigation class. As for the average of posttest score was 61.11 for the Inquiry Lab class and 54.01 for the Group Investigation class. The average of N-gain for the Inquiry Lab class was 0.49 and the average of N-gain for the Group Investigation class was 0.426. Pre-requisite test performed on the data pretest through Kolmogorov-Smirnov test for normality test and Homogeneity of Varians for homogeneity test. The results show that the pretest data was not normal and not homogenous. Then we calculated the posttest data statistically. Based on the calculation, we found that the posttest data was not was not normal and not homogenous. To determine the effect of students' quantitative literacy, we continued to non-parametric hypothesis test by using Mann-Withney U test.

Table 2. The result of U Mann Withney test on students' quantitative literacy posttest score

U Mann-Whitney	Z	Asymp. Sig. (2-tailed)
299.000	-1.143	.253

When finding in Table 2 are taken into consideration (2-tailed = .253 > p= 0.05), it can be said that two research groups did not have significant differences, this indicates that there is no difference between the posttest of Inquiry Lab class and the posttest of Group Investigation class. It can be concluded that Inquiry Lab and Group Investigation were equal to give effect of students' quantitative literacy. Although there is no statistically differences between those learning models, the students in both classes has been developed their literacy quantitative skill because both of classes have the same final competences which the average posttest score was higher than the pretest score.

3.2. *Analysis of students' quantitative literacy achievement*

Quantitative literation dimension divided into six important skills including interpretation, representation, calculation, application/analysis, assumptions, and communication, each defined in terms of proficiency level [11]. The elaboration of quantitative literacydimensions can be seen from the following table:

Table 3. Comparison of students' quantitative literacy dimensions

Group	Quantitative Literacy Dimension (%)					
	Interpretation	Representation	Calculation	Application	Assumption	Communication
IL	81,5	63	40,7	29,6	57,4	50
GI	81,5	42,6	51	55,6	50	38,9

The result shows that the highest achievement of quantitative literacy skills in Inquiry Lab class and Group Investigation class was the dimension of interpretation. In both groups, the students has the ability to explain the information of the mathematic form as well. The lowest achievement of quantitative literacy skills in Inquiry Lab class was dimension of application and lowest achievement of quantitative literacy skills Group Investigation class was dimension of communication.

4. Discussion

Based on the pretest analysis, the data shown there was no difference Inquiry Lab class and Group Investigation class on pretest achievement. The mean of Inquiry Lab class was 32,84 (22,67%) and the mean of the Group Investigation class was 31,73 (21,60%). The achievement after learning process of the Inquiry Lab class through Inquiry Lab showed the higher results than the Group Investigation class through Group Investigation. Achievement from Inquiry Lab class was 61,11 (61,11%) and the Group Investigation class was 54,01 (58,92%). Learning through Inquiry Lab tends to get higher mean score than students who learn through Group Investigation. Inquiry Lab and Group Investigation were equal to give effect of students' quantitative literacy. Although based on statistical calculation shows that there is no differences between those learning models, the students in both classes has been developed their literacy quantitative skill because both of classes have the same final competences which the average posttest score was higher than the pretest score.

Both of the experimental groups adequate to create an active learning environment for the students. Through the active learning, students could train their quantitative literacy. By giving factual quantitative data sets continuously during learning process, students encouraged to construct their new knowledge and develop their quantitative literacy. During the process of learning through Inquiry Lab, students were observed to have an enthusiastic learning. According to [12] inquiry-based learning has a positive impact on student learning achievement. Through this learning model, students and teachers can be more motivated to fill science learning process. Students are conditioned to find out the "something" new by the nature phenomenon directly. Some students claim that the learning process was very fun and it was their first experience in this semester. The results of this study indicate that modified Inquiry Lab learning models by Vee Diagrams metacognitive aspects influenced the students' quantitative literacy. Students' trained to develop their quantitative literacy skill: interpretation, representation, calculation, application/analysis, assumptions, and communication based on the numerical sets data. This finding mostly corroborates the other finding of the studies of [8] according to his research, the long term inquiry-based learning significantly could affect the quantitative literacy of students.

Good achievement result also shown by the students of Group Investigation experimental class. The involvement of students in learning process, could increase the ability of higher level thinking and metacognitive skills and also could create the meaningful learning during the class. The modified Group Investigation learning model could influence the students' quantitative literacy. Students worked with a set of quantitative data that must be investigate. Students are required to make an interpretation, representation, assumptions, calculations, applications, and communications based on their own quantitative data.

Reasoning skills has been developed through this learning activity. This is consistent with the findings of [13] in which the Group investigative learning model can improve students' reasoning ability. Reasoning itself is quite closely related to the students' quantitative literacy. Students' reasoning is the process of making conclusions based on information. It involves complex cognitive processes that need the strategy to organizing all knowledge aspects [14]. Quantitative literacy does not only consist of one aspect but involves several aspects including interpretation, representation, assumptions, calculations, applications, and communications, so it needs a strategy to achieve an integration of all quantitative literacy aspects. To achieve this, students are required to have reasoning skill. With that assumption, good reasoning capability can deliver students to good quantitative literacy.

The highest percentage of the Inquiry Lab class has indicated by the dimension of interpretation, followed by representation, assumptions, communications, and applications. While the highest percentage in the Group Investigation class was also interpretation dimension, followed by application skill, calculations, assumptions, representations, communication. Based on quantitative literacy analysis results of students' posttest, the highest percentage is interpretation dimension. This could happen because the students consider the interpretation of quantitative data, either from graphs or tables are considered easier than other dimensions. Study by [15], shows that the ability of representations, such as create a new chart requires was need more competence than interpretation skill. [16] describe a similar finding, where the result of students' performance on the tasks of interpreting the data was better than the performance of students in representation process. Other dimensions such as calculations, assumptions, applications and communications require more complex cognitive processes.

5. Conclusions

In both of the experimental groups showed an increase in students' quantitative literacy. Nevertheless, There was no significant difference between learning with metacognitive strategies based Vee Diagram through Inquiry Lab and Group Investigation toward the improvement of students' quantitative literacy. Generally, the student has been able to get high percentage of interpretation skill. This means that student has the ability to explain the information of the mathematic form as well after took learning process.

6. References

[1] Speth E B, Momsen J L, Moyerbrailean G A, Ebert M D, Long T M, Wyse S and Linton D 2010 *Sciences Education* **9** 323-332
[2] Richardson R and Callum M W G 2002 The Quantitative Literacy:Why Numeracy Matters for Schools and Colleges (United State: The National Council on Education and the Disciplines)
[3] National Council on Education and the Disciplines 2001 *Quantitative Literacy: Why Numeracy Matters For Schools And Colleges* (United State: The National Council on Education and the Disciplines)
[4] Frith V and Gunston G 2011 African J. of Health Professions Education 3 1 19-23
[5] Wardhani, Sri 2011 Instrumen Penilaian Hasil Belajar Matematika SMP: Belajar dari PISA dan TIMSS (Jakarta: Kemendiknas)
[6] Costa L A 1985 Developing Mind-A Resource Book For Teaching Thinking (Virginia: Association for Supervision and Curriculum Development)
[7] Wulansari B, Iskandar S M and Suryadharma I B 2013 Thesis, Universitas Negeri Malang
[8] Mc.Craight A M 2012 J.of the Scholarship of Teaching and Learning 12 4 86-102
[9] Santyasa I W 2007 Thesis, Universitas Pendidikan Ganesha
[10] Slavin R E 2005 Cooperative Learning (Boston: Allyn and Bacon)
[11] AAC&U 2009 Quantitative Literacy Value Rubric (United State: Association of America Colleges and Universities)
[12] UNESCO 2010 Current Challenges in Basic Science Education UNESCO (France: Education Sector 7)
[13] Noperman F 2012 Thesis, Universitas Pendidikan Indonesia.
[14] SüB H M 2002 Intelligence 30 261–88
[15] Leinhardt G, Zaslavsky O, and Stein M 1990 Review of Educational Research 60 (1) 1-64
[16] Uzun M S, Sezen N and Bulbul A 2012 Procedia-Social and Behavioral Sciences 46 2942-2946

Acknowledgments
Special thanks to headmaster and science teachers team in SMPN 2 Jatiwangi. Thanks also to all those who have helped our project, so this research can be completed properly. I extend my gratitude to my students for their enthusiastic participation in my research project.

IOP Conf. Series: Journal of Physics: Conf. Series **812** (2017) 012097 doi:10.1088/1742-6596/812/1/012097

The Effectiveness of Problem-Based Learning Approach Based on Multiple Intelligences in Terms of Student's Achievement, Mathematical Connection Ability, and Self-Esteem

A Kartikasari[1], D B Widjajanti[1]

[1]Universitas Negeri Yogyakarta, Indonesia
Email: ariskartikasari25@gmail.com

Abstract. The aim of this study is to explore the effectiveness of learning approach using problem-based learning based on multiple intelligences in developing student's achievement, mathematical connection ability, and self-esteem. This study is experimental research with research sample was 30 of Grade X students of MIA III MAN Yogyakarta III. Learning materials that were implemented consisting of trigonometry and geometry. For the purpose of this study, researchers designed an achievement test made up of 44 multiple choice questions with respectively 24 questions on the concept of trigonometry and 20 questions for geometry. The researcher also designed a connection mathematical test and self-esteem questionnaire that consisted of 7 essay questions on mathematical connection test and 30 items of self-esteem questionnaire. The learning approach said that to be effective if the proportion of students who achieved KKM on achievement test, the proportion of students who achieved a minimum score of high category on the results of both mathematical connection test and self-esteem questionnaire were greater than or equal to 70%. Based on the hypothesis testing at the significance level of 5%, it can be concluded that the learning approach using problem-based learning based on multiple intelligences was effective in terms of student's achievement, mathematical connection ability, and self-esteem.

1. Introduction

The aim of teaching and learning mathematics mostly focused to derive mathematical concept and skills, while developing the student's abiliy to recognize and apply mathematical concept in daily life situation was not a priority [1] [2]. The ability to recognize and applying mathematical concept was very important in learning mathematics, in terms of mathematics was a comprehensive and interconnected science which every concept, principle, and procedure are connected each other [3].

The connection in mathematics gives the meaning of learning and provide the reason why students learn mathematics [4]. Therefore, the ability of mathematical connetion was very important in learn mathematics.

Build the connection inter ideas will strengthened the students understanding and reduce the pressure of students to memorize and reduce their worrying of forget the materials they have gained [5] [6]. Mathematical connection was students ability to relate the mathematical knowledge and problems they gainded in mathematical class with the problems or situations they face in daily life [7]. Mathematical connection was the most important ability to learn mathematics and as one of the process standard of NCTM and one of the aim of learning mathematics in Indonesia[5].

Although the mathematical connection ability is very important, but sometimes school mathematics is seen as something that is not interconnected. This is demonstrated by the textbook, which presents mathematics as a separate and unrelated materials or topics, and yet unfamiliar students in making connections in learning. This fact was causing the ability students's mathematical connection still low.

The study of Anita [8] found that out of 72 middle school students, only 14 students or 19,44% achieve high score of mathematical conection test, 43 or 59,72% of studets have score on medium category of mathematical connection test, and 15 or 20,83% of students achieve low score of mathematical connection test. This result shows mostly students get mathematical connection score under high category.

The study of Nurasyiyah [9] showed that the high school students mathematical connection ability to connect some concepts or principless with the problems with other diciplines context or in daily life context only 13,5% of students. Total 35% of students able to applying the mathematical concept to solve the problems in another disciplines context, or daily life context. Instead, the result of prasurvey in January 2016 in MAN Yogyakarta III, showed that less than 50% of students achieve minimum high score in mathematical connection ability test.

Low ability of mathematical connection causes the lack of meaning and interest of students to learn mathematics. It can be seen from a great fear and the lack of attention of students in learning mathematics.

Mostly, students in all levesl, from elementary to high school perceive mathematics as a difficult and unattractive subject. The higher of the education level, the more complex of materials of mathematics and requires a process of abstraction. Therefore, more students in high school do not like mathematics.

The fact of lack students interest in mathematics, causes low mathematics achievement of students. Lack of achievement demonstrated by the results of the absorption of the National Examination (UN). One of the material tested in the national examination and has a low absorption is geometry and trigonometry. Based on data from the National Education Standards in 2013, 2014 and 2015, the absorption of students nationwide in the two materials is lower than other materials. One of the schools that have absorption in both the material and lower than the absorption of Sleman, Yogyakarta Special Province (DIY), as well as national level absorptive capacity is MAN Yogyakarta III.

The low of student achievement in MAN Yogyakarta III also demonstrated by the results of Deuteronomy Central Semester (UTS) in 2015/2016. The result of UTS showed that mostly students still have not reached the minimum completeness criteria (KKM), which is 75.

Instead the mathematical connection capabilities as one of the important ability and influence on student achievement in learning mathematics, other aspects that affect student achievement in mathematics is the affective aspect. One of the important affective aspects of the students is self-esteem.

Self-esteem is related to how students see themselves or individual assessments of their capabilities. Self-esteem is the acceptance of individuals against her and that she can vote, deserve to be appreciated, deserves to be successful, and valuable [10] [11] [12]. Self-esteem of the students different from one another. Self-esteem of students can be high, but can also be lower. Tendency, the students with high self-esteem will be more confident in social situations and also in dealing with the tasks assigned by the teacher, while students with low self-esteem tend to prefer punished or may be viewed as a hero by his friends than look stupid [13] [14] [15]. Therefore, it is very important for teachers to pay attention to the self-esteem of students and facilitate students to develop their self-esteem so that students will be more enthusiastic about learning mathematics and student's achievement will be increased.

Self-esteem is closely related to students' attitudes toward mathematics. One indicator of students' attitudes toward mathematics is confidence in the ability of students to learn mathematics. Based Mullis, et al [16], the confidence of students in ability to learn mathematics can be seen from the students' self confidence and self-concept of the student on the ability of students to learn mathematics. In this case the self-concept is closely related to self-esteem. This is corroborated by the opinion of Lawrence15, that self-concept can be seen as the term "umbrella" in which there is the self-esteem of an individual.

Based on the results of TIMSS, the scores obtained by Indonesian students related to confidence in the ability of learning mathematics is 9.7, the difference is only 0.3 points from the lowest score, namely 9.4 [16]. The result of TIMSS shows that the self-esteem of students Indonesia relating to confidence in the ability in mathematics is still low. Low self-esteem of students in learning mathematics, especially in MAN Yogyakarta III also obtained from the prasurvey conducted by investigators in January 2016. Based on the results of prasurvey, only 3,3% of students in very high category of self-esteem, 43,3% in the high category, 50% in the medium category, and the remaining 3,3% are in the low category. Therefore, teachers need to build community in the classroom, where each student will recognize and be actively involved in learning process.

Based on the description that has been described, the teacher must design learning that can develop the achievements, the mathematical connection ability and self-esteem of students in the learning mathematics. One approach of learning that can be used is the Problem-based learning (PBL). One of the characteristics of PBL is to provide a problem that close to real life and may occur in real life, so that PBL is a solution to facilitate the students to perform connection process between the material to real life. PBL allows students acquire new knowledge or solve problems tightly everyday life [17].

The advantages PBL in developing students' mathematical connection ability is also supported by the results of study. Based on the study conducted by Ratnaningsih [18], it showed that the students who received problem-based learning tend to btain a significant increase in the ability of connections compared to conventional learning. In addition based on the results of study, PBL chosen because it has some of the following characteristics: 1) presents a problem that close to real life and may occur in real life, 2) encourage the students are engaged in learning activities, 3) encourage the use of various approaches, 4) allow students to make a choice about how and what will be learned, 5) to encourage collaborative learning, and 6) help achieve a quality education [17].

PBL is a constructivist learning approach with many advantages may still have a problem for some students who do not have an interest in mathematics. Teachers need to consider the potential of the students, one of the potential is intelligence. According to Gardner [18] each individual has all the capacity and intelligence with different combinations. Gardner stated that there are nine types of intelligence, such that intelligence linguistic, musical, logical-mathematical, spatial, kinesthetic, intrapersonal, interpersonal, naturalits, and existential. Thus intelligences was called multiple intelligences intelligences. The theory of multiple intelligences believe that every student is unique, and at least have one intelligences [19] [20].

The diversity of intelligence of students does not mean that teachers should conduct learning activities individually, but the diversity of intelligence of students should be used as capital for a teacher to develop methods of learning and develop learning tools in order to facilitate the students through multiple intelligence possessed by the student [21]. By implementing multiple intelligences in the learning process will make students feel recognized in the classroom and students will better appreciate its intelligence [22]. Learning mathematics attention will make students feel recognized in the classroom and students will appreciate the intelligence they have and make the students realize how smart they are, that is where it can make the students develop self-esteem and self-confidence of students [23] [24]. In addition, the results of research which arecarried out by Al-Zyoud and Nemrawi [25] that the class that implements the intelligences multilpe in the learning process gives better results in the development of self-concept than students in classes that do not use multiple intelligences.

Based on the description that has been presented, learning with PBL approach by taking into account the multiple intelligences of students are expected to develop the achievements, the mathematical connection capability, and self-esteem of students in the learning of mathematics. Therefore, it is necessary to study on the effectiveness of problem-based learning approach based on multiple intelligences in terms of achievement, the mathematical connection ability, and self-esteem of students.

IOP Conf. Series: Journal of Physics: Conf. Series **812** (2017) 012097 doi:10.1088/1742-6596/812/1/012097

2. Research Method

2.1. Type, Time, Place, and Research Sample

This study was a quasi experiment. This study conducted in MAN Yogyakarta III from April until May 2016. The population of this study was all students of grade X of MAN Yogyakarta III. The sample of this study was grade X students of MIA 3 of MAN Yogyakarta III.

2.2. ResearchProcedure

This study began by providing a self-esteem questionnaire before learning process. In the implementation of learning through problen-based learning approach based on Gardner's multiple intelligences used the learning kits, consist of lesson plans and worksheets. Before preparing lesson plans and worksheets, researchers first identified the tendency of intelligence of students by distributing questionnaires tendency of intelligence, where the results are used for several things, including: preparing lesson plans and worksheets, arrange some activities based on multiple intelligences, and arrange group of students with heterogeneous intelligence. Lesson plans using PBL approach based on Gardner's multiple intelligences having learning steps as follows: (1) Teacher orient students to the problems in diverse contexts to encourage all students actively involve in learning process; (2) Teachers organize students to learn; (3) Students identify the problems and analyze the data individually or in a heterogeneous group; (4) The students presented the results in different ways; (5) The teacher and the student are evaluate the problem solving process. While the worksheets have a variety activities based on multiple intelligences, also there are some connection problems to help students to develop their connection ability, and some "quote of the day" to help students develop self-esteem. The developing teaching kit fulfill the validation criteria based on the results of expert judgement which is respectively obtaining average score 4.54 and 4.43 out of a maximum score of 5.

The instrument used in this study was a self-esteem questionnaire, achievement test and mathematical connection ability test. Based on the validity of the content, all three instruments are valid.

Instrument reliability of students' self-esteem questionnaire is 0.901 with SEM 1.38. The estimation results of the reliability of test achievement on trigonometry and geometry of the material can be seen in Table 1 below.

Table1. The reliability of Achievement Test

No	Instruments	Reliability	SEM
1	Achievement test of Trigonometri	0,784	1,807
2	Achievement test of Geometry	0,847	1,835

2.3. Data Analysis Technique

Data analysis techniques consisted of descriptive analysis and statistical analysis. Descriptive data are translated based on the average, maximum score, and minimum score. The learning approach satisfied effectiveness criteria if the percentage of students who get minimum criteria of achievement test, and achieve minimum category of mathematical connection test and also in self-esteem questionnaire are greater than or equal to 70%. Statistical analysis using the proportion test of the result of achievement test, mathematical connection test, and self-esteem questionnaire. The learning approach effective if at the significance level of 5%, the proportion of students who achieve minimum criteria of achievement test, also achieve minimum category of connection test and self-esteem questionnaire are greater than or equal to 70%.

3. Result and Discussion

The result of achievement test are presented in Table 2 below.

Table2. The Description of Achievement Test Result

No	Description	Score of Achievement Test
1	Average	77
2	Percentage of Students Completeness (%)	86,7

Tabel 3 and Table 4 respectively present the proportion of students of the result of mathematical connection test and self-esteem quistionnaire.

Table3. Mathematical Connection Test Result

No	Criterion	Number of Students	Precentage (%)
1	Very High	21	70
2	High	4	13,3
3	Medium	5	16,7
4	Low	0	0
5	Very Low	0	0

Table 4. Self-Esteem Questionnaire Result

No	Criterion	Number of Students	Precentage (%)
1	Very High	4	13,3
2	High	22	73,3
3	Medium	4	13,3
4	Low	0	0
5	Very Low	0	0

Based on Table 2, was obtained that based on achievement test result, the percentage of students who achieve score in minimum KKM was 86,6%. Table 3 and Table 4 respectively shows that the percentage of students who achieve score in minimum high cathegory of mathematical connection test and self-esteem questionnaire was 83,3% and 86,6%. The data obtained are the data that came from normal population based on normality test result. The percentage of each test used to test the proportion to find how significant the result of each test.

The next phase was performing the proportion test of student's achievement test, mathematical connection test, and self-esteem result. Based on the proportion test of achievement test obtained the Z-value 2,092 Z-table, which $Z_{0,05} = 1,645$, it concluded that H_0 rejected. At the sifnificant level 5%, it can be concluded that the problem-based learning approach based on Gardner's multiple intelligences was effective in terms of learning achievement.

This result shows that problem-based learning based on Gardner's multiple intelligences could develop the students achievement. The result appropriate with the study of Al-Zyoud and Nemrawi[25]about the effectiveness of multiple intelligences theory and the result shows that there is the significance difference between the experiment class and the control class in terms of students achievement.

Even the result was good, still there are some problems which achieve under 70% students who make correct answers. One of the problem of achievement test was number 11 on trigonometry, which achieve 20%. The problem shows in figure 1 below.

11. Seorang pengendara sepeda *downhill* menuruni sebuah bukit yang memiliki ketinggian h. Jika kemiringan bukit adalah 20,2° dan pengendara tersebut membutuhkan waktu 10 menit sampai ke bawah dengan kecepatan tetap 60 km/jam, Rumus trigonometri yang sesuai untuk mencari tinggi bukit adalah ....

A. $10 \times \cos 20{,}2°$ D. $600 \times \cos 20{,}2°$

B. $10 \times \sin 20{,}2°$ E. $600 \times \sin 20{,}2°$

C. $10 \times \tan 20{,}2°$

Figure 1. Achievement Test of Trigonometry

The result of students work on problem number 11, 13.3% of students answered A, 20% of students answered B, 6.7% of students answered C, 6.7% answered D, and 53.3% answered E. Mostly students choose E, namely $600 \times \sin 20{,}2°$. Most likely the students answered E due to an error in determining the distance rider. The distance calculation used should $60km\ /\ h\ \times\ 10\ =\ 600$. The correct calculation is first change the number of minutes in an hour, so $10\ minutes\ = \frac{1}{6} hour$, so that the correct distance is $60\ km/h \times \frac{1}{6} h = 10$ km. The correct answer to question number 11 was B.

Furthermore, based on the proportion test of mathematical connection test results, obtained Z-value was 1,697 where the value is greater than from Z table, which is $Z_{0,05} = 1.645$, so H_0 rejected. Then, at a significance level of 5%, it can be conclude that probleim-based learning approach based on Gardner's multiple intelligences effective in terms of mathematical connection test results.

These results correspond with the results of study of Sumarmo and Permana[18]. The results showed superiority of experimental group, which is group with problem-based learning approach than the control class. On the implementation of learning by problem-based learning approach based on Gardner's multiple intelligences of students accustomed to practice solving the problems associated with other topics, everyday life as well as those related to other disciplines in diverse contexts, so that students are accustomed to make a connection. In addition, student learning is given a few questions to choose the procedure used or concepts used in solving the problem. It is appropriate to[26], that in order to assist students in making connections and help the to develop their thinking, judging conclusions, strategies and procedures used. Although the mathematical connection test results have been good, but there is a number that have low achievement, which is a connection problem of number 2. In which from 30 students, only five students who answered correctly. Based on the interview, mostly students find the problem at number 2 was too difficult.

Furthermore, the proportion test of the result of self-esteem questionnaire obtained Z = 2.092 greater than $Z_{0,05}$= 1.645, so that H_0 rejected. At the 5% significance level, learning by problem-based learning approach based on Gardner's multiple intelligences effective in terms of students self-esteem. This is consistent with the notion that multiple intelligences based learning makes students realize how smart they are, that is where it can make the students develop self-esteem[17].

These results are also consistent with the results of Rafianti study[26] that students who join in the learning based on Gardner's multiple intelligences a high self confidence. This study is relevant because the self-esteem and self-confidence are interconnected. belief or confidence (self-confidence) that the person is capable of performing the task can raise self-esteem or self-esteem[10]. Moreover, the results indicate that self esteem scores for all items increased, and there are three item questionnaire highest increase, ie item numbers 7, 8, and 17. Where the item number 7 and 17 refer to the indicator of self-esteem about showing confidence of being acceptable by others, while the item number 8 respect to the indicators of self-esteem about understanding the self. The increase was made possible due to the application of the theory of multiple intelligences, which gives students the chance to learn in a variety of different ways, and the activities of the group in which one group is composed of some intelligence, so that each student can be actively engaged in learning, and feel valued.

4. Conclusion and Suggestion

At the 5% significance level, learning by problem-based learning approach based on Gardner's multiple intelligences effective in terms of student's achievement, the mathematical connection capability, and self-esteem of students. However, although there are some problems have been effective on achievement tests and mathematical connection test result which have low achievement. While some of the items on the self-esteem questionnaire having high improvement, which is possible because the learning that take account of students' diversity as one of the characteristics of the multiple intelligences theory that led the students felt more appreciated in learning.

Therefore, it is advisable for teachers to use problem-based learning approach based on Gardner's multiple intelligences in the learning process of mathematics. For the next researcher is expected to consider the difficulty level of the problems in order to obtain better results.

5. References

[1] Kennedy LM, Tipps S and Johnson A2008 *Guiding Children's Learning of Mathematics (11th Ed.)*(Belmont, CA: Thomson Wadsworth)
[2] Gainsburg J2008*Journal of Mathematics Teacher Education***11** 199-219
[3] Brumbaugh D K, Moch P L and Winkinson ME 2005 *Mathematics Content for Elementary Teachers* (London: Lawrence Erlbaum Associates)
[4] Johnson E B 2009*Contextual Teaching and Learning: What Is And Why It's Here To Stay.*(Thousand Oaks, CA: Sage Publications)
[5] NCTM 2000 *Principles and standards for school mathematics* (Reston, VA: NCTM)
[6] The Report of the Expert Panel on Student Success in Ontario 2004 *Leading Math Success: Mathematical Literacy, Grades 7–12*(Ontario: Ontario Education)
[7] Jaisook S, Chitmongkol S, Thongthew S 2013 *Journal of Social Sciences, Humanities, and Arts, Silpakorn University***13** 271-294
[8] Anita I W2011 *Prosiding Seminar Nasional Matematika STKIP Siliwangi Bandung***3**186-193.
[9] Nurasyiyah D A 2014 *Jurnal Ilmiah Matematika dan Pendidikan Matematika***6** 115-125.
[10] Schunk D H 2012 *Learning Theories: An Educational Perspective (6th Ed.)*(Boston, MA: Pearson Education)
[11] Heatherton T F and Wyland C L 2003Assessing self-esteem*Assessing Positive Psychology*ed S. Lopez and R. Snyder (Washington, DC: APA) pp 219-233
[12] Branden 1994 *The Six Pillars of Self-Esteem* (New York: Bantam Book)
[13] Alpay E2008 *Self-concept and self-esteem*(London: Department of Chemical Engineering and Chemical Technology, Imperial College of Science)
[14] Lawrence D 2006 *Enhancing Self-Esteem in the Classroom (3rd Ed.)* (London: Chapman Publishing)
[15] Happy N and Widjajanti D B 2014 *Jurnal Riset Pendidikan Matematika* **1** 48-57
[16] Mullis IV S, Martin M O, Robitaille D F and Foy P 2012 *TIMSS 2011 International Results in Mathematics* (Boston, MA: TIMSS and PIRLS)
[17] Delisle R 1997 *How to Use Problem-Based Learning in The Classroom* (New York: Association for Supervision and Curriculum Development)
[18] Permana Y and Sumarmo U 2007 *Jurnal Educationist***1** pp116-123
[19] Gardner H 2011 *Frames of Minds: The Theory of Multiple Intelligences* (New York, NY: Basic Books)
[20] Hoerr T R 2000*Becoming A Multiple Intelligences School* (Alexandria, VA: Association for Supervision and Curriculum Development)
[21] Chatib M 2009 *Sekolahnya Manusia: Sekolah Berbasis Multiple Intelligence di Indonesia* (Bandung: Khaifa)
[22] Widjajanti D B 2012*Prosiding Seminar Nasional Penelitian, PendidikandanPenerapan MIPA*(Yogyakarta: FMIPA UNY)
[23] Kohar A W and Rosyidi AH2011 *SEA-DR Proc. Int. Conf.*(Palembang: Universitas Sriwijaya) pp 20-31
[24] Fleetham M 2006 *Multiple Intelligences in Practice: Enhancing Self-Esteem and Learning in The Classroom* (Stafford: Network Continuum Education)
[25] Al-Zyoud N F and Nemrawi Z M 2015 *American International Journal of Social Science***4** 171-180
[26] Rafianti I 2013 *Penerapan Model Pembelajaran Matematika Berbasis Multiple Intelligences Terhadap Kemampuan Pemahaman Konsep, Penalaran Matematis, dan Self Confidence Siswa MTs.* (Bandung: TesisUniversitasPendidikan Indonesia)

IOP Conf. Series: Journal of Physics: Conf. Series **812** (2017) 012098 doi:10.1088/1742-6596/812/1/012098

Design and Development Computer-Based E-Learning Teaching Material for Improving Mathematical Understanding Ability and Spatial Sense of Junior High School Students

Nurjanah[1], J A Dahlan[1], Y Wibisono[1]

[1]Department of Mathematics Education, Universitas Pendidikan Indonesia, Bandung 40154, Indonesia

E-mail: nurjanahfattah@yahoo.co.id

Abstract. This paper aims to make a design and development computer-based e-learning teaching material for improving mathematical understanding ability and spatial sense of junior high school students. Furthermore, the particular aims are (1) getting teaching material design, evaluation model, and intrument to measure mathematical understanding ability and spatial sense of junior high school students; (2) conducting trials computer-based e-learning teaching material model, asessment, and instrument to develop mathematical understanding ability and spatial sense of junior high school students; (3) completing teaching material models of computer-based e-learning, assessment, and develop mathematical understanding ability and spatial sense of junior high school students; (4) resulting research product is teaching materials of computer-based e-learning. Furthermore, the product is an interactive learning disc. The research method is used of this study is developmental research which is conducted by thought experiment and instruction experiment. The result showed that teaching materials could be used very well. This is based on the validation of computer-based e-learning teaching materials, which is validated by 5 multimedia experts. The judgement result of face and content validity of 5 validator shows that the same judgement result to the face and content validity of each item test of mathematical understanding ability and spatial sense. The reliability test of mathematical understanding ability and spatial sense are 0,929 and 0,939. This reliability test is very high. While the validity of both tests have a high and very high criteria.

1. Introduction

Among the various branches of mathematics, geometry of position of the most concern. There are still many students who had difficulty in understanding the geometry, particularly the geometry of space [1] [2] [3]. Similarly this condition found in Indonesia on the level of primary and secondary education [4] [5]. Factors that play a role in learning geometry that is the understanding of the concept and spatial sense [6] [7] [8]. Based on the above description, then the necessary efforts to develop and improve the ability of mathematical understanding and spatial sense. Therefore researchers propose a study with the title "Design and Development Computer-Based E-Learning Teaching Material for Improving Mathematical Understanding Ability and Spatial Sense of Junior High School Students".

Design and development computer-based e-learning teaching material referred to in this study is a strategy of learning by the use of computer media for delivering learning materials all or part with the help of the media learning shaped interactive multimedia software. As for the problems in this research is "How to design and development computer-based e-learning teaching material for improving mathematical understanding ability and spatial sense of junior high school students?"

2. Method

This research method used of this study is developmental research which is conducted by thought experiment and instruction experiment, and concludes with a study of experiments for the purposes of validation of the learning model developed.

Step of research are reflected in the diagram below:

Stage	Forms of Study	Types of Methods	Research Steps
I	Theoritical	Study of The Development	Design and development computer-based e-learning teaching material for improving mathematical understanding ability and spatial sense of junior high school students
	The Empirical	The Study Implementation	Getting teaching material design, evaluation model, and instrument to measure mathematical understanding ability and spatial sense of junior high school students
		The Study Implementation	Conducting trials computer-based e-learning teaching material model, asessment, and instrument to develop mathematical understanding ability and spatial sense of junior high school students
	Theoritical	Study Documentation	Completing teaching material models of computer-based e-learning, assessment, and develop mathematical understanding ability and spatial sense of junior high school students

Figure 1. Research Flowchart

3. Results And Discussion

3.1 The development of learning materials

Computer-based e-learning teaching material used in the study is organized in the form of computer-based multimedia learning using Macromedia Flash software program and 3D Max. The preparation of computer-based multimedia considering students ability in operating computer designed to increase learning ability in understanding mathematical and spatial sense on geometry with flat-sided

The procedure of making computer-based multimedia learning according to Rusman [9] is (1) manufacturing of implementation plan of learning; (2) the creation of computer-based learning programme planning, including the introduction (the title of the program, the purpose of presentation, instructions), goals, learning experiences, and treatment; (3) the creation of computer-based learning, flowchart in accordance with the model used; (4) the making of a multimedia learning computer-based storyboards. Further development of learning materials is done through the following steps.

- Analysis of computer programs according to the presentation of images and animation in the concepts of mathematics. At this stage the researcher using Macromedia Flash with 3D Max that has any purpose provider tool making animations or dynamic and interactive writing.

Furthermore the technology choice of the resulting animations can be run in multiple platforms and a wide range of computers.

- Analysis of the subject matter of mathematics junior high school class VIII. At this stage the subject was taught on a semester in progress to inventory and further prepared to become subject to published materials. The fundamental thing of taking the topic of learning materials is the essence and the suitability of the presentation with computer media. After selected topics that meet these criteria, then made public its computer program design. In this stage the author chose the material geometry with flat sided is presented in the form of interactive learning materials.
- Manufacture of model design learning materials in the form of interactive computer programs. First thing to do after choosing an appropriate discussion topic presented with computer media is designing a computer program Flowchart, and then continued with the storyboarded. Story board this is the reference framework, making the program more. In addition to visualize ideas story board also presents the contents of the display of interactive multimedia.
- The process of making an interactive multimedia program. After the making of the Story Board, then the next stage is the process of making an interactive multimedia program.

Development of computer-based e-learning teaching material prior to use validated by expert people multimedia 5. As for the given sheet of multimedia considerations include: (1) Learning materials which are made in accordance with the principles of interactive multimedia computer-based; (2) Learning materials through interactive multimedia can help students develop the ability of understanding the mathematical and spatial sense; (3) Guidance in the learning materials in accordance with the level of development of junior high school students; (4) Organizing systematic learning materials; (5) Materials through interactive multimedia can help students establish the concepts/principles of mathematics with their own abilities; (6) The language used is clear/precise; (7) Images or representations mathematically used clearly/precisely. Based on the results of consideration to 5 validator retrieved the conclusion that learning materials can be used very well. As for the instruments used in this research are mathematical comprehension tests and tests the ability of spatial sense. A test of the ability of mathematical understanding and spatial sense ability tests developed by researchers based on a scoring rubric refers to Facione [10].

Instruments validated the theoretical validity of face and contents. Validity was done by 5 judgements people who are considered experts in mathematics education. The purpose of doing validation is expected to generate a valid instrument and reliability. As for the validity of the contents weighed are associated with: (1) The given subject matter; (2) Indicators of achievement of the learning results; (3) The aspect of the ability of mathematical understanding and spatial sense measured; (4) The difficulty level for junior high school students of class VIII. While the validity of advance which weighed include: (1) The clarity of the language or editorial; (2) The clarity of the picture; (3) The language is standard.

After the validity of the face and contents by 5 judgements done, then done little revision and testing to students who have received material learning materials to be used in research activities. Tryout to know the reliability, validity of the distinguishing power grain of matter, and the difficulty level of each grain of matter. the whole things is measured on the test input problems on test instruments and statements on non test is used or not.

Below will be described from each of the research instrument:

3.2 Test the ability of Mathematical Understanding

The results of the consideration of the validity of the advance and the content of the five penimbang with the hypothesis tested is as follows.

H_o : All judgements give the same considerations

H_1 : All judgements gives the consideration is not the same (at least there are two different judgements give consideration)

The criteria used is the test: If the value of the probability (sig) greater than $\alpha = 0.05$, then H_o accepted and in other circumstances H_o denied.

The results of the calculation of validity and contents of mathematical comprehension tests by using the Q-statistic Cochran presented in table 1.

Table 1. Test Results of Consideration of The Validity of Advance Mathematical Understanding Ability Test

N	Cochran's Q	Df	Asymp. Sig.
13	2.476[a]	4	.649

Based on table 1 above seen that value the Q-statistic Cochran to face validity was 2.476 asimtotis 0.65 significance with numbers greater than 0.05, so Ho accepted. So all judgements give the same reasoning against the validity of advance at any granule tests. As for the validity of the content of the assessment with a score of judgements give the same i.e. 1 so that it can be concluded all judgements give same consideration against the validity of the contents of each grain tests MU.

Next to test the reliability and validity of the grain problem by using reliability test values obtained ANATES 0.929 belongs to very high. Based on the calculation of the accrued ANATES 8 grains of matter, 2 grains of matter have validity with high criteria and 6 grains of matter has the validity criteria are very high. The results of the analysis showed that the mathematically understanding ability tests can be used for research.

3.3 Test the ability of Spatial Sense

The results of the consideration of the validity of the advance and the content of the five judgements with the hypothesis tested is as follows.

H_o : All judgements give the same considerations

H_1 : All judgements gives the consideration is not the same (at least there are two different judgements give consideration)

The criteria used is the test: If the value of the probability (sig) greater than $\alpha = 0.05$, then Ho accepted and in other circumstances Ho denied.

The results of the calculation of validity of the advance and the contents of the test spatial sense by using Q-statistic Cochran presented in table 2 and table 3.

Table 2. Test Results of Consideration of The Validity of The Advance Tests The Ability of Spatial Sense

N	Cochran's Q	Df	Asymp. Sig.
16	6.435[a]	4	.169

Table 3. Test Results of Consideration of The Validity of The Content of The Tests The Ability of Spatial Sense

N	Cochran's Q	Df	Asymp. Sig.
16	4.000[a]	4	.406

Based on table 2 above seen that value the Q-statistic Cochran to face validity is 6.4 with number asimtotis 0.17 significance greater than 0.05, so H_o accepted. From table 3 is also seen that the value the Q-statistic Cochran for the validity of the content is 4 by number asimtotis 0.41 significance greater than 0.05, so H_o accepted so the judgements give the same reasoning against the validity of advance and the contents of each grain of ability test spatial sense.

Next to test the reliability and validity of the grain problem by using reliability test values obtained ANATES 0.939 belongs to very high. Based on the calculation of the accrued ANATES 12 grains of matter, 6 grain question has validity with high criteria and 6 grains of matter has the validity criteria are very high. The results of the analysis show that tests the ability of spatial sense can be used for research.

4. Discussion

Learning materials computer-based e-learning that is generated is interactive multimedia learning materials-based computer by using several models of learning (model tutorial, simulation, discovery, drill and practice) in the form of a media text, images, sound, animation, and interaction to convey the overall learning materials packed in computer software (CD study).

Based on the results of the validation of materials against computer-based e-learning that is validated by an expert multimedia, retrieved the conclusion that learning materials can be used very well. As for pieces of interactive multimedia computer-based considerations are given include: (1) Learning materials which are made in accordance with the principles of interactive multimedia computer-based; (2) Learning materials through interactive multimedia can help students develop the ability of understanding the concept and spatial sense; (3) Guidance in the learning materials in accordance with the level of development of junior high school students; (4) Organizing systematic learning materials; (5) Materials through interactive multimedia can help students establish the concepts/principles of Mathematics with their own abilities; (6) The language used is clear/precise; (7) Images or representations mathematically used clearly/precisely.

The following is a printscreen from one of the videos on an interactive multimedia program about the shape of a flat wake looks up, looks ahead and looks right in the wake of its spatial model.

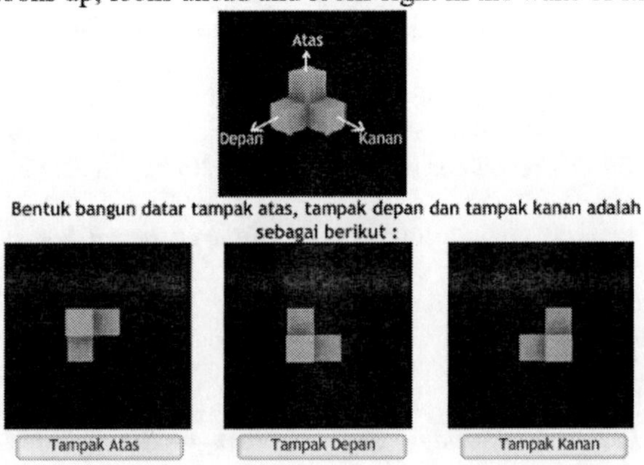

Figure 2. Video Footage on Learning

Based on an interactive multimedia learning materials are developed, it is possible to have a spatial sense ability students increased, because students were given activities that can develop their spatial sense. This is in line with Van den Heuvel & Buys [11] which States that the activities of the geometry can stimulate the child's ability to refine and help in developing spatial sense as well as the skill of thinking of the child. This is in line with the opinion of the Bruni & Seidenstein [12] and Liedtke [13] which declare that spatial sense can be improved through specific activities forms in geometry.

5. Conclusion

The results showed that the learning materials computer-based e-learning that is validated by expert people multimedia 5 can be used very well. The results of the consideration of the validity of the advance and the contents of the 5 judgements shows that the judgements gives the same reasoning against the validity of advance and the contents of each grain of mathematical understanding ability test and spatial sense. Reliability ability test mathematical understanding and spatial sense of 0.929 and 0.939 belongs to very high. While the validity of both the test has high and very high criteria. The results of the analysis showed that the mathematically understanding ability test instruments and spatial sense can be used.

6. References

[1] Suwaji U T 2008 *Paket Fasilitasi Pemberdayaan KKG/MGMP Matematika. Permasalahan Pembelajaran Geometri Ruang SMP dan Alternatif Pemecahannya.* (Yogyakarta: Pusat Pengembangan dan Pemberdayaan Pendidik dan Tenaga Kependidikan Matematika)

[2] Ben-Chaim D, Lappan G and Houang R 1989 *Focus on Learning Problems in Mathematics* **11** 49-60

[3] Orton A 1992 *Learning Mathematics Issue, Theory and Classroom Practice* (New York: Cassel)

[4] Soedjadi 1991 *Wajah Pendidikan Matematika di Sekolah Dasar Kita. (Beberapa Hasil Pengamatan Lapangan sebagai upaya Perbaikan di Masa Depan)* (Jakarta: Makalah Penataran penyiapan Calon Penatar Dosen PGSD-DII Guru Kelas)

[5] Fauzan A 1996 *Penelusuran Kemampuan Persepsi Ruang Siswa Kelas I SMU di Propinsi Sumatera Barat.* (Surabaya: Tesis Program Pascasarjana IKIP Surabaya)

[6] Soedjono 1983 *Diagnosis Kesulitan Belajar dan Pengajaran Remedial Matematika* (Jakarta: P2LPTK)

[7] Del Grande J 1990 *Arithmetic Teacher* **37** 14-20

[8] Wheatley G H and Reynolds A M 1999 *Developing Spatial Sense* [online: https://my.nctm.org/eresources/article_summary.asp?URI=TCM1999-02-374a]

[9] Rusman, Kurniawan D and Riyana C 2012 *Pembelajaran Berbasis Teknologi Informasi dan Komunikasi : Mengembangkan Profesionalitas Guru* (Jakarta: Rajawali Pers)

[10] Facione P A 1994 *Holistic Critical Thinking Scoring Rubric* [online: www.temple.Edu/tlc/resources/handouts/holistic20Critical20Thinking20Scoring%20Rubric.v2.pdf]

[11] Van den Heuvel M and Buys K 2005 *Young Children Learn Measurement and Geometry. A Learning-Teaching Trajectory with Intermediate Attainment Targets for the Lower Grades in Primary School.* (Utrecht: Freudenthal Institute, Utrecht University)

[12] Bruni J and Seidenstein R 1991 Geometric concepts and spatial sense *Mathematics for The Young Child* ed J N Payne (NCTM: Reston) pp 202-227

[13] Nurjanah and Subardja S 2003 *Suatu upaya Menumbuhkan Spatial Sense dalam Pembelajaran Geometri di SLTP melalui Model Pengajaran dengan Alat Peraga.* (LPTK: DIKTI)

Utilization of Lesson Analysis as Teacher Self Reflection to Improve the Lesson Design on Chemical Equation Topic

E. A. Edyani[1, a)], A. Supriatna[2], Kurnia[2], L. Komalasari[3]

[1]Sekolah PascaSarjana, UniversitasPendidikan Indonesia, Jl. Dr. Setiabudi No. 229, Bandung 40154, Indonesia.
[2]Departemen Pendidikan Kimia, Universitas Pendidikan Indonesia.
[3]SMA Laboratorium Percontohan, Universitas Pendidikan Indonesia.
[a)]icha.edyani@gmail.com

Abstract. The research is aimed to investigate how lesson analysis as teacher's self-reflection changes the teacher's lesson design on chemical equation topic. Lesson Analysis has been used as part of teacher training programs to improve teacher's ability in analyzing their own lesson. The method used in this research is a qualitative method. The research starts from build lesson design, implementation lesson design to senior high school student, utilize lesson analysis to get information about the lesson, and revise lesson design. The revised lesson design from the first implementation applied to the second implementation, resulting in better design. This research use lesson analysis Hendayana&Hidayat framework. Video tapped and transcript are employed on each lesson. After first implementation, lesson analysis result shows that teacher-centered still dominating the learning because students are less active in discussion, so the part of lesson design must be revised. After second implementation, lesson analysis result shows that the learning already student-centered. Students are very active in discussion. But some part of learning design still must be revised. In general, lesson analysis was effective for teacher to reflect the lessons. Teacher can utilize lesson analysis any time to improve the next lesson design.

1. Intorduction

Teachers as an important pillar in the world of education should have four competencies: pedagogic competency, personal competency, social competency, and professional competency. (Permendiknas, 2007). These competencies are crucial for the success of an education. Although wearing a good curriculum, but without balanced with the ability of the teacher then all in vain. Therefore, the Government demanded that all teachers to increase the competency of their self. One way teachers can do in enhancing the roles and responsibilities of the profession is to always do a self-reflection (Rahman, 2014). Self-reflection is a process of thinking back to interpret the experience for better planning in the future (Alwasilah, 2011). Through self-reflection teachers review the learning process that has been done, see advantages and disadvantages, evaluate, then apply it in further learning. This is expected to increase the teacher's self-competence. Especially in the designing process of learning (lesson design).

Lesson analysis is a way to see, hear, describe, discuss, and understand the interactions among teacher and students that comprise a lesson (Romagnano, Evans, & Gilmore, 2008). Many expert develop lesson analysis framework i.e. Fernandez, Kuno, Matsubara, and Hidayat&Hendayana. Hidayat&Hendayana framework already adapted to the process of education and learners in Indonesia. In this framework, teachers can see whether the learning process already student center or still teacher center, teachers can also view the interactions that occur between teachers with students and students

IOP Conf. Series: Journal of Physics: Conf. Series **812** (2017) 012099 doi:10.1088/1742-6596/812/1/012099

with students. Previous research that has been done by S (2015) shows that the teacher can generate a lesson of good design through self-reflection process with the help of lesson analysis on simple organic and inorganic compounds nomenclature topics. The next question is, how utilizing a lesson analysis as teacher self-reflection can improve the lesson design on chemical equation topic? In this study, main focus is the analysis of the effects of self-reflection process with the help of lesson analysis towards the lesson design designed by teachers.

2. Method

Qualitative method was used in this research. All data shown as descriptive form. The data collected is in the form of words of picture rather than number (Sugiyono, 2013). The research starts from build a chemical equation lesson design, implementation lesson design to senior high school student, utilize lesson analysis to get information about the lesson, and revise lesson design. The revised lesson design from the first implementation applied to the second implementation, resulting in better design for second implementation. The lesson process in the second implementation also analyzed by using lesson analysis. Then the findings from the analysis will be used to revise the last lesson design.

This research use lesson analysis Hendayana&Hidayat framework (2013) that have been adapted with student characteristic and learning condition in Indonesia. During the lesson process, all students and teachers activity are recorded using voice recorder and video recorder. Those records then transcripted, and applied to Hendayana&Hidayat lesson analysis framework. This lesson analysis format consist of conversation codes that occurs during lesson process. Each student-student and student-teacher conversation will be coded. The pattern that formed from the coding result will shows whether the lesson is teacher-centered or student-centered. Teacher then may use that pattern to reflect the lesson and improve the next lesson process. Flow of the research can be seen on Figure 1.

Figure 1.Research cycle

3. Results And Discussion

3.1. Build a Chemical Equation Lesson Design

Lesson Design that made in this research is different from Lesson design in general. The design consist of lesson activity or deductive situation created by the teacher, prediction of student's response, and teacher's anticipation. Objective of the lesson is to make the student be able to writes and equalizes chemical equation. So that the learning objectives achieved, the student's response prediction and teacher's anticipation should also be included into lesson design.

Lesson design that has been designed consists of two events, classical discussions and group. Both of these activities are designed to be student-centered learning through teacher's directives. The learning process of chemical equation starts from classical session. At classical session, teacher demonstrate reaction of nail with hydrochloric acid. This demonstration was observed by the students.Students and teachers then jointly determine the changes and write the equation.Then the students assisted by the teacher determine how to equalize the coefficients of equation.

Once the students gain a way to write and equalizes the equation, the learning process switched to a group. Students were divided into 8 groups, each consisting of four. Students work on five questions given by the teachers through discussion with a friend in the group. Each question will be reconfirmed in the classical style with the help of teachers.

3.2. First Lesson Analysis and Teacher Self Reflection

Following data obtained after transcripting and applying the recorded sound and video into lesson analysis format:

3.2.1. Classical Session. At this session, there are 184 classical dialogue/step which consists of 170 steps monologue of teacher's question-student's responses, and 10 steps student's initiative-teacher's response. Table 1 shows the categories of monologue teacher's question-student responses appear on a classical session in the first implementation. Based on existing data, it can be concluded that the learning process of a classical session in the first meeting is almost student-centered. This is because many students respond or argue when the teacher asked. Students do not just answer the close-question or resume fragments, but students often provide longer answers. As at the time of observing the demonstration, students observe and propose changes. There is also a dialogue initiative of the student-teacher response in the form of questions. This dialogue is there, although only slightly.

Table 1. Step number appears during classical discussions in the first implementation

No.	Category		Step Number Appears
	Teacher's Question	**Student's Response**	
1	Ask Yes/No	Yes/No	7
2	Ask resuming fragments	Resuming fragments	11
3	Ask definition	Answer by remembering notes	5
4	Ask with limited answer	Choose answer with no reason	12
5	Ask reason (contend)	Answer by reading text-book	39
		Answer by remembering notes	41
		Answer using own words	50
6	Paraphrase question	Answer using own words	6

3.2.2. Group Discussion Session. In the group sessions there are 1093 steps student's dialog. Number of dialogue with the involvement of teachers are 438 steps, while the dialog without the involvement of teachers are 655 steps. Based on these data it can be seen that the student-student dialog more dominant than the dialog between teacher-student. This indicates that students are active in group discussions. But in the group discussion of this first meeting there still some interruptions or teacher's involvement during the discussion process. Data of student's dialog with teacher's involvement is shown in Figure 2, and without teacher's involvement is shown in figure 3.

Based on figure 2, dialogs that occur the most are TTS-RIP and TPS-RPI. TPS-RPI dialog is a dialog in which student asks for teacher's approval regarding their assignment, and teacher lead them to find correct answer through question. TTS-RPI is a dialog in which student asks a solution of unresolved assignment, and teacher answers by giving question so the student can find their own answer. Based on this data, it can be concluded that there still many teacher's involvements in group discussion session. Teachers should let students discuss more with friends in the group. But teachers can anticipate questions from students well. Teachers did not directly answer student's questions, but first lead students by asking questions. Through these questions students are encouraged to think and obtain their own answers.

Figure 3 shows the kind of student's dialogue that appears during the student group discussion. The dialog that appears most in every group is EKB and EKM. These data indicate that students in each group do a question and answer to complete the tasks assigned by the teacher. Besides question and answer dialog, the students also argue about a problem. It can be seen from the number of agree-expressions (EKS) and disagree-expressions (EKT) showed by the students. In addition there is also the expression of "AHA" indicating that students achieve or obtain something. This further confirms that the students work together in solving the problem given by the teacher in the group.

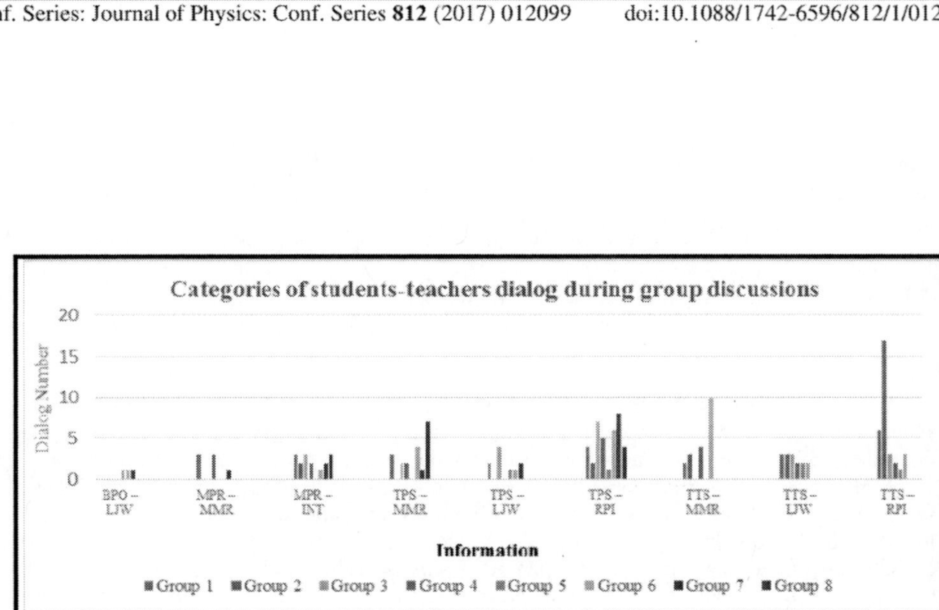

BPO – LJW	:	Student ask procedure Teacher answer student's question directly	TTS – MMR	:	Student ask unresolved solution Teacher instructs the student to do something
MPR – MMR	:	Student report group progress Teacher instructs the student to do something	TTS – LJW	:	Student ask unresolved solution Teacher answer student's question directly
MPR – INT	:	Student report group progress Teacheraskresult/progress/notification	TTS – RPI	:	Student ask unresolved solution Teacheranswers using scientific question
TPS – MMR	:	Student ask teacher approval Teacher instructs the student to do something	TPS – LJW	:	Student ask teacher approval Teacher answer student's question directly
			TPS – RPI	:	Student ask teacher approval Teacheranswers using scientific question

Figure 2. Categories of students-teachers dialog during group discussions in the first implementation

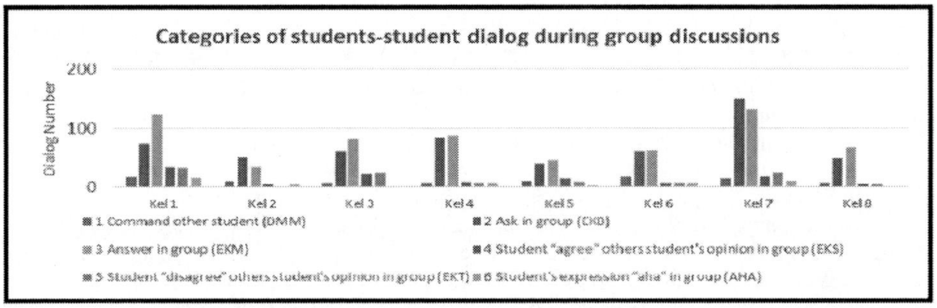

Figure 3. Categories of students-student dialog during group discussions in the first implementation

Based on the data that has been described above, it can be concluded that in the implementation of first chemical equation lesson design, the learning process has been student center, although there are still many interruptions from the teacher. Besides the data described above, it can be seen how the teacher guides the learning process, whether the learning process goes according to lesson design or not. Based on the results of the lesson analysis, teacher self-reflection results obtained through interviews as follows:

G : *In classical session, students have a great curiosity. Can be seen from the number of students coming to see the changes that occur from the reaction between nail and HCl. Students actively answer teacher's question about writing and equalizing chemical equation. Even in classical session students rarely ask, but all of the students involved in these activities.*

G : *Sharing that occurred in the group has been very good. Students are divided into groups in which there are students with high cognitive ability, medium and low. Then one group consisted of two men and two women. Therefore, the discussion in the group will be effective. Although there is still many question to teacher.*

G : *Almost all the group can solve the jumping problem. This happens because of the assistance given by the teacher that rice contains carbohydrates. In this session, the students are very enthusiastic to search for information, from open books and search the internet. Students discover the formula of carbohydrates but are still confused by the combustion reaction and write energy is formed. Overall, students actively discuss, to find out, keen to communicate its findings, as well as appreciate the friend who gives some opinions.*

G : *In terms of attitude, there are many students who changes at the time of learning, some students who had just silence before is being active and even willing to move forward, like Rafli and Harpa. Students who previously lacking focus and work on other things while learning, on learning yesterday they were focus and discuss in the group. In terms of cognitive, students who previously could only write chemical equation of a simple matter, be able to write and equalizes chemical equation of a problem or application.*

G : *Most students already know how to equalize the equation. Students are less able to write down chemical equation caused by lack of knowing about nomenclature compound.*

G : *There are many groups, the total are 8 groups. Very tired when describing or response student's question in each group. In addition there no enough time for confirming at the end, so the confirmation wasn't effective.*

G : *In general what is planned has been implemented. But it takes so much time in first classical session, so the confirmation about the activities and the final confirmation becomes not effective.*

G : *In terms of lesson design, the first classical activities should not be too long to allow more time to confirm all the questions well. Then a final confirmation really has to be effective so that the students fully understand. Change the demonstration to a more observable, such as the burning of Mg. Besides the nomenclature of the compounds will be emphasized so that they can write the equation. In terms of technical implementation, at the time of the classical all students have to actually rotate the seat facing forward so that everyone is in focus. During the discussion the teachers reduce interruptions and answer student questions. Teachers should direct them to discuss.*

3.3. Revise First Lesson Design

Based on the results of teacher self-reflection, it can be concluded that the interruption of teachers in the next learning should be reduced in order to maximize the student's ability to complete the task by discussing in groups. Demonstration of nails and hydrochloric acid is less attracted to the students and students hard to write the reaction, therefore, the demonstration should be replaced with a more interesting reaction, burning Mg. In addition, teachers must review the rules of nomenclature of compounds to make students more easily write down the equation. Then, answer's confirmation is also less effective so that many students hesitate and many students ask answers and confirmation to teachers.

3.4. Second Lesson Analysis and Teacher Self Reflection

After the lesson design was revised and implemented in a different class, the teacher re-analyze and do self-reflection. A*t classical* session, there are 320 steps that consist of 253 steps monolog of teacher's question-student's response and 13 steps monolog of student's initiative-teacher's response. Table 2 shows the categories monologue of teacher's question-student's response that appear in classical session at the second meeting. Based on these data it can be seen that the teacher has been directing the learning process in becoming a classical session centered on students. It can be seen from the reduced number of questions with short answers and the growing number of questions with open answers. Student's re*sponse was also nice, shown in t*able 2 many students answered teacher's questions using their own sentences.

Table 2. Step number appears during classical discussions in the second implementation

No.	Category		Step Number Appears
	Teacher's Question	**Student's Response**	
1	Ask Yes/No	Yes/No	5
2	Askresuming fragments	Resuming fragments	3
3	Askdefinition	Answer by remembering notes	3
4	Ask with limited answer	Choose answer with no reason	29
5	Askreason (contend)	Answer by reading text-book	24
		Answerwith student's own argument	149
6	Paraphrase question	Answer by remembering notes	8
		Answer using own words	26

Based on table 2 can be seen that in the second implementation, the student-teacher dialogs were much reduced. This means that teacher already reduces interruptions during the group discussions. Data of student's conversation during group discussions with teacher's involvement is shown in figure 4, and without the teacher's involvement is shown in figure 5.

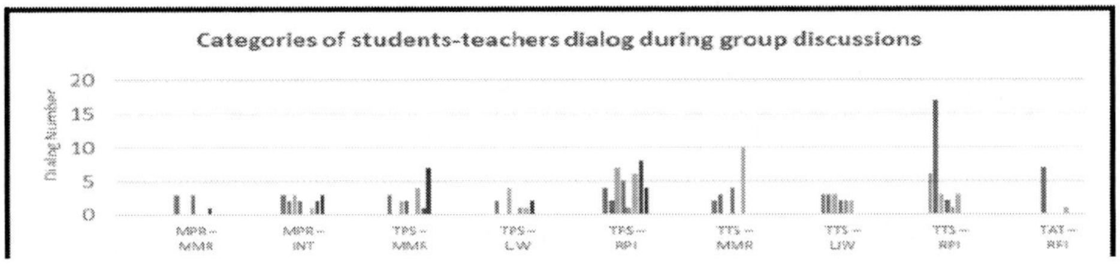

BPO – LJW	:	Student ask procedure Teacher answer student's question directly
MPR – MMR	:	Studentreport group progress Teacher instructs the student to do something
MPR – INT	:	Studentreport group progress Teacheraskresult/progress/notification
TPS – MMR	:	Student ask teacher approval Teacher instructs the student to do something
TTS – MMR	:	Student askunresolved solution Teacher instructs the student to do somethin
TTS – LJW	:	Student askunresolved solution Teacher answer student's question directly
TTS – RPI	:	Student askunresolved solution Teacheranswers using scientific question
TPS – LJW	:	Student ask teacher approval Teacher answer student's question directly
TPS – RPI	:	Student ask teacher approval Teacheranswers using scientific question

Figure 4. Categories of students-teachers dialog during group discussions in the second implementation

Based on the figure 4, a category of dialog that appears the most TTS-RPI, RPI-TPS, and TTS-MMR. TTS-RPI is a category of dialog in which students asking the solution of the problems they face, and the teacher directs them find the solution by giving questions. This category appears because most students confused when writing the chemical formula in the equation, other than that the students are also confused in equalizing the coefficients of the equation. TPS-RPI is a category in which students requesting approval of teachers and the teacher did not directly answer, but to give students directive question and gain confidence in their answers. Students often still have a doubt about the chemical formula which he wrote, in addition students are also often hesitant when equalize the equation. TTS-MMR is a category of dialog in which students ask about the solution of an unresolved problem. Teachers answered these questions by giving steps so that the students can obtain the solution of the problem. It happen while there some confusion occurs when students equalizes equation.

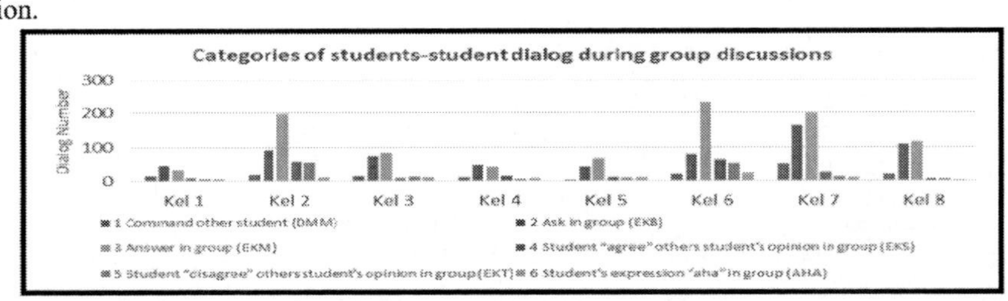

Figure 5. Categories of students-student dialog during group discussions in the second implementation

In group discussions, category of dialog most commonly found are answering (EKM) and asking (EKB) in the group, as well as affirmative and negative expressions (figure 5). This shows that students discuss well in the group. Group 6 and group 7 has the most amount of asking and answering dialog, this group also had the most AHA expression. These groups discussed so well that they found many new facts and find solutions to their problems (expression AHA).

Based on the above data, it can be concluded the learning process has been student-center, the interruption of the teachers have been greatly reduced. Based on the results of the lesson analysis, teacher self-reflection results obtained through interviews as follows:

G : *The learning process at this second meeting is already much better than in the first meeting. Last week's demonstrations about HCl and nails reaction are not too observed. This week has been replaced with Mg combustion demonstration. The changes in the reaction is very clear. Students are very interested on the flame and can observe the process when the ash was being formed. My interruption on classical sessions and group sessions has been significantly reduced, so that students can be more dominant in the learning process. Students seemed to focus and enthusiastic in discussions to solve problems. They want to find out the answers by opening the book and discuss, and are willing to come forward to communicate the results of their discussion. Teacher is not too significant in answering student's questions directly, but students are guided to find the answers themselves.*

G : *In classical session students are required to rotate the chair, so that students concentrate fully forward and pay attention to the demonstration. Demonstrations in the classical session encourages students to know more. It can be seen from the number of students that asked to see Mg ribbon, and see the changes that occur from the combustion reaction Mg. Students are also active when asked by the teacher to explain what happened before and after the reaction. In addition, students are also actively answered teacher's questions on how to write and equalize the equations.*

G : *During the group discussion, some students was silent and did not contribute at first. As the conversation progresses the students become interested and actively involved. Even these students came forward to present the answers of their group. Students who changes the most during the lesson process are David and Anhal.*

G : *Because some things are already pointed out such as confirmation of coefficients and indexes, how to write chemical formulas, and how to equalize the reaction coefficient, the student becomes better understood. By the time working on the problems the students do not really need help from the teacher.*

G : *There are 5 problems given to the student. It turns out they need more time to discuss these problems.*

G : *In terms of lesson design, exercises should be reduced, due to lack of time to discuss the problems. Classical activities mainly concentrated more to students. Students are more directed to be more active. In terms of technical implementation, teachers are significantly reduced interruptions than the first meeting, but it would be better if it reduces again. Teachers should lead students to discuss.*

3.5. Revise Second Lesson Design

Based on the results of teacher self-reflection, lesson design had to be revised again. Exercises should be reduced because of the lack of time available. Confirmation about the activities also require additional time for more effectiveness. In addition, teachers should be working to make learning more student-centered. Teacher's effort can be further reduced interruptions when students discuss and provide students the opportunity to be asked for more.

4. Conclusion

Lesson analysis can be used as a tool to reflect on the learning process by the teacher. Lesson analysis gives teachers a variety of views to observation, reflection, and the study of learning in the classroom. Through self-reflection activities that assisted with lesson analysis, the teacher can be aware of how the learning process has been done, how the teacher interaction with students or students with the students during the learning process. In addition, teachers can also find out the problems that occur during the learning process, either in terms of material or on how to guide the class. After realizing the problems, teacher can correct or seek solutions to these problems so that the learning process is getting better in the future.

References

[1] Alwasilah, A. C. *2011Pokoknya Action Research*. (Bandung: Kiblat Buku Utama)

[2] Hendayana S., & Hidayat A. 2013 *Developing tools for analyzing of classroom interaction: Does it student-centered or teacher-centered lesson* (Bandung: PPT Seminar international MSCEIS. UPI)

[3] Kemendikbud 2007 *Peraturan Menteri Pendidikan Nasional Republik Indonesia nomor 41 tahun 2007*

[4] Rahman, B. 2014 *Paedagogia. Universitas Lampung*. Retrieved July 11, 2016, from dosen.fkip.unila.ac.id/data/index.php?prodi=P19&id=1960031502

[5] Romagnano L., Evans, B., Gilmore, D *2008 AMTEMonograph* (4), pp. 103-15

[6] Saputra, B 2015 *Desain didaktis pada pembelajaran tatanama senyawa anorganik dan organik sederhana berbantuan lesson analysis sebagai self-reflection guru di sekolah menengah atas*. (Thesis, Universitas Pendidikan Indonesia, 2015, Unpublished).

[7] Sugiyono 2013 *Metode Penelitian Kuantitatif, Kualitatif dan Kombinasi (Mixed Methods)* (Bandung : ALFABETA)

IOP Conf. Series: Journal of Physics: Conf. Series **812** (2017) 012100 doi:10.1088/1742-6596/812/1/012100

Developing Instructional Design to Improve Mathematical Higher Order Thinking Skills of Students

E Apino, H Retnawati
Universitas Negeri Yogyakarta Jln. Colombo No. 1, Karangmalang, Sleman, DI Yogyakarta, Indonesia
Email: apinoezi@gmail.com

Abstract. This study aimed to describe the instructional design to improve the Higher Order Thinking Skills (HOTS) of students in learning mathematics. This research is design research involving teachers and students of class X MIPA 1 MAN Yigyakarta III, Special Region of Yogyakarta, Indonesia. Data collected through focus group discussions and tests. Data analyzed by quantitativedescriptive. The results showed that the instructional design developed is effective to improving students' HOTS in learning mathematics. Instructional design developed generally include three main components: (1) involve students in the activities non-routineproblem solving; (2) facilitating students to develop the ability to analyze and evaluate (critical thinking) and the ability to create (creative thinking); and (3) encourage students to construct their own knowledge.

1. Introduction

Education has an important role in improving the quality of human resources in order to compete in the era of globalization and modernization. The law of the Republic of Indonesia Number 20 Year 2003 about National Education System [1] states that the national education serves to develop the ability and character development as well as a dignified nation's civilization in the context of the intellectual life of the nation. One of the efforts to achieve these goals is through the learning of mathematics.

Permendikbud Number 59 Year 2014 about Curriculum SMA/MA [2] mentioned that the mathematics courses should be offered to all learners from primary schools, to equip learners with the ability to think logically, analytical, systematic, critical, innovative and creative, as well as the ability to cooperate. In order to achieve these objectives, it is important in learning mathematics developed various aspects of skills, one of which is a higher order thinking skills (HOTS). This is in line with the views expressed by Yen & Halili [3] that HOTS is the skills that are needed by every individual in the educational environment.

Experts define HOTS with different approaches and viewpoints. Resnick [4] argues that HOTS is hard to define, but easily recognizable by its characteristics. Further, Resnick [4] revealed some of the characteristics of HOTS as follows: (1) non-algorithmic, meaning that the action steps can not be fully determined at the beginning; (2) tends to be complex, meaning that steps can not be seen or predictable directly from a certain perspective; (3) often produces a lot of solutions rather than a single solution; (4) involve disagreements (nuanced judgment) and different interpretations; (5) involves the application of multiple criteria, which are sometimes mutually contradictory; (6) often involve uncertainty; (7) involving self-regulationin the process of thinking; (8) involving imposing meaning, such as discovered the structure of the irregularity; and (9) requires effortfull. If examined closely, the general characteristics of HOTS above demonstrates the need for unusual thought processes or thinking that is more complex and requires an unusual effort anyway.

Thomas & Thorne [5] state that "higher level thinking is thinking on a level that is higher than memorizing facts or telling something back to someone exactly the way it was told to you". Furthermore Thomas & Thorne [5] state that "higher level thinking requires that we do something with the facts. We must understand them, infer from them, connect them to other facts and concepts, categorize them, manipulate them, put them together in new or novel ways, and apply them as we seek new solutions to new problems". In line with these opinions, Lewis & Smith [6] states that "higher order thinking occurs when a person takes new information and information stored in memory and interrelates and/or rearranges and extends this information to achieve a purpose or find possible answers in perplexing situations". From the opinions that have been raised before it can be concluded that HOTS demand for more complex thinking in dealing with situations or solve a problem.

In connection with the framework of thinking skills, HOTS is often described as a cognitive activity or thinking skills at a high level. In this case, the term is usually contrasted with the term LOTS (Lower Order Thinking Skills). Saido, Siraj, Nordin, & Al-Amedy [7] states that the LOTS only require routine applications of information obtained previously as registering the information from memory beforehand and enter the numbers into the formula that has been studied before, while HOTS requires students to interpret, analyze, or manipulate information.

Some experts, including Fisher [8] classifies the cognitive process analysis, synthesis, and evaluation in Bloom's taxonomy as HOTS, while cognitive process knowledge, understanding, and application as LOTS. While in the cognitive process by Krulik & Rudnick [9], explained that the critical thinking and creative thinking as HOTS, while recall and basic included in LOTS. As if it is associated with Bloom's taxonomy revisions proposed by Anderson & Krathwohl [10], Liu [11] argues that the dimension of the cognitive process analyze, evaluate, and create as HOTS.

Based on some of these opinions can be concluded that HOTS is thinking skills that are more complex in solving various non algorhitmic problems that include critical thinking and creative thinking, As if it is associated with cognitive processes in Bloom's taxonomy, HOTS involve students' skills to analyze, synthesize, and evaluate, whereas if it is associated with cognitive processes in Bloom's taxonomy revision HOTS involves the ability to analyze, evaluate, and create.

Facts in the field indicate that HOTS of students in Indonesia is still [12]. Low HOTS of students in mathematics can be seen from the results of the study Trends in International Mathematics and Science Study (TIMSS) organized by the International Association for the Evaluation of Educational Achievement (IEA). TIMSS data in 2011 [13] shows that on cognitive dimension, Indonesia was ranked 38 out of 42 participating countries.

Low students' HOTS in Indonesia is caused by many factors. Results prasurvei the author of the 10 high school math teachers showed that only 2 teachers (20%) who have applied mathematics learning oriented to development students' HOTS. Besides other findings also show that teachers' understanding about HOTS is also lacking, even exist high school math teacher who is unfamiliar with the term of HOTS. Based on these findings, it is necessary to develop an instructional design to improve students' HOTS in mathematics learning. This article aims to descriptionan instructional design to improve students' HOTS in mathematics.

Method

This study using design research approach aimed to design and develop an intervention (such as programs, teaching-learning strategies and materials, products andsystems) as a solution to a complex educational problem as well as to advance our knowledge about the characteristics of these interventions and the processes to design and develop them, or alternatively to design and develop educational interventions (about for example, learning processes, learning environments and the like) with the purpose to develop or validate theories [14].The subject of research involving teachers and students of class X MIPA 1 MAN Yogyakarta III, Special Region of Yogyakarta, Indonesia. Data collection was done through Focus Group Discussions (FGD), instructional design trials and tests. The effectiveness of instructional design was analyzed by quantitative descriptive, by seeing an increase in scores pretest-posttest of students.

2. Result and Discussion

Based on the review of the theory and FGD, instructional design to improve student HOTS as follows: (1) learning objectives lead to the development of critical thinking skills (characterized by the ability to analyze and evaluate) and creative thinking are characterized by their ability to create; (2) the inducement activities of teachers by asking the question "why" and "how", which serves as a stimulus for students to develop HOTS; (3) using the problems that generate many answers or many ways of solving (4) content in worksheet have connectivity between the material and concepts; (5) worksheets that are used to build the concept leads to the discovery and investigation activities adjusted to the characteristics of the material; (6) worksheets that are used for the application of the concept using creative problem that demands a lot of answer and or a lot of ways to solve the problems; (7) problems associated with the real world context and have irregular structure (ill-structure); and (8) There are the reflection to find the reasons why students choose or use a particular method or concept in solving the problem.

Saido, Siraj, Nordin, & Al-Amedy [7] stated that the strategy can be used to improve students' HOTS is through problem solving activities. It is become a demand for math teachers to have the skills to provide the problems that can develop students' HOTS. Here is one example of the use of problem oriented students' HOTS in mathematics quoted from PISA [15].

Problem 1:

"A pizzeria serves two round pizzas of the same thickness in different size. The smaller one has a diameter of 30 cm and costs 30 zeds. The larger one has a diameter of 40 cm and costs 40 zeds. Which pizza is better value for money? Show your reasoning."

The discussion of the problem is as follows:

Discussion:

Smaller Pizza (diameter 30 cm):

$$Area = \pi \times \frac{30}{2} \times \frac{30}{2} = 225\pi\ cm^2$$

$$Area\ of\ each\ one\ Zed = \frac{225\pi}{30} = 7{,}5\pi\ cm^2$$

Larger Pizza (diameter 40 cm):

$$Area = \pi \times \frac{40}{2} \times \frac{40}{2} = 400\pi\ cm^2$$

$$Area\ of\ each\ one\ Zed = \frac{400\pi}{40} = 10\pi\ cm^2$$

Based on the calculation above shows that with 1 Zed earned $7{,}5\pi\ cm^2$ smaller pizza and $10\pi\ cm^2$ larger pizza. It can be concluded that a larger pizza (diameter 40 cm) cheaper than the smaller pizza (diameter 30 cm)

In the above problem, the student is required to understand the meaning of the problem, then able to calculate the extent or magnitude of the pizza, pizza magnitude obtained with 1 Zed price or price per cm2 pizza in units Zed. Furthermore, students are required to conclude the pizza which are cheaper and why. For students who are accustomed given routine matter will be difficulties in translating the intent of the question. As if the problem is converted into a regular math problems would read as follows:

Problem 2:

Routine mathematics mathematics:

A pizzeria provides two choices of pizza with the same thickness and different sizes. The smaller pizza has a diameter 30 cm is sold at a price of 30 Zed. Pizza with a larger size that has a diameter 40 cm for 40 Zed. Determine the price to pay if you buy two smaller pizza and theree larger pizza.

The discussion of the problem is as follows:
Discussion:
Known:
Smaller Pizza (diameter 30 cm) = 30 Zed;
Larger Pizza (diameter 40 cm) = 40 Zed
Asked:How the price of 2 smaller pizza + 3 larger pizza?
Solution: 2 (30 Zed) + 3(40 Zed) = 180 Zed.

Problem 2 shows that to solve the problem is given, students simply apply the concept of multiplication and addition. Forms of problem like this clearly do not require students to use HOTS.

As other examples of problems that can train students to think critically is as follows.

Problem 3 [16]

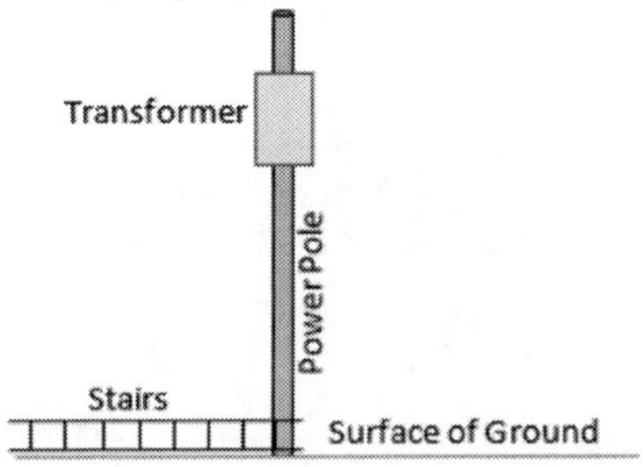

A technician will repair the transformer contained in one of the power poles using astairs, as shown in the illustration above. If the height of the transformer is 5 meters from the ground, what is the minimum required length of stairs so that technicians can repair the transformer?

To solve problem 3, students should know the concept of a minimum of what is needed to solve the problem, facts or information known, and what alternative solution to the problem. The above problem is not a problem that only requires the application of a concept, but students are asked to analyze in advance the concept of what it takes to solve the problems. The discussion of problem 3 is as follows.

If the height of the transformer is exemplified by y and the length of the stair is exemplified by r, it will be obtained illustrated the problem as follows.

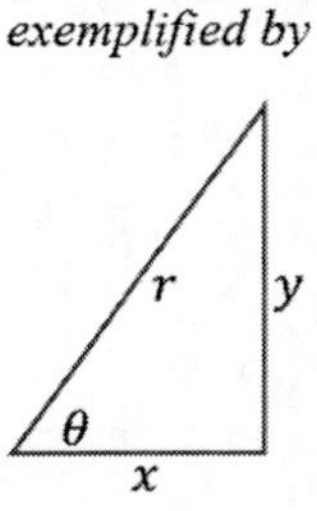

In beside picture, it appears that the problem can be solved by using a comparison trigonometry in a right triangle, where x is distance of end stair with a power pole. The problem is to determine the length r, then it takes a great corner between the stairs to ground level (θ), but on the problem the angle size θ is unknown.

Based on the above illustration angle size θ is between 0° to 90°. In this case the alternatives that can be chosen by selecting special angle which are quadrant I.

If $\theta = 30°$, *then* $\sin 30° = \frac{y}{r} \Leftrightarrow \frac{1}{2} = \frac{5}{r} \Leftrightarrow r = 10$ *meters*

If $\theta = 45°$, *then* $\sin 45° = \frac{y}{r} \Leftrightarrow \frac{1}{2}\sqrt{2} = \frac{5}{r} \Leftrightarrow r = 5\sqrt{2}$ *meters*

If $\theta = 60°$, *then* $\sin 60° = \frac{y}{r} \Leftrightarrow \frac{1}{2}\sqrt{3} = \frac{5}{r} \Leftrightarrow r = \frac{10}{3}\sqrt{3}$ *meters*

All of the above alternatives is correct, but it should be considered whether the angle chosen will produce the safest slope of the stair. Based on these three alternatives, the safest slope of the stair is when θ = 60°, so the length of the stair is required is $\frac{10}{3}\sqrt{3} \approx 5{,}77$ *meters.*

Based on discussion of problem 3, it appears that students are required to be able to make an illustration of the problem, it indicates that it needed the ability to create. In addition, students are also required to be able to identify important facts and information that is relevant to formulate the desired solutions, in this case the student is required the students skill to analyze. On the other hand, to find the best solution or the most reasonable solution of the above problems, it is necessary their students' ability to evaluate of each solution were found.

3.1 The effectiveness of instructional design

Before the instructional design applied in the classroom, students are given a pretest that aims to identify early students' HOTS. The next after implementation instructional design that has been developed, students are given a posttest to determine the impact of the implementation of the instructional design. Pretest and posttest mean score of students is presented in Figure 1.

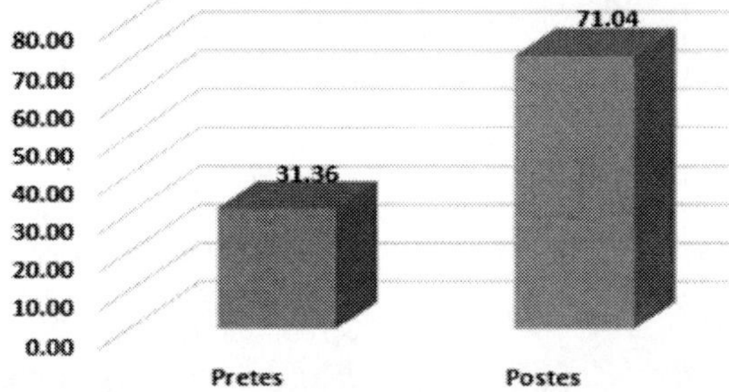

Figure 1. The Enhancementof Mean Score of Students' HOTS

Figure 1 shows that the score of students' HOTS has increased significantly. This shows that the implementation of instructional design provides a positive impact on the enhancement of students' HOTS. The performance score HOTS each student is presented in Figure 2.

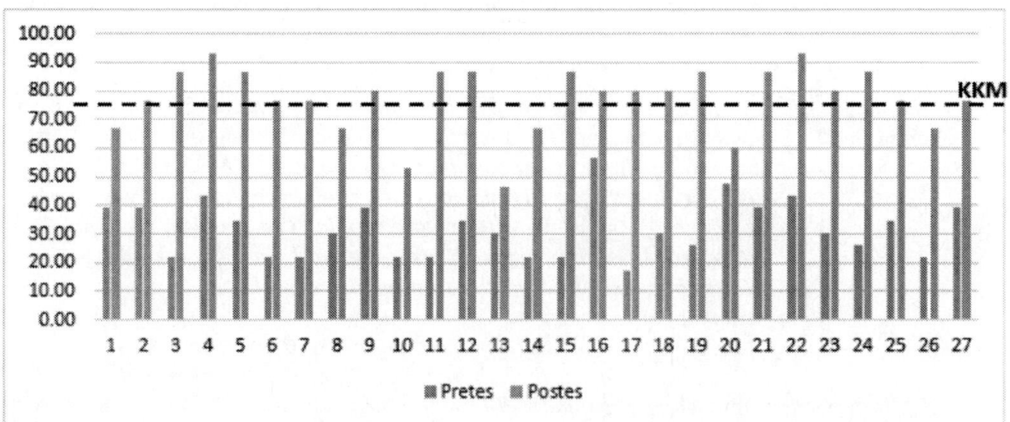

Figure 2. Graph of Enhancement of Students' HOTS

Based on Figure 2 shows that each score HOTS students when the pretest increased after the implementation of instructional design developed. Although not all students achieved a Minimum Completeness Criteria score, these achievements clearly indicate that the developed instructional design can improve each students' HOTS. Thus the implementation of instructional design that has been developed effective to improve student HOTS.

3.2 Discussion

The results of data analysis showed that the instructional design developed is effective for improving students' HOTS. This is due to the filing of an open question (creative problem) through worksheet make students more challenged to explore the various possibilities of ideas that can be used to solve the problem. Loewen [17] suggested that with the filing of creative problems can generate awareness for students that not all the problems have only one correct solution. This is what can trigger and train the students' creativity in mathematics learning, and creativity are part of HOTS.

Another factor which led to increased student HOTS is presence of meaningful learning activities, where students are actively involved in the discussion process to construct knowledge and utilize various relevant sources to explore the knowledge desired. This is in accordance with the opinion raised by Bohan & Bohan [18] that the learning process that involves the active participation of students to solve various problems can present meaningful learning activities for students.In addition, students become enthusiastic because they feel challenged to a given problem. This of course could improve students' motivation to learn. This is in line with the opinion of Loewen [17]which states that the presentation of creative problems challenging can increase students' interest and motivation to learn. Thus the increased interest and motivation to learn students become a determining factor for the increase students' HOTS.

3. Conclusion

Based on the result and discussion can be concluded that HOTS in mathematics learning is important. It aims to develop students' ability to analyze, evaluate, and create, so that students have the critical power and creativity that can be used to solve problems in everyday life. Some efforts should be made to improve students' HOTS in mathematics, namely: (1) engaging students in the activities of non-routine problem solving; (2) facilitate the students to develop the ability to analyze and evaluate (critical thinking) and the ability to create (creative thinking); and (3) encourage students to construct their own knowledge, so that learning becomes meaningful for students.

Based on the findings in this study suggested for teachers to apply instructional design that has been developed in this research in mathematics learning. To ensure that the instructional design can be implemented properly, teachers should provide various non-routine problems close to the everyday life and train students the ability to ask "why" and "how".To assist in making problem-oriented HOTS, teachers can adapt and modify the questions TIMSS and PISA

4. References

[1] Republik Indonesia 2003 The Law of Republic of Indonesia Number 20 Year 2003 about National Education System

[2] Mendikbud 2014 *Regulation of the Minister of Education and Culture Number 59 Year 2013 about Curriculum 2013 SMA/Madrasah Aliyah.*

[3] Yen T S and Halili S H 2015*The Online Journal of Distance Education and e-Learning***3** 41-47

[4] Resnick L B1987*Educational and Learning to Think* (Washington DC: National Academy Press)

[5] Thomas A and Thorne G2009 *How to Increase Higher Order Thinking*(Los Angeles: Center for Development and Learning) [online: http://www.cdl.org/resource-library/articles/HOT.php?type=subject&id=18]

[6] Lewis Aand Smith D 1993 *Theory into Practice* **32**131-137.

[7] Saido G A M, Siraj S, Nordin A B and Al-Amedy O S2015 *Malaysian Online Journal of Educational Management***3** 16-30.

[8] Fisher R2010 Thinking skill*Learning to teach in the primary school (2nd ed.)*ed J Arthur and T Cremin (New York, NY:Routledge)

[9] Krulik S and Rudnick J A 1999 Innovative Tasks to Improve Critical andCreative Thinking Skills. In Stiff, Lee V. & Curcio, Frances R. (Ed)Developing Mathematical Reasoning in Grades K-12. 1999 Yearbook.Reston, Virginia: NCTM.

[10] Anderson L W and Krathwohl D R2001*A Taxonomy for Learning, Teaching, And assessing: A Revision of Bloom's Taxonomy of Educational Objectives* (New York: Addison Wesley Longman)
[11] Liu X2010 *Essentials of Sciences Classroom Assessment* (Los Angeles: SAGE Publication)
[12] Susanti E, Kusumah Y S, Sabandar J andDarhim2014 *J. of Education and Practice***5** 51-58.
[13] Mullis I V S, Martin M O, Foy P, and Arora A2012 *TIMSS 2011 International Results in Mathematics*(Chestnut Hill, MA: IEA)
[14] Plomp T 2013 Educational design research: An introduction*Educational Design Research* ed J Van den Akker et al (Enschede, Netherland: National Institute for Curriculum Development) pp. 10-51
[15] Wardhani S and Rumiati2011 *Instrumen Penilaian Hasil Belajar Matematika SMP: Belajar dari PISA dan TIMSS.*(Yogyakarta: P4TK)
[16] Apino E 2016 *Pengembangan Perangkat Pembelajaran Matematika SMA Kelas X Semester Genap Menggunakan Model Creative Problem Solving Berorientasi pada Higher Order Thinking Skills Siswa* (Yogyakarta: Tesis Universitas Negeri Yogyakarta)
[17] Loewen A C 1995 *Teaching Children Mathematics* **2** 96-99
[18] Bohan H and Bohan S 1993 *The Arithmetic Teacher***41** 83-87

Character Education in Islamic Boarding School and The Implication to Students' Attitude and Critical Thinking Skills on Biodiversity Learning

R K Hayah*

Departement of Biology, Faculty of Mathematics and Natural Sciences, Universitas Indonesia, Depok, 16424, West Java, Indonesia
*Corresponding author: karimatulhayah@yahoo.co.id

Abstract. In this globalization and modernization era, Indonesia as an expanded country is trying to improve in various field, technology and information development in this era has caused changes in various field like economy, politic, social and also culture. Be sides caused the positives impact, this development also caused negatives impact. For example technology development in internet has caused some negative impact like sarcasm, insulting, materialism and hardness. The education process whether formally, informally, or nonformally, become a pillar to bore a new generation of Indonesia with the strong character. This strong character marked by moral capacity like sincerity, individual quality which differentiated with others and obstinancy to face the hardness. This research used a descriptive research. The subject is students from MA based Islamic boarding school at class X MIA. Sample in this research used purposive sampling tehnique.The result of this research express that students' atitude and critical thinking on biodiversity learning have average score whether character educated students in islamic boarding and also in school. This is proved that the strong character can be obtained from a brilliant education system and not just emphasize intelectual intellegence but education based on belief in God and fear and also able to give the creation which useful for religion, country and people. Then education that contain the two main element, that is the superiority in academic and also nonacademic (include moral and spiritual superiority) is needed. Students' critical thinking on biodiversity learning also have average score whether character educated students in islamic boarding school.

1. Introduction

Developments perceived as a result of globalization today, more positive impact that causes the rapid development of civilization in this country, but in addition to the positive impact of these developments had a negative impact. An example is the development of technology in the Internet, the negative impact a lot of things that happen like cynicism, harassment, materialism, promiscuity, and violence. The process of education whether formal, informal and non-formal, became the foundation for new Indonesian with a strong character. The robust character is characterized by the capacity of one's morals, such as honesty, the peculiarities of the quality of someone, which distinguishes itself from others, as well as the strength to face difficulties [1].

In psychology, the purpose of education is to form the character embodied in the essential unity of the subject with the behavior and attitude of its life. According to figures from the German character education, FW Foerster, character is something that qualify a person. Karate became identity that transcends the ever-changing experience kontinguen [2].

Thus, character education is a process that is integrated with education widely and gradually, of education in the family, educational institutions (both formal, informal and non-formal), until in community life. Character education is also reaching process of planting the value of religion, culture, customs, and aesthetics. In other words, character education is an effort to make the students familiar, caring and internalize the values so that they can berprilau as insal Kamil [3].

Description of the background of the above formulation of the problem of this research is how school-based character education in schools and the implications for biology student attitudes to learning? Therefore, it is expected to do the research formulation of research problems can be missed.

2. Research Method

This research is a descriptive research. Descriptive research is research that is intended to gather information about the status of an existing symptoms, the symptoms according to what their circumstances at the time of the study [4]. The subjects of this study were Madrasah Aliyah students who staying in pesantren in the class X MIA. Selection of the samples in this study using *purposive sampling* technique. Samples are all concerned with the focus of research is the principal, curriculum vice principal, vice principal of student, biology teachers, religious teachers, counseling teachers, pesantren religious teacher, pesantren educators, parents and students.

Table 1 Research Instrument

No	Type Instruments	Data Taken
1	Questionnaire attitude scale (Likert)	Students' attitudes
2	Draft interview	Character education curriculum of schools and boarding schools
3	Items (essay)	Critical thinking skills
4	Observation sheet	Activities biology learning

Data obtained by interviewing the relevant sample, giving the questionnaire and essay to the students. Then the data is processed by descriptive results so as to produce an answer from the formulation of research problems.

3. Result

3.1. Overview Implementation of School-Based Character Education at Pesantren

Based on the findings in the form of observations and interviews conducted with research samples totaling 16 respondents consisting of a chairman of the foundation, principals, teachers biologist, teacher pesantren, vice principal of curriculum and vice principal of student and five student daughter and 5 male students living in boarding schools, researchers formulate some aspects to see how the implementation of character education programs in high school Bidayatul Faizin, as follows:

1. Policy and administrative support to the implementation of character education schools
2. School environment in support of the implementation of character education
3. Teacher's knowledge of the implementation of character education
4. Increased competence of teachers in implementing the learning pendidian character
5. Used school curriculum content

The findings of some aspects of the above is as follows:

Table 2. Policy and support of the school administration to the implementation of character education in learning in high school Bidayatul Faizin

Statement	alternative Answers			
	Yes		No	
	F	%	F	%
1. Schools have the vision and mission of character education	13	81.25	3	18.75
2. To socialize continuously to the students / students and parents of students	12	75	4	25
3. Data or important documents regarding character education programs in learning for students	10	62.5	6	37.5
4. Show specific ways that school administrators are very concerned goes on character education	14	87.5	2	12.5
5. Have data regarding the enforceability of the development program of character education in schools	12	75	4	25
6. Provide flexibility to teachers to make teaching methods are creative, innovative in helping children learn	15	93.75	1	6.25
7. Having a support mechanism, supervision and monitoring effective for every person to be able to participate	14	87.5	2	12.5

Table 3. Environmental conditions in support character education schools

Statement	Alternative Answers			
	Yes		No	
	F	%	F	%
8. Have a facility that meets the needs of learners to develop character education	15	93.75	1	6.25
9. Having a clean environment, a healthy and open	12	75	4	25
10. Having a team as an evaluator keberjalanan character education program	12	75	4	25
11. Having ordinances and procedures appropriate to help teachers, faculty, parents and children to work together in developing a character education	15	93.75	1	6.25

Table 4. Knowledge of teachers on the implementation of character education lesson

Statement	Alternative Answers			
	There Is		There Is No	
	F	%	F	%
12. Can explain the meaning of character education and implement a character education lesson	12	75	4	25
13. Knowing how to make lesson plans sound character education	10	62.5	6	37.5
14. Involved in making the learning device sound character education	13	81.25	3	18.75
15. Knowing about how to create a syllabus and RPP sound character education	13	81.25	3	18.75
16. Engaged in the manufacture of the syllabus and RPP sound character education	13	81.25	3	18.75
17. Having a teaching materials that assist the learning of character education	13	81.25	3	18.75

18. Know the procedures for the implementation of character education lesson	13	81.25	3	18.75
19. Make three stages in its core activities, namely exploration, elaboration and confirmation	13	81.25	3	18.75
20. Can explain the components in character education	12	75	4	25
21. Knowing aspects of the value contained in character education	11	68.75	5	31.25
22. Knowing the learning goals of character education	15	93.75	1	6.25
23. Adapting the curriculum and school activities to the needs of learners	13	81.25	3	18.75
24. Being able to modify the child's learning in beragai way to properly and in accordance with the abilities and needs of the child as well as the character values memnunjukkan	12	75	4	25
25. Reflecting and terb u ka towards learning and change	10	62.5	6	37.5

Table 5. Increased competence of teachers in implementing character education

Statement	Alternative Answers			
	There Is		There Is No	
	F	%	F	%
26. Provide explanations to other teachers, parents and community members about the learning development of character education	12	75	4	25
27. Increase their knowledge in understanding the course content andintegrate into character values	12	75	4	25
28. Improve teacher knowledge to develop materials	13	81.25	3	18.75
29. Having a work space so that they can prepare lesson materials and share ideas on teaching character education	13	81.25	3	18.75
30. Implement seminar or briefing on the implementation of character education lesson	14	87.5	2	12.5

Table 6. The contents of the curriculum used school

Statement	Alternative Answers			
	There Is		There Is No	
	F	%	F	%
31. The curriculum introduces methods of learning and different learning styles, such as discussions, games, or play a role	13	81.25	3	18.75
32. Curriculum content includes daily experience of all learners in schools to instill character values in schools	13	81.25	3	18.75
33. The curriculum integrates subjects with ni lai-value karak t er	14	87.5	2	12.5
34. The curriculum promotes attitudes such as mutual respect, tolerance and knowledge of the entire value of character haris achieved	13	81.25	3	18.75

Based on the table for the average percentage of the various aspects of the measure, based on the results of interviews found that:

a. Policy and administrative support to the implementation of character education schools 80.4% of respondents answered no, and 19.6% answered no support and policy administration school
b. School environment in support of character education, 84.4% of respondents answered that the conditions of the school environment to support the implementation of character education, but by 14.6% more there are still some conditions that do not supported characters education.

c. Teacher's knowledge of the implementation of the teaching of character education, is 71.4% of respondents stated that teachers have a good knowledge about character education.
d. Increased competence of teachers in implementing character education, 80% of respondents stated that the competence of teachers in implementing character education increased.
e. The contents of the school curriculum used, the respondents said that 82.81% responded that the content of the school curriculum supports the enforceability of character education.

3.2. Implications of Character Education Students Against Attitude
Based on the results of an assessment of the attitude scale questionnaire that was distributed to the students obtained the following data:

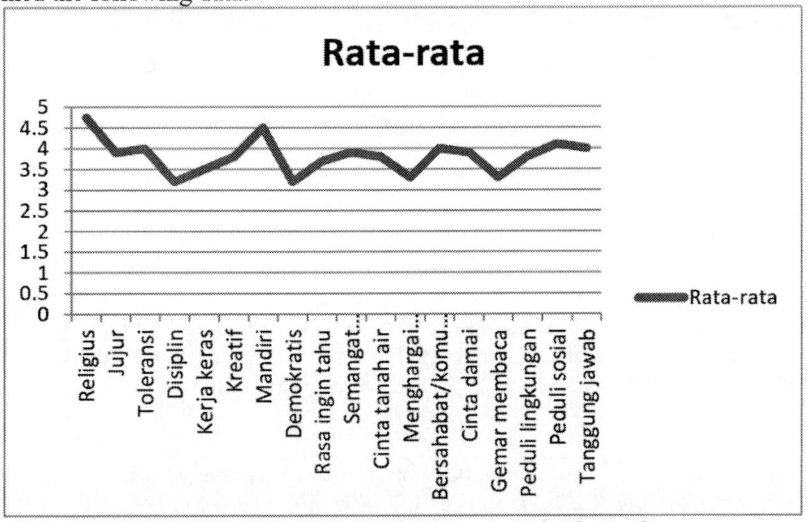

Figure 1. Graph Average yield score Attitude Students

From Figure 1 it is known that the average acquisition score varied for each aspect of the character, for the average acquisition *religious* score reaches 4.75 with very good category. *Honestly* had an average of 3.9 in both categories. *Tolerance* has a score of 4.0 with both categories. the discipline has a score of 3.2 with the category quite well. *Work Hard* had a score of 3.5 in both categories. Creative has a score of 3.8 in both categories. *Mandiri* has a score of 4.5 with a very good category. *Democratic* scores 3.2 with a category quite well. *Curiosity* has a score of 3.7 in both categories. *Spirit of ethnicity* has a score of 3.9 in both categories. *Love the homeland* has a score of 3.8 in both categories. Rewarding achievement has a score of 3.3 by category quite well. Bersahabat / communicative has a score of 4.0 in both categories. *Love peace* has a score of 3.9 in both categories. Joy of Reading has a score of 3.3 with the category quite well. *Care for the environment* has a score of 3.8 in both categories. S*ocial Care* has a score of 4.1 in both categories. And *responsibility* has a score of 4.0 in both categories.

3.3. Implications of Character Education Students Against Critical Thinking Skills
Based on the results of an assessment of the attitude scale questionnaire that was distributed to the students obtained the following data:

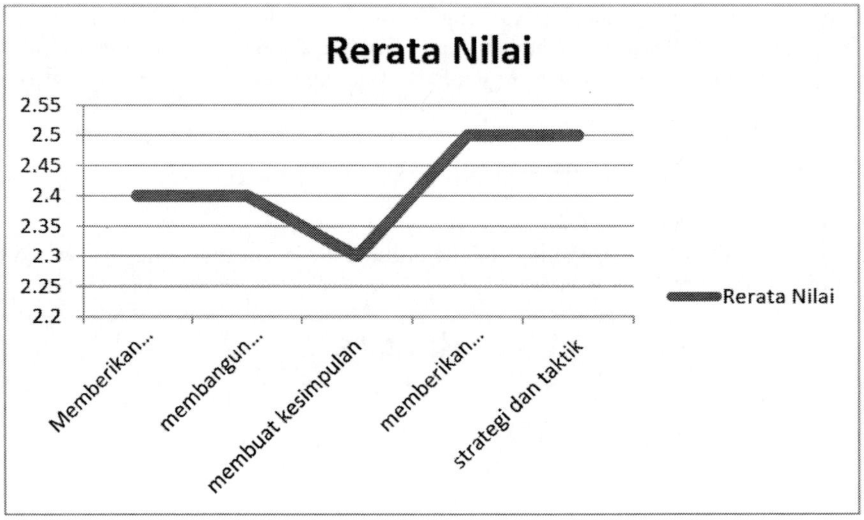

Figure 2 Graph The mean value of critical thinking skills (maximum value 3.0)

Based on Figure 2 is known that to give a simple explanation average value reached 2.4 of the maximum value of 3.0, to build basic skills rerta value maximum value 3.0 2.4 ari, to make conclusions mean value of 2.3 of the maximum value 3 , 0, to provide further explanation of the mean value is 2.5, while for the strategy and tactics of the average value of 2.5.

4. Discussion

Character education is moral education plus, namely involving aspects of knowledge *(cognitive)*, feeling *(feeling)*, and action *(fashion)*. Character education is a system of cultivation of character values to the school community, which includes knowledge, awareness or volition, and actions to implement these values, both against the Almighty God, ourselves, others, the environment, and nationality to become a human better.

Based on the results of research conducted in MA Fauzaniyyah, in the form of observations and interviews with teachers, principals, chaplain schools and pesantren leaders on the implementation of character education and the distribution of a questionnaire scale student attitudes and essay critical thinking skills, the result that before students get gemblengan with character education programs in schools and the school, the students have a character that can be said is not good, as yet the realization of politeness, discipline, responsibility, good cooperation, and a sense of concern among fellow, even follow the spirit of learning in schools any less, it this was seen when researchers conducted preliminary observations at the school.

Implementation karakterdi high school education Bidayatul Faizin done in a variety of activities in schools and boarding schools. In the field of character education curriculum into the learning process, it is known from the implementation plan of learning in which the learning process implied by integrating learning with spiritual values. Based on observations and interviews it managed to increase the students' attitude and enthusiasm of student learning.

Character values can also be implanted in many ways in addition to learning, in MA Fauzaniyyah held spiritual guidance on a regular basis in the form of study is directly supervised by the chairman of the foundation as well pesantren leaders, besides the students at each level are required to complete the terms of skills ubudiyah in each semester. Based on the observations of the character education program at MA Fauzaniyyah have managed to improve some students' character and improve the attitudes of students in learning. This is consistent with the statement of [3] that character education is a process that is integrated with education widely and gradually, of education in the family, educational institutions (both formal, informal and non-formal), until in community life. Character education is also reaching process of planting the value of religion, culture, customs, and aesthetics. In other words, character education is an effort to make the students familiar, caring and internalize the values so that they can berprilau as insal Kamil.

According [6] character education has the goal to instill values in self-learners and governance reform more respectful coexistence of individual freedom. Besides the goal also improve the quality of education in schools and the results towards achieving the formation of character and noble character of learners as a whole, integrated, and balanced in accordance with competency standards [6].

In character education in schools, all of the components *(stakeholders)* should be involved include educational components itself, namely the content of the curriculum, learning and assessment, quality of relationships, handling or management subjects, school management, implementation of programs and activities, empowerment means and infrastructure, financing and the work ethic of all citizens and the environment and also a boarding school.

5. Conclussion

Implementation of education karakterdi MA Fauzaniyyah done in a variety of activities in schools and boarding schools. In the field of character education curriculum into the learning process, it is known from the implementation plan of learning in which the learning process implied by integrating learning with spiritual values. Based on observations and interviews it managed to increase the students' attitude and enthusiasm of student learning.

Character values can also be implanted in many ways in addition to learning, in MA Fauzaniyyah held spiritual guidance on a regular basis in the form of study is directly supervised by the chairman of the foundation as well pesantren leaders, besides the students at each level are required to complete the terms of skills ubudiyah in each semester. Based on the observations of the character education program at MA Fauzaniyyah have managed to improve some students' character and improve the attitudes of students in learning, and students who earn gemblengan character education in schools and boarding schools also have good critical thinking skills.

6. Refferences

[1] Suhardi D 2012 J. Pendidikan Karakter, TahunII Nomor 3
[2] Koesoema, A 2007 *Pendidikan Karakter, Strategi Mendidik Anak Zaman Global* (Jakarta: Grasindo)
[3] Syafaruddin 2012 *Pendidikan dan Pemberdayaan Masyarakat* (Medan: Perdana Publishing)
[4] Sudjana, N 2004 *Penilaian Hasil Proses Belajar mengajar* (Bandung: PT. Rosdakarya)
[5] Setiawan dkk 1997 *Teknologi Pengajaran* (Bandung : Sinar Baru Algesindo)
[6] Asmani S andYudianto 2011 *Dimensi Pendidikan Karakter/Nilai Dalam Model Sains-Biologi Untuk Pembelajaran Manusia Sebagai Upaya Mengatasi Krisis Nilai dan Moral Bangsa.* Bandung: Pidato Pengukuhan Guru Besar Universitas Pendidikan Indonesia

Acknowledgments

The author express her thanks to Dr. H. Riandi M.Si and Prof. Dr. H. Suroso Adi Yudianto for great contribution and support to this study, and to reviewers for comments and help.

IOP Conf. Series: Journal of Physics: Conf. Series **812** (2017) 012102 doi:10.1088/1742-6596/812/1/012102

Recostructing the Physics Teaching Didactic based on Marzano's Learning Dimension on Training the Scientific Literacies

S Karim[1], E C Prima[2*], S Utari[1], D Saepuzaman[1] and M G Nugaha[1]

[1]Department of Physics Education, Universitas Pendidikan Indonesia, Jl. Dr. Setiabudhi No. 229, Bandung 40154 Indonesia
[2]Department of Science Education, Universitas Pendidikan Indonesia, Jl. Dr. Setiabudhi No. 229, Bandung 40154 Indonesia

[*]Corresponding author's email : ekacahyaprima@upi.edu

Abstract. Scientific literacy is currently considered as an important aspect supporting an useful citizenship ability for civilians inhabiting highly developed countries as well as developing countries. Consequently, certain countries recommended this scientific literacy to be applied at a national curricula. The PISA study showed the Indonesian scientific literacy level of 1, which means as just simple science phenomenon that could be exactly descibed by a student. This condition indicates that common science teachings do not optimally facillitate students to guide the scientific literacy. By proposing this research, the science didactic reconstruction will be offered in order to gain the students' scientific literacy evaluated from the qualitative analysis of the action research and the students' respons during learning science. The qualitative evaluation was developed based on the Marzano's learning dimension about the scientific literacy. This research, involving 29 students as participants, analyzed the improved physics teaching didactic as described in the following sentences. The teaching reconstruction concerned a high attention to the development of the structural knowledge. The knowledge was acquired from a real phenomenon followed by giving the instructed questions as the second learning dimension. The third dimension of leearning reconstruction aimed to provide the knowledge repetition on an appropriate science context. At the fourth dimension, the reconstruction should be improved in order to find the best treatment for the students. Hopefully, they can control the physical parameter and evaluate the result of their investigation related to the given science problems. It can be concluded that most of the students were interested in learning science. However, the productive learning didn't accompany students to the Marzano's second, third, and fourth learning dimensions.

1. Introduction

Nowdays, the science literacy is a purpose of the science teaching [1].Dewey (1904) stated that learning science guided with the direct student's inquiry experience is important to be prepared in order to improve their science literacy [2]. Moreover, the National Science Education Standard initiates the science literacy including (a) the scientific concept and the scientific process for physics and biology, (b) the inquiry scientific method, (c) the application of science in daily life, (d) the social and environment implications of science to the technology development [3]. In 1990, the brenchmarks for science literacy developed by the National Research Council for the National Science Education Standarexplainedabout how the students delvelop their science literacy and how the literacy can be improved to achieve the standardized level. This council also focused on the science literacy development to be applied at the elementary and middle school levels. According to Five (2014), science literacy is identified as the ability to comprehend the scientific process and the meaningful activity involvement acquired from the scientific information presented in a daily life [4]. Futhermore, the National Research Council (1996) stated the scientific literacy as the knowledge and the understanding of scientific concepts and processes required for personal decision making, participation in civic and cultural affairs, and economic productivity [3]. Therefore, the science literacy can be defined as the ability to comprehend the knowledge to be applied in a daily life, and it is required for the personal decision making and the active civil participation. Science literacy gives high attention to the use of science effectively with respect to the nature phenomena [4].

Science literacy as the knowledge and the competence is important to be attained by civils of a developing country as well as a developed country in order to build the national curricula implemented at the elementary and middle schools [5]. This program can support students the ability to solve their problems happening in a daily life. Consequently, science literacy is the students' right to be realized from a high quality science teaching [6]. Since 2005, the survey about students' comprehension on science literacy in the middle school has been taken involving 628 students from 5 different schools. The result showed that the students' science literacy must be improved including their abilities in explaining the science phenomena, contructing and evaluating the experiment, and interfering the scientific evidences [6]. The previous stategies to improve the students' science literacy has been performed by applying the level of inquiry and using the scientific approach. The finding presented that students' science literacy could be slightly improved, but the students tend to accomplish the literacy using sporadic ways. Hence in this research, the consecutive approach will be carried out to discover the ways in exercising students' science literacy.

The research is conducted in order to improve the quality of science teaching using Marzano's dimension approach [9]. This approach consists of first dimension D1 for positive perception and attitude, second dimension D2 for knowledge building, third dimension D3 for knowledge refining, fourth dimension D4 for knowledge broadening, fifth dimension D5 for habit of mind. The first and fifth dimensions of D1 and D5 are closely related to the pedagogy foundation in training the science literacy. The figure 1 illustrate the relationship among the Marzano's dimensions.

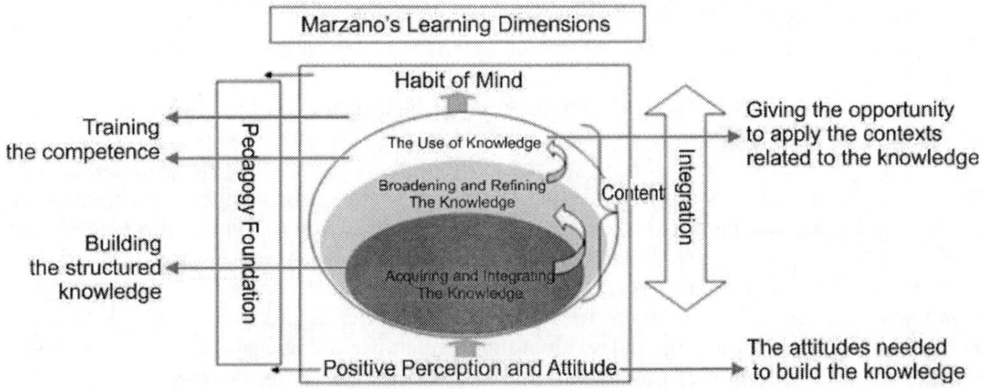

Figure 1. The relationship among the Marzano's dimensions in order to train the science literacy

According to the Marzano's learning dimensions, a teacher firstly accompanies the students to possess the positive perception and attitude in all of the processes of acquiring, broadening, and applying the knowledge. Theteacher engagement is needed to simplify the problems so that the students will not be frustrated in handling their difficulties. The engagement will also contribute to the improvement of the meaningful students' attention to acquire a good perception. On the other hand, a teacher should facillitate the students to think productively called as the habit of mind. This habit is required during acquiring, integrating, broadening, refining, and applying the knowledge. The scientifc questioning ability of a teacher will obviously assist the students in habituating their mind[6].

In the case of students' ability to acquire the concept of heat transfer, some misconceptions were found, and some students felt difficult to understand some concepts of heat transfer such as the heat is comprehended as the temperature transfer, the heat is the internal energy, and the heat is also influenced by the specific heat and the mass. To make a meaningful teacing as well as to give a good motivation to students, a teacher have to provide some applications related to the heat concept in a daily life called as a contextual base approach [10, 11]. A good student's perception and attitude needs certain activities involving the knowledge acquiring. Some science demonstrations can be shown to students in order to engage the students' interest and to discover the students' misconception through giving the cognitive conflicts [12]. Moreover, the contextual problem may also be presented to students with the intention of the students' interest [13]. Finally, a teaching should be directed to construct good scientific attitudes, and it is better if a relation between science and their future career could be introduced to students [14].

Some science literacy problems had by the Indonesian students were found that they don't have enough knowledge to discuss with a teacher, so the new knowledge couldn't be acquired [15]. If a teacher do not provide an appropriate aperception, students will not accomplish a desirable knowledge constructivism because the students' knowledge is not inexpedient with the new introduced knowledge. Therefore, a teacher needs to be patient to elaborate the knowledge so that the students will ready to attain a new knowledge. The infusion reading strategy might be proposed to a teacher with the purpose of handing the gap between students' knowledge and a new knowledge. For a students who not be accustomed to accept a new information, a teacher might provide a suitable article regarding the basic knowledge required to pursue some scientific questions and answers. Thus, the discussion can be well cunducted [16].

A teacher commonly uses an available textbook to teach the concept of heat transfer without considering the knowledge cunstructing structure, whereas this structure needs to be elaborated by a teacher through analyzing contextual science evidences which are easy to be observed by students in constructing their factual knowledge [17].In this research, the concept of heat transfer is constructed using the factual knowledge. It can be considered that the heat is not the internal energy of an object, but it is an energy transferred due to the two different object temperatures while the temperature is a physical parameter possessed by two different systems when they reach the termal equilibrium. Actually, these definitions must be constructed using some facts given through open-ended inquiry elaborated by a teacher [18]. A science concept might be constructed to students using the inquiry process causing the habit of mind.

In the students' knowledge refining as the third dimension of D3, the process is related to comparing, the inductive reasoning, the deductive reasoning, the error analysis, constructing support, abstracting, and analyzing the science phenomena. Fourth dimension of D4 about the perspective and the use of knowledge is related to making a decision, science investigation, conducting the experiment, solving the problems, seeking the new invention, and analyzing the system [9]. Those third and fourth dimensions contribute to facillitating the science literacy regarding describing the science phenomena, constructing and evaluating a scientific investigation, interpreting data and scientific evidences. Moreover, broadening the knowledge needs to be engaged to students by questioning science phenomena. This process is also necessary to improve the students' productive thinking. The engagement can be elaborated by explaining the scientific information, and showing the experiment resultsincluding datum and graphs. This work will provide students the opportunity in interpreting data and scientific evidences. Finally, a teacher should evaluate the students' work [7].

2. Methods

This research aims to qualitatively analyze the physics teaching reconstruction on the topic of heat transfer based on Marzano's learning dimensions in order to train the science literacy [19]. The description of teaching reconstruction is obtained from the students' response during learning science, the students' worksheet answers, and the profile of science literacy. The concept of heat transfer will be presented to students with the aim of training the science literacy. The learning process will be profiled to understand its treatment to indonesian students. The research is participated by 29 students in a middle school Bandung categorized as an ordinary school cluster. The students are grouped into 5 teams. To obtain an obvious observation data, the students' activities are monitored using a video recorder. Moreover, the students' worksheet answers indicating the students' responses are interpreted qualitatively as tabulated in tabel 1.

Tabel 1. Interpretation of students' response based on the students' worksheet answers [20]

Percentage (%)	Interpretations
100	all
76-99	almost all
51-75	mainly
	half
50	almost half
26-49	rarely
1-25	nothing
0	

3. Results And Discussion

According to the students' worksheet and the direct class observation recorded by the video, the description of students' learning achievement on the topic of heat transfer are shown in table 2.

Tabel 2. The profile of students' learning achievement based on their answered worksheets and the direct observation

No	Learning Activities	Findings	Observation results and recommendations
1	Defining the temperature	(71%) The students are mainly able to define the temperature. Some students were unable to define the temperature as two or more systems attaining the termal equilibrium.	Some students were difficult to define the temperature comprehended from the direct observation. Consequently, the demonstration of defining the temperature should be engaged by questioning science structurally. The students' observation might also be followed by discussing students' finding in defining the temperate.
2	Describing the heat	(57%) The students are mainly able to describe the heat as the energy transfer. Some students explained the heat as the energy transfer, which was not affected by the temperature difference of two object.	Some students considered the heat as the internal energy possessed by an object. It could be demonstrated that the sporadic particles' movement in the high temperature fluid will disperse quicker than the sporadic particles' movement in the low temperature fluid, so the temperature is reated to the internal energy of certain objects.
3	Explaining the effect of the heat absorbed by an object	(63%) The students frequently comprehend the heat, which is able to increase or decrease the object temperature. It is interesting that the	The phenomenon of an heated object should be demonstrate. When an object is being heated, the termal expansion will occur. It should be considered during the discussion about the physical and chemical changes to an objecy during the heat treatment.

Tabel 2. The profile of students' learning achievement based on their answered worksheets and the direct observation

No	Learning Activities	Findings	Observation results and recommendations
		heat was able to change the shape of an object although this phenomenon was not shown directly.	
4	Defining the termal expansion	(86%) Almost all of the students are able to define the termal expansion. However, (14%) the students rarely state the termal expansion as a reversible process.	The description should be dirrected by showing a related phenomenon. For example, a teacher demonstrates a baloon, expanded by a heated air, showing a reversible process.
5	Elaborating the inquiry questions	(86%) Almost all of the students are able to provide the inquiry questions.	A teacher shows a video regarding the termal expansion on two different metal rods. Afterwards, he guides the students making a graph L=f(T). Based on the graph, the students provide inquiry questions helped by a teacher. At the beginning session, a teacher has a significant role to direct students on elaborating inquiry questions.
6	Determining the dependent, independent, and controlled parameters	(29%)few students who enable to determine the dependent, independent, and controlled parameters	The students haven't recognized the dependent, independent, and controlled parameters obtained from the previous experiment. Consequently, a graph showing the relationship among the dependent, independent, and controlled parameters cannot be considered by the students. Therefore, through conducting the termal expansion experiment, a teacher must demonstrate which one is dependent parameter, independent parameter, and controlled parameter.
7	Constructing the experiment procedure	(49%) Almost half students are able to construct the experiment procedure, but only 29% students who completely compose the experiment procedure.	Although the students were able to take the data, they couldn't contruct the experiment procedure systematically. Thus, a teacher needs to guide the students writing an experiment procedure on the available worksheet.
8	Making a table	(43%) Almost half students are able to tabulate the datum taken from their observation.	A teacher should emphasize the parameters measured by the students and he also should discuss about drawing an efficient graph.
9	Making a graph using excell program	(14%) Few students enable to make a graph using excell program.	Although all of the groups were equipped by the computers, almost all students didn't have the initiative using excell in making a graph. This condition happens because each

Tabel 2. The profile of students' learning achievement based on their answered worksheets and the direct observation

No	Learning Activities	Findings	Observation results and recommendations
			group could not take the data during conducting the experiment, so it is recommended to ensure that this task is communicated before running the experiment.
10	Making a graph manually	(28%) Few students enable compose a graph manually recorded from experiment result	The students unusually draw a scientific graph. As a result, some groups fell difficult, unconfident, and unabled in creating a graph. It is recommended to a teacher how a good graph is composed both using excell program and a manual graph. In that case, the seconder data is needed to handle the problem.
11	Comparing the prediction to the experiment findings	Only 28% students who enable to analyze the experiment findings	It is recommended to instruct how to analyze a graph and compare it to the prediction or a teoretical framework. A teacher might employ the epistemic knowledge in investigating a graph and making a conclusion.
12	Determining the termal expansion coefficient based on the own graph	(14%) Few students are able to determine the termal expansion coefficient based on the own graph	It could be firstly analyzed that the students didn't have the experience in deriving a graph into a mathematics equation. A teacher requires to explain other related concepts even though the material is presented for a junior high school student.
13	Concluding the findings	(28%) Few students are able to conclude the experiment findings based on their own data	It is needed to model on how the experiment datum are compiled to arrange a conclusion using an epistemic knowledge.
14	Critisizing the results	None of the students enable to critisize their result or give the suggestion based on the experiment conducted	It requires to demonstrate how the findings are being evaluated and how the suggestion is arranged based on their findings.

According to our findings, we suggest to reconstruct the teaching didactic on the topic of heat transfer using Marzano's learning dimensions in order to train the students' science literacy as presented in the following sentences.

1. a teacher should consider the time required to present the learning materials to be adjusted with the available time given by a school. The required information for middle school students or other supporting information should be provided before conducting the experiment. In this case, the information on how to create a scientific graph based on data must be readby students from a module before they enter the class. This module is constructed in order to correct the students' misconceptions using multirepresentative approach equipped with the scientifically directional questions. This step is needed to construct the new knowledge required during conducting the experiment.

2. A teacher requires to analyze the students' characteristics. The students' learning capability will determine the role of a teacher conducting the learning in a class. A teacher will dominantly give the instruction in a class if the class consists of more students with their low abilities in accomplishing the learning materials. Consequently, it should be ensured that the students could encounter their difficulties during learning science, and the tasks have been done by the students so that the students will have the positive perceptions and attitudes. For example, a science-directional question has been demonstrated by a teacher to be solved correctly in order to improve the students' attention. Finally, the students will dominantly take a role in offering their scientific investigation questions.
3. A teacher must firmly conduct the science learning so that the science teaching could be fully dirrected. The concept repetition can be elaborated by a teacher to make sure that the students could accomplish the science learning well. In a group discussion, all students must contribute to their collaboration and cooperative learning.
4. A science demonstration should be well constructed to clearly show the science phenomena. In this research, the tools required to show the termal expansion phenomena on the metal rods during the heat treatment are equipped using microcontroller so that the students will have the factual information about the relation of $\Delta l = f(\Delta T)$. The demonstration is followed by offering the sequentially scientific questions with the purpose of the good acquiring and integrating the knowledge.
5. The learning scenario is developed to ensure that each step of broadening the knowledge could achieve the habit of mind. The questions have been elaborated such as why does an object's volume expand when it is being heated?, explain your prediction to the molecular system during termal expansion!, do the objects expand with the same amount?, why do you explain this prediction?. In order to apply those concepts, the students' habit of mind should be well trained by giving the scientifically related questions as follows: how does the working principle of the bimetallic strip?,can all materials be used as the bimetallic strip? what are the scientific requirements to show that phenomena?.
6. It should be ensured that the learning strategies is well constructed in order to train the science literacy including explaining the science phenomena, constructing and evaluating the experiment (comprehending the dependent, the independent, and the controlled parameters), and interpreting the data obtained from the scientific evidence.

4. Conclusions

According to the analysis based on the science learning process on the topic of heat transfer, the reconstruction of physics teaching didactic has been employed using Marzano's learning dimensions which emphasize the certain findings as narrated in the following sentences. It is required to prepare the scientific supporting information that should be comprehended by students before attending the class in order to handle their difficulties to construct the new knowledge. The directional questioning can be elaborated by a teacher in order to acquire the structurally subsequent knowledge while the concept repetition is important to ensure that students have understood the precise context scientifically. Moreover, a teacher must firmly conduct the science learning so that the science teaching could be fully dirrected. Finally, the teaching scenario must be well constructed to ensure that learning strategies could train the science literacy and the productive thinking.

5. References

[1] Wenning C J and Wenning R 2006 *Journal of Physics Teacher Education Online* **3** pp 3-14
[2] Dewey J 1904 *The Relation of Theory to Practice in Education," in Third Yearbook of the National Society for the Scientific Study of Education*, ed. N. L. Zimpher (Chicago: University of Chicago Press) pp 9-30
[3] National Research Council 1996 *National science education standards* (Washington, DC: National Academy Press)

[4] Fives H, Huebner W, Birnbaum A S and Nicolich M 2014 *Science Education* **98** pp 549-580
[5] Hobson A *Physics Education* 2003 **38** (109)
[6] Eisenkraft A 2010 *American Journal of Physics* **78** pp 328-37
[7] Utari S, Karim S, Setiawan A,Nugraha M G, Saepuzaman D and Prima E C, "Designing Science Learning for Training Students' Science Literacies at Junior High School Level," in Int. Conf. on Mathematics, Science, and Education 2015, edited by H. D. Barke et al. (Universitas Negeri Semarang, Semarang, 2015). pp.SE 1-6.C. D. Smith and E. F. Jones, "Load-cycling in cubic press," in Shock Compression of Condensed Matter-2001, AIP Conference Proceedings 620, edited by M. D. Furnish et al. (American Institute of Physics, Melville, NY, 2002), pp. 651–654.
[8] Marzano R Jand Pickering DJ 2006 *Demensions of learning Teacher's manual* (Virginia: Association for Supervision and Curriculum Development) pp. 1-212
[9] Gutwill-Wise J P 2001 *Journal of Chemical Education* **78**, 684
[10] King D T and Ritchie S M *International Journal of Science Education* 2013 **35** pp 1159-82
[11] Madu B C and Orji E 2015 *SAGE Open* **5** 2158244015594662
[12] Hagger M S and Chatzisarantis N L 2016 *Review of educational research* **86** pp 360-407
[13] Said Z 2016 *Eurasia Journal of Mathematics, Science & Technology Education* **12** pp 2253-65
[14] Akpan J P and Beard L A 2016 *Universal Journal of Educational Research* **4** pp 392-8
[15] Fang ZandWei Y 2010 *The Journal of Educational Research* **103** pp 262-73
[16] Kurnaz M A and Çalık M 2008 *Journal of Physics Teacher Education Online* **5** pp3-6
[17] Tanahoung C, Chitaree Rand Soankwan C 2010 *Eurasian Journal of Physics and Chemistry* Ed.**2** pp 82-94
[18] Creswell J W2013 *Sage publications, Los Angeles* pp 1-273.
[19] Koentjaraningrat 2013 *Metode Metode Penelitian Masyarakat Edisi Ketiga* (Jakarta : Gramedia) pp 1-55

Acknowledgments
We acknowledge DIKTI for the financial supports of Hibah Bersaing Dikti 2016.

Investigating Pre-Service Science Teachers (PSTs)' Technological Pedagogical Content Knowledge Through Extended Content Representation (CoRe)

R R Agustin [1,2] and L Liliasari[1,3]

[1] School of Postgraduate Studies, Universitas Pendidikan Indonesia, Jl. Dr. Setiabudhi No. 229, Bandung 40154, Indonesia
[2] International Program in Science Education, Universitas Pendidikan Indonesia, Jl. Dr. Setiabudhi No. 229, Bandung 40154, Indonesia
[3] Department of Chemistry Education, Universitas Pendidikan Indonesia, Jl. Dr. Setiabudhi No. 229, Bandung 40154, Indonesia
Email: liliasari@upi.edu

Abstract. The purpose of this study was to attain an insight into pre-service science teachers' technological pedagogical content knowledge (TPACK) as an integrative competency that is addressed by 21st century skills. The methods used in the study was descriptive. Nineteen pre-service science teachers (PSTs) of an educational university in Indonesia were involved in a semester long school science course. The course mainly develop students' pedagogical content knowledge (PCK) by utilizing content representation (CoRe) template. Furthermore an infusion of technological knowledge (TK) analysis led to the study of their TPACK by extending the template with a question in line to TK. The extended CoRe and self-reported survey were employed as instruments. The analysis of data used were quantitative and qualitative technique to obtain the insight into PSTs' PCK and TK. The results shows contrary value of PCK and TK identified by CoRe template to those measured by self-reported survey. However, the PSTs perceive their TPACK much higher, that, is 74.74%. Further investigation regarding PSTs ability to compose lesson plan was recommended for further research to capture more comprehensive insight into PSTs' TPACK.

1. Introduction

Technological Pedagogical Content Knowledge (TPACK) has emerged to unpack how teachers integrate technology into their teaching as well as to improve the quality of teaching with ICT. The framework provides a new way of thinking about preparing teachers to teach and learn with technology [1]. Furthermore, TPACK has been widely used as a tool to investigate the effectiveness of technology use in science classrooms because it is considered to help teachers make connections among technology, subject matter of science, and pedagogical knowledge [2] .

TPACK comprises seven constructs which are *technological knowledge (TK), pedagogical knowledge (PK), content knowledge (CK), technological pedagogical knowledge (TPK), technological content knowledge (TCK), pedagogical content knowledge (PCK) and technological pedagogical content knowledge (TPACK)*. TK refers to knowledge about various technology that can be implemented in everyday life as well as in teaching and learning activities. PK is the knowledge about teaching and learning strategies. CK is knowledge about the actual subject matter to be learned. TPK, TCK and PCK refers to knowledge about interplay between the original constructs. Thus, TPK is knowledge relates to how technology can be use appropriately for particular teaching and learning strategy. TCK relates to how specific technology play importance role in delivering certain content so as to it is being conveyed successfully. Meanwhile, PCK relates to how certain teaching and learning strategy fits subject matter delivery [1]. Constructs of the framework including the intersections is shown in figure 1.

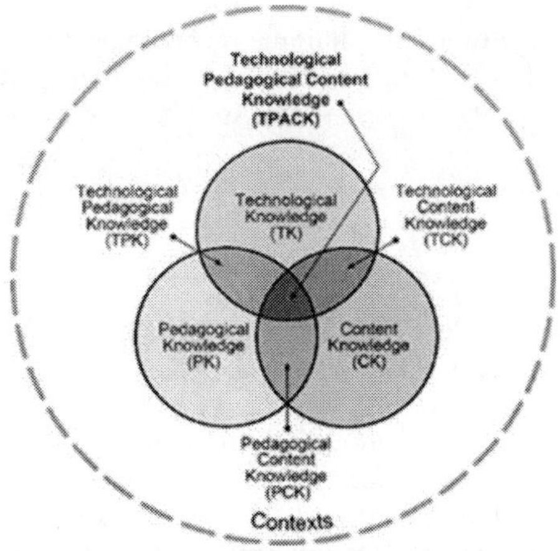

Figure 1. Technological pedagogical content knowledge (Retrieved from www.tpack.org)

Several studies have been conducted to capture insights into pre-service teachers' TPACK. Pre-service science teachers involved in 2-year master of teaching program infused technologies that fit to the content and context that aimed at facilitating non-experimental and experimental inquiry experiences [2]. Another aspect that was found to influence pre-service teacher competence in integrating ICT was ICT course experience [3]. However, pre-service teachers from diverse major found difficulties in understanding pedagogical knowledge when they involved in a technology integration course employing instructional design model [4]. On the other hands, proto-TPK was found as a keystone to set a starting point for interconnected development of TPK with pre-service teachers in their study at university [5].

The studies mentioned above commonly employ TPACK self-reported survey to captures TPACK constructs. However, there has no one that investigate TPACK constructs by involving Content representation (CoRe)- the origin of PCK instrument. Thus, this study aims at investigating Pre-Service Science Teachers (PSTs)' Technological Pedagogical Content Knowledge through Extended Content Representation (CoRe).

2. Research method

Descriptive method was utilized in this research to portray pre-service science teacher TPACK linked to their PCK. This study was the final part of a sixteen meetings 'school science' course aimed at preparing pre-service science teachers' competence in planning to teach science at a schooling year 7 to 9 (12-15 years old). At the beginning of the course, the pre-service science teachers were trained to construct CoRe in peer for several science content area. In this session, students' presentation and discussion were conducted to triangulate their PCK. Constructing CoRe for the topic being investigated was then conducted after they fluently constructing CoRe. The investigation was focused on the content area for 'Human Digestive System'. Meanwhile, investigating TPACK was carried out by employing two extending CoRe questions as well as self-reported survey in order to depict pre-service science teachers' perception on their TPACK.

Nineteen pre-service science teachers (PSTs) at a teacher education university in Indonesia were involved in this study. Moreover, instrument of the study consists of CoRe template [7] and TPACK self-reported survey that was adopted. In addition, a question linked to pre-service teachers' TPACK were added to the template. Those questions were asking technology preference in teaching certain idea and the way to compensate the absence of preferred technology.

Data were analysed qualitatively and quantitatively. Qualitative analysis was employed to capture pre-service science teachers' PCK on the basis of broadened CoRe template. The answers to CoRe template questions were identified and coding into big ideas emerged and teaching difficulties might pre-service science teachers confront. In addition, technological knowledge was investigated by questions no. 9. Addressed to technology preference in teaching particular idea. In order to grasp participants' TPACK, descriptive statistics was used to recap pre-service science teachers' perception on their TPACK.

3. Results and discussion

1.1. Pre-service science teachers' PCK

By implementing the extended CoRe, this research have indentified PSTs' PCK. Moreover, for 'Human Digestive System' it is captured six big ideas which are: *1) Human needs a food containing enough essential nutrition to be healthy, 2) Food digestion is helped by the digestive organs and accessory digestive glands, 3) Ingestion, digestion, absorption and excretion are the process of food digestion in human body, 4) Digestive deficiency can be caused by internal and external factor, 5)Food processing in digestive system occur in several stages which involves chemical and mechanical digestion, and, 6) Human's digestive system begins from mouth and ends at anus.*

Those big ideas are considered as an important means of depicting pre-service science teachers' PCK. It leads the investigation to the way pre-service science teachers' build the limitation of the topic, and can be viewed as the main ideas that they view as worth way to help students comprehend the whole topic. The big ideas are then explored by the second row of the CoRe asking pre-service science teachers to state the reason behind the importance for students to know these big ideas. The answers to these question delineate their reasons for developing this topic in this way. Regarding big idea 1, one PSTs viewed that it is important to the students to learn big idea 1 because learning this idea is supposed to lead the students to choose better food, so they will stay healthy. By stating the big idea and the reason behind emerging it, pre-service science teachers start to direct the learning into the stuff of content area and deliver it by means that clearly aimed. It is line that as science teachers begin to uncover their content knowledge they then focus on what matters in a content area and teach the matter for particular purpose and focus in developing a conceptualization of the subject area [7].

Pre-service science teachers' difficulties is captured by the fourth row of the CoRe template. For example, a participant views that there are various kind of vitamins, notably vitamin B. Students might confuse about it and thus the teacher should find appropriate technique to teach this content very well. Another difficulty that is confronted is for big idea 6. A participant views that it will be difficult to the students to learn about enzyme since it is a new concept for them. By this case it is found that the deeper content knowledge is needed to identify the depth of certain concept since the concept of enzyme for junior high school should be appropriately convey in accordance to the level. Thus, this questions help the researchers to arrest teacher insight through 'Human Digestive System' content area. Teachers begin to build and respond to insights they obtain about difficulties when teaching certain topic, and, PCK start to appear.

1.2. Pre-service science teachers' TK captured by extended core

Technological knowledge (TK) was captured by an additional question regarding technology pre-service science use to overcome the difficulties of teaching particular idea. The preference of technology for big idea 1 is shown in Table 1.

Table 1. PSTs' preference of technology usage.

Hardware	Frequency	Software	Frequency
Laptop	8	Flash simulation	1
LCD Projector	5	Digital images	1
Audio device	1	Power Point^{TM}	3
		Web learning resource	2
		Game software	1
		Video	1
		Audio software	1

Based on Table 1, it is shown that there are three kinds of hardware considered to support the learning of big idea 1, that are laptop, LCD projector and audio device. Meanwhile, seven software types chosen are flash simulation, digital images, power point, web learning resource, game software, video and audio software. This finding can be interpreted as TK posed by PSTs since TK can be represented by typical hardware and software [6]. This finding reveals the importance of preparing teachers educator to more emphasize technology involvement in regard to 21st century skills.

In order to validate PSTs PCK and TK the self-reported survey then being analyzed. PSTs' perception of their seven TPACK constructs is shown in Figure 2.

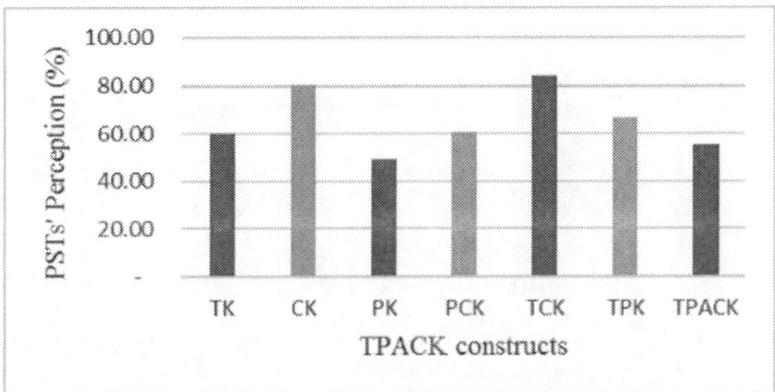

Figure 2. PSTs' perception on TPACK seven constructs.

Based on Figure 2, it is shown that PSTs perceive their TK as many as 60.15%, 80.26% for CK, 48.87 for PK, 60.53% for PCK, 84.21% for TCK, 75.94% for TPK and 74.74% for TPACK. This finding indicates that PSTs' PCK reach low value which is 60.53%. In contrast, in CoRe template PSTs state the importance of teaching particular idea as well as choosing the way appropriate to conceptualize it. Likely, TK reach low value as well which is only 60.15%. This finding is proving that PSTs TK are still low. This is validated by the preference of technology devices and software that is captured in Extended CoRe template. This extended CoRe template can be used to capture PSTs' TK. However, PCK is found to be contrary between that in CoRe template and self-reported survey. With that, it is known that PCK cannot be totally extended into any ways. Rather, it is more complex than merely about knowledge of students' conceptions and knowledge of instructional strategies [7]. Moreover the only extended CoRe template cannot be used to investigate PSTs constructs of TPACK. Comprehensively, another proxy is needed to complete the investigation. This finding is inline with several previous study that imply the need of lesson plan to capture teachers' TPACK in practical context [6]. Thus, further research may implement more various instrument to validate PSTs TPACK, such as, PSTs' lesson plan.

4. Conclusion

Extended CoRe template can be used to capture PSTs PCK and TK. PCK was represented in big ideas arisen and difficulties might confronted by PSTs in teaching the ideas. TK appeared in term of various kind of typical technology that are chosen to support teaching the ideas. However, the results of extended CoRe is not line with that in self-reported survey. Despite PSTs perform their PCK in extended CoRe, they still perceive it low in self-repoted survey. To better understand seven constructs of TPACK, it is needed another lens which is PSTs' lesson plan.

5. References

[1] Mishra P and Koehler M J 2006 *Teachers College Record* **108** 1017-54
[2] Maeng J L *et al* 2013 *J. Sci. Education Technology* **22** 838-57
[3] Yurdakul I K and Coklar A N 2014 *J. Computer Assisted Learning* **30** 363-76
[4] Lee C J and Kim C M 2014 *Educational Technology Research and Development* **62** 437-60
[5] Kontkanen S *et al* 2014 *Education and Information Technologies* **21** 919-43
[6] Pringle R M, Dawson K and Ritzhaupt 2015 *J. Sci. Education Technology* **24** 684-62
[7] McNeill, K L *et al* 2016 *Journal of Research in Science Teaching* **53** 261-90

Acknowledgments

We acknowledge Dr. Phil Ari Widodo, M.Ed. as the head of school science course and Ikmanda Nugraha, M.Pd. as a lecturer team of the course.

Increasing Self-Confidence of Indonesian Low Ability Student with Green's Motivational Strategies

I Siregar[1,2], Darhim[1], and E C M Asih[1]

[1]Department of Mathematics Education, Universitas Pendidikan Indonesia
Jl. Dr. Setiabudi No. 229, Bandung 40154, Indonesia
[2]STKIP Sebelas April Sumedang, Indonesia
Email: endangcahya@upi.edu

Abstract. Self-confidence is the important factor of mathematical learning, But in the reality, many 8th grade students do not have good self-confidence. To increasing the self-confidence is used Green's motivational strategies. So the purpose of the research is to know whether the Green's motivational strategies can increase the self-confidence of 8th grade students. The research focus on Indonesian junior high school student with lower ability. The research used qualitative research method, with basic qualitative research approach design. The activity included teaching material development and interview with student. From all conversation on the interview with students, students have satisfied feel. Students's self-confidence has changed better. All students show up that they can handle all matematical problems.

1. Introduction

Self-confidence is an important factor of mathematical learning. According to [6], the mathematics achievement depends on phsycological factors and the most important factor is self-confidence. But in the reality, many 8th grade students do not have good self-confidence. TIMSS 2011 report gave some facts that only 14% of 8th grade students which have good confidence on mathematics, 45% of 8th grade students is somewhat confidence and 41% of 8th grade students is not confidence[3]. Bad self-confidence will disturb the mathematics achievement on matematical learning.

Özbey [5] defines self-confidence as attitude to belief at self ability and always keep the emotion. Beside that, [1] self-confidence is belief can controling all situation by self. In other side, Oxford dictionaries [2] defines self confidence as belief to self ability, quality,and judgment. The conclution about self confidence based of tree definition is some one who have good self confidence will have good belief to his/her ability.

On this research, Green's motivational strategies is used to increase the self-confidence of 8th grade students. Green's motivational strategies were created by Green and developed by rudhumbu [4]. The purpose of the research is want to know are Green's motivational strategies can increase the self-confidence of 8th grade students.

Green's motivational strategies have four categories, that is conveying confidence, conveying high aspirations, giving comments, and valuing of tasks [4]. Conveying confidence is bassic category and most important. On conveying confidence, students is gave stimulate by teacher to build they belief at they ability on mathematics [4]. The stimulae is began with giving some motivation words is like "come on.., you can do it!", or "come on, you will get it!" [4]. Another stimulate is create situation and condition which make student success every find the mathematical problem [4]. The situation and the condition is like began the practice with easy mathematical problem. Student perception about success will influence they motivation (Middleton and Spanias, 1999). Student must belief that they will success.

Conveying high aspirations and giving comments are category which make Green's motivational strategies more precious and suitable on matematical teaching and learning. On the second category, conveying high aspirations, students is gave challenge to find more difficult mathematics problem [4]. Do not forget to completing the situation with propocatif words like " who was brave handle this problem?" or " come on be more creative!" [4]. On the third category, giving comments, students is gave some positive comments for everything which have done. The comments is like " good job!" or " exelent!!" or " goog, you can do better". The both Conveying high aspirations and giving comments will make the process matematical teching and learning not boring coused easy problem and help students on increasing they ability happily.

On the fourth category, valuing of tasks, students are explained the meaning of the task. Students are not only handle mathematical problem happily, but they are understand that the exercise is so important for them. Valuing of tasks have some sub categoty, that is Emphasizing the usefulness of a task, Emphasizing the importance of a task, Emphasizing enjoyment, offering rewards, teacher modeling enjoyment, connecting tasks, mentioning the rationale for a task, and offering choice[4].

2. Experimental Method

This research used qualitative research method, with a basic qualitative research approach design. The activity included the development of teaching material based on GMS and interview with student about their feelings after learning by teaching material developed by GMS. The research focus on Indonesian junior high school student with lower ability. Two schools were chosen and from each school was chosen two class from 8th grade. The name of two class at the first school in this paper are SMP 1 A and SMP 1 B. The names of two class at the second school are SMP 2 A and SMP 2 B.

3. Result and Discussion

3.1. Teaching Material Development

The teaching material development is based on Green's motivational strategies (GMS). The explanation of The teaching material development can be seen in Table 1 and Table 2. The example of teaching material can be seen in Figure 1.

Table 1. The teaching material development product base on GMS

Category of GMS	Teaching material development
Conveying confidence	Prepare some easy mathematical problem
Conveying high aspirations	Prepare some more difficult mathematical problem.
	All problem conected with prime subject.

Table 2. Teaching and learning proses product base on GMS

Category of GMS	Teaching material development
Conveying confidence	Give some easy mathematical problem at the beginning.
Conveying high aspirations	Continue with more difficult mathematical problem.
	All mathematical problem is gave in competition model. Student must finishing the problem on limited time (This proses is called warming up, don't more of 20 minutes).
	When we enter at prime subject, we can start with problem as challenge. And then we can give some explanation and example.
	Second and third example can gived as challenge again.
	The next exercise must begine with easy problem and continue with more difficult problem.
Giving commentsandvaluing of tasks	Give verbal motivation base on situation.

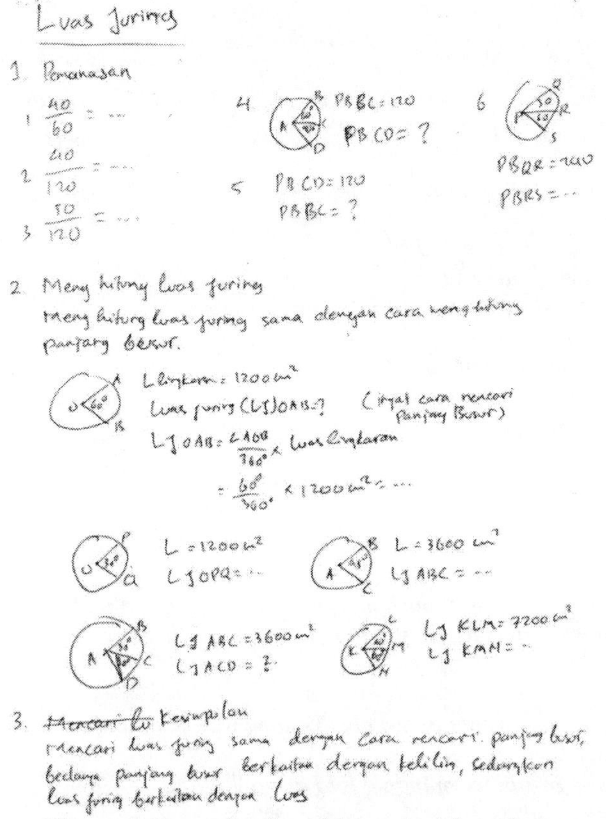

Figure 1. Example of teaching material product

Findings on the research:

- Mentioning the rationale for a task will be better is gave at the last or when students ask.
- The motivation words is like "good job!" or "wonderful!!" and teacher signatur is more valuable for student than candy or any gift els on offering rewards.
- The lesson, usually is began with apersepsi and motivation speech, but now apersepsi and motivasi can be melted on learning prosess. So, students can get more time for learning.

3.2. Interview With Student

The conversation in all interview at this paper is done with different students in different class from different school. All student's name is hidden and be replaced with simbol.

Symbol Explanation of conversation in interview script:

R = Researcher
M1stS = Male first Student
M2ndS = Male second Student
M3thS = Male third Student
F1stS = Female first Student
F2ndS = Female second Student
F3thS = Female third Student

SMP 1A

R	*: I am have some question, how your filling about my lesson? Maybe you are have some trace with my way, compared with the last teacher*
F1st &2ndS	*: Hehehe...*
R	*: its just question, if you think my lesson is different you can say different, but if not different, you can say it. Is my lesson different?*
F1stS	*: No, Not different*
R	*: How about you? (F2nd2)*
F2ndS	*: Comfortable*
R	*: why you said comfortable?*
F2ndS	*: I can understand easily*
R	*: what parts which make you can understand easily*
F2ndS	*: when I am learning*
R	*: when You are learnning?*
F2ndS	*: doing exercise*
R	*: ooo.. the exercise make you understand easily*
F2ndS	*: yes..*
R	*: ooo.. the exercise at the beginning lesson, warming up?)*
F2ndS	*: yes..*
R	*: at the last study, you get warming up at the start of learning?*
F2ndS	*: no..*
R	*: so which one do you think better, use warming up or not?*
F2nd&3thS	*: use warming up is better !!*

SMP 1B

R	*: when the lesson is begin, we always start with warming up. Do you like it or not?*
All student	*: We like it!!*
R	*: which one better, warming up first or not?*
All student	*: warming up first..*
M3thS	*: Its like sport mr, if we not doing warming up, we can get injury*
R	*: which one injury?*
M3thS	*: the foot..*
M3thS	*:if the lesson not use warming up, the lesson always difficult*
Other stud	*: that is right..*
R	*:hahaha.. have you fill scared with mathematics?*
M2nd&3thS	*: its oke..*
M1stS	*: (shouting) for me.. when I at first grad, mathematics were called death lesson*
R	*: death ?*
M1stS	*: yes mr, it make our head is explode.. becouse over there not use warning up..*
M2ndS	*: death... haha...*
M1stS	*:at Sindangbarang*
R	*: ooh.. piooh.. you are transfer student from Sindangbarang?*
M1stS	*: and I get transfer again to Tangeung. At Tangeung the lesson is better.. when in Sindangbarang, al students always says.. wooy... death lesson!!!...)*
R	*: hahaha... now if we comparing at Tangeung and in here, which one better?*
M1stS	*:in here better... over there always get explanation but saldom practice.. not fun.*
R	*: in here alway practice..*
M1stS	*:always practice, the explanation is few butunderstandable..)*
...	
R	*: so.. you not had prablem with mathematics, now how you filling?*
All student	*: more fun...*
R	*:because?*

M1st S : *fun.. becouse we are is like competition on handle the problem*
Other stud : *thats right..*

SMP 2A
R : *what is the benefit of warming up?*
F1st S : *so..*
F2nd S : *if we are forget,we can remember again*
F1st S : *helfull on exercise*
R : *ohh.. remember the lost memory by warming up.., are the warming up help you on exercise?)*
F1st S : *yes.. the exercise become more easy..*
R :*oh.. more easy.. how about you?*
F2nd S : *my answer is like her answer*
R : *warming up is fun*
F2nd S : *yes.. hehe..*
...
...
R :*ya, from the start I continue give you some problems*
F2nd S : *that were fun.. become challenge..*
F1st S : *ya.. that were fun.. go to infront of class.. again and again... I had perception before other studen I must be the first go to the infront of the class)*
R : *oh.. your perception is like that..*
F2nd S : *yes.. fast and more fast..*
R :*you want to get competition*
F1st S : *I must be the first..*
R : *have you ever learn like this before?*
F2nd S : *not yet*
R : *not yet, so it is the first time.. fun or not?*
F1st S : *fun*

SMP 2B
R : *the different, the past and present, what is the different?*
F1st S : *the learning proses become more fun*
R : *do you like exercise before?*
F1st S : *bored..*
R : *Hah..*
F1st S : *bored..*
R : *what your mean with bored?*
F1st S : *ya.. just bored*
R : *that is past, how about now?*
F1st S : *no..*
R : *why?*
F1st S : *Was fun...at the past the females grop saldom go to the infront of the class, but now they want to go to the infront of the class)*
R : *why?*
F1st S : *challenge*
R : *challenge.. any else?*
F2nd S : *I belife can handle the problem*
R : *you belife can handle the problem.. ooo.. can you do it before?*
F2nd S : *I can, but always confuse*

IOP Conf. Series: Journal of Physics: Conf. Series **812** (2017) 012104 doi:10.1088/1742-6596/812/1/012104

The conversation is done with different students in different class from different school, but all conversations is show same expretion. From all conversation students have satisfied filling. Students's self-confidence is change better. All students is show up that they belief can handle all matematical problem.

4. Conclusion

Green's motivational strategies can increase self-confidence of 8th grade students, especially, for Indonesian students with lower ability. From the interview results, students have satisfied feeling. Students's self-confidence has changed better. The researh get some findings, specially on applying Green's motivational strategies at Indonesian junior high school with lower ability. That is mentioning e rationale for a task will be better when students ask.The motivation words, such as "good job!" or "wonderful!!" and teacher signatur are more valuable for students than candy or any gift else on offering rewards.The lesson, usually is began with motivation speech, but now apperception and motivation can be integrated on the learning prosess. So, students can get more time for learning.

The research is too simple, but we belief that it can make much alternation. We need more profesional research with better instruments and more number of samples.

5. References

[1] Cambridge Dictionary [Online: http://dictionary.cambridge.org/dictionary/american-english/self confident?q=self+confidence]
[2] Roxana G, Terry L, and Joel S 2013 *Journal Obstet Gynaecol* **35** pp 355–361
[3] Ina V S M, *et al* 2012 *TIMSS 2011 International Research in Mathematics* (Amsterdam: TIMSS & PIRLS International Study Center)
[4] Norman R 2014 *International Journal of Education Learning and Development* **2** pp 76-103
[5] Şara A H, *et al* 2010 *Procedia Social and Behavioral Sciences* **5** pp 1205–1209
[6] Iskandar W, *et al* 2014 *International Journal for Innovation Education and Research* **2** p 10

Analyzing Students' Level of Understanding on Kinetic Theory of Gases

T Nurhuda*, D Rusdiana and W Setiawan

Departemen Pendidikan Fisika, Universitas Pendidikan Indonesia, Jl. Dr.Setiabudhi No. 229 Bandung, 40154 Indonesia

Email: tiaranurhuda@student.upi.edu

Abstract. The purpose of this research is to analysis students' level of understanding on gas kinetic theory. The method used is descriptive analytic with 32 students at the 11th grade of one high school in Bandung city as a sample. The sample was taken using random sampling technique. Data collection tool used is an essay test with 23 questions. The instrument was used to identify students' level of understanding and was judged by four expert judges before it was employed, from 27 questions become to 23 questions, for data collection. Questions used are the conceptual understanding including the competence to explain, extrapolate, translate and interpret. Kinetic theory of gases section that was tested includes ideal gas law, kinetic molecular theory and equipartition of energy. The result shows from 0-4 level of understanding, 19% of the students have partial understanding on the 3th level and 81% of them have partial understanding with a specific misconception on 2th level. For the future research, it is suggested to overcome these conceptual understanding with an Interactive Lecture Demonstrations teaching model and coupled with some teaching materials based on multi-visualization because kinetic theory of gases is a microscopic concept.

1. Introductions

In a teaching learning process, teachers always expect instructional goals that have been set can be achieved. Each of them sometimes feels that they have struggled in teaching students, but in fact many students are not able to achieve the expected instructional goals. During the teaching process, sometimes teachers are difficult to determine how many students have achieved the objectives and how many students have learning difficulties. The students' difficulties need to be detected as early as possible by the teachers and the results of this difficulty analysis are further used as a consideration in planning the next instructional programs. Students' learning difficulties are indeed varied, either in terms of scope or cause. Some of the indications that show a student with learning difficulties are, for example: acquisitions show lower educational outcomes, unequal results to the efforts that have been made and tardiness in performing tasks and learning activities. Based on these indications, an effective way to detect students' learning difficulties, especially in large lecture classes, is by measuring learning outcomes, for example, by employing diagnostic tests. A diagnostic test need to be valid, that is, able to measure what is being measured, and consistent when it is used to measure on another occasion. However, in a case with the need for accountability in higher education, standardized outcome assessments have been widely used to evaluate learning and inform policy [1].

The instructional process of physics should facilitate the students to discover concepts by themselves. Considering this situation, exploring and developing students' ability to understand concepts should become the teachers' commitment. Comprehension is the ability to understand or comprehend anything once they are known and remembered [2]. Students are supposed to get a lot of opportunities to use their logical reasoning abilities, practice, formulate concepts, and participate in solving complex problems that require their great effort. Then, the students are encouraged to reflect their thoughts in drawing an accurate conclusion.

Kinetic Theory of Gases is one of the key concepts used to understand other scientific concepts. In this research comprehension ability is divided into explaining, interpreting, inferring, exampling and comparing [3]. This concept is important to be understood by the students beforehand to explain other laws of thermodynamics. Besides, the most important is that this theory specifically discusses the behavior of gas particles microscopically and its relation to macroscopic quantities. This study is focused on the exposure profile of students' concept understanding, which is further broken down into the level of student understanding [4]. The results of this study can be used for further research to determine the appropriate treatment for improving students' concept understanding of the topic of gas kinetic theory.

2. Methodology

This research uses descriptive analysis to describe the level of students' understanding of the gas kinetic theory. The test instrument used was an essay test that had been content-validated by 4 experts, and this validated instrument was tested on 32 eleventh grade students of one senior high school in Bandung. The results of the test were then processed and analysed to see the validity, reliability, discriminating power, and level of difficulties [5]. Reliability was defined by researchers and statisticians as an indicator that provides information about the uniformity of a test when repeated measures are conducted. Reliability is also defined as the consistency of measurements used to measure the same thing [6]. Based on these criteria, the instrument is reduced to the one that can be trusted to describe the level of students' understanding which are categorized in Table 1 [4]

Table 1. Evaluation Scheme for Concept Test Items

Numerical score and degree of understanding	Criteria for scoring
0-No Understanding (NU)	Blank, repeats question, irrelevant or unclear response, no explanation given for choice answer
1-Specific misconception (SM)	Scientifically incorrect responses
2-Partial understanding with a specific misconception (PUSM)	Responses that show understanding of the concept, but that also contain a misconception
3-P3-Partial Understanding (PU)	Responses that contain a part of the scientifically accepted concept
4-Sound Understanding (SU)	Responses that contain all parts of the scientifically accepted concept

3. Results and Discussions

Table 2 shows content validity results of the instrument validated by 4 judges. On average, the judges decide that 98% of instrument can be used. From 27 validated items, the questions are reduced to 23 items that fit the criteria, including the suitability of the questions with learning indicators, aspects of concept understanding, and with the answers. The reduced questions are those which do not meet all of the three criteria. Of the 23 questions, there are 15 questions that were revised in terms of indicators. Most of the revised sub-sections are about gas kinetic energy and average speed of gases.

Table 2. Validation content result

Judges	Percentage	Criteria
Judge I	92%	Excellent
Judge II	88%	Good
Judge III	92%	Excellent
Judge IV	85 %	Good

Table 3 shows the results of the instrument validation after being content-validated that has been reduced to 23 questions. Based on the instrument validation, the instrument (23 essay test questions) can measure what is being measured and is consistent when tested in different samples. George and Mallery (2003) provide the following rules of thumb: "_ > .9 – Excellent, _ > .8 – Good, _ > .7 – Acceptable, _ > .6 – Questionable, _ > .5 – Poor, and_ < .5 – Unacceptable" [8]. Based on the 23 questions, an overview of students' level of understanding of the topic about the kinetic theory of gases can be seen, as shown in Table 4.

Table 3 Results of Instrument Validation

Results	Value	Criteria
Validity	0,73	Acceptable
Reliability	0,85	Good
Average	52,94	
Standard deviation	7,87	

Table 4 shows an overview of students' level of understanding on each sub material. The highest level of understanding, that is at level 4, occurs at 'the ideal gas law' section (33%), while the lowest level of understanding or not understanding, that is at level 0, occurs at 'pressure' and 'kinetic energy of gases' section (14%).

Table 4 Percentage of understanding levels on kinetic theory of gases

Level of understanding	Ideal gas law (%)	Pressure gas and kinetic energy gases (%)	Equipartition theorem (%)
SU	33%	4%	27%
PU	36%	26%	40%
PUSM	17%	35%	12%
SM	6%	21%	8 %
NU	8%	14%	13%

From further analysis, it is found that a few students have misconceptions in a specific way. The greatest misconception is on the question number 19 and 20, in the aspect of interpreting the questions from verbal words into mathematical equations on the 2nd level of understanding. The question number 19 is *"A gas cylinder with a certain volume contains ideal gas with the pressure P, the square root of the average squared-speed of the gas molecules is called V_{RMS}. If the tube is pumped with the similar gas, so the pressure becomes 2P, while the temperature is made constant, determine v_{RMS} now!"* In this case, most of the students answer $v_{RMS}=\sqrt{\frac{3P}{\rho}}$ ed, which is supposed to be

RMS= $\sqrt{\frac{3RT}{M}}$ v ; therefore, $vRMS \sim \sqrt{T}$ and because the temperature is constant, then *Vrms* is also constant. Hence, the present *vrms* is the same as the original *vrms*.

At the question number 20 some students answered on the 3rd level of understanding. The question is "*the effective speed or root mean square (RMS)of ideal gas molecules at a particular temperature is C, then the temperature of the gas is converted at the constant volume, so that the pressure becomes a quarter of the original. Estimate effective speed after the gas temperature is changed*!" Most students answered directly using $vRMS = \sqrt{\frac{3P}{\rho}}$ and $vRMS = \sqrt{\frac{3RT}{M}}$ equations. Whereas, they actually need to use Gay-Lussac equation first

$$\frac{P_2V_2}{T_2} = \frac{P_1V_1}{T_1} \quad (1)$$

$$\frac{\frac{1}{4}P_1}{T_2} = \frac{P_1}{T_1}$$

$$T_2 = \frac{1}{4}T_1$$

$$v_{RMS} = \sqrt{\frac{3RT}{M}} \quad (2)$$

$$v_{RMS} \sim \sqrt{T}$$

$$\frac{v_2}{v_1} = \sqrt{\frac{T_2}{T_1}}$$

$$\frac{v_2}{v_1} = \sqrt{\frac{\frac{1}{4}T_1}{T_1}}$$

$$\frac{v_2}{v_1} = \frac{1}{2}$$

$$v_2 = \frac{1}{2}v_1$$

At the question number 9, in the sub-material 'ideal gas law', most students answered with the 1st level of understanding. The question is *"If the volume of gas that is located in a closed vessel is kept constant 0.001 m ^{3}', there is a relationship between gas pressure (P) and the absolute temperature (T), as presented in the following table*[9]*:*

Table 5. Question number 9

Experiment	T (K)	P(N/m^2)
1	170	8,5
2	190	9,5
3	210	10,5

Draw a pattern graph of the relationship between pressure (P) and temperature (T), if the volume is more than 0,001 m ^{3} and less than (V) = 0,001 m ^{3}' ". Most students answered by directly modifying the table into a graphic form of temperature (T) against the volume by using a linear chart V = 0.001

m^3 and the pressure (T) against the volume (V) = 0,001 m^3, so the pattern graphs are constant. Whereas, there should be three chart pattern lines, namely V_1 <0,001 m^3, V_2 = 0,001 m^3, and V_3> 0,001 m^3 , as shown in **Figure 1**.

$$PV = nRT$$

$$V = \frac{nRT}{P}$$

$$V \sim \frac{T}{P}$$

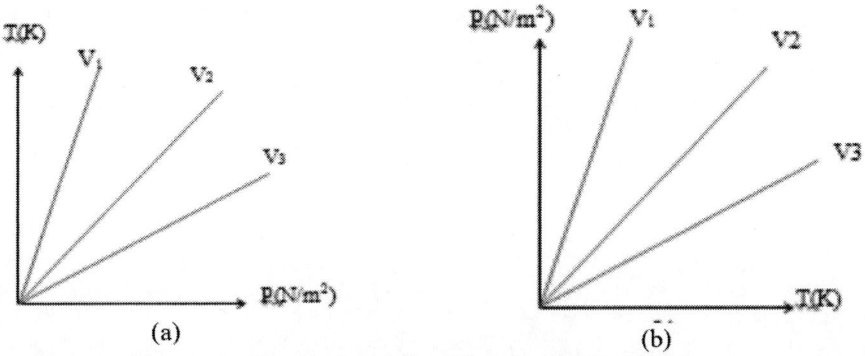

Figure 1. Graph of pressure (P) and temperature (T)

Furthermore, at the question number 5 most students answered on the 4th level of understanding. The question is *"During the summer, children are playing football and the ball is made of plastic. However, the ball gets dented. How do you fix this dented ball? Explain usingthe relationship between volume and temperature."* Most of the students answered correctly that the dented ball will return to its normal state when it is heated by putting it into a container of boiling water. This happens because when the ball is heated along with water, the temperature of the ball will increase, therefore the volume on the ball will expand and the pressure will be constant. This is in relation to *Charles' law* which states that if the gas pressure is constant, the gas volume (V) is directly proportional to its absolute temperature (T). The longer the surface of the ball is heated, the flatter it becomes because the volume inside the gas increases until the ball is restored to its original state.

4. Conclusions

Based on the elaborated discussion, it can be concluded that at the 'ideal gas law' section, 33% of the students have understood fully. However, 14 % of them have a lot of misunderstanding on the sub-material 'pressure and kinetic energy of gas'. Essentially, the concept of gas pressure and gas kinetic energy is a microscopic concept and cannot be directly observed. It is also associated with Newton's laws of motion; considered from its microscopic aspects, those are speed, collisions, pressure gas, kinetic energy and density.

5. References

[1] Lydia O L 2012 Educational Researcher **41** (9) 352–62
[2] Bloom, Benjamin S, etc 1956 *Taxonomy of Educational Objectives: The Classification of Educational Goals, Handbook I Cognitive Domain* (New York : Longmans, Green and Co)

[3] Anderson L W and Krathwohl R 2001 *A Taxonomy for Learning, Teaching and Assesing;A Revision of Bloom's Taxonomy of Education Objectives* (New York : Addion Wesley Lonman Inc.)
[4] Abraham, MR *Journal of Research in Science Teaching* **31**(2) 147-65
[5] Cronbach L J 1951 *Coefficient alpha and the internal structure of tests. Psychometrika,* **16**(3) 297-334.
[6] Baltes PB, Reese W, Nesselroade JR 1977 *Life-span developmental psychology: Introduction toresearch methods*. (Monterey, CA: Brooks/Cole)
[7] Gullford, JP 1956 *Fundamental Education Statistics in Psychology* (New York : Mc Graww-Hill Book Co. Ind)
[8] George D, & Mallery, P 2003 *SPSS for Windows step by step: A simple guide and reference. 11.0 update 4th ed* (Boston: Allyn & Bacon)
[9] Tipler P A 1991 *Physics for Scientist and Engineer, 3th Edition* (Worth Publisher, Inc)

Developing Science Virtual Test to Measure Students' Critical Thinking on Living Things and Environmental Sustainability Theme

M N Akbar, H Firman and L Rusyati

International Program on Science Education-Universitas Pendidikan Indonesia, Jl. Dr. Setiabudi No. 229, Bandung 40154, Indonesia
Email : muhamad.nur@student.upi.edu

Abstract. Critical thinking is skill and ability to use of risk taking creativity to make a decision and knowledge as a result, analysis and synthesis that, evaluation, to acquire, information search, to develop thinking, as an individual aware of his or her own thinking. The aim of this study is to develop the science virtual test to measure students' critical thinking on living things and environmental sustainability theme. The research method that is used in this research was descriptive research. The development of science virtual test item consist of five steps: (1) content analysis; (2) constructing the instrument (multiple choice) based on the elements of critical thinking by Inch; (3) validity judgment of the instrument by the expert; (4) legibility test of the instrument; (5) conducting the large field test. On the large field test was gained the results of validity and reliability of the test, difficulty index, discriminating power, and quality of distractor. The subjects of research were 8th grade students at International Junior High School in Bandung with 125 total of respondents. The coefficient alpha (α) was 0.747, the reliability of the test was categorized as 'high' and value of RXY correlation was 0.63 which mean that the validity of the test was categorized as 'high'. These means that science virtual test can be used to measure student's critical thinking with a good consistency. It is expected for other researcher to take this description as one of the basic information to be considered in developing science virtual test for improving students' critical thinking by various kind of topic.

1. Introduction

Modern technology offers educators a variety of new tools that can be used in the classroom. Technology can help teachers track and assess their students' as well as their own performance in the classroom. This research introduces a virtual test to measure the critical thinking of the students. Critical thinking is the most important skill for problem solving, inquiry and discovery. The ideal critical thinker is habitually inquisitive, well-informed, trustful of reason, open-minded, flexible, fair-minded in evaluation, honest in facing personal biases, prudent in making judgments, willing to reconsider, clear about issues, orderly in complex matters, diligent in seeking relevant information, reasonable in the selection of criteria, focused on inquiry, and persistent in seeking results which are as precise as the subject and the circumstances the inquiry permit[1].

Various examination methods are used in higher education institutions to assess the academic progress, for example paper-pencil-based examinations, assignments, and presentations[2]. More than fifty varied techniques used within higher education for assessment purposes; the most commonly method used are examinations[3]. The rapid advancement of Information and Communication Technologies (ICT) in teaching and learning has shifted the paradigm from paper-pencil-based to computer-based system of examinations which are usually termed as Computer Assisted Testing, Computer Based Testing (CBT), Computer Based Assessment (CBA), Online Assessment, E-Assessment and Web-Based assessment. This test uses computer based test method known as CBT[4].

Nowadays knowledge always develop, that's why the test should be developed because one of the test purpose is to develop students' achievement, ability and skill. The item tests should be selected because it has the highest fidelity with the target domain of task representing an important ability, such as reading, writing, or mathematical problem solving and it should be developed in consistently with the most rigorous guidelines[5]. The quality of test questions or items can be measured with the help of "Item Analysis Process". The major concepts related to item analysis including validity, reliability, item difficulty, and item discrimination[6]. As item development is a major step in test development, validity can be greatly affected by a sound, comprehensive effort to develop and validate test items[5]. The test item should be fieldly tested, and test takers should be appropriately prepared, then they should know what is the expected of item. Item analysis uses statistics and expert judgment to evaluate tests based on the quality of individual items, item sets, and entire sets of items, as well as the relationship of each item to other items[6]. Therefore, the quality of test will be increased.

2. Research methodology

The research method that used in this research was descriptive research. According to best, descriptive research is concerned with how what is or what exists is related to some preceding event that has influenced or affected a present s condition or event[7]. Furthermore, the descriptive studies describe a given state of affairs as fully and carefully as possible[8]. In this method the research focus to develop inch critical thinking multiple choice question from 8th grade on living things and environmental sustainability. It was consist of 26 items with 4 options each number.

The subjects in this research were 8th grade students from International Junior High School in Bandung. The sample was drawn from 125 respondents of the 8th grade students from three schools of International Junior High School in Bandung. The sampling technique that used for this research was a purposive sampling. For purposive sampling the sample has been chosen for a specific purpose[7]. The consideration was because the researcher choose sample based on the requirement such as the school implemented the curriculum 2013 or student has been learned about the concept that will be tested, have a good multimedia facilities e.g. personal computer, and located in Bandung, due to limited time and resources.

3. Result and discussion

As the result of the research there are three part that has been founded: (1) The general characteristic of the virtual test to measure students' critical thinking, (2) Legibility of virtual test by the students, and (3) Test item analysis such as validity, reliability, difficulty level, discriminating power, and distractor.

3.1 Critical thinking test

The designing of science virtual test to measure students' critical thinking on living things and environmental sustainability consist of three steps: (1) content analysis; (2) constructing the instrument (multiple choice test); (3) validity judgment of the instrument by expert. Before constructing the test item, content analysis was needed to determine the content that used to measure students' critical thinking on Junior High School. The theme of this research about the living things and environmental sustainability in 8th grade that limited by Core Competency and Basic Competence based on 2013 Curriculum. Then, there were five major topic that appropriate with the theme used such as structure and function of plant, reproduction system, developing of citizen and the impact, sun radiation, and climate change.

The instrument that has been constructed are in multiple choice form. It was constructed to measure the students' critical thinking based on the elements of critical thinking by Inch. The detailing the eight elements of critical thinking is a function that is interconnected, as follows: purpose, question at issue, assumptions, point of view, information, concepts, interpretation and inference, implication and consequences[9]. It is contains 26 sub elements, which Paul and Elder used as a guide for student in the critical thinking[16]. Then, author adjust it for constructing the instrument of science virtual test based on that criteria. In order to test the validity of the instrument the questions that have been constructed then judged by the 3 experts of education, 3 experts of content, and 3 experts of media. The science virtual test made by Adobe Flash 9.0. The example appearance of science virtual test can be seen on Figure 1.

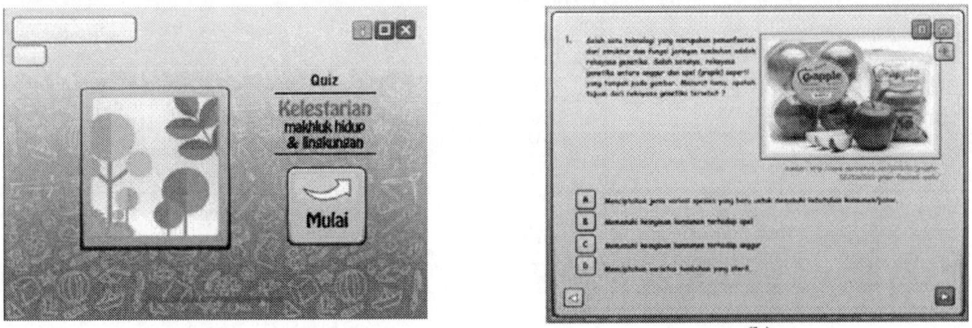

(a) (b)

Figure 1. Appearance of Science Virtual Test. (a) Opening and (b) Test Item.

3.2. Legibility test of science virtual Test

Legibility is ease and accuracy of reading of written or printed material[10]. The test item that has been made based on the sub-element of critical thinking by Inch, then the researcher do the legibility test of science virtual test. This test was done by five students of 8th grade in private junior high school. There are five criteria that has been test such as: (1) the description of article / picture / comic / video / tables easily understandable, (2) question easily understandable, (3) the option of the answer easily understandable, (4) questions and answers are interrelated, and (5) there are no words or terms that are difficult to understand or give rise to ambiguous meanings.

3.3. Test item analysis

The data obtained from the test were used in the statistical analysis. The analysis of the result was using SPSS Program Version 16 and ANATES Program Version 4.1.0 with the result show the reliability and validity of the test, item difficulty, item discrimination, and distractor analysis.

Table 1. Result of the reliability by SPSS 16.

Cronbach's Alpha	N of Items
.747	26

Based on the SPSS Program Version 16 the result of the reliability of the test $\alpha = 0.747$. Interpretation reliability coefficient of the test was categorized as 'high'[11]. Then, the result of the validity (using ANATES version 4.1.0) correlation of Rxy = 0.63 means that the validity of the instrument was categorized as 'high' according to interpretation value of Rxy (correlation)[11]. For item difficulty, item discrimination, and distractor analysis of each item have been measured by the ANATES version 4.1.0 analysis. The result of analysis per each item are:

IOP Conf. Series: Journal of Physics: Conf. Series **812** (2017) 012106 doi:10.1088/1742-6596/812/1/012106

3.3.1 Item difficulty (difficulty index). The indices of item difficulty level of each item are presented in Table 2. Test items were classified into three categories in terms of level of difficulty[12] as indicated in the Table 3.

Table 2. Item difficulty of comprehensive test items.

Item No	Difficulty index (p)	Item No	Difficulty index (p)	Item No	Difficulty index (p)
1	0.824	10	0.784	19	0.696
2	0.768	11	0.744	20	0.784
3	0.464	12	0.616	21	0.320
4	0.808	13	0.472	22	0.832
5	0.496	14	0.728	23	0.600
6	0.368	15	0.688	24	0.816
7	0.256	16	0.544	25	0.720
8	0.168	17	0.888	26	0.440
9	0.680	18	0.576		

Table 3. Classification of difficulty level.

Value of Difficulty Index (P)	Classification
0.00 – 0.30	Difficult
0.31 – 0.70	Moderate
0.71 – 1.00	Easy

3.3.2 Item discrimination. The indices of item discrimination of each item are presented in Table 4. The items were classified accordingly to their discrimination index[12] on Table 5. If the value of the discriminating power is negative, the item test should be rejected[12].

Table 4. Item discrimination of comprehensive test items.

Item No	DP	Item No	DP	Item No	DP
1	0.260	10	0.176	19	0.588
2	0.529	11	0.411	20	0.353
3	0.441	12	0.206	21	0.206
4	0.294	13	0.558	22	0.353
5	0.353	14	0.500	23	0.588
6	0.176	15	0.323	24	0.617
7	0.530	16	0.529	25	0.382
8	0.530	17	0.206	26	0.500
9	0.265	18	0.706		

Table 5. The Classification of Discriminating Power (DP).

Discriminating Power (DP)	Interpretation
0.00 – 0.20	Poor
0.21 – 0.40	Satisfactory
0.41 – 0.70	Good
0.71 – 1.00	Excellent

3.3.3 Distractor analysis

In the section on item analysis, the choice of distractors is critically important. Studies have shown that it is rare to find items for which more than three or four distractors operate efficiently. The distractor is usable if the distractor was chosen by 5% of all the respondent[12]. Result analysis shown in the Table 6.

Table 6. Distractor analysis of comprehensive test items.

Item Number	Option			
	A	B	C	D
1	Answer	1	0	21
2	8	7	13	Answer
3	Answer	55	10	1
4	Answer	4	9	11
5	5	45	Answer	11
6	Answer	25	2	32
7	66	13	Answer	14
8	Answer	22	65	17
9	5	6	27	Answer
10	Answer	9	2	16
11	Answer	18	4	10
12	29	Answer	8	9
13	Answer	29	24	12
14	20	3	Answer	5
15	6	Answer	10	22
16	13	39	Answer	4
17	8	Answer	1	3
18	5	45	Answer	11
19	Answer	30	4	3
20	10	Answer	9	8
21	Answer	63	3	19
22	3	15	Answer	2
23	13	Answer	29	8
24	Answer	6	2	15
25	Answer	14	13	8
26	Answer	28	32	7

Science virtual test was constructed to measure students' critical thinking on living things and environmental sustainability. It is a new way to test the level of student's critical thinking by Inch especially for 8th grade students in junior high school. Advantages of virtual science test are: (1) it was in the form of multiple choice, in contrast to the type of test used to measure students' critical thinking is commonly in essay form. With this test, measurements made are more objective than the essay form and it easy to know the profile or level of the students' critical thinking based on the students answer. Multiple choice questions are quick and easy to score, it is can be written so that they test a wide range of higher order thinking skills, and cover lots of content on a single test[13]; (2) compared to the base paper test, science virtual test can use multimedia which is cannot used in paper test, such as video, animation, and music. In this science virtual test is provided the interesting video that given an information that give more visualization to student, then student can see the phenomena clearly; (3) the result of students answer directly shown in the last session of test, so student will know the score that students gained; (4) paper less and it is very simple to use.

The item analysis conducted in this study implicates three statistics to assists in analyzing the effectiveness of each of the comprehensive test questions specifically validity and reliability of test, item difficulty, item discrimination and distractor analysis. The statistical analysis was done by SPSS version 16 and ANATES version 4.1.0 in the following:

- Reliability and validity of the test, based on the SPSS version 16 the number of Cronbach's alpha or coefficient alpha was used to determine the reliability of the test. If alpha is high, then we might assume that the reliability of the test is acceptable. Reliability coefficient was employed to evaluate the performance of the test as a whole. According to the result of the coefficient alpha (α) is 0.747. One should attempt to generate a KR20 reliability coefficient of .70 and above to acquire reliable score[8]. So, the reliability of the 26 question of the test was categorized as 'high', means that the question has a good consistency and accuracy to measure students' critical thinking. The number of the validity of whole of the test based on the ANATES version 4.1.0 from value of XY correlation is 0.63 which mean that the overall question or item was valid with significant result that categorized as 'high'. Beside that result, the validity that used is from the expert judgment with the result that all of the question can be used by some revision.
- Difficulty index, difficulty index in one of important aspect to determine that the instrument of test is feasible or not. The difficulty index is important because it helps to know how well students have acquired the knowledge they are supposed to[14]. Difficulty index of this test is resulted from the analysis by ANATES version 4.1.0. Test items were classified into three categories in terms of level of difficulty as indicated in the Table 3 such as easy, moderate, and difficult. Based on the result the item number of test which are classified as easy level are number 1, 2, 4, 10, 11, 14, 17, 20, 22, 24, 25. Item number which classified as moderate level are 3, 5, 6, 9, 12, 13, 15, 16, 18, 19, 21, 23, 26 and classified as difficult are 7 and 8. The total number which classified as easy 11 items, moderate 13 items, and difficult 2 items. From the result most of the item number have item difficulty in categorized as 'moderate'. While, two items which are categorized as 'difficult'. Analysis from the researcher that the one of factor that influence to the difficulty index is the level of students understanding to the question of the items and the information that provided. Another factor is when the instrument is tested in the class, there are students that still confuse with the term that used in the question and the information that has provided.
- Item discrimination, item discrimination as a measure used to discriminate between students in the top with that of the low group who obtained the correct responses. A highly discriminating item indicates that the students who had high tests scores (high achiever) got the item correct whereas students who had low test scores (low achiever) got the item incorrect[15]. The item number that has a categorized by poor items are number 6 and 10, satisfactory items are number 1, 4, 5, 9, 12, 15, 17, 20, 21, 22, 25, good items are number 2, 3, 7, 8, 11, 13, 14, 16, 19, 23, 24, 26, and excellent item is number 18. While, the other item such as item number 6 and 10 are categorized by poor item. From the result the number of item that has a good quality to differentiate between high respondent or achiever and low respondent or low achiever are 24 test items or 92% of test items.
- Distractor analysis, **a** distractor analysis assists in distinguishing plausible distractors from implausible ones. One way to study responses to distractors is with a frequency table that tells the proportion of students who selected a given distractor. Remove or replace distractors selected by a few or no students because students find them to be implausible. Distractors are classified as the

incorrect answer in a multiple-choice question. According to Instructional Assessment Resources (IAR), student performance in an exam item are very much influence by the quality of the given distractors[14]. The distractor should be considered an important part of the item test. A total of distractor that used are 104 from 26 items. Based on the result the total distractor that not usable are 21 because it was less than 5% chosen by total of respondent. For the other distractor is useable but there are some of the distractor have a poor distractor because the total number of respondent who choose that distractor near from the total number of respondent that choose the correct option or answer such as the distractor of item number 3. The distractor of number 3 option B, correct answer option A chosen by 58 respondent and option B as distractor chosen by 55 respondents. It means that the distractor of option B is similar with the correct answer. For another example is item number 7, option A as distractor the number respondent who choose the option A is more than the correct answer (option C). It means that the respondent think that the correct answer is option A. In general the distractor that should be change or revise is 25% from overall of the distractor.

4. Conclusion

Based on the research results it is acquired some conclusions that are: (1) in general the development of science virtual test item to measure students' critical thinking on living things and environmental sustainability consist of five steps: content analysis, constructing the instrument (multiple choice), validity of the instrument by the expert, legibility test by students, and conducting the large test; (2) science virtual test item constructed to measure the students' critical thinking based on the elements of critical thinking by Inch. Detailing the eight elements of critical thinking is a function that is interconnected, as follows: purpose, question at issue, assumptions, point of view, information, concepts, interpretation and inference, implication and consequences; (3) there were five criteria that have been tested on legibilty test such as the description of information, question, the option, interelation of questions and answers, and terms that have been provided; (4) based on the large test field is gained the results of validity and reliability of the test, difficulty index, discriminating power, and quality of distractor. The coefficient alpha (α) was 0.747, the reliability of the test was categorized as 'high' and value of XY correlation was 0.63 which mean that the validity of the test was categorized as 'high'. These means that science virtual test item can be measure a students' critical thinking with good consistency.

5. References

[1] Zhang L F 2003 *The Journal of Psychology* **137** 517-44
[2] Jamil M, Tariq R H, Shami P A and Zakariys B 2012 *TOJET: The Turkish Online Journal of Educational Technology* **11**
[3] Sim G, Holifield P and Brown M 2004 *Research in Learning Technology* **12**
[4] Aliyu S, Ibrahim U E, Junaidu S B and Kana D F 2015 *International Journal of Advanced Research in Computer Science and Software Engineering* **5** 25-8
[5] Haladyna T M and Rodriguez M C 2013 *Developing and Validating Test Items* (New York: Routledge)
[6] McCowan R J and McCowan S C 1999 *Item Analysis for Criterion-Referenced Tests* (New York: Buffalo State College)
[7] Cohen L, Manion L and Morrison K 2007 *Research Methods in Education* ed 6 (Companion Website)
[8] Fraenkel J R, Wallen N E and Hyun H H 2011 *How to Design and Evaluate Research in Education* ed 8 (New York: McGraw-Hill)

[9] Inch E S, Warnich B and Endres D 2006 *Critical Thinking and Communication: The Use of Reason in Argument* (Boston: Allyn and Bacon)

[10] Page G T and Thomas J B 1979 *International Dictionary of Education* (London: Pitman)

[11] Jacobs L C and Chase C I 1992 *Development and Using Test Effectively* (San Fransisco: Jossey – Bass Publisher)

[12] Arikunto S 2012 *Prosedur Penelitian Suatu Pendekatan Praktik* (Jakarta: PT. Rineka Cipta)

[13] Weimer M 2015 Advantages and Disadvantages of Different Type of Test Questions *Faculty Focus* (Mumbai: Magna Publication)

[14] Adeleke A A and Joshua E O 2015 *Journal of Education and Practice* **6** 28-42

[15] Zimmaro D M 2004 *Writing Good Multiple-Choice Exams* (The Division of Instructional and Assessment, University of Texas)

[16] Paul R and Elder L 2006 *The Miniature Guide to Critical Thinking: Concepts and Tools* (The Critical Thinking Community)

The Effect of Teaching Model 'Learning Cycles 5E' toward Students' Achievement in Learning Mathematic at X Years Class SMA Negeri 1 Banuhampu 2013/2014 Academic Year

N Yeni[1,2], E P Suryabayu[2], T Handayani[3]

[1]Departement of Mathematics Education, School of Postgraduate, Universitas Pendidikan Indonesia, Jl. Dr. Setiabudi No. 229. Bandung 40154, West Java-Indonesia
[2]Departement of Mathematics Education, SekolahTinggiKeguruanIlmuPendidikanAhlussunnahBukittinggi, Jl. Diponegoro no. 8, Bukittinggi 26117, West-Sumatera, Indonesia
[3]SekolahMenengahAtasNegeri 1 Banuhampu, Jl. PakanSinayan, West-Sumatera, Indonesia
Email: nelfitriyeni@student.upi.edu

Abstract. Based on the survey showed that mathematics teacher still dominated in teaching and learning process. The process of learning is centered on the teacher while the students only work based on instructions provided by the teacher without any creativity and activities that stimulate students to explore their potential. Realized the problem above the writer interested in finding the solution by applying teaching model 'Learning Cycles 5E'.The purpose of his research is to know whether teaching model 'Learning Cycles 5E' is better than conventional teaching in teaching mathematic. The type of the research is quasi experiment by Randomized Control test Group Only Design. The population in this research were all X years class students. The sample is chosen randomly after doing normality, homogeneity test and average level of students' achievement. As the sample of this research was X.7's class as experiment class used teaching model learning cycles 5E and X.8's class as control class used conventional teaching.The result showed us that the students achievement in the class that used teaching model 'Learning Cycles 5E' is better than the class which did not use the model.

1. Introduction

Learning is the most basic activities in the whole process of education at schools. This means that the success or failure of education goals depends on how the learning process experienced by the students in the classroom. The learning activities at schools also require students to learn optimally. It is expected that learning goals in gaining minimum competency achievement that must be full filled by the student.Mathematics is an abstract and deductive science as well as its usefuness in daily life. Sometime people think that mathematic is abstract in science and learn a lot of formulas that lead to difficulties students understand the concepts of math. So that students are less excited when learning time takes place. In addition, the learning process is not as effective as less precise methods used, and the lack of learning resources can also lead them not interested in learning. Consequently, the result the score of the learning outcomes decreased.

Problems of decreasingstudent learning outcomes is not much different from the observations done during the Practice Field Experience at SMAN 1Banuhampu. Based on these observations, the researchersnoticed that the learning process still a one-way. The process of learning is centered on the teacher while the students only work based on instructions provided by the teacher without any creativity and activities that stimulate students to explore their potential. The usual steps in learning involved; teachers prepare students to learn, explain the material, giving the example of the problems, then provide practice questions, students asked to do the exercises and ends with the provision of homework. The lack of variety of learning such the example above caused boredom for students. So that when the teacher explains the lesson only some of the studentspay attention while the others are busy with other activities. When the teacher asked them to solve problems (do the test) only a few students done it seriously, most of them just cheated on his friend's work, and some did not do at all.

Some problems can be as source of poor behavior of student in learning mathematic such as monotonous method, lack of the opportunity for the students to explore their own capabilities, lack of understand of the teacher's explanation, and the classroom atmosphere less conducive. These conditions indicated that learning is still dominated by teachers so that students are less active in the learning process, majority of students have not been able to construct their own knowledge.Addressing problems in teaching mathematics above. It is important to give the opportunity for the students to construct their own understanding through learning activity. Constructivism learning theory stated that learning is an active activity in which students construct their own knowledge and find the meaning of something they learned. Teachers act as facilitators, helping student activity in the form of knowledge that learning is an active process of the student. One of the learning activities isbased on constructivism theory is a 'Learning Cycle 5E' model.

Teaching model 'learning cycle 5E' is part of the learning based on theory of constructivism [1]. Model learning cycle 5E introduced by Robert Karplus in Science Curriculum Improvement Study/SCIS [2]. According [3]is a model of learning cyclethat enables students to find their own concept orconsolidate concepts learned, prevent misconceptions, andprovides opportunities for students to apply the concepts that have beenlearned in new situations. Teaching model 'learning cycle 5E'is a learning model that is centered on students which consists of five phases. They are engagement, exploration, explanation, elaboration, and evaluation; [4]; [5]; [6]; [7].

In engagement phase. Students are invited to correlate their knowledge to the material that will be learned. The teachers should facilitate the students to relate their own knowledge to the teaching material based on their experience in daily life. The teachers should be able to awaken and develop the students interest and curiosity about the topics that will be discussed. This is done by asking some questions about the factual processes in everyday life (related topic). Thus, students will give a response. Then, students can be used as a foothold for teachers to know students prior knowledge on the subject. In addition teachers need to identify the presence of conceptual errors on the students. In this case the teacher must build aengagement between the everyday experiences of students with learning topics to be discussed. Exploration phase is the second step. In the this phase teacher formed small groups of 2-4 students, and then give them the opportunity to work together in their groups without direct instruction from the teacher (teacher distributed learning material in the form of student's activities sheets). In this group, students are encouraged to test hypotheses or create new hypotheses, try alternative solutions in the group. At the time teacher conduct and record observations and ideas or opinions that developed in the discussion conducted by the students. In this phase, the teacher acts as a facilitator and motivator. Basically the goal of this stage is to check the knowledge of the students if it is correct, it is still wrong, or perhaps partly wrong, partly right. Explanation phase. At this stage, teachers are required to encourage students to explain a concept with a sentence / thought itself, ask for clarification on the evidence and explain other students, and each other think critically. After doing the discussion, teacher gives explanations of concepts discussed before, use former student explanation as a basis for discussion. The next is Elaboration. At this stage students tried to apply the concepts and skills they have learned in a new situation or a different context. Thus, students will be able to learn significantly, because it has been able to implement/apply the newly learned concepts in new situations. At this stage it's hoped that teachers and students' motivation will increase. Increasing student motivation can certainly boost student learning outcomes. The last steps isEvaluation. At this phase teachers can observe students' knowledge or understanding in applying new concepts. Students can do a self-evaluation by asking open-ended questions and seek answers through observation, evidence, and the teachers as an evaluation of the implementation process of learning cycle 5E model . The results of this evaluation can be made by the teacher to know whether it has gone very well, fairly well, or still less. Similarly, through self-evaluation, students will be able to know the lack or progress in the learning process which has been done.

In previous studies conducted by [8] states that teaching model learning cycle 5E is one alternative of learning models that can improve student learning outcomes. [9] suggests the influence of learning cycle can improve student achievement in terms of initial ability of students. [10] stated that the application of the teaching model learning cycle 5Ecan improve students' mathematical problem solving abilities. Based on these results, the alleged application of learning cycle 5E can improve students' mathematics learning outcomes.

Based on the explanation above it can be concluded that the teaching model 'learning cycle 5E' can facilitate students to construct their own knowledge. Before the study began, the teacher tried to arouse interest and curiosity of students to the material to be learned. Then the students build their math knowledge with members of his group followed by communicating and discussing the knowledge they gained from solving problems . After that, students are given the opportunity to apply new knowledge. These five stages are the things that must be done in implementing the teaching model 'learning cycle 5E'. In this research, students were grouped based on academic ability of students. A highly capable students will join to each group. It is aimed to help the group in discussing the material given. Because some of them involved in low ability student.

Student learning outcomes showed by the evaluation process or test, then the results of the test assessed by the teacher.[11] states that there are three aspects of the assessment of learning, namely: a) cognitive sphere, in terms of thinking skills, including the ability to memorize, understand, apply, analyze, synthesize, and evaluate; b) affective sphere, includes character behaviors such as feelings, interests, attitudes, emotions, and values;c) The sphere of psychomotor, include imitation, manipulation, precision, articulation, and naturalization. Based on the above description learning outcomes in learning mathematic by using teaching model 'learning cycle 5E' is the result of learning on the cognitive obtained after carrying out tests.Based on these description above the researcher intends to try applying teaching model 'learning cycle 5E' on the result of teaching mathematic at X years class at SMAN 1Banuhampu in the 2013/2014 academic year.The main purpose of this study was to determine whether learning outcomes of students who take mathematics learning by using teaching model 'learning cycle 5E' is better than the students who take the learning by using conventional learning models in class X SMA N 1 Banuhampu in 2013/2014 academic year.

2. Method

This research was conducted by using a quasi-experimental methods.[12] argues that the quasi-experimental research is experimental research is not successfully pursue or realize the things that are required on a purely experimental research (True Experiment). The design used randomized control group only design.

This experimental study has a population of all students in grade X SMAN 1 Banuhampu enrolled in the 2013/2014 academic year. The samples in this study were taken randomly (random is done only on the selection of the class).It means researchers took two classes at random from all one class as an experiment and another as a control class.Every member of the population gets equal opportunity to become members of the sample. Because in this study population members are distributed to the whole class, then the sample selection is done on these classes. This is done because it is not possible to random system against these individuals. In this research, researcher implemented quasi-experimental research by giving treatment in the experimental class by applyingteaching model 'learning cycle 5E' and model conventional learning in the control class.

3. Results and Discussion

This study was conducted to see the effect of the teaching model 'learning cycle 5E' on learning outcomes. Implementation of learning by using Teaching model 'learning cycle 5E'which was implemented in the experimental class X.7 gave the better result than learning from ordinary class at X.8.The stages are implemented are first, engagement that aims to arouse students' interest and curiosity of the material to be studied.

First stage, the results showed that the process of engagement have not maximized yet because of the curiosity of students toward the material is low. It may be caused by lack of active students in initial discussions to identify the material provided. On the second stage; exploration, students are grouped by ability evenly and then given the opportunity to discuss the matter within each group the aim is for students to discover / comprehend the material provided. [13], argues that cooperation is an important tool in helping to achieve the solution of social problems, helping individuals to adjust to other people. In Exploration phase [6] students are given the opportunity to work both independently and in groups, this phase gives students concrete experiences that help students begin to build concepts and develop skills without instruction or guidance directly from the teacher. At this stage the student activity looks quite active in group discussions, this indicates that the purpose of the learning cycle 5E to activate students in the learning process can be done well, students do not look shy to express opinions in front of his friends though sometimes wrong. In explanation phase, at this stage the student should be able to explain a concept by using his own words before the other groups. At this stage showed that students brave enough to explain their findings in front of his friends. In elaboration phase, student should be able to apply concepts and skills learned in different contexts. In this study, the researchers observed that students were able to apply through other matters relating to the material being studied. In evaluation phase, at this stage the students are expected to evaluate themselves by asking several questions aimed to solidify the understanding of the material that has been given. In this phase showed that the students asked some questions if they were less familiar with the material provided. It means they realized their weaknesses in learning process and ask forthe teacher's help to give more explanation.

From the findings above it can be concluded that the learning outcomes of students increased from one meeting to another. This is evidenced by the results of the average value of student work sheets or LKS is increased from the first meeting until the fifth meeting. At the five phases applied to the learning model It showed that the exploration stage indicated students more active in the learning process, they are able to focus on understanding the material in their own group. After testing the hypothesis with t test results are obtained as follows.

Table 1. Final Tests Hypothesis Testing

Elements	Class Experiment	Control class
N	28	29
$\bar{X}$	81	67
S	12,55	14,97
S^2	167,5025	224,1009

Based on Table 1. It can be conclude that $t_{hitung} = 21{,}34 > t_{tabel} = 1{,}6725$, mean$H_0$rejected, in other words the learning outcomes of students who used teaching model 'learning cycle 5E' is better than the usual learning.It shows that the application of teaching model 'learning cycle 5E' can improve student learning outcomes. This is supported by studies Marek[14], which states that a teacher can not accommodate the concept for students, but through the questions, teachers can help students construct meaning from the concept of their own experience.

Learning cycle 5E model gave a positive impact on student learning outcomes. Studentslook more active in learning, they discover new ideas within the group, and not ashamed to express their opinion, although sometimes it's wrong or different with other friends. In the implementation teaching model 'learning cycle 5E' also found weaknesses. The weaknesses found suck as the limited time allocated, some students feel bored sit in group work, it may be caused by a group of their friends which cannot cooperate one to another, consequently they do not want to finish their activities on their LKS. But the researchers sought for the purpose of this study can be achieved well.That is the way to maximize time and made it clear to students that learning in groups will train students to learn actively, independently work on the problems and is responsible for their own work. The average value of 81 experimental class, while the class only controls 67. It can proved by the results of the final tests on the experimental class is better than the control class.

4. Conclusion

Based on the results of research and discussion can be taken conclusion that mathematics learning outcomes of students who follow the teaching model 'learning cycle 5E' is better than the students who followed conventional teaching in grade X SMAN 1 Banuhampu 2013/2014 academic year. It needed other strategies to use the time allocatedeffectively in applying this model. As a whole students can be involved in learning process by activated them in each of the phases. As the result it can reduce the role of the teacher in teaching and learning process. Students also can showed their ability to solve the problem in learning mathematic by sharing the idea to their friend in the work group.

5. References

[1] Made W2012*Contemporary Innovative Learning Strategies(* Jakarta: Earth Literacy)p 170
[2] Trowbridge LW dan Bybee RW 1996*Teaching Secondary School science: Strategies for Developing Scientific (6'h Ed.)* (New Jersey: Prentice Hall)
[3] Soebagio S& Soetarno S 2009 *Media Komunikasi Kimia***5** 50
[4] Lorsbach A W 2002 *The learning Cycleas A tool for Planning Science Instruction*[Online: http://www.coe.ilstu.edu/scienceed/lorsbach/257lrcy.htm]
[5] Bybee Rodger W 2006 The Biological Sciences Curriculum Study 5E Instructional Model: Origins Effectiveness and Applications [Online:http://www.bscs.org/.pdf/bscs5eexemummary.pdf]
[6] Madu B C & Amaechi C C 2012 Effect of five-step learning cycle model on students' understanding of concept related to elasticity *Journal of education and Practice***3** 173-181
[7] Ergin I 2012 *J Phys Educ***6** 14-20
[8] Ari Wibowo dkk 2009*Penerapan model pembelajaran siklus belajar (learning cycle) 5E dalam meningkatkan hasil belajar siswa pada mata pelajaran Teknologi Informasi dan Komunikasi* (Bali:Thesis Universitas Pendidikan Ganesha)
[9] Purwanto 2008*Pengaruh pembelajaran menggunakn learning cycle pada pokok bahasan bangun ruang sisi lengkung terhadap prestasi belajar matematika ditinjau dari kemampuan awal siswa kelas VII SMP Negeri 2 Grobogan*(Bali: Thesis Universitas Pendidikan Ganesha)
[10] Apriyani 2009*Penerapan model learning cycle "5E" dalam upaya meningkatkan kemampuan pemecahan masalah matematika siswa SMP N 2 Sanden kelas VIII pada pokok bahasan prisma dan limas*(Bali: Thesis Universitas Pendidikan Ganesha)
[11] Kunandar 2007*Teachers Implementation Unit Level Curriculum (SBC) and Success in Teacher Certification*(Jakarta: King Grafindo Persada) p 385
[12] SuharsimiA 2005 *Management Research*(Jakarta: Rineka Reserved)p 209
[13] Fatih T2011*Education Sciences: Theory & Practice***11** 273-277
[14] Marek Edmund A 2014 Challenges in a Physics Course: Introducing Student-Centred Activities for Increased Learning*Journal of University Teaching & Learning Practice* vol **11** no 2 article 8

Acknowledgments
Thank youvery much for the suggestion and idea from the supervisor and colleagues. It hope that this study will be one solution for any problems in teaching mathematics specially for the students of senior high school level.

Development of Reasoning Test Instruments Based on TIMSS Framework for Measuring Reasoning Ability of Senior High School Student on the Physics Concept

Muslim*, ASuhandi and M G Nugraha

Physics Education Department,Universitas Pendidikan Indonesia, Jl. Dr.Setiabudhi 229 Bandung, Indonesia, 40154

*Corresponding author's email: muslim@upi.edu

Abstract. The purposes of this study are to determine the quality of reasoning test instruments that follow the framework of Trends in International Mathematics and Science Study (TIMSS) as a development results and to analyse the profile of reasoning skill of senior high school students on physics materials. This research used research and development method (R&D), furthermore the subject were 104 students at three senior high schools in Bandung selected by random sampling technique. Reasoning test instruments are constructed following the TIMSS framework in multiple choice forms in 30 questions that cover five subject matters i.e. parabolic motion and circular motion, Newton's law of gravity, work and energy, harmonic oscillation, as well as the momentum and impulse. The quality of reasoning tests were analysed using the Content Validity Ratio (CVR) and classic test analysis include the validity of item, level of difficulty, discriminating power, reliability and Ferguson's delta. As for the students' reasoning skills profiles were analysed by the average score of achievements on eight aspects of the reasoning TIMSS framework. The results showed that reasoning test have a good quality as instruments to measure reasoning skills of senior high school students on five matters physics which developed and able to explore the reasoning of students on all aspects of reasoning based on TIMSS framework.

1. Introduction

One of the goals of physics subjects in the curriculum of 2013 is that the students have the competency to develop the ability of reasoning skill within inductive and deductive analysis using the concepts and principles of physics to explain natural events and solve the problems, both qualitatively and quantitatively [1]. This purpose is in line with one of the cognitive domains in TIMSS, namely reasoning dimension.Reasoning can be defined as the ability to think logically to solve the problem using scientific methods [2,3,4], while scientific reasoning skills are the competencies needed to develop critical thinking skills students [5]. Therefore, reasoning skills need to be trained in physics teaching and assessed using an assessment instrument test with good test construction.

Assessment is one important component and it cannot be separated from the learning activities.Assessment was conducted to determine the extent of students' ability to understand and mastering the concept.In Permendikbud number 23 in 2016, the scope of assessment includes a competence attitudes, knowledge, and skills [1].Assessment includes all aspects of competencies using appropriate techniques, to monitor the development of the learners ability.

Results of an international test into the cognitive abilities of Indonesian students from *Trends in International Mathematics and Social Study* (TIMSS) conducted by the IEA *(International Association for the Evaluation of Educational Achievement)* are in low category.This result due to the ability of Indonesian students are currently have the knowing ability and not accustomed to solve the application and reasoning problems.This Indonesian student difficulties caused by students' reading strategy is still very limited so that the level of reasoning is still low, linear, and fragmented (not comprehensive).Learning process should be encourage students to find out from various sources of observation, be able to formulate a problem (queries) not only solve the problem, to train students think analytically (decision-making) and not thinking mechanistic (routine) and have capability to cooperate and collaborate in solving the problem.

Mullis *et al*(2014) stated that in TIMSS 2015, the cognitive dimension are divided into three domains which describe the thought process that can be used to complete the science developed item [6]. The first domain is knowing, discusses the student's ability to remember, recognize and explain the facts, concepts, and procedures as the foundation in science.The second domain is applying, focusing on the use of knowledge to produce explanations and solve practical problems.The third domain is reasoning, including the use of evidence and scientific understanding to analyse, synthesize, and generalize, often in unfamiliar situations and complex context.Aspects of knowing, application, and reasoning in the cognitive abilities as implemented in TIMSS can be used to show the students' thinking skills profile.Of these three aspects, knowing and application aspects are basic thinking skills, while reasoning skills is high-level thinking skill.TIMSS 2015 classifies reasoning ability *(reasoning)* into eight categories: analyse, integrate/synthesize, hypothesize/predict, design, draw conclusions, generalize, evaluate, and justify [7].

Based on analysis result of the Physics national examination (UN),found that most of the problems of physics are tested only measure low-level cognitive abilities that only measures cognitive ability to remember, understanding and applying.The tendency of the assessment construction is more mechanistic, primarily on the use of formulas of physics.While question that can measure high-level thinking skills are rare. Based on aspects of the reasoning, the questions of physics tested yet measure reasoning ability, especially when referring to the TIMSS reasoning skills framework. In line with Wasis (2014) research based on analysis of the Physics national examination in 2013 which showed that the highest proportion of matter is measure applying cognitive processes with the conceptual knowledge dimensions [8].

Considering there are linkages between assessment and learning process and the importance to measure the reasoning ability, as well as national examination in 2016 that based on reasoning which refers to the TIMSS framework, it is necessary to develop the instrument test to measure reasoning skills, so it can serve as a guide for physics teachers in teaching physics at schools to equip students reasoning skills.It is expected that the students will be able to develop the reasoning skill so that they can compete with other students both nationally and internationally. Mardapi (2008) states that the best test instrument can improve the quality of assessment that describe students' ability profiles [9]. The qualityof instrument tests can be determined by analysing the validity, reliability, distinguishing features and the level of test difficulties [10].

2. Research Methodology

This research used research and development (R&D) design. The research focused on the development of instruments test to measure reasoning skills in high school physics material.The development step using 4D stage, which includes: 1) define, 2) design, 3) develop, and 4) disseminate [11], but in this research the development step only from define to develop stage.Research subjects were 104 students at three high schools in Bandung.Sampling was conducted by random sampling technique.Reasoning instrument test is constructed following the TIMSS framework and it has 30 multiple choice questions (MCQs) that cover parabolic and circular motion, Newton's gravity law, work and energy, harmonic oscilation, momentum and impulse.The quality of reasoning skill tests were analysed using the Content Validity Ratio (CVR) based on assessment results by validator (expert judgment) to obtain the content validity and construct validity items, use the formula:

$$CVR = \frac{n_e - \frac{N}{2}}{\frac{N}{2}} \tag{1}$$

n_e is number of validator agreeing on item test,N is total number of validator. The Category of CVR according to Lawshe shown in Table 1 [12].

Table 1. Index of CVR

Index CVR	Criteria
0,00 – 0,33	Not suitable
0,34 – 0,67	Suitable
0,68 – 1,00	Very suitable

Furthermore, analysis conducted to the instrument test, including: the validity of the item, level of difficulty, discriminating power, and reliability of the test. Instrumen test that have been constructed and analysed, used to get a profile of students' reasoning skills, which is based on the achievement of the average score on all reasoning skills aspects.

3. Results and Discussion

Reasoning skills instrument test was developed based on the basic competencies of physics of the 2013 curriculum andreferring to the eight categories of reasoning skills by TIMSS 2015 frameworks, which are analyse (A), integrate/synthesize (I), hypothesize/predict (H), design (D), draw conclusions (DC), generalize (G), evaluate (E), and justify (J).Based on the physics matterial and reasoning skills, matrix for distributing items was created as presented in Table 2.

Table 2. Distribution of reasoning skills test items

No.	Physics subject	Reasoning skills category and number of item test								Total item tes
		A	I	H	D	DC	G	E	J	
1	Parabolic circular motion	1	2	-	3	4	5	-	-	5
2	Newton's gravity law	6	-	-	7	8	9	10	-	5
3	Work and energy	-	11	12	-	13	14	15	16	6
4	Simple harmonic oscillation	17	18	19	20	21	-	22	23	7
5	Momentum and impulse	24	25	26	-	27	28	29	30	7
Total item test		4	4	3	3	5	4	4	3	30

CVR test results based on five validators to 30 items of the instrument tests showed that almost all of the questions have a CVR value between of 1 and 0.60 which means that all items meet the suitable criteria.Although the results of the validation have met the criteria, there are some suggestions for improvement given by validator that are associated with changes in answer options and the category of reasoning skills because it does not appropriate, and improvement in writing the questions.

1.1. Characteristics of reasoning skills instrument test

Characteristics of the reasoning instrument tests obtained after limited testing and revision then performed extensive testing with a sample of 104 students in three high schools in Bandung.Based on analysis using the Anates V4 software, the value of validity of the item, level of difficulty and discriminating power presented in Figure 1.

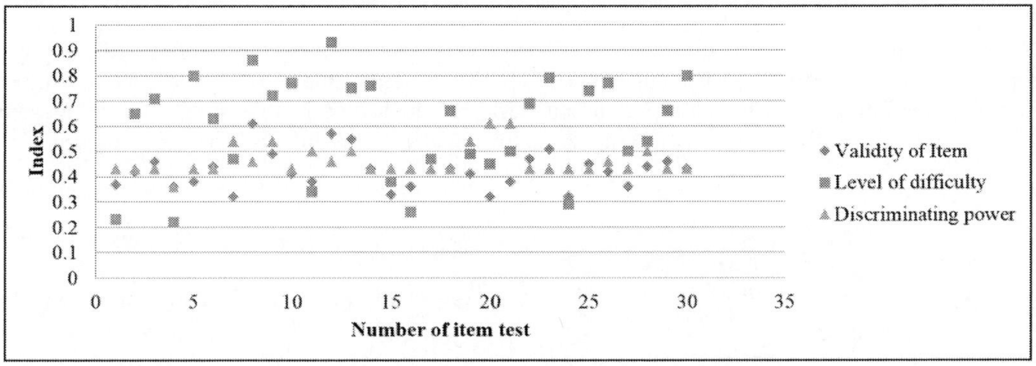

Figure 1. Analysis of the reasoning skills items instrument test quality

Base on the analysis of the 30 items shown that validity: 11 items low category, 18 items sufficient category, and one item high-category.Thus, for discriminating power: 1 item in enough category and 29 items good category, for level of difficulty: 9 items easy category, 19 items medium category, and 2 items difficult category. Furthermore, analysis result found thatthe value of reliability test is 0.82, which mean very high category that indicates the reasoning developed test is reliable for use in assessing the students [13].Then, it can be seen that the reasoning developed instrument test has good quality as a test instrument to measure reasoning skills.

1.2. Profile of student reasoning skills
Students' reasoning skills profile based on the results of implementation test by using prototype and depeloved reasoning test are shown in Figure 2.

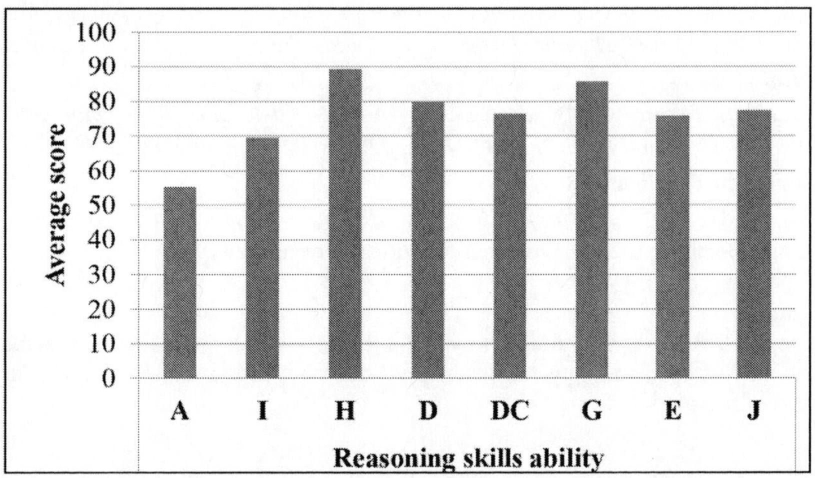

Figure 2. Distribution of students' reasoning skills profile

Based on Figure 2, it can be seen that the achievements of the students' reasoning skills is good enough with average score of 76.2 and none of the students' reasoning component that is less than 50. The ability of the students achieved the highest reasoning in the category of formulate hypotheses while analyse ability was lowest components.

4. Conclusion

Characteristics of reasoning skills instruments test reffering to TIMSS framework in physics at high school developed based on quality tests results has good quality to measure reasoning skills in physics teaching materials:parabolic and circular motion, Newton's gravity law, work and energy, harmonic oscilation, momentum and impulse. Reasoning skill instrument test is able to explore the reasoning skills of students on all aspects of reasoning based on TIMSS framework with enough category.

5. References

[1] Kemendikbud 2016 *Peraturan Menteri Pendidikan dan Kebudayaan Nomor 23 Tahun 2016 tentang Standar Penilaian Pendidikan*

[2] Ammer 2006 *Basic Proficiencis in Scientific Reasoning* Forsburg (United States : Forsburg State University)

[3] Zimmerman C 2007 *The Development of Scientific Reasoning Skill: What Psychologist Contribute to an Understanding of Elementary Science Learning.* (United States : Illinious State University)

[4] Koenig K2012 *Explicitly Targeting Presevice Teacher Scientific Reasoning Abilities and Understanding of Nature of Science Throught an Introductory Science Course* **21**(2) 1-9

[5] Fischer F, Kollara I, Uferb S, Sodiana B, et al. 2014 *Scientific Reasoning and Argumentation: Advancing an Interdisciplinary Research Agenda in EducationJournal of Frontline Learning Research* **5** 28-45

[6] Mullis2014 *TIMSS Advanced 2015 Assessment Frameworks*. USA : Boston College

[7] Mullis I V S & Martin, M O2013 *TIMSS 2015 Assessment Frameworks* (United States : TIMSS & PIRLS International Study Center)

[8] Wasis 2014 *Proceeding of International Conference On Research, Implementation, and Education of Mathematics and Science* **8** (1) 147-54

[9] Mardapi, D 2008 *Teknik Penyusunan Instrumen Tes dan Nontes*(Yogyakarta : Mitra Cendikia)

[10] Matlock, S & Hetzel 1997 *Annual Meeting of the Southwest Educational Research Association, Austin*

[11] Thiagarajan S, Semmel D S and Semmel M 1974 *Instructional Development for Training Teachers of Exeptional ChildrenSource Book* (Bloominton: Center for Innovation on teaching the Handicapped)

[12] Lawshe C H 1975 *A Conf. Held at Bowling Green State University* **28** 563-75

[13] Ding L and Beichner R 2009 *Physics Education Reseach* **5** 20103

Acknowledgments

The author would like to thank validators who have provided assessment for the improvement of reasoning instruments test and physics teacher in the city of Bandung who has assisted in the implementation of this study.

Analogy Mapping Development for Learning Programming

R A Sukamto, H W Prabawa and S Kurniawati

Universitas Pendidikan Indonesia, Jl. Setiabudhi No. 229, Bandung 40154, Indonesia
E-mail: rosa.ariani@upi.edu

Abstract. Programming skill is an important skill for computer science students, whereas nowadays, there many computer science students are lack of skills and information technology knowledges in Indonesia. This is contrary with the implementation of the ASEAN Economic Community (AEC) since the end of 2015 which is the qualified worker needed. This study provided an effort for nailing programming skills by mapping program code to visual analogies as learning media. The developed media was based on state machine and compiler principle and was implemented in C programming language. The state of every basic condition in programming were successful determined as analogy visualization.

1. Introduction

The media developed in this study maps many lines of computer program code to image as an analogy. The analogies used to provide assistance to students in representing the concept of programming that requires a process of abstraction into concrete form. The researcher already made observation for many years how teaching students using analogy as visualization of computer program code. The ideas using the analogies for teaching basic programming have been used for many years by first author. Therefore, the media was made for implementing the analogies ideas. There are many researchers using visual simulation for learning such as research which use simulator to make a program in programmable controller [1], reseach about using state machine for checking complex programmable logic devices (CLPD) design [2], and research about a ModelSim tool for designing electronic circuits in a machine using the state machine principle [3], animated interactive analogies for understanding programming variables [4]. Thus the visualization method is one of the solution to help novice learners understand the concept.

The media used state machine and compiler principles. The state machine is a sequence of analysis condition as processes or ideas [5]. The state machine principle was used for visualizing sequence of the program code process as sequence of image of analogy. The state was also considered using the compiler process concept. CodeMirror was used for tokenizing the program code as a token which CodeMirror is a library for javascript [6]. This media in purpose is one effort to implement basic programming analogy in order to make the students more easy understanding programming than before.

2. Why using analogies?

The analogies can be understood as a comparison of the similarity of the two concepts. The familiar concept is called the analog concept and the unfamiliar one the target concept. Both the analog and the target have features (also called attributes) which shown in Figure 1. If the analog and the target share similar features, an analogy can be drawn between them. A systematic comparison, verbally or visually, between the features of the analog and the target is called a mapping [7][8].

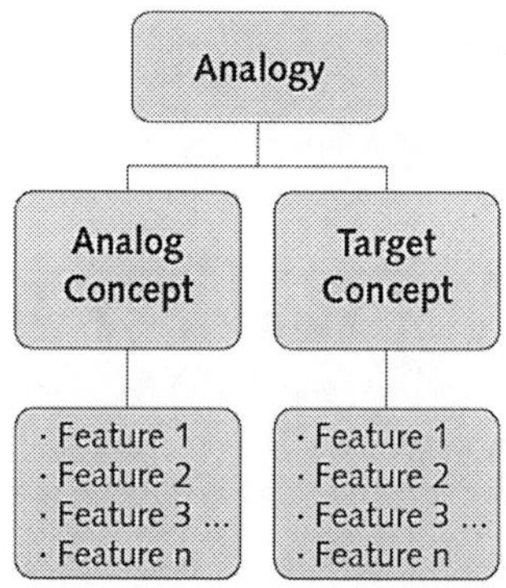

Figure 1. A conceptual representation of an analogy [8]

History proves that the analogy has contributed not a simple scientific discoveries. not by evidence, but rather as an inspiration. Analogy has also given its role in providing an explanation for a particular discoveries [8]. An analogy used in the classroom, textbooks and web-based learning should be designed to be activated in the elaboration, the cognitive processes that establish a relationship between what is already known and what will be known [8] including in programming [9][10][11]. Therefore analogy can be used as an effective tool in learning, teachers need to have enough knowledge regarding the pedagogical aspects. In its most basic form, at least the teachers have the ability to [12]:

- defines correspondence between analogy chosen by students as well as mapping targets: which concepts are obtained directly by teachers and which concepts developed by the students as a result of generalization analogy
- understand that analogy can not provide all of the aspects of learning that is required by students. It can be interpreted that the more the analogy is used to represent a concept, it would be better
- understand that not all students can understand the linkage between the concepts taught by analogy given.

Studies have shown that when used effectively, the analogy is a valuable pedagogical tool for teachers and media training for students in understanding the concepts.

3. Analogy mapping development

The student's learning style can categorized as follows:

- visual learners which understand better by seeing some image or imagination image in mind,
- auditory learners which understand better by hearing,
- kinesthetic learners which understand better by seeing real things that touched or moved [13].

According to this, there are two parts using images or things surrounding as visualizing something. A media can help students for recording easier in their mind the things by using images. Thus, the students can conceive any text easier by using image if the text is converted as an image as an analogy. This research was developed a media for mapping every line of program code as an analogy which based on the concept can help students to understand programming. The analogies used in this research can be seen in the Table 1. Analogy Mapping is written using Indonesia language.

IOP Conf. Series: Journal of Physics: Conf. Series **812** (2017) 012109 doi:10.1088/1742-6596/812/1/012109

Table 1. Analogy mapping screen shoots examples.

Program Code	Analogy Image	Description

`typedef struct{`
`int a;`
`int b;`
`int a1;`
`int b1;`
`int a2;`
`int b2;`
`int a3;`
`int b3;`
`}bungkusan;`
	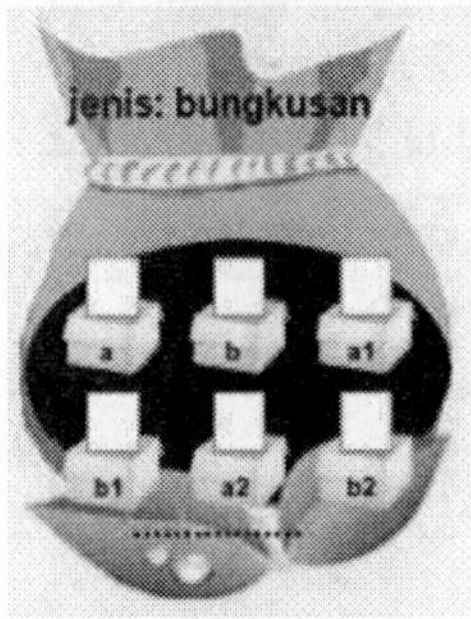
	
		a structured type declaration, a bud is a symbol for early declaring (*jenis* = type, *bungkusan* = type name)

`void proc(int a, int b){`
 	
`}`

		a procedure declaration (*pabrik* = factory, proc = factory name, *bahan baku* = raw material)

`int func(int a, int b){` `}`	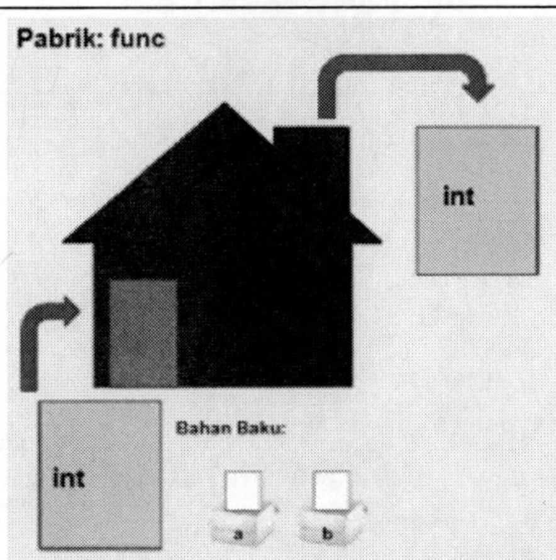 	a function declaration (*pabrik* = factory, func = factory name, *bahan baku* = raw material)
`int a, b = 7, c, i, tarr[9], yarr[1][2];`	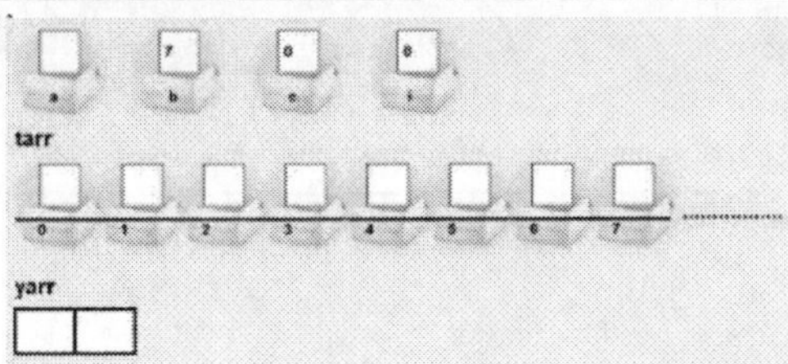 	variables and arrays declaration in one line

Every analogy was developed by using the state machine principle manually. State machine principle more popular with finite automata which can model many state conditions in software or hardware. There are many state machine model examples used in this research as shown in the Table 2. The result analogies of sequence instruction in program code are shown in the Figure 2.

Table 2. State machine examples.

<table>
<tr><th>Description</th><th>State Machine</th></tr>
<tr><td>A variable declaration
A variable usage</td><td>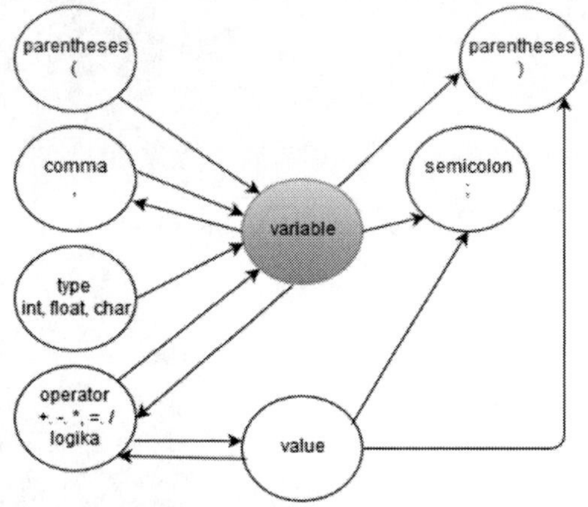
</td></tr>
<tr><td>If condition</td><td>
</td></tr>
<tr><td>Looping for</td><td>
</td></tr>
<tr><td>A one dimension array</td><td>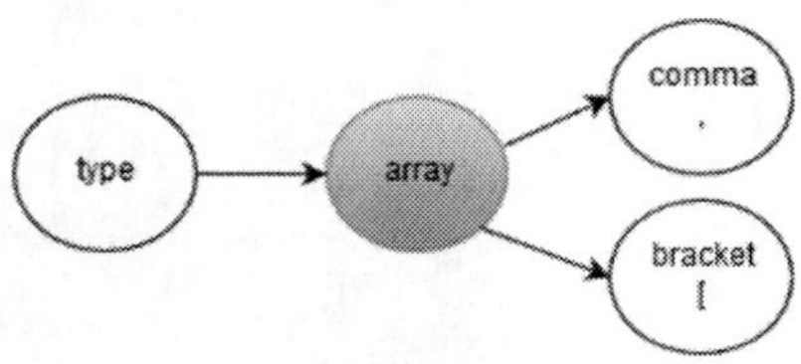
</td></tr>
<tr><td>A structured type declaration</td><td>
</td></tr>
<tr><td>A procedur declaration
A function declaration</td><td>
</td></tr>
</table>

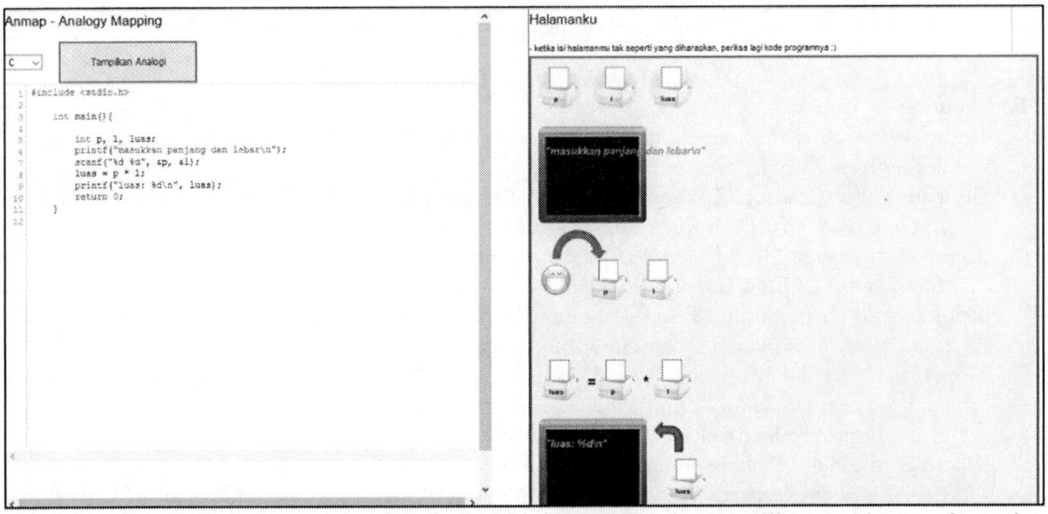

Figure 2. Analogies of sequence instruction in program code (tampilkan *analogi* = show the analogies, *halamanku* = my yard, *ketika isi halamanmu tak seperti yang diharapkan, periksa lagi kode programnya* = if your yard picture not as your expectation, please check your code).

Analogy Mapping was tested by 10 students that registered as first algorithm and programming subject students. They made many source codes to solve several problems and those source codes were used as input of Analogy Mapping. The result of the students tested was shown in Table 3.

Table 3. Analogy mapping testing result

Students	Source Codes	Lines	Correct Analogies	Wrong Analogies
Student 1	7	527	521	6
Student 2	9	263	257	6
Student 3	8	455	449	6
Student 4	8	498	495	3
Student 5	6	116	99	17
Student 6	5	115	113	2
Student 7	4	105	103	2
Student 8	3	47	44	3
Student 9	4	357	357	0
Student 10	3	79	77	2

4. Discussion

There are still appeared wrong analogies because there were several state undefined yet. The researcher will use the media to teach the students in order to get data as evaluation of the advantages of the media using the minimalist method in visual analogy learning model.

5. Conclusion

According to the result, all basic programming concept can be converted into an image as analogy using state machine and compiler principles, but the tested result showed that 47 analogies were wrong analogies (1.835938 %) and 2513 analogies were correct analogies (98.16406%). The media was supported by CodeMirror library as part of the process which converts program code into token and its label.

6. References

[1] Bathelt, Jens 2004 *Programming a Programmable Logic Controller* (Swiss: Federal Instituts of Technology Zurich)
[2] Simeonov, Simeonov I, Kukenska V S and Ilarionov R T 2005 *Realization of The Final State Machine with CPLD Device* (Bulgaria: Electronics)
[3] Altera Corporation 2011 Using *ModelSim to Simulate Logic Circuits for Altera FPGA Devices.* (San Jose: ALTERA Corporation-University Program)
[4] Doukakis D, Grigoriadou M and Tsaganou G 2007 *Understanding The Programming Variable Concept with Animated Interactive Analogies*
[5] Hopcroft, John E, Motwani R and Ullman J D 2001 *Introduction to Automata Theory, Languages, and Computation: Second Edition* (Boston: Addison-Wesley)
[6] _______, https://codemirror.net/(accessed 20th June 2016).
[7] Glynn S M 1994 Teaching science with analogies : a strategy for teachers and textbook author. 1994 *Reading Research Report No.15* (National Reading Research Project of University of Georgia and University of Mariland)
[8] Glynn, S M 2008 *Making Science Concept Meaningful to Students : Teaching with Analogies.*
[9] Forisek M and Steinova M 2012 *Proc. of the 43rd ACM technical symposium on Computer Science Education (Raleigh)* pp 15-20
[10] Gorgina, Moape T, Ojo S O and Lall M 2015 *Proc. of the EDSIG Conference on Information Systems and Computing Education (Wilmington)* pp 1-9
[11] Levy, Omer, Markovitch S 2012 *Proc. of the Twenty-Sixth AAAI Conference on Artificial Intelligence (Ontario)* pp 991-7
[12] Harrison, Allan G, Treagust and David F 2005 Metaphor and Analogy in Science Education. (Berlin: Springer Science and Business Media)
[13] Gardner Howard 2010 *Multiple Intelligences: New Horizons in Theory and Practice* (New York: Basic Books)

Optimisation of Deacetylation Process for Chitosan Production from Red Snapper (*Lutjanus* sp.) Scale Wastes

N D Takarina[1*], A B Indah[1], A A Nasrul[1], A Nurmarina[1], A Saefumillah[2], A A Fanani[3] and K D P Loka[1]

[1] Departement of Biology, Faculty of Mathematics and Natural Sciences, Universitas Indonesia, Depok, 16424, West Java, Indonesia
[2] Department of Chemistry, Faculty of Mathematics and Natural Sciences, Universitas Indonesia, Depok, 16424, West Java, Indonesia
[3] Marine Science Postgraduate Study Program, Department of Biology, Faculty of Mathematics and Natural Sciences, Universitas Indonesia, Depok, 16424, West Java, Indonesia
Corresponding author: noverita.dian@sci.ui.ac.id

Abstract. Red snapper (*Lutjanus* sp.) is common tropical fish that known as important source of marine product in particular Indonesia. This research aimed to optimise the chitosan synthesis from the red snapper scale waste through deacetylation process. Method in this research was divided into three stages which were deproteination, demineralization, and deacetylation. Deproteination stage was done with solution containing 4.2% w/v NaOH and heated at 60° C for 5 hours and followed by the demineralization stage with solution containing 52% v/v 2 N HCl at room temperature for 6 hours. The comparison between fish scales and solutions was 1: 6. After that, process continued with the deacetylation. Several treatment during the deacetylation process were taken into consideration to determine the effective concentration for yielding optimum chitosan output. Chitosan produced were having moisture content of 2.88%, ash content of 1.10%, and nitrogen content of 0.0136%. Optimal Degree of Deacetylation (DDA) was up to 90.83% that obtained by heating treatment at a temperature of 110° C with solution containing 80% NaOH for 4 hours, and comparison between chitin : solution was 1 : 3. This result indicated that chitosan extracted from red snapper scale is very potential and can be applied to industry.

1. Introduction

Crustacea shell waste is known as chitosan source that important for commercial use due to its high content and availability. Chitosan usually is used in medical and pharmaceutical industries. Chitin has close relation to protein, mineral, lipid and pigment [1]. Chitin is natural polysacharide with chemical structure (poli [β- (1 → 4) -2- acetamido-2-deoxy-glucose]), mainly found in shell of crustacea namely crab, shrimp, insect cuticle and fungi cell wall. Chitin is also biopolymer with chemical structure resemble to cellulose and it is the primary component of exoskeleton. Beside from crabs and shrimp shell, chitin has been successfully isolated from carp (*Cyprinus carpio* L.) scale [2]. Furthermore, chitin and chitosan has been extracted from tilapia (*Tilapia nilotica*) scale [3,4]. It was also confirmed that chitin and chitosan can be extracted from scale of *Labeo rohita*, likewise extraction of chitosan and carboxymetil chitosan from scale of same species [5,6].

From above studies, chitosan and chiton are most likely extracted from freshwater fishes. Due to limited information about extraction similar compound on saltwater fishes in particular red snapper, investigation on method for extracting and characterizing chitosan is urgent. Therefore, this paper aims to investigate the optimum deacetylation process for producing chitosan from red snapper (*Lutjanus* sp.).

2. Experimental Method

2.1. Preparation of Fish Scale

Scales were washed using running water to remove wastes and contaminants from scales. Contaminants will affect the chitin and chitosan quality. After been cleaned and washed, scales were dried out under natural sunlight. Using of furnace in drying out will cause discoloration on fish scale sand cause scales turn yelow and will affect chitin and chitosan contents. Dry scales then kept in the watertight container to keep the moisture.

2.2. Deproteinase and Demineralisation

During the deproteination stage, scales were dissolved in the NaOH 4,2% (w/v) with ratio of scale to NaOH was 1:6 . The solution was stirred for 5 hours at 60 °C and sieved for obtaining residu. This residu then washed with aquadest to obtain neutral pH . During the demineralisation, the washed solution then dissolved in solution containing 52% v/v 2 N HCl at room temperature for 6 hours with ratio of 1:6. The solution then washed with aquadest to obtain neutral pH. The result from deproteination and demineralization of fish scales was chitin.

2.3. Deacetylation

This stage aimed to produce chitosan. The deacetylation process were conducted by using treatments combine the concentration of NaOH (60, 70, and 80%), temperature (110, 120, an 130 °C) and duration (4 and 6 hours) difference (Table 1). The purpose of combining those treatment was to find which treatment can produce chitosan with high DDA value (> 75%). In these treatments, the chitin was added by NaOH (ranged from 60 to 80 %) and then boiled at temperature 110-130 °C from for 4 to 6 hours.

Table 1. Experimental design on deactylation stage

List of treatment					
1) 110 °C; 60%; 4 hours	2) 110 °C; 60%; 6 hours	3) 110 °C; 70%; 4 hours	4) 110 °C; 70%; 6 hours	5) 110 °C; 80%; 4 hours	6) 110 °C; 80%; 6 hours
7) 120 °C; 60%; 4 hours	8) 120 °C; 60%; 6 hours	9) 120 °C; 70%; 4 hours	10) 120 °C; 70%; 6 hours	11) 120 °C; 80%; 4 hours	12) 120 °C; 60%; 6 hours
13) 130 °C; 60%; 4 hours	14) 130 °C; 60%; 6 hours	15) 130 °C; 70%; 4 hours	16) 130 °C; 60%; 6 hours	17) 130 °C; 80%; 4 hours	18) 130 °C; 80%; 6 hours

2.4. Measurement of ash content, nitrogent content, and DDA

Quality of chitosan can be seen from several parameters. Some parameters like ash content, nitrogen content and DDA are important characters in determining chitosan wheter it is good or not [7]. Ash and nitrogent content were analysed using proximate analysis while DDA was using FTIR. For ash content, chitosan were placed on the petri dish and heated inside the furnace at 100-105 °C for 30 minutes. Ash content from chitosan can be calculated using formula below

$$\%\ \text{ash} = \frac{W_3 - W_1}{W_2 - W_1} \times 100\ \% \qquad (1)$$

where, W_3 is weight of chitosan ash and petri dish, W_2 is weight of chitosan and petri dish, and W_1 is weight of petri dish only.
Nitrogen content was can be calculated based on formula

$$\%\ \text{nitrogen} = \frac{(Va - Vb) \times N \times 14.007}{W \times 1000} \times 100\ \% \qquad (2)$$

where, Va is sample titration volume of HCl , Vb is void sample volume of HCl, N is normality of HCl, and W is chitosan weight.

Degree of deacetylation (DDA) influences the physical, chemical and biological properties of chitosan, such as acid base and electrostatic characteristics, biodegradability, self aggregation, sorption properties, and the ability to chelate metal ions [8]. DDA was measured by using the equation below [9].

$$\% \text{ DDA}=100 - \left(\frac{A_{1655}}{A_{3450}} \times \frac{100}{1.33}\right) \quad (3)$$

where, A_{1655} is chitosan absorbance at 1655/cm wavelength and A_{3450} is chitosan absorbance at 3450/cm wavelength.

3. Result and Discussion

3.1. Ash content on chitosan extracted from red snapper scale waste

Ash content is one of important parameters that determine the quality of chitosan. The lower ash content then the higher chitosan quality. The ash content determined by the deactylation process and the chitosan raw material. Based on the measurement, the ash content of chitosan synthesized from red snapper scale has percentage of ash ranged from 1,1 % to 1,88 % (Figure 1).

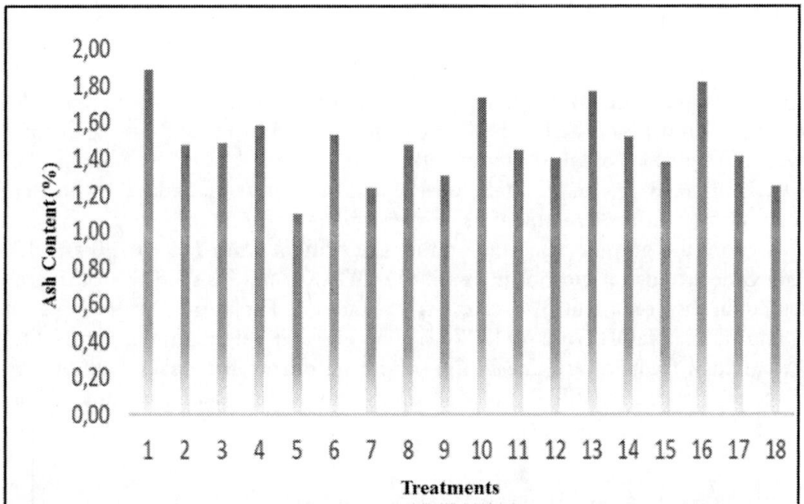

Figure 1. Ash content on chitosan extracted from red snapper scale waste

The lowest ash content which is 1,1 % found in the deacetylation process that added by NaOH 80 % boiled at temperature 110 °C for 4 hours (treatment 5). Likewise, the highest ash content which is 1.88 % found in the deacetylation process that added by NaOH 60 % boiled at temperature 110 °C for 4 hours (treatment 1). From the graph it can be drawn from treatment 10, 13 and 16 that higher temperature combined with the extensive boiling time will increase the ash content and lower the chitosan quality. Ash content of chitosan from red snapper has fulfilled the ash content maximum quality standard which is equal to 5 % because red snapper scales has ash content below 5% [7]. This result also contrast to some researches which showed higher content. Content of ash on chitosan extrcacted from Nigerian shrimp was 6.41% while in mussel shell was 36.87% [11,12]

3.2. Nitrogent content on chitosan extracted from red snapper scale waste

In general, nitrogen content of chitosan from red snapper were very low (Figure 2). The lowest nitrogen content which is 0,0136 % found in the deacetylation process that added by NaOH 80 % boiled at temperature 110 °C for 4 hours. Likewise, the highest ash content which is 0,0278 % found in the deacetylation process that added by NaOH 80 % boiled at temperature 110 °C for 6 hours.

Nitrogen content of chitosan from red snapper below the minimum quality standard which is equal to 5 % [7]. Nitrogen content on this research were also contrast to others. In chitosan extracted from Nigerian shrimp content of nitrogen was 2.71% while in mussel shell was 2.29% [11,12]

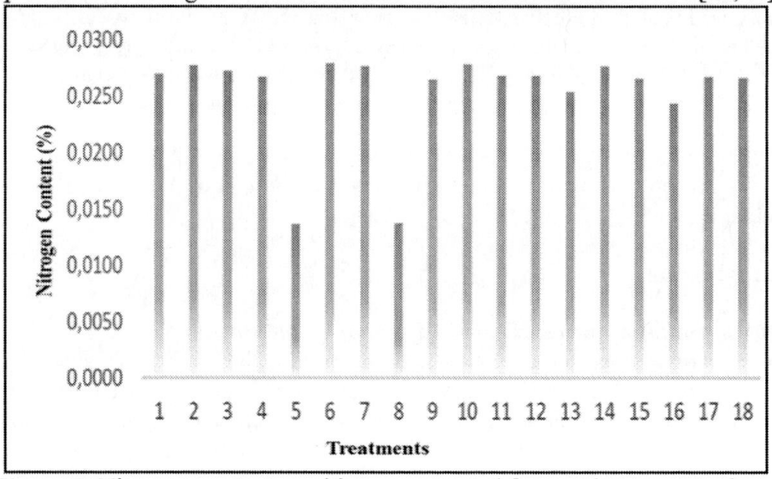

Figure 2. Nitrogen content on chitosan extracted from red snapper scale waste

3.3. DDA of chitosan extracted from red snapper scale waste
Based on the measurement, the DDA of chitosan synthesized from red snapper has DDA ranged from 79,32 % to 90,83 % (Figure 3). The lowest DDA which is 79,32 % found in the deacetylation process that added by NaOH 60 % boiled at temperature 110 °C for 4 hours (treatment 1). Likewise, the highest DDA which is 90,83 % found in the deacetylation process that added by NaOH 80 % boiled at temperature 110 °C for 4 hours (treatment 5). However, the boiling temperature and duration were not affect the DDA. From the graphic, the temperature increement from 110 to 120 and 130 °C combined with the boiling time extension did not increase the DDA. It can be seen from the graphic, there is 6 peaks represented in the treatment 5, 6, 10, 11, 15 and 17. Those treatments have similarity which influenced by adding the NaOH from 60 to 80%. Therefore, it can be drawn that NaOH concentration is more important than temperature and boiling duration during the deacetylation process. DDA of chitosan from red snapper has fulfilled the DDA minimum quality standard which is equal to 75 % [7].

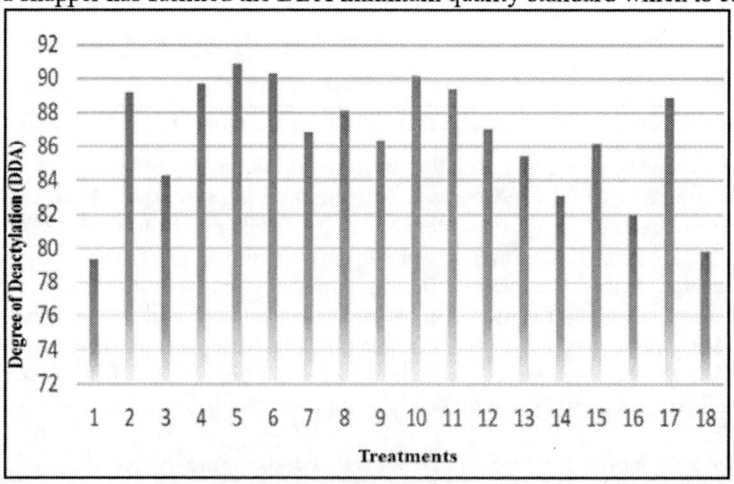

Figure 3. DDA on chitosan extracted from red snapper scale waste

Content of ash and protein among all treatments performed decreased with increasing of DDA (see treatment 5). Reaction with NaOH can decrease protein level, resulting purer chitosan [10].

4. Conclusion

Chitosan synthesized from red snapper scale has high DDA and low ash contents. Deacetylation process by adding chitosan with NaOH 80 % and boiled at temperature 110 °C for 4 hours will increase the DDA and reduce the ash content.

5. References

[1] Park B K and Kim M M 2010 *Int. J. Mol. Sci.* **11** 5152
[2] Zaku S G, Emmanuel S A, Aguzue O C and Thomas S A 2011 *Afr. J. Food Sci.* **5** 478
[3] Uawonggul N, Kongsri S and Chantai S 2011 *IJPAC* **6** 139
[4] Weeraphan W Chitosan extraction from fish scale of tilapia (*Tilapia nilotica*) for edible film production 2011 Thesis Chang Mai University
[5] Kumari S and Rath P K 2014 *Procedia Mater. Sci.* **6** 482
[6] Muslim T, Rahman M H, Begum H A and Rahman Md A 2013 *Dhaka Univ. J. Sci.* **61** 145
[7] National Standardization Agency of Indonesia 2013 Letter Number 298/KEP/BSN/12/2013 SNI 7949
[8] Li Q, Dunn E T, Grandmaison E W and Goosen M F A 1992 *J. Bioac. Compat. Polym.* **7** 370
[9] Domszy J G and Robert G A F 1984 *J. Macromol. Sci., Pure Appl. Chem.* **186**, 1671
[10] Yuan Y, Chesnutt B M, Haggard W O and Bumgardner J O 2011 *Materials* **4** 1399
[11] Isa M T, Ameh A O, Tijjani M and Adama K K 2012 *Int. J. Biol. Chem. Sci.* **6** 446
[12] Abdulkarim A, Isa M T, Abdulsalam S, Muhammad A J and Ameh A O 2013 *Civil Environ. Res.* **3** 2222

Acknowledgments

We acknowledge Ministry of Research, Technology and Higher Education also Directorate of Research and Community Engagement Universitas Indonesia for its PITTA Grant Fiscal Year 2016 with contract number: 2047/UN2.R12/HKP.05.00/2016.

IOP Conf. Series: Journal of Physics: Conf. Series **812** (2017) 012111 doi:10.1088/1742-6596/812/1/012111

Modelling Scientific Argumentation in the Classroom : Teachers perception and practice

R M Probosari[1*] , Sajidan[1], Suranto[1], B A Prayitno[1], and F Widyastuti[1]

[1]Science Education Doctoral Program, Faculty of Teacher Training and Education, Universitas Sebelas Maret, Jl. Ir. Sutami No. 36A Surakarta, Indonesia
[*]Corresponding author : riezkymaya@fkip.uns.ac.id

Abstract. The purposes of this study were to investigate teacher's perception about scientific argumentation and how they practice it in their classroom. Thirty biology teachers in high school participated in this study and illustrated their perception of scientific argumentation through a questionnaire. This survey research was developed to measure teachers' understanding of scientific argumentation, what they know about scientific argumentation, the differentiation between argument and reasoning, how they plan teaching strategies in order to make students' scientific argumentation better and the obstacles in teaching scientific argumentation. The result conclude that generally, teachers modified various representation to accommodate student's active participation, but most of them assume that argument and reasoning are similar. Less motivation, tools and limited science's knowledge were considered as obstacles in teaching argumentation. The findings can be helpful to improving students' abilities of doing scientific argumentation as a part of inquiry.

1. Introduction

Lately a lot of educational research that examines the importance of argumentation in learning, both on the subjects of social and exact sciences. Facts show that many students who have the skills to argue because he was not accustomed to doing so. In learning science, scientific arguments are part of a form of scientific communication which is an integral part of the nature of science. Low ability scientific argument indicates the limited ability of students in scientific thinking and understanding of science as a whole, both the content and context, but learn not only assumed as absorbing or transfer of knowledge, but rather build or reorganize knowledge [1][2]. Several studies have shown that the weakness of the arguments do not only occur in primary or secondary education, but also on higher education. Most students only partially understand the concept of argumentation alone and this is mainly due to a lack of knowledge about the importance of the argument to define their own position in the academic discussion [3]. The purposes of this study were to investigate teacher's perception about scientific argumentation and how they practice it in their classroom.

1.1. Deliberate Discourse in Scientific Argumentation

Classroom discourse plays an important role for the achievement of learning goals in science, especially those relating to scientific communication. This mainly refers to the use of language used by teachers and students in classroom interaction. In relation to science learning, classroom discourse is focused on the use of scientific language and how the situation after a practicum class discussion or lesson. Scientific discussion involving the participation of students in the form of a question, answer, assertion and argument to clarify the resolution of a case or problem. Argumentation in science learning does not end merely positioned as brainstorming, but led to an innovation, either in the form of a change of attitude, perspective or knowledge. In this case the students can work individually or in collaboration with others. Some things to note in this regard is considerable discussion time, place and situation conducive to stimulate students to exchange ideas and thoughts without making others feel inferior or cornered [4]. Deliberative discourse in the scientific arguments made in the form of participatory communication and collaborative cooperation focused on solving the problem on a particular topic. In deliberative discourse, the debate in the discussion would make the students know how a problem is seen from different views with reasons and arguments respectively.

1.2. Argumentation and Scientific Explanation
Many people who claim that argumentation and scientific explanation are basically similar, assuming both are equally in need of a statement with the reasons. In fact, the arguments have different standards of adequacy with the scientific explanation. Scientific explanation is considered good if it can be clearly understood, convincing, and relevant. The argument requires different standards is strong evidence and anything that can make the explanation can be explained both by the context and the content so that students can see the explanation more than just facts and challenges them to build an explanation for addressing problems [5].However, explanation and argumentation together complete the practice of science through the construction process knowledge in the scientific community. Both of these components work in synergy and complement. Scientific explanation was obtained through the efforts of individuals, whereas others involve arguments that need to be assigned their position, whether it supports or opposes the idea it conveys. The success of the arguments requires their statements, evidence and convincing explanation [6]. The argument is needed to validate the scientific explanation because it is often found several different explanations for the same phenomenon [7].

1.3. Justifying Scientific Argumentation in the Classroom
In practice, the scientific justification of argumentation in science classes include two things, the truth is the truth content and structure. Structurally, the argument can be measured with Toulmin's Argument Pattern (TAP), which contains four aspects of the argument, namely the claim, data, warrants and backing. Data, claims, and warrant a fundamental element of the argument. Claim is a statement obtained through argument. Data are facts that support the claim. Warrant used to explain the claim through the evidence obtained. Backing consolidate warrant and describes the relationship between data, warrants and claims [8]. The structure of the argument can be different, depending on the context. In many cases, often encountered the confusion between the arguments that are structurally appropriate but the content is wrong, or vice versa [9].

2. Methods
The study involved 15 participants' biology teacher in the scope of Central Java, which were selected through purposive random sampling taking into account interest and their interest towards learning arguments in class. All participants are biology teachers who have taught for at least five years in high school between the ages of 30-45 years. Methods of data collection is done by filling questionnaire survey regarding the views of the argument, what distinguishes the argument in science classes with others, why the scientific arguments essential for science learning and the extent to which they accommodate students to argue. Furthermore, researchers and the teachers do a focus group to get recommendations on the appropriate learning models to stimulate and enhance the ability of the student's argument. As a supporter, made the observation at a grade one teacher who gets the best ratings to see how the implementation of learning that are thought to increase the motivation and skills of the students argued. The study analysed qualitatively to obtain an overview of the arguments of teachers in science classroom.

3. Findings and discussion

3.1. Teachers Perception of Argumentation in Learning Science
Observations and structured interviews with teachers show many differences of opinion regarding the scientific argumentation teacher. Most teachers still consider that the same scientific argumentation with reasoning and explanation.

Table 1. Teachers perception of scientific argumentation

Name	Institution	Perception of scientific argumentation	Teaching argumentation in the class
A	State high school	The argument is the student's ability to give a reason	Yes
B	Private high school	The argument must be accompanied by tangible evidence	Yes
C	State high school	The argument must be through discussion	Yes
D	State high school	The argument is identical to the factual truth	Sometimes
E	State high school	Arguments part of the scientific method	Yes
F	Private high school	The same arguments with reasoning	Yes
G	State high school	Scientific argument is difficult because not all students are actively participating	Yes
H	State high school	Arguments = reason with proof	Yes
I	State high school	The argument is the ability to convince others	No
J	State high school	Arguments need data and facts	Sometimes
K	Private high school	Scientific argument is part of the discussion to find the truth	Yes
L	State high school	Arguments = grounds with a solid foundation	Yes
M	Private high school	Arguments = reasoning	No
O	Private high school	Arguments = an argument	Yes

The table above shows that most teachers feel that they have done the learning skills of argumentation accommodate students, although many still do not understand the meaning, structure and how scientific argumentation is formed. The results of the interview further mentioned that most of the activities carried out in the form of argumentation class discussions and presentations.

3.2. Obstacles with scientific argumentation

Some difficulties in teaching argumentation in science classes are presented as follows:

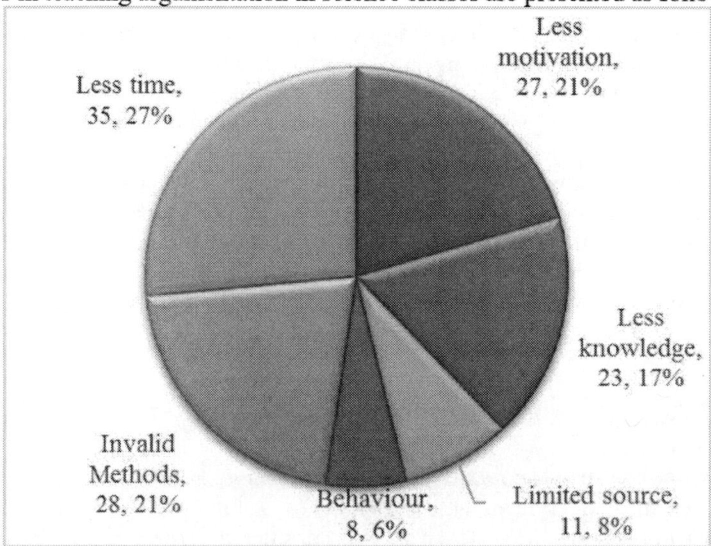

Figure 1. Teachers obstacles with teaching argumentation

Figure 1 shows that the difficulty in teaching the scientific arguments in school mainly due to circumstances that are not conducive. The results of the investigation with the teacher mentioned that some of the obstacles are : the limited time, less student motivation, lack of knowledge, limited learning resources, the environment is not supportive, and less accuracy of learning methods. Some teachers still consider that the application of argumentation in science learning takes more time, because it should be done through open discussion, but on the other side, the time schedule of their lesson plan is running.

3.3. Teachers activities daily notes in teaching argumentation
Some examples of daily records of teachers in science learning is presented below :

3.3.1. Hydrilla sp and fishes. are the components of abiotic, while water is a abiotic components in pond ecosystem.Student 1 = True, because Hydrilla sp. and animals are living things, and be able to breathe, so it can meet the biotic component categories. While the water is not a living being so that said abiotic components (A good argument because there are some proper and appropriate reason) Student 2, Abiotic: physical and chemical components in the ecosystem. Biotic: all living things (The argument is not appropriate for the wrong reasons)

3.3.2. Each trophi. levels in the food chain could eat or be eaten for one organism or more in the other trophic levels. Students 3 = Correct, because the food web consists of some of the food chain and an organism that occupy trophic levels in the food chain not only eat one type of food, nor eaten by one type of organism, such as cattle and buffalo grass eaten. (A good argument because there are some proper and appropriate reason) Students 4 = Right, because the food web is not only one trophic. (The argument is not appropriate for the wrong reasons)

The samples above indicates that the teacher gives the student activity assessment argument based on their reasons and their relevance to the question asked, not referring to the standard argument intact. Teachers reported that the debriefing and discussion activities students are also still dominated by many students who say "do not want" and "agree with ......" but still they can not able to provide the strong and logical reasons. The standard argument is the simplest such claims, evidence and reasoning [10] has not been seen here.

3.4. Limitation of modelling argumentation in the classroom
The argument is crucial to develop critical thinking and give more deep understanding of ideas and opinions, referring to the proposition with supporting evidence and reasoning [11]. Students perform the process of thinking and social interaction to build and evaluate the arguments of others so that they understand how they think a true scientist [12].It is a major concern in the learning of argumentation in science classes is how to condition conducive learning environment for students so that students really enjoy the excitement of finding new ideas or even disappointment if there are inaccuracies in giving argument. Internal factors such as motivation and the feeling is very influential here so in practice, teachers have to be smart to choose a model and learning strategies that can stimulate students to win the argument without feeling burdened. In practice, the teacher considers the argument hardly taught because of the assumption that the ability to argue, pull out the ideas, the natural talent of the students. In addition there is the assumption that as long as the students can understand the concepts being taught and to answer exam questions are given, meaning the learning objectives have been achieved, and scientific arguments are less important to students because it is not all about the type of open-ended question or require detailed reasons. It should be understood, how teachers teach students to understand that the simplest argument is more than just giving a reason, but must meet certain components, such as from the claim, evidence, and reasoning [13]. Claim is a statement to answer a question or problem. Truth claims are determined by evidence and reasoning used. Claim is considered correct when it was able to answer questions, along with supporting evidence and reasoning that provide justification or excuse. Evidence is evidence or scientific data supporting the claim, such as measurement data, results of research, observation, and others. Reasoning contain scientific principles that explain the relationship evidence to claim higher levels of the argument made,

the evidence presented in more detail. Reasoning is the justification or reason for linking evidence used by the claim [10].

Learning strategies to improve students' skills of argumentation should be directed to the investigation or inquiry as well as how they communicate during the investigations carried out [14].The argument is very useful to support the scientific communication so that students can see an invention with different viewpoints and support their scientific attitude.

4. Conclusion

This study investigated teacher's perception about scientific argumentation and how they practice it in their classroom. Based on the results, some conclusions are made as follows : generally, teachers modified various representation to accommodate student's active participation, but most of them assume that argument and reasoning are similar. Less motivation, tools and limited science's knowledge were considered as obstacles in teaching argumentation. The findings can be helpful to improving students' abilities of doing scientific argumentation as a part of inquiry.

5. References

[1] Park Y S 2008 *J. Korean Association for Research in Science Education* **28** 211
[2] Park J, Park Y, Park J and Jin S 2014 *J. Baltic Science Education* **13** 259
[3] Wingate U 2012 *J. English for Academic Purposes* **12** 45
[4] Asterhan C S and Babichenko M 2015 *J. Educational Psychology* **107** 740
[5] Brigandt I 2016 *Science and Education* **25** 251
[6] Berland L K and McNeill K 2012 *Science Education* **96** 808
[7] Osborne J and Patterson A 2011 *Science Education* 95 627
[8] Toulmin S E 2003 *The Layout of Argument* (New York: Cambridge University Press)
[9] Acar O and Patton R 2012 *Proc. Social and Behavioral Sciences* vol 46 (Barcelona: Spain/Elsevier Ltd) p 4756
[10] McNeill K, Singer R, González-Howard M and Loper S 2016 *Int. J. Science Education* **38** 1
[11] Deane P and Song Y 2014 *Psicología Educativa* **20** 99
[12] Jaber L and Hammer D 2016 *Science Education* **100** 189
[13] McNeill K 2011 *J. Research in Science Teaching* **48** 793
[14] Duschl R and Osborne O 2002 *Studies in Science Education* **38** 39

IOP Conf. Series: Journal of Physics: Conf. Series **812** (2017) 012112 doi:10.1088/1742-6596/812/1/012112

Analizing Student Biology Education Misconception And Scientific Argumentation Ability Using Diagnostic Question Clusters (Dqcs) Of Molecular Genetic Concept

L Nurlaila*, S Sriyati and Riandi

Departement of Biology Education, Universitas Pendidikan Indonesia, Jl. Dr. Setiabudi no 229, Bandung 40154, Indonesia

*Corresponding author: nurlailalela29@gmail.com

Abstract. The purpose of this research is to describe the profile of misconceptions and scientific argumentation ability using Diagnostic Question Cluster (DQCs) of molecular genetics concept. This research use descriptive research method and biology education students as a research subject. The Instrument that used in this research are DQCs, sheets interviews, observations, and field notes. The DQCs tested by writing and oral that used to analyze misconceptions and scientific argumentation ability. Sheets interviews, observations and field notes, are used to analyze the possible factors causing misconceptions and scientific argumentation ability. The results showed that misconception of molecular genetics are: DNA (23.75%), genes (18.75%) of chromosomes (15%) and protein synthesis (5.5%). The pattern of the highest misconceptions owned Misconception-Understand Partial. The average scientific argumentation ability is 55% and still categorized warrant (W). The pattern of the scientific argumentation abilities formed is level 2 to level 2 that consists of the arguments in the form of a claim with a counter claim that accompanied by data, collateral (warrant) or support (backing) but does not contain a disclaimer (rebutal).

1. Introduction

Biology is one of science subject matter that includes concepts that are concrete and abstract concepts. Biological materials abstract be one of the sources of difficulty in understanding the concepts students are causing misconceptions. One of biological materials that are difficult to understand by most students is the genetic material concept [1]. The misconception is not just happening in the students of primary and secondary school level, but the level of high school students until the students were told there students who experienced misconception especially for concepts that require deep understanding. Just as in biological materials, which have the abstract concepts that can't be explained in real terms.

Based on the results of an initial interview with a lecturer of the course [2] note that the genetic material that is highly vulnerable to any misconceptions or fail to understand the concept is in subconcepts cell reproduction, genetic control (protein synthesis and gene expression), as well as the basic concepts of relationship genes, DNA and chromosomes. In addition, according [3] adds that the genetic material included in the material a tedious and tiring. It can happen one of them is that students tend to memorize concepts compared to apply concepts, besides genetics are biological materials that constitute the conceptual knowledge that contains some complex concepts [4]. The results of the research to the understanding of genetics concepts made by [5] suggests that approximately 15% of the students are unable to provide an explanation of the chromosomes, and 70% of students give incorrect explanation of the process of mitosis and meiosis. In addition, based on research results [6] at the upper secondary level school students in the city of Palu in mind that students have misconceptions on the concept of chromosomes (76.1%), genes (75.0%) and DNA (76.5%) and protein synthesis (63.1%). Understanding the concept of the genetic material is essential as a basis for understanding other concepts in other biological concepts. It can be said that understanding the concept of genetics is a prerequisite for understanding other concepts such as biotechnology [7].

[8] suggest that the science learning should promote the teaching for understading, but in reality their science lessons in college now synonymous with the provision of comprehensive science course content. Basically, the provision of comprehensive content in college is important to convey to the students, but to ensure that students understand the whole content should also be analyzed using the capabilities of higher order learning schemes include the ability of reasoning or in other words is the ability of the scientific argument. Over the last two decades the ability of argumentation becomes very interesting in science education in university education [9]

Based on the research results [10] argued that the ability of high school students in Bandung on learning biology is still not well developed. It can be seen from the percentage of the level of ability that decreases from the lowest to highest. Decrease in the ability of these arguments are influenced by the methods and strategies of teaching teachers. Through these studies it is known that high school students in the city of Bandung is still not able to develop the ability of the argument. So it is very interesting, especially what arguments the ability of students in college, whether the result will be the same as the argument ability of high school students or it will be better.

Based on the interview notes that the lecturer has struggled to develop questions or questions that can stimulate students to think of using the theory / concept of relevant science that can explain the data obtained as the basis for a claim scientific argument, so if the ability of this argument is not developed, the students' answers to the question is feared will have misconceptions. So that measure students' conceptual understanding has become increasingly important to the biology faculty members were involved in evaluating and improving departmental programs [11]. It needed the assessment tool to identify and detect the extent of scientific argumentation ability of students to questions about the concept of genetically so that it can detect the tendency of their misconceptions. Assessment tools that can be used as an alternative in identifying the ability of scientific arguments (scientific reasoning) and the tendency of their misconceptions include Diagnostic Question Clusters (DQCs).

Diagnostic Question Clusters (DQCs) is one concept conceptual inventory or assessment in biology [12]. According [13] that the diagnostic test is useful to know the learning difficulties faced by the students, including the misunderstanding of the concept. The results of the diagnostic tests provide information about the concepts that have not been understood and that was understood. Through the use of group diagnosis in a test question can provide the opportunity for students to be able to think like a scientist biology "thinking like a biologists'. The important aspect to think like scientists in biology is always apply the basic principles of scientific argument (reasoning). Besides this, there are several studies also use diagnostic tests to determine students' understanding of the molecular and cell biology [13]. Some diagnostic tests may be used to determine the understanding of concepts and misconceptions which are diagnostic question clusters (DQCs) and interviews.

According [14] that this type of test is used to reduce the student can guess the answer, because they are required to give arguments (reasoning) so it can know the extent of their understanding of a concept. In addition, based on research results Hartley et al (2011) also explained that the function of the Diagnostic Question Clusters (DQCs) is to measure comprehension through the ability of argumentation based on scientific principles, as an assessment tool that can inform learning difficulties students and their misconceptions learners to a concept.

Based on the background of the above problems, the authors are interested to do research with the title Analysis of misconception and Capabilities argument Student Science Biology Education by Using Diagnostic Question Clusters (DQCs) on the concept of Molecular Genetics with a problem of this research is "What is the profile of misconceptions and the ability of scientific argumentation biology education students using Diagnostic instrument Cluster Question (DQCs) on the concept of molecular genetics? ". The research objective to be achieved in general is to analyze misconceptions and scientific argumentation ability of biology education students using the instrument cluster diagnostic question (DQCs) on the concept of genetics. In particular the research objectives to be achieved are as follows: 1. Analyze misconceptions biology education students on the concept of genetics 2. Analyzing the ability of scientific arguments biology education students on the concept of genetics 3. Uncover the tendency pattern of misconceptions and scientific argumentation ability of biology education students on the concept of genetics

The results of this study are expected to provide benefits as preliminary information for the lecturers of the misconceptions and the ability of scientific argumentation experienced by students of biology education who take courses in genetics, and as study materials to anticipate the possibility of misconceptions in students when the learning process and informing them of alternative instruments upon ability in analyzing the scientific arguments and misconceptions. Results are expected to be the basis for further research, especially in developing diagnostic question cluster (DQC) as an instrument in analyzing misconceptions and scientific argumentation ability of biology education students, as well as preliminary information on further action to anticipate any misconceptions in students.

2. Method

The method used descriptive research. This research method used to analyze the data in ways that describe or depict the data that has been collected by the instrument DQCs as it is. The population of this study were all students of biology in one of public university in Bandung who take the classes genetics academic year 2015/2016. The sampling technique used in this research is purposive sampling. This technique is a technique of sampling data source with a certain considerations. The sample of this research is a biology education students who follow courses of Genetics in the academic year 2014/2015 of 40 people. The independent variables in this study were analyzed misconceptions aspect of the students' answers on DQCs given the lead to the genetic material and the use of assessment rubrics [16]

Table 1. Misconception rubric from [16]

Critera	Description
Understand (U)	The answer to each concept statement shows mastery of concepts
Partial Understand (PU)	Answer demonstrate mastery of concepts that marked their partial (at least one) feature or element answers represent mastery of concepts
Partial Understand with Misconception (PUM)	Answers indicate that understood the concept but contains statements that misconception
Misconception (M)	Answer that contains information that is not logical or appropriate
Don't understand (DU)	Answer indicate: (1) Repetition question / statement (2) The answer is of no relevance to the statement (3) The answer "I do not know / understand"; (4) No answer or not filled

The ability of a scientific argument as independent variables in this study were analyzed based on the quality of student writing arguments. The dependent variable in this study is the Problem Question Diagnostic Clusters (DQCs).

Scientific Argument Ability in this study is a written response (claim, warrant, backing, and rebuttal) by [16] that is given by using questions DQCs developed into a two-tier DQCs type of test or multiple choice questions about the type of reasoned and essays. Observation sheet used as an instrument to measure the adherence to the learning process in the classroom during the study. This observation sheet aims to identify the learning process that occurs in the classroom so that it can support data capabilities misconceptions and scientific arguments that have been obtained. Notes field is used as a supporting instrument in this study in order to get the field data during the study process. Notes field will be obtained through field data collection tool such as stationery, recorders, cameras, and others.

3. Result And Discussion

3.1 Analysis misconception of genetic molecular concept

Conception is the ability to understand the concept of good obtained through interaction with the environment as well as from formal education [18].A misconception is a conclusion that's wrong because it's based on faulty thinking or facts that are wrong about concept. Based on the result about written and oral conception, there is a data :

3.2.1 Written and oral conception. Most of the conception of the students were in the category of understanding of the concept . However , there are some students who still have misconceptions and did not even understand the concept. Based on the data it is known that sub highest concept is understood correctly by students is sub concept of chromosomal abnormalities by the number of students who correctly answered is 27 students or 68 % of the total number of students . While the concept of the lowest sub misreading by the students is sub concept prokaryotic protein synthesis in the number of students who answered with these categories is 20 students or 50 % of the overall student population . The sub concepts that led to a lot of misconceptions are sub concepts chemical structure of DNA by the number of students who have misconceptions are 9 people or 23 % of the overall student population.

The average confidence level of all students answer is as much as 81.5 % . It shows that a written test used may indicate a confidence level of students as much as 18.5 % of the answers to questions that have been given through the provision of reasons.

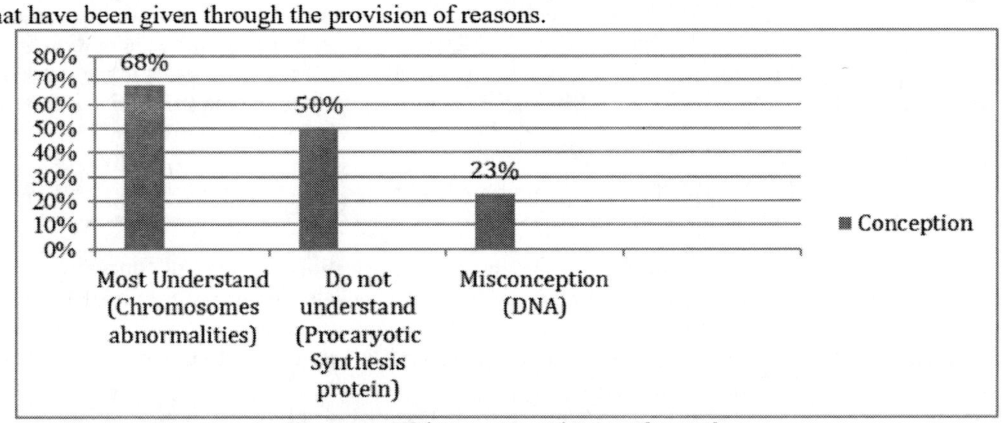

Figure 1. Written conception result graph

Sub highest concept is understood correctly by students is sub concept of chromosomal abnormalities by the number of students who correctly answered is 25 students or 63 % of the total number of students. While the concept of the lowest sub misreading by the students is sub concept prokaryotic protein synthesis in the number of students who answered with these categories is 15 students or 38 % of the overall student population. The sub concepts that led to a lot of misconceptions are sub concepts chemical structure of DNA by the number of students who have misconceptions are 10 people or 25 % of the overall student population.

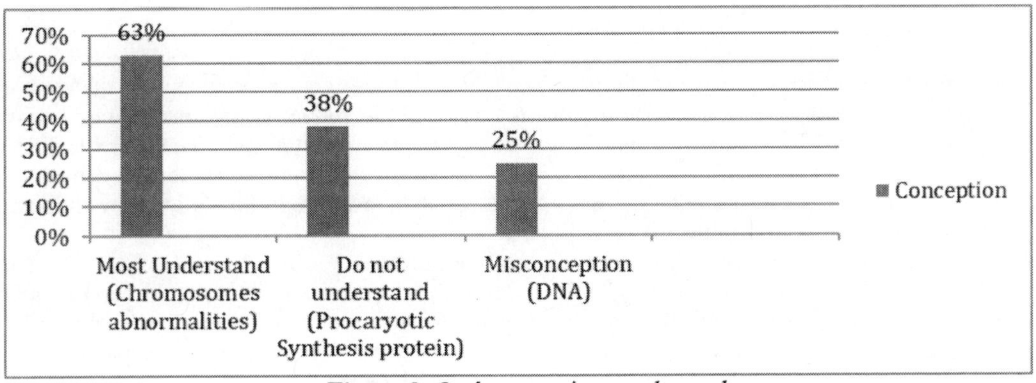

Figure 2. Oral conception result graph

The average confidence level of all students answer is as much as 82 % . It shows that a written test used may indicate a confidence level of students as much as 12 % of the answers to questions that have been given through the provision of reasons

3.2.2 Misconception of genetic moleculer

The results showed that misconceptions subconcepts highest in the chemical structure of DNA with a percentage of 23,75 % of the total number of students . It shows that there is a misconception that occur and must be investigated causes of the misconceptions. The results showed that misconception of molecular genetics are: DNA (23.75%), genes (18.75%) of chromosomes (15%) and protein synthesis (5.5%).

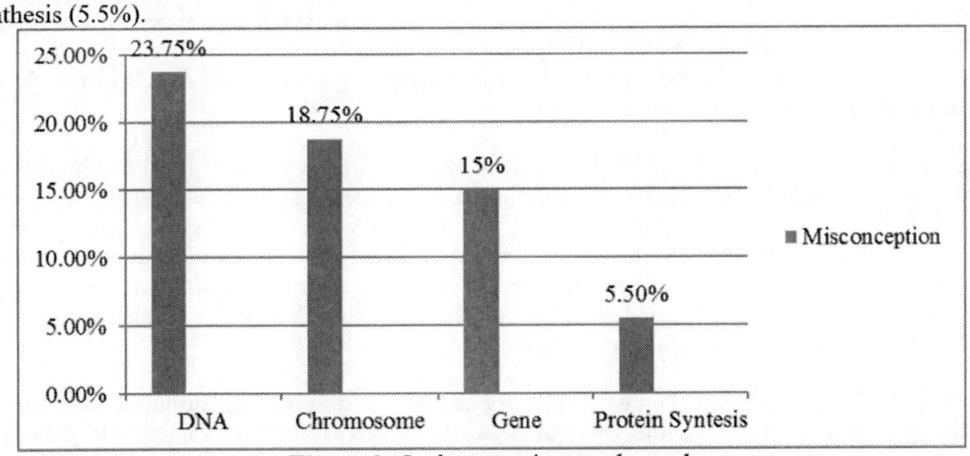

Figure 3. Oral conception result graph

Results of the analysis showed that most students still do not understand the true concept of the chemical structure of DNA and its characteristics . While the concept of differences in DNA and RNA that almost all students can understand . This is in contrast to the results of research which states that the concept of DNA better than other concepts in genetics.

3.2.3 Misconception pattern

The highest misconceptions pattern formed in a changing conception is *Misconception (MC) – Understand Partially* (UP) as much as 35 % on the concept of DNA and RNA . While the pattern formed is the highest argument argumentation level 2 - level arguments 2. It shows that there is a change in the conception of the misconceptions students have come to understand the concept of majority. The pattern of the argument, unchanged. That means there is no credible form or the learning process to develop the student 's argument.

3.3 Analysis student scientific argumentation ability

3.3.1 Reality of student scientific argumentation. The ability of the scientific arguments in question in this research is the ability , based on the argument that Toulmin model of argumentation. Ability arguments are analyzed Claim (C), Warrant (W), Backing (B) and Rebuttal (R) . The four categories are analyzed to produce scientific arguments profile capabilities biology education students who attended the lectures of genetics and cell biology . The results showed that the ability of written and oral arguments have different results.

The reality of general scientific argumentation abilities of students based on questions given in writing or orally in mind that the ability of the warrant (W) is the most prominent ability compared with the ability of other arguments in the following order: 50 % of the students have the ability argumentation warrant categories; 22 % only mengclaim; 18 % appears categories backing and rebuttal as much as 3 % of the total number of students. The facts are not in accordance with the theory of Inch and [18] states that a person who makes a claim is expected to provide support by using evidence and reason in accordance with the facts on the ground.

3.3.2 The Quality of Argumentations. As for the quality of the arguments that were analyzed using the argument Toulmin quality showed that the students were in the highest quality of written arguments on the quality of conceptual category descriptors I with an average percentage of 22% and the lowest in the category descriptors V with an average percentage of 18%. Similarly, the quality of the arguments orally or verbally with the highest percentage in the descriptor I with an average percentage of 25% and the lowest in the descriptor V with a percentage of 19%.

This shows that the quality of the arguments the highest student is at the level proposed claim causal framework theoretical specific and still low in unifying the proper reference when developing data. This is consistent with the results of [20] through the sheet questionnaires and interviews show that low descriptors V due to the difficulties faced by students in the argument is the ability to develop data / backing in support of their claim.

3.3.3 Level of Argumentation. In addition to quality-based argument is conceptually and epistemic , the quality of the arguments analyzed through arguments based on the level of [21]. The results showed that the level of argumentation students in writing and orally dominant sequence is the level of argument 2, 1 , and 3 , which means that the students are still at the level of argument in the form of a claim with a counter claim and be accompanied by the data , collateral (warrant) or support (backing) but little or no rebuttal. This is in contrast to the results of [20] which explains that the quality of the arguments of individual students are at level three.

3.3.4 Pattern of Argumentation. The pattern of argumentation abilities formed based on changes in the level of argument, based on the level of argumentation [23] Of the 25 possible patterns of argumentation is found that the highest argument is owned by the pattern of argumentation level 2 - level arguments 2. It shows that there is no change in the ability of argument are formed. This means that the ability of the student's argument remained at level 2 or argument in the form of a claim with kounterklaim containing the data, guarantees, and there is no objection. No changes in the ability of the argument can be caused by several factors, namely 1) there is no learning activities that develop student argumentation; 2) students have little opportunity in berpendapatPola ability argumentation formed based on changes in the level of argument, based on the level of argumentation [23]. Of the 25 possible patterns of argumentation is found that the highest argument is owned by the pattern of argumentation level 2 - level arguments 2. It shows that there is no change in the ability of argument are formed. This means that the ability of the student's argument remained at level 2 or argument in the form of a claim with kounterklaim containing the data, guarantees, and there is no objection. No changes in the ability of the argument can be caused by several factors, namely 1) there is no learning activities that develop student argumentation; 2) students have little opportunity to argue.

4. Conclussion

The concept of molecular genetics is one of the concepts in biology pembelajaaran prone with their misconceptions. The misconception is to misunderstand the concept that a person either students or students. Miskonsepi research results of students on the concept of molecular genetics with DQCs shows that students still have misconceptions about the concept of molecular genetics. The subconcepts with the highest percentage rate misconception is on subconcepts DNA & RNA as much (23.75%), genes (18.75%), chromosomes (15%) and protein synthesis (5.5%). It shows that the concept of DNA has the highest level of vulnerability compared with other concepts of molecular genetics. In addition to the analysis of misconceptions, this study also analyzes the ability of the scientific arguments that are owned by students.

The ability of the scientific arguments that are owned by the student is still in the category of the warrant (W) as much as 50% of all students. While the quality of the argument is still dominated by the quality of the conceptual arguments that are in the first or lowest descriptors owned by the descriptor V on the quality of epistemic argument. As for the level of argumentation that is formed through written tests and oral tests known that argument level 2, 1, and 3 most prominent than the level of the other arguments. Ability argument dianalsis also form a pattern similar to the pattern of misconceptions.

The highest misconceptions pattern formed in a changing conception is Misconception-Understand Partiallty as much as 35% on the concept of DNA and RNA. While the pattern formed is the highest argument argumentation level 2 - level arguments 2. It shows that there is a change in the conception of the misconceptions students have come to understand the concept of majority. The pattern of the argument, unchanged. That means there is no credible form or the learning process to develop the student's argument.

5. References

[1] Meilinda 2009 E-modul interaktif berbasis kontruktifisme pada materi Genetika untuk meningkatkan kompetensi guru Biologi SLTP (*Tesis of Universitas Pendidikan Indonesia*)
[2] Benabdallah N, Bhatia S , Williamson I and Bickmore W 2015 *European J. of Genetics* **23** 1 Planary Lecture
[3] Venvile G, Gribble S and Donovan J 2002 *Science Education* **89** 614–633
[4] Urey and Calik 2008 *Asia pasific forum on science learning and teaching* **9** 2 12:1-15
[5] Infante, Malachias M E 2010 *Electronic J. of Science Education* **9** 3 657-668
[6] Wangintowe T 2008 Miskonsepsi Siswa SMA pada Mata Pelajaran Biologi dan Faktor-faktor Penyebabnya. *Disertation of Universitas Negeri Malang*
[7] Tekkaya 2006 *J. of Hacettepe University Education Faculty* 23 259-266
[8] National Research Council 1996 *National science education standards* (Washington DC: National Academy Press)
[9] Kuhn D 2010 *Science Education 94* 810–824
[10] Kurniadi E 2015 *J.Pendidikan* **14** 1-11
[11] Fisher K M, Williams K S and Lineback J E 2011 *CBE-Life science education* **10** 418–429
[12] Suwirto 2013 *Pengembangan Tes Diagnostik dalam Pembelajaran* (Yogyakarta:Pustaka Pelajar)
[13] Shi D, Xu Y, Hopkinson B M, François M, and Morel M 2010 *Science* **327** 5 February 2010
[14] Treagust D F 2007 J. of Research in Science Teaching **44** 2 205–235
[15] Morgil, İnci and Nuray Y 2006 *J. of Turkish Science Education* **3** 1
[16] Toulmin S 2003 *The uses of argument, updated* (Cambridge: Cambridge University Press)
[17] Suparno 2005 *Filsafat Konstruktivisme dalam Pendidikan* (Yogyakarta: Kanisius)
[18] Warnick B K, Thompson G W and Tarpley R S 2006 *Proc.* of the Thirty-third Annual National *Agricultural Education Research Conference* **33** 601-603
[19] Erduran S andMaria P J 2004 *Argumentation in science education* (London: Springer)
[20] Herlanti Y 2012 *J. Pendidikan IPA Indonesia* **1** 2 168-177
[21] Erduran S, Simon and Osborne J 2004 *Science Education* **88** 915-933

IOP Conf. Series: Journal of Physics: Conf. Series **812** (2017) 012113 doi:10.1088/1742-6596/812/1/012113

Gesture Recognition using Latent-Dynamic based Conditional Random Fields and Scalar Features

I N Yulita[1*], M I Fanany[2] and A M Arymurthy[2]

[1]Departemen Ilmu Komputer, Universitas Padjadjaran, Jl. Raya Bandung – Sumedang Km. 21, Sumedang 45363, Indonesia
[2]Fakultas Ilmu Komputer, Universitas Indonesia, Kampus UI Depok, Depok16424, Indonesia

Email: intan.nurma@unpad.ac.id

Abstract. The need for segmentation and labeling of sequence data appears in several fields. The use of the conditional models such as Conditional Random Fields is widely used to solve this problem. In the pattern recognition, Conditional Random Fields specify the possibilities of a sequence label. This method constructs its full label sequence to be a probabilistic graphical model based on its observation. However, Conditional Random Fields can not capture the internal structure so that Latent-based Dynamic Conditional Random Fields is developed without leaving external dynamics of inter-label. This study proposes the use of Latent-Dynamic Conditional Random Fields for Gesture Recognition and comparison between both methods. Besides, this study also proposes the use of a scalar features to gesture recognition. The results show that performance of Latent-dynamic based Conditional Random Fields is not better than the Conditional Random Fields, and scalar features are effective for both methods are in gesture recognition. Therefore, it recommends implementing Conditional Random Fields and scalar features in gesture recognition for better performance

1. Introduction

The need for segmentation and labeling of sequence data appears in several fields [1]. Many methods were developed to solve this problem. One of them is Conditional Random Fields [2]. This technique has been implemented for the recognition of the object [3], speech [4], and relational learning [5]. The advantage of this technique is its ability in processing the knowledge through the use of a series of sequences to form a model which can overcome the bias of the data. If the data has a bias that is the resemblance between several labels, then Conditional Random Fields will pay attention to the output/label sequence.

However, Conditional Random Fields did not use intrinsic structure from its observation so that further developed state Hidden-state Conditional Random Fields [6,7]. But Hidden-state Conditional Random Fields would eliminate the ability to capture the dynamics of extrinsic between their labels by CRF [8]. Thus, it developed into Latent-Dynamic Conditional Random Fields to merge the two mechanisms [9].

On the other hand, the recording of data gesture is usually a point of articulation which is every point consist of several coordinate axes. Therefore, it is important to use only important features for its modelling. This process is can be done by using feature extraction. Thus, this paper offers the use of scalar features to improve the accuracy of Latent-Dynamic Conditional Random Fields on gesture recognition. Its use reduces the number of features so make modeling time will be reduced.

2. Experimental Method

In this chapter will describe methods that will be used in gesture recognition is Conditional Random Fields, Latent-Dynamic Conditional Random Fields, and Scalar Features. Also, it will be explained the used dataset and the flowchart of the proposed method.

2.1. Method

Features on the same position with different coordinate axes will be changed to a scalar value by using Euclidean distance. The calculation is as follows:

$$s_{ij} = \sqrt{x_{ij}^{2} + y_{ij}^{2} + z_{ij}^{2}} \quad (1)$$

Where
x_{ij}, y_{ij}, z_{ij} are the three values of the coordinates of the data for the same position
s_{ij} is a scalar value at a specific position

Illustration of the scalar features is described in Figure 1. If the data have coordinates in the x, y, and z coordinates, then the scalar features are the calculation of Euclidean distance.

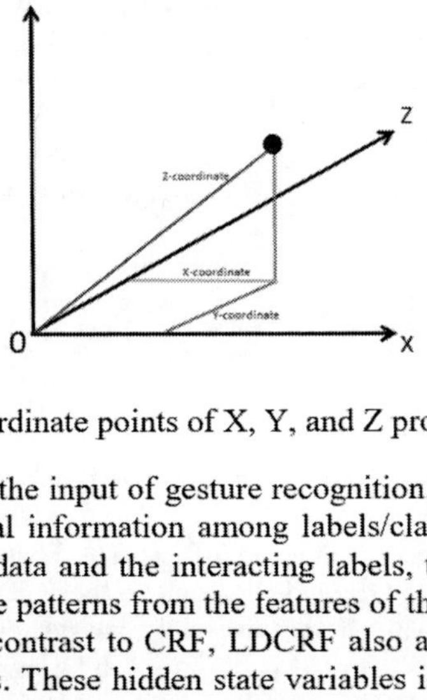

Figure 1. Three coordinate points of X, Y, and Z produce a scalar feature

These scalar features become the input of gesture recognition. In this study, the recognition is also processed by using the sequential information among labels/classes of the dataset. Thus, to capture both factors: the features of the data and the interacting labels, this study proposes LDCRF which it can capture both internal structure patterns from the features of the data, and internal structure patterns of the interacting labels[10]. In contrast to CRF, LDCRF also adds some hidden state variables that models internal structure patterns. These hidden state variables is placed in latent sub-structure layer. This layer serves to filter the input sequence. The other hand, HCRF also uses this layer but it ignores the external structure of the interaction of the label. So the lack of CRF and HCRF are resolved by building of the model of LDCRF.This method is a combination of CRF and HCRF. The modelingof LDCRF is presented in Figure 2,the architecture used is based on research conducted by Fabio Tamburini, Chiara Bertini, Pier Marco Bertinetto for prosodic Prominence detection [11]

LDCRF is one type Probabilistic Graphical Models that represent the model through a probability distribution [12]. The label is determined based on the maximum value of the probability as follows:

$$p(y|x) = \frac{\partial}{\sum_{h'} \partial} \quad (2)$$

$$\partial = \exp\left[\sum_{a=1}^{b} \sum_{c=1}^{d} w_a f_a(x, c, h_c, h_{c-1})\right] \quad (3)$$

Where
$f_a$is feature function
$w_a$is weight of feature function
$h_c$is hidden state

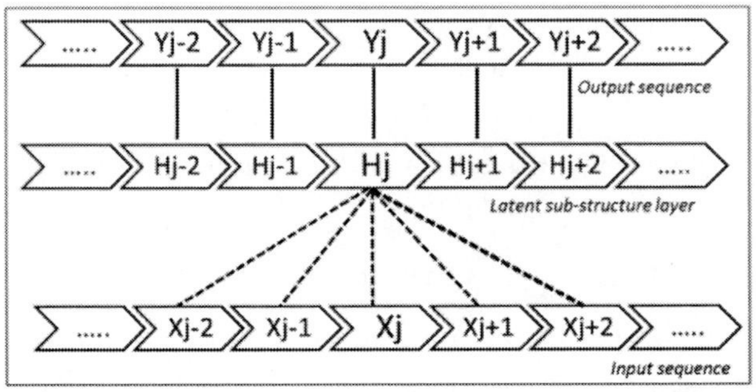

Figure 2. The architecture of Latent-Dynamic Conditional Random Fields

The weights of the feature function are obtained from the training data using optimization techniques. Some widely used optimization technique is Conjugative Gradient and Broyden-Fletcher-Goldfarb-Shanno. While the feature function is obtained by latent sub-structure layer. A feature function defines a value (0 or 1) of the sequence x, sequence c of two hidden state being observed. For example the value of feature function is 1 if the the position is 3 and also the hidden state 3rd and 4th are the representation of the label "rest" and "preparation".

2.2. Flowchart of the Recognition

The recognition process of this research is based on scheme that it will be illustrated in Figure 3.

Figure 3. The Proposed Method

In this study, thedata used aresegmentation of gesture phase from some videos, which was developed by the University of Sao Paulo [13]. In each of these segmentation presented six points of articulation that is left hand, right hand, left wrist, right wrist, head, and spine, which eachposition have coordinates of x, y, and z, as illustrated in Figure 4. The purpose of gesture recognition is the prediction of gesture phase (rest, preparation, stroke, hold, and retraction).

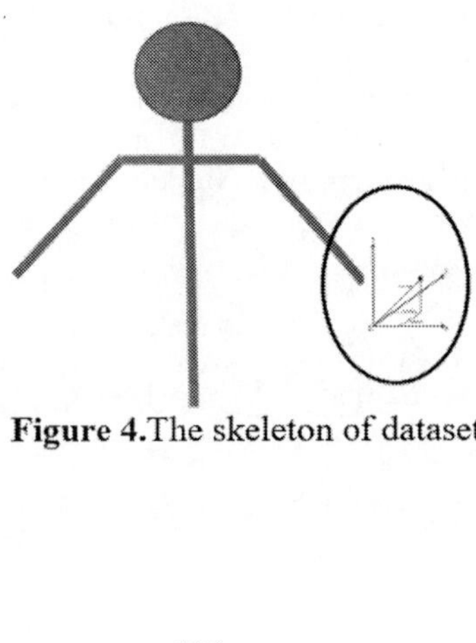

Figure 4. The skeleton of dataset

Then the dataset is preprocessed in the first step. This preprocessing includes feature selection. Through preprocessing, only the relevant features are used for building modeling. From the dataset, they have the feature of timestamp. The preprocessing will eliminate this feature because this feature is not helpful to build the proposed modelling.

The second step is feature extraction. The obtained features are six scalar features of the 18 features. This process is done by using the calculation of scalar features. Then these x scalar features are modeled using Latent-Dynamic Conditional Random Fields. From these resulting models, the maximum probability of data to the label will be the final label.

3. Result and Discussion

Some experiment tests have been done using the dataset in this research. The evaluation scheme isbased on 3-cross-validation and calculation of the accuracy and execution time for both methods. The test results are described in the next section.

2.3. Accuracy

Effectiveness in the process of recognition will be tested based on the percentage of the accuracy so that it can evaluate the performance of the resulting model to the data being tested.The accuracy is presented by %.

Table 1. The Accuracy of this gesture recognition

	Without Scalar Features		With Scalar Features	
Accuracy	CRF	LDCRF	CRF	LDCRF
First Iteration	50.14	54.14	57.75	54.61
Second Iteration	50.00	50.00	78.08	71.08
Third Iteration	64.14	50.07	64.70	68.06
Average	54.85	51.40	66.84	64.59

From Table 1, it can be seen that the LDCRF implementation without scalar features of the first iteration gets a higher accuracy than CRF. But accuracy is lower in the second and third iterations so that the average accuracy is also lower than the CRF. While the implementation of the scalar features, LDCRF the first and second iteration is lower than the CRF. But in the third iteration, higher accuracy. Similarly, the implementation without the scalar features, accuracy LDCRF on average lower than CRF. This indicates that the addition of hidden layer LDCRF no more effective for the gesture recognition even worsens the performance of the external structure.

Also, the table 1 is known that the use of scalar features can boost recognition accuracy of both CRF and LDCRF movement. In the first iteration, the accuracy of CRF increased 7.61% and amounted 0.47 LDCRF%. hile in the second iteration, respectively 28.08% and 21.08% and in the third iteration of 0.56% and 7.99%. So that the average increase was 11.99% and 13.19%. With the increasing accuracy, it can be seen that the effective use of scalar features for gesture recognition.

2.4. Running Time

Another evaluation parameter is therunning time in order to know how long it takes to produce a model in the recognition process. The execution time is presented by seconds.

Table 2. The running time of this gesture recognition

Running Time	Without Scalar Feature		With Scalar Feature	
	CRF	LDCRF	CRF	LDCRF
First Iteration	115.06	1671.98	48.67	969.42
Second Iteration	49.00	1891.90	68.23	1106.89
Third Iteration	52.71	3035.23	66.82	1154.95

Average	72.26	2199.70	61.24	1077.09

In Table 2, it can be seen that the implementation without the scalar features, running time of CRF and LDCRF are 115.06 and 1671.98 seconds. The running time is decreased by more than half in the second iteration of CRF but a slight increase on the third iteration. This is in contrast with LDCRF for the second iteration; there is an increase almost be doubled in the third iteration. Based on the calculation of average it was found that LDCRF takes three times greater than the CRF.

With the use of scalar features, the running time of CRF and LDCRF in the first iteration have decreased by about 50%. But in the second and third iteration, CRF increased slightly, but LDCRF sharp decline. From the calculation of averages, CRF and LDCRF decrease running time when used the scalar features.

4. Conclusion

From the results of experiments that have been done, it is known that CRF is superior to LDCRF. This indicates that the use of hidden state will not necessarily increase the accuracy of labeling. Regarding the running time, the LDCRF takes than the CRF. For faster labeling, the CRF is selected for our gesture recognition. The use of scalar features proven to be effective for our gesture recognition as shown by higher accuracy and lower running time

5. References

[1] Lafferty J D, McCallum A and Pereira F C N 2001 *Conditional random fields: Probabilistic models for segmenting and labeling sequence data* ICML (San Francisco, CA, USA: Morgan Kaufmann Publishers Inc.)

[2] Sutton C and McCallum A 2011 *An Introduction to Conditional Random Fields* (Foundations and Trends in Machine Learning)

[3] Quattoni A,Collins M and Darrell T 2004 *Conditional random fields for object recognition* (MIT Press) pp. 1097–104

[4] Charles S and Andrew M 2006 *An Introduction to Conditional Random Fields for Relational Learning* (MIT Press) chapter 4 p 93–128

[5] Zweig G and NguyenP 2009 *A segmental CRF approach to large vocabulary continuous speech recognition* (IEEE ASRU Workshop)

[6] Gunawardana A, Mahajan M,Acero A, and Platt J C 2005 *Hidden conditional random fields for phone classification* (InInterspeech) p 1117–20

[7] Quattoni A, Wang S, Morency L P, Collins M and Darrell T 2007 *Hidden conditional random fieldsIEEE Trans. Pattern Anal. Mach. Intell.* **29** 1848–52

[8] Tong Y,Chen R and Gao J 2015 *Hidden State Conditional Random Field for Abnormal Activity Recognition in Smart Homes*

[9] Boonsuk S, Suchato A, PunyabukkanaP C, Wutiwiwatchai N and Thatphithakkul 2014 *Mathematical Problems in Engineering*

[10] Wittner C, Schauerte B and Stiefelhagen R 2015*What's the point? Frame-wise Pointing Gesture Recognition with Latent-Dynamic Conditional Random Fields*

[11] Rahimi A M, Ruschel R, and Manjunath B S 2016 *UAV Sensor Fusion with Latent-Dynamic Conditional Random Fields in Coronal Plane Estimation CVPR*

[12] Tamburini F, Bertini C and Bertinetto P M 2014 *In Proceedings of Speech Prosody* (SP-2014) 285–9

[13] MadeoR C B, Wagner P K and Peres S M 2013 *International Journal of Computer Science and Information Tech.* **5** 1-20

Acknowledgments
The Author thanks to the Indonesian Endowment Fund for Education (LPDP) and Machine Learning and Computer Vision Laboratory, Universitas Indonesia that contributed and supported the study.

Creative Mathematical Games: The Enhancement of Number Sense of Kindergarten Children Through Fun Activities

Mirawati[1]

[1]UniversitasMuhammadiyahTasikmalaya, Indonesia
Email: mirapaud@gmail.com

Abstract. The research departed from an issue found regarding the number sense of kindergarten children and as a solution to this problem, the research proposes the use of creative mathematical games in the teaching and learning. Departing from the issue and the offered solution, the following problems are about Children's ability of number sense before and after the implementation of creative mathematical games; the forms of creative mathematical games in improving children's number sense; the implementation of creative mathematical games in improving children's number sense; and the factors possibly affecting the implementation of creative mathematical games. This study use action research method. The data were collected through observation, interview, and documentation and then qualitativelyanalysed using thematic analysis technique. The findings show that children respond positively to the creative mathematical games. They demonstrate fairly high enthusiasm and are able to understand number as well as its meaning in various ways. Children's number sense has also improved in terms of one-on-one correspondence and mentioning and comparing many objects. The factors possibly affecting the implementation of these creative mathematical games are the media and the stages of teaching and learning that should be in accordance with the level of kindergarten children's number sense.

1. Introduction

The important role makes mathematics to become a science which should be introduced and it has to be taught to every individual starts from pre-school to high school level [21].Furthermore, [21]stated that mathematics learning for young learners is aimed to stimulate the thinking ability of the children so they are ready to learn mathematics for the next step and the children are able to master various mathematical science and skill which enable them to solve problems in their daily life.Based on this, it is implied that mathematics learning for young learners is a medium to develop their simple thinking skill which can help the children to understand several concrete concepts, yet it is not intended to allow the children to have academic skill as soon as possible. The learning process is more likely to the ability of solving problems that exist in their surroundings. Another important aspect that needs to be developed and facilitates the children in mathematics learning is *number sense*or the sensitivity of numbers[6].

Number sense is an intuition about numbers and matters related to numbers.It can be broadly defined as understanding of the meaning of numbers and the relationship between numbers [10]. Some aspects of *number sense* skill possessed by the children are the ability to mention the sound of numbers, mention the sound of number sequences, do one to one correspondence, connect sound to the symbol of numbers, count concrete object, compare the amount of objects, do the addition by combining group of objects and do the subtraction by separating the group of objects[3]. Those *number sense*skills are a part of the children's thinking skill.A child with a good *number sense*, one of which, will have the ability to do correspondence between the sound of numbers and concrete objects so it can bring meaning to the numbers[6].

Number sense problem of the children which is frequently found is related to quantity and counting [2]. A research conducted by[19]also shows that one of the problems faced by the kindergarten students is the ability to count and recognize the sequence of numbers. The*number sense* problem is frequently happened in several organizations of Early Childhood Education Program. Based on the observation conducted by the writer, some children seem to have difficulties in understanding the changes in number of objects and the sequence of numbers. For example, the children may not be able to count in sequence from larger numbers such as 10, 9, 8and so on. The children also experience confusing in distinguishing ordinal concept like the order of number such as first, second, third and so on with cardinal concept to mention the amounts of objects such as one object, two objects, three objects, and so on. The *number sense*problem is not a negligible problem since several researches state that *number sense* is very important in developing mathematics learning for early childhood students. Jordan [9]says that difficulties in mathematics learning at kindergarten level is not on the children's problem in memorizing mathematics concepts, but it is *more* likely on the lack stimulation of *number sense*possessed by the children. Another research conducted by[1] shows that *number sense*can help facilitate the child in the operational activities of quantity and number systems. Some statements from those researches imply that*number sense*has a great influence to the increasing of the students' understanding in mathematics learning.So, it can be said that the*number sense*skill is an early foundation of the skill and the mastery concepts of mathematics at higher level.On the other hand, the*number sense*skill of every child might be different because of the different surroundings and cultural background. Adults around the children can provide appropriate stimulation for the development of the children's*number sense*so that the children are able to develop it optimally.

One of the promoted activities to learn the mathematics concepts for young learners is creative game, because learning by playing games can help the children to understand various concepts of mathematics through fun and meaningful activities [17]. Creative mathematics game is a creation of mathematics learning through playing ([12]; [15]). According to [16]the activities of playing, including games, are able to encourage the children to develop their logical thinking skill and understand mathematical process such as addition, subtraction and other concepts.The description above implies that the creative mathematics games can help the children to understand the concepts of mathematics including the children's number sense aspect.Based on the description, the writer intends to make the creative mathematics game as a solution to the problem of*number sense*in Kindergarten.

2. Experimental Method

The research ocation of this study is Preschool UPI and the method is action research by Kemmis and MC Taggart [14]. Kemmis and MC Taggart's research design consists of four components as follows:

- Planning. The researcher made coordination with the teachers about implementing the creative math games to improve the children's number sense in kindergarten.
- Implementation. The researcher implemented the creative math games in the learning process.
- Observation. The next phase was observation to figure out how far the children's number sense develops as well to control the implementation of the creative math games.
- Reflection. The reflection phase was performed by the researcher to discuss and evaluate the results of the activities that had been carried out.

The data collection technique in this research employed three techniques, namely observation in the forms of field notes, interview, and documentation. The data obtained were analyzed through a qualitative approach by using thematic analysis.

3. Result and Discussion

3.1. The Initial Condition of the Number Sense Ability of Children

Findings in this research is advanced on the Number sense ability of the children prior to the implementation of creative math games in Preschool UPIis in the state of mentioning numbers and the sequence of numbers. The findings show that some of the students experienced difficulties when doing one to one correspondence, in which they had not been able to perform correspondence of sounding numbers and things that represent the number values, thereby being an obstacle to them in understanding the meanings of some particular numbers. The students also experienced difficulties in comparing the numbers of objects and administering simple summation and subtraction regarding the objects.

The difficulties found in the findings are typically the problem with number sense that was experienced by the students. Theoretically, children at the age of 4 to 6 years old should have been able to do one to one correspondence and carry out simple numeration [3]. Deheane[6] suggests that the potential of number sense is already owned by children since birth and infancy. It is in line with the statement of [18] that every child has the potential of number sense though could be in different state, which one of the causes is due to the different stimulation they get. A good stimulation could enhance the development of number sense ability in children in as well a good way, and vice versa. The problem of number sense that was experienced by the students in Preschool UPI is partially caused by lack of efforts made by the school in developing their children's number sense ability neither through the forms of programs nor the other activities. The math learning is sometimes carried out in the forms of student worksheets whereas, according to [21] the math learning process for young children should better be presented in the activities that attract the children's attention and through the use of authentic media. In different notes, [13] also notes that math learning process presented to children regardless of their needs and interests will probably be less meaningful or even meaningless for the children.

3.2. Creative Mathematical Games to Enhance the Number Sense of Kindergarten Students

The selection of math games to be carried out in the research did not take any direct implementation for granted but going through the discussion and workshop conducted by the researcher with the school and teachers. The games should be in corresponding with psychological principles of children in terms of the game rules, players, media selection, as well as the length of time needed to perform the games ([5]; [7]; [11]). The brief description of the four games is as follows:

- Fishing For Fun. Fishing for fun game is created by the group of Green Apple Lessons [23] which aims to help children be able to enhance numerical literacy or numeracy, reading numbers, correspondence, and memorizing. The media used in this game are ten paired picture cards of cats and fishes. The game takes place for about 10 to 15 minutes with pretty simple rules, that children pair the picture cards of cats and fishes with the same numbers. Once the children found the pairing cards, then they may take both of the cards, but if not, then they are asked to put the cards back into the original place.
- Five Little Monkeys. Five little monkeys game is created by Eileen Christelowwhich aims at helping children enhance their skills in numeracy, reading numbers, correspondence, and practicing simple numeral operation [3]. The media used in this game are five little monkeys board, beam or pin, and dice. The game is carried out by storing the beam or pin into the boxed on the board that has been provided with the number on the dice that came up.
- Dominoes Modification. Dominoes modification game is developed by [19] with the aims to help children in counting, corresponding, and comparing various objects as well as in simple numerical operation. The media used in this game are ten domino cards that have been modified in the forms and the dot illustrations on the cards. The dots are tailored to the theme taken, which in this research is the illustration of donuts.
- My Numerical Carpet. The initial idea of making this numerical carpet is to introduce the concept of numeric, its symbols, and simple numerical operation by using objects. My numerical carpet is created by [15]. The media used in this game are numerical carpet with

2x3 meters in size and stuffed dice. The rules of this game are the children to roll the dice then move according to the number on the dice that just came up.

The four creative math games as presented above are redesigned in order to be able to provide stimulation in various aspects of number sense on children gradually. It aims to equip children with optimal number sense ability and thereby the children could pass every stage required. It is as noted by[20]that math learning for very young learners should consider the stages of concept they could attain from the beginning to the end, so there will be no missing stages or phases and could help them understand the concept thoroughly. A similar conception is noted by [4]that young children learn math from the very simple things around them.

3.3. The Implementation of Creative Mathematical Games to Enhance Number Sense of Children in Kindergarten

The implementation of creative math games in improving number sense of Preschool UPIstudents is conducted into five cycles. The theme administered in this research is accustomed to the students' needs and interests, namely animals, foods, and space. The Figures 1 and 2 are the documentation of the implementation of creative math games in improving students' number sense in Preschool UPI.

(a)

(b)

Figure 1. The young learners playing Fishing for Fun and Five Little Monkey game.(a) They took pairing cards (both the cards have the same number of fishes), and they counting fish in the cards. (b) The young learners playing Five Little Monkeys by rolling the stuffed dice, mentioning the number the dice showing, and placing the monkey pin on the column appropriate with the number on dice

(a)

(b)

Figure2. The young learners playing Dominoes Modification and My Numerical Carpet.(a) Two young learners arranging dominoes from the smallest number, one to ten, and making domino effect. (b) The young learners with their teacher observing My Numerical Carpet.

The implementation of each activity as figured above was conducted in pairs or in groups. This supposed that the other aspects of children development such as social skills could also be stimulated since one of the principles of early childhood education is the comprehensive and thorough [22]. The stages of number sense acquisition become the standards in planning the learning scenario thorough creative math games therefore the children attain all the aspects. Besides, the learning activities conducted consist of various games and plays. Those various activities are meant to support the learning process thereby not being boring and monotonous for the children, as noted by [3] that every child sometimes needs different activities to establish and develop their number sense ability.Another important point to note as reference in the implementation of creative math games is the repetition of activities that leads to the accomplishment of number sense aspect in each game.The repetition is conducted as well to recall the prior concepts therefore the children could understand more [13].According to the discussion above, the criteria of the implementation of creative math games in enhancing number sense of children are as follows:

- Creative math games are packed in the forms of attractive thematic learning activities for the children.
- The games are aimed at developing and enhancing number sense of children but still could stimulate other development aspects.
- The games are designed to be performed in pairs or in groups.
- Learning scenario through creative math games is designed accustomed to the phases of number sense aspect the children own.
- Creative math games employed in this study consist of various kinds of plays and games.
- There are repetition of activities which aims to advance the number sense of children more in every activity of creative math games

3.4. The Factors Affecting the Implementation of Creative Math Games

Some factors that probably affect the implementation of creative math games are found when the research conducted, they are summarized as follows:

- Media.The media used in the learning activities are designed accustomed to psychological principles of children, namely containing interesting pictures and forms, eye-catching with various colors, proper in size, having soft texture, and made from materials that are safe for children ([7] 2009; [11]).
- The learning stages focuses on children's stages of number sense.
- Teacher Behavior. The results of study show that active and fully attentive behavior of the teacher could enhance enthusiasm in the children when about to do the learning activities.
- PeerTeaching. One of the other factors that indirectly influence the success of creative math games implementation is the help and support from colleges (peer teaching). The support and help from peers are gained by the children through the learning process that is designed properly to give them opportunity to work or do activities in groups, so they could interact with each other.

3.5. Number Sense Ability of Children after the Implementation of Creative Math Games

Based on the results of study related to number sense aspect of children in Preschool UPI after the implementation of creative math games the number sense skills of the children increased. Some of the number sense ability grow after the stimulation given through creative math games such as one to one correspondence, connecting the numerical sound to its symbol, the ability to mention more objects, comparing the number of objects, and establishing simple numerical operation, either summation or subtraction with the help of objects.From the results of the study, it suggests that creative math games generate benefits that can enhance number sense of children. This is corresponding with the purposes of the creative math games for children as noted by experts that games could encourage children to advance and strengthen their mathematical understanding, find out the alternatives of problem solving spontaneously, and give opportunity to the children to communicate mathematical process as well as encourage their mathematical knowledge through some entertaining activities ([16]; [17]).

4. Conclusion

Based on the findings and discussion, it can be concluded that creative math games could stimulate number sense of children in Preschool UPI. The number sense ability of the children that is originally not optimal grows to develop after the implementation of creative math games, so in the other words, the increase of number sense ability of the children represents the success of the implementation of creative math games.

5. References

[1] Aunio, *et al*2006*Routledge:Sandivanian Journal of Educational Research***50**pp 483-502.
[2] California Department ofEducationChild Development Division2010 *Desired ResultsDevelopmental Profile*(Sacramento: California Department of EducationChild Development Division)
[3] Confer C 2005 *Teaching Number sense in Kindergarten*(United States of America: MathSolutions Publication)
[4] Copley J V 2001 *The young child and mathematics*(Virginia: National Council of Teachers of Mathematics)
[5] Coughlin P2000 *Menciptakankelas yang berpusatpadaanak*(Jakarta: Children Resources International Inc.)
[6] Deheane S 2011 *The Number sense, How The Mind Creates Mathematics*(United Kingdom:Oxford University Press, Inc.)
[7] Eliyawati C 2000 *Media Pembelajaran Anak Usia Dini* (Bandung: Pustaka Utama)
[8] Jacobsen D,*et al*2009 *Methods for teaching: Metode-metodepengajaransiswa TK-SMA*. (Yogyakarta: PenerbitPustakaBelajar)
[9] Jordan N C 2007*ASCD: Educational Leadership***65** pp 63-65.
[10] Malofeeva E, Day J, Saco X, Young L, andCiancio D2004*Journal of Educational Psychology***96**pp 648-659.
[11] Mariyana R2009 *PengelolaanLingkunganBelajar*(Bandung: UPI Press)
[12] Mayeski M 2002 *Creative Activities for Young Children 7th Edition*(United States: Delmar)
[13] McGrath C2010 *Supporting Early Mathematical Development: Practical Approaches to Play Based Learning*(New York: Routledge)
[14] McNiffJandWhiteheadJ 2002*Action Research:PrinciplesAnd Practice, Second Edition*. (USA: Routledge Falmer)
[15] Mirawati2014*Proc. Konferensi Nasional Pendidikan Dasar SPs UPI 2014(Bandung)* (Bandung: Rizki Press) pp 672-678
[16] Özdogan E2011*Elsevier: Procedia Social and Behavioral Sciences***15** pp3118–3120
[17] Perry B and Dockett S 2010*Proc. of the 33rd annual conference of the Mathematics Education Research Group of Australasia*(Fremantle: MERGA)
[18] PilmerD2008 *Number sense* (Canada: Nova Scotia School for Adult Learning, Department ofLabour and Workforce Development)
[19] Rahayu W A 2013 *Meningkatkan Kemampuan Mengenal Konsep Bilangan Anak Melalui Penggunaan Media Kartu Domino Bergambar di TK Al-Hidayah* (Bandung: Skripsi UniversitasPendidikanIndonesia)
[20] Sarama J and Cements DH 2009 *Early Childhood Mathematics Education Research: Learning Trajectories for Young Children*(New York: Routledge)
[21] Sriningsih N 2008 *Pembelajaran Matematika Terpadu untuk Anak Usia Dini* (Bandung: Pustaka Media)
[22] Suyadi and Ulfah 2013 *Konsep dasar PAUD* (Bandung: PT. Remaja Rosdakarya)
[23] TPT 2013 *Mathematic for Kinderarten in Teacher Pays Teacher*. [Online:http://www.teacherspayteachers.com/]

Acknowledgments
Thank you for Prof. Dr. Didi Suryadi, M.Ed from UPI, who provided insight and expertise that greatly assisted the researchand for comments that greatly improved the manuscript in this research.

IOP Conf. Series: Journal of Physics: Conf. Series **812** (2017) 012115 doi:10.1088/1742-6596/812/1/012115

Meaning of Fractions

D A K Dewi[1], D Suryadi[1], T Suratno[2], E Mulyana[1], and H Kurniawan[3]

[1]Departement of Mathematics Education, Universitas Pendidikan Indonesia, Jl. Dr. Setiabudi No. 229, Bandung 40154, Indonesia
[2] Department of Primary School Teacher Education, KampusSerang, UniversitasPendidikan Indonesia, Jl. Ciracas No. 18, Serang 42116, Indonesia
[3] SDN Cibaduyut 3, Jl. Cibaduyut Raya Gg. Maeja, Bandung 40236, Indonesia
Email: diazdiniasri@student.upi.edu

Abstract. Introducing fractions is identical to divide an object. Suppose we divide the apple into two parts. One divided into two parts, the question arises whether one part can be called a half or not. Based on this activity, how can students give meaning to fractions. This study aims at designing a different fractions lesson by applying Didactical Design Research. In doing so, we undertook several research phases: 1) thinking what is fractions and why students should learn this concept; 2) designing didactical situation based on identified learning obstacles; and 3) reflecting retrospectively on the lesson design and its implementation as to redesign the fractions lesson. Our analysis revealed that most students held epistemological obstacles in giving meaning of fractions because they only know fractions as numbers that have numerator and denominator. By positioning ourselves as students, we discuss the ideal design to help students in constructing the meaning of fractions.

1. Introduction

Reference teachers usually teach the curriculum and textbooks. The curriculum used as a reference for decision making related to the election materials, learning objectives, and measures of learning. Textbooks used as a reference in the selection of materials and the plot. In practice, the teacher usually first learn the material content of the textbook. Then, the next morning he was to transfer the knowledge to the students, so that students tend to learn in one direction. When the teacher out of the classroom, it ended its activities.

In elementary school, students begin to learn fractions ranging from grade 3 to grade 6. The curriculum has been mapped this topic into several sections according to the capacity of students in each level. In grade 6, fractional topic focuses on fractional multiplication and division operations. Our initial analysis of the students' learning experience previously indicated likely students are still not getting the meaning of fractions. This is a result of the learning experience of students who merely accept it, without making the process of getting something. If learning imposed on the operation of multiplication and division of fractions, then the students will experience learning obstacles. To anticipate the learning obstacles, it is important for us to establish the meaning of fractions in the classroom with students as a first step [2].

From the life history class, our teacher explained that students accustomed to dependence on others when in difficulty. They tend to ask the teacher directly without trying to find out the answer to friends who can. As a result, the teacher asks the students could visit a friend who has not been able to provide assistance. It turned out that this dependence as a result of the habits of the students who learn imitative. They often do not get the room to explore their capabilities, but only receive, accept, and accept. That is, there are other things to consider besides the material in the learning process, namely class life.

This paper is not intended to debate whether things above are true or false, but is intended to distribute a collaborative learning experience that we have implemented. Here we will discuss how we jointly collaborate in creating a learning that is rich in meaning, ranging from design, implementation, up to reflection. The goal is to build the care and concern of teachers to students' development as well as life in society as a decision-making basis professionalism [4].

2. Experimental Method

This study applies the method didactical Design Research through three stages of analysis. The first reflection for action made prior to the implementation of learning in which the main focus Repersonalisasi prospective analysis through the process and re-contextualization of teaching materials underlying the design of didactic hypotheses included the anticipation. Second, reflection in action undertaken during the learning process by applying the Metapedadidactic analysis (unity, flexibility, and coherence.Third, reflection of action that is performed after learning through a retrospective analysis comparing the situation with the hypothesis didactic Metapedadidaktik analysis [5].

The main focus of the study were to assess the learning needs of students and to develop a didactic design based learning obstacle and didactic learning trajectory so that the design is expected to improve and develop the learning process towards better and be able to overcome learning obstacles experienced by students. Subjects of this study is a 6th grade elementary school students in one school in Bandung.

3. Result and Discussion

The thought process of a teacher [3] occurred in three phases, namely before learning (reflection for action), when learning takes place (reflection in action), and after learning (reflection on action). All three reflect the thought process of curriculum development framework at the level of practical implementation (enacted curriculum) in the classroom (learning). This will be the basis of discussion of our findings.

3.1. First Phases: Being Student

In this phase, we are exploring a series of questions about 'the content of what material should be taught?' And 'how should the content of the material taught?'. The reason is to guide teachers when choosing the content of the materials and methods relating to the characteristics of a number of students were taught to the period of time and a certain place. As for the question 'why' and 'what' that helped develop a variety of reasons and the purpose behind the teacher's choice. The last question that is important to be explored for more leads and strengthen teachers' belief that certain material content is more important and valuable to students than others [4]

The questions that eventually led us into more specific questions, namely: 1) What is the meaning of fractions for ourselves ?; 2) How do I introduce fractions to students so that they acquire meaning ?; and 3) How is the response of students to learning situations we design? Through discussions deep enough, we interpret the fractional part of the whole.

Once we are able to construct meaning for ourselves fractions, the next challenge is how to introduce fractions to students. The way varied, some use apples, cake, paper, chalk, and so forth. First, we try to apples. Suppose an apple we divided into two parts to introduce ½. After the apples are cut, we are not confident to say that 1 parts is ½ because the concept of fractions said every part of it must be the same. Same here can be just as much, the same length, and another that has a standard size. While the results of the apple pieces can be seen from the side of the weighing also its size. This can lead to ambiguity in the minds of students. If using some other foods like cake and pizza is also the case will be the same as apples. The second idea that emerged was to use chalk. Chalk split into two with broken. Given the nature of chalk that is easily shattered when broken, we think if the pieces are recombined, the result will not necessarily be returned intact into one. Can be seen, nor tubular chalk so that the balance is not the same on the right and left. Another option is to use paper. Paper is more easily divided into several equal parts by means of folded or cut, even can also shaded [1].

Fractional didactic design construct meaning we finally decided to use paper, paper precisely terraced. Design, teachers will encourage students to shading terraced paper representing a value of $\frac{1}{2}, \frac{1}{3},$ and $\frac{1}{4}$. Typically, the value of $\frac{1}{2}, \frac{1}{3},$ and $\frac{1}{4}$ introduced as shown in Figure 1. That's the simplest way of introduction.

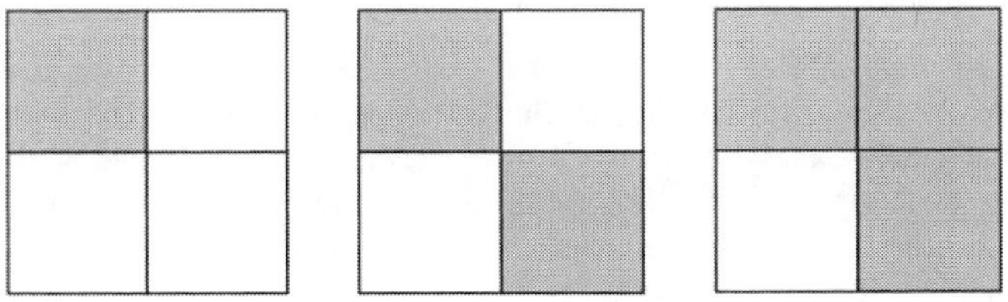

Figure 1. Figure 1 show that the simplest way of introduction fraction, such as $\frac{1}{4}, \frac{2}{4},$ and $\frac{3}{4}$

This time we want to provide more challenge to the students, because they are the 6th grade. The terraced paper that should be shaded students will be provided in a size of 12 x 2 as shown in Figure 2 and must be shaded in five different ways.

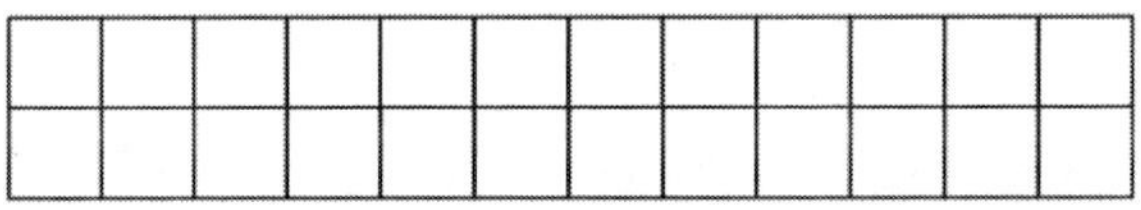

Figure 2. Figure 2 show that the terraced paper that should be shaded students for representing a value of $\frac{1}{2}, \frac{1}{3},$ and $\frac{1}{4}$ in five different ways

Interest ask students to shading with five different ways in which the first is to lure cooperation. We plan to make students sit in groups, each consisting of four students. Each group will be given 20 sheets of paper's puzzle (each 5 to work on $\frac{1}{2}, \frac{1}{3},$ and $\frac{1}{4}$, and 5 for backup). Secondly, we want to spark the creative power of students, because they should not be shading in the same way. Thirdly, we hope will appear among children in the class collaboration in completing the challenge. So the media is not just facilitate the students in solving the problems, but also to develop the potential and social life. More away, the nature of child dependency can be reduced.

To find out the possibilities of the ideas of the students, we try to direct shading in the terraced paper along with several other fellow teachers. An important point of this activity is that we can feel what the student. If before we think of the student's perspective, so this time we move as a student. The benefit is the teachers will be more confident with the election-election with still no grace period to repair if there is something lacking. Momentum like this can make the design more and more mature.

3.2. Second Phases: Learn from Students

The learning objectives of this day is that students can make sense of fractions, since their foundation in learning is next. There are some findings that we gained in the classroom when the implementation takes place. First, students confused when asked shading ½ in five ways. This is in accordance with our expectations. The confusion is a result of everyday students who are accustomed to know one answer. They are also not used to try to directly before the given example. Anticipation that teachers do when all students give a hint of confusion is simple, does not give an example.

The second discovery occurred in the group Marni and Siti. Siti trouble shading $\frac{1}{3}$. Siti invited the teacher asked the group Marni. Siti has not been accustomed to ask, so he accompanied the teacher. When finally Siti asked, Marni reluctant to answer. Chances are there are two, Marni doubt about the answer or is still not able to share the results of his efforts for finding a way shading 1/3 was quite difficult. But with the guidance of teachers, Marni finally answered questions Siti taste.

Findings third occurred in the group Eneng and Lilis. Eneng groups who have difficulty shading ¼ invited the teacher asked the group managed to get guidance Lilis good problem solving. Lilis groups answer questions not to show the answer, but the guiding Eneng by systematically asking questions, such as: 1) How many box ?; 2) If you want shading $\frac{1}{4}$, how the results of 24 divided by 4 ?; 3) So how much should you hatched? Seeing the incident, Marni learn from Lilis humble, so that finally they changed for the better. Another difficulty the student group was eventually by itself going to ask her. Teachers successfully played rhythm of new life in the classroom. Gradually, the atmosphere of collaboration between students in the class formed gently.

Eventually it was revealed that the learning objectives that the teacher wants the students can actually collaborate in the classroom, ranging from attempted when in difficulty. He was deliberately invites students who struggle to ask the students who are able to be in the habit of changing for the better class. Usually, the teacher could ask students to tell the students who are struggling. This time, he wants to play behind them. The formation of this new culture cannot be separated from a challenge. Therefore, when the teacher asked us to ask, he was setting up the guidelines, namely: 1) Try to ask your difficulties before; 2) Try to help her friend not trouble again; 3) What did you tell your friends? Do you understand? If yes, please tell a friend in your group.

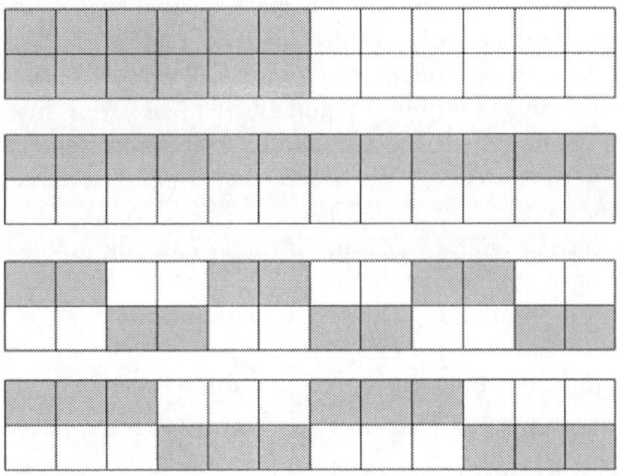

Figure 3. Figure 3 show that are some answers to students who appear in representing a value of $\frac{1}{2}$.

3.3. Third Phases: Learning from Students, Learning from Friends

When reflection, students become our primary focus. We each shared our findings on the activities of students in the class, which we call the learning of students. One interesting finding we are turned out Eneng a very quiet student. Everyday she only interact with her close friends. But that day we both saw her successfully demonstrated the courage to ask the group will be invited Lilis then distributed to friends of the results of their study group. This is a remarkable achievement Eneng.

Findings like that has always been a great conversation when reflections. We all know that not all students are monitored at the same time by the teacher. There is just something that will be missed by us as educators. This is where the function of reflection, sharing and fill in the blanks. As well as a finding that we learned from one observer about the way of thinking a boy named Gio. For shading

$\frac{1}{2}$, Gio took two boxes and shading in one box, took another two boxes and shading 1 box, and so on (Figure 4). The same way Gio did for shading $\frac{1}{3}$, and $\frac{1}{4}$. But unfortunately, the way he was not understood by his friends and did not previously anticipated by the teacher. In fact, the very way shows the meaning of fractions.

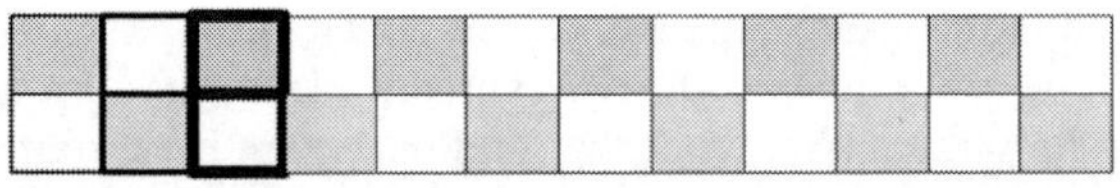

Figure 4. Figure 4 show that how Gio thinking in shading terraced paper for representing a value of $\frac{1}{2}$

How Gio think not anticipated earlier by us. In class, the teacher saw Gio shading it that way and say that it was wrong. Being aware that there are innocent teacher, the teacher admitted that he was wrong and apologize immediately to Gio. It also becomes important lessons for us that it is not permissible for us as educators justify to students. Today we find students who are considered 'less intelligent' apparently showing the mindset is unique, so unique that be had not considered before. It may be that during this time he was considered 'less intelligent' does not mean it can not be anything, but he has not facilitated their potential. Guilt teacher today is proof that the learning of the students is equal to self-correct without having patronized.

4. Conclusion

The process of learning mathematics is nothing perfect. In it there is always the challenge: how to design meaningful learning of mathematics so that students understand it as a whole. From the series we have done, with the main focus of students, we found a few keys: 1) to position themselves into the student when the design process; 2) learning of students; 3) learn from my students and colleagues. If such activity is carried out continuously, then gradually shift the way we can all think of the imitative to more independent way of thinking. We hope that what we have done is a form of real manifestations to education.

The series of activities carried on not to assert its right or wrong in what he has done. But how can we put ourselves to learn from students and learning to hear the opinion of friends. Because in fact, learn from children make us more mature. The proof, learning that day has made students and teachers each become more mature. So here too we interpret that in a collaboration no caste, just as humans are created by God, not caste, just different course. And the differences that make us increasingly rich, not making us disunited. Its the same thing as $\frac{1}{2}$to $\frac{2}{4}$, $\frac{3}{6}$, $\frac{4}{8}$, and $\frac{12}{24}$have a numerator and denominator are different but have the same value that is $\frac{1}{2}$.

What we write here only a didactic situation of the whole learning process. In fact, in the learning process we have 3 pieces of this activity and the first activity. More than that, we're here together trying to be a skilled educator learning, not teaching experts.

5. References

[1] Indikka K 2016 Pecahan *Think Math: Rancangan Pembelajaran Matematika* ed D. Suryadi,*et al*(Bandung: Rizqi Press)
[2] Kurniawan H 2016 Pecahan *Think Math: Rancangan Pembelajaran Matematika* ed D. Suryadi,*et al* (Bandung: Rizqi Press)
[3] Suratno T dan Suryadi D 2013 *Proc. Peran Lesson Study dalam Menyongsong Implementasi Kurikulum 2013* ed M Anggaryani and M S Prastiwi (Surabaya: FPMIPA Universitas Negeri Surabaya)
[4] Suratno T 2016 Didaktik dan didactical design research. *Monograf Didactical Design Research* ed D. Suryadi,*et al*(Bandung: Rizqi Press)

[5] Suryadi D (2013) *Far East Journal of Mathematical Education* **10** 91-107.

Acknowledgments
We thank teachers at SDN Cibaduyut 3 who have already worked to build a new culture for the nation's children for more brilliant life. Also, we thank students who actively participated in this study.

Professionality of Junior High School (SMP) Science Teacher in Preparing Instructional Design of Earth and Space Sciences (IPBA)

L Marlina[1*], Liliasari[2], B Tjasyono[3] and S Hendayana[2]

[1]Lecturer of Physics Education, Sriwijaya of University, Jl. Raya Palembang-Prabumulih KM 32 Ogan Ilir 30662, Indonesia.
[2]Lecturer at Departement of Science Education, Postgraduate School, Universitas Pendidikan Indonesia. Jl.Setiabudhi 229,Bandung 40154, Indonesia.
[3]Lecturer at Departement of Geoscience and Technology Faculty, Bandung of Institute Technology. Jl.Ganesa 10, Bandung 40154, Indonesia.

*Corresponding author's email : leni_fisika@yahoo.co.id

Abstract. The teacher is one important factor in the provision of education in schools. Therefore, improving the quality of education means we need to enhance the quality and the professionalism of teachers. We offer a solution through education and training of junior high school science teachers in developing the instructional design of Earth and Space Sciences (IPBA). IPBA is part of the science subjects which is given to students from elementary school to college. This research is a preliminary study of junior high school science teacher professionalism in creating instructional design IPBA. *Mixed method design* is used to design the research. Preliminary studies conducted on junior high school science teacher in one MGMPs in South Sumatera, and the respondent are 18 teachers from 13 schools. The educational background of science teachers who teach IPBA not only from physical education but also biology and agriculture. The result of preliminary study showed that the ratio of teachers who teach IPBA are 56% from physic education, 39% from biology, and 5% from agriculture. The subjects of IPBA that considered difficult by teachers are the distribution of sun, moon, and satellite motion; specific processes in lithosphere and atmosphere; and the correlation between lithosphere and atmosphere with the environment. The teachers also face difficulty in preparing media, choosing the right methods in teaching IPBA.

1. Introduction

Teacher, as a professional, have a function, role, and position which is very important in achieving the vision of education in 2025, to create an intelligent and competitive Indonesian people. In UU No. 20 of 2003 about National Education System article 39 (2) states that teachers are professionals. As professionals, teachers should have an academic qualifications and competency as a learning agent (Depdiknas, 2005). Competency as learning agents include four aspects: pedagogical competence, personal competence, professional competence and social competence.

Professional development that prepares teachers to design learning experiences for students has not been widely studied. The current study appears in the program that prepares teachers to play an important role in designing a series of lectures for students using available materials and to use the lessons they developed (Penuel, *et al.*, 2011). Professional teachers are those who have the expertise, skills, or skills that meet certain quality standards or norms and require professional education. They must have academic qualifications and competence (Depdiknas, 2005).

Several studies have been conducted in a teacher professional development program. Trautman & Makinster (2010), stating that the model of adaptive and flexible professional development can improve the perception of teachers in teaching using geospatial technology to improve students' critical thinking skills. Professional development to assist teachers in teaching instructional design leading to enhanced science learning (Penuel, *et al.*, 2011).

Science including Earth and Space Sciences (IPBA), developed in order to understand natural phenomena (National Research Council, 1996; Brotosiswoyo, 2000). According to the National Science Education Standards (National Research Council [NRC], 1996), IPBA is one of the primary content on science education that should be given in each grade, level or stage. Unfortunately, the teacher's ability to transfer the knowledge to the students is considered low. Poor ability science teacher / IPBA can not be separated from the teacher preparation process itself by the Institute of Education Personnel (LPTK) (Pujani *et al.*, 2011). As stated by McDermot (1990), one of the important factors that affect the poor performance is the lack of science education teachers are well prepared.

The efforts to improve the quality of teachers start from early education, continuing education, and education and training of teachers in the office (*in-service training education*). According to Musset (2010) in Sulaeman (2016), stated that the ongoing training is a means to develop skills needed for teachers so that students can achieve higher learning outcomes. Meanwhile, according to Bayrakci (2009) in Sulaeman (2016), until now, in-service teacher training is accepted as an effective method for improving the knowledge, skills, and confidence of teachers. Training for science teachers has been seing as a set of academic activity that attended by teachers actively, fully creative, and teaching practices how to face students (Brito, et al., 2005; Trova and Trna, 2014 in Sulaeman, 2016).

Interviews with the head of MGMP in Ogan Ilir, South Sumatera, show that the science teacher who teach IPBA has different educational backgrounds, not only physics, but also chemistry, biology, and agriculture. This leads to mismatch both in terms of qualified teachers and in mastery of the material. The results of further observations of the material scope IPBA in middle and high school curriculum, it is found that there is shifting in curriculum about IPBA contents. It is indicated from the decreasing in material scope of IPBA in physics curriculum of the competency-based curriculum (KBK) in 2004 to KTSP 2006, as well as on Curriculum 2013.

Based on the review, it is necessary to develop education and training programs to improve the professionalism of junior high school science teacher in one MGMP South Sumatera province in the preparation of instructional design of Earth and Space Sciences (IPBA).

2. Research Methods

Mixed method design is used to do the research with a sample of 18 junior high school science teachers from 13 schools. The data is collected through questionnaires.

3. Results and Discussion

3.1 Background Science Teacher Education

The initial study, shows that educational degree of science teachers in Ogan Ilir mostly bachelor degree. It can be seen that their educational background is 56% physics education, 30% biology education, and 5% agriculture (shown in Figure 1). The education background has high influence in mastery of some specific subject of Earth and Space Sciences (IPBA). Furthermore, teacher is demanded to be more innovative in delivering the material to the students. Thus, teaching and learning in the classroom well conducted.

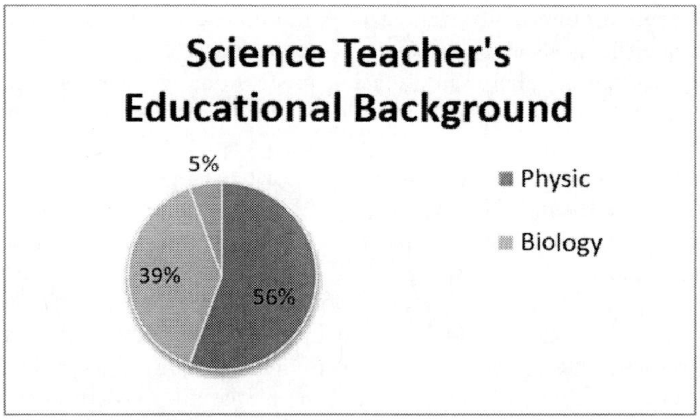

Figure 1. Science Teacher's Educational Background

3.2 Science Teacher Teaching Experience

The teachers' teaching experience in Ogan Ilir can be classified, as follows:

1. 0% has 0-5 years of experience,
2. 67% has 6-10 years of experience,
3. 16% has 11-15 years of experience, and
4. 17% has 16-20 years of experience

Based on the data, their experience is still relatively young. An experience teacher is different from the inexperienced one. Experience for a science teacher is important to determine the success of the learning process.

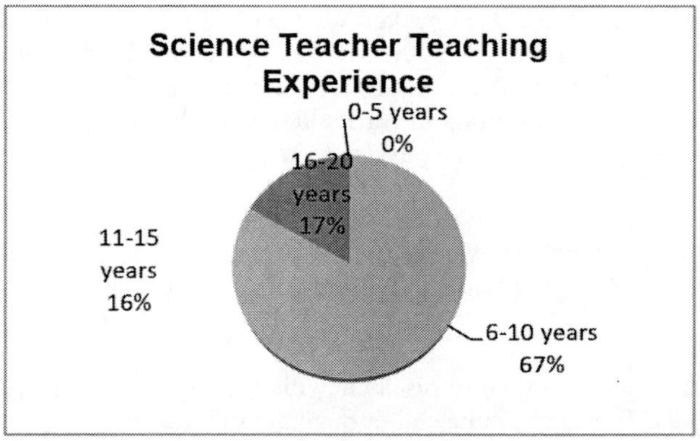

Figure 2. Science Teacher's Experience in Teaching

3.3 Difficulties in understanding the material of Earth and Space Science (IPBA).

In the term of understading the concept, the data shows that:

1. On concept of the solar system, 21% of teachers said it is easy, 72% is moderate, 7% is difficult, and 0% is very difficult.
2. On the concept of the sun as a star and the earth as a planet, 13% of teachers said easy, 74% is moderate, 13% is difficult, and 0% is very difficult.
3. On the concept of distribution of motion of the sun, moon, and satellite; 0% easy, 71% moderate, 29% is difficult, and 0% is very difficult.
4. On the concept of special processes in the lithosphere and atmosphere, 0% easy, 57% moderate, 43% is difficult, and 0% is very difficult.
5. On the concept of relationship processes in the lithosphere and atmosphere with the environment,

0% easy, 79% moderate, 21% is difficult, and 0% is very difficult.

In general, it can be concluded that there are still some science teachers in Ogan Ilir have difficulty in understanding the concept of IPBA.

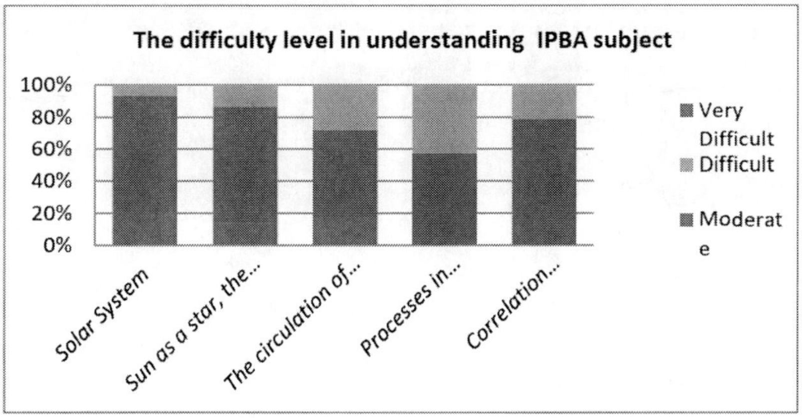

Figure 3. The difficulty level in understanding IPBA subject.

3.4 Difficulties in Creating the Instructional Design of IPBA

The questionnare shows that science teachers have difficulty in developing instructional design IPBA, for instance in making lesson plans, providing teaching materials, choosing a rigth media, methods, and assessment about motion of the sun, the earth, and satellite; specific processes in the lithosphere and atmosphere; and the correlation process in the lithosphere and atmosphere with the environment.

Figure 4 (a) shows that teachers have moderate difficulty in making the lesson plan, while in choosing media, method and making assessment is more difficult. Further, they face difficulty in subject about process in the lithosphere and atmosphere.

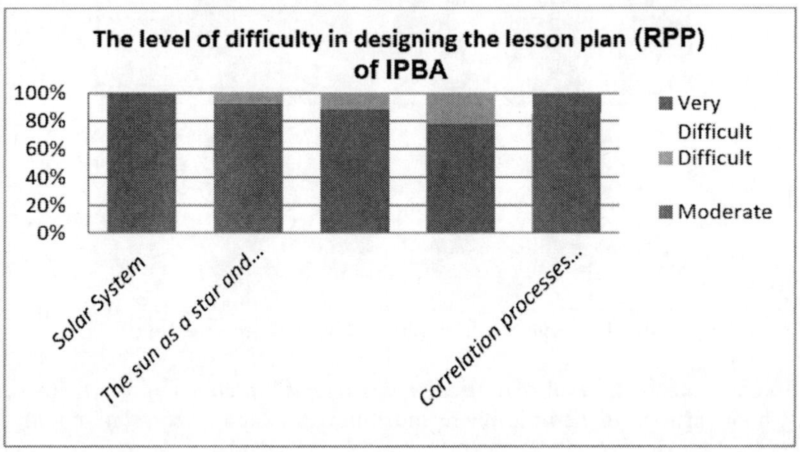

Figure 4a. The level of difficulty in designing the lesson plan (RPP) of IPBA

Figure 4 (b) describe that the level of difficulty is moderate in providing the teaching material, but they face more difficulty about process in the lithosphere and atmosphere.

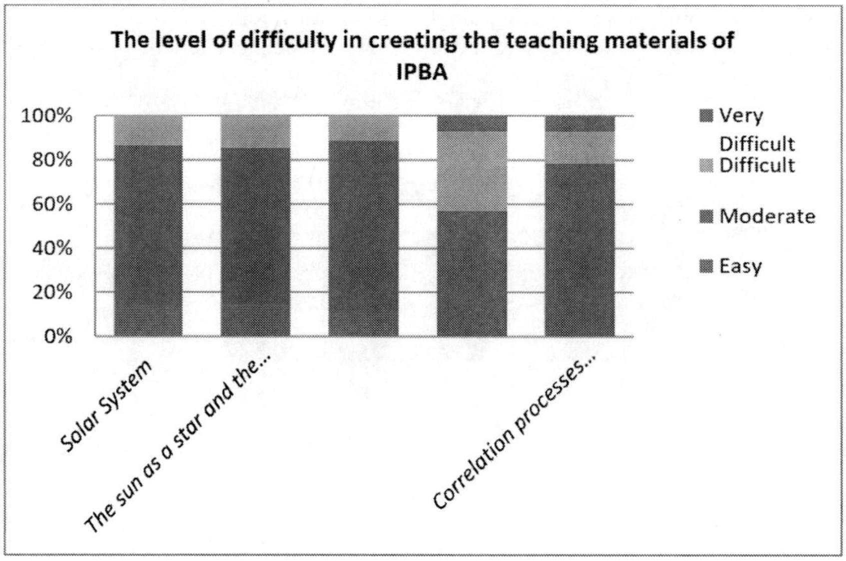

Figure 4b. The level of difficulty in creating the teaching materials of IPBA

Figure 4 (c) describe that the level of difficulty is moderate in choosing the media, but they face more difficulty about process in the lithosphere and atmosphere.

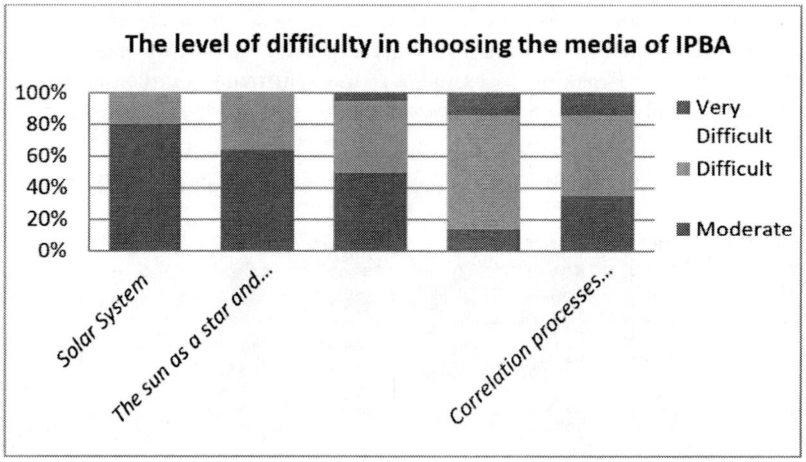

Figure 4c. The level of difficulty in choosing the media of IPBA

Figure 4 (d) shows that the level of difficulty is moderate in choosing the method, but they face more difficulty about process in the lithosphere and atmosphere and its correlation with environment.

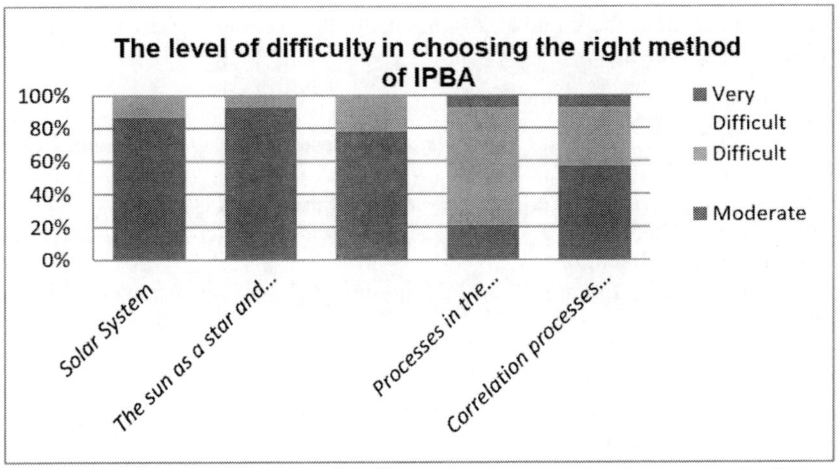

Figure 4d. The level of difficulty in choosing the right method of IPBA

Figure 4 (e) shows that the level of difficulty is moderate in designing the assesment, but they face more difficulty about process in the lithosphere and atmosphere and its correlation with environment.

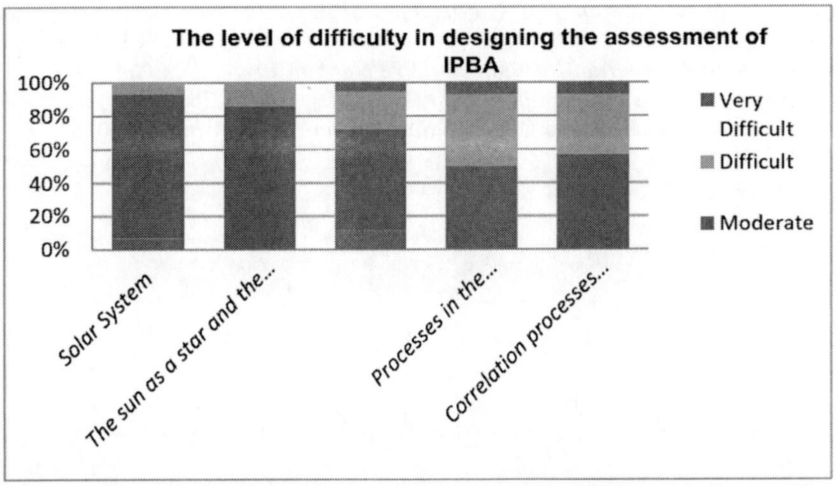

Figure 4e. The level of difficulty in designing Instructional IPBA: (a). Lesson Plan; (b). Teaching materials; (c). Media; (d). Method; and (e). Assessment.

The data can be concluded that they still facing difficulty, primarily in the last 3 subjects: the circulation of the sun, earth, and satelite; processes in litosphere and atmospher; and correlation process betweeen litosphere and atmosphere with environment. The results show their opinion, not the reality in the schools.

4. Conclusion

Teacher is one of important factors in the provision of education in schools. Therefore, improving the quality of education means we need to enhance the quality of teachers. Through education and training about preparation of instructional design IPBA we can improve the quality. Preliminary study through a questionnaire given to 18 science teachers from 13 schools, the educational background of the teachers are varies. While, their experience in teaching is considered poor. Furthermore, they have difficulty in creating the instructional design of IPBA, mainly about motion of the sun, earth, and satellite; processes in the lithosphere and atmosphere; and the correlation between the lithosphere and atmosphere with the environment.

5. References

[1] Brotosiswoyo B S 2000 *Hakekat Pembelajaran Fisika di Perguruan Tinggi dalam Hakekat Pembelajaran MIPA dan Kiat Pembelajaran Fisika di Perguruan Tinggi* (Jakarta : Proyek Pengembangan Universitas Terbuka Departemen Pendidikan Nasional)

[2] Creswell J W and Clark V L P 2007 *Designing and Conducting Mixed Method Research.* California : Sage Publication)

[3] Depdiknas *2003Undang-undang RI No. 20 tahun 2003 tentang Sistem Pendidikan Nasional.* (Jakarta : Depdiknas)

[4] Depdiknas 2005 *Undang-Undang RI Nomor 14 Tahun 2005 tentang Guru dan Dosen*. Jakarta: Fokus Media

[5] McDermot, L C 1990 *American Journal Physics* **58** (8) pp 734-42

[6] National Research Council (NRC) 1996 *National Science Education Standards* (Washington DC : National Press)

[7] Penuel W R, Gallagher P L and Moorthy S 2011*Am Educ Res J.* **96** pp.996-1025

[8] Pujani M N 2011 *Pembekalan Keterampilan Laboratorium IPBA Berbasis Kemampuan Generik Sains Bagi Calon Guru*. Disertasi. SPs Upi. Tidak diterbitkan.

[9] Pujani M N, Liliasari and Herdiwijaya D 2011. *Prosiding Seminar Nasional Penelitian, Pendidikan dan Penerapan MIPA, Fakultas MIPA, Universitas Negeri Yogyakarta, 14 Mei 2011*

[10] Sulaeman A A 2016 *Program Pendidikan dan Pelatihan Kreativitas Guru Biologi SMA dalam Pembelajaran Berbasis Komoditas Hayati Unggulan Lokal*. SPs UPI. Tidak diterbitkan.

[11] Trautmann N M and MaKinster J G 2010 *J Sci Teacher Educ.* **21** pp 351-70

Development of *Two-Tier* Diagnostic Test Pictorial-Based for Identifying High School Students Misconceptions on the Mole Concept

W Siswaningsih, H Firman, Zackiyah and A Khoirunnisa

Department of Chemical Education, Universitas Pendidikan Indonesia, Jl. Dr. Setiabudi No. 229, Bandung 40154, Indonesia
E mail: wiwi2450@gmail.com

Abstract. The aim of this study was to develop the two-tier pictorial-based diagnostic test for identifying student misconceptions on mole concept. The method of this study is used development and validation. The development of the test Obtained through four phases, development of any items, validation, determination key, and application test. Test was developed in the form of pictorial consisting of two tier, the first tier Consist of four possible answers and the second tier Consist of four possible reasons. Based on the results of content validity of 20 items using the CVR (Content Validity Ratio), a number of 18 items declared valid. Based on the results of the reliability test using SPSS, Obtained 17 items with Cronbach's Alpha value of 0703, the which means that items have accepted. A total of 10 items was conducted to 35 students of senior high school students who have studied the mole concept on one of the high schools in Cimahi. Based on the results of the application test, student misconceptions were identified in each label concept in mole concept with the percentage of misconceptions on the label concept of mole (60.15%), Avogadro's number (34.28%), relative atomic mass (62 , 84%), relative molecule mass (77.08%), molar mass (68.53%), molar volume of gas (57.11%), molarity (71.32%), chemical equation (82.77%), limiting reactants (91.40%), and molecular formula (77.13%).

1. Preliminary

Chemistry became subjects difficult to understand for some students because they contain abstract concepts such as moles, molecules and particles [1]. In fact, there are some subject of chemistry that can be seen, for example, a solution, or demonstrated, for example, the combustion reaction, however, there are some subject of chemistry that can not easily be demonstrated so that students can not see directly, such as atoms and electrons [2]. According to Johnstone, chemistry consists of three levels of representation [3]. Three levels of representation are: (1) the representation of the macroscopic ie representations that describe the bulk properties of the phenomenon and is seen in the daily experience of students, (2) the representation of submicroscopic ie representations that provide explanations on the particulate levels are described as atoms, molecules, or ions , (3) the symbolic representation is the representation that involves the use of chemical symbols, formulas, equations, diagrams, models, and animations [4].

The ability to master the three-level representation makes students have difficulties in understanding chemical concepts. According Nakhlekh explained that one reason for the difficulties that students link the properties of matter at the level of representation used to illustrate and explain the chemical phenomena [5]. Many basic concepts of chemistry involving submikroskopis and symbolic representations are quite difficult to study by students [6].

The results of the study Steiner to the chemistry teacher stated that the concept of the mole is the toughest material as the initial concept in studying chemistry [7]. In this matter, the students do not have the ability to transfer meaning between the macroscopic level with submikroskopis [8]. This relates to the concept of matter mole pretty abstract so it took a visualization in describing the phenomenon. Mole concept important to learn because as a basis for further study material. The concept of the mole is also important for other chemical concepts more complex such as stoichiometry [9]. Research study on the material concept of the mole has been done by Furio *et al* [8], the results found that students have difficulty in mastering the concept of the mole.

Students tend to build their own concepts in the brain that lead to concepts or visual images that they have become one, so that it can interfere with the ability of students to understand the true concept [10]. The inability of the students to understand the concept is right on the concept of the mole may result in conception is not intact.Incomplete conception can lead to misconceptions. The misconception is defined as the conceptions of different students with the conception of the experts [11]. One cause of the misconceptions that students often do not see the relationship between the three levels of the chemical representation [12].

The diagnostic test multiple-choice *two-tier* can be used in particular to identify misconceptions experienced by students on specific content [4]. Excess use of multiple choice tests *two-tier* compared to interview or a map of concepts such as planning and assessment test is easy to do [13]. A diagnostic test multiple-choice *two-tier* depicted as a diagnostic instrument with *the* first *tier* on each item consists of questions with answer choices of two to four options. *tier* two of each item containing reasons which refers to the first *tier* [14]

According to Edens and Potter, in presenting scientific concepts visually through images, incorporating text is elaborative process that includes three levels of representation [15]. Context display can become more effective tools and an important tool to be applied to activate the relevant scheme which required students to understand the text [16]. Therefore, the test involved the pictorial form. According Tavassoli *et al* , the pictorial comes from the word *"picture"* which can be defined as everything that is poured in the form of images [16]. Pictorial representations submikroskopis emphasis on problem-solving chemistry [17].

Research on the misconception diagnostic test on the material concept of the mole has been done by Grace (2013, p. 41), but still a test narrative. Questions can be tiring long narrative that students in understanding the crux of the matter in question [18]. According to Hanson, misconceptions students may be examined through pictorial representations [19]. The study focused on mendiagnostik misconceptions with good reliability and validity have an important role in science education research [20].

Based on the exposure, researchers conducted a study with the title "Development of *Two-Tier* Diagnostic Tests to Identify Miskonsepi pictorial Based High School Students at Material Concept Mol". *Two-tier* diagnostic tests based on the material pictorial concept of the mole hereinafter referred TDMKM.

The purpose of this study is to determine the structure of a test developed in diagnostic tests *two-tier-based* pictorial on the material concept of the mole, produce diagnostic tests*two-tier-based* pictorial in the material concept of the mole who meet the criteria of content validity and reliability, and identify misconceptions students on the material concept mol through *two-tier* diagnostic test developed pictorial-based.

2. Method

The method used in this study is based on method development and validation conducted by Adams and Wieman [21]Development of an activity increases one thing to perform certain steps. Validation is the process of an investigation carried out in describing the items, so that each item can measure what is to be measured [22]. Validation is an important part of evaluation that aims to assess a feasible thing or not according to the criteria tertentu.Penelitian TDMKM development carried out through four stages, namely (1) the development stage items, (2) the validation phase, (3) the stage of determining determination key, and (4) the test phase of development aplikasi.Tahap items carried by adapting and modifying the development stages of *the two-tier* diagnostic test performed by Chandrasegaran *et al* [4] with reference to the proposed Treagust stages [14]. These stages, namely, (1) the determination of the purpose and scope of the test, (2) the prediction data collecting student misconceptions, and (3) development TDMKM items.

In defining the purpose and scope of tests done through a literature review about the material concept of the mole which further developed into the concept map material concept of the mole. In a concept map consists of a number of labeling concepts used to develop essays. Essay aims to collect data prediction misconceptions experienced by students. Essays given to high school students who have studied the concept of the mole and the material is then compiled in the form of a matrix table misconceptions. Items are developed in the form of *a two-tier* test with each consisting of four possible answers. In the first *tier,* involved *stem* pictorial form. Problem on the first *tier* is based on a literature review of research journal teksdan book. The second *tier* items developed four possible answers with a matrix based on misconceptions obtained from essay. Distractor on the matter which was developed based on identified students 'conceptions of the students' answers on the test essay [4].

The validation phase was conducted in order to determine the quality TDMKM instruments are developed in accordance with certain criteria. Phase validation is performed to test the validity of the content. The validity of the content is based on *judgment* or consideration by experts in the field kimia.Hasil *judgment* of the validators on each item on developed, processed using techniques CVR *(Content Vallidity Ratio)* is expressed by Lawshe [23] by using the formula (1):

$$\text{CVR} = \frac{(n-N/2)}{N/2} \qquad (1)$$

Information:
CVR = ratio of content validity
= Number of validators declared "valid"
= Number of validator

CVR acquisition value is then compared with the minimum value of the CVR (Lawshe, 1975, p. 568). Items considered valid in terms of content validity if the value of the CVR is equal to or greater than 0.99 (CVR ≥ 0.99), whereas if the items have a CVR value less than 0.99 (CVR <0.99), the items were declared invalid.

Items that have been considered and declared valid by the validator, corrected based on the suggestions given validator, and then tested on a number of students who have studied the material concept of the mole. The results of the students' answers acquisition assessed by giving a score of 1 (one) when a student answers correctly on the second *tier,* while a score of 0 (zero) if only one or both *tier* one. This is done to determine reliability. Reliability in this study is used to calculate the Cronbach alpha values. To obtain a Cronbach alpha values, the results of students' answers on each item on processed using SPSS *(Statistical Package for Social Science)* version 20.0. Criteria Cronbach alpha values are presented in Table 1 below [24].

.**Table 1.** Criteria cronbach alpha value

Cronbach Alpha Criteria	Classification
a ≥ 0.9	Very good
0.8 ≤ a <0.9	Good
0.7 ≤ a <0.8	Be accepted
0.6 ≤ a <0.7	Doubtful
0.5 ≤ a <0.6	Bad
a <0.5	Not acceptable

Determination key is determined by the pattern of responses the students' answers to each item on. The test was developed in the form of multiple choice diagnostic tests *two-tier* thus obtained 16 possible answers students' response patterns. The pattern of response to the students' answers are presented in Table 2 below.

Table 2. Response patterns answer students in each item problem TDMKM.

T ier second The first *tier*	(1)	(2)	(3)	(4)
a	a (1)	a (2)	a (3)	a (4)
b	b (1)	b (2)	b (3)	b (4)
c	c (1)	c (2)	c (3)	c (4)
d	d (1)	d (2)	d (3)	d (4)

Questions have been declared valid and reliable applied to the students to be identified miskonsepsinya. The students' answers were analyzed based on the predetermined determination key. Revealed misconceptions can be seen from the students' answers on the first *tier* and second *tier*. Answer students are classified into several categories. Here is presented Table 3 on the classification of students' understanding [25].

Table 3. Classification of understanding students.

Combination Answers First Tier Second Tier	Classification
True - True	understand the concept
True - False	Misconceptions
False – True	misconceptions
False - False (related)	specific misconceptions
False - False (not related)	Do not understand the concept

Each possible answer students are calculated as a percentage by the formula (2):

$$\mathrm{KNP} = \frac{X}{N}\ 100\% \qquad (2)$$

Information:
KNP =% criteria
X = the number of students who answered
N = the total number of students

3. Results and discussion

3.1. Grain development problem

Development of items TDMKM on *the* first *tier* is based on text books and research journals regarding the material concept of the mole. *Stem* in *the* first *tier* in the form of pictorial with consisted of four answer options, while the second *tier* in the form of the reasons which refers to the answer to *the* first *tier* with consisted of four possible answers. The reason the second *tier* is a basic concept in the material concept of the mole. Each *tier* consists of a selection of the correct answer and three distractors. Correct answers obtained from literature review based on the book teksdan research journal. Three distractor on the second *tier* is obtained from answers to students based on predictions of student misconceptions in the material concept of the mole.

3.2. Quality Tests

Items considered valid in terms of content validity if the value of the CVR is equal to or greater than 0.99 (CVR $\geq$ 0.99), whereas if the items have a CVR value less than 0.99 (CVR <0.99), the items were declared invalid. A total of 20 items validated, the result is a total of 18 items declared valid by the CVR value is equal to one. After that, each item on revised based on suggestions validator.

A total of 18 items are declared valid, then tested to 34 high school students of class X that has been studying the concept mol.Berdasarkan material calculations, the value alpha Cronbach of 0.659 which is in the range 0.6 $\leq \alpha$ <0.7. The range is included in the category of "doubtful" that provisions are about to be obtained Cronbach alpha values $\geq$ 0.7 were included in the category of "acceptable". As many as one problem aside, judging by the item total correlation is low, at about fifteen numbers on the concept of limiting reagent, so the Cronbach alpha value becomes 0,703. So the total TDMKM items that have a valid and accepted that as many as 17 items.

Based on the consideration to reduce errors in identification of misconceptions students' answers on the test results of the application, then the election items. Selection of items is based on the representation of each concept in test and item total correlation on each item on, in order to obtain 10 questions with a Cronbach alpha value of 0.707 which remain in the category of "acceptable".

A total of 10 items that have been declared valid and acceptable to the test content validity and reliability then compiled into a determination key based student response patterns.From this response pattern can be seen in the mindset of the students answered each item on that misconception experienced by students can be identified.

3.3. Students misconceptions on limiting reagent material

Items that have been declared good and acceptable validity and reliability were then applied to 35 high school students of class X science in high school who has studied the material concept of the mole. According to Peterson [11], misconceptions considered significant if it is found at least 10% of the total sample of students.

Table 4. Answers students at Grain Boundary problem concepts reagents number nine.

Butir Soal Nomor Sembilan	Pola respon	Jumlah Siswa	Persen-tase (%)
Note the following reaction illustration.	a(1)	9	25,71
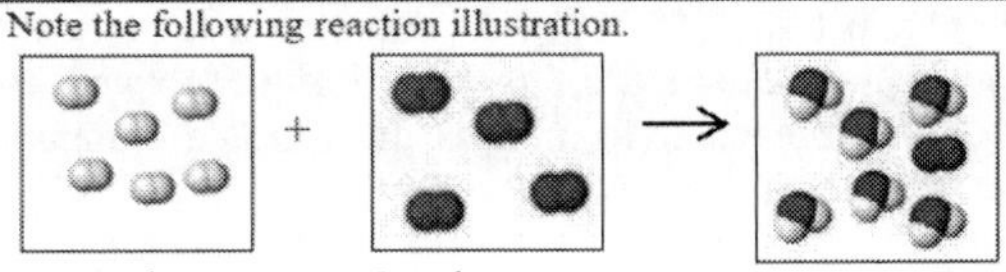	a(4)	2	5,71
3 mol 2 mol	b(1)	5	14,28
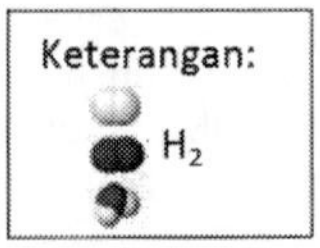	b(3)	12	34,28
	b(4)	5	14,28
	c(1)	1	2,85
Reactant molecule that acts as a barrier is .... a. H_2 b. O_2 c. H_2O d. H_2 and O_2 The reason is .... (1) reagent is the limiting reagent smallest coefficient (2) the limiting reagent is the mass of the smallest reagent (3) the limiting reagent is a reagent that most small amount molnya (4) limiting reactant is completely reacted first reactant	d(4)	1	2,85

A response pattern (1) with a total percentage of 25.71% stated that the molecule that acts as a limiting reagent is H_2 because the limiting reagent is a reagent that most small coefficient. The pattern of response b (1) with a total percentage of 14.28% stated that the molecule that acts as a limiting reagent is O_2 as reactants limiting reagent coefficient is the smallest. The pattern of response b (3) with a total percentage of 34.28% stated that the molecule that acts as a limiting reagent is an O_2 for limiting reagent is the reagent the smallest amount molnya. The pattern of response b (4) with the total percentage of 14.28% stated that the molecule that acts as a limiting reagent is the limiting reagent O_2 as the first reactant is completely reacted. Percentage who have misconceptions response pattern that is equal to 91.40%, to understand the concept as much as 5.71%, while the do not understand the concept as much as 2.85%. From these data, we can see a comparison of three categories of students' understanding shown in Figure 1 below.

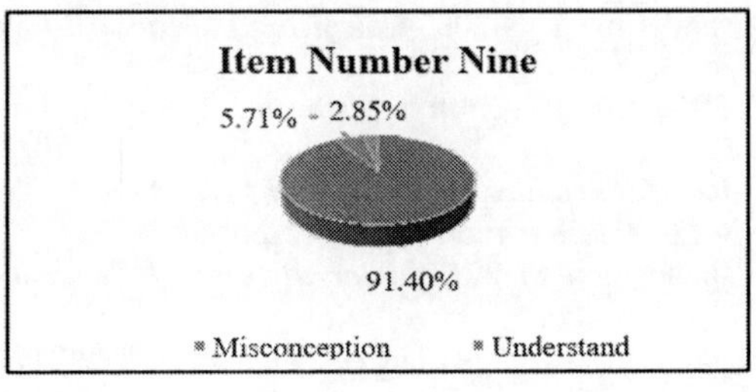

Figure 1. Percentage of students on item response pattern problem number nine concepts limiting reagent.

Based on the test results of the application, the student misconceptions identified in all the label concept. The misconception that many experienced students on each label concept, namely: the amount of 1 mole of oxygen a different structure, namely O_3 and O_2, each consisting of one molecule for 1 mol of O_2 and O_3 each contain a number of substances produced from the calculation results in an element or compound (45.71%); one mole of sulfur molecules have a number of particles 6:02 x 10^{23} for 6.02 x 10^{23} is a number that becomes the benchmark in calculating the amount of the element / compound (25.71%); relative atomic mass is the atomic mass ratio of the mass of atoms of other elements (34.28%); molar mass is mass divided by the number of moles of these substances (45.71%); molar volume of 1 mole of gas (A) was 22.4 L because it has a number of 22.4 liters of gas at a particular mole, occurs in conditions of 25 ° C and 1 atm (45.71%); the equation is the reaction of a compound of formula which is formed of two or more atoms (22.85%); the limiting reagent is the number of moles of reagents smallest (34.28%); the molecular formula is the formula compound / molecular multiple of the empirical formula (34.28%).

4. Conclusion

Two-tier diagnostic tests based on the material pictorial concept of the mole, hereinafter referred TDMKM developed with *stem* on the first *tier* in the form of pictorial with four possible answers and the second *tier* in the form of reason with four possible answers. Results validation TDMKM 20 items declared valid by as much as 18 about six validator. A total of 18 items tested reliability and alpha values obtained Cronbach 0.659 so that no one items that must be set aside. It improves the Cronbach alpha value becomes 0,703. Based on the test content validity and reliability, as many as 17 items have been categorized as good and acceptable. Misconceptions students identified on the labels on the material concept with the concept of the mole concept of the mole percentage of (60.15%), the Avogadro constant (34.28%), relative atomic mass (62.84%), a period of relative molecular (77.08%), molar mass (68.53%), molar volume of gas (57.11%), molarity (71.32%), equation (82.77%), the limiting reagent (91.40%), the molecular formula (77.13%).

5. References

[1] Uce M 2009 *ProQuest Education Journals* **129** 683-91
[2] Taber K S 2009 *Educació Quimica EduQ* **4** 13-20
[3] Johnstone A H 2000 *Chemistry Education Research and Practice in Europe* **1** 9-15
[4] Chandrasegaran A L, Treagust D F and Mocerino M 2007 *Chemistry Education Research and Practice* **8** 293-300
[5] Nakhleh M B and Krajcik J S 1994 *Journal of Research in Science Teaching* **31** 1077-96
[6] Ahiakwo M J 2013 *African Journal of Chemical Education* **3** 29-38
[7] Steiner R P 1986 *Journal of Chemical Education* **62** 1048
[8] Furio C, Azcona R and Guisasola J 2002 *Chemistry Education Research and Practice in Europe* **3** 277-92
[9] Staver J R 1993 *Journal of Reasearvh in Science Teaching* **30** 321-37
[10] Modic A L 2011 *Student Misconception-Identifying and Reforming What They Bring to The Table Chemistry* (Bozeman: Tesis, Master of Science, University of Montana)
[11] Tan K C D, Taber K S, Goh N K and Chia L S 2005 *Chemistry Education Research and Practice* **6** 180-197
[12] Barke H D, Hazari A and Yitbarek S 2009 *Misconception in Chemsitry* (Berlin: Springer)
[13] Tüysüz C 2009 *Academic Journa* **4** 626-31
[14] Treagust D 2006 *Journal of Science and Mathematics Education Centre* **23** 1-9
[15] Edens K M and Potter E F 2001 *Studies in Art Education-Spring* **42** 214-33
[16] Tavassoli A, Jahandar S and Khodabandehlou M 2013 *Indian Journal of Fundamental and Applied Life Science* **3** 553-65
[17] Noh T and Scharmann L C 1998 *Journal of Research in Science Teaching* **34** 199-217
[18] Halakova Z and Proksa M 2007 *Journal of Chemical Education* **84** 172-4
[19] Hanson R 2015 *International Journal of Innovative Research & Development* **4** 115-22
[20] Gurel D K 2015 *Eurasia Journal of Mathematics, Science & Technology Education* **11** 989-1008
[21] Adams W K and Wieman C E 2010 *International Journal of Science Education* **33** 1-24

[22] Haladyana T M and Rodriguez M C 2013 *Developing and validating test items* [Online] (Newyork: Routledge)
[23] Lawshe C H 1975 *Personnel Psychology* **28** 563-73
[24] Gliem J A and Gliem R R 2003 *Midwest Research to Practice Conference in Adult, Continuing, and Community Education (Columbus)* (Indianapolis: ScholarWorks) pp 82-8
[25] Tarakci M, Hatipoglu S, Tekkaya C and Ozden M Y 1999 *Journal of Education* **15** 84-93

Physical Characteristics, Chemical Composition, Organoleptic Test And The Number Of Microbes In The Biscuits With Addition Of Flour Banana Peels

Hernawati*, **A Aryani, R Shintawati**

Biology Education Program, FPMIPA Universitas Pendidikan Indonesia,
Jl.Dr.Setiabudi No. 229, Bandung 40154, Indonesia
Corresponding author: hernawati@upi.edu

Abstract. The purpose of this study to analyze the physical characteristics, chemical composition and organoleptic test of biscuit flour with the addition of flour banana peel. Materials used are banana peels Kepok. Kepok banana peel has been found to contain high fiber food. Biscuit-making stage includes the formation of cream, adding flour and wheat flour dietary fiber from banana peels to concentrations of 0% as control, 25%, 50% and 75% of 100 grams of wheat flour; mixing; molding; baking in the oven for 20-25 minutes with a temperature of 180°C. Parameters to be measured, namely the physical characteristics include: hardness, softness, consistency, crispness. Furthermore, the biscuits were tested by chemical analysis (proximate). Organoleptic test include: aroma, taste, mouthfeel, aftertaste. Data were analyzed statistically using SAS computing programs. Physical and organoleptic test results biscuits with the addition of flour banana peels has sufficient level of preference between like-liked. Based on the results of the proximate analysis of biscuits with the addition of flour banana peels has generally been in accordance with the National Standards of Indonesia (SNI). Conclusion of the study that the addition of flour banana peels in biscuits has the potential to become functional foods that contain high fiber.

1. Introduction

At this time Indonesia is in full swing to build towards a developed and prosperous country. The most important resource for the course of development is human resources. In this case the man not only as objects of development, but as subjects important role in the development process. However, this time there has been a shift or change in lifestyle, which causes mortality and morbidity in the community increased, characterized by changing patterns of infectious diseases become degenerative and metabolic diseases. Therefore, maintaining health is very important. Health is one of the major factors that support each person can carry out the processes of life, such as to grow and develop, and carry out all the activities of life are socially and economically productive.

Various ways can be done in order to obtain optimal health status, one of them with a set menu of food with added fiber foods. Dietary fiber consists of total dietary fiber, which includes soluble dietary fiber and insoluble dietary fiber [1]. Fiber is not soluble in water consisting of cellulose, hemicellulose, and lignin, which can be found in bran, cereal, nuts, vegetables and fruits; whereas is water-soluble pectin, agar-agar, carrageenan, musilase, and gum [2][3]. Dietary fiber derived from plants can be used as functional materials [4] [5], because the fiber can interact physiologically to provide many benefits to support the setting in digestion , hyperlipidemia and hypercholesterolemia disease, can be prevented by improving diet high in dietary fiber.

Dietary fiber can provide physiological effects, such as (i) an increase in gastrointestinal health; (ii) improved glucose tolerance and insulin response; (iii) reduction in the risk of cancer; and (iv) setting and the digestion of fat and body weight [6]. Type, source, and amount of fiber affect bowel function in a different way. In general, the fiber that is resistant to fermentation in the colon such as wheat bran, largely increasing the volume of intestinal contents [8], thus helping to facilitate the stools. One source that can be used as a dietary fiber that is a banana peels.

Indonesia is one of the centers of origin of various types of bananas. It is therefore not surprising that in Indonesia the number of types of bananas are quite abundant. bananas are commonly cultivated today are the descendants of Moses acuminta Colla and Musa balbisiana Colla which has a lot of diversity in Thailand, Malaysia, Indonesia and Papua New Guinea [9]. However, these plants typically grouped into seedless banana or banana cultivation and wild bananas have seeds [10].

Peels bananas is part of the agricultural waste which is generally discarded as trash or used as animal feed. This has an impact on the increase of solid waste dumped into the environment. One effort is utilizing banana peels overcome by means of shaping it into nutritious food such as biscuits and certainly healthy. Results of previous studies show that the flour banana peels is a functional food ingredient containing high fiber [11]. Biscuits with the addition of flour banana peels is a product that needs to be developed and the potential to meet their food needs. Based on the above studies have been conducted to examine the physical characteristics, chemical composition, and organoleptic test in the biscuit with the addition of a banana peels.

2. Experimental Method

Materials used are Kepok banana peels. Kepok banana peels dried in the oven [11]. Once dried banana peels smoothed by using a blender. Phase manufacture of biscuits includes the formation of cream (creaming method), adding flour and dietary fiber flour banana peels with a concentration of 0% (not added flour banana peels as a control), 25%, 50% and 75% of 100 grams of wheat flour, mixing (mixing), printing, roasting in the oven for 30 minutes at a temperature of 150°C.

Parameters to be measured, namely the physical characteristics include: hardness, softness, consistency and crispness. Furthermore, the biscuits were tested by chemical analysis (proximate) which aims to see the magnitude of the nutritional value content, especially the content of dietary fiber of biscuit products are added to flour banana peels. In addition it aims also to see the trend change that occurred with the increasing flour banana peels are added. Organoleptic test include: flacour, color, mouthfeel, aftertaste.

Physical and organoleptic test conducted by test hedonic scale 1-4 (1 = less liked, 2 = quite likes, 3=likes, 4=love). Number of panelists to test the cake flour banana peels as many as 48 people. Parameters that have been observed to be analyzed using a statistical test by using analysis of variance (ANOVA) followed by Duncan's Multiple Range Test test [12]. The overall analysis performed with SAS software with SAS software 9.1.3.

3. Result and Discussion

Physical testing (texture) conducted on biscuits with the addition of flour banana peels of 0%, 25%, 50%, 75% gave results that were not significantly different in terms of hardness, softness, consistency and crispness at the 95% confidence level ($\alpha = 5\%$). Panelists passions levels of the physical condition of biscuits with the addition of flour banana peels were on a scale from 1.98 to 3.13 (less like- like) of four scale used. It shows that the formulation does not affect the general acceptance of the biscuit formulation with the addition of flour banana peels. Biscuit flour formulation with the addition of banana peels can be used because it has the same level of acceptance. Panelists passions levels Against Physical Appearance Flour Biscuits With the addition of Flour Banana Peels can be seen in Table 1.

Texture is an important aspect of the quality of food, sometimes more important than the taste, flavor and color. Texture most important influence food quality soft and crunchy. Characteristic textures are most often referred to is the hardness, cohesiveness and water content. The fiber content on a banana peels does not affect the hardness of the biscuits. That is because prior treatment mashed banana peels first. Biscuits produced tend not hollow (solid) but easily broken.

Table 1. Panelist passion levels against physical appearance flour biscuits with the addition of banana peels

Biscuit Code	Hardness	Softness	Consistency	Crispness
KP0%	2,94	2,91	3,10	3,10
KP25%	2,67	2,35	3,02	2,94
KP50%	2,50	2,71	2,92	2,87
KP75%	1,98	3,02	2,71	2,83

Annotation : Skala 1= less likely; 2= enough like, 3= like, 4= really like

Some materials used in the manufacture of biscuits with the addition of flour banana peels, among others, wheat, eggs, sugar flour, margarine and butter, skim milk. These materials are used to produce biscuits with the taste, texture and consistency were good. The addition of egg yolk in the manufacture of biscuits with the addition of flour banana peels produce biscuits with a soft texture. The use of egg yolks will produce a more tender biscuit than using whole eggs [13]. This is due to lecithin in egg yolk has emulsifying power. The presence of emulsifier makes eggs can improve the texture, increase the volume and increase the protein content. The role of the functional properties of the protein in the egg depends on the type of product to be made. Functional properties of proteins in eggs play a role determining the quality of the final product in the food industry [14].

In the manufacture of biscuit with the addition flour banana peels is used the sugar for generate taste and structure biscuit who good. Sugar in the manufacture biscuit serves to give a sense a sweet, softens and smoothes the texture as well as the make the colors crust biscuit become brown attractive [15]. In addition it, added also milk skim. Milk is functioning hold back absorption of water and for increase the value the nutritional of the product biscuit. The addition of milk skim can provide a sense, flavor, appearance the final product, regulate the density of dough, dissolving and propagate dough [13]. The addition of milk skim generate consistency of biscuit flour banana peels good enough.

The addition of flour banana peels does not affect the level of crispness biscuit produced. The level of crispness biscuit determined by the type of flour used, the higher the protein content of the flour, biscuits produced less crisp. This is because the high protein flour has a high gluten content. Conversely, the use of flour with a low protein content will produce a crunchy biscuit. This is in accordance with the opinion [16] that the flour is good for the manufacture of biscuits are wheat flour which has a low protein.

Biscuits with the addition of a banana peels flour 0%, 25%, 50%, and 75% effect on water, ash, protein, crude fiber, crude fat, carbohydrates, gross energy, and minerals can be seen in Table 2. Proximate test results showed that the biscuits with the addition of flour banana peels has a good nutrient content in accordance with the Indonesian National Standard (SNI 01-2973-1992) [17]. Water content values obtained from the biscuit with the code KP0% (1.96%), KP25% (1.51%), KP50% (1.31%), and KP75% (1.08%). The percentage of water content of biscuits with the addition of a banana peels flour is not more than 5% so in accordance with the Indonesian National Standard (01-2973-1992) biscuit is a maximum of 5%. Biscuit with a water content below 5% is expected to maintain the shelf life of biscuits, because the water content of less than 5% will be free of damage and of harmful microbes. Microbes will easily grow on media containing moisture or high water. The water content in food will determine acceptance, freshness and durability of the food. In the roasting process biscuit, a process of heating and the reduction of water content. The water content in the biscuits will affect consumer acceptance, especially on the attributes of texture (crispness). Biscuits with high water content tends not crisp so that the texture is less desirable and have short storage durability.

Table 2. Proximate result of biscuits were added flour banana peels

Parameters	SNI	KP0%	KP25%	KP50%	KP75%
Water (%)	Maks 5	1,96	1,51	1,31	1,08
Ash (%)	Maks 1,6	1,11	1,55	2,39	3,06
Protein (%)	Min 9	8,33	8,05	9,00	9,21
Crude Fiber (%)	Maks 0,5	2,83	1,92	2,47	3,13
Crude Fat (%)	Min 9,5	14,85	20,70	17,83	17,73
Carbohydrates (%)	Min 70	72,88	66,57	68,31	68,08
Gross Energy (Kkal/kg)	4000	4196	4596	4342	4451
Calsium (Ca) (%)		0,13	0,24	0,31	0,29
Phosfor (P) (%)		0,02	0,049	0,19	0,16

Annotation: KP0% = biscuit were not added flour banana peels; KP25%= biscuits were added flour banana peels amount 25%; KP50%= biscuit were added flour banana peels amount 50% KP75%= biscuit were added flour banana peels amount 75%

Results obtained proximate to the ash content biscuit code KP0% (1.11%), KP 25% (1.55%), KP50% (2.39%), and KP75% (3.06%). According to the Indonesian National Standard (01-2973-1992) ash content of the biscuit is a maximum of 1.5%. Biscuits with the addition of flour banana peels as much as 25%, 50%, 75% have ash content exceeds provision made SNI (1992). It can be explained that the mineral content found in flour banana peels is high enough so that the test results also high ash content. High levels of ash in biscuits allegedly because of biological minerals in a banana peels. The ash content is mineral elements as the residue left after the material was burnt to free the element carbon. The ash content can also be interpreted as a non-volatile components, remained in combustion and annealed organic compounds. The ash content of a food indicates residual inorganic material that remains after the organic material in the food broken. The ash content is not always equivalent to mineral material, because there are some minerals lost during volatilization or interactions between constituents.

The protein content flour biscuits with the addition of a banana peels is in conformity with the provisions of the Indonesian National Standard (01-2973-1992) at least 9%. It can be seen in 50% and 75%, namely biscuits protein content flour biscuits with the addition of a banana peels as much as 50% at 9.00% and the addition of flour banana peels as much as 75% of 9.21. This is due to banana peels flour contains low protein. The addition of protein from eggs, margarine and skim milk does not significantly raise the level of proteins. The results showed that with the addition of a banana peels flour biscuits protein levels can meet the standard and suitable for consumption.

Fiber content of flour biscuits with the addition of a banana peels is higher than the Indonesian National Standard maximum of 0.5%. Fiber content biscuits with the addition of flour banana peels 25%, 50%, 75% respectively was 1.92%, 2.47% and 3.13%. Along with the addition of flour biscuits banana peels made a little hard, rough and crumbly consistency. This is presumably because the influence of the fiber content of flour banana peels. Biscuit made seemed a little harsh, rough, crumbly consistency, and less crisp. This is presumably because the influence of the fiber content of flour banana peels is high. Kepok the banana peels in the oven containing 14.04% crude fiber and dry drying of 16.14% [11]. The high fiber content of flour biscuits with the addition of the potential banana peels to become functional foods. This biscuit is expected to be a food that can lower blood cholesterol levels in patients with hypercholesterolemia or can maintain the balance of lipid levels and blood sugar levels.

According to the Indonesian National Standard (01-2973-1992) fat content of the biscuit is a minimum of 9.5%. The fat content of biscuits with the addition of flour banana peels 25%, 50%, and 75% higher than the required standards. Height biscuits can be caused by the materials used such as butter, margarine, milk powder is quite high in fat. In addition, such materials, high levels of fat biscuits with the addition of flour banana peels because the kepok banana due also used to have a fairly high fat content 14.63 (oven dry) and 15.29 (dry drying) [18]. The high fiber content of flour biscuits with the addition of the potential banana peels to become functional foods. This biscuit is

IOP Conf. Series: Journal of Physics: Conf. Series **812** (2017) 012118 doi:10.1088/1742-6596/812/1/012118

expected to be a food that can lower blood cholesterol levels in patients with hypercholesterolemia or can maintain the balance of lipid levels and blood sugar levels.

Many carbohydrates contained in foodstuffs in the form of starch, sugar and crude fiber. According to the Indonesian National Standard (01-2973-1992) carbohydrate content in biscuits is at least 70%. Based on the amount of carbohydrates, biscuits with the addition of a banana peels starch still under SNI, this was due to the fat content of biscuits with the addition of a banana peels flour is quite high. The higher the fat content of biscuits contain a lower carbohydrate content. The low carbohydrate content allegedly due to reduction of wheat flour is replaced with flour banana peels low in protein and carbohydrates.

According to the Indonesian National Standard (01-2973-1992) biscuits calorie value is at least 400 Kal in 100g of material. Overall biscuit made without flour banana peels and floue banana peels with the addition of caloric value meets the requirements of SNI. that is a biscuit KP0% (4196 Kcal/kg, biscuits KP25% (4596 Kcal/kg), biscuits KP50% (4342 Kcal/kg) and biscuits KP75% (4451 Kcal/ kg). High levels of energy on a biscuit base material is due to banana peels kapok have high energy levels, as well as several other ingredients such as flour, sugar and skim milk powder. The higher the addition of flour banana peels higher energy levels produced biscuits. High levels of energy biscuits with the addition of a banana peels flour showed that the biscuits produced has good nutritional value and can be used as an energy source.

Testing organoleptic conducted on biscuit with addition of flour banana peels with a composition 0%, 25%, 50%, 75% gave results that were not significantly different in terms of flavor, color, mouthfeel, aftertaste at the 95% confidence level ($\alpha = 5\%$) (Table 3). Scores were obtained for flavor ranged from 2.71 to 3.21 (enough like-like), the color ranges from 2.19 to 3.27 (enough like- like), moutfeel ranged from 2.23 to 3.08 (enough like-like), aftertaste ranging from 2.23 to 2.77 (enough like). A panelist level of the physical condition of biscuits with the addition of flour banana peels can be seen in the Table 3.

Table 3. Organoleptic test panelists to biscuit flour flour with the addition of a banana peel

Biscuit Code	Flavor	Color	Mouthfeel	Aftertaste
KP0%	3,21	3,27	3,08	2,77
KP25%	2,85	2,56	2,63	2,46
KP50%	2,71	2,31	2,23	2,42
KP75%	2,79	2,19	2,25	2,23

Annotation : Skala 1= less likely; 2= enough like, 3= like, 4= really like

In addition to protein, fat is one of the important components in the manufacture of biscuits because it serves as an aroma enhancer and produce products that crunchy texture. In the manufacture of biscuits with the addition of a banana peels flour also added oils, margarine and metega. The use of margarine in the manufacture of biscuits are based on the characterisctic plastically. The characterisctic of plastically is important in producing a crunchy biscuit [13]. Biscuits with the addition of flour banana peels has a low protein content and high fat, so that the resulting products fairly crisp pastry. The panelists judged that the level of crispness of biscuits with the addition of flour banana peels is favored.

Flavors of food can be judged based on the smell, taste, and oral stimulation. Organoleptic test results in terms of aroma, color, mouthfeel and aftertaste flour biscuits with the addition of flour banana peels was not significantly different at the 95% confidence level ($\alpha = 5\%$). A panelist on the second level of biscuit products include enough like to like (2.19 to 3.27). This result is quite good because the flavor is an important factor in determining the quality of a food ingredient [18]. The addition of flour banana peels still give a distinctive flavor, so most panelists still feel the particularities flavor of banana peels. Flavor associated with the sense of smell. Oven or warming process allegedly able to increase the Maillard reaction, which is important as a source of flavor [19].

Similarly, the taste when chewed (mouthfeel) and after ingestion (aftertaste) flour biscuits with the addition of flour banana peels still feels a little bitter. One of the compounds that are suspected as the cause bitter or bitter taste in biscuits with the addition of flour banana peels is tannin. The presence of

IOP Conf. Series: Journal of Physics: Conf. Series **812** (2017) 012118 doi:10.1088/1742-6596/812/1/012118

tannins in foodstuffs are expected to cause a bitter taste in biscuits with the addition of flour banana peels. Allegedly excessive bitter taste associated with lipid and protein breakdown process. Oxidation phosphatidylcholine, amino acids, and peptides known to cause a bitter taste [19].

Biscuit brown color caused by the addition of banana flour brown. The brown color is due to the phytochemical compounds owned banana peels. In addition, the wet heating process will enhance the yellow color component and lower white [20].

4. Coclusion

Physical and organoleptic test results biscuits with the addition of flour banana peels has sufficient level of preference between like-liked. Based on the results of the proximate analysis of biscuits with the addition of flour banana peels has generally been in accordance with the National Standards of Indonesia (SNI 01-2973-1992). Based on proximate test fiber content of flour biscuits with the addition of flour banana peels potential as functional foods for health.

5. References

[1] Wang J, Rosell CC and Benedito de Barber C 2002 *Food Chemical* **79** 221–226.
[2] Carvalho AFU, Portela MCC, Sousa MB, Martins FS, Rocha FC, Farias DF and Feitosa JPA 2009 *Ulva fasciata* Delile. *Braz J Bio.* **69** 3 969–977
[3] Viuda M, Marcos M L, Fern´andez L, Sendra E, Lo´pez V and Pe´rez A 2010 *Comprehensive Reviews in Food Science and Food Safety* **9** 240-258
[4] Fern´andez L´ J, Viuda M M, Sendra E, Sayas B E, Navarro C and P´erez A J A 2007 *Europe Food Research Technologyl.* **226**(1–2) 1–6
[5] Sendra E, Fayos P, Lario Y, Fernandez L J, Sayas B E and Perez A JA 2008 *Food Microbiololgy* **25**(1) 13–21
[6] Lunn J and Buttriss JL 2007 *Nutrition Bulletin* **32** 2 1–64
[7] Border´ıas AJ, S´anchez A I and P´erez M M 2005 *Trends Food Science Technology* **16** 458–465
[8] Sastrapradja S,Lubis S H A, Sukma E D, Soetarno H and Lubis I 1977 *Sayur-sayuran* (Bogor: Proyek Sumber Daya Ekonomi, LIPI)
[9] Nasution R E andYamada I 2001 *Pisang-pisang Liar di Indonesia (*Bogor: LIPI)
[10] Hernawati and Aryani A 2008 Biosantifika **11** 1-12
[11] Steel R G D and Torrie J H 1993 *Prinsip dan prosedur statistik* (Jakarta : PT. Gramedia)
[12] Claudia R, Estiasih T, Ningtyas D W and Widyastuti E 2015 *J. Pangan dan Agroindustri, 3* 4 1589-1595
[13] Aini N 2009 *Lebih Jauh Tentang Sifat Fungsional Telur.* http://kulinologi.biz/index.php. (August 12, 2014)
[14] Ikrawan Y 2006 Biskuit, Makanan Pengganti Saat Lapar http://pikiranrakyat.com/Cetak/2006/ 122006/28/Cakrawala/Lain05.Htm (Juni 9 2014)
[15] Whitely PR 1991 *Biskuit Manufacture* (London: Applied Science Publishing, Ltd)
[16] Badan Standarisasi Nasional Indonesia 1992 Standar Mutu Biskuit (Jakarta: Badan Standarisasi Nasional)
[17] Khomsan A, Anwar F, Sukandar D, Riyadi H and Mudjajanto E S 2007 *Studi Implementasi Progam Gizi: Pemanfaatan, Cakupan, dan Dampak Terhadap Status Gizi* (Bogor: Institut Pertanian Bogor)
[18] Sarbini D, Rahmawaty S and Kurnia P 2009 *J. Penelitian Sains & Teknologi* **10** 1 18 – 26
[19] Damayanthi E, Tiong T L and Arbianto L 2007 *Rice Bran* (Penebar Swadaya, Jakarta)

Acknowledgments

We are gratefull to Ministry of Research, Technology and Higher Education of the Republic of Indonesia and Universitas Pendidikan Indonesia, which has funded this research with competitive grants scheme, DIPA funds DRPM 2016, corresponding assignment agreement implementation of research programs Number: 058 / SP2H/PL/DIT.LITAMBAS/ II / 2015. Date Desember 31, 2015.

Teacher Pedagogical Content Knowledge (PCK) and Students' Reasoning and Wellbeing

A Widodo

Faculty of Mathematics and Science Education, Universitas Pendidikan Indonesia, Jl. Dr. Setiabudi No. 229, Bandung 40154, Indonesia
Email: widodo@upi.edu

Abstract. This paper summarizes findings of a study on efforts to improve teachers Pedagogical Content Knowledge and how it affects students' reasoning and wellbeing. It was found that improvement of teachers' PCK was not very strong but we managed to develop strategies to facilitate their developments. In the second year, the research was focused on identifying students' reasoning skills both informal reasoning and formal reasoning. Data showed that students reasoning is relatively low (level 2 of five levels) and they could not construct highly coherence arguments. In addition alternative strategies to promote students' reasoning were explored. Attempts to support teachers to conduct lessons that facilitate students' reasoning found that teachers need intensive and continuous support. The study also identifies students' wellbeing as the impact of improvement of lessons and other activities designed to improve students' wellbeing. Research on students' wellbeing is not yet given attention in Indonesian schools although it plays very important roles in students' academic and nonacademic achievements.

1. Introduction

Indonesian government is concerned about the level of students' achievement as published in international benchmarking exercises such as TIMSS and PISA[1]. Teachers quality was identified one of crucial factors attributed to the situation. As one of the responses, the Indonesia government launched a program called "Teacher Certification". The program was designed as a tool to improve teacher quality through three strategies, i.e. attraction channel (increasing the enrolments of pre-service teachers), upgrading (improving teacher qualifications), and the behavioral (changing teachers' teaching practice). To obtain the certificate a teacher as to pass a certification qualification test conducted at the end of a training program especially designed for the purpose.

A study on the impact of teacher certification, however, suggest that teacher certification does not lead to improvement of teachers' teaching performance, teachers' behavior and teachers' participation in professional development [1]. The study also finds that there is no performance difference between certified and uncertified teachers. Indeed, data on students' enrolment shows that there is a growing interest to be a teacher, but not impact is yet to identify in the classrooms

Other criticism on the quality of education is a claim that students are not meeting employer expectations. Employers said that fresh graduates lack of competencies needed for the jobs. Some researchers suggest that argumentation skills and well-structured explanations can support the development of student abilities to meet these criticisms because this strategy can promote high-level thinking [2]. As suggested by OECD (2013) future jobs demands for high-level cognitive skills, interpersonal skills, and some generic skills such as interpersonal communication, self-management, and ability to learn. Reasoning skills play the role as a base for developing high-level cognitive skills.

This paper summarizes findings of a three-year research project on teacher PCK, reasoning and students' wellbeing. In the first year a focus on improving teacher Pedagogical Content Knowledge (PCK) was done to identify teachers PCK and alternatives to improve teachers PCK. In the second year, the focus was on strategies to improve students' reasoning. In third year, the study moved further to how teachers' PCK may affect students' reasoning and wellbeing.

2. The important of teachers PCK

Since its introduction by Shulman (1987), PCK has been growing to be a special area in teacher education. Numerous research studies on different strategies to measure PCK have been conducted [3],[4]. Despite significant research conducted on PCK, however, it is not easy to draw a general understanding of PCK. Instead there are variations of PCK that emerge as different researchers may develop different interpretation of PCK. For some, PCK is considered more or less as teachers' knowledge and skills [5],[6] while other consider PCK as teachers' values and judgements [7]. Therefore, it is understandable that critiques on PCK find confusion not only about the nature of PCK but also the contribution of PCK to teaching practice [8].

Loughran suggests that viewing PCK as transformation can be more productive since PCK itself "...is concerned with how subject matter knowledge can be transformed through pedagogical situations so that the particular subject matter under consideration is better understood by students" [7]. It means that PCK should be viewed from both the perspective of teachers and students. Although PCK may not directly relate to students' learning, improvement of teachers' PCK may contribute to deeper understanding of effective lessons to facilitate students' learning.

Pre-service period is a critical and yet an important phase in teachers' career. During the period student teachers develop not only knowledge and skill needed to be a teacher but also views and values they will bring throughout their career. Numerous studies on PCK were conducted on pre-service teachers [5],[9],[10],[11],[12]. Many studies on PCK did not include how pre-service teachers implement their PCK in the classroom. Although the same lessons conducted by two teachers do not necessarily suggest that the two teachers have the same PCK, but observing lessons help to understand why such a teacher develop such a PCK. Every experience in conducting lesson should provide feedback that contributes the development of a teacher's PCK.

As summarizes by van Driel *et al*, studies on teachers' PCK identifies the importance of teachers' knowledge of subject matter, teaching experience of the specific topic, knowledge of students' conceptions and learning difficulties, and participation in workshops on teaching [4]. These suggest that teachers with more teaching experience may develop higher level of PCK compare to inexperience teachers.

Since the development in early 2006, CoRes has generally been recognized as a valid instrument in articulating science teachers' PCK in the science education literature Participants, in the initial development of the CoRes must consider the "Big Ideas" associated with a topic. These are ideas that participants see as being crucial in helping students understand the topic [3]. CoRes were intended to capture instances of teachers' PCK and explicitly reveal their specialist knowledge of practice. In so doing, their PCK becomes meaningful, powerful, recognized and valued as part of their professional knowledge, and provides a new discourse of teacher professional language. A study has confirmed the effectivenessof CoResin engaging pre-service science teachers with PCK and providing them with greater academic efficacy and confidence in their preparedness for teaching [3].

Our previous study on strategies to improve teachers PCK showed that video based coaching [13] and intensive guidance during teaching practice [14] contribute to the development of teachers PCK. Our study suggests that a deliberate and well-designed program is needed to develop a teacher PCK. Teachers with strong PCK will be in a better position to present lessons more effectively and promote students' learning.

A study conducted on PCK of pre-service teachers who are graduated from biology major found that pre-service teachers' PCK is very limited [15]. Many of the respondents were unable to relate content and pedagogy. The lack of opportunities for them to integrate their content knowledge and their pedagogy knowledge led them to poor PCK. This suggests that teaching experience is an essential component in the development of teachers PCK, although, it is not enough to facilitate the development.

Although PCK is not teaching skills but PCK influences teachers' teaching practice. Teachers' lack of PCK may in turn lead to poor teaching and learning process in the classroom. It is argued that there is a nexus between teachers' pedagogical content knowledge (PCK) – the teachers' knowledge of how to teach particular content effectively – and student learning and students' reasoning skills.

3. Enhancing learning

When the focusof education is to enhance students' learning, then the quality of a teacher's teaching does play a significant role. Teaching capability must be cultivated so that students can develop the competencies and dispositions that will prepare them to be creative and collaborative life-long problem solvers. Comparatively, there is much more emphasis on what the teacher is doing and relatively little on what and how the student is achieving. Hence, there is a need to understand how teacher practices can address student learning.

One of the dominant theories of learning in the last decades is constructivism [16]. Constructivism suggests that learners actively construct and re-construct their own sets of meanings or understandings. They do not simply receive packets of knowledge from their teachers or other information sources; instead they actively construct and re-construct their own conceptions based on their existing conceptions. Knowledge cannot simply be transferred. Giving verbal explanations, for instance, does not necessarily lead to understanding unless the listeners themselves actively construct the knowledge. It is important to note here that learning experiences from the constructivist point of view do not necessarily need to be the ones that make the students physically active as claimed by critics [17]. Hands on activities may support learning but from the constructivist point of view, mental activities are what matter.

Since learners have developed pre-instructional conceptions that are not (totally) in accordance with the concepts to be learned, learning has to be viewed as a change of these students' pre-conceptions. Students' pre-conceptions not only serve as the starting points of the learning process, but also as a guide for learning [16]. Students' pre-conceptions show both the students' current position of their understanding and the way they view and act on certain events. Research on conceptual change reveals that changing students' conceptions is not easy and that students' new conceptions very often are not the ones expected by the teacher.

In their early theory of conceptual change, Posner *et al* suggest that there are four conditions required for conceptual change to occur: dissatisfaction with the existing conceptions, intelligibility, plausibility and fruitfulness of the new conceptions [18]. Later it is acknowledged that these are not sufficient to facilitate conceptual change. In their revised version of this theory, Strike and Posner [19] suggest that conceptual change is affected not only by "cognitive factors" but also by "affective factors", e.g. motives and goals as well as institutional and social resources. In a nutshell, conceptual change is not only facilitated by logic and scientific findings (cold rationality), rather it is a complex process that involves non rational factors [20]. Furthermore, Pintrich (1999) suggests that conceptual change can be facilitated by adoption of a mastery goal orientation, adoption of more constructivist epistemological beliefs, embracing higher levels of personal importance, value and interest, adoption of higher levels of self-efficacy for learning and adoption of a belief in personal control of learning [21].

Duit, Treagust and Widodo (2013) identified different foci of research on conceptual change that include epistemological, ontological, affective, intentional and multidimensional [16]. Clearly research on conceptual change suggests that conceptual change approach is an effective strategy to facilitate conceptual change but more evidence is needed.One of recommendations on how to facilitate conceptual change that deserve more attention is the use of mental model [22],[23,[24]. Mental model is closely relate to how students reason about the concept. Therefore, in the next part I will discuss how to develop students' reasoning.

4. Developing reasoning

Reasoning is largely believed as important skills to be developed by students but it is not yet received sufficient attention. Reasoning is not just a result of learning but it helps students to learn more effectively and meaningfully. Our studies on students reasoning found that there is improvement of students' rational reasoning from primary school to senior high school but there is a high prevalence of intuitive and emotive reasoning [25]. This suggests that many students keep on using informal reasoning. Our other study shows that there is no significant change of students' formal reasoning from primary school to senior high school [26]. Findings of our studies clearly show that reasoning skills need to be given more attention in our school. Reasoning is as important as students' understanding of the subject and therefore it should be given as equal attention.

Some reasoning skills, however, do not show a clear pattern of development. This suggests that to develop students' reasoning, lessons should be carefully designed. Reasoning skills is not a side effect of lessons; rather it is an important skill that should be given priority. It seems, however, teachers do not aware the importance of developing students' reasoning. Moreover, teachers also need supports on how to plan and to conduct such lessons.

Reasoning skills may naturally develop as people mature. However, full development of reasoning skills would not occur without practice and application [27]. There are ways improve students' reasoning, such as using Toulmin' argumentation firework (1958) that require students to provide data, warrant, qualifier, and rebuttal to every claim they made [28]. Other way to promote students' reasoning is using representation [29]. In this strategy students are encouraged to make explicit their thinking by representing it into different representations.

There are a number of views as to how people reason. Hand *et al*, Kuhn and Osborne focused on the structure and the process of argumentation [2],[27],[30]. These researchers are interested in how argumentation leads to a final conclusion. The use of logic is central to the argument.

Other, such as Tytler, Prain, Hubber & Waldrip (2013) used representation [29]. Studies on representation suggested that when students were encouraged to express and justify their ideas using visual representations, they tended to develop a stronger content knowledge, and learn subject-specific ways of representing scientific ideas (Waldrip & Prain, 2011; Waldrip, Prain & Sellings, 2013). Logical reasoning skill is a significant component for deep scientific understanding because it facilitates learners to connect what one is learning with what is already explored effectively [31]. When the teacher focuses on students' thinking and reasoning with a series of representational challenges, teachers are actually supported students for quality conceptual learning [32],[33],[34].

Based on the arguments presented, I believe that effective teaching for conceptual change can be facilitated when teachers has a solid PCK that they have good understanding of the content as well as strategies to present it.

Drawing from constructivist theory, itis important that teachers explore students' initial understandings prior to presenting the concepts. Students can be encouraged to draw their representation and re-consider it throughout the lessons. Teachers may also use discussion between students and with their teacher to facilitate students to evaluate their representation. Reasoning processes should align with student learning experiences, enabling these experiences to function as an induction into this domain and its discursive purposes and resources.

Tytler *et al* argued that there are a number of components to the reasoning process. These components involve the following steps [31].

- Making a claim. Learners need to make their claim explicit. This claim making includes exploring the phenomenon and asking an opened question
- Checking claim with the evidence. Learners need to judge their claims in light of the evidence. Teachers may use this process intervention strategy to guide learners to the revise their clams

- Refining the claim and exploring the robustness of the claim. Learners should be encouraged to refine their claims. It may include the limitation of the claims or the applicability of the claims to new settings.
- Coordinating explanations and evidence to explain and clarify ideas. Learners should be encouraged to check and accommodate their claims based the new evidence.
- Publicly justifying and communicating ideas, evidence and exploring problem-solving related situations. Finally, learners should communicate their new claims and applied them to new context.

5. Improving students wellbeing

Students' wellbeing has been given an increasing attention in education despite the lack of consensus on the term. Based on a number of literature [35] summarizes that youth wellbeing covers factors related to physical and mental health, character strengths and virtues, environmental integrity, as well as involvements such as extra-curricular or civic engagement. A more specific indicator of wellbeing in the school context is supposed to comprises two main aspects, i.e. personal and social wellbeing. Personal wellbeing consists of emotional wellbeing (positive feelings and absence of negative feelings), satisfying life, vitality, resilience and self-esteem, and positive functioning (autonomy, competence, engagement, and meaning and purpose). Social wellbeing is subdivided into supportive relationships, and trust and belonging.

Reasoning and wellbeing links together and may even have a reciprocal relationship. Student with good reasoning capabilities may have an advantage in learning so that they can learn in more effective and efficient ways that at the end contribute to her/his wellbeing. Alternatively, students with who are in wellbeing stage can reason better that help them to learn and understand the lesson. Clearly there reasoning and wellbeing are two areas that still need more research. Our initial research on wellbeing and reasoning indicates that lessons that focus on promoting students reasoning results in improvement of students' wellbeing.

6. Conclusion

This study reiterates the importance of teachers PCK to teachers teaching practice. Although PCK may not directly relate to teaching practice, however, PCK shapes teachers their practice. Deliberate and systematics efforts are needed to facilitate the improvement of teachers PCK. Although teachers acknowledge the importance of reasoning for the students, however, lessons may not sufficiently facilitate students' reasoning because such lessons require teachers to use new teaching strategies and create a potential risk to the teachers. Teachers aware about the importance students' wellbeing, however, support is needed to help teachers in creating lessons that promote students' wellbeing.

7. References

[1] Chang M C, Shaeffer S, Al-Samarrai S, Ragatz A B, de Ree J and Stevenson R 2014 *Teacher reform in Indonesia: The roles of politics and evidence in policy making* (Washington DC:: The World Bank)
[2] Hand B, Norton-Meier L, Staker J and Bintz J 2009 *Negotiating science. Thecritical role of argument in student inquiry*. Portsmouth, NH: Heinemann
[3] Loughran J J, Berry A and Mulhall P 2012 *Understanding and Developing Science Teachers' Pedagogical Content Knowledge* 2nd edition (Rotterdam: Sense Publishers)
[4] vanDriel J H, De Jong O and Verloop N 2002 *Science Education* **86** 572-90
[5] Beyer C J and Davis E A 2011 *Science Education* **96** 130-57
[6] Magnusson S, Krajcik J and Borko H 1999 Nature, sources and development of pedagogical content knowledge for science teaching. In J Gess-Newsome and N G Lederman (Eds.) *PCK and Science Education* (Netherlands: Kluwer Academic Publishers) pp 95-132
[7] Loughran J 2013 *Curriculum Inquiry* **43** 119-41
[8] Settlage J 2013 *Journal of Science Teacher Education* **24** 1-12

[9] Brown P, Friedrichsen P and Abell S 2013 *Journal of Science Teacher Education* **24** 133-55
[10] Hume A and Berry A 2011 *Research in Science Education* **41** 341-55
[11] Hume A and Berry A 2013 *Research in Science Education* **43** 2107-36
[12] Nilsson P and Loughran J 2012 *Journal of Science Teacher Education* **42** 1-23
[13] Rusmana N E, Widodo A, Riandi and Rochintaniawati D 2014 Kemampuan P*edagogical Content Knowledge* (PCK) Guru Peserta Musyawarah Guru Matapelajaran (MGMP) IPA di Kota Sumedang (Paper Presented at UPI-UPSI Conference)
[14] Herlina L, Widodo A, Rochintaniawati D and Riandi 2015 Perkembangan Pedagogical Content Knowledge MahasiswaCalon Guru BiologiPeserta Program PengalamanLapangan. *Prosiding Seminar Nasional IPA VI.* Semarang: Jurusan IPA TerpaduUniversitasNegeri Semarang.
[15] Anwar Y, Rustaman N, Widodo A and Redjeki S 2012 *Proc. Seminar Nasional IPA III (Semarang)* (Semarang: Universitas Negeri Semarang)
[16] Duit R, Treagust D and Widodo A 2013 Teaching science for conceptual change, In S Vosniadou (Ed) *International Handbook of Research on Conceptual Change* 2nd edition (New York: Routledge)
[17] Fox R 2001 *Oxford Review of Education* **27** 23-35
[18] Posner G J, Strike K A, Hewson P W and Gertzog W A 1982 *Science Education* **66** 211-27
[19] Strike K A and Posner G J 1992 A revisionist theory of conceptual change. In R A Duschl & R A Hamilton (Eds.) *Philosophy of Science, Cognitive Psychology, and Educational Theory and Practice* (State University of New York: Albany)
[20] Pintrich P R, Marx R W and Boyle R A 1993 *Review of Educational Research* **63** 167-99
[21] Pintrich P R 1999 Motivational beliefs as resources for and constraints on conceptual change, In W Schnotz, S Vosniadou & M Carretero (Eds) *New Perspectives on Conceptual Change* (Amsterdam: Pergamon) pp 33-49
[22] Clement J 2013 Roles of explanatory models and analogies in conceptual change. In S Vosniadou (Ed) *International Handbook of Research on Conceptual Change* 2nd edition (New York: Routledge)
[23] Nersessian N J 2013 Mental model in conceptual change, In S Vosniadou (Ed) *International Handbook of Research on Conceptual Change* 2nd edition (New York: Routledge)
[24] Jonassen D and Easter M A 2013 Model building for conceptual change, In S Vosniadou (Ed). *International Handbook of Research on Conceptual Change* second edition (New York: Routledge)
[25] Saptarini D, Widodo A, Riandi and Rochintaniawati D 2015 *Proceedings of International Seminar on Mathematics, Science and Computer Science Education (Bandung)* (Bandung: Faculty of Mathematics and Science Education, UniversitasPendidikan Indonesia)
[26] Amalia N F, Widodo A, Riandi and Rochintaniawati D 2015 *Proc. of International Seminar on Mathematics, Science and Computer Science Education (Bandung)* (Bandung: Faculty of Mathematics and Science Education, UniversitasPendidikan Indonesia)
[27] Kuhn D, Katz J B, and Dean D 2004 *Thinking & Reasoning* **10** 197–219
[28] Toulmin S 1958 *The Uses of Argument* (Cambridge: Cambridge University Press)
[29] Tytler R, Prain V, Hubber P and Waldrip B 2013 *Constructing Representations to Learn in Science* (Dordrecht: Sense Publishers)
[30] Osborne J 2010 *Science* **328** 463-6
[31] Tytler R, Prain V, Hubber P and Haslam F 2013 Resoning in Science through Representation, In R Tytler, V Prain, P Hubber & B Waldrip (eds) *Constructing Representations to Learn in Science* (Rotterdam: Sense Publishers)
[32] Waldrip B, Prain V and Carolan J 2010 *Research in Science* Education **40** 65-80
[33] Waldrip B and Prain V 2006 *Teaching Science* **54** 17-21
[34] Prain V and Waldrip B 2006 *International Journal of Science Education* **28** 1843-66

[35] Soutter A K, O'steen B and Gilmore A 2012 *Curriculum Journal of Curriculum Studies* **44** 111–42

The Effect of MnO_2 Content and Sintering Atmosphere on The Electrical Properties of Iron Titanium Oxide NTC Thermistors using Yarosite

Wiendartun[1*], **and Dani Gustaman Syarif**[2]

[1]Departmen Pendidikan Fisika, Universitas Pendidikan Indonesia, Jl. Dr. Setiabudi No. 229, Bandung 40514, Indonesia
[2] PSTNT-BATAN, Jl Tamansari 71, Bandung, Indonesia

Email: wien@upi.edu

Abstract. The effect of MnO_2 content and sintering atmosphere on the characteristics of Fe_2TiO_5 ceramics for Negative Thermal Coefficient (NTC) thermistors by using Fe_2O_3 derived from yarosite has been studied. The ceramics were produced by pressing a homogeneous mixture of Fe_2O_3, TiO_2 and MnO_2 (0-2.0 w/o) powders in appropriate proportions to produce Fe_2TiO_5 based ceramics and sintering the pressed powder at 1100-1200°C for 3 hours in air, O_2 and N_2 gas. Electrical characterization was done by measuring electrical resistivity of the sintered ceramics at various temperatures from 30°C to 200°C. Microstructure and structural analyses were also carried out by using an scanning electron microscope (SEM) and x-ray diffraction (XRD). The XRD data showed that the pellets crystallize in orthorhombic. The presence of second phase could not be identified from the XRD analyses. The SEM images showed that the grain size of pellet ceramics increase with increasing of MnO_2 addition, and the grains size of the ceramic sintered in oxygen gas is smaller than sintered in nitrogen gas. Electrical data showed that the value of room temperature resistance (R_{RT}) tend to decrease with respect to the increasing of MnO_2 addition and the pellet ceramics sintered in oxygen gas had the largest thermistor constant (B), activation energy (E_a), sensitivity (α) and room temperature resistance (R_{RT}), compared to the sintered in nitrogen gas. From the electrical characteristics data, it was known that the electrical characteristics of the Fe_2TiO_5 pellet ceramics followed the NTC characteristic. The fabricated Fe_2TiO_5 ceramics have thermistor constants (B = 2207-7145K). This can be applied as temperature sensor, and will fulfill the market requirement.

1. Introduction

Negative Temperature Coefficient (NTC) thermistor has been widely used around the world today, due to its capability used in various fields of electronics, such as thermometer, electric current limiter, water flow sensor, and pressure sensor [1-2]. The NTC thermistor is generally made of ceramic having structure of spinel of AB_2O_4 where A is the ion occupies tetrahedral position and B is the ion occupies octahedral position [3-4]. Fe_2TiO_5 ceramic is one of some ceramics that can be applied for NTC thermistor. The thermistor may be produced in the form of disk/pellet and thick film. Fe_2TiO_5 is one of semiconductor ceramics used as based material for main components fabrication of NTC thermistor as temperature sensor. The composition of mineral Fe_2TiO_5 is belong to pseudobrokyte group where the general formula of this compound is X_2YO_5 with octahedral in both side, X and Y [5]. The

Fe_2TiO_5 ceramic has capability of being NTC thermistors, potentially, due to its semi conductive property. However, its characteristics can be still improved by addition of additive and sintering atmosphere condition. Since the MnO_2 added into the Fe_2TiO_5 ceramic, the following conditions may happen. First, the MnO_2 will dissolve in the Fe_2TiO_5 ceramic by substituting a part of Fe^{3+} ions and/or Ti^{4+} ions. The second, the MnO_2 does not dissolve but segregates at grain boundaries of the Fe_2TiO_5 ceramic. Since the first condition happens, the Fe_2TiO_5 ceramic may have a lower electrical resistivity when the substitution of Fe^{3+} and /or Ti^{4+} creating free electron in the conduction band. For the second condition, the electrical resistivity may be higher because the segregated MnO_2 may change the microstructure. Thermistor constant *B* is a quantity which determine typical characteristic of thermistor corresponding to electrical resistance changes with temperature. The larger thermistor constant lead to better thermistor quality. The addition of MnO_2 dopant into the Fe_2TiO_5 ceramic with sintering atmosphere variation may increase the thermistor constant which then improves the performance of the thermistor. Here, the effect of MnO_2 content and sintering conditions on the electrical properties of iron titanium oxide ceramic using yarosite as raw material for NTC thermistor based on the above hypothesis were studied. In addition, the possibility of local material utilization was also studied in order to step up the added value of the local material especially the material contains Fe_2O_3.

2. Material and Method

Fe_2TiO_5 thermistor ceramic was prepared by using powder of Fe_3O_3 derived from a yarosite mineral deposit and TiO_2. The extraction was done by using a precipitation process. The Fe2O3 was chemically analyzed to know the composition. The chemical composition of the yarosite powder was depicted in Table 1 [3, 6]. Powders of Fe_2O_3, TiO_2 and MnO_2 were weighed in appropriate proportions to fabricate MnO_2 added-Fe_2TiO_5 ceramics, where the MnO_2 were 0, 0.5, 1.0, 1.5 and 2.0 w/o (weight %), was calcinated at 700°C for 2 hour. In order to form pellets, pressed powder, the homogeneous mixture of Fe_2O_3 and TiO_2 was pressed at 4.10^7 kg/m^2. The raw pellets were sintered at a temperature of 1200°C for 3 hours in air atmosphere and 2.0 w/o MnO_2 dopant was sintered at 1100°C for 3 hours in oxygen (O_2) and nitrogen gas (N_2). Sintering in N_2 was intended to determine the resistivity of the pellets can be decreased.

Table 1. Chemical composition of Fe_2O_3 powder derived from yarosite

No	Compound	Concentration (%)
1	Fe_2O_3	93.80
2	SiO_2	1.02
3	MgO	0.09
4	CaO	0.19
5	TiO_2	1.15
6	MnO	0.12
7	Na_2O	0.59
8	K_2O	0.50

The crystal structure of the sintered pellets was analyzed with x-ray diffraction (XRD) using Kα radiation at 40 kV in voltage and 25 mA in current. The microstructure of fractured pellets was investigated by a Scanning Electron Microscope (SEM). The opposite-side surfaces of the sintered pellets were coated with Ag paste. After the paste was dried at room temperature, the Ag coated-pellets were heated at 600°C for 10 minutes. The resistance of the pellets was measured at various temperatures from 30 to 200°C in steps of 10°C using a digital multimeter. Thermistor constant (B) was derived from Ln resistivity versus $1/T$ curve where B is the gradient of the curve based on (1) [6-7] :

$$R = R_{o.}\, exp\ (B/T) \tag{1}$$

Where, R is the electrical resistance, R_o is a constant or the resistant at the infinite temperature, B is the thermistor constant and T is the temperature in Kelvin and k is the Boltzmann constant. Room temperature resistance (R_{RT}) was determined as the electrical resistance at room temperature (25°C).

From the value of B, the activation energy (E_a) and sensitivity (α) were calculated using equation 2 and 3 [8-9].

$$E_a = B.k \tag{2}$$

$$\alpha = B/T^2 \tag{3}$$

3. Result and Discussion

The XRD profile of Fe_2TiO_5 without doping MnO_2, sintered at 1200°C for 3 hours in air atmosphere is shown in Figure1A and XRD profiles of 2.0 % mole MnO_2 doped Fe_2TiO_5 pellet ceramics sintered at 1100°C for 3 hours with sintering atmosphere of Oxygen and Nitrogen gas respectively are shown in Figure 1b and Figure 1c.

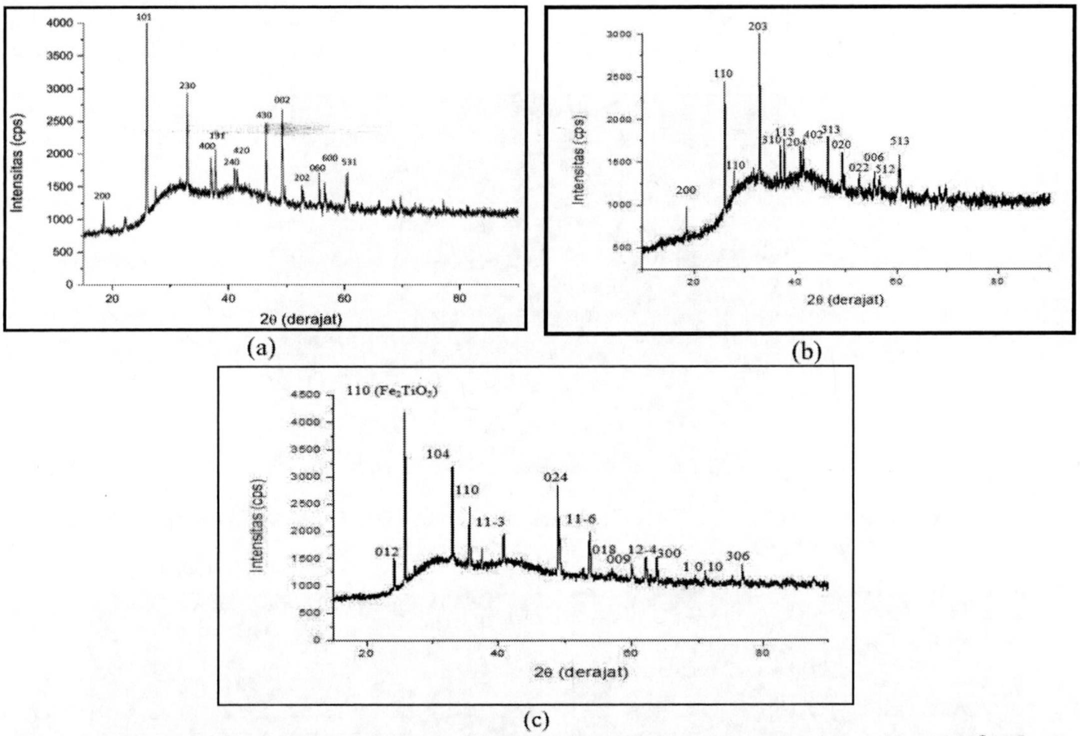

Figure 1. XRD profile of the (a) 0.0 w/o MnO_2 added- Fe_2TiO_5 Ceramic sintered at 1200°C for 3 hours in air, (b) 2.0 % mole MnO_2 doped Fe_2TiO_5 ceramic sintered in Nitrogen gasand (c) Oxygen gas.

As shown in the figure 1 the profiles are generally similar. The XRD profiles show that the structure of the pellet ceramics is orthorhombic after being compared to the XRD analyses program: X Powder aplication vers.2004.04.46 Pro. No peaks from second phases observed. It may be due to the small concentration of impurities which is smaller than the precision limit of the x-ray diffractometer used. The XRD data of Figure1 indicates that the synthesis of the Fe_2TiO_5 pellets has been well prepared from Fe_2O_3 (Yarosite) and TiO_2 powder with MnO_2 content and sintering atmosphere condition.

Microstructures of the Fe_2TiO_5 ceramic added with 0, 1.5 and 2.0 w/o MnO_2 respectively, sintered at 1200°C in air are represented in Figure2. Microstructures of the Fe_2TiO_5 pellet ceramic 2.0 % mole MnO_2 doped Fe_2TiO_5 pellet ceramics using Fe_3O_3 from yarosite sintered at 1100°C for 3 hours with sintering atmosphere of oxygen and nitrogen gas respectively, are showed in figure 2.

(a) (b)

(c)

Figure 2. Microstructure of the Fe_2TiO_5 Ceramics sintered at 1200°C/3h/Air Doped -0.0 w/0 (A), 1.5 w/o (B) and 2.0 w/o MnO_2(C).

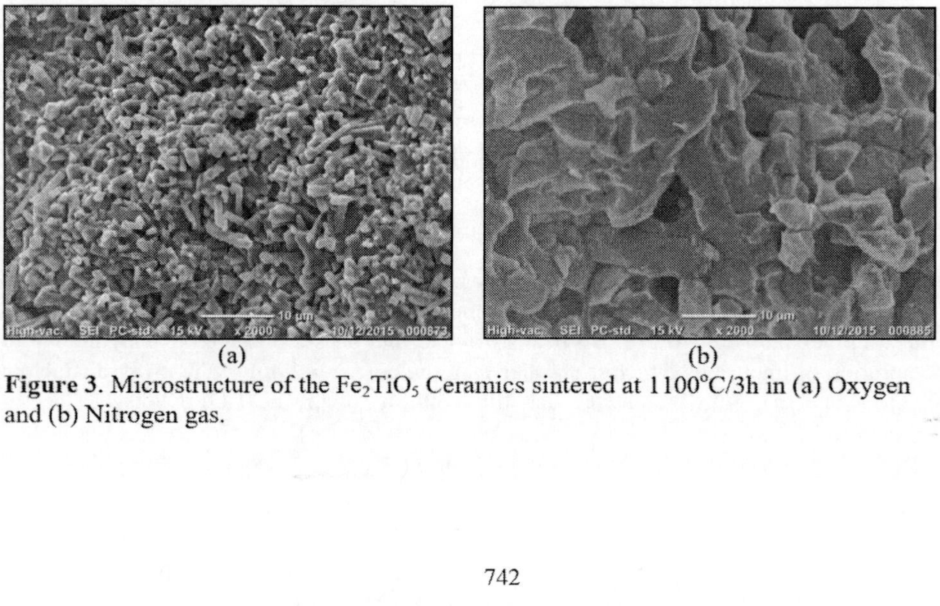

(a) (b)

Figure 3. Microstructure of the Fe_2TiO_5 Ceramics sintered at 1100°C/3h in (a) Oxygen and (b) Nitrogen gas.

The samples could be well synthesized with air, nitrogen and oxygen gas sintering atmospheres. All of the pellets are characterized in porous structure with different grain size depending on the MnO_2 addition and sintering atmosphere condition. From Figure 2, it is seen that grains of the ceramic tend to become larger as the increase of MnO_2 addition. The increase of the grain size may be caused by the segregation of the added MnO_2. The segregated MnO_2 inhibits the grain growth during sintering [10-11] The grain size calculated by using of the intercept method is found to be 11.10 μm, 11.50 μm and 11.67 μm, respectively. From Figure 3, it is seen that the grain size of the ceramic sintered in oxygen gas is much smaller than that of the ceramic sintered in nitrogen gas. This situation can be explained as follow, a relatively poor oxygen for sample sintered in nitrogen gas, make the ceramic could not be well synthesized. Some of Fe_2O_3 and TiO_2 segregated at grain boundaries and inhibited grain growth, producing small grains. The grain size calculated by using of the intercept method is found to be 2.84μm and 8.08 μm for sintered in oxygen and nitrogen gas atmospheres, respectively. Sintering gas atmosphere also influence of the electrical properties of the pellets Fe_2TiO_5. In an atmosphere in which oxygen-rich grain growth does not work properly causing small grain size. In the otherwise, in an nitrogen gas atmosphere grains growth better [12-13].

The electrical data of the MnO_2 added-Fe_2TiO_5 ceramics and sintering atmosphere variation are shown in Figure 4 and Table 2-3. The electrical data of Figure 4 show that the Ln resistance increases linearly as the $1/T$ increases, indicating that the electrical characteristics of the ceramics follows the NTC tendency expressed by equation (1). As shown in Table 1, the increase of the MnO_2 added-Fe_2TiO_5 ceramics tend to decreases the thermistor constant (B), activation energy (E_a), sensitivity (α) and room temperature resistance (R_{RT}). As shown in Table 2, electricity resistance and thermistor constant of Fe_2TiO_5 ceramic in oxygen gas is larger than in nitrogen gas. The ceramics sintered in oxygen gas have relatively smaller oxygen vacancy and means smaller number of electron, so the resistance of this ceramic is larger. The small value of the activation energy exhibits the extrinsic property of the ceramics. The value of B and E_a of our ceramics is large enough and fits the requirement [14-15].

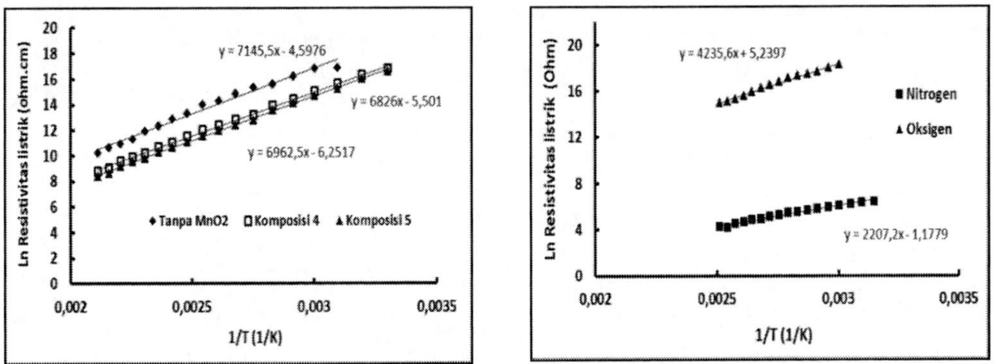

(a) (b)

Figure 4. The relation between Ln resistance (R) vs $1/T$ of (a) MnO_2 added-Fe_2TiO_5 ceramics and (b) Oxygen-Nitrogen gas sintering atmospheres

Table 2. Electrical characteristics of the MnO_2 added-Fe_2TiO_5 ceramics 1200°C/3h/air

Content of MnO_2 (w/o)	B(K)	E_a(eV)	α(%/K)	R_{RT}(MΩ)
0	7145	0.62	8.0	261.14
1.5	6826	0.59	7.7	36.21
2.0	6962	0.60	7.8	27.02

Table 3. Electrical characteristics of the Fe_2TiO_5 doped 2.0 % MnO_2 Sintered at 1100°C for 3hrs with sintering conditions.

Atmosphere	B (K)	E_a (eV)	A (%/K)	R_{RT} (MΩ)
O_2 gas	4235	0.36	4.8	280.79
N_2 Gas	2207	0.19	2.5	5.07

4. Conclusion

Fe_2TiO_5 pellet ceramics using Fe_2O_3 from yarosite have been well produced with MnO_2 content and sintering atmosphere conditions. All of the pellets crystallize in orthorhombic structure. The grain size of the Fe_2TiO_5 ceramics tends to be larger by addition of MnO_2. The electrical characteristics of the Fe_2TiO_5 base ceramics followed the NTC characteristic. The value of B and R_{RT} of the produced Fe_2TiO_5 ceramics namely B = 2207-7145 K and R_{RT} = 5.07-280.79 MΩ, fitted market requirement and the Fe_2TiO_5 ceramic can be applied as NTC thermistor.

5. References

[1] Metz R 2000 *Journal of Materials Science* **35** 4705-11
[2] Vakiv M, Shpotyuk O, Mrooz O and Hadzaman I 2001 *Journal of the European Ceramic Society* **1** 1783-85
[3] Wiendartun,. Syarif D G 2012 *Journal of Materials Science Research* **1** 70-5
[4] A. Feltz, W. Polzl 2000 *Journal of the European Ceramics Society* **20** 2352-66
[5] Csete de Gyorgyfalva G D C, Reaney I M 2001 *Journal of the European Ceramics Society* **21** 2145-48
[6] Wiendartun, Waslaluddin, Syarif D G 2013 *Journal of the Australian Ceramic Society* **49** 141-7
[7] Wiendartun, Risdiana, Fitrilawati, Siregar R E 2015 *Materials Science Forum* **827** 262-5
[8] Jadhav R N, Mathad S N, Puri V 2012 *Ceramic International* **38** 6481-86
[9] Park K, Lee J K, 2009 *Journal Alloy and Compound* **475** 513-17
[10] Ming Long Liu, DeYang, Yuan-Fang Qu 2010 *Journal of Alloys and Compdound* **508** 559-64
[11] Zhang H, Chang A, Changwen P 2011 *Microelectron Enginering* **88** 2934-40
[12] Veres A, Noudem J G, Perez O, Fourrez S, Bailleul G 2007 *Journal European.Ceramic Society* **27** 3873–76
[13] Feteira A 2009 *Journal American. Ceramic Society* **92** 967-73
[14] Ma C, Liu Y, Lu Y, Qian H2015 *Journal.Alloy Compound.* **650** 931-35
[15] Wang Z, Chun Wang, Yongfei Wen & Liangliang Zhang 2016 *Ferroelectrics* **492** 126-33

Acknowledgments
The authors wish to acknowledge their deep gratitude to Directorate General of Higher Education (DIKTI), Ministry of National Education of Indonesian Government for financial support under HIBAH BERSAING program with contract No 305/SP2H/LT/DPRM/II/2016 Tanggal 10 Maret 2016. It is a pleasure to Koike Laboratory, Department of Applied Physics of Engineering Tohoku University Japan.